Sunscreens

COSMETIC SCIENCE AND TECHNOLOGY

Series Editor

Eric Jungermann
Jungermann Associates, Inc.
Phoenix, Arizona

1. Cosmetic and Drug Preservation: Principles and Practice, *edited by Jon J. Kabara*
2. The Cosmetic Industry: Scientific and Regulatory Foundations, *edited by Norman F. Estrin*
3. Cosmetic Product Testing: A Modern Psychophysical Approach, *Howard R. Moskowitz*
4. Cosmetic Analysis: Selective Methods and Techniques, *edited by P. Boré*
5. Cosmetic Safety: A Primer for Cosmetic Scientists, *edited by James H. Whittam*
6. Oral Hygiene Products and Practice, *Morton Pader*
7. Antiperspirants and Deodorants, *edited by Karl Laden and Carl B. Felger*
8. Clinical Safety and Efficacy Testing of Cosmetics, *edited by William C. Waggoner*
9. Methods for Cutaneous Investigation, *edited by Robert L. Rietschel and Thomas S. Spencer*
10. Sunscreens: Development, Evaluation, and Regulatory Aspects, *edited by Nicholas J. Lowe and Nadim A. Shaath*
11. Glycerine: A Key Cosmetic Ingredient, *edited by Eric Jungermann and Norman O. V. Sonntag*
12. Handbook of Cosmetic Microbiology, *Donald S. Orth*
13. Rheological Properties of Cosmetics and Toiletries, *edited by Dennis Laba*
14. Consumer Testing and Evaluation of Personal Care Products, *Howard R. Moskowitz*
15. Sunscreens: Development, Evaluation, and Regulatory Aspects. Second Edition, Revised and Expanded, *edited by Nicholas J. Lowe, Nadim A. Shaath, and Madhu A. Pathak*
16. Preservative-Free and Self-Preserving Cosmetics and Drugs: Principles and Practice, *edited by Jon J. Kabara and Donald S. Orth*

ADDITIONAL VOLUMES IN PREPARATION

Hair and Hair Care, *edited by Dale H. Johnson*

Sunscreens

Development, Evaluation, and Regulatory Aspects

Second Edition, Revised and Expanded

edited by

Nicholas J. Lowe
Skin Research Foundation of California
Santa Monica, California

UCLA School of Medicine
Los Angeles, California

Nadim A. Shaath
KATO Worldwide, Ltd./Nickstadt-Moeller
Mt. Vernon, New York

Madhu A. Pathak
Harvard Medical School
and Massachusetts General Hospital
Boston, Massachusetts

Marcel Dekker, Inc. New York • Basel • Hong Kong

ISBN: 0-8247-9306-4

The publisher offers discounts on this book when ordered in bulk quantities. For more information, write to Special Sales/Professional Marketing at the address below.

This book is printed on acid-free paper.

Marcel Dekker, Inc.
270 Madison Avenue, New York, New York 10016

Current printing (last digit):
10 9 8 7 6 5 4 3 2 1

PRINTED IN THE UNITED STATES OF AMERICA

About the Series

The Cosmetic Science and Technology series was conceived to permit discussion of a broad range of current knowledge and theories of cosmetic science and technology. The series is made up of books written by one or more authors or edited volumes with a number of contributors. Authorities from industry, academia, and the government are participating in writing these books.

The aim of this Series is to cover the many facets of cosmetic science and technology. Topics are drawn from a wide spectrum of disciplines ranging from chemistry, physics, biochemistry, analytical and consumer evaluations to safety, efficacy, toxicity, and regulatory questions. Organic, inorganic, physical and polymer chemistry, emulsion technology, microbiology, dermatology, and toxicology all play a role in cosmetic science.

There is little commonality in the scientific methods, processes, or formulations required for the wide variety of cosmetics and toiletries manufactured. Products range from hair care, oral care, and skin care preparations to lipsticks, nail polishes and extenders, deodorants, body powders and aerosols to over-the-counter products, such as antiperspirants, dandruff treatments, antimicrobial soaps, and acne and sunscreen products.

Cosmetics and toiletries represent a highly diversified field with many subsections of science and "art." Indeed, even in these days of high technology, "art" and intuition continue to play an important part in the development and evaluation of formulations, and the selection of raw materials. There is a move

Preface

Increasing awareness of the damaging effects of sunlight has led to a significant demand for improved photoprotection by topically applied sunscreening agents. Several studies have shown that sunscreens are able not only to protect against ultraviolet-induced erythema in human and animal skin but also to inhibit photocarcinogenesis in animal skins. More recently, knowledge of the harmful effects of UVA augmentating those of UVB, the direct dermal effects of UVA (even with low levels of radiation) as a potential cause of skin aging, and the clinical problems of UVA-induced photosensitivity reactions has led to a quest for improved UVA-absorbing sunscreens.

In recent years, the most important assay of sunscreen evaluation has been the sun protection factor (SPF). This is predominantly a measure of the ability of a sunscreen to protect against sunburn wavelengths within the UVB range of the ultraviolet (290–320 nm). Several protocols have recently been described in the literature addressing the issue of UVA testing and the degree to which sunscreen cosmetic products protect in that region of the ultraviolet spectrum.

The regulation of sunscreen cosmetic products in the United States, in particular, has remained a major topic of discussion. Issues relating to safety, evaluation, and compliance have captured the attention of many involved in this industry. The safety of PABA and its derivatives, as well as the benzophenone class of sunscreening agents, has recently been questioned. SPF testing procedures recommended by the Food and Drug Administration (FDA) in the United

States are now very similar to those recommended by the European COLIPA standards. Other issues currently being debated worldwide include specific protocols for UVA testing, the lack of criteria for UVA protection factor labeling instructions in the United States, the difficulties in introducing new active sunscreening agents, and antiaging claims by sunscreen cosmetic manufacturers.

Other important challenges have included the evolution of new vehicles possessing superior substantive qualities that reduce the problems of sunscreen removal by water, the effect of solvents and other ingredients in improving or reducing the effectiveness of sunscreen active ingredients, and the photochemical stability of different sunscreens.

This book attempts to provide the reader with current knowledge about sunscreen development, evaluation, formulation, and regulatory issues. It combines a variety of chapters written by prominent scientists and practitioners from appropriately varied disciplines, including academics, industry, the medical community, marketing, and regulatory agencies.

The book is organized into six major sections:

1. Introductory chapters on the evolution of sunscreens and on the need for photoprotection.
2. The photobiological aspects of sunscreens, with some underlying physics.
3. The state of the industry in the United States, Europe, Japan, and Australia—past, present, and future—as well as the regulatory aspects in these countries.
4. The chemistry of sunscreen ingredients, tanning accelerators, and cosmetic formulations.
5. Human and in vitro evaluation of sunscreen products.
6. Chemical/physical evaluation of sunscreen products.

We believe that this second edition continues to reflect the concerns and ideas of those involved in sunscreens and photoprotection. It is a reference guide for all.

Nicholas J. Lowe
Nadim A. Shaath
Madhu A. Pathak

I would like to thank my wife, Pamela, and daughters, Nichola and Philippa, for their continued patience, love, and understanding in allowing me the time to work on this second edition.

—NJL

To my wife, Laila, and my daughter, Mona, "I love you" and thanks for your patience, love and guidance. I would also like to thank my parents for inspiration, and my colleagues at KATO Worldwide Ltd./Nickstadt-Moeller for their encouragement and support at the most critical juncture of the growth and establishment of our company.

—NAS

Contents

Contributors

Patricia P. Agin, Ph.D. Schering-Plough Health Care Products, Memphis, Tennessee

Gideon I. Andemicael KATO Worldwide, Ltd./Nickstadt-Moeller, Mount Vernon, New York

Michael W. Anderson, B.Sc., Ph.D., M.B.A. Tioxide Specialties, Ltd., Durham, England

O. Binet Fondation A. de Rothschild, Paris, France

Homer S. Black, Ph.D. Department of Dermatology, Veterans Affairs Medical Center and Department of Dermatology, Baylor College of Medicine, Houston, Texas

Teresa D. Bourget Skin Research Foundation of California, Santa Monica, California

Alain Chardon L'Oréal, Clichy, France

Miles R. Chedekel, Ph.D. Mel-Co®, Orland, California

Lisa E. Paloympis KATO Worldwide, Ltd./Nickstadt-Moeller, Mt. Vernon, New York

Madhu A. Pathak, B.Sc. (Hons.), B.Sc. (Tech.), M.B., M.Sc. (Tech.), M.S., Ph.D. Department of Dermatology, Harvard Medical School and Massachusetts General Hospital, Boston, Massachusetts

Sheldon R. Pinnell, M.D. Department of Medicine/Dermatology, Duke University Medical Center, Durham, North Carolina

Betty Rajan, M.D. Department of Dermatology, Texas Tech University Health Science Center, Lubbock, Texas

Nadim A. Shaath, B.Sc. (Hons.), Ph.D. KATO Worldwide, Ltd./Nickstadt-Moeller, Mount Vernon, New York

Robert M. Sayre, Ph.D. Department of Medicine, Division of Dermatology, University of Tennessee Center for Health Sciences, Memphis, Tennessee

Ian R. Scott Unilever Research U.S., Edgewater, New Jersey

Stephen R. Spruce, B.Sc. (Hons.) Tioxide Specialties, Ltd., Durham, England

Joseph W. Stanfield, M.S. Associate Research Director, Product Evaluation Laboratory, Schering-Plough Health Care Products, Memphis, Tennessee

James M. Wilmott Mary Kay Cosmetics, Inc., Dallas, Texas

Antony R. Young United Medical and Dental Schools of Guy's and St. Thomas's Hospital, University of London, London, England

Lisa Zeise, Ph.D. Mel-Co ®, Orland, California

Alexander P. Znaiden Chesebrough-Ponds U.S.A., Trumbull, Connecticut

Sunscreens

I

INTRODUCTION

1

Evolution of Modern Sunscreen Chemicals

Nadim A. Shaath
KATO Worldwide, Ltd./Nickstadt-Moeller, Mt. Vernon, New York

I. HISTORICAL

The first reported use of sunscreens in the world was in 1928, in the United States, with the commercial introduction of an emulsion containing two sunscreening chemicals, benzyl salicylate and benzyl cinnamate (1). In the early 1930s, a product containing 10% Salol (phenyl salicylate) appeared on the Australian market (2). In France, in 1936, a chemist named E. Schueller introduced the first commercial sunscreening product (3). In the United States lotions containing quinine oleate and quinine bisulfate appeared in 1935 (4). *p*-Aminobenzoic acid (PABA) was first patented in 1943, leading the way for the incorporation of several *p*-aminobenzoate derivatives in sunscreen formulations (4). During World War II red petrolatum was used by the U.S. military, which led to the extensive use of both physical and chemical ultraviolet filters after the war. The U.S. military specifications (5) issued on July 10, 1951, listed approved sunscreening compounds and their recommended concentrations, namely glyceryl PABA (3%), Escalol 75A (5%), 2-ethylhexyl salicylate (Sunarome WMO; 5%), digalloyl trioleate (3%), homomenthyl salicylate (8%), and dipropylene glycol salicylate (4%).

This chapter deals with the evolution of modern ultraviolet (UV) filters or sunscreen chemicals from early reported uses of salicylates and PABA derivatives to the current approved list of sunscreen chemicals in the United States,

Europe, Japan, South America, Australia, and New Zealand (6). The design of the current, as well as future, ultraviolet filters is an excellent example of classic structure–activity relationship (SAR) studies.

II. STRUCTURE-ACTIVITY RELATIONSHIP

The relation of chemical structure and efficacy of ultraviolet filters is clearly evident, as will be documented in this chapter. The evolution of the currently approved sunscreen agents was not necessarily through concerted and well-planned design, for contributors of these chemicals have been as varied as the chemicals themselves. Sunscreens have originated from both academic and industrial research laboratories with completely diverse uses. The cost, safety, and marketability of the new filters have had a dramatic influence on the evolution of the current approved list of sunscreen chemicals, regardless of their efficacy, degree, and nature of their protection.

In the United States, the so-called category I approved list of sunscreen chemicals (7) lists 21 UV filters, 19 of which are chemical absorbers and 2 of which are physical blockers/absorbers. This list reflects the state-of-the-art design of sunscreen chemicals dating back to its presence in any U.S. commercial product before 1972. Currently, any sunscreen supplier wishing to introduce a new UV filter must either go through a costly and time-consuming New Drug Application (NDA) to the U.S. Food and Drug Administration (FDA) or sit back hoping for the reopening of the monograph for additional sunscreen approvals before its anticipated adoption at the turn of the new century. Unlike medical drugs, for which the return on investment may be in the hundreds of millions of dollars, sunscreening chemicals do not afford substantial returns to the companies producing them. This substantial cost, estimated to be several million dollars, and a waiting period exceeding 3 years are prohibitive for most manufacturers embarking on both research designed to produce new and innovative UV filters, as well as the costly endeavor in obtaining an NDA. In Europe, the process is simpler, because European Economic Community (EEC) member countries have established a body called COLIPA (8) that effectively regulates the sunscreen industry. The cost in both time and the production of the necessary safety toxicological data is reasonable, allowing for more new introductions and innovations. Japan (9) and Australia (2) have similar, less costly measures for the adoption and introduction of new sunscreening agents.

The need for new and better sunscreen agents (responding to the recent medical advances and expanded demands of both the industry and the consumers ultimately using these products) is well documented (10). The more than half a million new cases of skin cancer reported each year in the United States alone (11), along with the new emerging evidence of the damaging effects of UVA rays (12) and the depletion of the ozone layer through the use of chlorofluorohydrocarbons (13), as well as for demographic considerations and the popularity of modern leisure outdoor lifestyles, are a few of the reasons the need for photoprotection (14) and the product of new, safe, more selective, specific, and effective UV filters is paramount.

The chemical UV filters in use throughout the world can be classified as derivatives of the following:

PABA and *p*-aminobenzoates
Salicylates
Cinnamates
Benzophenones
Anthranilates
Dibenzoylmethanes
Camphor derivatives
Miscellaneous chemicals

The compounds listed in this chapter to illustrate the evolution and the design process of sunscreening chemicals may or may not be commercially available or approved for use in cosmetic products. Chapters 11–14 list the regulatory status of any UV filter in question for use in the United States, Europe, Australia, and Japan.

A. PABA and *p*-Aminobenzoate Derivatives

p-Aminobenzoic acid (PABA), the first sunscreen patented from this family of UV filters, was a popular sunscreen chemical in the 1950s and 1960s. It has an absorption maxima at 296 nm and a moderate molar extinction coefficient of 13,600. Its chemical structure (**1**) reveals the presence of two reactive functional groups; namely, amino and carboxylic acid moieties, substituted in a *para* orientation on the benzene nucleus.

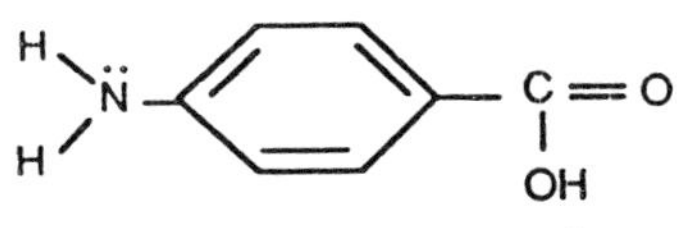

1

This particular configuration of an electron-releasing group (—NH_2) *para* to an electron-acceptor group (—COOH) allows the smooth electron delocalization shown in **2**:

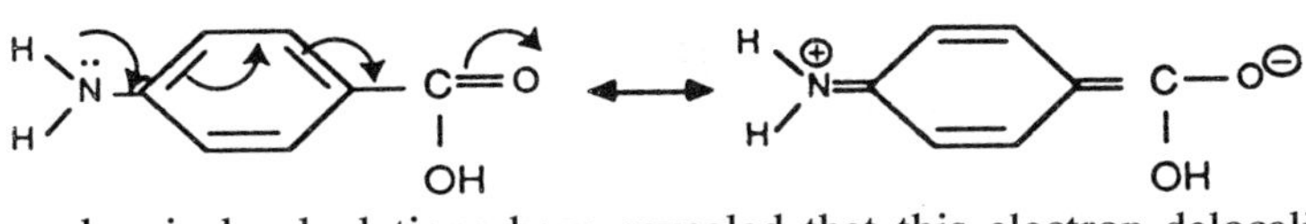

2

Quantum chemical calculations have revealed that this electron delocalization energy corresponds to the electronic transitions associated with the UVB region of the solar spectrum (15).

Unfortunately, the presence of these two extremely polar groups, the amine and the carboxylic acid, situated away (*para*) from one another contributes a number of problems that render the use of this product commercially less appealing, namely:

1. Free amines tend to oxidize rapidly in air and thereby produce off-colors.
2. Amines and acids are extremely polar groups that tend to hydrogen bond intermolecularly as shown by **3**:

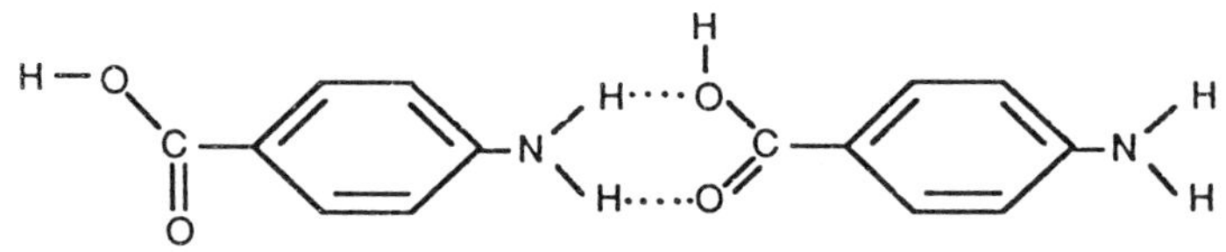

3

The intermolecular hydrogen bonding leads to the increased association of the molecules, thereby producing a crystalline physical state. This crystal structure poses various constraints on the liberal use of the product in cosmetic formulations. A suitable solvent will be required to ensure the rapid and continuous dissolution of PABA in the formulation. Improper dissolvation may yield gritty and crystalline finished formulations, which are both unappealing and prevent the formation of continuous films on the skin, a necessary condition for good sun protection.

3. The presence of both the polar amine and carboxylic acid group promotes the water solubility of the sunscreen chemical in the finished cosmetic formulation owing to the excessive hydrogen bonding with solvents (**4**).

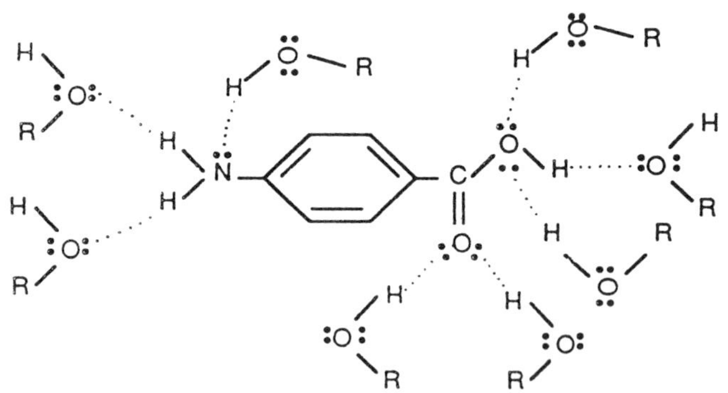

4

where R = —H (water), —C_2H_5 (ethanol), glycols, etc.

4. Because of excessive hydrogen bonding with solvents, a dramatic solvent effect is observed (16). This solvent effect will shift the λ_{max} by 27 nm from 293 nm in nonpolar solvents to 266 nm in polar solvents. Such solvent effects have an influence on the efficacy of the sunscreen agent in cosmetic formulations.

5. Carboxylic acids and amines are sensitive to pH changes.

In addition to the foregoing chemical limitations of PABA, several recent reports have cast doubt on its safety as a UV filter (17); hence, its worldwide decline as a sunscreen agent.

Researchers in the field of sunscreens responded to the consumer's need for better UV filters, and several sunscreens based on the strengths of the PABA moiety emerged, which include chemicals that address some of the problems just outlined.

1. A sunscreen was designed to protect the carboxylic acid grouping from pH changes and potential chemical reactions. Glyceryl PABA (**5**), a creamy white powder, is currently approved in both the United States and Europe (18).

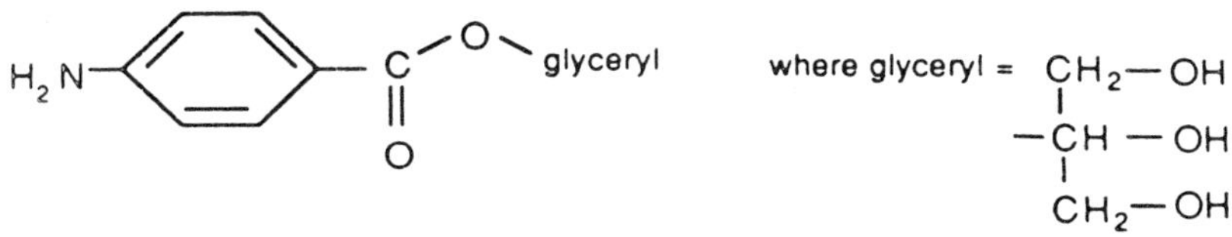

5

Unfortunately, even though the carboxylic proton is protected, the amine group has not been blocked. The compound is still a solid and more water-soluble (owing to the presence of the three hydroxyl groups in glycerol) than the parent PABA compound. Also, two possible isomers emerge from the reaction of glycerin with PABA (**6**). The ester linkage may be on the terminal carbon (major isomer) or the central carbon.

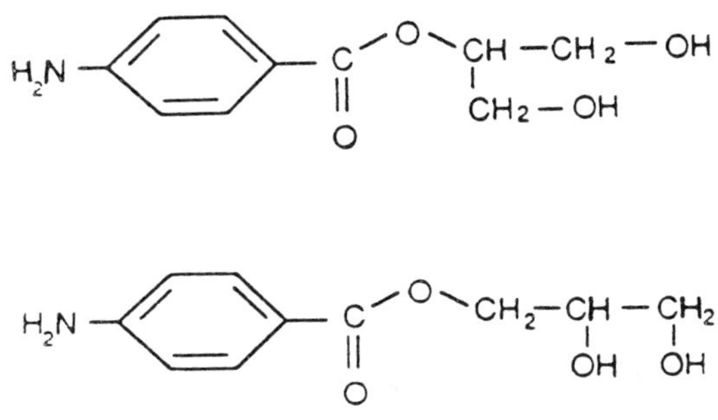

6

Other impurities found in this compound include glyceryl-di-*p*-aminobenzoate and traces of PABA and glyceryl tri-*p*-aminobenzoate.

2. A sunscreen was designed to eliminate both the protons in the carboxylic acid and the amine groupings. Ethyl-4-bis(hydroxypropyl)aminobenzoate is approved for use in both the United States and Europe (#2.1-COLIPA #S.2) and is sold under the trade name of Amerscreen P. The carboxylic

proton was protected with a two-carbon chain (ethyl) and the two amine protons were substituted with two hydroxypropyl groups (**7**).

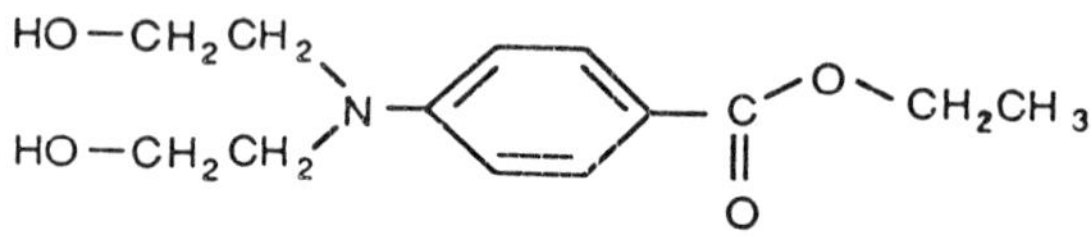

7

The drawbacks of this compound are several; namely, their partial water solubility (owing to the two added hydroxyl groups) and its relative insolubility in cosmetic formulations, which is primarily due to their isomeric impurity. Masse and co-workers (19) have reported the presence of four

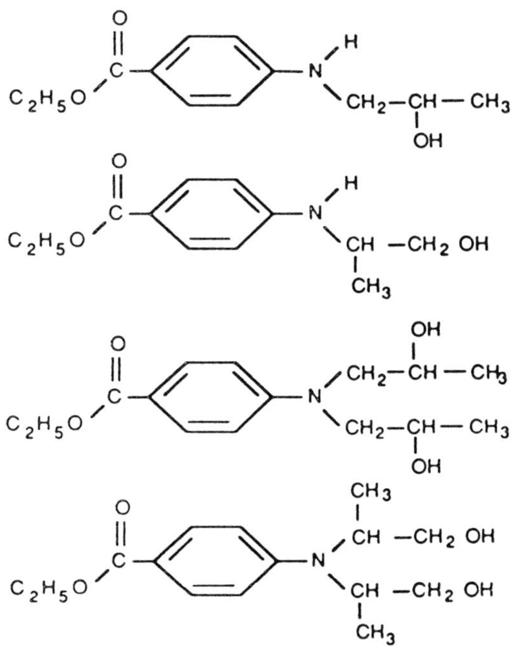

8

isomers (**8**) in the Amerscreen P product identified by nuclear magnetic resonance (NMR) spectroscopy and quantitatively measured by high-performance liquid chromatography (HPLC) in sunscreens.

3. Several sunscreens have successfully addressed the problem of protecting both the amino and carboxylic group (**9**):
   a. *N,N*-dimethyl PABA ethyl ester ($R{=}C_2H_5$)
   b. *N,N*-dimethyl PABA butyl ester ($R{=}C_4H_9$)
   c. *N,N*-dimethyl PABA amyl ester ($R{=}C_5H_{11}$)
   d. *N,N*-dimethyl PABA octyl ester ($R{=}C_8H_{17}$)

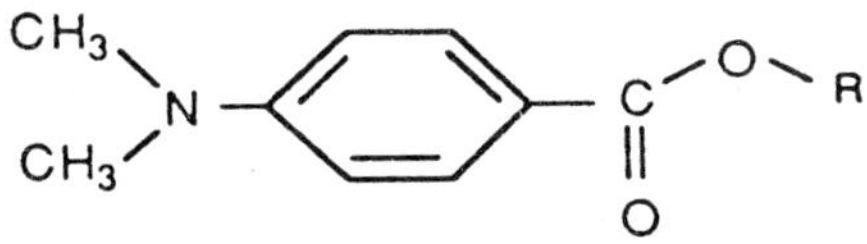

9

where $R = C_2H_5, C_4H_9, C_5H_{11}, C_8H_{17}$

The four compounds represent the best molecules that can capitalize on the parent PABA molecules and its ideal UVB sunscreening capability, yet diminish many of the problems associated with PABA itself. In fact, the fourth compound, also known as padimate-O, represents the ultimate in sunscreen design as the intermolecular association was decreased. This change in structure resulted in a UV filter that is a liquid instead of a crystalline solid, and the problems associated with the protons on the amine and the carboxylic acid group are eliminated. Its molar extinction coefficient is one of the highest found in a UVB filter, reaching 28,400 in polar solvents. The extinction coefficient is more than double that of PABA. Even though it is still subject to solvent effects, the shifts reported are from 300 nm in nonpolar solvents to 316 nm in polar solvents, values that are well within the UVB range. This chemical has enjoyed tremendous commercial exposure as reported in an FDA report on frequency of use in suntan preparations in the United States (20).

B. Salicylates

Salicylates were the first UV filters used in sunscreening preparations (1). Several of these derivatives have enjoyed substantial sales worldwide. These include (**10**) benzyl salicylate (COLIPA No. S-15), octyl salicylate (S-13), homomenthyl salicylate (S-12), triethanolamine salicylate (S-9), potassium salicylate (S-9), amyl salicylate (S-10), *p*-isopropylphenyl salicylate (S-15), and 4-isopropylbenzyl salicylate (S-16)

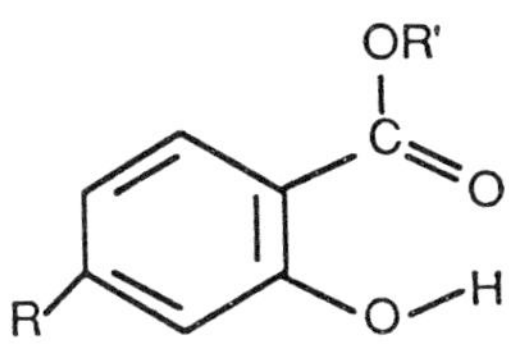

where R= H and R'= $-CH_2-C_6H_5$ *(S-15)*,
$-C_8H_{17}$ *(S-13)*,
-Homomenthyl *(S-12)*,
$-N^+(CH_2CH_2OH)_3$ *(S-9)*,
-K *(S-9)*, **10**
$-C_5H_{11}$ *(S-10)*,
$-CH_3-CH(OH)-CH_2-C_6H_5$ *(S-15a)*, and

R= $CH(CH_3)_2$ and R'= $-CH_2C_6H_5$ *(S-16)*

The salicylates, as a group, are *ortho*-disubstituted compounds, with a spatial arrangement permitting internal hydrogen bonding within the molecule itself (**11**) allowing a UV absorbance of about 300 nm.

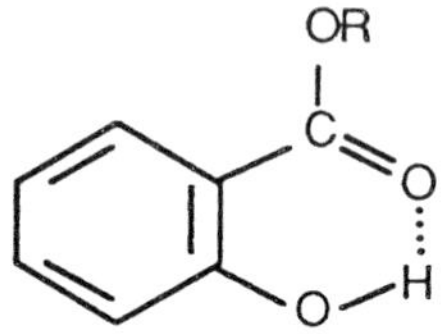

11

Normally, hydroxybenzoates in the *para*- or *meta*-position will exhibit a λ_{max} well below the 300-nm range, closer to 270 nm. Thus the parabens (**12**) have absorbances in about the 270-nm range.

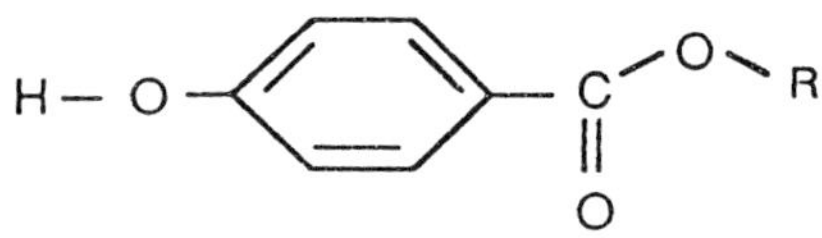

12

The reason for the unusually large λ_{max} difference of 30 nm or more in comparing the *ortho* compound (salicylates) with the *para* compound (parabens) is due to the previously mentioned intramolecular hydrogen bonding. The hydrogen bonding possible in the salicylates lowers the energy requirements for the compound to be promoted to its excited state (**13**):

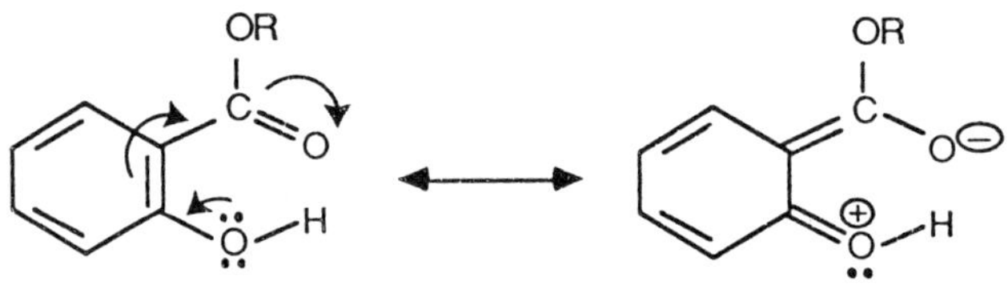

13

The *para*-disubstituted compound can achieve this electronic transition only through internal resonance delocalization unassisted further by the spatial proximity that is found in the salicylates and not in the parabens (**14**):

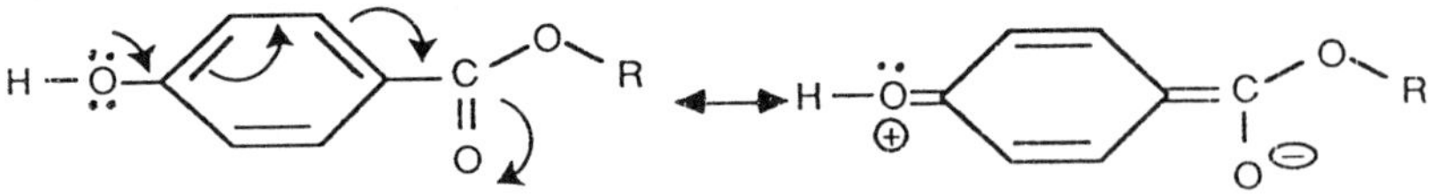

14

Thus, even though the salicylates have the ideal UVB sunscreen range of 300–310 nm, nevertheless, for precisely the same reason, namely, the *ortho* relationship, they have a much lower extinction coefficient. The *ortho* relationship of the phenolic group to the bulky carboxylic ester grouping causes crowding and strain on the molecule as a whole. To counterbalance this steric strain, the two groups deviate ever so slightly from planarity. Any minor deviation from planarity of the molecule causes a lowering of the extinction coefficient because the symmetry will dictate whether an electronic transition is allowed or forbidden. A discussion of such rules is beyond the scope of this chapter; however, the interested reader is referred to the many excellent monographs on the topic (21).

That there are eight salicylates considered for use in Europe and three on the U.S. FDA's category I list attests to the versatility of these compounds. Although they are relatively weak UV absorbers, they have an excellent safety record, and several salicylates, notably octyl salicylate, are used in many formulations owing to their ease of incorporation into cosmetic formulations, aesthetics, stability, emoliency, and non–water-solubility. Therefore, salicylates are excellent solubilizers of other traditionally nonsoluble cosmetic ingredients, such as the benzophenone class of UV absorbers. An explanation for their mildness, their stability, and their excellent safety record, despite their continued use for more than 50 years, stems from their *ortho*-disubstituted chemical structure. The two active groups, the hydroxyl and the carboxyl ester groups, are intramolecularly hydrogen bound to one another; hence, their electrons are less available for interaction with other ingredients or with biological substrates found on the skin.

Similarly, salicylates are not subject to solvent effects, owing to the unavailability of the hydroxyl and carbonyl groups to interact with the surrounding solvent molecules. Thus, the behavior of salicylates in all media, polar or nonpolar, is extremely stable and, consequently, is predictable.

Of all the salicylates on the market today, the one that enjoys the most exposure and is found in numerous formulations is 2-ethylhexyl salicylate, also called octyl salicylate. This UV filter incorporates the 2-ethylhexyl grouping found in several of the leading sunscreens: 2-ethylhexyl *p*-methoxycinnamate, 2-ethylhexyl dimethyl *p*-aminobenzoate, and 2-ethylhexyl-2-cyano-3,3-diphenyl acrylate. The presence of this eight-carbon unit ensures its insolubility in water and other aqueous formulations.

Homomenthyl salicylate, also called homosalate or HMS, is also extremely popular in sunscreen formulations in the United States. This compound has a nine-carbon unit attached to the carboxylic acid moiety of the salicylate molecule. However, this nine-carbon unit, called homomenthyl or 3,3,5-trimethylcyclohexyl, can exist in two possible isomeric forms, the *cis* and the *trans*, **15**.

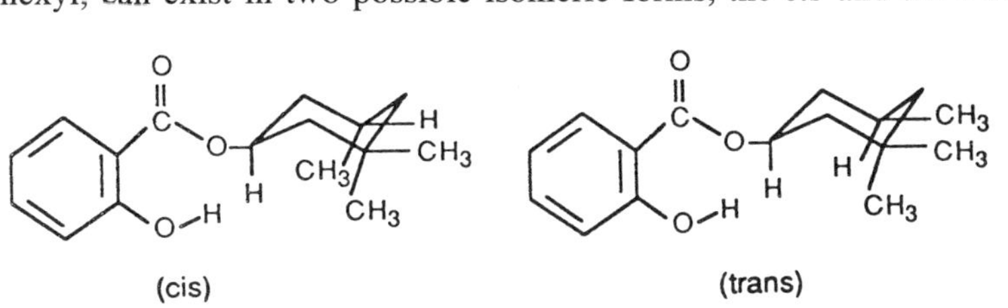

15

Hence, most commercially available homosalates are mixtures of two isomers, the *cis* and the *trans*. Typical mixtures contain 15% *cis* and 85% *trans*, although other mixtures of 40% *cis* and 60% *trans* were also reported.

In fact, the ethyl or methyl esters have a degree of water solubility that renders them less reliable in waterproof formulations. The presence of the isopropyl groupings, in both the *ortho-* and *para-*positions, is certainly a step in the right direction, as alkyl groups [—$CH(CH_3)_2$] are slightly electron-releasing through hyperconjugation. Electron-releasing groups in the *ortho-* or *para-*position assist in the electron delocalization necessary for the UV absorption of the harmful rays centered at about 305 nm. Compounds 6, 7, and 8 (see Table 1) all are used in water-soluble applications with the diethanolamine salt (No. 8) enjoying appreciable sales, mostly for use in hair preparations. Compounds 9–14 are all attempts to decrease the water solubility. They all contain the electron-releasing group OCH_3 in the proper *para-*position; hence, they are all superior to the unsubstituted counterparts. As in the PABA derivatives, the ideal sunscreen is the octyl derivative, with the cyclohexyl derivative being a distant second in its effectiveness. To my knowledge, none of these derivatives are being used to any degree, with the exception of the octyl derivative.

The last three cinnamate derivatives, 15, 16, and 17 (see Table 1), are all cyano-substituted cinnamates. None enjoys any significant exposure, with the exception of No. 16, which has the added advantage of COLIPA approved and U.S. FDA category I status. More recently developed sunscreen formulations have incorporated ethylhexyl-α-cyano-β-phenyl cinnamate, also known as octocrylene, in their product.

D. Benzophenones

Unlike other sunscreening agents available, the benzophenones are the only class of compounds that belong to the aromatic ketone category (24). Dibenzoylmethane derivatives, even though they are arylalkyl ketones, exhibit their UV-filtering effect through a keto–enol tautomerism not possible in benzophenones. Resonance delocalization, as in all the other classes of compounds discussed earlier, is aided by the presence of an electron-releasing group in either the *ortho-* or *para-*position or both. The electron-accepting group, in this case, the carbonyl group itself, participates in the resonance delocalization process, **17**:

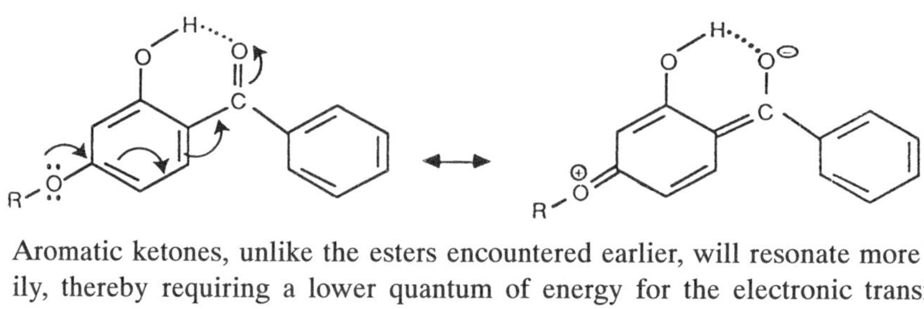

17

Aromatic ketones, unlike the esters encountered earlier, will resonate more easily, thereby requiring a lower quantum of energy for the electronic transition required. Since energy (ε) is inversely proportional to the wavelength (λ), as derived from the following familiar energy relationship:

$$E = \frac{hc}{\lambda}$$

it then follows that a longer wavelength is required for such electronic transitions; hence, the values exceed 320 nm. This can be seen in the following diagram contrasting a typical aromatic ester with an aromatic ketone with similar substituents (**18**):

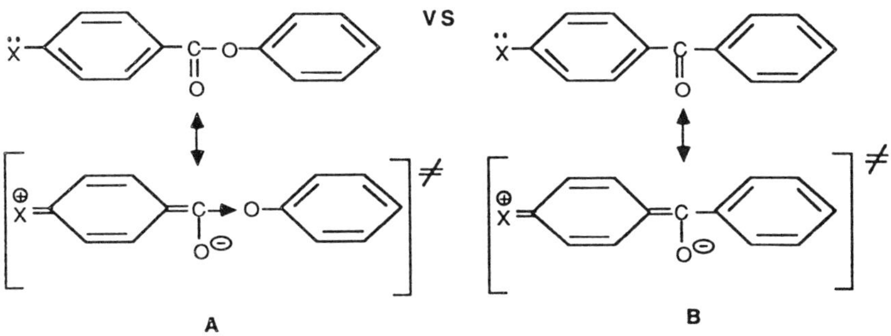

18

If you observe the transition state (A) for esters, you will note the presence of an oxygen atom that is destabilizing this transition, owing to its electron-withdrawing capability. To overcome this destabilization, extra energy is required which, in turn, means a lower wavelength (λ_{max}); hence, most esters have a λ_{max} of less than 310 nm, unless exceptional situations prevail in the molecule, as occurs in anthranilates, which are discussed later in this chapter.

The drawback in using benzophenones as UV filters is due to various factors. (1) Aromatic ketones are chemically different from esters. Metabolically, esters can be hydrolyzed in vivo, rendering by-products that the body can metabolize (a detoxification mechanism). It has been reported that patients developed statistically more allergic reactions to oxybenzone than PABA (25). Aromatic ketones, on the other hand, do not have as many handles that accelerate the detoxification process. (2) These products are, without exception, solids that are generally difficult to handle and to solubilize in cosmetic formulations. Benzophenone-8, for example, is notorious for its poor solubility in standard cream or lotion formulations. (3) Their maximum absorption generally is less than 330 nm, which is a value barely into the UVA region—the main reason for the use of benzophenones. (4) Some of these derivatives exhibit unusually large λ_{max} shifts in different solvents. For example, dioxybenzone (benzophenone-8) exhibits a λ_{max} shift of 26 nm, with a value of 326 nm in polar solvents versus a value of 352 nm in nonpolar solvents reported (16).

Of the 12 benzophenones with European COLIPA numbers (18), only three are approved in the U.S.-FDA category I listing; namely, oxybenzone, dioxybenzone, and sulisobenzone. Oxybenzone (S-38) is on the EEC Council Directive of permitted UV filters. Three other benzophenones are on the provisionally accepted list in Europe (8); they are sulisobenzone (S-40), mexenone (S-39), and 2-ethylhexyl-4-phenyl benzophenone,2-carboxylate (S-42).

The structures of the five benzophenones that are currently approved for use in the United States or Europe are shown in **19**.

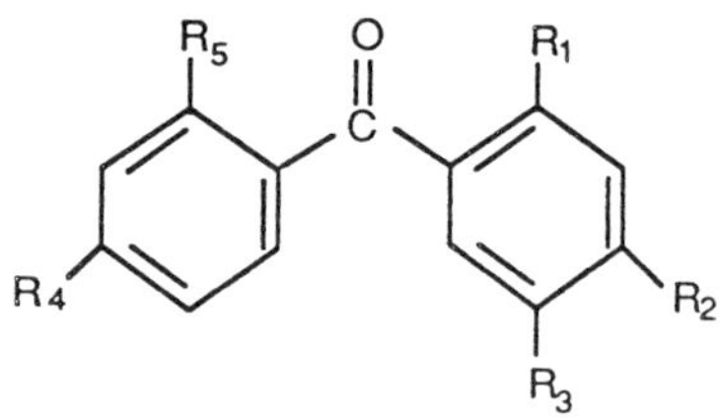

19

where:

1. oxybenzone (S-38); R_1=OH, R_2=OCH_3, R_3=R_4=R_5=H
2. dioxybenzone; R_1=OH, R_2=OCH_3, R_3=R_4=H, R_5=OH
3. sulisobenzone (S-40); R_1=OH, R_2=OCH_3, R_3=SO_3H, R_4=R_5=H
4. mexenone (S-39); R_1=OH, R_2=OCH_3, R_3=H, R_4=CH_3, R_5=H
5. 2-ethylhexyl-2-(4-phenyl benzoyl)benzoate (S-42); R_1=OC_8H_{17}(iso), R_2=R_3=R_4=H, R_5=-C_6H_5

E. Anthranilates

Anthranilates, or *o*-aminobenzoates, are an interesting class of UV filters. Two such compounds are used in commerce today, homomenthyl-*N*-acetylanthranilate, approved in Europe (EEC No. 2.7 and COLIPA No. 5.18), and menthyl anthranilate, which is on the U.S.-FDA category I listing (see **20** for their chemical structures).

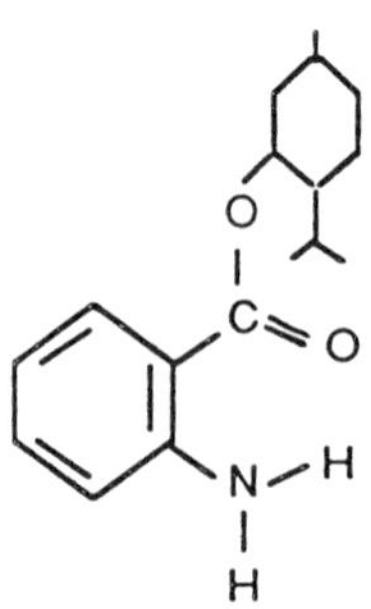

menthyl anthranilate

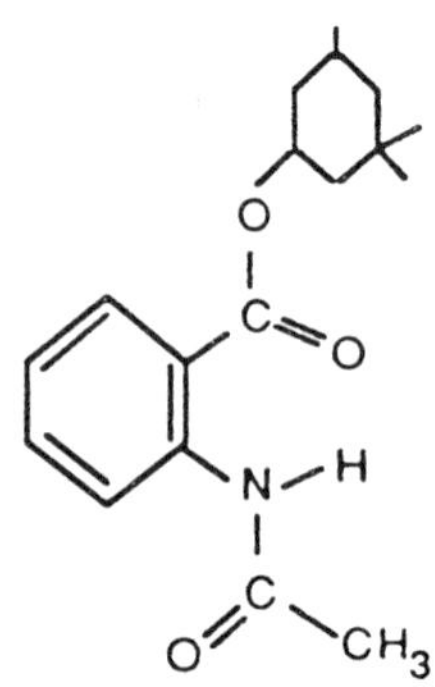

homomenthyl-n-acetyl anthranilate

20

This class of compounds offers an elegant example on the effect of chemical structure on UV absorbance characteristics. This effect, termed here the *ortho effect*, has been observed in numerous organic compounds, as is illustrated in Table 2.

Menthylanthranilate, an *ortho*-disubstituted aminobenzoate, has a λ_{max} of 336 nm, whereas octyl PABA, a *para*-disubstituted aminobenzoate has a λ_{max} of only 298 nm, **21**.

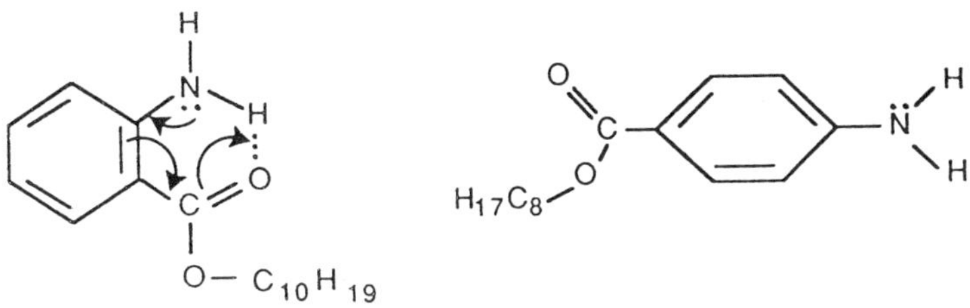

21

λ max 336 nm　　　　λ max 298 nm

This dramatic 38-nm shift in the maximum absorption is clearly due to the ease in electron delocalization (recall, lower energy translates into a longer wavelength) in the *ortho*-disubstituted compounds for which the geometry allows for this "through space" assistance. This situation is analogous to that for salicylates, for which the *ortho*-disubstituted phenolbenzoate has a λ_{max} of 308 nm and parabens, the *para*-disubstituted phenolbenzoate has a λ_{max} of 260 nm.

Unfortunately, the same situation is also true here; namely, that the molar extinction coefficient of the anthranilates is much lower than that of the *p*-aminobenzoate. Again, the steric crowding in the *ortho*-disubstituted compound causes the molecule to deviate from coplanarity, thereby reducing the intensity of the absorbance.

Anthranilates, as with salicylates, are stable and safe compounds to use owing to the *ortho*-disubstituted relationship and, as in salicylates, do not exhibit any significant solvent effects in cosmetic formulations (16).

Table 2 The λ_{max} of *ortho*- and *para*-Disubstituted Compounds

ortho-Disubstituted compounds	Wavelength (nm)	*para*-Disubstituted compounds	Wavelength (nm)
o-Nitrophenol	351	*p*-Nitrophenol	317
o-Nitroaniline	412	*p*-Nitroaniline	381
o-Hydroxybenzaldehyde	325	*p*-Hyroxybenzaldehyde	282
o-Hydroxyacetophenone	329	*p*-Hydroxyacetophenone	276
o-Hydroxybenzoic acid	301	*p*-Hydroxybenzoic acid	251
o-Aminobenzoic acid	327	*p*-Aminobenzoic acid	284

F. Dibenzoylmethanes

Dibenzoylmethanes, or substituted diketones, are a relatively new class of UV filters. Only one has now received limited approval for use in the United States, whereas three are approved for use in Europe. At least two others are available, but do not have approvals anywhere in the world as yet. The five dibenzoylmethanes, their chemical structures, and their λ_{max} are listed, **22**.

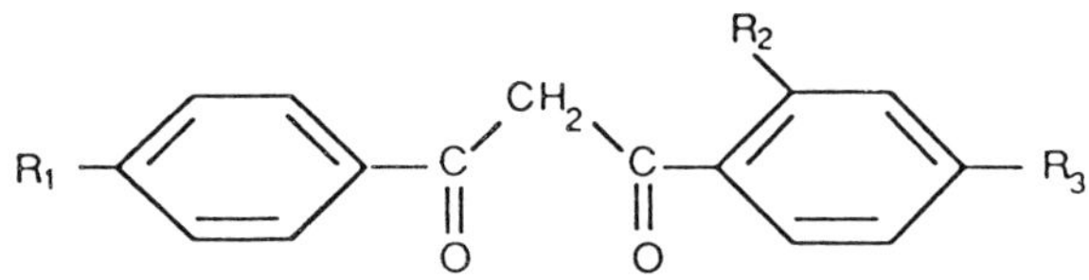

where: **22**

1. R_1= t-C_4H_9, R_2=H, R_3=OCH_3 ; (S-66); λ max=355nm
2. R_1=i-C_3H_7, R_2=H, R_3=H ; (S-64); λ max=345nm
3. R_1=R_3=OCH_3, R_2=H; (S-52); λ max=363nm
4. R_1=t-C_4H_9, R_2=OH, R_3=H; λ max=370nm
5. R_1=t-C_4H_9, R_2=OH, R_3=OCH_3; λ max=373nm

This group of UV filters exhibits properties resulting from a keto–enol tautomerism. The keto form of these compounds actually has a λ_{max} of about 260 nm. However, the enol form has the observed λ_{max} with values exceeding 345 nm, making them suitable candidates for UVA filters.

This keto–enol tautomerism is shown in structure **23**.

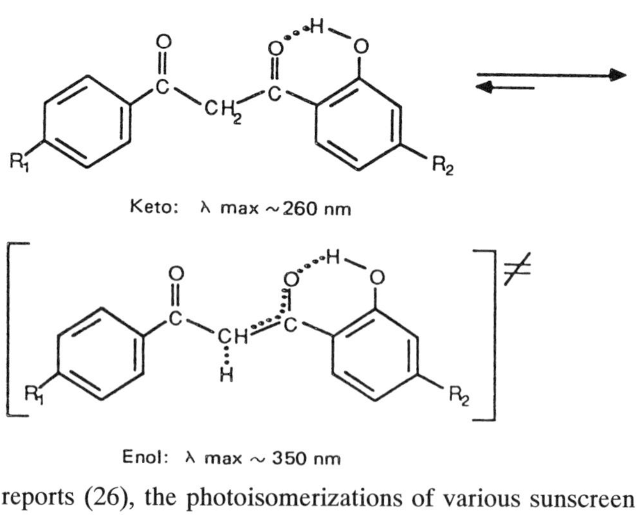

23

Enol: λ max ~ 350 nm

In several reports (26), the photoisomerizations of various sunscreen chemicals have been listed. The three approved dibenzoylmethane UVA filters had excessively high photoisomerizable compounds from an irreversible disappearance, which results in a significant loss of protective power. Thus, compound 1, also known as Parsol 1789, had a percentage irreversible loss of 36%; compound 2, also known as Eusolex 8020, had a loss of over 40%; and compound 3, also known as Parsol DAM, had a loss of 26%. Interestingly, compounds 4 and 5 had irreversible losses of only 3.6 and 5%, respectively.

Even though no explanation was offered, it is obviously due to the presence of the hydroxyl group in the *ortho*-position to the carbonyl side chain (analogous to the salicylate situation). The *ortho* relationship has two effects. The first is undesirable, since it yields a compound with a λ_{max} of more than 370 nm, a value too high to afford any significant protection from the more damaging lower UVA rays (320–350 nm; 12). The second is extremely helpful in reducing the percentage loss caused by the irreversible photoisomerization of these types of molecules. The *ortho*-hydroxyl group will shift the equilibrium of this keto–enol tautomerism clearly away from the keto form, which is mostly responsible for this irreversible loss of activity.

This *ortho*-effect in the dibenzoyl methane derivatives is shown in **24**.

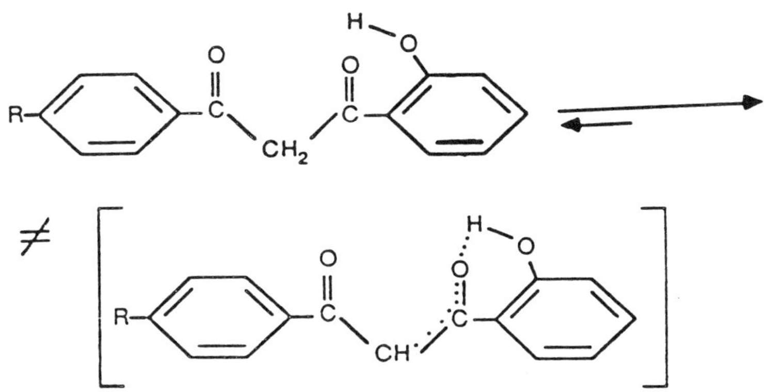

24

The dibenzoylmethane derivatives have exceptionally high molar extinction coefficients (>30,000). However, they suffer from possessing low photostability. Designing compounds to lower their photostability has resulted in dibenzoylmethane derivatives with higher than desired wavelength protection (over 370 nm).

G. Camphor Derivatives

Six bicyclic compounds (**25**) are approved for use in the EEC member countries. None are approved in the United States.

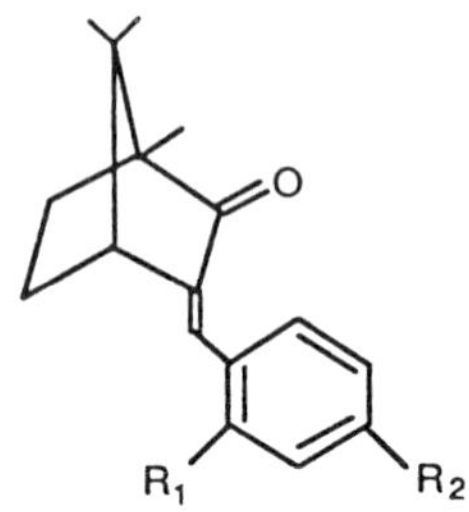

where: 25

1. α - (2-oxoborn-3-ylidene)-p-xylene-2-sulfonic acid (2.23, S-58); R_1=SO_3H, R_2=CH_3; λ max= 296 nm
2. α - (2-oxoborn-3-ylidene)-toluene-2-sulfonic acid (2.23, S-59); R_1=H, R_2=SO_3H; λ max= 296 nm
3. 3-(4-methylbenzylidene)-bornan-2-one (2.25, S-60); R_1=H, R_2=CH_3; λ max= 300 nm
4. 3-Benzylidenebornan-2-one (2.26,S-61); R_1=H, R_2=H; λ max= 290 nm

The two other bicyclic derivatives approved for use in Europe are:

5. *N,N,N*-Trimethyl-4-(2-oxoborn-3-ylidenemethyl) anilinium methyl sulfate (**26**; No. 1.2, S-57)

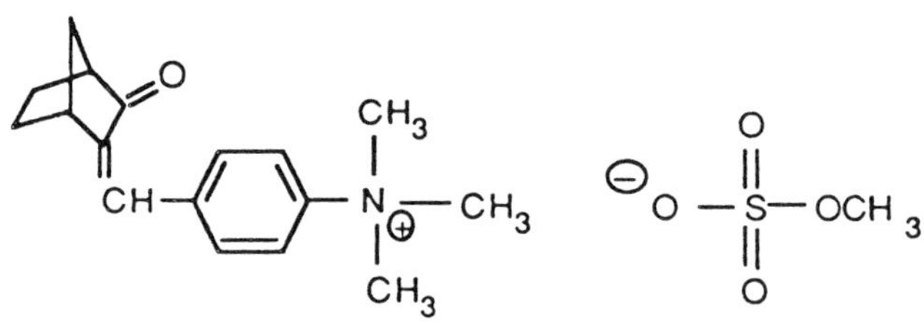

26

λ max = 288 nm

6. 5-(3,3-Dimethyl-2-norbornylidene)-3-pentene-2-one (**27**; No. 2.2, S.56)

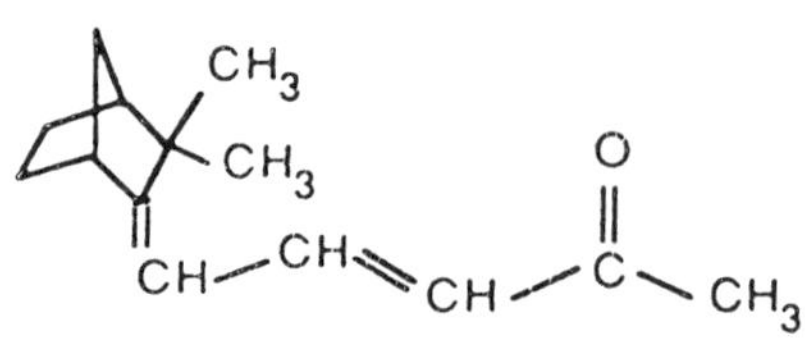

27

λ max = 300 nm

All of the foregoing bicyclic derivatives are solids, with the exception of No. 6, which is an oily liquid. These compounds have a high molar extinction coefficient generally above 20,000 and absorb in the UVB range of 290–300 nm. They all owe their photoabsorptivity to this resonance delocalization, **28**;

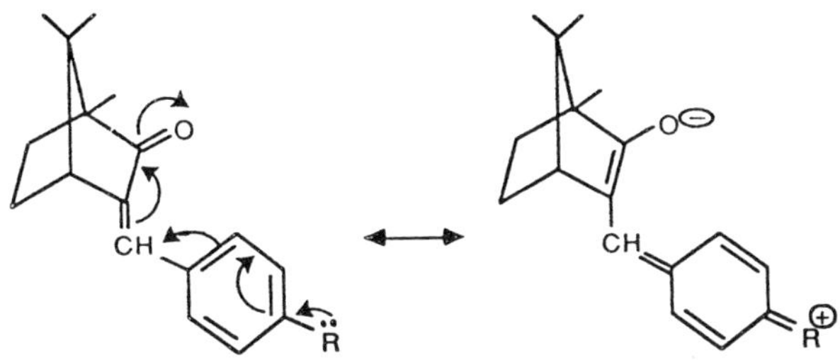

28

Beck et al. (23) have reported on the photostability of the benzylidene camphor derivatives and have found them to be extremely stable to photodegradation with losses generally below 2.5%.

H. Miscellaneous Compounds

Under this category several interesting sunscreening chemicals are found that were not listed under the previous classifications:

U.S.A FDA Category I Listing of Approved Sunscreens

1. Digalloyl trioleate, **29**

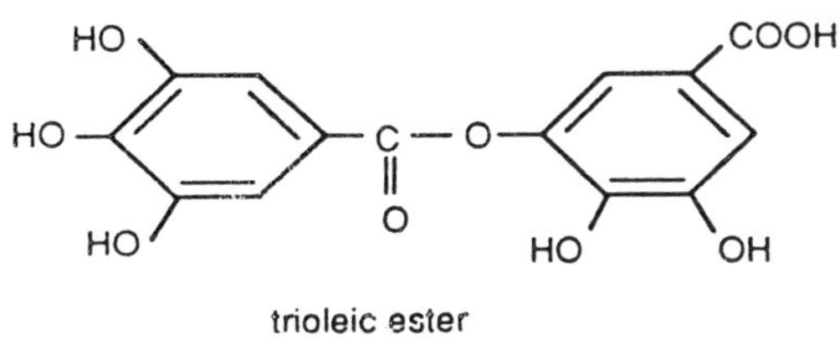

29

   This chemical has had no major exposure in the United States. It is approved at a dosage level of 2–5%.
2. Lawsone with dihydroxyacetone, **30**

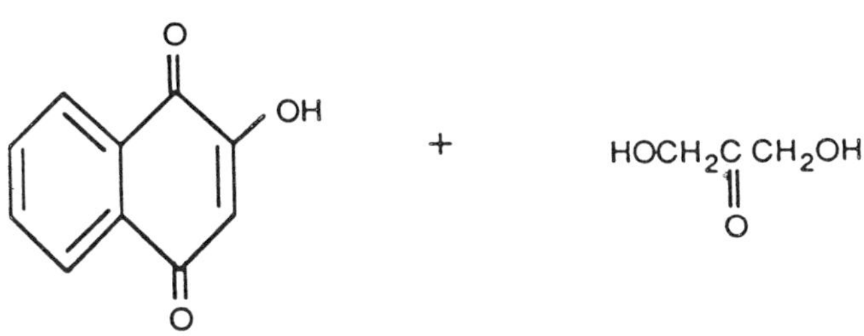

30

Dihydroxyacetone (DHA) has been used as an artificial tanning accelerator, since it is reported to impart a dark, yellowish color to the skin. Lawsone is in this mixture to act as a barrier to UV rays. This mixture is rarely used anymore.

3. 2-Phenylbenzimidazole-5-sulfonic acid, **31**

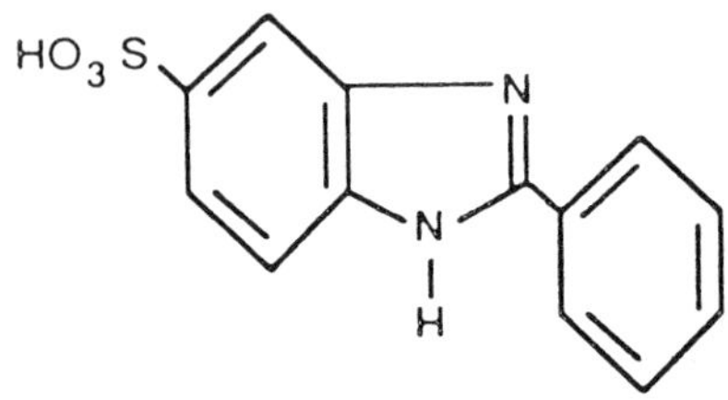

31

This chemical has limited usefulness in the United States. It is an extremely high-melting white powder, with water solubility. It is approved in the United States at a dosage level of 1–4%. It is also on the EEC list of permitted UV filters. (No. 1.6, COLIPA No. S.45.) Its λ_{max} is about 308 nm.

The EEC List of Approved UV Filters

1. 2-Phenylbenzimidazole-5-sulfonic acid (No. 1.6, S-45) discussed earlier.
2. 3-Imidazol-4-ylacrylic acid and ethyl ester (No. 1.5, S-46), also called urocanic acid (**32**). It has a low λ_{max} of about 285 nm.

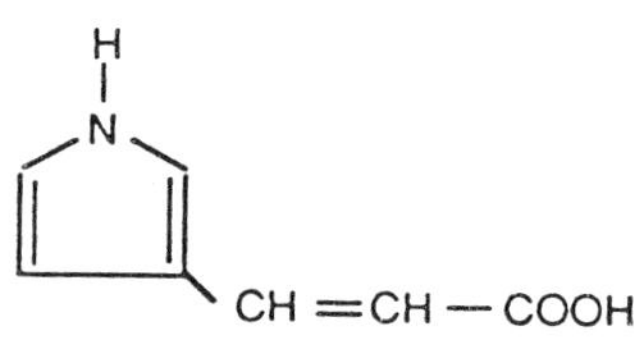

32

The EEC Provisionally Accepted UV Filters

1. 5-Methyl-2-phenylbenzoxazole (**33**; No. 2.19, S-47)

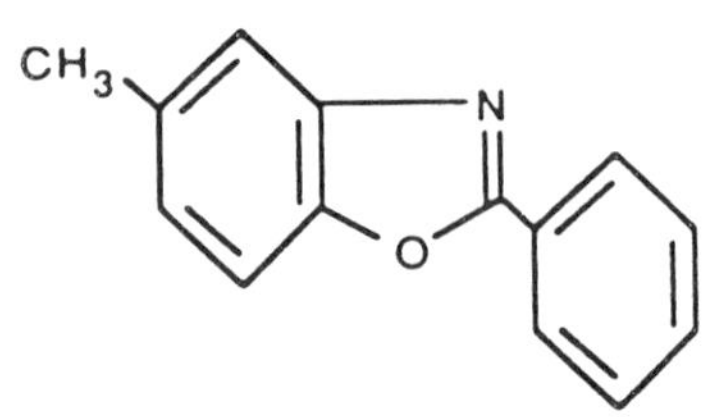

33

It is a solid, insoluble in water, with a λ_{max} of about 305 nm and allowed at a maximum concentration of 4%.

2. Sodium 3,4-dimethoxyphenylglyoxylate (**34**; No. 2.20, S-50)

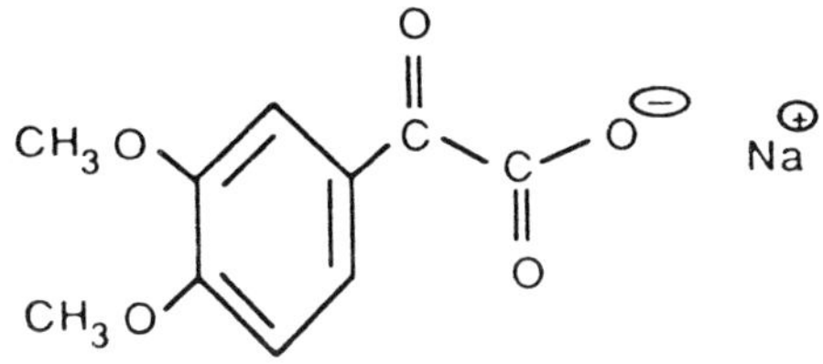

34

It is a solid, highly soluble in water, with a λ_{max} of about 275 nm and a weaker absorption at 305 nm and is allowed at a maximum concentration of 5%.

Other Chemicals Originally Considered by COLIPA

1. 2,2′-Hydroxy-5-methylphenyl benzotriazole (**35**)

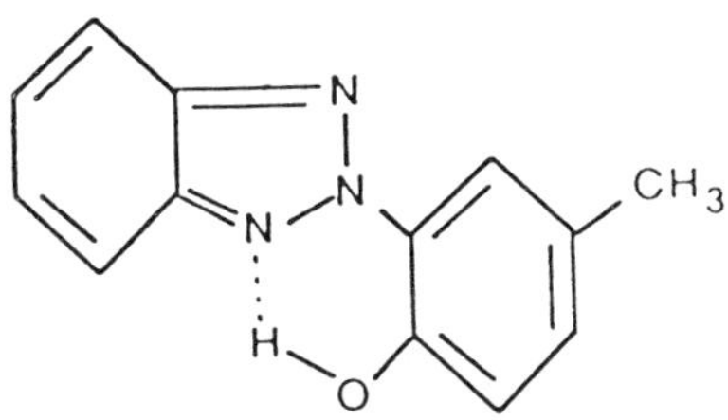

35

This chemical is currently sold for noncosmetic uses. It is a crystalline powder that is insoluble in water and has a λ_{max} of about 340 nm and 295 nm. Note the interesting *ortho* effect in this molecule. The *t*-octyl derivative in the 5-position is also available as a noncosmetic sunscreen.

2. Dibenzalazine (**36**)

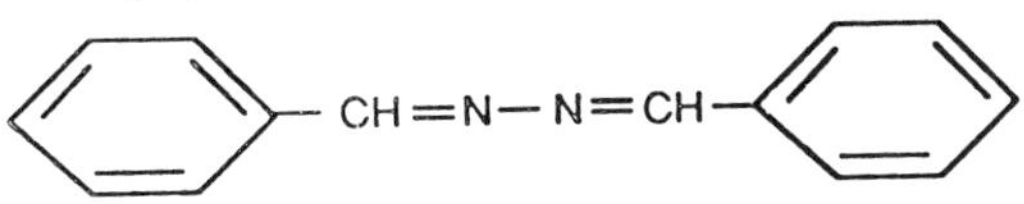

36

This compound, also called benzalazin or dibenzaldehydrazine, has a λ_{max} of about 302 nm. Note that the extended conjugation throughout the molecule causes this chemical to serve as a UVB filter.

III. NEW SUNSCREEN CHEMICALS

In reviewing the literature of the last 3 years, several new UV filters have been reported. The following review is not intended to be an exhaustive listing of work completed or in progress, rather it is meant to report on the more important discoveries made today, which may be the sunscreens of the future.

The sunscreening compositions reported in the literature over the last 3 years can be subdivided into new chemical UV filters and new forms of physical blockers.

A. New Chemical UV Filters

Several patents for modifications of the benzophenone molecule for use as UV filters have appeared and these include three Japanese patents by the Kao Corporation. The first patent (27), issued on June 22, 1987, includes the following derivatives (**37**):

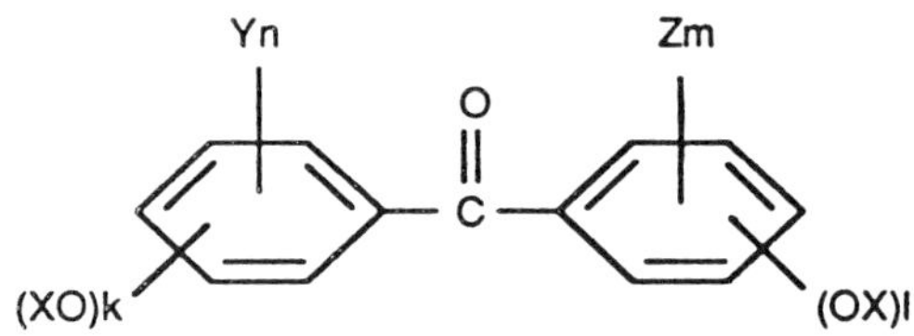

37

where

X = alkali metal ion, ammonium ion, or amino acid organic cation; Y and Z = OH, C_{1-24} alkyl, alkoxy, sulfonic acid, or alkali metal salt; n and m = 0–3; k + 1 = 1–4.

The second patent (28), issued on June 22, 1987, includes the phenolic derivatives of the foregoing compound.

The third patent (29), issued on August 8, 1987, describes the same compositions bound to montmorillonite-type clay mineral–metal oxide complex.

The Sogo Yakuko Co. in Japan has received a patent for their thiazoline derivatives (**38**) that absorb UV radiation with medium and long wavelengths and also inhibit tyrosinase activity involved in the formation of melanins from tyrosine (30).

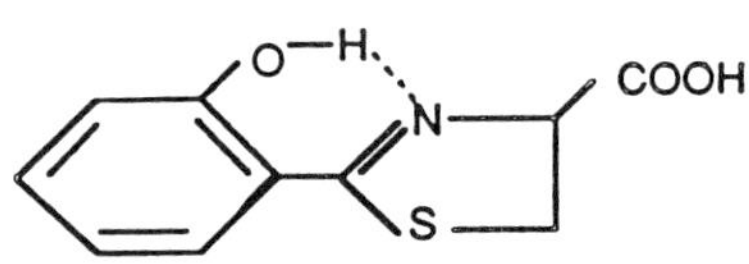

38

A German patent, issued to the Bayer A.-G. company describes the use of several indoline derivatives (**39**) useful as UVA filters, with λ_{max} ranging from 330 to 345 nm (31).

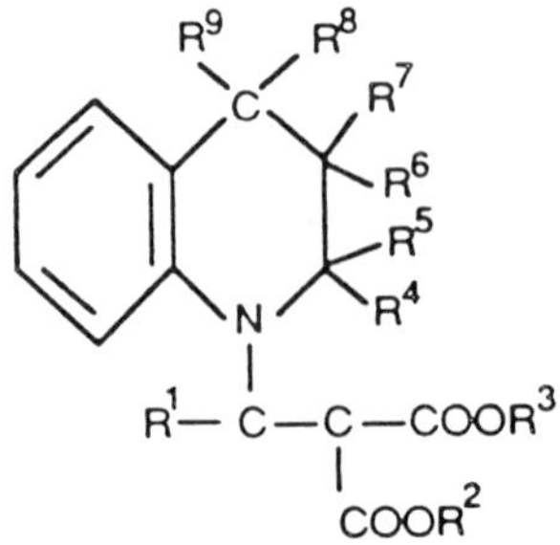

39

where:

$R' = H$, C_{1-8}-alkyl; R_2, R_3 = C_{1-8}-alkyl; R_4–R_9 = H, C_{1-8} alkyl, C_{5-6} cyclo alkyl, phenyl-C_{1-4}-alkyl; R_5 R_6 may complete another C_5–C_6 ring; R_{10} = H, ≥1 C_{1-8} alkyl, alkoxy, C_{5-6} cyclo alkyl, phenyl-C_{14}-alkyl, carb-C_{1-8}-alkoxy ester, halo; n = 0, 1.

Another German patent, issued to the Kao Corporation, offers several chalcone derivatives (**40**) useful as UVA filters (max absorption 350 nm) (32).

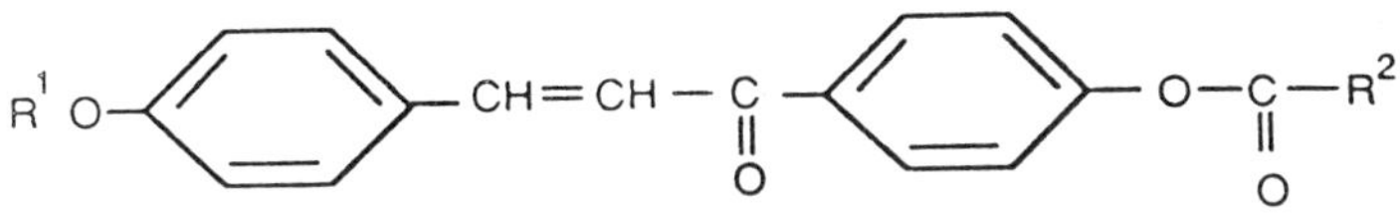

40

where:

R_1 = H, CH_3

R_2 = C_{2-24} aliphatic hydrocarbyl derivatives

A more recent patent, issued to Oreal S.A., lists various chalcone derivatives that are useful for the protection of skin and hair against UV radiation (33). A sunscreen containing chalcone, flavone, or coumarin derivatives, or a mixture thereof, offering broad-spectrum protection was the subject of a Japanese patent (**34**).

L'Oreal has reported on a new UVA sunscreen agent that is presumed to have excellent photostability and photochemical inertness (22). This compound is of the benzylidene camphor type and has a λ_{max} of 345 nm. The chemical is 3,3′-(1,4-phenylenedimethylidene) bis [7,7-dimethyl-2-oxo-bicyclo (2.2.1) heptane-1-methane sulfonic acid] sodium salt (**41**).

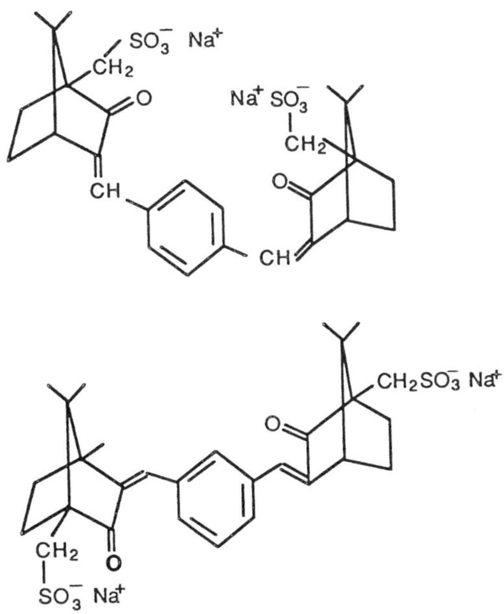

41

Sunscreens containing acrylamide polymers of benzylidene camphor, benzotriazole, or dibenzoylmethane derivatives (**42**) were assigned to Oreal in a German patent (35).

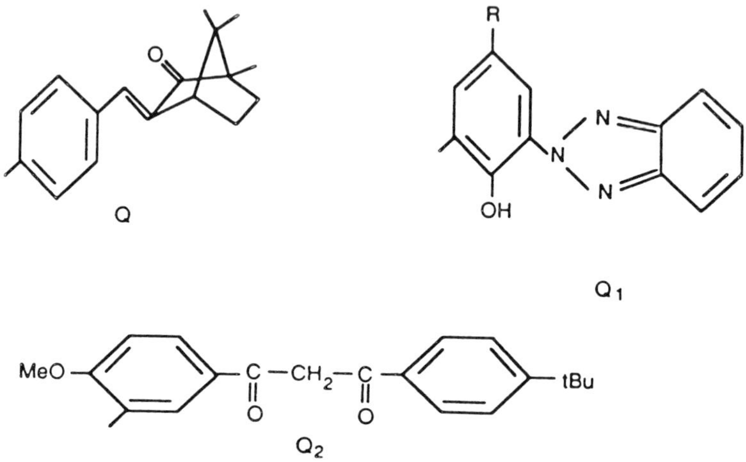

42

The sunscreen contains the repeating units [$CH_2CH(CONHCH_2X)$] where:

X = Q, Q_1, Q_2; and R = CH_3, *tert*-octyl

The title compounds were polymerized in isopropanol in the presence of H_2O_2 and ascorbic acid.

Italian scientists from the University of Pisa have reported on two series of quaternary ammonium UVB sunscreens for the hair (**43**).

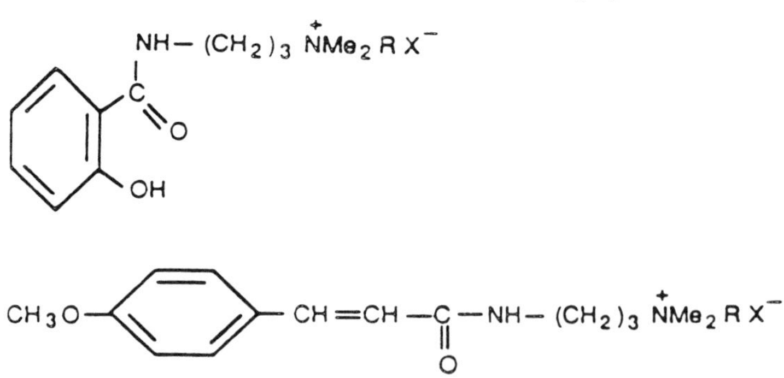

43

where R = ethyl, C_8H_{17}, $C_{12}H_{25}$, or $C_{16}H_{33}$, and X = Cl, Br, or I.

The compounds were subjected to substantivity tests on animal wool keratin and microbiological tests. The presence of a quaternary ammonium ion confers on the sunscreen molecule the capacity to adhere to and be sorbed by a keratin substitute (36).

α-Dehydroamino acid derivatives are disclosed as UV filters for cosmetics in a Japanese patent assigned to Ajinomoto Co., Inc. (37).

In a patent assigned to Takasago, the claim is made that 3,4-methylenedioxyphenol or its esters are tyrosinase inhibitors and that cosmetics containing this material inhibit melanin formation and prevent sunburn and darkening of the skin (38).

A number of patents on derivatives of dibenzoylmethane have been issued. Two granted to the Kao Corporation involve the avid (39) and amide (40) derivatives of the parent compound (**44**):

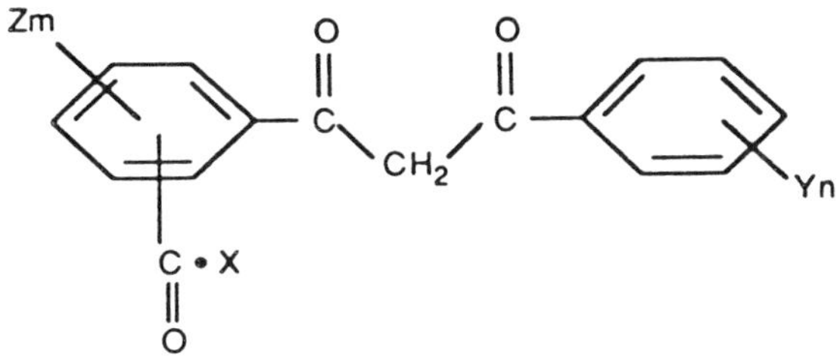

44

where

$X = OH, NR^1R^2$
$Y, 2m = OH, C_{1-24}$, alkyl, alkoxy, or polyoxyalkyene
$n, m = O\text{-}3$
$R^2, R^2 = C_{1-24}$ linear or branched aliphatic hydrocarbonyl

A patent issued to Shiseido Co., Ltd. (41) reports on sunscreens containing 4-(1,1-dimethylethyl)-4′-methoxydibenzoylmethane and long-chain aliphatic acid esters.

A Japanese patent issued to Pola Chemical Industries, Inc. describes the use of ferulic acid amides (**45**) as sunscreening agents (42), and claims the compound inhibits inflammation and darkening of the skin,

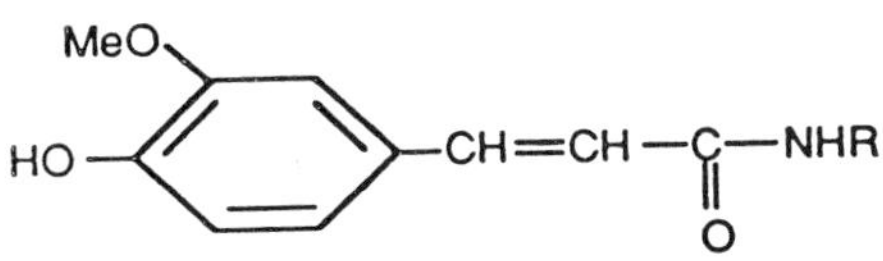

45

where R = H, or linear, branched alkyl or alkenyl.

The preceding compound is claimed to inhibit skin inflammation and darkening.

5-Alkylidene-3-pyrrolin-2-one (**46**) was evaluated for its UVA-filtering capability, as it has a λ_{max} of 346 nm (43).

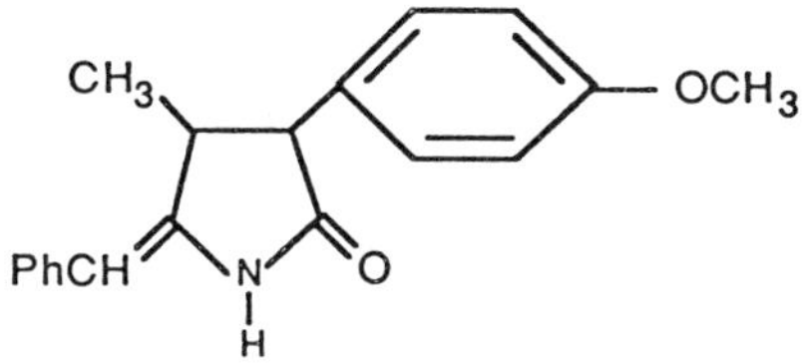

46

A European patent describing the sunscreening properties of dimethylaminocinnamate derivatives (**47**) was issued to Graesser Laboratories, Ltd. (20, 44).

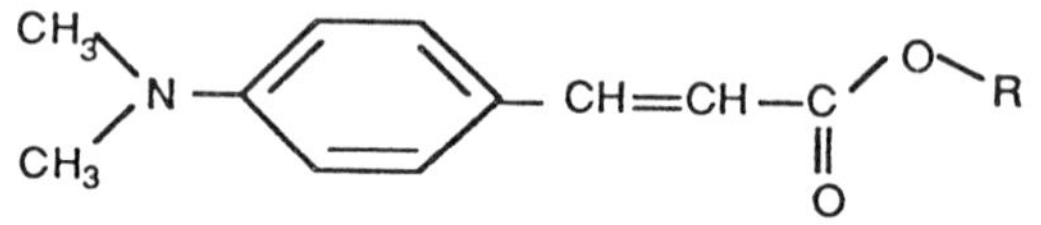

47

where R= 2-ethyl hexyl

Amide derivatives of PABA, hydroxybenzoic acids, cinnamic acids, urocanic acids, and benzimidazoles have UV-absorbing properties (21). The compounds were reported to have stimulated melanogenesis in the Keeman tumor line IGR-37, in vitro.

A series of patents describing various polyvalent metal salts of PABA, *p*-dimethylaminobenzoic acid, and 4-methoxycinnamic acids were assigned to Kanebo (46–54). Sunscreen pressed powders containing zinc, calcium, magnesium, aluminum, and zirconium salts of the foregoing compound were reported. A German patent issued to Oreal, S.A. describes a two-component artificial suntan agent containing mesotartaric acid and hydroxyquinones (55).

A U.S. patent assigned to Johnson and Johnson describes a series of vinylogous amides (**48**) as sunscreen agents (56).

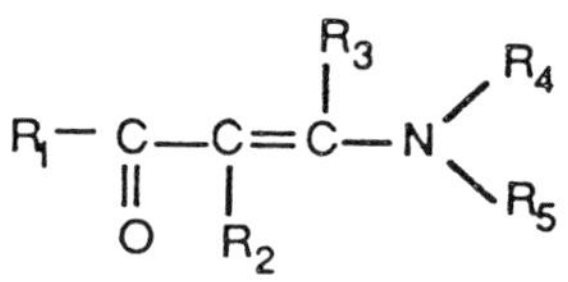

48

where

R = alkyl, cycloalkyl, alkenyl, cycloalkenyl, aryl, substituted aryl, or alkaryl
R_2, R_3 = H, alkyl, alkenyl, cycloalkyl, cycloalkenyl
R_4, R_5 = H, alkyl, alkenyl, cycloalkyl, aryl, etc.

Natural or biosynthesized raw materials have been described in several recent patents or articles.

A 1988 German patent describes the use of DL-α-tocopherol as a sunscreen that absorbs the UV light at 280–320 nm (57). Advanced Polymer Systems Inc. announced the development of a sunscreen product that incorporates genetically produced melanin entrapped in a polymer-based delivery system, which mimics the body's natural means for distributing melanin through the skin (see Chap. 20). The use of melanoidins was described as being responsible for the prevention of rashes caused by UV irradiation and the formation of dark melanin in the skin (58). "Cosmetics containing methyl hesperidin as a UV absorber," was the title of a recent Japanese patent (59). Methyl hesperidin is a mixture of 3′-

methyl-7-(rhamnosyl-2-methyl glycosyl)hesperitin as a major component (≥90%), 3,6′-dimethylhesperidin chalcone, 3′-monomethylhesperitin, 3,6′-dimethyl-4′-(rhamnosyl-2-methylglucosyl)hesperitin chalcone. Finally, thiamin or thiamin esters such as thiamin pyrophosphate and thiamin monophosphate were reported as effective sunscreen compositions.

Skin-lightening cosmetics, containing hydroquinone and sugar condensates were reported by Shiseido (61). Topical cosmetics contain alkyl hydroquinone–sugar (**49**) condensates. (I:R_1 = pentose residue, hexose residue, amino sugar residue, uronic acid residue, or these residues methylated; R_2 = C_{1-4} alkyl).

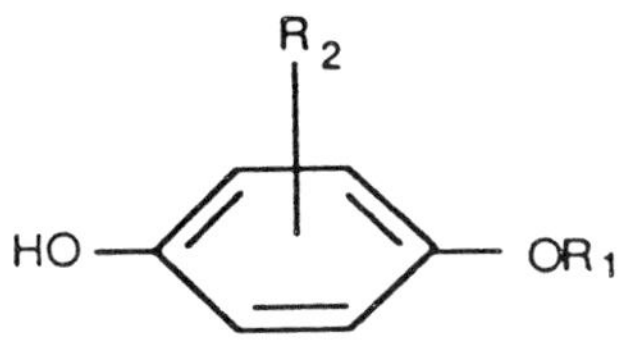

49

B. New Physical Blockers

Several forms of titanium dioxide, zinc oxide, and other metal oxides were reported as possessing superior qualities to those already available on the sunscreen market (see Chaps. 17 and 18).

A recent Japanese patent discloses the use of sunscreens containing metal oxides as an absorber of ultraviolet light (62). The transparent ultramicrometallic oxides, with an average diameter smaller than 300 Å include titanium dioxide, zinc oxide, chromium, cobalt, and tin oxides. Another Japanese patent reveals the use of hydrophobic titanium dioxide microparticle-coated spheric organic powders for cosmetics (63).

A sunscreen containing microparticles of hydroxylated titanium (rutile type), having a surface area of 60–250 m^2/g have been reported (64). Siloxane-treated titanium dioxide particles, with an average diameter of 40–70 μm, are the subject of another Japanese patent (65).

Sunscreens containing 1–30% ZnO ultrafine powder (maximum diameter 0–1 μm) that are colorless and protect the skin from UV irradiation have been the subject of a patent issued for the Kobayashi Kose Co. (66).

A European patent describing cosmetics containing infrared radiation blockers was issued to the University of Miami (67). The reflecting material may be a semiconductor (e.g., Sn oxide, In oxide, and Cd orthostannate), doped with F, P, or Sn, or a noble metal, especially Au, Cu, or Al. A layer of the material may be sandwiched between a layer of dielectric material (e.g., TiO_2 or Zn sulfide), with the bottom layer superimposed over the substrate.

A number of additional innovations include the use of plastics, resin emulsions or particles, as well as other reflecting material for the blocking of UV rays. For example, a European patent (68) discloses the use of UV-scattering lamina that comprises a laminar substance as matrix and a finely divided metal dispersed in the matrix where the difference in refractive index between the laminar substance and fine compound is >0.1. Sunscreens containing powders coated with UV light absorbers have been reported (69). Ultraviolet absorbers coated on the powder surface have less undesirable effects on the skin than noncoated absorbers.

A resin emulsion containing UV absorbers prepared by polymerizing monomers in the presence of UV absorbers was the subject of two Japanese patents issued recently. The first uses benzoguanamine resin particles and is reported to be both transparent and safe to the skin (70).

The second uses a mixture of methylmethacrylate, ethyl methacrylate, divinyl benzene, and acrylic acid in the presence of a UV absorber (padimate O) to produce the polymers, which are reported to absorb UV light effectively and to protect the skin from the harmful radiation (71).

IV. CONCLUSION

The relationship of chemical structure and efficacy of ultraviolet filters was clearly demonstrated through the review of all the currently available UV filters, as well as the recent reports of new chemicals possessing unique sunscreening capabilities.

The evolution of modern sunscreen chemicals, although still not complete, has nevertheless produced an impressive array of ultraviolet filters that have been incorporated into a multitude of products used in our daily lives, such as moisturizers, creams, lotions, and shampoos.

The future will witness not only new applications in which sunscreens may be used, but will require more effective substantive sunscreen chemicals that have fewer drawbacks and limitations. This calls for additional extensive research and resources devoted to this effort by chemical and cosmetic concerns, newer more effective testing procedures to ensure the safety of UV filters at the anticipated expanded usage levels, and the reasonable relaxation of regulations by governmental and regulatory agencies to allow the speedier adoption of novel sunscreens. The current situation in the United States, for example, is intolerable. Of the 21 sunscreen chemicals allowed, only 2 are physical reflectors (e.g., zinc oxide is not on the Category I listing), only 4 are UVA absorbers (with an additional derivative, a dibenzoylmethane, receiving limited approval in one sunscreen formulation), and 15 UVB absorbers. Of these 15 UVB sunscreens, only 9 are used to any extent commercially, and with only 5 having any waterproofing capabilities. In fact, with the recent, not fully justified trend of ''PABA-free'' products, this list diminishes even further. In the December 19, 1988 issue of the *Cosmetic Insiders Report* a suggestion was made that ''benzophenone-free'' sunscreen products may have at least as much legitimacy in preventing allergy as does ''PABA-free.'' If this suggestion is adopted, consumers will only have menthyl anthranilate and, in a few formulations, methoxy-*t*-butyldibenzoylmethane as the only UVA sunscreens allowed in the United States. This scenario is by no means ''far-fetched'' or unlikely, thereby demonstrating the need for new UVA and UVB sunscreens, as well as UV physical blockers worldwide.

REFERENCES

1. Patini, G. Perfluoropolyethers in sunscreens, *Drug Cosmet. Ind., 143*:42 (1988).
2. Groves, G. The sunscreen industry in Australia: Past, present, and future, *Sunscreens* (N. J. Lowe and N. A. Shaath, eds.), Marcel Dekker, New York, 1997, Chap. 13.
3. Rebut, D. The sunscreen industry in Europe: Past, present, and future, *Sunscreens* (N. J. Lowe and N. A. Shaath, eds.), Marcel Dekker, New York 1997, Chap. 12.
4. Safer and more successful suntanning, *Consumers Guide*, Wallaby Pocketbooks, New York, 1979, pp. 31–33.
5. Kumler, W. D. Action of sunscreen compounds, *Perfumer Essential Oil Rev., 12*: 427 (1952).
6. Shaath, N. The encyclopedia of UV absorbers for sunscreen products, *Cosmet. Toilet., 3*:21 (1987).
7. *Federal Register, 43*:266,38206–38269 (Aug. 25, 1978).
8. Janousek, A. *Cosmet. Toilet., 3*:79 (1983).
9. Fukuda, M. and Naganuma, M. The sunscreen industry in Japan: Past, present and future, *Sunscreens* (N. J. Lowe and N. A. Shaath, eds.), Marcel Dekker, New York, 1997, Chap. 14.
10. Butts, J. T. Sunscreen agents no longer confined to sun care products, *Soap Cosmet. Chem. Specialties*, Oct. (1984).
11. Robbins, P. *The Skin Cancer Foundation Journal*, New York, 1985.
12. Parrish, J. A., Anderson, R. R., Urbach, F., and Pitts, D. The spectrum of electromagnetic radiation: UVA in perspective, *UV-A: Biological Effects of Ultraviolet Radiation with Emphasis on Human Responses to Long-wave Ultraviolet*, Plenum Press, New York, 1978.
13. Scientists warn 1 in 7 Americans will develop skin cancer, *The News Journal*, Tuesday, March 10, 1987: 3A
14. Lowe, N. UVA photoprotection, *Sunscreens* (N. J. Lowe and N. A. Shaath, eds.), Marcel Dekker, New York, 1997, Chap. 26.
15. Shaath, N. A. On the theory of ultraviolet absorption by sunscreen chemicals, *J. Soc. Cosmet. Chem., 82*:193 (1987).
16. Agrapidis-Paloympis, L. E., Nash, R. A., and Shaath, N. A. The effect of solvents on the ultraviolet absorbance of sunscreens, *J. Soc. Cosmet. Chem., 38*:209 (1987).
17. Kligman, A. M. The identification of contact allergens by human assay: III. The maximization test: A procedure for sunscreening and rating contact sensitizers, *J. Invest. Dermatol., 47*:393–409 (1966).

18. Liem, D. H. and Hilderink, L. T. H. UV absorbers in suncosmetics 1978, *Int. J. Cosmet. Sci.*, *1*:341–361 (1979).
19. Masse, M. O., Herpol-Borremans, M., Grimee, R., and Gleviczky, S. *Int. J. Cosmet. Sci.*, *4*:235–246 (1982).
20. Editor. Sunscreen use in cosmetic formulations, *Cosmet. Toilet.*, *102*:67 (1987).
21. Jaffe, H. H. and Orchin, M. *Theory and Application of Ultraviolet Spectroscopy*, John Wiley & Sons, New York, 1964, p. 111.
22. Deflander, A. and Lang, G. Photoisomerization of benzylidene camphor and derivatives, *15th IFSCC International Congress, Cosmetic Science Preprints/Platform Presentations*, Vol. 1, Sept. 27, 1988, London, 1988, p. 49.
23. Beck, I., Deflander, A., Lang, G., Arnaud, R., and Lemaire, J. Study of the photochemical behaviour of sunscreens benzylidene camphor and derivatives II: Photosensitized isomerization by aromatic ketones and deactivation of the 8-methoxy psoralin triplet state, *J. Photochem.*, *30*:215 (1985).
24. Dibenzoyl methane derivatives, even though they are arylakyl ketones exhibit their UV-filtering effect through a keto/enol tautomerism which is not possible in benzophenones.
25. Davis, D. *Cosmet. Insiders Rep.*, *7*:1–2 (1988).
26. Beck, I., Deflander, A., Lang, G., Arnaud, R., and Lemaire, J. Study of the photochemical behaviour of sunscreens benzylidene camphor and derivatives, *Int. J. Cosmet. Sci.*, *3*:139–152 (1981).
26a. Saettone, M. F., Giannaccini, B., Morganti, C., Persi, A., and Cipriani, C. Substantivity of sunscreens: An appraisal of some quaternary ammonium sunscreens, *Int. J. Cosmet. Sci.*, *8*(1):9–25 (1986).
27. Kao Corp. Japanese patent 62,138,422 (1987).
28. Kao Corp. Japanese patent 62,138,423 (1987).
29. Kao Corp. Japanese patent 62,181,213 (1987).
30. Sogo Yakuko Co., Ltd. Japanese patent 62,22,711 (1987).
31. Bayer A-G. German patent 3,519,926 (1985).
32. Kao Corp. German patent 3,411,636 (1983).
33. Oreal S.A. German patent 3,742,690 (1988).
34. Matsuura Yakugyo Co., Ltd. Japanese patent 63,96,120 (1986).
35. Oreal S.A. German patent 3,700,531 (1986).
36. Fox, C. Cosmetic raw materials literature and patent review 1985–1987, *Cosmet. Toilet.*, *103*:31 (1988).
37. Ajinomoto Co., Inc., Japanese patent 62,36,484 (1987).
38. Saettone, M. F., Giannaccini, B., Checchi, M., Anselmi, C., Bartalini, P., and Andreassi, L. 5-Alkylidene-3-pyrrolln-2-ones as potential UV-A filters, *Med. Biol. Environ.*, *13*:111–114 (1985).
39. Kao Corp. German patent 3,543,496 (1984).
40. Shiseido Co. Ltd. Japanese patent 61,215,316 (1985).
41. Kao Corp Japanese patent 62,56,418 (1987).
42. Pola Chemical Industries Inc. Japanese patent 62,249,918 (1986).
43. Editor. Biosynthesized melanin as sun protection, *Cosmet. Toilet.*, *10*:19 (1988).
44. Graesser Laboratories Ltd. European patent 153,089 (1985).

45. Dominique, R. and Jung, L. European patent 235,064 (1986).
46. Kanebo Ltd. Japanese patent 60,130,653 (1983).
47. Kanebo Ltd. Japanese patent 60,130,655 (1983).
48. Kanebo Ltd. Japanese patent 60,130,654 (1983).
49. Kanebo Ltd. Japanese patent 60,149,515 (1985).
50. Kanebo Ltd. Japanese patent 60,149,516 (1985).
51. Kanebo Ltd. Japanese patent 60,149,517 (1985).
52. Kanebo Ltd. Japanese patent 60,174,710 (1985).
53. Kanebo Ltd. Japanese patent 60,174,711 (1985).
54. Kanebo Ltd. Japanese patent 60,174,709 (1985).
55. Oreal S.A. German patent 3,642,216 (1985).
56. Johnson and Johnson Baby Products Co. U.S. patent 4,663,156 (1987).
57. Fomin, A. German patent 3,620,989 (1988).
58. Ichimaru Farukosu K. K. Oriza Yuka Co., Ltd. Japanese patent 60,215,609 (1985).
59. Momotani Juntenkan, Ltd. Japanese patent 63,66,110 (1986).
60. Segawa, H. Japanese patent 62,258,313 (1986).
61. Shiseido Co., Ltd. Japanese patent 62,89,608 (1985).
62. Okamura Oil Mill, Ltd. Japanese patent 62,84,017 (1987).
63. Pola Chemical Industries, Inc. Japanese patent 61,257,907 (1986).
64. Mitsubishi Metal Corp. Japanese patent 62,281,812 (1987).
65. Kao Corp. Japanese patent 62,145,011 (1987).
66. Kobayashi Kose Co., Ltd. Japanese patent 60.231,607 (1985).
67. University of Miami, European patent 260,751 (1988).
68. Sumitomo Chemical Co., Ltd. European patent 268,938 (1988).
69. Noevir Co., Ltd. Japanese patent 62,185,775 (1987).
70. Shiseido Co., Ltd. Japanese patent 60,243,005 (1984).
71. Dainippon Ink and Chemicals, Inc. Japanese patent 62,93,220 (1985).

2

Sunscreens: Rationale for Use to Reduce Photodamage and Phototoxicity

Nicholas J. Lowe
Skin Research Foundation of California, Santa Monica, and UCLA School of Medicine, Los Angeles, California

Josia Friedlander
Skin Research Foundation of California, Santa Monica, California

I. INTRODUCTION

The incidence of sunlight-induced skin aging and skin cancers has been increasing in many parts of the world. In particular, the incidence of melanoma skin cancer has shown a well-documented increase in several continents over the last several years (1).

Many authorities are recommending primary prevention programs to reduce this incidence of cutaneous photodamage and skin carcinogenesis. An integral component of these programs is the use of protective clothing (2) as well as hats and effective sunscreens (3,4).

Most modern sunscreens have highly efficient absorption or reflecting capabilities throughout the ultraviolet B (UVB), partly the UVA, and in some instances, infrared wavelengths. Over the last several years, more efficient sunscreening ingredients have been developed for improved protection of the population (5–8).

More recently, direct evidence has been derived, from a large population study in Australia, of the effectiveness of sunscreen in their ability to reduce

---

Part of this manuscript has been reprinted with permission from Friedlander, J. and Lowe, N.J. Sunburns and sunscreens, *Management of Wilderness and Environmental Emergencies* (Auerbach P., ed.), St. Louis, Mosby, 1995.

the incidence of solar keratosis. This was a particularly important observation, as solar keratoses have been established as risk factors for basal cell carcinoma, melanoma, and are known precursers for squamous cell carcinoma (4).

II. SOLAR RADIATION

A. The Solar Spectrum

Solar radiation encompasses the electromagnetic spectrum from short, high-energy cosmic, gamma and x-rays, to the longer, lower-energy ultraviolet (UV), visible, infrared (IR), microwaves, and radio waves. Nearly 30–40% of solar radiation, including most of the harmful portions, is absorbed in the atmosphere by the ozone layer at 15–35 km above sea level (9). High-energy radiation (wavelength < 10 nm) displaces electrons from incident tissue to form ions. Therefore, radiation in this category, including gamma and x-rays, is termed *ionizing radiation.* Because they lack the energy to do this, UV, visible, and shorter IR are categorized as *nonionizing irradiation.*

The radiation striking the earth is approximately 50% visible (wavelength 400–800 nm), 40% IR (wavelength 1300–1700 nm), and 10% UV (10–400 nm). The UV spectrum is divided for convenience into UVA, from 320 to 400 nm; UVB, from 290 to 320 nm; UVC, from 100 to 290 nm; and vacuum UV, from 10–100 nm. The human eye cannot detect wavelengths other than those in the visible range. Vacuum UV, UVC, and the shortest UVB are screened by the ozone layer, which eliminates essentially all UV radiation (UVR) shorter than 290 nm. Little UVA is filtered in the atmosphere. The remaining UV that reaches the ground is about 10% UVB and 90% UVA at midday. The UVB intensity declines from the noontime apex, but UVA intensity remains relatively constant throughout the day (9).

B. Human Skin and Acute Solar Overexposure (Sunburn)

Erythema, the most familiar manifestation of UVR exposure, occurs in a biphasic manner (10). The early portion of this reaction, also known as immediate pigment darkening (IPD), is mediated by UVA and lasts 13–30 min. The UVA-induced erythema is marked in some individuals, whereas it is undetectable in others. Delayed erythema, a function primarily of UVB dosages, begins 2–8 h after exposure and reaches a maximum in 24–36 h, with erythema, pruritus, and pain in the sun-exposed areas (11). This reaction resolves over a 3- to 5-day period and initiates increased melanogenesis, which reaches an apex at 2–3 weeks (10).

The *minimal erythema dose* (MED) is defined as the amount of UVB exposure after which minimally perceptible erythema can be detected (11): UVB at 307 nm is the most efficient wavelength for erythema. The dosage of UVA required to elicit erythema, 20–70 J/cm^2, is 600–1000 times that required for UVB, which ranges from 20 to 1000 mJ/cm^2. Although the intensity of UVA at noon is typically ten times that of UVB, UVB radiation is responsible for 98–99% of delayed erythema development.

A typical MED value for an individual who burns easily and tans slightly (skin type II) is 20 min at midday in midsummer in Southern California (Lowe, N. J., personal observation). An exposure of 5 MED produces a painful sunburn. At 10 MED, edema, vesiculation, and bullae formulation often occur. Skin necrosis, leading to scarring and uneven pigmentation, may result. Depending on the area involved, severe sunburn can cause the systemic symptoms of "sun poisoning" including fever, nausea, vomiting, severe headache, prostration, and even shock (9).

Microscopically, changes are detectable as early as 30 min after UVR exposure. Epidermal changes include intracellular edema, vacuolization and swelling of melanocytes, and the development of characteristic "sunburn cells" (12). In the dermis, UVR initially leads to interstitial edema and endothelial cell swelling. Later, there is perivenular edema, degranulation, and loss of mast cells, decrease in number of Langerhans cells, neutrophil infiltration, and erythrocyte extravasation. This reaction reaches its apex at 24 h resolving over 3–5 days (9,10).

Sunburn cells are dyskeratotic keratinocytes with pyknotic nuclei and homogeneous eosinophilic cytoplasm (9). Sunburn cell formation is maximal at 18–24 h after exposure and resolves in 3–7 days. These cells occur in proportion to the amount of skin's UVB exposure; UVA is less effective in the development of sunburn cells (5).

Later changes include hyperkeratosis (increased scale), acanthosis (epidermal thickening), disorganization and misalignment of keratinocytes, dermal vascular ectasia, and mononuclear perivascular infiltration (13).

C. Prolonged Effects of Solar Radiation on the Skin

There are two major adaptive responses in the skin initiated by UVR exposure: melanogenesis (tanning) and thickening of the stratum corneum. Tanning occurs as a result of de novo production of melanin in melanocytes. The UVB rays enhance the binding of circulating melanocyte-stimulating hormone (MSH) to melanocytes, leading to melanocyte proliferation, dendritic arborization, and pigment production. Interestingly, UVB radiation actually inducts melanocyte proliferation in both exposed and covered areas of the body (13).

Melanin is produced from tyrosine through several enzymatically controlled reactions. The initial, rate-limiting step is the conversion of tyrosine into dihydroxyphenylalanine (DOPA). The melanin, incorporated into organelles called melanosomes is then distributed to surrounding keratinocytes. Some melanosomes remain intact in the keratinocytes as they keratinize and migrate to become the stratum corneum. However, others are degraded enzymatically into an amorphous form that deposits in the interstitium. This amorphous melanin undergoes oxidation in the "immediate pigment-darkening" reaction when exposed to UVA radiation (14).

Melanin absorbs, reflects, and scatters UVR and also acts as a free-radical trap. Although persons of all skin types have approximately the same number of melanocytes, black-skinned persons have approximately 400 melanosomes per basal epidermal cell, a fourfold increase over the typical pale Caucasian. This difference reduces the UVA and UVB penetration into the dermis by a factor of five in blacks as compared with Caucasians and is responsible for the 30-fold increase in the typical MED value for blacks (13).

Tanning induces a sunprotection factor (SPF) increase between 2 and 4 against further UVR exposure. Even with a "protective tanning base," Caucasians are susceptible to a significant amount of UVR-induced damage.

The skin surface is composed of a layer of densely keratinized cells, called the stratum corneum. In response to prolonged UVB exposure, the stratum corneum can hypertrophy to six times its original thickness by increased synthesis of keratin by basal keratinocytes. This response is independent of pigmentary changes, evidenced by the act that it occurs in albinos. The stratum corneum is particularly hypertrophied on the palms and soles, which explains why these areas are so resistant to sun damage. Thickened stratum corneum absorbs or reflects 90–95% of incident UVB so that just 5–10% reaches the basal keratinocytes, melanocytes, and superficial dermal vascular system (14). Interestingly, extended UVA exposure does not induce a similar proliferation of the stratum corneum.

Although UVA at high dosage (such as in a tanning salon) can produce erythema and melanogenesis, it does not provide the same degree of protection as a naturally acquired UVB suntan. This is due to at least two different factors. The UVA does not induce thickening of the stratum corneum, an important protective factor induced by UVB. Also, the pigmentary changes induced by UVA exposure occur primarily in the basal layer, whereas those initiated by UVB are distributed throughout the entire epidermis. Recent evidence points to the aging effects of relatively low levels of UV exposure of human skin (see Chap. 5).

D. Solar Radiation and Photoaging

Human skin changes in grossly and microscopically detectable ways over the lifetime of an individual. However, exposure to excessive amounts of solar radiation greatly increases the rate at which skin ages. Dermatoheliosis (*photoaging*) is the wrinkled, leathery skin seen in persons repeatedly exposed to the sun. Photoaging is induced by prolonged exposure to all portions of the solar spectrum, including UVA, UVB, and infrared (15).

Grossly photodamaged skin appears wrinkled and atrophic, with mottled pigmentation. Superimposed are hypopigmented macules, as well as hyperpigmented lentigines (''liver spots'' or ''sun spots''), telangiectasias, and raised, roughened precancerous actinic keratoses.

Microscopically, as skin ages normally, the epidermis atrophies, the dermis becomes hypocellular, the vasculature remains intact, and collagen forms a stable, increasingly cross-linked matrix. By contrast, the epidermis of photoaged skin becomes acanthotic (thickened), the dermis hypercellular (owing to fibroblast and mast cell proliferation), and the vessels tortuous and dilated. Elastic fibers degenerate into a thickened, amorphous mass displacing the normal collagen in a process termed *elastosis*. These changes are detectable in most Caucasians before the age of 30. Shorter UVA, in the 320–340 nm range, is probably responsible for the bulk of UVA-induced photodamage (15).

Prolonged exposure to solar radiation also has dramatic effects on the life span of fibroblasts and keratinocytes. Human fibroblasts in vitro will divide only a limited number of times before they die. Although the mechanism remains unclear, the number of times they do replicate correlates with the age of the subject from which the fibroblasts are obtained. More recent studies have demonstrated that fibroblasts and keratinocytes from a given individual will replicate more times if obtained from an unexposed area of the body, as compared with those derived from more repeatedly sun-exposed tissue (16). This implies that solar radiation exposure decreases the regenerative capacity of human skin and, thus, hastens skin aging.

E. Solar Radiation and Cancer

Solar radiation-induced local immune suppression may induce or promote nonmelanoma skin cancer (NSMC); UVR definitely induces NMSC in mice. Transformation of normal cells into malignant ones probably occurs as a result of UVR-initiated DNA mutation. Patients with xeroderma pigmentosum (XP), who lack the usual reparative enzymes for damaged DNA, have a greatly increased risk of developing NMSC (16). Although cumulative UVB appears to be more closely linked to NMSC development, UVA also plays a role. Both can induce thymine dimer formation, which may be an initiating step in cell transformation. Also, UVA augments the development of UVB-induced NMSC in hairless mice (13).

Epidemiological studies of cumulative sun exposure and the incidence of NMSC in normal humans are also extremely strong. The risk of basal cell carcinoma in Atlanta, Georgia, is 50% greater than that seen in Minneapolis, Minnesota. The age-adjusted risk for squamous cell carcinoma in Atlanta is four times that seen in Minneapolis. The mutagenicity of 1 J/cm^2 of UVB at 300 nm is 1000 times that of 1 J/cm^2 of UVA at 330 nm. The carcinogenic effects likely correlate with the mutagenicity; therefore, UVB can be assumed to be manyfold more carcinogenic than UVA (3).

The data on malignant melanoma (MM) are not clear-cut. Nevertheless, MM is probably induced or promoted in some way by UV exposure. Recent trends have shown an alarming rise in development of MM. Risk factors for the development of MM are complex and do not correlate with cumulative UVB exposure. However, intermittent, intense sunlight exposure may correlate with the development of melanoma (17). As mentioned earlier, UVB actually induces melanocyte proliferation in both exposed and covered areas of the body. This may explain why MM sometimes occur in areas without significant previous sun exposure (13).

Garland et al. (17) postulate that the broader use of sunscreens that protect primarily against UVB wavelengths encourages susceptible individuals to spend more time in the sun. These persons received a larger cumulative dose of solar radiation outside the UVB range. Increased exposure to non-UVB solar radiation has been suggested to promote the development of melanoma skin cancer. Further scientific study of this question is needed to justify this hypothesis.

III. SUNSCREENS

A. History and Development

In 1928, the world's first commercial use of a sunscreen, an emulsion of benzyl salicylate and benzyl cinnamate, was developed in the United States. In the 1930s, a 10% solution of phenyl salicylate appeared on the Australian market. In 1935, lotions of quinine oleate and quinine bisulfate were available in the United States. In 1943, *p*-aminobenzoic acid (PABA) was patented, introducing this popular agent and subsequently its derivatives. The U.S. military used red petrolatum, a physical blocking agent, as a sunblock during World War II. Their efforts also led to popularization of other UV-filtering agents, including glycerol-PABA, 2-ethylhexyl salicylate, digalloyl trioleate, homomenthyl salicylate, and dipropylene glycol salicylate (18). Most of the early agents were directed toward UVB, the portion of the sun's spectrum responsible for the most palpable and distressing effect, the familiar sunburn.

Parasol 1789 (avobenzone or 4-*t*-butyl-4′-methoxydibenzoylmethane), an agent broadly effective in the UVA spectrum is now marketed in combination with UVB-blocking agents. These formulations provide broad-spectrum chemical sun protection to the U.S. consumer (19,20).

A new subclass of physical blockers, micronized reflecting powders, have more recently been made available from a variety of manufacturers. Unlike traditional physical blockers, micronized reflecting powders are less visible, yet provide broad-spectrum protection against UVR. These should prove useful in UVR-sensitive patients resistant to older physical blockers for cosmetic reasons. An additional benefit is that they do not cause photosensitization.

Future trends in the development of sunscreens include products with efficacy against UVC (which may become more important with further erosion of the ozone layer) and agents directed against infrared (IR), because of the concerns about IR-induced photoaging.

B. Evaluation of Sunscreening Agents

Evaluation of Sun Protection Factor

The U.S. Food and Drug Administration (FDA) approved techniques to assess the efficacy of a sunscreen in the UVB portion of the spectrum is determination of the sunscreen protection factor (SPF) (11). The *SPF* is defined as the ratio of the time of UVR exposure necessary to produce minimally detectable erythema in sunscreen-protected skin to that time for unprotected skin.

A typical testing protocol is as follows: Skin in a non–skin-exposed area, such as the buttocks or lower back, is covered with light-proof adhesive foil; 1-cm^2 areas of foil are removed sequentially so that each area receives a defined dose of UVB. The following day, the patient returns to be examined and areas are assessed for erythema. The amount of exposure necessary to generate barely detectable erythema is determined.

The following day, sunscreen is applied uniformly with a pipette at 2 mg/cm^2 or 2 $\mu l/cm^2$. The sunscreen is allowed to dry for 15 min, and then the subject is irradiated for incrementally increasing times, based on the estimated SPF, which is usually determined spectrophotometrically. The volunteer is asked to return again 24 h later, at which time the skin is reexamined. The SPF is then calculated from the ratio of the protected MED to the unprotected MED (11).

There is some controversy about whether SPF thus defined accurately reflects the action spectra for induction of skin cancer in humans, or the application spectra for induction of human skin cancer. Also, application of 2 mg/cm^3 correlates with using at least 29.6 ml (1 oz.) over the body's surface. Since studies have revealed that most individuals apply about 22.2 ml (0.75 oz.; 1.5 mg/cm^2 or 1.5 $\mu l/cm^2$), the usage SPF is sometimes significantly less than the listed in vivo SPF (21).

It is also important to recognize that the SPF is a measure of UVB plus UVA protection, providing a suitably filtered xenon arc solar simulator is used. We use a 1000-W solar simulator (Oriel) with different filters. For, SPF assays, we routinely use this machine with Schott WG 320-nm, 2-mm thick, plus UG 11 filter.

Evaluation of Protection Against Ultraviolet-A

Currently, there is no consensus of an ideal in vivo assay to assess protection against UVA. The UVB spectrum readily causes erythema that can be used for the SPF assay; UVA can cause erythema, but only after a much larger dose. However, the action spectra for other UVA-induced phenomena (e.g., photosensitivity dermatitis, photoaging, development of melanoma and nonmelanoma skin cancer) are not specifically defined. A variety of different assays have been described for evaluation protection against UVA.

Phototoxic Protection Factor The phototoxic protection factor (PPF) was developed to try to decrease the UVA dose necessary to achieve a detectable difference in erythema (19). This involves sensitizing the subject's skin with 8-methoxypsoralen (8-MOP) either topically or orally, and then comparing the MED, as done for UVB sunscreen testing. It is a reliable assay, with clearly defined visual erythema, but can result in long-term pigmentation of skin sites (19). This has a minimal risk for the volunteers because their skin is irradiated only once.

Ultraviolet-A Erythema Protection Factor The UVA erythema protection factor (APF) is defined analogously to SPF and assesses the time to induce erythema in UVA-exposed skin. The reliability of such testing is of some question because of the inadequacy of current filters and long deviation of irradiance with many solar simulators. Recently, however, Cole and Van Fossen (22) described a high-intensity UVA source that they contend is a more effective method to assess UVA protection based on either erythema or tanning in an unsensitized individual.

We have also evaluated this assay using a 1000-W xenon arc solar simulator (Oriel) filtered with WG 345-nm, 2-mm plus UG 11 filter. We find the assay to be reproducible with this apparatus. It can also be used for evaluating water-resistant sun protection against UVA (Lowe, N.J. et al., unpublished data, 1993).

Pigment-Darkening Protection Factor Immediate pigment-darkening (IPD), a UVA and visible spectrum-mediated transient oxidation of preexisting melanin, has also been used to evaluate UVA sunscreen efficacy (23). This method has been criticized because it must be performed on individuals with darker skin, because pale-skinned individuals who are the most in need of protection do not demonstrate easily measurable IPD (24).

The delayed pigment-darkening response evaluated 2 h after UVA radiation may be more reproducible. This has been well described (25). We have also investigated this assay. It requires more UVA energy than for IPD, but it is more reproducible, and is achieved with a greater variety of skin phototypes than the IPD assay (Lowe, N.J., et al., unpublished data, 1993).

In Vitro Transmission Protection Factor The *transmission protection factor* is defined as the ratio of the photocurrent measured by a spectroradiometer through Transpore TM tape (3M Company, St. Paul MN) to the photocurrent measured through sunscreen-coated tape at a given wavelength. According to Diffey and Robson (26), this in vitro method provides close agreement with in vivo recorded SPF data. Other techniques include spectrophotometrically comparing sunscreen applied to either a quartz crystal (27) or excised, treated epidermis (28), with control as the untreated surface.

Sunburn Cell Protection Factor Finally, the inhibition by sunscreen of sunburn cell formation in 8-methoxypsoralen (8-MOP)-sensitized skin in both mice (27) and in humans (28) has been used to assess UVA sun protection. The inciting wavelength for sunburn cells is in the UVB range, but this is shifted into UVA by sensitization with 8-MOP (12). The limitations of these final two techniques are obvious in light of the difficulty in assuring a correlation between these in vitro tests and the human response.

C. Substantivity

Substantivity is the characteristic of a sunscreen that reflects how effectively the advertised degree of protection is maintained under adverse conditions, including repeated water exposure or sweating. According to the U.S. FDA, a sunscreen is declared water-resistant if it can maintain its original SPF after two 20-min immersions. A sunscreen is very water-resistant if it retains its protective integrity after four 20-min immersions (29). Substantivity is of enormous importance, because sunscreens are used outdoors in settings where abundant sweating and repeated immersion in water is common.

D. Active Sunscreen Ingredients (Table 1)

Protectants Against Ultraviolet-B

p-*Aminobenzoic Acid* *p*-Amino benzoic acid (PABA) has been available since 1943 and was popular in the 1950s and 1960s. Nowadays, PABA is used infrequently as a sunscreen for a variety of reasons. Its absorption peak at 296 nm (18) is relatively far from the UVB-induced erythema peak at 307 nm. It is poorly soluble in water, and must be used as a 5–15% solution in alcohol. After application, PABA penetrates the stratum corneum effectively, where it is trapped and remains bonded by hydrogen bonding to epidermal proteins. This greatly enhances it substantivity (30), but also increases the risk for contact or photocontact dermatitis. Sensitivity of this sort is seen in up to 4% of the population (31). It can cause a stinging sensation when applied and stains both cotton and synthetic fabrics. After photooxidation, this can leave a permanent yellow discoloration. PABA is no longer commonly used in sunscreens.

The PABA Derivatives The PABA esters are created by addition of hydrocarbon groups to the PABA molecule. Many of these molecules are an improve-

Table 1 FDA-Approved Sunscreen Ingredients in the United States (1993)

Chemical	Approved (%)
UVA absorbers	
Oxybenzone	2–6
Sulisobenzone	5–10
Dioxybenzone	3
Methyl anthralinate	3.5–5
Avobenzone	3[a]
UVB absorbers	5–15
Aminobenzoic acid (PABA)	1–5
Amyldimethyl PABA (padimate A)	1–3
2-Ethoxyethyl-*p*-methoxycinnamate	8–10
Digalloyl trioleate	2–5
Ethyl 4-bis(hydroxypropyl)aminobenzoate	1–5
2-Ethylhexyl-2-cyano-3,3-diphenylacrylate	7–10
Ethylhexyl-*p*-methoxycinnamate	2–7.5
2-Ethyhexyl salicylate	3–5
Glyceryl aminobenzoate (Glyceryl PABA)	2–3
Homomenthyl salicylate	4–25
Lawsone with dihydroxyacetate	0.25% with 3%
Octyldimethyl PABA (Padimate O)	1.4–8
2-Phenylbenzimidazole-5-sulfonic acid	1.4
Thethanolamine salicylate	5–12
Physical blockers	
Red petrolatum	30–100
Titanium dioxide	2–25

Available in USA (1993) in Shade UVA Guard sunscreens.

ment over PABA, in that they are water-soluble and do not penetrate the stratum corneum. The most widely used of the PABA derivatives is padimate O, or octyldimethyl PABA. Its absorption peak is more desirable at 300 nm in nonpolar solvents, ranging to 316 nm in polar solvents (18). Padimate A (amyldimethyl PABA) is similar to padimate O and is also used in combination with other agents. Neither ester stains clothing although both can cause stinging when applied.

Another derivative, glycerol PABA, has frequently been implicated as the etiologic agent in both contact and photocontact dermatitis. However, after careful testing it appears that contaminating impurities, including benzocaine, may have actually been responsible (32).

Salicylates The salicylates are *ortho*-distributed aromatic compounds with a peak absorption of about 300 nm. Two compounds of this type, octyl salicylate and homomenthyl salicylate (homosalate) are currently approved in the United States. Although not very effective as sunscreens, they have the benefit of being exceptionally stable, essentially nonsensitizing, and water-insoluble, leading to high substantivity. They are also useful as solubilizers of other poorly soluble sunscreen ingredients, such as the benzophenones (18). They are used commonly in "PABA-free" products (32).

Cinnamates 2-Ethylhexyl-*p*-methoxycinnamate (Parasol MCX; absorption maximum 310–311 nm), 2-ethoxyethyl-*p*-methoxycinnamate and octylmethoxycinnamate are available in the United States. These are effective in blocking UVB, but have poor substantivity and are generally found in combination with other agents (8). Cinoxate, the cinnamate most often seen with other cinnamates, is found in balsam of Peru, balsam of Tolu, coca leaves, and cinnamon oil (32).

Protectants Against Ultraviolet-A

Benzophenones Benzophenones are aromatic ketones that absorb predominantly in the UVA portion of the spectrum, between 320 and 350 nm. For example, oxybenzone has an absorption maximum of 326 nm in polar solvents, compared with 352 nm in nonpolar solvents. Benzophenone-3 (sulisobenzone) and dioxybenzone are also approved for the U.S. market (18). Oxybenzone is frequently implicated as the etiologic agent in photocontact allergy, although reactions have also been reported with dioxybenzone (32).

Dibenzoylmethanes Dibenzoylmethanes are substituted diketones that undergo keto–enol tautomerism on absorption of UVR. The keto form of the dibenzoylmethanes has a UV absorption maximum of 260 nm, whereas the enol form absorbs above 345 nm. Parasol 1978 (avobenzone or 4-*t*-butyl-4′-methoxydibenzolymethane) with an absorption maximum at 355 nm is the only agent of this class available in the United States. Although these compounds are capable of a high degree of UV absorption (19), they are unstable and can undergo photoisomerization to compounds that are not protective. Parsol 1789 has a loss of protective power as high as 36% by photodegradation (18). Even Parsol 1789 does not provide significant protection against UVA radiation that is close to 400 nm (6).

E. Physical Blockers

Physical blocking agents or reflecting agents, such as zinc oxide, titanium dioxide, iron oxide, kaolin, ichthammol (Ichthymall) (33), red veterinary petrolatum, talc ($MgSiO_x$), and calamine (FeO_2) are composed of particles of a size that scatter, reflect, or absorb solar radiation in the UV, visible, and even infrared ranges. Twenty percent zinc oxide, 20% titanium dioxide, as well as 1% iron oxide have been spectrophotometrically demonstrated to reduce transmittance in the UVA and visible ranges to a maximum of approximately 20%. However, a combination of the zinc and iron oxides together is synergistic, effectively reducing transmittance in the UVA and visible ranges to as low as 1.5%. These data were confirmed in vivo in a hematoporphyrin derivative-sensitized guinea pig model (6).

Older physical blockers had the disadvantage that they were comedogenic, had to be applied in a relatively thick layer, and melted in the sun, staining clothing. They were opaque and, therefore, visible, making them cosmetically undesirable for many individuals. These formulations found a market in young people who apply brightly pigmented products in limited areas. Because of their efficacy and broad-spectrum coverage, physical blockers are potentially important for persons with certain photosensitivity disorders.

Recently developed micronized preparations now available in the United States provide an excellent option within this class of sunscreens. Micronized physical blockers are suspensions of finely ground material, such as titanium dioxide, that reflect at wavelengths shorter than the visible spectrum. Because they do not reflect in the visible spectrum, they are invisible; thus, more cosmetically acceptable. Micronized titanium dioxide is chemically stable and does not cause any photoallergic or contact dermatitis.

Micronized sunscreens are more effective at the shorter UV wavelengths. A major difficulty in formulating micronized sunscreens is in preventing agglomeration of the particles. If this occurs, the portion of the spectrum reflected will shift into the visible range and the product will have characteristics of traditional opaque physical blockers.

F. Vehicle

The choice of vehicle for a sunscreen is important for several reasons. First, the proper vehicle can enhance a sunscreen's substantivity or ability to remain on the skin and be effective. Second, the wrong vehicle can act as a skin irritant, induce a phototoxic or photoallergic reaction, or may be comedogenic. Finally, a solvent may modify the sunscreening agent because of its polarity and, thereby, dramatically shift the absorption spectrum of the agent toward or away from the desired range (7).

G. Systemic Photoprotection

Over the years, a variety of systemic agents have been investigated as oral agents for sunscreening purposes. The appeal is threefold. First, oral agents are convenient. Second, they provide coverage for the entire body. Finally, oral agents are likely to eliminate the concern over substantivity so critical for topically applied products. Included in the list of these products are PABA, antihistamines, aspirin, indomethacin, retinol, ascorbic acid, α-tocopherols (e.g., vitamins A, C, and E), corticosteroids, psoralens, β-carotene, and antimalarials. The last three are used in some individuals afflicted with certain photosensitivity dermatoses. Otherwise, these agents have been disappointing for persons with normal skin (16).

H. Sun-Protective Clothing

Although ultraviolet radiation is partially reflected by clothing, some lightweight fabrics typically worn in the summertime do not even provide an SPF of 15. Once damp from sweat or water, the SPF of some fabrics falls. The weave of some fabrics does appear to affect the flux of solar radiation through garments. Solumbra TM, a maker of sun-protective wear, manufactures clothing that has been demonstrated in vivo to provide an SPF of greater than 30, regardless of textile, color, or moisture content (2). Such special fabrics are useful for some photosensitivity sufferers. Practitioners should be aware that some fabrics provide protection by an applied coating, which may lose efficacy over time. Formal protection standards for sunprotective clothing need to be developed.

I. Other Barriers

Window glass effectively blocks UVR below 320 nm, thus providing solid protection against UVB. Tinted windows and the plastic interleaf found in auto windshields effectively block nearly all UVA (34). However, front and side window tinting sufficient to protect individuals with UVA-sensitive skin is illegal in many areas because of concerns about reduced driver visibility and limited ability for law enforcement officers to see what is happening inside an automobile with tinted windows.

IV. SUN PROTECTION

A. Sun Protection for Normal Skin

It is well established that prolonged exposure to UVB leads to deleterious effects on human skin, including photoaging and nonmelanoma skin cancers. Application of a UVB-blocking sunscreen does decrease the risk for development of nonmelanoma skin cancer (9). Therefore, we recommend that patients use sunscreens, depending on their skin phototype as described in Table 2.

Additionally, all individuals should be encouraged to stay indoors or seek shade during the peak hours of solar radiation flux from 10 AM to 2 PM. A hat or a sun visor is a useful addition. Patients with pale complexions should be reassured that "fair" skin is attractive. Those who insist on darker skin should use a self-tanning lotion containing dihydroxyacetone (DHA). Tanning salons should be avoided as intense UVA exposure provides limited protection and induces photoaging and photoallergic responses.

Table 2 Skin Types and Suggested Sunscreen Protection Factors

Type	Characteristic	Examples	Suggested SPF: Routine day	Suggested SPF: Outdoor activity
I	Always burns easily, never tans	Celtic or Irish extraction often blue eyes, red hair, freckles	15	25–30 (waterproof)
II	Burns easily, tans slightly	"Fair skinned" individuals; often have blond hair	12–15	25–30 (waterproof)
III	Sometimes burns, then tans gradually and moderately	Most Caucasians	8–10	15 (waterproof)
IV	Burns minimally, always tans well	Hispanics and Asians	6–8	15 (waterproof)
V	Burns rarely, tans deeply	Middle Easterners, Indians	6–8	15 (waterproof)
VI	Almost never burns, deeply pigmented	Blacks	6–8	15 (waterproof)

Persons who spend a significant amount of time outdoors, particularly in areas of high solar flux, should be cautious about the immediate and long-term risks from UV exposure and should be advised about how to properly protect themselves. Included in this list of individuals are naturalists, skiers, hikers, bicyclists, fishermen, mountain climbers, gardeners, as well as members of certain professions (e.g., farmers and farm workers, lifeguards, and postal delivery persons).

Finally, this information should be disseminated to children at an early age. Childhood is typically the time of life when maximal sun exposure occurs. Childhood sunburns are implicated in increasing risk for malignant melanoma and NMSC and thus should assiduously be avoided. It has been calculated biostatistically that if children consistently used an SPF 15 sunscreen through the age of 18, the occurrence of nonmelanoma skin cancer could be reduced by 78%. If an SPF 7.5 product were used throughout life, the risk of NMSC would decrease 97% (41).

B. Evidence for Sunscreen Reduction of Actinic Keratoses

In previous animal studies, sunscreens possessing an SPF of 15 or greater, were capable of reducing the incidence of cutaneous preneoplasia and neoplasia. In addition, when sunscreens were used and irradiation continued, there was evidence in the reduction in the degree of photoaging skin changes (15).

In a recent investigation of a large population of Australians, the daily use of a sunscreen that contained a combination of 8% 2-ethylhexide-*p*-methoxy cinnamate and 2% avobenzone (which had an SPF of 17) resulted in a significant reduction in the number of new actinic keratoses appearing in the sunscreen-protected population, compared with the placebo cream control population (4). This was an important study as it was conducted in 588 individuals over one Australian summer and showed the importance of daily sunscreen protection and the ability of human skin to undergo apparent repair during sunscreen protection. The reversal of epidermal dysplasia suggests that the daily use of effective sunscreens should be an important part of any community program for reduction of UV-induced skin carcinogenesis and aging.

C. Photosensitive Skin Reactions

There is a special need for photoprotective measures, including sunscreens and sun-protective clothing. Because the action spectra for many photosensitive reactions or photosensitive disease lie within the UVA, sunscreens are needed that efficiently protect individuals from these wavelengths of UV. Table 3 lists drugs that can precipitate a photosensitive skin reaction.

D. Phototoxicity

Phototoxicity can occur in any individual exposed to photosensitive agents and radiant energy of sufficient intensity. The action spectra usually involves UVA, and it is wise to use a broad-spectrum sunscreen as well as removing the offending phototoxic agent.

Many plants elaborate photoactive substances, such as furocoumarins (particularly psoralens), that can induce phototoxic reactions (Table 4). This particular category of phototoxic reactions is termed *phytophotodermatitis*. It has been noted in vegetable pickers and other food handlers as well as naturalists. Flower leis given to visitors in Hawaii have also been implicated.

Phytophotodermatitis is commonly exploited by dermatologists as part of a therapeutic regimen to control psoriasis and to repigment the skin of persons with vitiligo (35). Patients are administered psoralens either topically or orally and are then exposed to controlled doses of UVA as photochemotherapy.

Psoralens is also incorporated into sunscreens as a suntan-enhancing agent in Europe. Use of PUVA can increase the risk of skin cancer, as it is documented that psoralens in combination with UVA increases the risk for development of NMSC. Therefore, it is important to protect patients with UVB- and UVA-protecting sunscreens following PUVA phototherapy sessions.

E. Photoallergy

Photoallergy involves interaction of the immune system and solar radiation. Photoallergic reactions are rare and usually triggered by UVA exposure. Indi-

Table 3 Agents That May Cause Photosensitivity

- Anticancer drugs
  - Dacarbazine (DTIC-Dome)
  - Fluorouracil (Fluoroplex; and others)
  - Methotrexate (Mexate; and others)
  - Procarbazine (Matulane)
  - Vinblastine (Velban)
- Antidepressants
  - Amitriptyline (Elavil; and others)
  - Amoxapine (Asendin)
  - Desipramine (Norpramin: Pertofrane)
  - Doxepin (Adapin; Sinequan)
  - Imipramine (Tofranil; and others)
  - Isocarboxazid (Marplan)
  - Maprotiline (Ludiomil)
  - Nortriptyline (Aventyl; Pamelor)
  - Protrityline (Vivactil)
  - Trimipramine (Surmontil)
- Antihistamines
  - Cyproheptadine (Periactin)
  - Diphenhydramine (Benadryl; and others)
- Antimicrobials
  - Antifungals (Fentichlor, Multifungin, Jadit)
  - Demeclocycline (Declomycin; and others)
  - Doxycycline (Vibramycin and others)
  - Griseofulvin (Fulvicin-U/F; and others)
  - Methacycline (Rondomycin)
  - Minocycline (Minocin)
  - Nalidxic acid (NegGram)
  - Oxytetracycline (Terramycin; and others)
  - Sulfacytine (Renoquid)
  - Sulfadoxine-pyrimethamine (Fansidar)
  - Sulfaguanidine
  - Sulfamethazine (Neotrizine; and others)
  - Sulfanilamide
  - Sulfapyridine
  - Sulfasalazine
  - Sulfathiazole
  - Sulfisoxazole (Gantrisin; and others)
  - Tetracycline (Achromycin; and others)
- Antiparasitic drugs
  - Bithionol (Bitrin)
  - Chloroquine (Aralen)
  - Hydroxychloroquine
  - Pyrvinium pamoate (Povan)
  - Quinine
  - Oxybenzone
  - PABA esters
  - *p*-Aminobenzoic acid
- Others
  - Amiodarone (Cordarone)
  - Bergamot oil, oils of citron, lavender, lime, sandalwood, cedar (used in many perfumes and cosmetics; also topical exposure to citrus rind oils)
  - Benzocaine
  - Captopril (Capoten)
  - Carbamazepine (Tegretol)
  - Chloradiazepoxide (Librium)
  - Coal tar and derivatives (containing acridine, anthracene, naphthalene, phenanthrene phenols, thiophene)
  - Contraceptives, oral (Norethynodrel)
  - Cyclamates (calcium cyclamate, sodium cyclohexylsulfamate)
- Antipsychotic drugs
  - Chlorpromazaine (Thorazine; and others)
  - Chlorprothixine (Taractan)
  - Fluphenazine (Pernitil: Prolixin)
  - Haloperidol (Haldol)
  - Perphenazine (Trilafon)

Table 3 Continued

- Sulfamethizole (Thiosulfil; and others)
- Sulfamethoxazole (Gantanol; and others)
- Sulfamethoxazole–trimethoprim (Bactrim, Septra; and others)
- Thiothixene (Navane)
- Trifluoperazine (Stelazine; and others)
- Triflupromazine (Vesprin)
- Trimeprazine (Temaril)

Diuretics
- Acetazolamide (Diamox)
- Amiloride (Midamor)
- Bendroflumethiazide (Naturetin; and others)
- Benzthiazide (Exna; and others)
- Chlorothiazide (Diuril; and others)
- Cyclothiazide (Anhydron)
- Furosemide (Lasix)
- Hydrochlorothiazide (HydroDIURIL; and others)
- Hydroflumethiazide (Diucardin; and others)
- Methyclothiazide (Aquatensen; Enduron)
- Metolazone (Diulo; Zaroxolyn)
- Polythiazide (Renese)
- Quinethazone (Hydromox)
- Trichlormethiazide (Methahydrin; and others)

Hypoglycemics
- Acetohexamide (Dymelor)
- Chlorpropamide (Diabinese; Insulase)
- Glipizide (Glucotrol)
- Glyburide (DiaBeta; Micronase)
- Tolzamide (Tolinase)
- Tolbutamide (Orinase; and others)

Nonsteroidal anti-inflammatory drugs
- Benoxaprofen (Oraflex)
- Ketoprofen (Orudis)
- Piperacetazine (Quide)
- Prochlorperazine (Compazine; and others)
- Promethazine (Phenegran; and others)
- Thioridazine (Mellaril)
- Naproxen (Naprosyn)
- Phenylbutazone (Butazolidin; and others)
- Piroxicam (Feldene)
- Sulindac (Clinoril)

Sunscreens
- 6-Acetoxy-2,4-dimethyl-*m*-dioxane (preservative in sunscreens)
- Benzophenones
- Cinnamates
- Diethystilbestrol
- Disopyramide (Norpace)
- Dyes (acridine, acriflavine, anthraquinone, eosin, erythrocine, fluorescein, methylene blue, methyl violet, orange red, rose bengal, toluidine blue, trypaflavin, trypan blue)
- Furocoumarins:psoralens (trioxsalen, methoxsalen, psoralen)
- Gold salts (Myochrysine; Solganal)
- Hexachloraphene (pHisoHex; and others)
- Isotretinoin (Accutane)
- 6-Methylcourmarin (used in perfumes, shaving lotions, and sunscreens)
- Mestranol
- Musk ambrette (used in perfumes)
- Quinidine sulfate and gluconate
- Saccharine
- Tattoo dye (red or yellow cadmium sulfide)

Table 4 Common Plants and Lichens Causing Photodermatitis

Common name	Botanical name	Family
Lime	*Citrus aurantifolia*	Rutaceae
Citron	*Citrus medica* (*C. acida*)	Rutaceae
Bitter orange	*Citrus aurantium*	Rutaceae
Lemon	*Citrus limon*	Rutaceae
Bergamot	*Citrus bergamia*	Rutaceae
Gas plant; burning bush	*Dictamnus albus* (*D. fraxinella*)	Rutaceae
Common rue	*Ruta graveolens*	Rutaceae
Persian lime (Tahitian)	*Citrus aurantifolio*, ''Persian''	Rutaceae
	Phebalium argenteum	
Cow parsley, wild chervil	*Anthriscus sylvestris*	Umbelliferae
Celery	*Apium graveolens*	Umbelliferae
Giant hogweed	*Heracleum mantegazzianum*	Umbelliferae
Parsnip (garden variety)	*Pastinaca sative* (*P. urens*)	Umbelliferae
Cow parsley	*Heracleum sphondylium*	Umbelliferae
Parsnip (wild parsnip)	*Heracleum giganteum*	Umbelliferae
Fennel	*Foeniculum vulgare*	Umbelliferae
Dill	*Anethum graveolens*	Umbelliferae
	Peucedanum ostruthium	Umbelliferae
Wild carrot, garden carrot	*Daucus carota*	Umbelliferae
Masterwort	*Peucedaum ostruthium*	Umbelliferae
	Ammin majus	Umbelliferae
Angelica	*Angelica archangelica*	Umbelliferae
Figs	*Ficus carcia*	Umbelliferae
Milfoil, yarrow	*Achillea millefolium*	Composiate
Stinking mayweed	*Anthemis cotula*	Composiate
Buttercup	*Ranunculus* spp.	Ranunculaceae
Mustard	*Brassica* spp.	Cruciferae
Bind weed	*Convolvulus arvensis*	Convolvulaceae
Agrimony	*Agrimonia eupatoria*	Rosaceae
Goose foot	*Chenopodium* spp.	Cheopodiaceae
Scurfy pea, bavchi	*Psoralea corylifolia*	Leguminosae
St. John's wort	*Hypericum perforatum*	Hypericaceae
	Hypericum crispum	Hypericaceae
	Schinopsis quebrachocolorado	
Red quebracho	*Schinopsia lorentzii*	Anacardiaceae
Lichens	*Parmelia* spp.	Lichen (symbiotic
	Hypogymnia spp.	association between
	Pseudovernia spp.	fungi and algae)
	Cladonia spp.	commonly grouped
	Platismatia spp.	with fungi
	Physcia spp.	
	Umbilicaria spp.	
	Cetrania spp.	

viduals with this disorder have previously been exposed to the responsible agent and, after reexposure to it and UVR, develop the dermatitis 24–48 h later. Because the UVA portion of the spectrum is most often responsible, it is important to use an effective UVA sunscreen product (6).

F. Photosensitivity Skin Disorders

Photodermatoses include several disorders. Since afflicted persons are extremely sensitive to the sun, it is incumbent to provide them with a highly efficacious sunscreen or properly tested sun-protective clothing.

Polymorphous Light Eruption (290–365 nm)

Polymorphous light eruption (PMLE) is a common, idiopathic skin eruption that occurs in susceptible individuals on exposure to more intense than usual solar irradiation. It affects 10–14% of the white population, most of them female, with onset usually in the first three decades of life. Typically, PMLE manifests as some combination of pruritic papules, macules, vesicles, plaques, or confluent erythema 1–2 days (2 h to 5 days) after sun exposure on areas of unprotected skin. It is seen most commonly at the beginning of summer or during a visit to a geographic clime with a higher solar flux. The reaction resolves over 7–10 days if additional sun exposure is avoided. As the name suggests, the reaction is polymorphic, although, in a given individual, it is usually a single morphology that remains constant.

Investigators have variously implicated UVA, UVB, UVC, visible radiation, x-rays, and even α-particles in the action spectrum. Often, PMLE flares can be prevented by use of newer broad-spectrum sunscreens. A recent study of an Inuit Indian population demonstrated that at least 18 or 30 were sensitive to UVA. According to the physician evaluators in this study, 86% of the 21 patients who completed the study obtained good to excellent results by using a sunscreen containing padimate O and avobenzone (36).

However, it must be kept in mind that sunscreens that provide modest UVA protection (e.g., benzopheones) are poorly effective for those who suffer with PMLE. Many patients benefit from prophylactic PUVA photochemotherapy, whereas others respond to oral agents, including antimalarials or β-carotene (16).

For some, it may be necessary to avoid the sun entirely and to wear effective sun-protective clothing.

Porphyrias (400–410 nm) Porphyrias are caused by inherited or acquired abnormality in the heme metabolic pathway. Substrates for the affected enzyme accumulate, leading to the clinical manifestations. There are a wide variety of porphyrias because of the complexity of heme biosynthesis.

Photosensitivity begins in childhood on sun-exposed areas, with the acute development of vesicles, bullae, and hyperpigmentation or hypopigmentation.

The skin is also more sensitive to trauma than is normal. Chronically, the skin scars atrophically, with mutilating deformities of exposed areas, including cicatrizing alopecia. Other manifestations are highly visible depending on the type of pophyria (37). The action spectrum is in the visible range from 400 to 410 nm (Soret band) and thus these unfortunate individuals must be protected by a physical blocking agent or by adequate clothing. Some sufferers do respond to oral β-carotene.

Solar Urticaria (290–515 nm) Solar urticaria is a rare, rapidly developing reaction to sun or UV exposure (20). Minutes after exposure, sensitive individuals develop pruritus, followed by the urticaria and erythema. The reaction quickly runs its course, rarely lasting longer than 24 h.

Persons with this syndrome encompass different subsets. Some produce a serum factor that after inoculation into a normal subject, transiently permits induction of solar urticaria in the recipient. This group appears to generate an immunoglobulin of the IgE class that causes the reaction. These subjects can rightfully be classified among those with photoallergies. Others cannot transfer the reaction passively, and the mechanism for the development of solar urticaria is unknown.

Antihistamines are effective at high doses in a few patients. Maintenance solar exposure and PUVA therapy have also been used successfully in some. Patients should be phototested to determine the wavelengths that promote their reaction and then provided with a protective sunscreen. Alternatively, or additionally, they may be instructed to avoid the sun whenever possible (38).

Chronic Actinic Dermatitis (290–360 nm) Chronic actinic dermatitis encompasses the gamut of disease previously described variously as chronic photosensitivity dermatitis and actinic reticuloid (39). Seen in older, preponderantly male patients, this dermatosis manifests initially as erythematous macules and plaques in sun-exposed areas, including the face, exposed areas of scalp, rim of the ears, back of the neck, forearms, and back of the hands. Over time, this disorder can evolve into the lichenified, infiltrated lesions of actinic reticuloid, with superimposed areas of edema and vesiculation during acute flares. Clinical evaluation in these patients should be conducted by phototesting across a broad portion of the UV and visible spectrum.

Histologically, the skin initially resembles a contact dermatitis, but can progress with actinic reticuloid to the pattern of a cutaneous T-cell lymphoma. The portion of the spectrum causing the reaction varies from one person to another. Thus, phototesting and careful selection of sunscreens or avoidance of the sun is mandated in the management of these patients. Treatment options include oral azathioprine, PUVA, with or without prednisone, etretinate, and cyclosporine (39).

Persistent Light Reaction (290–400 nm) Most patients who develop a chemical photosensitivity dermatitis note that their reaction resolves within a week or two if the offending agent is eliminated. A subset of individuals continue to overreact to the sun in spite of careful avoidance of the putative photosensitizer.

This condition is called persistent light reaction and sometimes resolves in months, although it may persist indefinitely. For some persons, the reaction is so severe it results in a generalized exfoliate dermatitis. Histologically, the tissue shows a dense perivascular round cell infiltrate.

In some cases, the mechanisms for this reaction is ongoing exposure to some chemical that is not correctly identified and eliminated. In other instances, the offending agent has bonded to dermal proteins in such a way that it cannot be eliminated. Often, the reactions is idiopathic. The action spectrum may be in the UVB and UVA is common. Phototesting to determine appropriate protective sunscreens and avoidance of the sun are usually effective in minimizing flares (38).

Lupus Erythematosus (290–330 nm) Cutaneous lupus erythematosus (LE) occurs in two broad forms, discoid LE and subacute cutaneous LE. The lesions of discoid LE are typically found on the head and neck and are well defined, variably sized scaly patches that heal leaving atrophy, scarring, and pigmentary changes. Subacute cutaneous LE manifests as papulosquamous or polycyclic lesions around the neck, the outer aspects of the arms, and on the trunk. These lesions are more mild than discoid LE and heal without scarring or atrophy (40). Although UVB appears to be more prominent in LE, UVA also plays a role in some individuals (34).

Interestingly, exacerbations of noncutaneous lupus (e.g., nephritis) are sometimes seen after sun exposure. Thus, it is necessary to provide these patients with broad-spectrum sunblock and encourage them to avoid the sun. Oral antimalarials, such as chloroquine, have been used with limited success. Recently, an open-label trial of a broad-spectrum sunscreen containing padimate O and avobenzone reduced clinical severity of cutaneous LE over a 4-week period (40).

Xeroderma Pigmentosum (290–340 nm) The autosomal recessive disorder xerodermic pigmentosum (XP) stems from a genetic defect in the ability to repair UVB-induced dimers. This leads to a high frequency of mutations and premature development of skin neoplasms, including malignant melanoma as well as basal cell and squamous cell carcinoma. One study noted the median age of skin cancer in XP patients was 8 years of age, in comparison with 60 in the general population. Other clinical manifestations include sun sensitivity, skin atrophy, and increased freckling. Photophobia and other ocular manifestations, including severe UV-induced conjunctivitis and keratitis, are common. Progressive neurological deterioration occurs in about 40%.

After diagnosis, treating physicians should aim for careful protection against the sun and vigilant detection of skin cancer. Eye protection should also be emphasized (16).

Albinism Oculocutaneous albinism is an autosomal recessive defect in the formation of melanosomes. Although melanocyts are present, these individuals are either hypomelanotic or amelanotic in the skin and hair. Patients living in areas of high solar flux uniformly develop premalignant (i.e., actinic keratoses) or malignant (i.e., nonmelanoma skin cancer) skin lesions by early adulthood if they are not careful to apply sunscreen and avoid unnecessary sun exposure. Melanoma is rare, although the reason for this is unclear (35).

V. CONCLUSIONS

Our understanding of the effects of sun on the skin continue to evolve. A relation between UVB exposure and the development of basal cell and squamous cell carcinoma has been documented. It appears likely that high-dose exposure to UVR increases the risk for the development of malignant melanoma. Prolonged exposure to either UVA or UVB increases the rate of skin aging.

Development of new, highly effective sunscreens of both the traditional chemical kind as well as newer micronized physical blockers continues. More effective UVB-blocking formulations may lead to increased cumulative sun exposure and, thus, UVA exposure. However, only time will delineate the effect this will have on the rates of skin cancer and skin aging. Meanwhile, physicians may benefit their patients most by discouraging sun exposure and encouraging prudent and frequent use of sunscreens.

Children as well as adults with a likelihood of increased sun exposure should be warned of the short and long-term consequences of UVR on the skin and should be instructed concerning sun protection. Any person suffering from one of the many photosensitivity dermatoses must receive counseling about the special requirement to protect his or her skin.

REFERENCES

1. Scotto, J., Fears, T. R., Fraumeni, J. F. Incidence of non-melanoma skin cancer in the United States, U.S. Dept. of Health and Human Services Publication, Washington, DC: 82-2433, 1983.
2. Sayre, R. M. and Lowe, N. J. Scientific poster presentation at the American Academy of Dermatology Meeting. December 1992.
3. Stern, R. S. Sunscreen use and non-melanoma skin cancer. *Sunscreens, Development, Evaluation and Regulatory Aspects* (N. J. Lowe and N. Shaath, eds.), Marcel Dekker, New York, 1990, pp. 85–92.
4. Thompson, S. C., Jolley, D., and Marks, R. Reduction of solar keratoses by regular sunscreen use, *N. Engl. J. Med.*, *329*:1147–1151 (1993).
5. Kaidbey, K. H. The photoprotective potential of the new superpotent sunscreens, *J. Am. Acad. Dermatol.*, *22*:449–452 (1990).

6. Roelandts, R. J. Which components in broad spectrum sunscreens are most necessary for adequate UVA protection? *Am. Acad. Dermatol.*, *25*:999–1004 (1991).
7. Lowe, N. J., Weingarten, D., and Wortzman, M. Sunscreens and phototesting, *Clin. Dermatol.*, *6*:40–49 (1988).
8. Luftman, D. B., Lowe, N. J., and Moy, R. L. Sunscreens: Update and review, *J. Dermatol. Surg. Oncol.*, *17*:744–746 (1991).
9. Roberts, J. Exposure to the sun. *Management of Wilderness and Environmental Emergencies*, 2nd ed. (Auerbach, P., ed.), Mosby, St. Louis, 1989.
10. Epstein, J. H. Biologic effect of sunlight, *Sunscreens, Development, Evaluation and Regulatory Aspects*. (N. J. Lowe and N. Shaath, eds.), Marcel Dekker, New York, 1990, pp. 43–54.
11. Lowe, N. J. Sun protection factor: Comparative techniques and selection of UV sources, *Sunscreens, Development, Evaluation and Regulatory Aspects* (N. J. Lowe and N. Shaath, eds.), Marcel Dekker, New York, 1990, pp. 379–394.
12. Garmyn, M. Sohrabvand, N., and Roelandts, R. Modification of sunburn cell in 8-MOP-sensitized mouse epidermis: Method of assessing UVA sunscreen efficacy. *J. Invest. Dermatol.*, *93*:642–645 (1989).
13. Kaplan, L. A. Suntan, sunburn and sun protection, *J. Wilderness Med.*, *3*:173–196 (1992).
14. Harber, L. C. DeLeo, V. A., and Prystowsky, J. H. Intrinsic and extrinsic photoprotection against UVB and UVA radiation, *Sunscreens, Development, Evaluation and Regulatory Aspects* (N. J. Lowe and N. Shaath, eds.), Marcel Dekker, New York, 1990, pp. 359–378.
15. Kligman, L. H. and Kligman, A. M. Ultraviolet radiation-induced skin aging, *Sunscreens, Development, Evaluation and Regulatory Aspects* (N. J. Lowe and N. Shaath, eds.), Marcel Dekker, New York, 1990, pp. 55–72.
16. Kraemer, K. H. Heritable diseases with increased sensitivity to cellular injury, *Dermatology in General Medicine*, 3rd ed. (T. B.Fitzpatrick, et al., eds.), McGraw-Hill, New York, 1987, pp. 1791–1811.
17. Garland, C. F., Garland, F. C., and Gorham, E. D. Rising trends in melanoma; an hypothesis concerning sunscreen effectiveness, AEP *3*:103–110 (1993).
18. Shaath, N. A. Evolution of modern sunscreen chemicals. *Sunscreens, Development, Evaluation and Regulatory Aspects* (N. J. Lowe and N. Shaath, eds.), Marcel Dekker, New York, 1990, pp. 3–36.
19. Lowe, N. J., Dromgoole, S. H., Sefton, J., et al. Indoor and outdoor efficacy testing of a broad spectrum sunscreen against ultraviolet. A radiation in psoralen-sensitized subjects, *J. Am. Acad. Dermatol.*, *17*:224–230 (1987).
20. Lowe, N. J. UVA photoprotection, *Sunscreens, Development, Evaluation and Regulatory Aspects* (N. J. Lowe and N. Shaath, eds.), Marcel Dekker, New York, 1990, pp. 459–468.
21. Gotlieb, A., Bourget, T., and Lowe, N. J. Sunscreens: Effects of amounts of application of sun protection factors. *Sunscreens, Development, Evaluation and Regulatory Aspects* (N. J. Lowe and N. Shaath, eds.), Marcel Dekker, New York, 1990, pp. 441–446.
22. Cole, C. and VanFossen R. Measurement of sunscreen UVA protection; an unsensitized human model, *J. Am. Acad. Dermatol*, *26*:178–84 (1992).

23. Kaidbey, K. H. and Barnes, A. Determination of UVA protection factors by means of immediate pigment darkening in normal skin, *J. Am. Acad. Dermatol.*, *25*:262–266 (1991).
24. Agin, P. P. and Stanfield, J. W. [letter], *J. Am. Acad. Dermatol.*, *27*:136–137 (1992).
25. Chardon, A. Immediate pigment darkening. Presented at 4th Congress of the European Society for Photobiology, Sept. 1–6, 1991.
26. Diffey, B. L. and Robson, J. J. A new substrate to measure sunscreen protection factors throughout the ultraviolet spectrum, *Soc. Cosmet. Chem.*, *40*:127–133 (1989).
27. Roelandts, R., Sohrbvand, N. and Garmyn, M. Evaluating the UVA protection of sunscreens. *J. Am. Acad. Dermatol.*, *21*:56–62 (1989).
28. Sayre, R. M. and Agin, P. P. A method for the determination of UVA protection for normal skin, *J. Am. Acad. Dermatol.*, *23*:429–440 (1990).
29. Kaidbey, K. H. Substantivity and water resistance of sunscreens, *Sunscreens, Development, Evaluation and Regulatory Aspects* (N. L. Lowe and N. Shaath, eds.), Marcel Dekker, New York, 1990, pp. 405–410.
30. Fisher, A. A. Sunscreen dermatitis: *para*-Aminobenzoic acid and its derivatives, *Cutis*, *50*:190–192 (1992).
31. Dromgoole, S. H. and Maibach, H. I. Contact sensitization and photocontact sensitization of sunscreening agents. *Sunscreens, Development, Evaluation and Regulatory Aspects* (N. L. Lowe and N. Shaath, eds.), Marcel Dekker, New York, 1990, pp. 313–340.
32. Dromgoole, S. H. and Maibach, H. I. Sunscreening agent intolerance: Contact and photocontact sensitization and contact urticaria, *J. Am. Acad. Dermatol.*, *22*:1068–1078 (1990).
33. Patel, N. H. Highton, A., and Moy, R. L. Properties of topical sunscreen formulations, *J. Dermatol. Surg. Oncol.*, *18*:316–320 (1992).
34. Johnson, J. A. and Fusaro, R. M. Broad spectrum photoprotection: The role of tinted auto windows, sunscreens and browning agents in the diagnosis and treatment of photosensitivity, *Dermatology*, *185*:237–241 (1992).
35. Mosher, D. B., Fitzpatrick, T. B., Ortonne, J. P., et al. Disorders of pigmentation, *Dermatology in General Medicine*, 3rd ed. (T. B. Fitzpatrick, et al., eds.), McGraw-Hill, New York, 1987.
36. Fusaro, R. M. and Johnson, J. A. Topical photoprotection for hereditary polymorphic light eruption of American Indians, *J. Am. Acad. Dermatol.*, *24*:744–746 (1991).
37. Bickers, D. B. and Pathak, M. A. The porphyrias, *Dermatology in General Medicine*, 3rd ed. (T. B. Fitzpatrick, et al., eds.), McGraw-Hill, New York, 1987, 1660–1715.
38. Bernhard, J. D. Pathak, M. A., Kochevar, I. E., et al. Abnormal reactions to ultraviolet radiation, *Dermatology in General Medicine*, 3rd ed. (T. B. Fitzpatrick, et al., eds.), McGraw-Hill, New York, 1987, 1481–1506.
39. Roelandts, R. J. Chronic actinic dermatitis, *J. Am. Acad. Dermatol.*, *28*:240–294 (1993).
40. Callen, J. P. Roth, D. E., McGrath, M. A., and Dromgoole, S. H. Safety and efficacy of a broad spectrum sunscreen in patients with discoid or subacute cutaneous lupus erythematosus, *Cutis*, *47*:130–136 (1991).
41. Stern, R. S., Weinstein, M. C., and Baker, S. G. Risk reduction for non-melanoma skin cancer with childhood sunscreen use, *Arch Dermatol.*, *122*:537 (1986).

3

Photoprotection Against Harmful Effects of Solar UVB and UVA Radiation: An Update

Madhu A. Pathak
Harvard Medical School and Massachusetts General Hospital, Boston, Massachusetts

I. INTRODUCTION

Sunlight is a part of our everyday life, and most persons accept it as ''good'' and beneficial to our health. With the advent of our knowledge in cutaneous photobiology and photodermatology within the past 50 years, the terrestrial solar radiation and particularly ultraviolet radiation (UVRs 290–400 nm) have become a major environmental factor deleterious to our health. Most of the effects (e.g., sunburn, suntanning, or pigmentation, photoaging (solar elastosis or dermatoheliosis), solar keratosis, skin cancer (nonmelanoma and melanoma), and immunosuppression, are attributed to ultraviolet radiation and, more particularly, to UVB radiation (290–320 nm). Topical sunscreens have become an essential armament for dermatologists and cosmeceutical companies in providing protection to human skin against acute and chronic adverse effects of solar radiation. In this short presentation, an attempt is made to present a brief updated overview of solar radiation and the general principles of protection with topical sunscreens for clinicians interested in science of photoprotection. This brief chapter will review (1) the solar radiation at the surface of our earth; (2) the proposed new definition of sunscreens in view of changing concepts in photoprotection; (3) the general effects of ultraviolet B (UVB) and ultraviolet A (UVA) radiation; particular emphasis is focused on the harmful effects of UVA radiation, with a view to modify and improve our approach to photoprotection not only against

UVB radiation, but more specifically against that of UVA wavelengths; and (4) photoprotective measures that are practical and useful for physicians to recommend to their patients who seek advice not only against sunburn reaction, but also against hyperpigmentation, photoprotection, photoaging, and other problems. New approaches on protection of skin against UVB and UVA radiation are emphasized, with specific recommendations for UVA protection. Additional information on this subject was recently reviewed (1), and it is recommended that the reader refer to this article, as well as others in this book on sunscreens.

II. SOLAR RADIATION

Although the sun emits all kinds of radiations, including x-ray, ionizing, ultraviolet, visible, infrared radiations, and radiowaves, for clinical and photoprotective considerations, the solar spectrum at the earth's surface (sea level) consists of wavelengths of electromagnetic energy between 290 and 3000 nm. The solar spectrum implicated in human skin reactions involves (1) ultraviolet radiation (290–400 nm); (2) visible radiation (400–760 nm); and (3) infrared radiation up to 1800 nm. Although the sun is known to emit radiation shorter than 290 nm, these wavelengths are not encountered at sea level, as they are selectively absorbed and filtered out by the ozone layer and by molecular oxygen of the stratosphere. For convenience, the ultraviolet spectrum from both sunlight and artificial sources is arbitrarily subdivided into three bands, from the longer to shorter wavelengths. These bands are referred to as UVA, UVB, and UVC regions (Tables 1 and 2). The UVA region extends from 320 to 400 nm. It is also referred to as longwave, or near-ultraviolet region. This spectrum causes immediate and delayed tanning reaction and several other effects, including photoaging (solar elastosis) and skin photosensitization (additional information about the effects of UVA radiation is outlined later in this chapter). The UVB band extends from 290 to 320 nm. It is also referred to as sunburn radiation,

Table 1 Wavelength Distribution of Solar Irradiance Before Attenuation by the Earth's Atmosphere

Wavelength region	Irradiance (W/m^2)	% or total
UVC (< 280 nm)	6.4	0.5
UVB (290–320 nm)	21.1	1.5
UVA (320–400 nm)	85.7	6.3
Visible (400–760 nm)	532	38.9
Infrared (>760–3000 nm)	722	52.8
Solar constant	1367	100.0

Table 2 Solar Irradiance at Sea Level in Spectral Bands and for Two Zenith Angles

Spectral band (nm)	Zenith angle = 0°	Zenith angle = 20°
290–320 (UVB)	0.5	0.3
320–400 (UVA)	6.3	5.0
400–760 (visible)	38.9	41.0
760–3000 (IR)	54.3	54.0

mid-UVR, or erythema band, and is the principal cause of sunburn, skin cancer, and many other effects, some of which are listed in Table 3. The UVC radiation (also referred to as germicidal, short UVR, or far UVR) comprises wavelengths shorter than 290 nm (from 200 to 290 nm). This spectral band is erythemogenic, mutagenic, and also carcinogenic, but is absent at the surface of the earth because it is absorbed and filtered by the ozone layer.

There is now a heightened concern for UVC and UVB radiation because of the depletion of the stratospheric ozone layer by chlorofluorocarbons, halons, nitric oxides, and so on. This will result in an increased irradiance level of both UVB and UVC radiation at the earth's surface that may eventually contribute to a higher incidence of skin cancer and other harmful effects to humans and to other forms of life on this planet (2). The UVC radiation from artificial light

Table 3 Risks and Benefits of Solar Radiation

Risks
- Sunburn
- Sun-tanning (immediate and delayed pigmentation)
- Photoaging (dermatoheliosis)
- Solar keratoses
- Skin cancer (basal and squamous cell carcinoma)
- Malignant melanoma
- Local and systemic immune suppression
- Photosensitivity diseases (polymorphous light eruption, lupus erythematosus)
- Drug-related phototoxic and photoallergic reactions
- Cataracts

Benefits
- Vitamin D synthesis
- Phototherapy and photochemotherapy
- Psychological comforts

sources (e.g., germicidal lamps, low-pressure mercury arc lamps, and high-pressure mercury arc lamps) can occasionally be encountered by humans; this may cause erythema as well as photokeratitis to our eyes.

Recently, Frederic (3) and Urbach (2) have presented some interesting data on solar radiation that are quite relevant to knowing the relative contributions of UVA and UVB radiation to sunburn reaction and other harmful effects. In Table 4, total UVB and UVA flux (watt per square meter; W/m^2) at sea level, latitude 30°N at noontime for the months of March, June, September, and December are presented (2,3). Also given in the table are the calculated contributions of solar UVB and UVA to the erythema reaction. These data indicate that approximately 15 minimal erythema doses (MEDs) of UVB and UVA radiation can be received in late June and early July at 30°N latitude at seal level during the day, assuming 1 MED = 250 J/m^2 or 25 mJ/cm^2. The total UV radiation flux, in terms of UVB and UVA radiation per square meter for different times of the year shown in Table 4 indicates that the effective UVB and UVA components capable of inducing sunburn reaction constitute approximately 85 and 15%, respectively, of erythemogenic radiation. For convenience the UVA spectrum is further subdivided into UVA-I (340–400 nm) and UVA-II (320–340 nm). The UVA-II is more damaging than UVA-I and, in many respects, behaves more like UVB radiation. It is erythemogenic, melanogenic, as well as carcinogenic.

It is difficult to provide data pertaining to solar radiation incident on various localities elsewhere in the world relative to solar altitude, irradiance values for UVB and UVA in terms of microwatt per square centimeter (μW/cm^2) and milliwatt per square centimeter (mW/cm^2) and the percentage values of UVA and UVB radiation based on latitude and seasonal variation. Some general statements can be made that are helpful for the physician in advising patients about the effects of solar UVA and UVB radiation.

Table 4 Total UVB and UVA Flux at Sea Level: Latitude 30°N, Noontime[a]

Date	Erythema flux (W/m^2)	MED/h	Erythemogenic radiation (%) UVB (290–400 nm)	UVA (320–400 nm)	UVA-I (340–400 nm)	UVA-II (320–340 nm)
Mar. 21	0.214	3.08	82.0	18.0	7.80	10.20
June 21	0.296	4.26	85.0	15.0	6.45	8.55
Sept. 21	0.230	3.31	83.6	16.4	7.14	9.27
Dec. 21	0.097	1.39	75.7	24.3	10.97	13.36

[a]Data based on 1 MED = 250 J/m^2 or 25 mJ/cm^2 for skin phototype II.
Source: Ref. 2.

1. In the Northern Hemisphere, the UV output of the sun is highest in summer months, particularly in the months of June and early July, and lowest in the winter months (December and January).

2. The flux of UVB radiation is maximum in a 2-h period close to solar noon (11 AM–1 PM; up to 60% of total UV radiation can be received between 11 AM and 3 PM.

3. The solar radiation incident in various localities on the earth's surface varies significantly with latitude, altitude, season (day of the year), time of the day, cloudiness, atmospheric aerosols, and so on. It is important to understand that UVB accounts about 4–6% of total terrestrial UVR; and UVA accounts for about 6–7% of total solar radiation. About 15 MED equivalent doses of UVB and about 4–5 MED equivalent dose of UVA can be received at the surface of the earth during a day. At high altitudes (e.g., 4600 m; about 15,000 ft), these values may still be higher (>18–19 MEDs), but for practical purposes, we should realize that, in summer months, we can face over 15 MEDs of sunburn radiation from sunrise to sunset at sea level in North America. For this reason, sunscreens with a sun protection factor (SPF) value of at least 15 are essential to achieve adequate protection against an average exposure of less than 4 h.

4. Since latitude determines the angle between a location and the sun, the latitude of every given location exerts a strong effect on the amount of radiation that is received. The equatorial belt (0–10°N, S) is known to receive the highest amount of UVR throughout the year. The angle between the location and the sun determines the amount of the atmosphere that photons have to pass through before reaching the earth. The lower the angle, the longer is the path that the photons have to pass through in the atmosphere (e.g., at a solar altitude of 90°, 60°, and 30°, the UVB flux is about 290, 230, 60 $\mu W/cm^2$, respectively). The flux of UVB (e.g., 295–300 nm) can be expected to vary from the equator to 70°N by a factor of over 100, whereas UVA flux (e.g., 330–340 nm) can be expected to vary by a factor of 5. Clearly the flux of UVB varies by latitude more than that of UVA.

5. The UVB radiation is the predominant cause of sunburn reaction during the months of March to September in the Northern Hemisphere. From personal observations on the determination of MED response between 11 AM and 1 PM, at varying solar altitudes (30°–80°), the percentage of UVB and UVA radiation contributing to erythema reaction varied from 70 to 80% and 30 to 20%, respectively.

6. Interestingly, the sunburn-producing capacity of UVB spectrum (290–320 nm) is approximately three orders of magnitude greater than that of UVA radiation. Thus, UVA is about 600–1000 times less effective than UVB radiation in inducing a sunburn reaction (e.g., the MED of UVB for a fair-skinned individual, who burns easily, but does not tan, is approximately 20–30 mJ/cm^2, the MED of UVA for the same skin-type individual is approximately 20–30 J/cm^2 (1).

III. DEFINITION OF SUNSCREEN (PRESENT AND FUTURE)

For medical, industrial, as well as legal explanations, there is a need for defining sunscreens that the cosmeceutical industry produces for photoprotection against solar radiation. Surprisingly, even the medical and leading English language dictionaries of United States do not define sunscreens. The conventional definition of a *sunscreen* that has gradually emerged since 1938, when the use of *p*-aminobenzoic acid (PABA) was first proposed as a sunscreen by a renowned dermatologist, the late Stephen Rothman, is to screen the sun and to protect or shelter viable cells of skin against potentially harmful effects of solar ultraviolet radiation (λ 290–400 nm), as exemplified by such reactions as sunburn, skin cancer, skin photosensitization, and so forth. The effects of UVR photons are targeted at the level of DNA, RNA, proteins, cell membrane lipids, enzymes, and organelles, that contribute to many harmful reactions. Protection against such harmful reactions is achieved by topical application of a sunscreen to the skin in the form of a thin film of a solution, lotion, cream, spray, or gel, containing a known quantity of UV-absorbing chemical (or chemicals) that singly or collectively are capable of impeding the penetration of impinging harmful UV rays into the skin. Until recently, photoprotection and the prevention of damaging effects of solar radiation have been achieved by an attenuation process, involving absorption, reflection, or scattering of UVR by chemicals that constitute sunscreen(s).

In view of the advances in biotechnology in percutaneous penetration and better modes of delivery of photoprotective molecules to the skin, either by topical or oral routes, at or near the target site of photons, this old conventional definition of sunscreens may have to be modified before the dawn of the 21st century to include new concepts or approaches of photoprotection. The absorbed UV energy can exert its effects in various forms. We know that free radicals, reactive forms of oxygen species ($^1O_2$, $O_2^{\cdot -}$, $^{\cdot}OH$, or other) and other reactive forms of photoexcited states, are generated by UV radiation, and they interact with target sites and exert specific harmful effects at cellular and molecular levels. The UV-induced free radicals can be quenched, the formation of reactive forms of oxygen species can be intercepted or inhibited, and the harmful products of interaction between photons and essential biological molecules of the skin can be prevented. Postinflammatory mediators released in response to UV exposure (e.g., prostaglandins, histamine, and other cytokines) can be effectively counteracted by specific pharmacological agents. The skin can be protected against free radical reactions and oxidative stress of reactive oxygen species or some other forms of photoactivated states of molecules. Accordingly, the future generation of photoprotectants (hence, sunscreens) may or may not absorb UV radiation, but may act by topical, oral, or injectable route at target sites as free radical scavengers or antioxidants, as cytokine inhibitors, or as quenchers of photoexcited states, and protect the skin cells against actinic damage. The regulatory agencies (e.g., U.S. FDA) will have to recognize such molecules as sunscreens. It will be imperative to demonstrate not only the photoprotective efficacy of such molecules under an in vitro system, but also under in vivo conditions after the delivery of photoprotective molecules to the skin at the site of photoreaction.

IV. EFFECTS OF UVB AND UVA RADIATION

In the first 75 years of the 20th century, sunlight and UV radiation from artificial sources was recognized for its tonic value, and sunbathing was widely advocated for health reasons, including the biosynthesis of vitamin D, essential for growth of strong bones, and for promoting the healthy appearance of skin with a golden brown tan. At the close of the 20th century, sunbathing still remains popular in Western Civilization and now persons from all over the world including Japan, China, Soviet Russia, and other continental countries, indulge in sunbathing, outdoor recreational activities, traveling, and visiting places with bright sun and warm climates. These persons are recognizing and realizing the potential harmful effects of sunlight. With the advent of our knowledge in cutaneous biology within the past 50 years, solar radiation has now become a major environmental factor deleterious to our health. Solar radiation, and particularly UV radiation (λ 290–400 nm), has been implicated in the induction of various acute and chronic harmful reactions in the human skin, briefly outlined in Table 1. Dermatologists worldwide are familiar with the effects of UVB radiation (290–320 nm) on human skin. Sunburn, suntanning (darkening of skin involving increased synthesis of melanin resulting from numerical increase in the population density of melanocytes); photoaging, or dermatoheliosis (a polymorphic response of the skin involving dermal connective tissue and the vascular system, causing wrinkling of skin and alterations in colagen, elastin fibers, and proteoglycans); development of nonmelanoma (basal and squamous cell) carcinomas, and malignant melanoma-type of skin cancers; immunosuppression in the form of local and systemic impairment of immune system, with altered cutaneous immune functions; photosensitivity problems of patients with polymorphous light eruptions, lupus erythematosus; as well as drug-induced or chemically induced phototoxic and photoallergic reactions, represent a few examples of growing concerns about the deleterious effects of solar radiation (290–320 nm) (1).

Unfortunately, very little attention is paid to the harmful effects of UVA radiation (320–400 nm), which is quite abundant in natural light. Indeed, until recently, the UVA region was considered to be innocuous and of little or no concern to dermatologists, except for an academic interest in understanding the nature of tanning reaction of skin. In 1974, with the development of psoralen and UVA (PUVA) in the photochemotherapy of psoriasis and other skin diseases (4), the interest of dermatologists and photobiologists in understanding the effects of UVA radiation became important and widespread. The availability of high-intensity UVA-emitting systems useful for PUVA therapy contributed to the growth of tanning salons. The tanning industry suddenly blossomed in the mid-1970s and early 1980s throughout European countries, the United States, and elsewhere, and it became increasingly evident that for cosmetic, recreational, and even health reasons, humans were exposing themselves to increasing amounts of UVA radiation, without realizing the potential acute and chronic harmful effects to their skin of solar UVA radiation, or UVA radiation from artificial light sources. It is now evident that UVA radiation is not safe, and many of the harmful effects produced by UVB radiation (see Table 1) are also induced by UVA radiation. The following facts about harmful effects of UVA, abstracted from various publications, are briefly summarized for the scientific interest of dermatologists and the cosmetic industry.

A. UVA Output

The UVA output by the sun is significantly greater than the UVB output throughout the year. The extraterrestrial UVA-irradiance, before attenuation by the earth's atmosphere, is approximately 85.7 W/m^2 (3). A major portion of this radiation after scattering by atmospheric molecules (e.g., N_2, O_2) reaches the earth's surface because the ozone in the outer atmosphere plays no specific role in filtration or diminution of UVA radiation. Table 2 provides approximate values of solar irradiance at sea level in spectral bands of UVB, UVA, visible, and such, for two zenith angles, expressed as the percentage of total irradiance. The major source of UVA radiation from artificial light sources is the emission from low-pressure mercury arc fluorescent lamps that are used by lighting industry in tanning salons, tanning beds, and phototherapy units. Most of these lamps emit high-intensity UVA radiation (13–20 mW/cm^2), and some industrial lamps may emit even higher fluences than these.

B. Effectiveness of UVA versus UVB

The UVB flux undergoes significant seasonal variation; the UVA flux, however, does not appreciably change seasonally (Table 5). The quantum efficiency of UVB photons, in terms of erythema reaction, is approximately three orders of magnitude greater than the quantum efficiency of UVA photons. If we determine the MED of these two spectral bands in the same individual (fair skin with no tan), and compare the erythemal effectiveness of UVB and UVA radiation, we shall find that UVB radiation is about 600–1000 times more effective in inducing a minimal erythema reaction than the dose required for a minimal ery-

Table 5 Erythema Effective Energy at 30°N (noontime)

Time	Erythema flux (W/m^2)	Total MED (Noon)	% UVB	% UVA
Mar. 21	0.214	3.1	82	18.0
June 21	0.296	4.3	85	15.0
Sept. 21	0.230	3.3	83.6	16.4
Dec. 21	0.097	1.4	76	24

thema reaction of UVA radiation (e.g., MED of UVB = 20–30 mJ/cm^2; MED of UVA = 30–35 J/cm^2).

C. Human Skin and UVA Transmission

The transmission of UVA into the dermis of an untanned fair-skinned individual is approximately 50% of the impinging flux. Even-pigmented persons with brown skin transmit a significant amount of UVA radiation (30–35%). The UVA radiation penetrates deeper into the skin (at a depth of about 160–250 μm) than UVB radiation (at about 17–49 μm). The UVA radiation can reach to the dermal papillae of hair in anagen phase when compared with UVB radiation that reaches only to the superficial papillary dermis. The UVA effects can occur to the vascular network in deeper layers of the reticular dermis. This radiation is known to induce structural and functional changes in keratinocytes, melanocytes, Langerhans cells, and fibroblasts. Children, teenagers, and young adults in their youthful life receive a significant portion of solar radiation before the age of 18. This exposure is usually uncontrolled and without photoprotection. This exposure contributes to many changes in the skin that results in premature aging of skin in later years of life.

D. UVA Radiation Causes Sunburn

The extent of contribution of UVA from sunlight for acute reactions such as sunburn, is about 18 to 20% (see Table 5). This percentage may vary between 15 and 25%, depending on the latitude, season (time of the year), time of day, and atmospheric conditions. The MED for UVA-induced sunburn is approximately 20–30 J/cm^2 for a fair-skinned individual who always burns, but does not tan well (1). Unlike the UVB burn that is visible between 4 and 6 h, with a peak response at 20 h, after exposure, the UVA burn, in a fair-skinned individual, is visible earlier than UVB burn and peaks between 6 and 10 h after exposure; it usually fades within 24–36 h, if the exposure dose is low (<2 MED). In persons who have tanned or brown skin, the UVA-induced burn is usually not visible unless the UVA exposure dose is greater than 50–60 J/cm^2.

E. UVA Effects Are Additive to the Effects of UVA

The UVA radiation penetrates deep into the skin; its effects are additive to the effects of UVB (5). The relative contribution of UVA and UVB radiation to sunburn reaction is 15–24% and 75–85%, respectively. Some years back, it was erroneously believed that UVA effects were augmentative and enhanced the effects of UVB. Only when exogenous photosensitizers are present (e.g., 8-methoxypsoralen or demethylchlortetracycline) can the effects of UVA be considered as augmentative owing to a photosensitization reaction that occurs at low fluences of UVA radiation.

F. UVA Activates Melanocytes and Causes Tanning of Skin

The immediate tanning reaction (also called as immediate pigment-darkening reaction, IPD) and delayed-tanning reaction (new melanogenesis) of human skin are caused by UVA radiation (1,5–7) and involve oxidation of preexisting melanin, followed by new and increased synthesis of melanin (7) that occurs within 48 h after exposure. This is accompanied by increased population density of melanocytes, with increased production of melanosomes and increased melanization of epidermis. This phenomenon, when studied under light and electron microscopes, indicates recognizable damage to functional melanocytes, and this process of increased melanogenesis is now quite well known to dermatologists and pigment cell biologists. People of skin phototypes II–VI (Caucasians, Orientals, Asiatic Indians, Hispanics, and blacks) become hyperpigmented very easily by UVA radiation in which molecular oxygen is known to play a prominent role. Without oxygen, the IPD reaction, involving oxidation of preexisting melanin in the skin, does not occur, and new melanogenesis or delayed-tanning reaction also fails to occur (5,6). The reactive forms of oxygen [e.g., superoxide anion ($O_2^{\cdot-}$) singlet oxygen ($^1O_2$), and hydroxy radicals ($^{\cdot}OH$) are readily and effectively induced by UVA irradiation; they play a distinct role in cellular damage (peroxidation of membrane lipids) and hyperproliferation of melanocytes. This is evidenced by observing facial hyperpigmentary problems in patients with melasma; the facial melasma of such patients is often aggravated by UVA as well as UVB radiation.

G. UVA Causes DNA Damage

Many lines of evidence indicate that the primary biological actions of UVA radiation involve DNA damage, mediated through absorption of UVA by other molecules, such as flavin mononucleotide (FMN), nicotinamide adenine dinucleotide NADH, NADPH (reduced form), certain quinones (ubiquinone), tryptophan, porphyrins, and riboflavin, present in the skin (8,9). These molecules act as chromophores to absorb UVA radiation. They are present in trace amounts in epidermal cells, and on photoexcitation by 320 to 400-nm radiation, they interact with molecular oxygen to produce reactive forms of oxygen, referred earlier to as $^1O_2$, $O_2^{\cdot-}$, $^{\cdot}OH$, and H_2O_2. These reactive oxygen species induce significant damage in the form of single-strand breaks, induction of C_4-cyclobutane-type dimers (e.g., thymine dimers), and DNA–protein cross-links. Thus, UVA radiation causes DNA damage and other alterations through non-DNA chromophores, as just briefly outlined (8–10). Indeed, there is good scientific evidence to indicate that reactive forms of oxygen, including free radicals, participate in acute and chronic responses of skin evoked by UVA radiation (e.g., sunburn, inflammation, skin pigmentation, and skin carcinogenesis).

H. UVA Causes Distinct Damage to Cell Membranes

The UVA damage is oxidative and is caused by UVA radiation interacting with oxygen to produce reactive forms of oxygen ($^1O_2$, $O_2^{\cdot-}$, $^{\cdot}OH$, and H_2O_2) (9–11) in epidermal cells. Cell membrane lipids of keratinocytes undergo lipid peroxidation reactions that contribute to an inflammatory response and the generation of inflammatory mediators (e.g., prostaglandins, such as PGE_2), inactivation of enzymes, denaturation of proteins, damage of cell organelles (involving mitochondria and impaired oxidative phosphorylation) that are essential for survival of cells. We must realize that our skin is profusely perfused with oxygenated blood; capillaries and venules are easy targets for the reactive oxygen species produced by UVA radiation.

I. UVA Radiation Causes Phototoxic and Photoallergic Reactions

Phototoxic and photoallergic reactions are invariably caused by UVA radiation. Certain drugs or chemicals often gain access to the skin, either after topical application, or after oral ingestion of certain drugs given for medicinal purposes (12). Examples of such drugs or chemicals include certain antibiotics (e.g., tetracyclines), sulfonamides, birth control pills, certain diuretics (Diuril) thiazides, certain antifungal agents, dyes, tranquilizers, furocoumarins (e.g., 4,5′,8-trimethylpsoralen, and 8-methoxypsoralen) that are often used for photochemotherapy of skin diseases, certain porphyrins (uro-, copro-, and proto-porphyrin), and others (9,12). Many of these drugs are photoactivated and become very reactive only in the presence of UVA radiation to cause DNA damage, cell membrane damage, and invariably induce skin phototoxic reactions such as erythema, edema, and vesiculation in many individuals and, occasionally, photoallergic reactions (eczematous dermatitis, puritus with papular and vesicular reactions) in a few individuals. Many diseases, such as polymorphous light eruptions, lupus erythematosus, and photosensitivity reactions in porphyrias (erythropoietic protoporphyria), are also aggravated by exposure to UVA radiation.

J. UVA Radiation Causes Alterations of the Immune System

In the past two decades, it has been well established that UVB radiation causes local and systemic alterations of immunological functions of epidermal cells, particularly of Langerhans cells and keratinocytes of skin (13,14). Langerhans cells are the principal (if not the sole) antigen-presenting cells in epidermis that play a key role in preventing the growth of UVB-induced skin cancer. The abrogation by UV radiation of the normal antigen-presenting function of Langerhans cells allows ''suppressor'' antigen-presenting cells in the epidermis to activate suppressor T cells and, hence, to increased susceptibility to skin cancer. Also, UVA radiation causes depletion of Langerhans cells and alterations of immunological functions of epidermal cells. Keratinocytes also secrete distinct immunosuppressive factors following exposure to UVA that lead to activation of suppressor T cells. This depletion of Langerhans cells and, subsequently, alterations in immunological functions of skin, leading to its susceptibility to cancer in the form of basal and squamous cell carcinomas, and possibly even to melanomas, is the major worry for dermatologists relative to UVA radiation.

K. UVA Radiation is Carcinogenic

In humans, most of the skin cancers occur on sun-exposed parts of the body (face, neck, lips, or hands). Laboratory studies in experimental animals (mice, rats, and guinea pigs) have revealed that UVB radiation is the most important cause for the induction of nonmelanoma skin cancers (basal and squamous cell carcinomas; 2,15). Exposure of animals to UVA radiation in combination with UVB radiation in a ratio proportional to that found in sunlight, gives rise to increased prevalence of skin cancer. It is increasingly evident that exposure of skin to solar UV radiation (UVB plus UVA) is a major risk factor that contributes to the development of human melanomas in light-skinned individuals. Although UVA radiation has not yet been demonstrated to induce malignant melanoma in laboratory animals, it activates and proliferates the melanocytic system of freckles, solar lentigenes, and nevi. The UVA radiation from tanning salons has been postulated and implicated in the pathogenesis of skin cancer, including melanoma.

L. UVA Causes Photoaging or Dermatoheliosis

Photoaging (actinic elastosis), with its unpleasant consequences of wrinkling, has become a problem of major concern to many young and adult persons who love sunshine, sunbathing, outdoor pleasures, and recreational activities in the sun, including sailing, golfing, playing tennis, and so forth. The skin shows tangled accretions and large masses of deranged elastic fibers, and altered collagen that shows denaturation and decreased solubility. In addition, there are increased levels of glycosaminoglycans, indicating progressive solar elastosis. Although UVB radiation contributes to photoaging, the deep, penetrating UVA also contributes and accelerates this process (9,16).

V. PHOTOPROTECTION AGAINST THE HARMFUL EFFECTS OF UVA AND UVB RADIATION

The photoprotective approaches to be discussed are applicable to normal individuals of various skin phototypes (SPT) who are fair-skinned (SPT I and II), or have tanned, light brown, dark brown, or black skin (SPT III, IV, V, and VI) and want to be protected against UVA and UVB radiation. It is assumed that such persons have a normal threshold of tolerance to UVA radiation and have no preexisting skin diseases that are recognized as "photosensitivity problems" manifesting abnormal responses of skin to sunlight (e.g., polymorphous photodermatitis, solar urticaria, lupus erythematosus, actinic prurigo, or vesicular eruptions in porphyria, or patients with melasma). Approaches of photoprotection for such light-sensitive patients are discussed elsewhere in this book.

The protection of skin of normal persons against UVA radiation alone is not easy, is often partial or incomplete because of excessive sun exposure, and in real life is not practical, because out in the sun a person is rarely exposed to pure UVA radiation unless he or she is receiving UVA exposures in a tanning salon or is being treated with phototherapy units or with PUVA therapy involving the oral use of psoralen and intentional exposure to UVA radiation. Such persons can be adequately protected by UVA-absorbing sunscreens discussed later on. The primary source of most intentional or nonintentional exposures to UVA and UVB radiation for most people is the sun. Any average person of any skin phototype can be adequately protected against the harmful effects of both the UVB and UVA radiation present in natural sunlight, as long as the duration of sun exposure is kept to less than 3 or 4 h.

A. Photoprotection (Principles and Practice)

Effective protection of human skin against solar radiation (exposure of less than 3–4 h) can be achieved by various methods (1), some of these are simple and obvious, others are a little bit complicated and require some understanding of avoiding the source of radiation before recommending the use of effective topical sunscreens. Any anticipated normal response to UVB or UVA radiation can be prevented by knowing and following four physical principles of photoprotection: (1) absorption, (2) scattering, (3) reflection, and (4) inactivation or quenching of free radicals and photoexcited states of molecules that are created by exposure of skin to solar radiation. (1) The *absorption* (hence, the filtration) of impinging ultraviolet radiation on skin can be achieved by providing a thin film, on the surface of skin, or a sunscreen lotion, cream, spray, or gel containing one or more chemical agents that selectively show high absorption spectrum of UVB and UVA radiation, and do not allow a significant portion (<95%) of the UV rays to penetrate to viable cells in the skin. (2) The *scattering* and (3) *reflection* are two additional ways to attenuate the impinging radiation on skin. This can be achieved by using a thin film or a lotion or cream containing particulate form of zinc oxide (ZnO) or titanium dioxide (TiO_2), or by a combination of both in a lotion (suspension) or cream form. These physical agents (ZnO and TiO_2) exhibit good refractive properties that allow the incident radiation (both UVB and UVA) to be deflected or scattered. Such a measure is very effective in protecting the skin (17). (4) The last approach involves inactivation or quenching of free radicals or reactive oxygen species generated in the skin by UV radiation. This can be achieved by using selective scavengers or quenchers of free radicals and reactive oxygen species. These free radical quenchers are antioxidant compounds that are required in high concentration (e.g., vitamin C, β-carotene, α-tocopherol, certain thiol or SH-rich compounds). Their effectiveness in protecting the skin against solar radiation is limited because of their poor ability to diffuse into the epidermis.

The following six or seven steps represent the recommended guidelines for achieving a comprehensive sun protection program against UVA as well as UVB radiation:

Avoidance

Avoid sunlight between 10 AM and 4 PM, especially in spring and summer months when sunlight is strong (from March to the end of September in the Northern Hemisphere). A certain attitudinal change must be stressed before encouraging and recommending that persons reschedule their outdoor activities to between 7 AM and 10 AM in the morning, and after 4 PM at the end of the day, during spring through the end of September. Avoidance of sun is easy to advocate, but difficult to implement for working class people and for children, teenagers, and adolescents, who are at risk of overexposure. Whenever possible, indoor activities should be encouraged. Persons should be advised to move to shaded areas under a tree or to remain in buildings and homes as much as possible. This attitudinal change for sun avoidance is certainly very helpful for fair-skinned individuals (SPT I–III) and in protecting the skin against UVA radiation and preventing problems of hyperpigmentation and premature aging of skin of Oriental and Asiatic people, Hispanic people, and people of the Middle-East region (SPT IV–VI). Fair-skinned persons, who burn easily and tan poorly, should avoid excessive sun exposure between 10 AM and 4 PM.

Block Out Sun's Rays

The direct effect of solar radiation can be easily avoided by using hats (broad-brimmed hats and baseball caps), parasols (umbrellas), and protective clothing (apparel) made of synthetic fabrics (nylon, polyester). The use of long-sleeve shirts, T-shirts, and pants is preferred. Most types of textiles, both natural (cotton) and synthetic (nylon, Dacron, or polyester), provide good protection against UVA and UVB radiation. Such protective clothes have SPF values that often exceed 20–30. Indeed, an apparel made of synthetic materials is superior in achieving protection against UVA radiation than the use of cotton fabrics. The density of the weave and the color (blue, red, green) of clothing appear to provide high SPF values and extend the duration of protection when sun exposure becomes prolonged and unavoidable. It is important to advise the person to protect the habitually exposed areas of face, neck, chest, and hands, that clothing cannot cover with topical sunscreens (SPF 15 or higher) and to shield these sites.

Chemical Sunscreens

The third and the most reliable approach *is to cover up the skin* by using a thin film of *topical sunscreen* formulations with SPF values of 15 or higher (up to 30). Topical sunscreens contain active chemical ingredients that absorb, scatter, or reflect the incident UV radiation, thereby reducing the direct penetration and effect of UV exposure. Commercially available chemical sunscreens can be arbitrarily classified into the following four categories that are sequentially presented and described: (1) Sunscreens of category one are those that block UVB radiation and contain one or more UVB-absorbing chemicals, such as PABA or esters of PABA, known as padimate O or Padimate A, or other chemicals, known as octyl-salicylate, cinnamates (octylmethoxycinnamate), homomenthylsalicylate, and so on. There are over 21 such chemicals that U.S. FDA has approved for protecting skin against UVB and UVA radiation (1, 17). These UVB-absorbing sunscreen chemicals and their recommended concentration for use are listed in this review article (1) as well as in the FDA monograph (17). In other countries (Germany, United Kingdom, France, and other countries of Europe, South America, Middle-East, Australia, Japan) similar types of UV-absorbing chemicals are available and generally used for over-the-counter sale of sunscreens. These UVB-absorbing chemicals are colorless and have high extinction values in the 290 to 320 nm range, and when applied uniformly to the skin in the form of a thin film of lotions, creams, or gels, can effectively block the penetration of UVB radiation into the skin. Most of these UVB-absorbing sunscreens, however, are ineffective in protecting the skin against UVA radiation, and such sunscreens are primarily formulated to minimize the discomforts of sunburn reaction and to optimize and facilitate the tanning reaction of skin that

Table 6 Sunscreen Chemicals That Absorb UVA Radiation and Protect the Skin

Chemical	Recommended concentration	Protection range (nm)	Side effects
Benzophenones			
Oxybenzone	2–6%	320–360	Partially effective and irritant to few
Sulisobenzone	5–10%	320–360	Partially effective and irritant to few
Dioxybenzone	3.0%	320–360	Partially effective and irritant to few
Other chemicals			
Menthylanthranilate	3%	300–340	Partially effective
Butylmethoxydibenzoylmethane (Parsol 1789)	3% or less	320–400	Effective up to 400 nm, but irritant
Red-Vet-petrolalum	>30%	320–370	Smarting and irritant
Titanium dioxide	2–25%	300–400	Effective, but cosmetically not acceptable
Zinc oxide	2–20%	300–400	Effective, but cosmetically not acceptable

is stimulated by UVA radiation. (2) Sunscreens of category two are called "combination sunscreens" that contain UVB- as well as UVA-absorbing chemicals in a suitable vehicle or base formulation. They were arbitrarily formulated to partially block or filter out UVA radiation to satisfy consumer's demands who wish to acquire a tan without the discomforts of sunburn reaction. In such combination sunscreens there are two or three UV-absorbing chemical ingredients. The primary sunscreen is the one that strongly absorbs in the UVB spectrum (e.g., PABA or esters of PABA known as padimate-O or padimate-A, octylsalicylate, octylmethoxycinnamate) and shows little or no absorption in the UVA spectrum. The formulation is then enriched or supplemented with a UVA-absorbing benzophenone compound such as oxybenzone, sulisobenzone or dioxybenzone or menthylanthranilate (see Table 6).

These second category types of combination sunscreens are partially effective in blocking UVB and UVA radiation (from 290 to 360 nm). Their effectiveness in blocking additional segment of UVA, from 360 to 400 nm radiation, is poor and often unsatisfactory in effectively preventing the penetration of 360 to 400 nm radiation because (1) the UVA-absorbing compound must be colorless, (2) it must show high and broad absorption spectrum in the 320 to 400 nm range,

(3) it must be nonsensitizing and nonirritant, and (4) colored compounds that show high absorption values in UVA spectrum (320–400 nm) are not acceptable to consumers, as well as to the industry. Because of these limitations, most sunscreens available on the market are not totally, but only partially, effective in blocking UVA radiation. The three benzophenones and menthylanthranilate (see Table 6) partially absorb UVA radiation up to 340–360 nm, whereas dibenzoylmethane (Parsol 1789) absorbs UVA radiation quite significantly and effectively up to 380 nm (see Table 6). Parsol 1789 is one of the best available agents to protect skin against UVA radiation. (3) Sunscreens of category three also belong to the combination category sunscreens that are good and quite effective in blocking UVA radiation. Parsol 1789 is one such effective UVA-blocking agent, and it is widely used in many European countries, as well as in the United States, Canada, Australia, and a few other countries. No UVA-blocking sunscreen can be used alone; it usually must be combined with a UVB-absorbing agent such as padimate O, a cinnamate derivative, or some other UVB-absorbing agent mentioned earlier. In European countries the combination of 2-ethylhexyl-*p*-methoxycinnamate and Parsol 1789 is popular in reducing the effects of UVA radiation. The effectiveness of such UVB- and UVA-absorbing combination sunscreens, containing 3–5% UVB-absorbing agent (e.g., padimate O or octylmethoxycinnamate), and 2–3% Parsol 1789, is usually rated as good to excellent, provided the sun exposure is not excessive (<3–4 h). Their SPF value in UVB spectrum may range from 15 to 30, and they provide reasonably good and satisfactory protection in the UVA spectrum (UVA SPF = 4–6). The problems associated with using high UVA SPF combination sunscreens are related to their irritancy and possibly low-grade photosensitizing properties. Benzophenones and Parsol 1789 are irritants and cause contact photosensitivity problems that are occasionally related to photoallergic reactions. Parsol 1789 is also photolabile (1).

The effectiveness of such UVA-absorbing sunscreens containing Parsol 1789 or benzophenones has been further enhanced by adding a 2–3% concentration of micronized zinc oxide and titanium dioxide. Such well-formulated sunscreens enriched with either ZnO or TiO_2, or both, protect the skin quite well in UVB and UVA spectral regions. In addition, such effective UVA-absorbing sunscreens are also formulated in a water-resistant base of acrylate polymer and can protect individuals after immersion of skin for 40–80 min of swimming in ocean water or chlorinated water of swimming pools.

Any recommended sunscreen against UVB and UVA radiation should have a minimum SPF value of 15, or a maximum SPF value of 30 (or may be as high as 45 for a sensitive person of SPT I), and should contain UVB- and UVA-absorbing chemicals discussed earlier. There is a limit to using increased SPF value for providing an extended duration of protection. The recommended sunscreen should be very water-resistant (waterproof), and substantive (i.e., adherent) to the skin. It would seem prudent, therefore, to encourage the regular use of topical sunscreens (SPF 15–30) that absorb optimally and provide effective protection throughout the UVB and UVA spectrum, and are very water-resistant.

Physical Sunscreens

So far we briefly discussed three categories of sunscreens. There is a fourth category of sunscreen that is devoid of chemical sunscreens (e.g., UVB or UVA-absorbing chemical agents discussed earlier), but generally contains only micronized particles of mineral oxides, such as ZnO and TiO_2 in a form of lotion or cream. The concentration ZnO or TiO_2 is usually high (>5–20%; 18). Sunscreens containing these particulate forms of mineral oxides are quite effective against UVB as well as UVA radiation. The most desirable particle size of ZnO or TiO_2 for optimal protection must be smaller than 100 nm (30–100 nm). Such particles, when applied to the skin, provide multiple surfaces for forward scattering and reflection of UVB as well as UVA radiation. These are inert chemicals that are nonirritant, nonsensitizing, and quite effectively protect the skin by processes involving absorption, reflection, and scattering of UVB as well as UVA radiation. These physical sunscreens also prevent tanning or hyperpigmentation of skin and other abnormal responses of skin to solar radiation in persons of brown or tanned skin color. They are good for children and for people who are allergic to sunscreen chemicals (e.g., PABA, benzophenones, Parsol 1789) or those who are constantly exposed to the sun because of the nature of their work (sailors, farmers, roofers, life guards). Consumers may not like and use such effective sunscreens that contain ZnO or TiO_2 because of their whiteness that imparts cosmetically unacceptable color or hue to the skin. Tinted physical sunscreens, containing iron oxides, talcum powder, ZnO, TiO_2, that approximately match the color of the skin in various shades are available and can be recommended; they are acceptable to persons with light brown or brown skin.

Dihydroxy Acetone

There is one additional way to protect the skin against UVA radiation. It involves the use of dihydroxy acetone (DHA at about 3–5% concentration) in sunless tanning lotion form, usually at an alkaline pH of 8 (19). The newly modified lotions of DHA manufactured by leading sunscreen companies are enriched with a UVB-absorbing chemical sunscreen (e.g., octylmethoxycinnamate, octylsalicylate, or padimate O) to provide broad-spectrum protection against UVB and UVA radiation. After three or four sequential topical applications of DHA-enriched lotion, DHA undergoes oxidation, and the skin develops an orange-brown color within 10–12 h, and this DHA-induced skin color is protective against UVA radiation, but not against UVB radiation. The color remains adherent to the stratum corneum and is not easily washed off. Although the color of the skin is not due to increased production of natural melanin, this water-resistant orange-brown color appears to simulate or mimic a suntanning reaction. Many fair-skinned people who prefer the skin to be tanned, but worry about the effects of sunlight, like such an artificially induced skin-tanning reaction without undergoing the harmful effects of sun exposure; such persons learn to acquire an esthetically pleasing color of their skin that is photoprotective against UVA radiation also. Patients who are sensitive to UVA radiation (e.g., patients with PMLE, or erythropoietic protoporphyria) can be helped and protected by such formulations (19).

Free Radical Scavengers

In recent years the cosmetic industry has promoted and advocated widespread use of topical antioxidants [e.g., α-tocopherol, β-carotene, vitamin C (ascorbic acid), and SH (thiol)-rich compounds] as photoprotective agents against harmful effects of UVB and UVA radiation, with the assumption that sun exposure of skin produces the generation of free radicals and reactive oxygen species ($^1O_2$, $O_2^{\cdot-}$, $^{\cdot}OH$) that damage DNA, cell membranes, and proteins (collagen, elastin, and enzymes), and thus contribute to premature photoaging of skin. Topical supplementation of skin with antioxidant-rich lotions should theoretically help to inactivate the free radicals and reactive oxygen species that damage the skin and, therefore, should protect the skin. The antioxidants are claimed by their manufacturers to prevent lipid peroxidation reaction in the skin and to inhibit free radicals and the resulting inflammatory response. Even the photoprotective properties of melanin (as a stable free radical quencher and an antioxidant) are claimed for justifying the use of topical melanin-rich creams for preventing the UV-induced free radical damage of the UV exposed skin (20). It is true that many of such antioxidant compounds (vitamin C, β-carotene, α-tocopherol, melanin, thiol, or SH-rich compounds) are free radical quenchers when they are evaluated under in vitro system. Their in vivo effectiveness to inactivate UV-induced free radical formation in human skin has not been adequately documented. To exert the inactivation of free radicals, the quencher has to be present in adequate concentration at the site of free radical generation. To be effective, these antioxidants have to be repeatedly applied because they do not easily penetrate and diffuse into the skin at the site of reaction to provide effective concentration in the epidermis to inactivate the UV-induced production of free radicals and reactive oxygen species. β-Carotene, α-tocopherol, vitamin C, and other free radical quenchers in reality are not sunscreens; they have very low SPF values in UVB and UVA spectral range, and unless they are supplemented with UVB-absorbing chemicals, such as padimate O, cinnoxate, or salicylates, the antioxidant property of these agents is clinically not at all satisfactory enough to rate them as "topical sunscreens" per se. Even vitamin C, which has been recently claimed to be anti-inflammatory (21) and helps control photoaging and the inflammatory response associated with sunburn, has no measurable sun protection factor (SPF). Any antioxidant-rich preparation containing vitamin C, α-tocopherol, or melanin, has to be used in conjunction with a sunscreen agent that has an SPF value of 15 or higher. Whether such antioxidant-rich formulations are helpful in preventing the photoaging of skin remains to be documented. Perhaps, with improved techniques for delivery of effective concentration of these free radical quenchers (e.g., the use of liposomes), it is conceivable and feasible to inactivate the harmful effects of UVA radiation involving free radicals and reactive oxygen species. Recently, we (22) have observed the topical and oral use of a plant extract, known as *Polypodium leucotomos*, to be effective in protecting individuals against UVB-induced sunburn reaction, and PUVA-induced phototoxic reaction in normal individuals undergoing psoralen plus UVA treatment. The extract, when formulated in a lotion form, acts as an antioxidant and a scavenger of superoxide anion ($O_2^{\cdot -}$). It protects the skin against UVB-induced sunburn reaction and PUVA-induced phototoxic reaction. Well-controlled, double-blind clinical studies are needed to accept this concept of photoprotection against the harmful effects of free radicals and reactive oxygen species.

REFERENCES

1. Pathak, M. A. and Fitzpatrick, T. B. Preventive treatment of sunburn, dermatoheliosis and skin cancer with sun protective agents, *Dermatology in General Medicine*, 4th ed. (T. B. Fitzpatrick, A. Z. Eisen, K. Wolff, I. M. Freedberg, and K. F. Austen, eds.), McGraw-Hill, New York, 1993, pp. 1689–1717.
2. Urbach, F. Introduction, *Biological Responses to Ultraviolet-A Radiation* (F. Urbach, ed.), Valdenmar Publishing, Overland Park, KS, 1992, pp. 1–6.
3. Frederick J. E. and Alberts, A. D. The natural UVA environment, *Biological Responses to Ultraviolet-A Radiation* (F. Urbach, ed.). Valdenmar Publishing, Overland Park, KS, 1992, pp. 7–18.
4. Parrish, J. A., et al. Photochemotherapy of psoriasis with oral methoxsalen and longwave ultraviolet light, *N. Engl. J. Med.*, *291*:1270 (1974).
5. Norris, P. G., Gange, R. W., and Hawk, J. L. Acute effects of ultraviolet radiation on the skin, *Dermatology in General Medicine* (T. B. Fitzpatrick, et al., eds.), McGraw-Hill, New York, 1993, pp. 1651–1657.
6. Auletta, M., et al. Effects of cutaneous hypoxia upon erythema and pigment response to UVA, UVB, and PUVA (8-MOP + UVA) in human skin, *J. Invest. Dermatol.*, *86*:649–652 (1986).
7. Pathak, M. A. Immediate and delayed pigmentary and other cutaneous responses of solar UVA radiation, *The Biological Effects of UVA Radiation* (F. Urbach and R. W. Gange, eds.), Praeger Publishers, New York, 1986, pp. 156–167.
8. Peak, M. J., Peak, J. G., and Churchill, M. E. Cellular and molecular effects of UVA radiation in mammalian cells, *Biological Response to Ultraviolet A Radiation* (F. Urbach, ed.), Valdenmar Publishing, Overland Park, KS, 1992, pp. 29–46.

9. Carbonare, M. D. and Pathak, M. A. Skin photosensitizing agents and the role of reactive oxygen species in photoaging, *J. Photochem. Photobiol.* (*B. Biol.*), *14*: 105–124 (1992).
10. Freeman, S. E., Grange, R. W., Sutherland, J. C., et al. Production of pyrimidine dimers in human skin exposed in situ to UVA radiation, *J. Invest. Dermatol.*, *91*: 349–352 (1987).
11. Carraro, C. and Pathak, M. A. Studies on the nature of in vitro and in vivo photosensitization reaction by psoralens and porphyrins, *J. Invest. Dermatol.*, *90*:267–275 (1988).
12. Elmets, C. A. Cutaneous phototoxicity, *Clinical Photomedicine* (H. W. Lim and N. A. Sober, eds.), Marcel Dekker, New York, 1992, pp. 207–226.
13. Kripke, M. L. Immunological unresponsiveness induced by ultraviolet radiation, *Immunol. Rev.*, *80*:87–102 (1984).
14. Cruz, P. D., Jr. and Bergstresser, P. R. Photoimmunology: Effects of UVB radiation on cutaneous photocarcinogenesis and allergic contact sensitivity, *Photomedicine for Clinical Dermatologists* (H. W. Lim and N. A. Soter, eds.), Marcel Dekker, New York, 1992, pp. 137–151.
15. Stevenborg, H. J. C. M. and van der Leun, J. C. Tumorigenesis by a long-wavelength source, *Photochem. Photobiol.*, *51*:325–330 (1990).
16. Kligman, L. H., Akin, F. J., and Kligman, A. M. The contribution of UVA and UVB to connective tissue damage in hairless mice, *J. Invest. Dermatol.*, *84*:272–276 (1985).
17. Federal Register: Sunscreen Drug Products for Over-the-Counter Human Drugs: Proposed Safety, Effectiveness, and Labeling Conditions. Washington, DC, Department of Health, Education, and Welfare, Food and Drug Administration. Vol. 43, No. 166, August 25, 1978, p. 38206.
18. Sayre, R. M., et al. Physical sunscreens, *J. Soc. Cosmet. Chem.*, *41*:103–109 (1990).
19. Follett, K. A., Johnson, J. A., and Fusaro, R. M. Protection of photosensitized rats against long ultraviolet radiation by topical application of compounds with structures similar to that of dihydroxy acetone. *Dermatologica*, *175*:58–63 (1987).
20. Pugliese, P. T., et al. UVA and UVB protection against lipid peroxidation by a melanin containing sunscreen, *Melanin*: *Its Role in Human Photoprotection* (L. Leise, M. R. Chedekel, and T. B. Fitzpatrick, eds.), Valdenmar Publishing, Overland Park, KS, 1994, pp. 271–276.
21. Darr, D., Combs, S., Dunston, S., et al. Topical vitamin C protects skin from ultraviolet radiation-induced damage, *Br. J. Dermatol.*, *127*:247–253 (1992).
22. Gonzalez, S. and Pathak, M. A. Inhibition of ultraviolet-induced formation of reactive oxygen species, lipid peroxidation, erythema, and skin photosensitization by *Polypodium leucotomos*, *Photodermatol. Photoimmunol. Photomed.* (in press, 1995).

II

PHOTOBIOLOGICAL ASPECTS

4

Biological Effects of Sunlight

John H. Epstein
University of California, San Francisco, California

I. INTRODUCTION

This volume is concerned with all aspects of sunscreens. The primary purpose for the existence of these agents is to protect human skin against the damaging effects of the sun's rays. This chapter is concerned with these adverse effects.

II. RADIATION

The electromagnetic spectrum emitted by the sun ranges from the very short cosmic rays to the very long radio waves and beyond. The ultraviolet (UV) rays, which comprise the shortest of the nonionizing rays, are responsible for most of the photocutaneous changes that occur (Table 1). They have been divided into three categories (1). The shortest UVC rays extend from 100 to 280 nm. No wavelengths shorter than 290 nm (288 nm) reach the earth's surface from the sun, primarily because of filtration by ozone. Thus, although UVC is quite photoactive, it does not constitute a risk to the general population. It can be somewhat hazardous in certain occupations, such as welding where considerable UVC is emitted from artificial sources (2). In contrast, some of the mid-UV or UVB rays (290–320 nm) penetrate the ozone layer and are responsible for most of the cutaneous photobiological events induced by exposure to the sun (3). These rays inhibit DNA, RNA, protein synthesis, and mitosis formation; labilize

Table 1 Nonionizing Radiation

Type	Wavelength (nm)
UVC	100–280 (290)
UVB	280 (290)–315 (320)
UVA	315 (320)–400
Visible	400–800
Infrared	800–1700

lysosomal and cellular membranes; are mutagenic; and induce early and delayed erythema responses. Prolonged UVB exposure damages dermal connective tissue and is the primary carcinogenic stimulus for nonmelanoma skin cancers. Although delayed erythema is usually considered a sunburn reaction, all of the immediate phototoxic events noted should be included in this concept. Beneficial effects include new pigment formation, which provides some protection, and initiation of the vitamin D cascade, which is important to calcium metabolism. The UVB effects are direct and do not require intermediate photosensitizers. As a rule, these rays do not pass through window glass. Thus, a sturdy piece of window glass will prevent all of these responses.

The long or near UVA rays (320–400 nm) do pass through window glass and produce a significant number of photobiological effects (4). In contrast with UVB, UVA induces its effects indirectly, requiring oxygen (5,6). Actually the UVA spectrum may have to be divided into at least three parts. The shorter rays between 315 or 320 and 327 nm appear to act directly on cell structures, similar to UVB. Between 327 and 347 nm both direct and indirect effects occur, being more direct toward the 327-nm and indirect toward the 347-nm end of the spectrum. At 335 nm, the contribution of each effect is equal. Between 347 and 400 nm the effects are indirect. Recently, several authors have been using the term UVI for the longer UVA rays (340–400 nm) and UVAII for the shorter UVA rays (315–340 nm). The UVAI spectrum has been reported to be therapeutically beneficial in atopic dermatitis (7) and systemic lupus erythematosus (8).

The UVA rays induce an immediate erythema that diminishes within 2 h and a delayed erythema response that reaches a peak at 6 h (9). This is in contrast with the UVB delayed erythema that tends to reach a peak in 12–24 h. These rays also induce an immediate pigment-darkening reaction and new melanin pigment formation (4,10). For persons who tan well (type IV skin) it requires more energy to induce an erythema than a tan (6). In fair-complexioned persons, about the same amount of energy is required for both the erythema and the new–pigment-formation responses. The UVA tan response is not associated with epidermal hyperplasia and provides a much less protective effect against a UVB-induced erythema than the UVB-induced tan (6). The UVA can also enhance or add to the immediate effects of UVB on human as well as experimental animal skin (4). In experimental animals, it induces dermal elastosis; although less efficiently than UVB (11); it augments UVB-induced carcinogenesis, and in large enough amounts UVA alone can induce cancer formation (4,12).

The visible rays are used primarily for vision, although they can induce reactions with certain chemical molecules. The infrared rays supply heat and can induce certain skin changes, such as cellular damage and erythema ab igne (13–15). The histological changes induced by prolonged exposure to infrared radiation are similar to those induced by chronic UV exposure, and in situ squamous cell carcinomas have been reported.

III. GOTTHUS DRAPER LAW

The Gotthus Draper law (16) is the primary law of photochemistry and photobiology. It states that nonionizing radiation must be absorbed to induce a photochemical and, thus, a subsequent photobiological reaction. As a corollary, the radiation-absorbing molecule, chromophore, photosensitizer, and such, must be present at the time of irradiation for a response to occur.

IV. DEFINITIONS

A. Absorption Spectrum

The rays absorbed by a molecule or a combination of molecules with a greater or lesser efficiency constitute the *absorption spectrum* of that molecule or those combinations of molecules (17).

B. Action Spectrum

The wavelengths that are absorbed by a molecule that induce a specific photochemical or photobiological reaction with greater or lesser efficiency constitute the *action spectrum* for that specific reaction. As a corollary, the action spectrum must be included in the absorption spectrum of the photosensitizing molecule, as the energy must be absorbed to produce any action (17).

C. Photosensitivity

Photosensitivity (18) is the broad term used for adverse responses to nonionizing radiation. These reactions may be phototoxic or photoallergic.

Phototoxicity

Phototoxicity is common. It will occur in everyone if enough energy, and in photosensitized reactions, enough of the photosensitizing molecules are present.

It is generally characterized by an erythema, usually delayed in onset, followed by hyperpigmentation and desquamation. Thus, it simulates the ordinary sunburn response. In fact, the sunburn reaction to UVB rays is the most common of all phototoxic responses.

The mechanisms of phototoxicity are somewhat complicated and may take one of several forms. As with UVB reactions they may result from specific damage to macromolecules, such as DNA, RNA, proteins, membranes, and the like (3). This is also true of photosensitization by certain molecules, such as 8-methoxypsoralen (8-MOP; 19). Such injury can occur with or without the presence of oxygen and the generation of free radicals and superoxides.

In other phototoxic events the presence of oxygen is essential, and membrane damage may play a most important role. These are photodynamic relations and are characteristic of photosensitization by porphyrin molecules, although 8-MOP can also induce such reactions (19).

Another mechanism of phototoxicity depends on the formation of toxic photoproducts. These, in turn, damage the tissues. This occurs, at least at times, with chemicals such as chlorpromazine (20).

Photoallergy

In contrast, photoallergy is uncommon. Like all allergies it is dependent on an acquired altered reactivity secondary to an immediate humoral or a delayed cell-mediated response. Clinically, it is characterized by an immediate wheal-and-flare reaction, or a delayed papular to eczematous reaction.

Photoimmunology

In addition to the photoallergic reactions, the UV rays, with or without exogenous sensitizers, can induce significant changes in the immune system. In vitro they damage lymphocytes (21); UVC is more toxic than UVB, which is more toxic than UVA. Psoralens plus UVA (PUVA) are also toxic to lymphocytes in vitro. Whole-body erythemogenic exposures to both UVB and PUVA reduce the number of circulating T cells in humans in vivo (22,23). However, the mechanisms of this effect appear to be different. Hersey and co-workers reported (24) that exposure to sunlight daily for a 2-week period induced a reduction in the total number of T lymphocytes, owing to a marked drop in T helper cells. Functionally, the T suppressor cell activity was increased, and natural killer cell activity was decreased. Subsequently, Hersey et al. (25) demonstrated suppression of natural killer cell activity by suntan solarium radiation depleted of UVB energy. Thus, although UVB is much more effective, UVA radiation does affect natural killer cells.

The UV modulation of contact hypersensitivity (CHS), delayed-type hypersensitivity (DTH), and cutaneous carcinogenesis has received a great deal of attention in recent years (26,27). Ultraviolet-induced immunosuppression caused by antigen-specific suppressor T cells has been demonstrated experimentally in all three of these responses.

Local photoimmunosuppression results from application of an antigen to a UV-irradiated skin site (28). This leads to the development of antigen-specific T suppressor cells and specific immunological tolerance to that antigen. UVB radiation has profound effects on Langerhans cells (LC), which are the antigen-presenting cells in the epidermis. These alterations in the LC are associated with activation of the suppressor T-cell pathway of the immune response (29). However antibodies to tumor necrosis factor (TNF)-α inhibit this local UV-induced immune suppression, suggesting that cytokines may play a role in this response.

Systemic photoimmunosuppression results from application of the antigen in a non–UV-irradiated site several days after high-dose UV irradiation of a distant skin site. This leads to inhibition of CHS, with the formation of antigen-specific suppressor T cells (27). Similarly, the DTH response is inhibited in the same manner. This is not an LC-driven process. Systemic photosuppression of CHS appears to be mediated by TNF-α (27); whereas the DTH photosuppression appears to be produced by interleukin-10 (IL-10), elaborated by UV-irradiated keratinocytes (30). Recent studies suggest that UV damage to DNA is the initial event in the induction of photoimmunosuppression (31,32).

The potential clinical importance of this photoimmunosuppression is suggested by the effects of UVB radiation before the inoculations of infectious agents into experimental animals and on photocarcinogenesis (26,27). Several infectious agents cause more extensive disease that is associated with UV-induced inhibition of DTH before inoculation. In addition, exposure of mice to subcarcinogenic amounts of UVB radiation induces tolerance to transplantation of highly antigenic UV-induced tumors, owing to the activation of antigen-specific suppressor T cells. Unfortunately, although UVB radiation does exacerbate or precipitate flares of certain infectious diseases, such as herpes simplex type 1 and type 2, the relation to photoimmunosuppression is unclear. This is also true of photocarcinogenesis in humans.

Photocarcinogenesis

Perhaps the most important detrimental effect of sun exposure is the induction of skin cancers. Nonmelanoma skin cancers—basal cells carcinomas (BCC) and squamous cell carcinomas (SCC)—are by far the most common cancers that occur in the United States each year (33). They make up 30 to 40% of all cancers (34). It has been estimated that over 700,000 new nonmelanoma skin cancers will occur in 1994 (35). Most of these will be BCC, but about 50,000 will be SCC and 2300 persons in the United States are expected to die of nonmelanoma skin cancer in 1994. The vast majority of these will die of SCC.

The incidence has been increasing at a rapid rate in the last several decades (34,36–38), so much so that it has been considered a quiet 20th century epidemic (36).

The role of the sun in the induction of nonmelanoma skin cancers was suggested by several astute clinical observations at about the turn of the 20th century and, subsequently, was confirmed by epidemiological and experimental studies (39–41). The primary action spectrum falls in the UVB range (41). However, UVA enhanced UVB carcinogenesis in experimental animals and, in large enough amounts, has induced cancers by itself (42–44). Thus, if we can extrapolate from the experimental data to the human experience, UVA rays may play a part in the development of nonmelanoma skin cancers in humans.

Malignant melanoma is the third most common skin cancer, which has been increasing in incidence at an even more alarming rate (37,38,45,46). Indeed, this incidence is increasing faster than any other reportable cancer in the United States (47). The UV rays, primarily UVB rays, from the sun are the primary cause of nonmelanoma skin cancers. This is most notable for SCC, which has an almost one on one relation to sun exposure (48). Although the sun's rays do appear to have an important influence on melanoma formation, this association is much less clear than for nonmelanoma skin cancers (46,49,50). How UV radiation induces skin cancers poses a most interesting question. Immune suppression, as occurs in kidney transplant patients, leads to a high incidence of skin cancers, mostly SCC, in sun-exposed skin (51,52). Subcarcinogenic exposures to UVB rays prevents rejection of highly antigenic UVB-induced transplanted cancers owing to the activation of specific suppressor T cells in experimental animals (27). As yet, no similar system has been described in humans. However, Yoshikawa et al. (53) reported on a group of patients in whom previous UVB irradiation inhibited sensitization to dinitrochlorobenzene (DNCB), a very potent contact allergen. When compared with another similar group, in whom previous UVB radiation did not prevent DNCB sensitization, they noted that 92% of those with skin cancers were in the former UVB-inhibited group. Thus, patients with skin cancers may not have the immunological ability to reject their sun-induced skin cancers.

Mechanistically, we must consider UV-induced DNA damage and repair, which are important issues when considering the role of the sun in skin cancer formation (54). Most of what we know of the functioning of this system in human skin derives from studies on the rare genetic disease xeroderma pigmentosum (55). These patients have a defective DNA repair enzyme system. In addition, they have an inordinate susceptibility to the carcinogenic effects of UV rays. Thus, it would seem reasonable to assume that the defects in DNA repair are responsible for this susceptibility. However, definitive proof of this relation remains to be established. Also, there is no specific evidence that defects in DNA repair are responsible for sun-induced cancers in the general population.

One of the most intriguing areas of recent interest concerns the relation to human carcinogenesis of oncogene activation and suppressor gene inactivation (56,57). These genes are part of the normal genetic wiring that directs growth and development of normal cells. However, they are susceptible to carcinogenic stimuli, such as UV radiation, ionizing radiation, chemicals, and age. Such exposures can activate oncogenes to produce more of their normal gene products or of aberrant gene products, or they may inactivate normal suppressor gene function. In addition, there is evidence that specific carcinogenic stimuli dictate which oncogene is activated, or which suppressor gene is inactivated (57,58). Along this line Brash et al. (59) reported specific UV-induced mutations in the p53 suppressor gene in over 50% of the invasive SCC that they studied (59). More recently, Campbell et al. (60) reported similar findings in Bowen's lesions and Nagano et al. (61) in actinic keratoses as well as SCC, indicating that these specifically UV-induced mutations are early events in sun-induced skin cancers.

Treatment of these sun-induced skin cancers depends on removal of the existing growths. However, prevention provides the best approach (62). The use of hats, with wide brims, can shield the face from 70% of the incident rays. The use of appropriate clothing can provide 100% protection of the clothed skin. Avoidance of the midday sun (10 AM to 3 or 4 PM) will reduce the UVB exposure by 60%, and the use of a sunscreen with an SPF of 15 will protect against 94% of these rays. In support of the value of sunscreens, Thompson and her co-workers (63) compared the effects of a good broad-spectrum sunscreen that absorbed 94% of the incident UVB rays and 90% of the UVA rays with the cream base on actinic keratosis (AK) formation. The 588 subjects were randomized and followed over the Australian summer between September 1991 and March 1992. There were significantly fewer new AK lesions and a significant regression of prior AK lesions in the sunscreen-protected subjects. Thus, the regular use of a potent broad-spectrum sunscreen significantly prevented the formation and persistence of these precancerous lesions. We hope that this will translate into a reduction in skin cancer development in the long term.

V. CLINICAL REACTIONS

There is an immense catalog of clinical photosensitivity reactions (18). In an attempt to bring some order to this complex situation I have divided the problems into three large categories: (1) Those reactions that are due to a loss or lack of protection; (2) problems that are due to the presence of a photosensitizing molecule; and (3) the catchall of the responses that are not due to a loss or lack of protection or the presence of a photosensitizer.

A. Loss or Lack of Protection

The primary protection the skin has against the damaging effects of the sun is melanin pigmentation. The primary problems in this category are related to the presence or absence of melanin. Included in this group are the following problems:

1. Occulocutaneous albinism (64): In this condition the melanocytes are present, but there is a complete or partial deficiency in the tyrosinase activity that is responsible for melanin formation.
2. Vitiligo (64): Vitiligo is characterized by a loss of melanocytes. Thus, the photosensitivity is due to the lack of cells to provide the pigmentary system.
3. Phenylketonuria (64,65) and Chediak–Higashi diseases (66): In both of these diseases the photosensitivity is the result of pigment dilution.
4. Xeroderma pigmentosum (XP; 67–69): These genetically inherited problems are due to enzymatic inabilities to repair ultraviolet (UV) damaged DNA, rather than a pigmentary defect.
5. Type I and II skin (70): This category also contains the most common of all the photosensitivity conditions that occur in the light-complexioned individual who sunburns readily and, with repeated such injury, develops chronic sun damage and skin cancers (71–74).
6. Action spectrum: The reactions in this group are phototoxic and the action spectrum is in the UVB range (72–74).
7. Therapy and prevention (70): Treatment in this category consists of removal of the precancerous and cancerous tumors that occur with the chronic damage. However, prevention can play a much more important role. This should include avoidance of the midday sun, between 10 AM and 3 PM, because of the concentration of the UVB rays during that time; the wearing of protective clothing, such as hats and tightly woven garments; and the use of the excellent sunscreens and blocks that are discussed elsewhere in this volume.

B. Photosensitization Reactions

The second clinical category consists of photosensitization reactions. The photosensitizer in this context will be considered a molecule that absorbs the sun's or artificially produced nonionizing radiation and, as a result of this process, induces an adverse response in the skin. The photosensitizer may be endogenous or exogenous in origin.

Porphyrins

The porphyrins are the only well-established photosensitizers manufactured by the human body. These are pyrrole ring structures that are essential to cellular metabolism throughout the body. Among other things, the heme part of hemoglobin is formed by these molecules. However, in the group of diseases named the porphyrias, large amounts of porphyrins are made in the erythropoietic (bone marrow) or the hepatic (liver) tissues (75). In these states, the porphyrins are not associated with iron and are potent photosensitizers. As such, they induce cutaneous changes, ranging from marked photodestruction of the skin and underlying tissues, to simple hyperpigmentation, depending on sun exposure and the amount of porphyrin molecules available. The cutaneous changes are due to photodynamic phototoxic responses, and the primary action spectrum is found between 400 and 401 nm.

Exogenous Photosensitizers

Exogenous photosensitizers may reach the skin by topical application (contact) or through the bloodstream (systemic).

Topical Chemicals Topical chemicals may cause photosensitivity reactions advertently, as occurs with tars and psoralen compounds in the treatment of psoriasis and vitiligo, or inadvertently, as occurs with the psoralen molecules in plants and perfumes, halogenated salicylanilides in deodorant soaps, sulfonamide and phenothiazine medications, and so on. In addition, contact photosensitizers are found in moisturizing creams, aftershave lotions, and industrial materials, such as polycyclic hydrocarbons contacted in the petrochemical industry, the making of tar products, contactants in agriculture and animal husbandry, and the like (18,19,76). The reactions are usually phototoxic, although occasionally, they are photoallergic. The phototoxic responses are characterized by erythema and edema, followed by hyperpigmentation and desquamation.

Photoallergic photocontact reactions are uncommon. As exceptions, the halogenated salicylanilides (77)—musk ambrette (78–80) and 6-methylcoumarin (81,82)—produce primarily photoallergic responses. These allergic photocontact reactions are all delayed hypersensitivity responses.

The action spectrum for these reactions, whether they are phototoxic or photoallergic, usually falls in the UVA range between 320 and 400 nm.

Systemically Administered Chemicals Systemically administered chemicals also may induce photosensitivity reactions advertently, as occurs with the use of psoralen compounds in the treatment of vitiligo and psoriasis, or inadvertently, which is the usual case (83,84). These inadvertent photoreactions are generally induced by commonly used medications including antibacterial sulfonamides, thiazide diuretics, sulfonylurea antidiabetic drugs, phenothiazines, certain broad-spectrum antibiotics, and the nonsteroidal anti-inflammatory drugs (NAIDS). Of the antibiotics, essentially all of the tetracyclines, but minocycline, are photosensitizers (85). However, doxycycline and demethylchlortetracycline are by far the most potent photosensitizers (84,85). Benoxaprofen was a very potent NAIDS photosensitizer (86–88), but it has been removed from the market. Several of the presently available NAIDS are photosensitizers and will induce acute reactions (89,90). Also, some of them have induced low-grade photosensitivity reactions, resulting in a pseudoporphyria clinical pattern in the skin. The thiazide diuretics are responsible for more photosensitivity reactions than any of the other commonly used drugs, because they are the most commonly used of these potent photosensitizers.

Amiodarone is an uncommonly used drug for cardiac stabilization. Over 50% of patients taking this drug have photosensitivity reactions (91). Also a significant number develop a blue-black discoloration of their skin. Apparently, this discoloration is due to phagocytosis of degraded cell membranes bound to lipid-soluble amiodarone (92). Photosensitivity may persist after the drug has been withdrawn because of tissue retention of the drug.

The vast majority of these clinical reactions are phototoxic, although occasionally delayed hypersensitivity responses may occur. The action spectrum for both phototoxic and photoallergic reactions usually includes long-wavelength UV radiation (19,76,84,93).

C. Normal Protection and No Known Photosensitizer

Photoreactions not due to deficient protection or the presence of known photosensitizers comprise a catchall of conditions.

Allergic and Autoimmune Conditions

Such immunological reactions include the apparent photoallergic diseases, the solar urticarias (94), the polymorphous light eruptions (95–97); and the autoimmune diseases such as lupus erythematosus (LE; 98,99), dermatomyositis (100), pemphigus, and pemphigoid (101).

The action spectrum for the solar urticarias ranges from UVC rays to infrared energy, depending on the specific disease process (94). The polymorphous light eruptions usually respond adversely to UVB rays (95). However, there is a subset of patients who react primarily to UVA radiation (96). The action spectrum for LE is both the UVA and UVB range (98,102), and fibroblasts from LE patients are abnormally sensitive to UVA (103). The reactivity of dermatomyositis patients appears to be UVB rays (100), and both pemphigus and pemphigoid are sensitive to UVB radiation (101). In these autoimmune diseases the abnormal reactions appear to follow phototoxic injuries.

Pellagra and Pellagralike Diseases

Such diseases include a dietary deficiency in nicotinic acid or tryptophan, which results in a cellular deficiency of nicotinic acid (104). Similar problems can be induced by competitive activity of certain medications, such as isoniazid, and metabolic deviation of nicotinic acid precursors, as may be caused by functioning carcinoid tumors. Also, certain disorders of amino acid metabolism, such as Hartnup disease (105), hydroxykynurenemia (106), and congenital tryptophanuria (107) may have abnormalities leading to deficiencies in nicotinic acid formation or utilization. These conditions are characterized by erythema and desquamation following sun exposure. This is followed by a dusky brownish-red discoloration. In chronic cases the skin is thickened, scaly, and may have a dusky magenta color (104). The action spectrum for this apparent phototoxic response has not been clarified.

Certain Recessive Degenerative Genodermatoses

The recessive degenerative genodermatoses may have photosensitivity components (108). In addition to xeroderma pigmentosum (XP), which was noted earlier in this discussion, Bloom's syndrome, Cockayne's syndrome, and poikiloderma congenitale (Rothmund-Thomson syndrome), all have photosensitivity problems, The responses appear to be phototoxic, but the action spectrum has not yet been clarified. Skin fibroblasts from patients with Bloom's syndrome have reduced survival after UVB radiation (109). In addition, cells from Cockayne patients are hypersensitive to UV radiation (110).

Miscellaneous Disorders

The miscellaneous group is the "catchall" of this catchall category. This group contains a number of diverse conditions including certain photodermatoses and other conditions that may be aggravated by sun exposure.

The Photodermatoses Including Hydroa Aestivale et Vacciniforme Patients with this syndrome respond adversely to UVA rays (111,112). Whether this is a phototoxic response has not been determined. Disseminated actinic porokeratosis (DSAO) appears to be a peculiar phototoxic response to UV radiation (113). However, the definition of the action spectrum has not been completely clarified.

Photoaggravated Diseases A number of photoaggravated diseases are included in this group. Darier's disease is adversely affected by UVB rays (114,115). This appears to be a phototoxic process. Familial benign chronic pemphigus does respond abnormallly to radiation throughout the UVB and UVA spectrum (116). This also appears to be a phototoxic response. Recent studies indicate that UVB can activate herpes simplex infections in humans (117). The action spectrum for reactions in photosensitive psoriasis, granuloma annulare, and acne rosacea remains undetermined. However, acne rosacea is one of the most common photosensitive diseases.

VI. SUMMARY

Adverse reactions to the sun's rays have become commonplace not only because of excessive sunbathing habits, but also because of the increasing number of photosensitizers in our environment. The sun is responsible for a significant amount of human morbidity, ranging from acute sunburn reactions to skin cancer formation.

Photodamage to cutaneous cellular macromolecules and membranes, and generation or release of inflammatory mediators are integral parts of this photoinjury. Endogenous and exogenous photosensitizing molecules in our personal and general environments supply further pathways for damage. In addition, there are several photodermatoses and ''photoaggravatable'' diseases that add to the potential morbidity induced by sun exposure. We hope that sophistication in the realm of photoprotection will add significantly to our preventive medicine armamentarium.

REFERENCES

1. Parrish, J. A. The scope of photomedicine, *The Science of Photomedicine* (J. D. Regan and J. A. Parrish, eds.), Plenum Press, New York, 1982, pp. 3–17.
2. Birmingham, D. J. Occupational dermatoses, *Prog. Dermatol.*, *3*:1–8 (1968).
3. Epstein, J. H. Photomedicine, *The Science of Photobiology, 2nd ed.* (K. C. Smith, ed.), CRC Press, Boca Raton, FL, 1989, pp. 155–192.
4. Urbach, F. and Gange, R. W., eds. *The Biological Effects of UVA Radiation*, Praeger, New York, 1986.
5. Peak, M. J. and Peak, J. G. Molecular photobiology of UVA, *The Biological Effects of UVA Radiation* (F. Urbach and R. W. Gange, eds.), Praeger, New York, 1986, pp. 42–56.
6. Gange, R. W., Park, Y.-K., Auletta, M., Kagetsu, N., Blackett, A. D., and Parrish, J. A. Action spectra for cutaneous responses to ultraviolet radiation, *The Biological Effects of UVA Radiation* (F. Urbach and R. W. Gange, eds.), Praeger, New York, 1986, pp. 57–67.
7. Krutmann, J., Czech, W., Dig-pgen, T., Niedner, R., Kapp, A., and Schopf, E. High dose UVA1 therapy in the treatment of patients with atopic dermatitis, *J. Am. Acad. Dermatol.*, *26*:225–230 (1992).
8. McGrath, H., Jr. UV-A1 light decreases disease activity and eliminates antibodies in patients with systemic lupus erythematosus, *Biological Responses to UVA Radiation* (F. Urbach, ed.), Voldenmar Publishing, Overland Park, KS, 1992, pp. 257–260.
9. Whitman, G. B., Leach, E. E., Deleo, V. A., and Harber, L. C. Comparative response to UVA radiation in humans and guinea pigs, *The Biological Effects of UVA Radiation* (F. Urbach and R. W. Gange, eds.), Praeger, New York, 1986, pp. 79–86.
10. Irwin, C., Barnes, A., Veres, D., and Kaibey, K. An ultraviolet radiation action spectrum for immediate pigment darkening, *Photochem. Photobiol.*, *57*:504–507 (1993).
11. Kligman, L. H. and Sayre, R. M. An action spectrum for ultraviolet induced elastosis in hairless mice: Quantification of elastosis by image analysis, *Photochem. Photobiol.*, *53*:237–242 (1991).

12. Sterenborg, H. J. C. M. and Van der Leun, J. C. Tumorigenesis by a long wavelength UV-A source, *Photochem. Photobiol.*, *51*:325–330 (1990).
13. Cross, F. On a turf (peat) fire cancer. Malignant change superimposed on erythema ab igne, *Proc. R. Soc. Med.*, *60*:1307–1308 (1967).
14. Arrington, J. H. III. Thermal keratoses and squamous cell carcinoma in situ associated with erythema ab igne, *Arch. Dermatol.*, *115*:1226–1228 (1979).
15. Kligman, L. H. Intensification of ultraviolet-induced dermal damage by infrared radiation, *Arch. Dermatol. Forschung*, *272*:229–238 (1982).
16. Harber, L. C. and Bickers, D. R. *Photosensitivity Diseases, Principles of Diagnosis and Treatment*, W. B. Saunders, Philadelphia, 1981, pp. 24–25.
17. Epstein, J. H. Photoallergy and photoimmunology, *Dermatologic Immunology and Allergy* (J. Stone, ed.), C. V. Mosby, St. Louis, 1985, pp. 629–640.
18. Epstein, J. H. Phototoxicity and photoallergy, *Sunlight and Man* (M. A. Pathak, et al., eds.), University of Tokyo Press, Tokyo, 1974, pp. 459–477.
19. Epstein, J. H. and Wintroub, B. U. Photosensitivity due to drugs, *Drugs*, *30*:42–57 (1985).
20. Kochevar, I. E. Phototoxicity mechanisms: Chlorpromazine photosensitized damage to DNA and cell membranes, *J. Invest. Dermatol.*, *77*:59–64 (1981).
21. Morison, W. L. Photoimmunology, *J. Invest. Dermatol.*, *77*:71–76 (1981).
22. Morison, W. L., Parrish, J. A., Bloch, K. J., and Krugler, J. I. In vivo effect of UVB on lymphocyte function, *Br. J. Dermatol.*, *101*:573–519 (1979).
23. Morison, W. L., Parrish, J. A., Bloch, K. J., and Krugler, J. I. In vivo effects of PUVA on lymphocyte function, *Br. J. Dermatol.*, *104*:405–413 (1981).
24. Hersey, P. M., Haran, E., Hasil, E., and Edwards, A. Alteration of T cell subsets and induction of suppressor T cell activity in normal subjects after exposure to sunlight, *J. Immunmol.*, *31*:171–174 (1983).
25. Hersey, P., MacDonald, M., Henderson, C., Schibeci, S., D'Alessandro, G., Pryer, M., and Wilkinson, F. J. Suppression of natural killer cell activity in humans by radiation from solarium lamps depleted of UVB, *J. Invest. Dermatol.*, *90*:305–310 (1988).
26. Morison, W. UV and immunity in humans, *Photochem. Photobiol.*, *50*:515–524 (1989).
27. Kripke, M. L. Immunosuppressive action of UV radiation, *The Dark Side of Sunlight* (F. R. deGruijl, ed.), Utrecht University, Utrecht, 1993, pp. 77–85.
28. Cruz, P. D. and Bergstresser, P. R. The influence of ultraviolet radiation and other physical and chemical agents on epidermal Langerhans cells, *Epidermal Langerhans Cells* (Schuler, G., ed.), CRC Press, Boca Raton FL, 1990, pp. 253–271.
29. Okamoto, H. and Kripke, M. L. Effector and suppressor circuits of the immune response are activated in vivo by different mechanisms, *Proc. Natl. Acad. Sci. USA*, *84*:3841–3845 (1987).
30. Rivas, J. M. and Ullrich, S. E. Systemic suppression of delayed-type hypersensitivity by supernatants from UV-irradiated keratinocytes. An essential role for keratinocyte-derived IL-10, *J. Immunol.*, *149*:3865–3871 (1992).
31. Applegate, L. A., Ley, R. D., Acalay, J., and Kripke, M. L. Identification of the molecular target for the suppression of contact hypersensitivity by ultraviolet radiation, *J. Exp. Med.*, *170*:1117–1131 (1989).

32. Kripke, M. L., Cox, P. A., Alas, L. G., and Yarosh, D. B. Pyrimidine dimers in DNA initiated systemic immunosuppression in UV-irradiated mice, *Proc. Natl. Acad. Sci. USA*, *89*:7516–7520 (1992).
33. Epstein, J. H. Nonmelanoma skin cancer overview 1992, *Photomed. Photobiol.*, *14*:49–57 (1992).
34. Scotto, J., Fears, T. R., and Fraumeni, J. F., Jr. Incidence of nonmelanoma skin cancer in the United States. *NIH Publication No. 83-2433*, US Department of Health and Human Services, Washington DC, 1983.
35. Boring, C. C., Squires, T. S., Tong, T., and Montgomery, S. Cancer statistics, 1994, *CA*, *44*:7–26 (1994).
36. Marks, R. Nonmelanotic skin cancer and solar keratoses: The quiet 20th century epidemic, *Int. J. Dermatol.*, *26*:201–205 (1987).
37. Glass, A. G. and Hoover, R. N. The emerging epidemic of melanoma and squamous cell carcinoma, *JAMA*, *262*:2097–2100 (1989).
38. Gallagher, R. P., Ma, B., McLean, D. I., Yang, C. P., Ho, V., Carruthers, J. A., and Warshawski, L. M. Trends in basal cell carcinoma, squamous cell carcinoma and melanoma of the skin from 1973 through 1987, *J. Am. Acad. Dermatol.*, *23*: 413–421 (1990).
39. Urbach, F. Man and ultraviolet radiation, *Human Exposure to Ultraviolet Radiation. Risks and Regulations* (W. F. Passchier and B. F. M. Boonjakovic, eds.), Excerpta Medica, New York, 1987, pp. 3–16.
40. Leong, G. K. P., Stone, J. L., Farmer, E. R., Scotto, J., Reizner, G. T., Burnett, T. S., and Elbern, D. J. Nonmelanoma skin cancer in Japanese residents of Kauai, Hawaii, *J. Am. Acad. Dermatol.*, *17*:233–238, 1987.
41. Epstein, J. H. Photocarcinogenesis, skin cancer and aging, *Aging and the Skin* (A. K. Balin and A. M. Kligman, eds.), Raven Press, New York, 1989, pp. 307–329.
42. Brodthagen, H. Interaction of UVA and UVB in photocarcinogenesis, *The Biological Effects of UVA Radiation* (F. Urbach and R. W. Gange, eds.), Praeger, Westport, CT, pp. 127–136.
43. van Weelden, H., deGruijl, F. R., and Van der Leun, J. C. Carcinogenesis by UVA, with an attempt to assess the carcinogenic risks of tanning with UVA and UVB, *The Biological Effects of UVA Radiation* (F. Urbach and R. W. Gange, eds.), Praeger, Westport, CT, 1986, pp. 137–146.
44. Sterenborg, H. J. C. and Van der Leun, J. C. Tumorigenesis by a long wavelength UV-A source, *Photochem. Photobiol.*, *51*:325–330 (1990).
45. Rigel, D. S., Kopf, A. W., and Friedman, R. J. Incidence and mortality of malignant melanoma in the US, *Melanoma Lett.*, *7*:1–2 (1989).
46. Epstein, J. H. Experimental models for primary melanoma, *Photodermatol. Photoimmunol. Photomed.*, *9*:91–98 (1992).
47. Ries, L. A., Hankey, B. S., and Edwards, B. K. Cancer statistics review 1973–1981. NIH Publication No. 902789. Division of Cancer Prevention and Control, National Cancer Institute, Bethesda, MD, 1990.
48. Johnson, T. M., Rowe, D. E., Nelson, B. R., and Swanson, N. A. Squamous cell carcinoma of the skin (excluding lip and oral mucosa), *J. Am. Acad. Dermatol.*, *26*:467–484 (1992).

49. Koh, H. K., Kligler, B. E., and Leu, R. A. Sunlight and cutaneous melanoma: Evidence for and against causation. Yearly review, *Photochem. Photobiol.*, *51*: 765–779 (1990).
50. Koh, H. K. and Lew, R. A. Sunscreens and melanoma: Implications for prevention [editorial], *JNCI*, *86*:78–79 (1994).
51. Boyle, J., Briggs, J. D., Mackie, R. M., Junor, B., and Aitchison, J. R. Cancer, warts and sunshine in renal transplant patients, *Lancet*, *1*:702–705 (1994).
52. Smith, S. E., Davis, I. C., Leshin, B., Fleischer, A. B., White, W. L., and Feldman, S. R. Absence of human papillomavirus in squamous cell carcinomas of nongenital skin from immunocompromised renal transplant patients, *Arch. Dermatol.*, *129*: 1585–1588 (1993).
53. Yoshikawa, T., Rae, V., Bruins-Slot, W., van den Berg, J. W., Taylor, J. R., and Streilein, J. W. Susceptibility to effects of UVB radiation on induction of contact sensitivity as a risk factor for skin cancer in humans, *J. Invest. Dermatol.*, *95*: 530–536 (1990).
54. Cleaver, J. E. Repair processes for photochemical damage in mammalian cells, *Adv. Radiat. Biol.*, *4*:1–75 (1974).
55. Kraemer, K. H. Twenty years of research on xeroderma pigmentosum at the National Institutes of Health, *Photobiology* (E. Riklis, ed.), Plenum Press, New York, 1991, pp. 211–221.
56. Bishop, M. J. The molecular genetics of cancer, *Science*, *235*:305–311 (1987).
57. Ananthaswamy, H. N. and Piercell, W. E. Molecular mechanisms of ultraviolet radiation carcinogenesis: Yearly review, *Photochem. Photobiol.*, *52*:1119–1136 (1990).
58. Elder, J. T. C-Ha-ras and UV photocarcinogenesis, *Arch. Dermatol.*, *126*:379–382 (1990).
59. Brash, D. E., Rudolph, J. A., Simon, J. A., Lin, A., McKenna, G. J., Baden, H. P., Halperin, A. J., and Ponten, J. A role for sunlight in skin cancer: UV-induced p53 mutations in squamous cell carcinoma, *Proc. Natl. Acad. Sci. USA*, *88*: 10124–10128 (1991).
60. Campbell, C., Quinn, A. G., Young-Suck, R., Angus, B., and Rees, J. L. p53 mutations are common and early events that precede tumor invasion in squamous cell neoplasia of the skin, *J. Invest. Dermatol.*, *100*:746–748 (1993).
61. Nagano, T., Ueda, M., and Ichihashi, M. Expression of p53 protein is an early event in ultraviolet-induced cutaneous squamous cell carcinogenesis, *Arch. Dermatol.*, *129*:1157–1161 (1993).
62. Marks, R. and Fraser, P. Ozone depletion and human health or fiction, *Dermatology: Progress and Perspectives* (W. H. C. Burgdorf and S. I. Katz, eds.), Proceedings of the 18th World Congress of Dermatology, Parthenon Publishing Group, New York, 1992, pp. 111–115.
63. Thompson, S. C., Jolley, D., and Marks, R. Reduction of solar keratoses by regular sunscreen use, *N. Engl. J. Med.*, *329*:1147–1151 (1993).
64. Bolognia, J. L. and Pawelek, J. M. Biology of hypopigmentation, *J. Am. Acad. Dermatol.*, *19*:217–255 (1988).
65. Knox, W. E. Phenylketonuria, *Metabolic Basis of Inherited Disease* (J. B. Stanbury, J. B. Wyngaarden, and D. S. Fredrickson, eds.), McGraw-Hill, New York, 1972, pp. 266–295.

66. Windhorst, D. B., Zelickson, A. S., and Good, R. A. A human pigmentary dilution based on a heritable subcellular structural defect—the Chediak–Higashi syndrome, *J. Invest. Dermatol.*, *50*:9–18 (1968).
67. Cleaver, J. E. Defective repair replication of DNA in xeroderma pigmentosum, *Nature*, *218*:652–653 (1968).
68. Epstein, J. H., Fukuyama, K., Reed, W. B., and Epstein, W. L. Defect in DNA synthesis in skin of patients with xeroderma pigmentosum demonstrated in vivo, *Science*, *168*:1477–1478 (1970).
69. Kraemer, K. A., Lee, M. M., and Scotto, J. Xeroderma pigmentosum: Cutaneous, ocular and neurologic abnormalities in 830 published cases, *Arch. Dermatol.*, *123*: 241–250 (1987).
70. Pathak, M. A., Fitzpatrick, T. B., and Parrish, J. A. Topical and systemic approaches to protection of human skin against harmful effects of solar radiation, *The Science of Photomedicine* (J. D. Regan and J. A. Parrish, eds.), Plenum Press, New York, 1982, pp. 441–473.
71. Scotto, J., Fears, T. R., and Fraumeni, J. F., Jr. Incidence of nonmelanoma skin cancer in the United States, U.S. Department of Health, Education and Welfare, Publication Number (NIH) 82-2433, 1981.
72. Urbach, F. Photocarcinoogenesis, *The Science of Photomedicine* (J. D. Regan and J. A. Parrish, eds.), Plenum Press, New York, 1982, pp. 261–292.
73. Epstein, J. H. Photocarcinogenesis, skin cancer and aging, *J. Am. Acad. Dermatol.*, *9*:487–502 (1983).
74. Van der Leun, J. UV carcinogenesis, *Photochem. Photobiol.*, *39*:861–868 (1984).
75. Poh-Fitzpatrick, M. B. Porphyrias, *Photomedicine*, *Vol. 1* (E. Ben-Hur and I. Rosenthal, eds.), CRC Press, Boca Raton FL, 1987, pp. 163–178.
76. Harber, L. C., Kochevar, I. E., and Shalita, A. R. Mechanisms of photosensitization to drugs in humans, *The Science of Photomedicine* (J. D. Regan and J. A. Parrish, eds.), Plenum Press, New York, 1982, pp. 323–347.
77. Herman, P. S. and Sams, W. M., Jr. *Soap Photodermatitis*, Charles C Thomas, Springfield IL, 1972.
78. Raug, G. J., Storrs, F. J., and Larsen, W. G. Photoallergic contact dermatitis to mens perfume, *Contact Dermatitis*, *5*:251–260 (1979).
79. Riovinazzo, V. J., Ichikawa, H., Kochevar, I. E., Armstrong, R. B., and Harber, L. C. Photoallergic contact dermatitis to musk ambrette: Action spectrum in guinea pigs and man, *Photochem. Photobiol.*, *33*:773–777 (1981).
80. Galosi, A. and Plewig, G. Photoallergic eczema caused by musk ambrette, *Hautarzt*, *33*:589–594 (1982).
81. Kaidbey, K. H. and Kligman, A. M. Contact photoallergy to 6-methylcoumarin in proprietary sunscreens, *Arch. Dermatol.*, *114*:1709–1710 (1978).
82. Jackson, R. T., Nesbit, L. T. and DeLeo, V. A. 6-Methylcoumarin photocontact dermatitis, *J. Am. Acad. Dermatol.*, *2*:124–127 (1980).
83. Epstein, J. H. Chemical phototoxicity in humans. *JNCI*, *69*:265–268 (1982).
84. Toback, A. C. and Anders, J. E. Phototoxicity from systemic agents, *Dermatologic Clinics*: *Photosensitivity Diseases* (V. A. De Leo, ed.), W. B. Saunders, Philadelphia, 1986, pp. 223–230.

85. Bjellerup, M., Kjellstrom, T., and Ljunggren, B. Influence of tetracycline phototoxicity on growth of cultured human fibroblasts, *J. Invest. Dermatol.*, *85*:573–574 (1985).
86. Ferguson, J., Addo, H. A., McGill, P. E., et al. A study of benoxaprofen-induced photosensitivity, *Br. J. Dermatol.*, *107*:429–442 (1982).
87. Halsey, J. Benoxaprofen side-effect profile in 300 patients, *Br. Med. J.*, *284*:1365 (1982).
88. Kochevar, I. E., Hoover, K. W., and Gawienowski, M. Benoxaprofen photosensitization of cell membrane disruption, *J. Invest. Dermatol.*, *82*:214–218 (1984).
89. Diffey, B. L. Daymond, T. J., and Fairgreaves, H. Phototoxic reactions to prixoicam, naproxen, and tiaprofenic acid, *Br. J. Rheumatol.*, *22*:239–242 (1983).
90. Stern, R. S. and Bigby, M. An expanded profile of cutaneous reactions to nonsteroidal anti-inflammatory drugs. Reports to a specialty-based system for spontaneous reporting of adverse reactions to drugs, *JAMA*, *252*:1433–1437 (1984).
91. Diffey, B. L., Chalmers, R. J. G., and Muston, N. L. Photobiology of amiodarone: Preliminary in vitro and in vivo studies, *Clin. Exp. Dermatol.*, *9*:248–255 (1984).
92. Zachary, C. B., Slater, D. N., Holt, D. W., Storey, G. C. A., and MacDonald, D. M. The pathogenesis of amiodarone-induced pigmentation and photosensitivity, *Br. J. Dermatol.*, *110*:451–456 (1984).
93. Elmets, C. Drug induced photoallergy, *Dermatology Clinics*: *Photosensitivity Diseases* (V. A. DeLeo, ed.), W. B. Saunders, Philadelphia, 1986, pp. 231–241.
94. Armstrong, R. B. Solar urticaria, *Dermatologic Clinics*: *Photosensitivity Diseases* (V. A. DeLeo, ed.), W. B. Saunders, Philadelphia, 1986, pp. 253–259.
95. Epstein, J. H. Polymorphous light eruption, *J. Am. Acad. Dermatol.*, *3*:329–343 (1980).
96. Holzle, E., Plewig, G., Hofmann, C., et al. Polymorphous light eruption: Experimental reproduction of lesions, *J. Am. Acad. Dermatol.*, *7*:111–125 (1982).
97. Epstein, J. H. Polymorphous light eruption, *Dermatologic Clinics*: *Photosensitivity Diseases* (V. A. DeLeo, ed.), W. B. Saunders, Philadelphia, 1986, pp. 243–251.
98. Tuffanelli, D. L. and Epstein, J. H. Chronic cutaneous (discoid) lupus erythematosus, *Dubois' Lupus Erythematosus*, 3rd ed. (D. J. Wallace and E. L. Dubois, eds.), Lea & Febiger, Philadelphia, 1987, pp. 283–301.
99. Hymes, S. R., Jordan, R. E., and Arnett, F. C. Lupus erythematosus, *Dermatologic Clinics*: *Photosensitivity Diseases* (V. A. DeLeo, ed.), W. B. Saunders, Philadelphia, 1986, pp. 267–276.
100. Caro, I. A dermatologist's view of dermatomyositis, *Clinics in Dermatology*: *Polymyositis/Dermatomyositis*, Vol. 2 (R. D. Sontheimer, ed.), J. B. Lippincott, Philadelphia, 1988, pp. 9–14.
101. Cram, D. L. and Fukuyama, K. The immunochemistry of ultraviolet-induced pemphigus and pemphigoid lesions, *Arch. Dermatol.*, *106*:819–824 (1972).
102. Lehman, P., Holzle, E., Kind, P., Goerz, G., and Plewig, G. Experimental reproduction of skin lesions in lupus erythematosus by UVA and UVB radiation, *J. Am. Acad. Dermatol.*, *22*:181–187 (1990).
103. Zamansky, G. B., Minka, D. F., Deal, C. L., et al. The in vitro photosensitivity of lupus erythematosus skin fibroblasts, *J. Immunol.*, *134*:1571–1576 (1985).

104. Stratigos, J. D. and Katsambas, A. Pellagra: A still existing disease, *Br. J. Dermatol.*, *96*:99–106 (1977).
105. Baron, D. N., Dent, C. E., Harris, H., Hartz, H., Hart, E. W., and Hepson, J. B. Hereditary pellagra-like skin rash with temporary cerebellar ataxia, constant renal aminoaciduria and other bizarre chemical features, *Lancet*, *2*:421–428 (1956).
106. Kromrower, G. M. and Westall, R. Hydroxykinureninuria, *Am. J. Dis. Child.*, *113*: 77–80 (1967).
107. Tada, K., Ito, H., Wada, Y., and Arakawa, T. Congenital tryptophanuria with dwarfism: Toboku, *J. Exp. Med.*, *80*:118–134 (1963).
108. Bligard, G. A. and Storer, J. S. Photosensitivity in infants and children, *Dermatologic Clinics*: *Photosensitivity Disease* (V. A. DeLeo, ed.), W. B. Saunders, Philadelphia, 1986, pp. 311–320.
109. Zbinden, I. and Cerutti, P. Near ultraviolet sensitivity of skin fibroblasts of patients with Bloom's syndrome, *Biophys. Res. Commun.*, *98*:579–587 (1981).
110. Lehmann, A. R., Francis, A. J., and Giannelli, F. Prenatal diagnosis of Cockayne's syndrome, *Lancet*, *1*:486–488 (1985).
111. Goldgeier, M. H., Nordland, H. H., and Lucky, A. W. Reproduction of hydroa vacciniforme with UVA, *J. Am. Acad. Dermatol.*, *9*:278–280 (1983).
112. Halasz, L. L. G., Leach, E. E., Walther, R. R., et al. Hydroa vacciniforme: Induction of lesions with ultraviolet A, *J. Am. Acad. Dermatol.*, *8*:171–176 (1983).
113. Chernosky, M. E. and Anderson, D. E. Disseminated superficial actinic porokeratosis, *Arch. Dermatol.*, *99*:401–407 (1969).
114. Shelley, W. B., Arthur, R. P., and Pillsbury, D. M. A view of keratosis follicularis (Darier's disease) as a neoplastic process, *Arch. Dermatol.*, *80*:332–338 (1959).
115. Penrod, J. M., Everett, M. A., and McCreight, W. G. Observations on keratosis follicularis, *Arch. Dermatol.*, *82*:367–370 (1960).
116. Cram, D. L., Muller, S. A., and Winkelmann, R. K. Ultraviolet-induced acantholysis in familial benign pemphigus, *Arch. Dermatol.*, *96*:636–641 (1967).
117. Perna, J. J., Mannix, M. L., Rooney, J. F., Notkins, A. L., and Straus, S. E. Reactivation of latent herpes simplex virus infection by ultraviolet light: A human model, *J. Am. Acad. Dermatol.*, *17*:473–478 (1987).

5

Exposure to Low Levels of Ultraviolet Light B or Ultraviolet Light A Induces Cutaneous Photodamage in Human Skin

David P. Meyers and Ian R. Scott
Unilever Research U.S., Edgewater, New Jersey

Nicholas J. Lowe
Skin Research Foundation of California, Santa Monica, and UCLA School of Medicine, Los Angeles, California

I. INTRODUCTION

There are significant data from both human epidemiological and experimental animal studies suggesting that the ultraviolet (UV) component of sunlight is responsible for the induction of photoaging or dermatoheliosis (1–8) and photocarcinogenesis (9–11). Despite the existence of a sizable database, little work has been conducted to investigate the repetitive effects of UV on human skin under experimentally controlled conditions. In addition, the amounts of UV delivered in most animal studies are much greater than those received by average members of the public in a normal day of activities (12–17). Usually, the average individual is exposed to relatively low levels of UVA. This is normally received filtered through either window glass or current sunscreen that provides primarily UVB protection. Those individuals receiving UVA tanning treatments would receive considerably greater amounts of UVA.

The study reported here investigated the effects of repetitive UV exposure, particularly UVA, on previously unexposed human skin. The study was designed to determine whether repetitive erythemogenic doses of UVA and solar-simulated UV are capable of inducing cutaneous photodamage. The effectiveness of repetitive suberythemogenic UVA in producing changes was also investigated in the same manner. The dosages selected are biologically relevant, as they are comparable with the solar exposures received by the public in their normal daily activities.

II. METHODS

A. Ultraviolet Irradiation Apparatus

An Oriel 1000-W xenon arc solar simulator was equipped for all irradiations with a dichroic mirror and a 1-mm UG11 filter. A 2-mm Schott WG-345 filter was used to deliver UVA radiation (320–400 nm) and a 1-mm WG-320 filter to deliver solar-simulated radiation (290–400 nm). The spectral output was determined using an Optronics spectral radiometer. The spectral characteristics were determined for both filter arrangements. The spectral analysis showed that the filter arrangements provided essentially UVA or solar-simulated radiation. The WG-345 and WG-320 effectively eliminated all relevant energy below 320 and 290 nm, respectively. This ensured that any possible synergism between UVB and UVA for UVA irradiations and UVB and UVC for solar-simulated irradiations was eliminated. Irradiation output was measured before each scheduled irradiation treatment period using an IL700 research radiometer (International Light Co.) with peak absorption at 305 nm for solar-simulated UV. For UVA irradiation, a broad-absorbing Dermalight radiometer, with a peak absorption of 365 nm, was used. Irradiance of the lamp during the 24 weeks of irradiations was generally 0.05 mW/cm^2 for solar-simulated UV dose group and 30 mW/cm^2 for the UVA dose groups.

B. Study Volunteers

Normal healthy skin phototype II and III individuals (mean age 30.9 $\pm$ 3.6), who had a history of skin that sunburned, but who eventually developed a mild to moderate tanning response, were recruited. A sufficient number of individuals were recruited to ensure that the data obtained were statistically relevant. All recruited individuals had no past history of skin disease, were not taking any photosensitizing drugs, and their buttock skin had no prior exposure to sunlight. Buttock skin of all potential study participants was examined to determine if any skin lesion or photodamage was present that could potentially accentuate the skin evaluation scores and, thereby, impinge on the conclusions drawn concerning the clinical effect of the UV treatments. Before enrollment in the study, each test subject reviewed and signed an informed consent form that described the investigation in detail.

C. Minimal Erythema Dose Determination

The minimal erythema doses (MEDs) for each test subject were determined for both solar-simulated UV and UVA on previously nonexposed buttock skin. Determinations were made using an opaque adhesive-backed foil template with 1-cm^2 openings. These openings were sequentially removed to deliver doses above and below the estimated MED for an average individual with type II skin. The sites were examined 24 h after irradiation, and the MED determined as the UV dose that induced minimal, but uniformly perceptible, erythema. These values were recorded, and the treatment regimen for each subject was individualized, based on their MEDs.

D. Experimental Protocol

Two sites were used on each buttock and were randomized to eliminate observer bias and any potential site-to-site variation. The irradiation regimen consisted of 1 MED of solar-simulated UV, 1 MED of UVA (high UVA), the amount of UVA in 1 MED of solar-simulated UV (low UVA), or no UV. The UV doses ranged from 0.03 to 0.07 J/cm^2 for solar-simulated UV, 25 to 35 J/cm^2 for high UVA, and 3 to 5 J/cm^2 for low UVA. The solar-simulated dose contains less energy than the low UVA dose owing to the UVB component, which possesses a high erythema effectiveness factor. Test sites were irradiated twice weekly with at least 3 days between each irradiation. The irradiation period was 24 weeks in duration. Clinical assessment of erythema, pigmentation, skin dryness, fine wrinkles, and skin atrophy (apparent progressive decrease in skin dermal components manifested as sagging or a sunken appearance) was conducted initially, and every 2 weeks thereafter, during the 24-week irradiation period and again 12 weeks after the last irradiation. Biopsy specimens (4 mm) were obtained at 12, 24, and 36 weeks after the initial irradiation. The 36-week specimens were obtained 12 weeks after the final irradiation. Biopsies were formalin-fixed, paraffin-embedded and sectioned at 5 μm, and stained with hematoxylin–eosin (H&E) for general tissue morphology, Gomori's aldehyde fuchsin for elastic tissue, and Lillies ferrous iron for melanin. Histopathological evaluations were conducted on coded samples to prevent observer bias. Evaluations of epidermal changes included epidermal thickness, granular layer thickness, stratum corneum thickness, dyskeratosis, and melanization. Dermal parameters included vascular dilation, perivascular infiltrate, and solar elastosis. Elastic fiber content was determined by computerized image analysis. The color-imaging system allows image processing to operate on color digital images. The system identifies elastic fibers, based on their color properties, as numerical values. The total area occupied by these specific values is determined and converted to grains, which are equivalent to the total number of pixels representing elastic tissue. Elastic tissue content is expressed as area fraction percentage (grains/area × 100) per field plus or minus the standard deviation.

III. RESULTS

A. Clinical Findings

The clinical changes observed after 24 weeks of treatment are shown in Figure 1. Detectable changes occurred in all the UV-treated groups, with the 1 MED of solar-simulated UV and high UVA treatments inducing changes of similar magnitude. Clinical changes were less severe in sites receiving the low dose of UVA.

Visible erythema was seen for the first 6 weeks in the sites receiving 1 MED of UVA and 1 MED for solar-simulated UV, after which its intensity gradually declined, presumably owing to a concomitant increase in epidermal pigmentation. Slight erythema was briefly present in the low UVA level, which was noticed after 6 weeks of irradiation and was absent at the next assessment period. The erythema observed on sites receiving the low dose of UVA suggests that lower amounts of UVA fail to produce pigmentation at a similar rate or quantity when compared with the other UV treatments, and the resulting lack of pigment was manifested as erythema. Although the high UVA and solar-simulated UV treatments produced considerable pigmentation, erythema was visible throughout

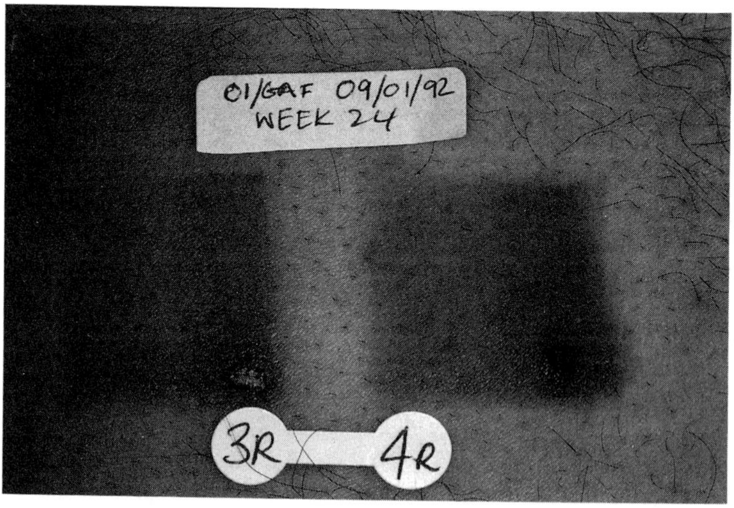

Figure 1 Clinical changes produced after 24 weeks of twice-weekly irradiation. The 1 MED of UVA produced the most profound pigmentation. The solar-simulated UV treatment induced a greater degree of change than the low UVA; however, irradiation with suberythemal UVA caused observable pigmentary changes; 1L, high UVA; 2L, control; 3R, 1 MED solar-simulated UV; and 4R, low UVA.

the study. The solar-simulated UV treatment resulted in greater epidermal thickening than was seen in either of the UVA groups.

Dryness is attributed to abnormalities in the stratum corneum. The high UVA treatment caused the highest skin dryness scores, with the solar-simulated UV and low UVA inducing similar dryness scores. A similar pattern of dyskeratosis was observed histologically for these treatments.

The most notable finding was the ability of UVA, delivered for 24 weeks, in twice weekly suberythemal doses, to cause observable changes to the skin. These changes, including pigmentation and fine wrinkling, were seen after only 6 weeks of treatment and persisted throughout the study, becoming more noticeable as the study progressed. The changes were still present and statistically relevant ($p < 0.05$) 12 weeks after the final irradiation.

Interestingly, (1) the high UVA treatment produced greater skin atrophy and fine wrinkles than the 1 MED solar-simulated UV treatment ($p < 0.05$), and (2) the low UVA treatment caused significant fine skin wrinkling ($p < 0.05$). These findings are important because it is believed that damage to the dermal components is responsible for the clinical manifestation of sagging and wrinkles. The clinical changes observed during this study, skin atrophy in particular, are consistent with dermal changes. These data suggest that even suberythemal doses of UVA are capable of damaging dermal components and that UVA may affect the dermis to a greater extent than UVB (15–17).

B. Histological Findings

Epidermal Parameters

The histological changes that occurred are graphically presented in Figures 2–5. Solar-simulated UV treatment caused an increase in total epidermal thickness, granular layer, and stratum corneum thickness. This increase was greatest after 24 weeks of irradiation ($p < 0.05$). Twelve weeks after the final irradiation (36 weeks), values were similar to those seen at 12 weeks, but still different from control ($p < 0.05$), suggesting that pretreatment morphology was returning, possibly owing to the presence of a more normal epidermal physiology. For the UV treatments, the high UVA group displayed an unexpected response in granular layer and total epidermal thickness. These values were increased at 12 weeks ($p < 0.05$), similar to the control at 24 weeks, and showed an increase at 36 weeks. The low UVA elicited the same response for epidermal thickness, except that a statistical difference ($p < 0.05$) was seen at 36 weeks. An increase in granular layer thickness was seen at all time points, but was significantly different from the control only at 12 weeks. Although both UVA groups produced increases in stratum corneum thickness at all time points, the increases in the low-dose UVA were not significant. Solar-simulated UV irradiation resulted in significant ($p < 0.05$) increases in stratum corneum thick-

(2)

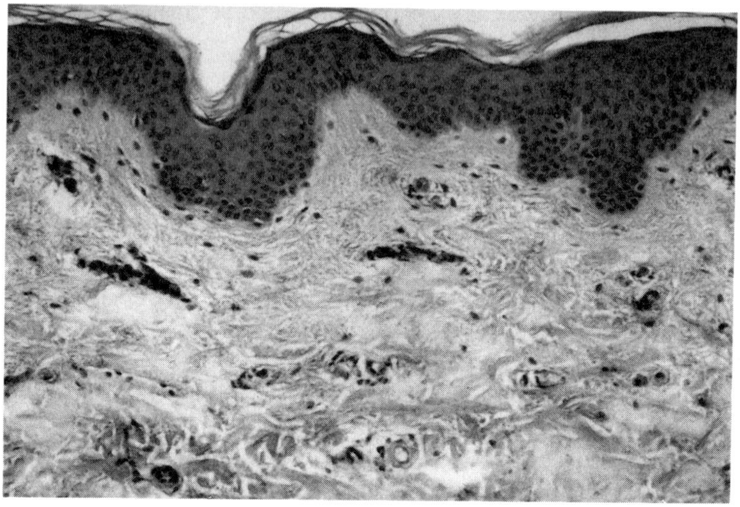

(3)

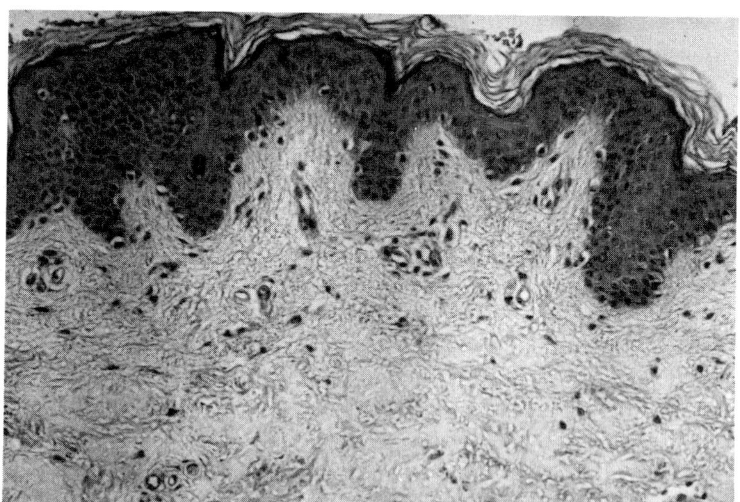

Figures 2–5 Photomicrographs demonstrating the histological changes observed after 24 weeks of twice-weekly irradiation. All UV treatments, including the suberythemal UVA, produced a thickened stratum corneum and epidermis and inflammation and vascular dilation in the dermis. (2) Control, (3) low UVA, (4) 1 MED solar-simulated UV, and (5) high UVA. Original magnification 250×.

(4)

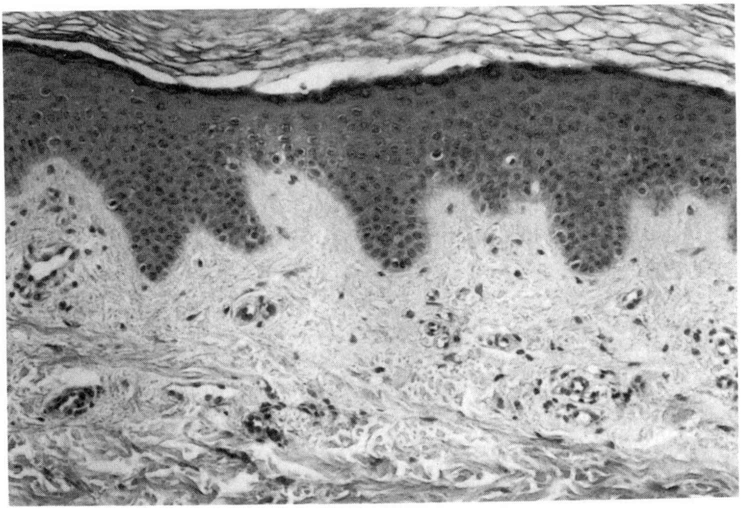

(5)

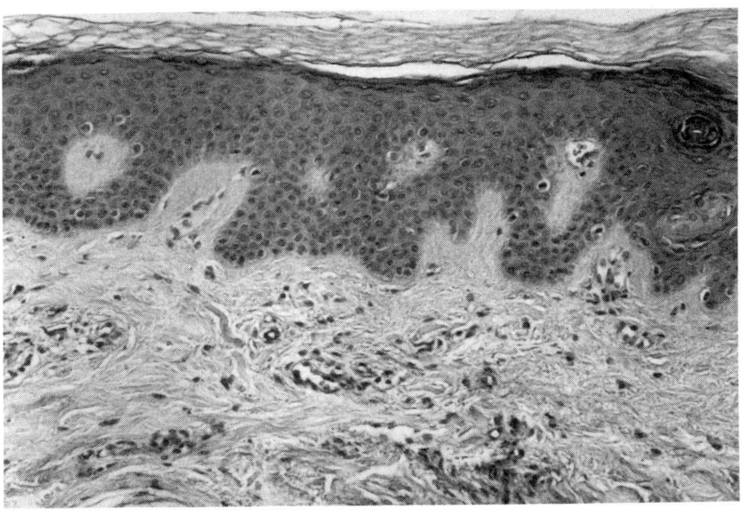

Figures 2–5 Continued

ness at all time points. All treatments produced increased numbers of dyskeratotic cells, although significant ($p < 0.05$) differences were noted only at 36 weeks in the solar-simulated UV and high UVA groups. Increased melanization was seen in all UV-treated groups with the high UVA being different from the control at all time points. The solar-simulated UV and low UVA produced increases in melanization, but no statistical differences were noted. This appears to be partly due to variability in test subject response to these treatments.

These results indicate that the damage produced by the UVA treatments persists longer than solar-simulated UV-induced epidermal damage. Most notable were the epidermal changes produced by suberythemal amounts of UVA. These alterations were present after only 12 weeks of twice-weekly irradiations. Although some improvement in morphology was seen 12 weeks after the final irradiation, marked changes were still quite evident. This suggests that even low doses of UVA have lasting effects on the epidermis.

C. Dermal Parameters

All treatment regimens induced observable inflammation, indicated by inflammatory cell infiltrate. The high UVA treatment resulted in significant ($p < 0.05$) inflammatory infiltrate at all time points. Interestingly, the low dose of UVA caused noticeable inflammation at all time points, showing a statistical difference at 24 weeks. Solar-simulated UV treatment also produced observable inflammation, with a statistical difference noted at 36 weeks, and significant ($p < 0.05$) vascular dilation at all biopsy intervals, which showed a gradual increase in severity with time. Irradiation with the low UVA dose resulted in vascular dilation that was significant ($p < 0.05$) at 24 and 36 weeks, whereas the high UVA dose caused an increase at 12 and 36 weeks. The most striking findings were the changes to dermal elastin assessed histopathologically, and observed in all treatment groups as early as 12 weeks ($p < 0.05$). At 36 weeks, elastosis was clearly evident in all treatment groups and was significant in the high UVA group. Morphological changes to dermis are shown in Figure 6.

Computerized image analysis quantification of elastic fiber content generated unexpected results (Table 1). Both UVA treatments produced a significant ($p < 0.05$) decrease in elastic fiber content after 24 weeks of irradiation and, in the high UVA group, remained significantly decreased 12 weeks after the final irradiation. Irradiation with solar-simulated UV produced a slight, but not significant, increase after only 12 weeks of treatment and remained increased at both 24 and 36 weeks.

As with the epidermis, all treatment regimens produced significant changes in the dermis and were noticeable after only 12 weeks of irradiation. These alterations increased in severity with treatment and showed little appreciable improvement 12 weeks after the final irradiation. Of greatest interest was the ability of the low-dose UVA to produce dermal change. The most notable was the elastosis observed after just 12 weeks of irradiation, which increased with treatment. The dermal changes seen in this group illustrate the capability that even low doses of repetitive UVA, delivered for as few as 12 weeks, have in producing dermal photodamage.

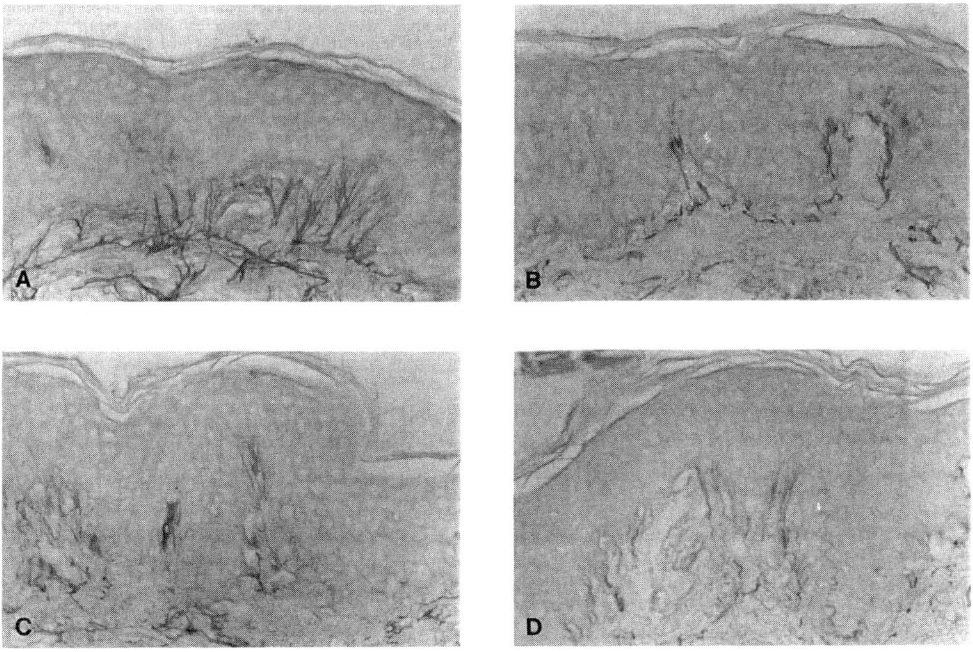

Figure 6 Photomicrographs illustrating the changes in elastic tissue after 24 weeks of twice-weekly irradiation. UVA irradiation resulted in a loss of elastic tissue, particularly the fine oxytalan fibers in the upper papillary dermis. Solar-simulated UV produced a slight increase, with clumping of the oxytalan fibers. (A) Control, (B) low UVA, (C) 1 MED solar-simulated UV, and (D) high UVA. Original magnification 400×.

Table 1 Effect of Repetitive Irradiation on Area Fraction Percentage of Elastic Fibers[a]

Treatment group	12-wk value[b] (n = 12)	24-wk value[b] (n = 12)	36-wk value[b] (n = 12)
1 MED SS	4.66 ± 0.29	4.65 ± 0.67	5.20 ± 0.49
Untreated control	4.53 ± 0.24	4.37 ± 0.72	4.49 ± 0.45
1 MED UVA	3.77 ± 0.29	2.87 ± 0.32	2.94 ± 0.44
UVA in solar-simulated	4.10 ± 0.23	3.14 ± 0.40	3.78 ± 0.68

[a]Human buttock skin irradiated twice weekly for 24 weeks. Biopsies obtained 12, 24, and 36 weeks after the initial irradiation and elastic tissue-like staining was quantified by image analysis.
[b]Data are mean ± SD.

IV. DISCUSSION

Previous investigations of repetitive UV irradiation have been conducted in animal models, in particular the hairless mouse (5–7). These models may not accurately reflect the effect of repetitive UV radiation on human skin for several reasons. First, the difference in histological characteristics between human and hairless mouse skin, in particular, the much thinner epidermis and relative lack of a epidermal pigmentation response to UV radiation, may make the mouse model more sensitive to UV effects than is human skin. In addition, previous studies performed on animals (5–7, 12–14) may have exaggerated the effects of UV on human skin owing to the large doses used. Therefore, animals may not serve as appropriate models for studying the effects of repetitive UV radiation on human skin.

Previous studies conducted on the effects of repetitive UVA and UVB irradiation on humans may be of little value owing to technical shortcomings. These studies used FS sunlamp bulbs that contain a significant amount of UVC, in addition to UVB and UVA (12–14). Therefore, in these studies, it remains unclear what changes UVA and UVB, without UVC, induce in human skin. In our study, care was taken to select and measure the spectral output of the solar simulator used. Full spectral characteristics were determined for the different filter arrangements to make certain that no UVC was being delivered and that UVA sites received only UVA without any contaminant UVB.

Brief exposure to UVA has been associated with a variety of histological changes (8). These include marked pigmentation and thickening of the epidermis and inflammation and elastosis in the dermis. The doses used in this previous work were significantly higher than the levels of UVA to which an average member of the public is exposed in normal daily activities. There has been little research reported documenting the effects of repetitive irradiation with small amounts of solar-simulated UV or UVA to human skin. We felt it was important to select doses that were comparable with those that average individuals would easily be exposed to in their normal activities. Also, we decided to conduct our investigation in human skin that had not been previously exposed to sunlight to permit an accurate assessment of the rate and magnitude of changes produced by the UV treatments on skin.

Skin phototype II individuals were selected for this study because this group is considered susceptible to photodamage from repetitive sun exposure (18,19). These skin types are common in the United States and Northern Europe. We carefully selected the amounts of UV delivered to be compatible with the exposure that is expected for an individual with an indoor job, with brief periods of outdoor activity (i.e., driving in their car or taking short walks). Individuals who are outdoors and protected with an efficient UVB sunscreen that provides little UVA protection would also be exposed to these amounts of UVA during a 3- to 4-h exposure to Southern California summer sun.

Both the clinical and histological data demonstrate the significant changes produced by all the UV treatments. Although we expected that the 1 MED solar-simulated UV and UVA treatments would result in cutaneous damage, the rate and extent to which changes occurred were not expected. These dramatic changes were noted both clinically and histologically after only 12 weeks of twice-weekly irradiation.

Treatment with 1 MED solar-simulated UV resulted in marked changes in the epidermis after 12 weeks of irradiation. Higher scores were recorded at 24 weeks, indicating that the proliferation and differentiation of keratinocytes were still increased. Following 12 weeks with no treatment, epidermal morphology showed noticeable improvement, suggesting a gradual recovery from the irradiation. Although the dermis responded in a similar manner during the 24-week irradiation period, 12 weeks after the last treatment, observable changes were still quite evident. This suggests that the dermal components are damaged by irradiation with solar-simulated UV and recover at a much slower rate than does the epidermis.

The high UVA treatment induced similar changes, but to a greater extent. At 24 weeks, epidermal changes showed improvement over the 12-week scores and increased again at 36 weeks. These findings are not yet understood. Conversely, dermal scores peaked at 24 weeks and showed a slight improvement at 36 weeks. It would be expected that greater epidermal thickening, caused by keratinocyte proliferation, would lead to increased pigmentation scores. This, however, was not true. The UVA-induced melanization displayed an inverted response curve with melanization scores at 24 weeks the lowest and greatly increased by 36 weeks. This suggests that lower levels of melanization, as seen between weeks 12 and 24, failed to protect the dermis against damage, as indicated by increased histology scores at week 24.

Of greatest interest were the unexpected changes observed in the skin irradiated with the low dose of UVA. Significant alterations were noted in both the epidermis and dermis after only 12 weeks of twice-weekly treatment. This dose was suberythemogenic and, as such, was not expected to produce significant clinical or histological changes. These findings significantly impinge on current thinking relative to the biological activity of repetitive exposure to low doses of UVA, since it is believed that UVA radiation, low doses in particular, have little effect on human skin.

The most notable finding in all UV treatments was the profound effect on dermal elastic tissue. Previous animal UV irradiation studies have shown changes in elastic fiber structure (7). In our study, the effects on elastic tissue were quite profound, as measured by both routine histology and computerized image analysis of elastic fiber content. Significant elastosis was observed in all UV-treated groups, including the low UVA, after only 12 weeks of irradiation.

Epidermal melanization appears to have played a minor photoprotective role relative to elastosis. Although all treatments produced pigmentation, significant elastosis was also noted. As melanin production in the solar-simulated UV group increased, the elastosis scores declined, suggesting a photoprotective effect. In both UVA groups, however, the results indicate that melanin production did not provide meaningful protection against elastosis. In all UV-treated groups, melanization displayed a sharp increase between 24 and 36 weeks. The data suggest that, even after 12 weeks with no irradiation, there is still an up-regulation of the melanization process.

The elastic fiber content, as measured by computerized image analysis, clearly shows a difference in qualitative effects of solar-simulated UV and UVA radiation. The methods used to measure dermal elastic fiber content have been reported previously (20,21) and have proved reliable in our studies. There was a slight increase in the dermal elastic fiber content with solar-simulated UV and a significant decrease in both the low and high UVA. Although there were no significant differences in the inflammatory responses elicited, differences because of greater or less proteolytic damage cannot be ruled out as a possible explanation and require further investigation. Another potential reason for the difference is that UVA may have a more direct effect on elastic tissue, leading to decreased quantities (12–14). The differing effects on elastin from different portions of the UV spectrum are not yet known, and further investigation is required. Although elastotic material has been investigated in previous studies of sun-damaged skin (22), the cause of this abnormal tissue is unclear.

There appears to be a close relation between elastic fibers and the lymphatic system (23,24). It has been demonstrated that the superficial lymphatics are surrounded by a network of elastic fibers, and this network is attached to other elastic fibers (25). These elastic fibers may provide not only structural support, but also may serve as prelymphatic pathways. The structural arrangement of the elastic fiber network is ideally situated to serve as a transport system for waste material moving downward from the epidermis to the initial lymphatics for removal (26). However, it remains unproved that elastic fibers influence lymphatic function (27,28). If there is a functional relation between elastic fibers and the lymphatics, the significant alterations to elastic fiber morphology seen in this study could have profound ramifications. If the fine oxytalan fibers, situated in the upper portion of the papillary dermis, act as prelymphatic drains, loss of or damage to these fibers could limit or prevent proper drainage into the initial lymphatics. Decreased skin drainage could lead to lymphoedema and alterations in normal immune function.

Langerhans cells leave the dermis through the lymphatics and there is an age-related reduction in Langerhans cell numbers (29,30). Ultraviolet radiation also results in decreased numbers of these cells, which is due to cell migration, rather than cell death (31). From these findings, the question arises of why Langerhans cells do not return to the skin. If a functional relation exists between elastic fibers and lymphatics, changes in elastic fiber morphology may have profound implications. It is possible that the alterations to elastic tissue, which occur in both intrinsically and actinically aged skin, in turn, may alter normal lymphatic function and impede Langerhans cells from reaching the lymph nodes. This impairment in motility may prevent an individual from mounting a normal immune response; an abnormal immune response is often seen with increasing age or sun exposure. Given the alterations in elastic fiber morphology observed in this study, it could be envisaged that normal Langerhans cell migration would be dramatically altered by UV radiation, resulting in impaired immune function. The relation between elastic fibers, Langerhans cell cycling, and the lymphatics remains unclear and requires further investigation in light of our findings.

V. CONCLUSIONS

Our results demonstrate that even small amounts of solar-simulated UV and UVA damage the skin. These results are particularly compelling; for even the lowest dose of UVA produced significant photodamage and elicited the cosmetic signs of photoaging. It is expected that low levels of UV radiation, delivered for even longer time periods, would lead to a greater photodamage and photoaging.

It is still unknown how quickly UV radiation induces histological changes in the skin. Our results indicate that these alterations were present considerably earlier than 12 weeks. It will be important to determine in future studies how soon UV-induced changes in skin morphology can be visualized.

Even when using many current sunscreens, which provide predominantly UVB protection, the public is exposed to UVA levels comparable with our low UVA dose. Our findings clearly demonstrate that there is a need for routine daily photoprotection with efficient UVB- and UVA-absorbing sunscreens and other methods of protection, such as photoprotective clothing. It remains unknown whether the skin is capable of recovering from UV-induced damage, particularly to the dermis. This underscores the importance of the regular use of sunscreens starting from early childhood.

ACKNOWLEDGMENT

The authors wish to thank Marjorie Lehman for her excellent histology contributions.

REFERENCES

1. Kligman, A. M. Early destructive effects of sunlight on human skin, *JAMA*, *210*: 2377–2380 (1969).

2. Lavker, R. M. Structural alterations in exposed and unexposed skin, *J. Invest. Dermatol.*, *73*:59–66 (1979).
3. Knox, J. M., Cockerell, E. G., and Freeman, E. G. Etiological factors and premature aging, *JAMA*, *179*:136–142 (1962).
4. Lavker, R. M. and Kligman, A. M. Chronic heliodermatitis. A morphologic evaluation of chronic actinic damage with emphasis on the role of mast cells, *J. Invest. Dermatol.*, *90*:325–330 (1988).
5. Kligman, L. H., Akin, F. J., and Kligman, A. M. The contributions of UVA and UVB to connective tissue damage in hairless mice, *J. Invest. Dermatol.*, *84*:272–276 (1985).
6. Kligman, L. H. Intensification of ultraviolet-induced dermal damage by infrared radiation, *Arch. Dermatol. Res.*, *252*:229–238 (1982).
7. Hirose, R. and Kligman, A. M. An ultrastructural study of ultraviolet-induced elastic fiber damage in hairless mouse skin, *J. Invest. Dermatol.*, *90*:697–702 (1988).
8. Gilchrest, B. A., Soter, N. A., Hawk, J. L., Barr, R. M., Block, A. K., Hensby, C. N., Mallet, A. J., Greaves, M. W., and Parris, J. A. Histologic changes associated with ultraviolet-A-induced erythema in normal human skin, *J. Am. Acad. Dermatol.*, *9*:213–219 (1983).
9. Lancaster, H. O. and Nelson, J. Sunlight as a cause of melanoma, *Med. J. Aust.*, *1*:452–456 (1957).
10. Kopf, A. W., Kripke, M. L., and Stern, R. S. Sun and malignant melanoma, *J. Am. Acad. Dermatol.*, *11*:674–684 (1984).
11. Lee, J. A. H. Melanoma and exposure to sunlight, *Epidemiol. Rev.*, *4*:110–136 (1982).
12. Bissett, D. L., Hannon, D. P., and Orr, T. V. An animal model of solar-aged skin; Histological, physical, and visible changes in UV-irradiated hairless mouse skin, *Photochem. Photobiol.*, *46*:367–378 (1987).
13. Nakamura, K. and Johnson, W. C. Ultraviolet light induced connective tissue changes in rat skin. A histopathologic and histochemical study, *J. Invest. Dermatol.*, *51*:253–258 (1968).
14. Poulsen, J. T., Staberg, B., Wulf, H. C., and Brodthagen, A. Dermal elastosis in hairless mice after UV-B and UV-A applied simultaneously, separately or sequentially, *Br. J. Dermatol.*, *110*:531–538 (1984).
15. Smith, J. G., Davidson, E. A., Sams, W. M., and Clark, R. D. Alterations in human dermal connective tissue with age and chronic sun damage, *J. Invest. Dermatol.*, *39*:347–350 (1962).
16. Kaidbey, K. H. and Kligman, A. M. The acute effects of longwave ultraviolet radiation on human skin, *J. Invest. Dermatol.*, *72*:253–256 (1979).
17. Schwartz, E., Cruickshank, F. A., Perlish, J. S., and Fleischmajer, R. Alterations in dermal collagen in ultraviolet irradiated hairless mice, *J. Invest. Dermatol.*, *93*:142–146 (1989).
18. Plastow, S. R., Lovell, C. R., and Young, A. R. UVB-induced collagen changes in the skin of hairless albino mouse, *J. Invest. Dermatol.*, *88*:145–148 (1987).
19. Stern, R. S. and Momtaz, T. K. Skin typing for assessment of skin cancer risk and acute response to UV-B and oral methoxsalen photochemotherapy, *Arch. Dermatol.*, *120*:869–873 (1984).

20. Fitzpatrick, T. B. The validity and practicality of sun-reactive skin types I through VI, *Arch. Dermatol.*, *124*:869–871 (1988).
21. Uitto, J., Paul, J. L., Brockley, K., Pearce, R. H., and Clark, J. G. Elastic fibers in human skin: Quantification of elastic fibers by computerized digital image analyses and determination of elastin by radioimmunoassay of desmosine, *Lab. Invest.*, *49*: 499–505 (1983).
22. Godeau, G., Gonnord, G., Jolivet, O., Schoevaert, D., Pompidou, A., Lagrue, G., and Robert, A. M. A selective histochemical method for the quantitative estimation of elastic fibers by computerized morphometric analysis. *Int. Acad. Cytol. Anal. Quant. Cytol. Histol.*, *8*:321–325 (1986).
23. Chen, V. L., Fleischmajer, R., Schwartz, E., Palia, M., and Timpl, R. Immunochemistry of elastotic material in sun-damaged skin, *J. Invest. Dermatol.*, *87*:334–337 (1986).
24. Mortimer, P. S., Cherry, G. W., Jones, R. L., Barnhill, R. L., and Ryan, T. J. The importance of elastic fibers in skin lymphatics, *Br. J. Dermatol.*, *108*:561–566 (1983).
25. Mortimer, P. S., Jones, R. L., and Ryan, T. J. Human skin in lymphatics: Regional variation and relationship to elastin, *Prog. Lymphol.*, *10*:59–64 (1985).
26. Ryan, T. J. and Curri, S. B. The structure of fat, *Clin. Dermatol.*, *7*:37–47 (1989).
27. Ryan, T. J. The lymphatics of the skin. *The Physiology and Pathophysiology of the Skin*, Vol. 5 (A. Jarrett, ed.), Academic Press, London, 1978.
28. Leak, L. V. The fine structure and function of the lymphatic vascular system. *Handuch der Allegeminen Pathologie* (H. Meesen, ed.), Springer Verlag, Berlin, 1972.
29. Unna, R. G. *The Histopathology of the Diseases of the Skin*, Macmillan, New York, 1986.
30. Silverberg-Sinakin, I. and Thorbecke, G. J. Contact hypersensitivity and Langerhans cells, *J. Invest. Dermatol.*, *75*:61–67 (1980).
31. Thiers, B. H., Maize, J. C., Spicer, S. S., and Cantor, A. B. The effect of aging and chronic sun exposure on human Langerhans cell populations, *J. Invest. Dermatol.*, *82*:223–226 (1984).
32. Gilchrest, B. A., Murphy, G. F., and Soter, N. A. Effect of chronologic aging and ultraviolet irradiation on Langerhans cells in human epidermis, *J. Invest. Dermatol.*, *79*:85–88 (1982).

6

Ultraviolet Radiation-Induced Skin Aging

Lorraine H. Kligman and Albert M. Kligman
University of Pennsylvania, Philadelphia, Pennsylvania

I. INTRODUCTION

Recognition of the environmental influences on the aging of human skin has followed a cyclic and often deviant path. In recent years, renewed emphasis has been placed on the major role of solar ultraviolet (UV) radiation in what we now call photoaging. Although a new term, photoaging is not a new concept. Dermatologists of the late 19th century, notably Unna (1) and Dubreuilh (2), described the devastating effects of sunlight on the skin of farmers and sailors, compared with the appearance of indoor workers. Appreciation of these findings was gradually lost as the pursuit of deeply tanned skin as an indication of health and well-being developed in Western Caucasoid cultures. For many years the consequences of overexposure to sunlight were ignored and misnamed. Because serious sunworshippers can look old, even at age 50, with wrinkled, leathery skin, it became customary to consider the worn facial appearance as "premature" aging. The changes were considered merely an acceleration of the natural or intrinsic aging process, differing only quantitatively (3). Often, descriptions of intrinsically aged skin failed to account for the effects of solar radiation on exposed areas of the body (4). To compound the difficulties, biopsies of aged skin were frequently taken from very old persons with various diseases and unknown amounts of sun exposure (5). The result was a blurred distinction between the characteristics of intrinsic aging and photoaging.

Recent experimental work has concentrated on clarifying past misconceptions. In addition, once again, clinical observations are reconfirming that octogenarians who have avoided sun exposure can still have smooth, unblemished skin, showing only a laxness and deepening of expression lines. On the other hand, photoaged skin displays a variety of neoplasms, deep furrows, extensive sagging, with a yellowed, nodular surface, marked by telangiectatic traceries. These drastic, visible aspects reflect profound structural dermal alterations that are quite different from those found in protected, intrinsically aged skin (6–8).

II. PHOTOAGING VERSUS INTRINSIC AGING

A. The Dermal Matrix

Although clinical evidence for photodamage can remain inapparent for many years, ultraviolet radiation-induced changes in elastic fibers can be detected, histologically, as early as the second decade (9). It begins as a marked hyperplasia of relatively normal fibers, leading to accumulations of massive quantities of thickened, tangled, degraded fibers that eventually degenerate into an amorphous mass (Fig. 1). Ultrastructurally, the normally homogeneous elastin matrix begins to appear granular and develops numerous electron-lucent areas that impart a motheaten appearance (10). As the photoaging process continues, the matrix disappears, leaving only tangled microfilaments. Elastosis to this degree never occurs in the protected skin of healthy older persons. Instead, intrinsically aged skin shows a modest increase in the number and thickness of elastic fibers (Fig. 2a) and some loss of the fine vertical fibers that, in young skin (see Fig. 2b), insert into the basement membrane (3,11). Ultrastructurally, there is deterioration characterized by the development of lacunae, which cause a loosening of the aggregated components of mature fibers. Frank loss of elastin matrix is not seen (10).

Collagen, the major component of dermal connective tissue, also undergoes UV-induced changes. In the past, few collagen-specific histochemical observations were made. Brief descriptions exist that note changes in staining characteristics (12), with emphasis on "basophilic degeneration" of collagen. This has proved to be a misnomer. The bluish-stained material in hematoxylin and eosin (H&E) sections has been shown, with differential histochemical and immunochemical stains, to be elastosis (13). A late change in photoaging is a substantial loss of collagen as it is replaced by the masses of elastosis (see Fig. 1).

Because few systematic studies have been done with humans, the early effects of UV radiation must be inferred from animal experiments. Initially, the dermis thickens, as fibroblasts are stimulated to synthesize collagen. With further UV exposure there is a progressive loss of collagen-specific stain in deep bands

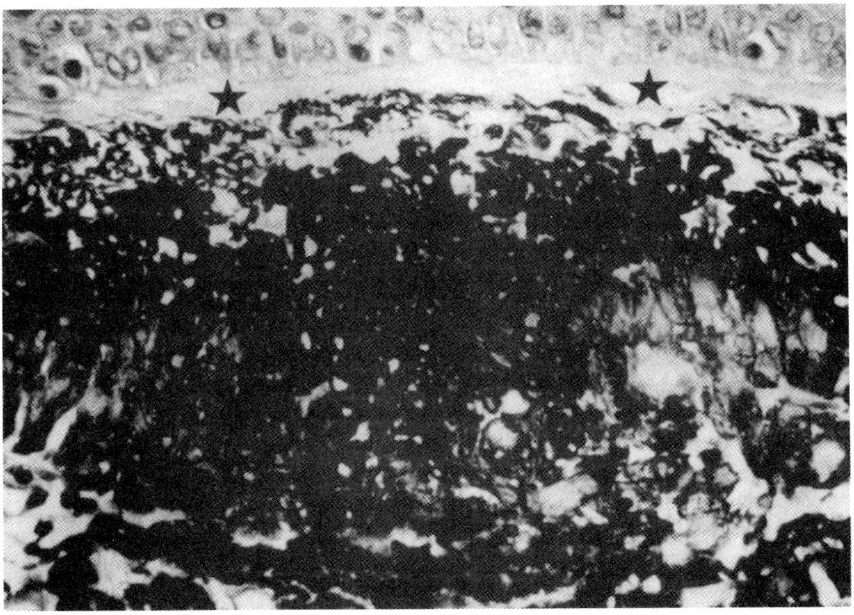

Figure 1 Advanced elastosis in human skin. Amorphous masses of degraded elastic fibers replace most of the collagen. Only a narrow Grenz zone remains (★). Luna's stain (×125).

across the dermis (14). Although there is no consensus on whether UV radiation has a direct effect on collagen (15,16), there is an abundance of experimental evidence suggesting that the UV-induced inflammatory infiltrate is at least partly responsible for collagen degradation (17–19). Recent work also indicates that dermal photodamage may be mediated by induction of signaling peptides in the epidermis. Cytokines, such as interleukin (IL-6) and tumor necrosis factor (TNF-α) are induced by UV radiation (20). TNF-α activates protease-containing neutrophils and macrophages, whereas IL-6 stimulates fibroblast-derived collagenase (21).

Early biochemical analysis of sun-damaged human skin showed a decrease in insoluble (mature) collagen and an increase in soluble (new) collagen (22). A more recent study, however, showed no changes in collagen in sun-exposed skin, when compared with unexposed sites of elderly subjects (23). Such discrepancies are very likely due to the degree of photodamage in the subjects and the time lapse since the subjects were undergoing prolonged UV exposure. In intrinsic aging, however, collagen apparently becomes more highly cross-linked,

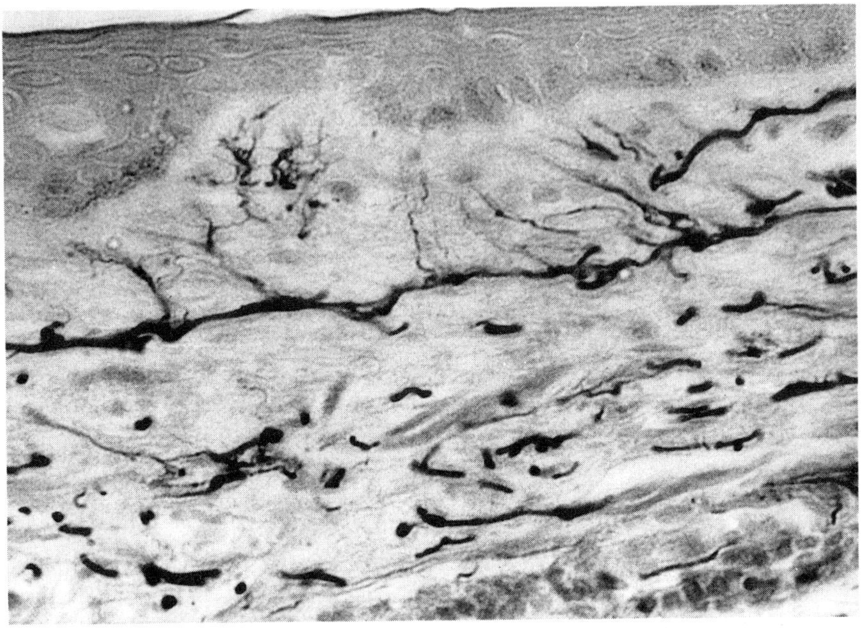

(a)

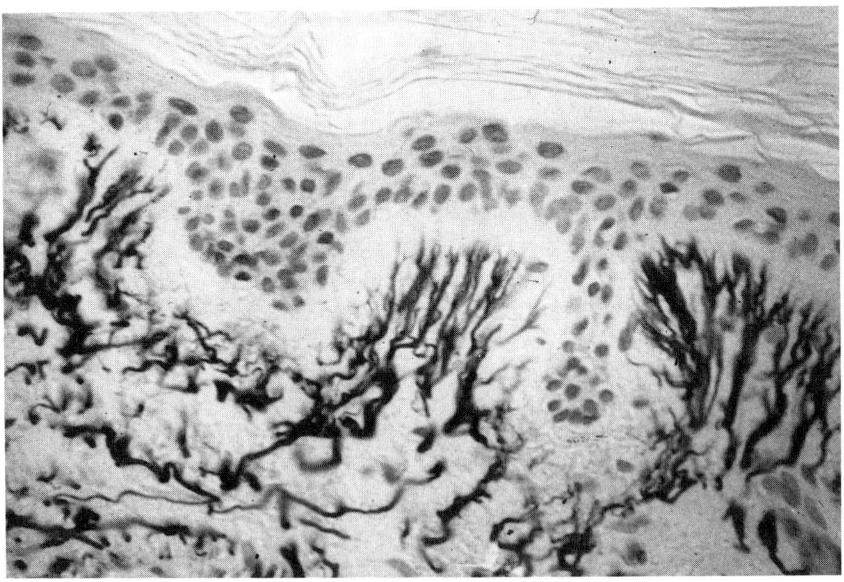

(b)

Figure 2 (a) Protected aged skin with slightly thickened elastic fibers and extensive loss of the vertical fibers that normally insert into the dermal–epidermal junction. Note the flattened junctional area and lack of rete ridges. Luna's stain (×380). (b) Young protected skin replete with fine candelabra-like vertical fibers. Orcein (×380).

with greater resistance to enzymatic degradation (24,25). Bundle deposition also becomes more dense (26).

Another component of the dermal matrix, the ground substance, is composed of the proteoglycans (PGs) dermatan sulfate and heparan sulfate, and the glycosaminoglycan (GAG) hyaluronic acid. Abundant in fetal skin, these decrease rapidly in early life. In protected skin, they remain low throughout most of adult life (22). In photoaged skin, ground substance drastically increases to levels almost matching that of fetal skin (22,27). Other molecules, such as fibronectin, increase in photoaged skin and bind to the elastotic masses (13).

B. Other Dermal Components

Fibroblasts, especially those of the upper dermis, in repeatedly sun-exposed skin are numerous and appear to be in a state of high metabolic activity, as judged by hypertrophy of the endoplasmic reticulum. The opposite occurs in intrinsic aging. There are fewer fibroblasts, and these are atrophic. Dermal thinning, over time, is characteristic of intrinsic aging, whereas repeatedly irradiated skin thickens as elastotic masses accumulate. In photodamaged skin, there is also a perivenular histiocytic–lymphocytic infiltrate in which numerous mast cells, often partially degranulated, are seen in close proximity to fibroblasts (18). In this chronic "heliodermatitis" (Fig. 3), it is likely that mast cell-produced mediators, in conjunction with proteolytic enzymes released by the infiltrating cells, may lead to the degradation of elastic and collagen fibers. In contrast, in chronologic aging, hypocellularity is the rule (28).

The microcirculation with its delicate capillary loops (Fig. 4a) also suffers from sun exposure. Vessels become dilated and tortuous (see Fig. 4b), seen on the surface as telangiectases. Finally, they become sparse, with massive loss of the capillary loops and disruption of the horizontal plexus (29). Ultrastructurally, walls of the vessels are greatly thickened by deposition of reduplicated basement membrane-like material that also contains reticulin fibers (30). In protected, aged skin, there is some loss of the subepidermal capillary loops, probably secondary to flattening of the dermal–epidermal junction. The vessels, however, are not dilated, and the horizontal plexus is largely undisturbed (29,31). Moderate thickening of the vessel walls is occasionally seen in intrinsic aging, but in some very old persons, the vessels often have abnormally thin walls (30).

Finally, benign, malignant, and premalignant lesions flourish in photoaged skin. Seborrheic and actinic keratoses, solar lentigos, keratoacanthomas, basel cell epitheliomas, and squamous cell carcinomas develop almost exclusively in sun-exposed areas (32). Senile or cherry angiomas are about the only growths that regularly appear in protected, aged skin.

In summary, contrary to the usual textbook statement that sun-damaged skin is thin and atrophic, ongoing photoaging is characterized by "more" not "less"

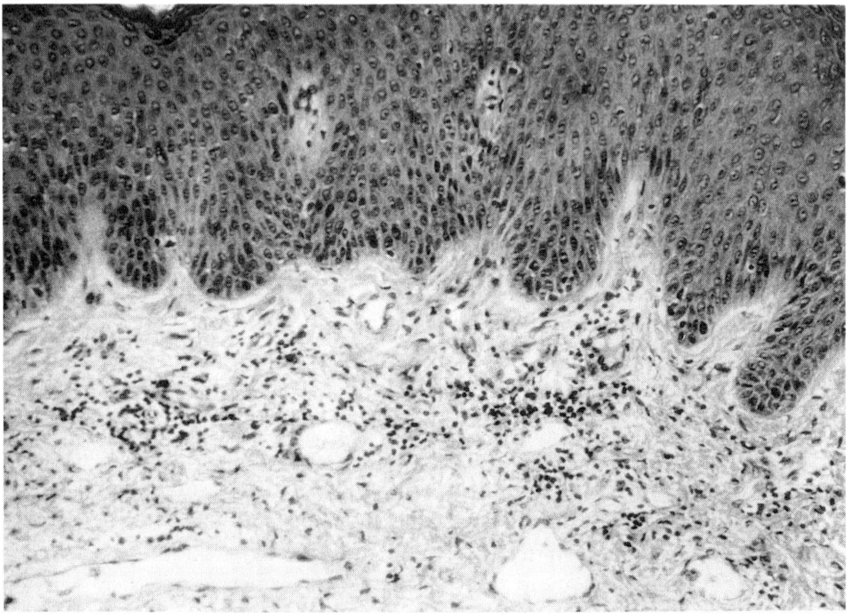

Figure 3 Heliodermatitis in repeatedly irradiated skin. Note the marked epidermal hyperplasia and severe cellular inflammatory infiltrate. H & E (×380).

until end-stage photodamage in the very elderly. Hypertrophy rules in photoaging: epidermis thickens and sprouts neoplasms, sebaceous glands enlarge (4,33), elastic fibers multiply, collagen synthesis is stimulated, ground substance increases, while inflammatory cells abound and vessel walls thicken.

II. EXPERIMENTAL STUDIES

Photoaging in humans is a slowly evolving process, taking decades to become clinically apparent and many more years to develop all of the manifestations, whether microscopic or biochemical. Furthermore, the UV dosimetry of human sun exposure is impossible to quantify with any degree of accuracy. Systematic studies require animal models for practical, as well as ethical, reasons. The hairless mouse has proved to be a relevant model for human photoaging (34). The UV-induced changes, ranging from immediate responses (35) to those associated with prolonged exposure, such as tumorigenesis (36) and connective tissue damage (37,38), are comparable with those in human skin.

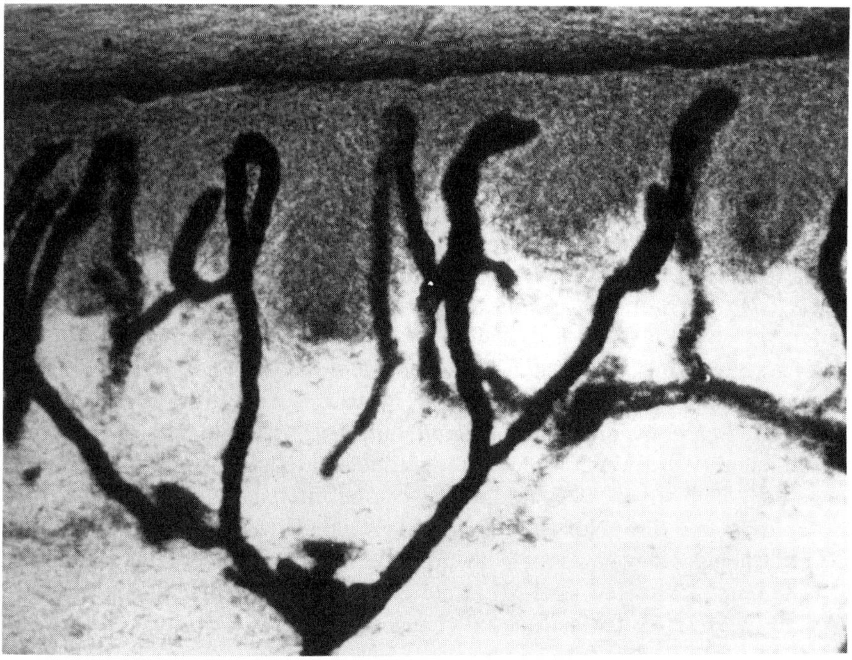

(a)

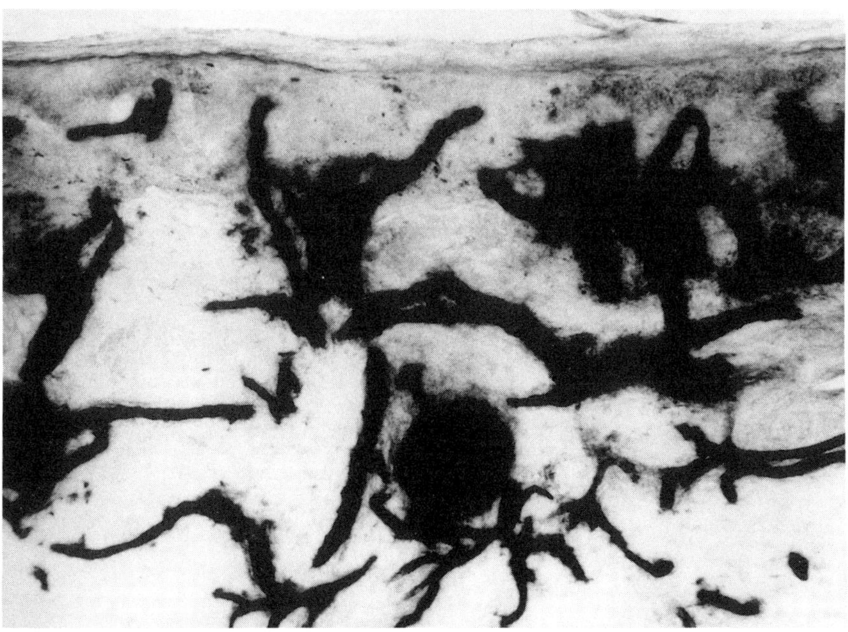

(b)

Figure 4 Superficial vasculature: (a) Protected young adult skin with well-defined capillary loops. (b) Dilated vessels with disrupted architecture in photodamaged skin. Alkaline phosphatase stain (×100).

A. Ultraviolet B and Photoaging

The ultraviolet B (UVB; 290–320 nm) waveband is responsible for erythema, DNA damage, and skin cancer. Some evidence was provided by Sams et al. (39) that it was also effective in damaging connective tissue. These authors produced a modest amount of elastic fiber hyperplasia in haired mice after a severely damaging dose of 30 to 50 minimal erythema doses (MED). Although their UV source emitted some UVC (200–290 nm), it was assumed that the effective waveband was the longer wavelength UVB, which is capable of penetrating into the dermis. With the hairless mouse, it has been possible to produce extensive tissue photodamage with far lower doses. A bank of FS Westinghouse ''sunlamps'' is a convenient high-energy source of predominantly UVB radiation. The quantity of UVA emitted is inconsequential relative to connective tissue damage (17). The small UVC component can be filtered out effectively with a cellulose triacetate film. Normally, the hairless mouse has very little elastic tissue. Nevertheless, it was possible to produce severe elastosis (Fig. 5) with a total of 5 J/cm^2, delivered as 2 MED per exposure, thrice weekly for approximately 30 weeks (17). Ultrastructural changes (Fig. 6) in the elastic fibers bear

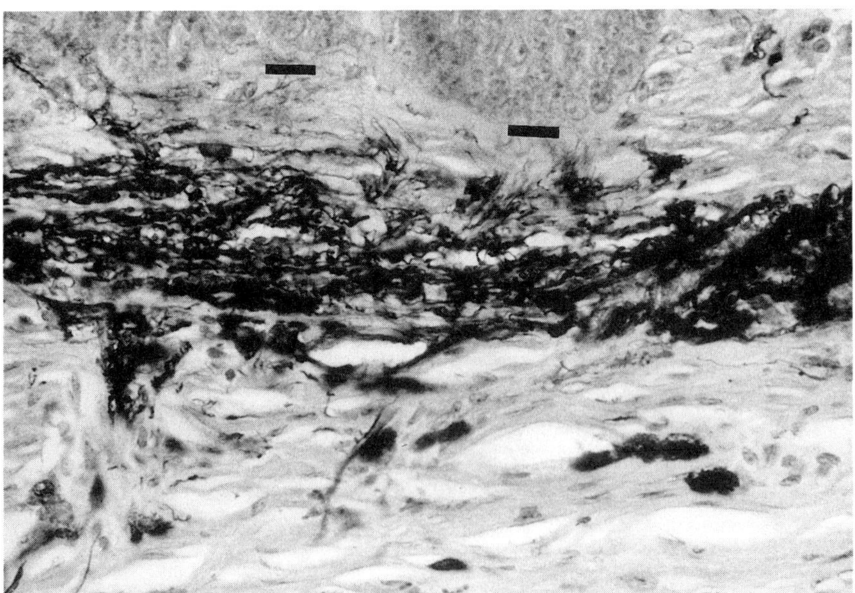

Figure 5 Mouse skin. Elastosis induced by repeated exposure to UVB. Luna's stain (×380). Dermal–epidermal junction (bars).

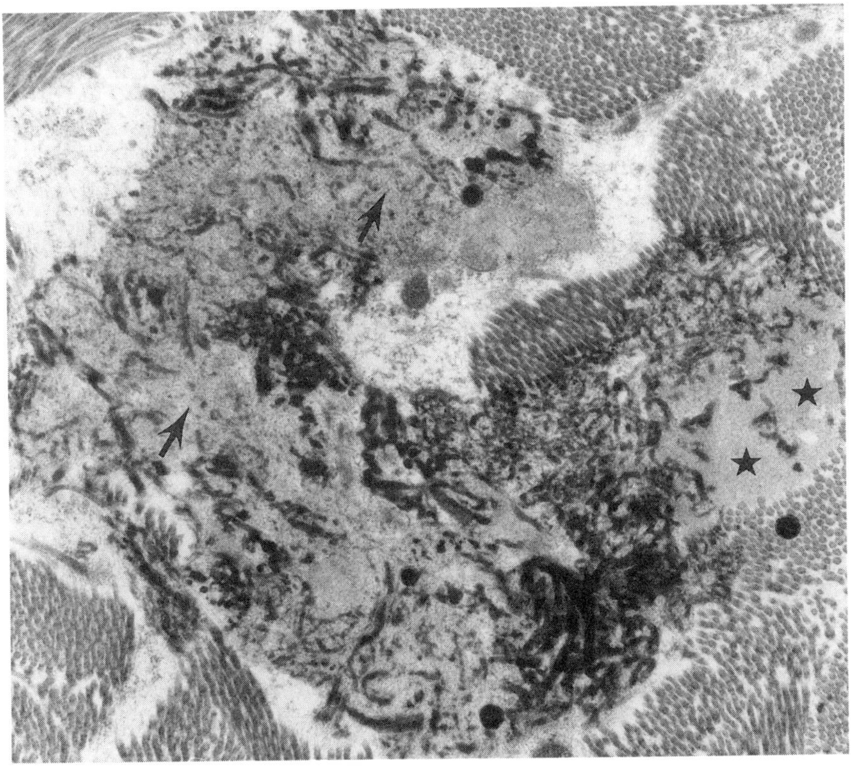

Figure 6 Elastotic clump of fibers in mouse skin irradiated for 30 weeks. Most of the electron-dense matrix is degraded, leaving granular (→) and amorphous (★) residues. Glutaraldehyde-osmium-tannic acid staining.

a remarkable similarity to those seen in photodamaged human skin (40). In irradiated animals, fibroblasts became more numerous. Ultrastructurally, they appeared to be metabolically active, producing increased quantities of collagen that contributed to a thickening of the dermis. Reticulin fibers, normally limited to the basement membrane zones, were prominent throughout the upper dermis, an indication of new collagen synthesis (37). A decreased affinity for van Gieson's stain suggested severe damage to mature collagen. The histological findings were confirmed by electron microscopy. In addition to a fraying and partial dissolution of large collagen fibers, there were numerous fibers of small diameters that may be newly synthesized collagen. The UVB-irradiated hairless mouse skin has been examined with biochemical techniques. In one study, using cyanogen bromide digests of whole skin, an early and progressive increase in the ratio of type III to total collagen was reported (41). It was implied that this was due to an increase in type III collagen. Another study, using pepsin digestion of isolated collagen, showed a progressive increase in total collagen during prolonged irradiation, but with a reduction to control levels after months of irradiation. There was no consistent increase in the ratio of type III to total collagen until the loss of collagen was detected late in the study (42). Schwartz et al. (43) also reported no change in the ratio in UVB-irradiated mouse skin, as did Oikarinen and Kallioinen (23) in photodamaged human skin. These results suggest that both types I and III collagen increase proportionally until end-stage photodamage, when there may be reduced synthesis of type I collagen.

A compilation of histochemical and biochemical damage suggests that photodamage to the collagen matrix is at least biphasic. Early on fibroblasts are stimulated to produce more collagen, with the new, soluble (less cross-linked) collagen having little affinity for van Gieson's stain, but appearing as "reticulin" fibers. After long-term repeated irradiation when mature (more highly cross-linked) collagen is lost at a faster rate than fibroblasts can replace it, the dermis thins, becoming atrophic in end-stage photoaging.

A third component of the mouse dermis, the GAGs and PGs of the ground substance, as in humans, increases greatly with repeated UV radiation. According to Margelin et al. (44), not all GAGs and PGs increase. On examining explants of UV-irradiated hairless mouse skin in vitro, synthesis of heparin and heparan sulfate was increased, whereas that of dermatan sulfate and hyaluronic acid was not. By contrast, our histochemical observations indicated large increases in hyaluronic acid (17).

B. Ultraviolet A and Photoaging

Brief exposure to UVA (320–400 nm) can, similar to UVB, produce erythema (45) and damage blood vessels (46). Because such effects require doses that may be hundreds of times greater than UVB, the role of UVA in photoaging was thought to be negligible. However, UVA is present in sunlight in amounts that can be 100–500 times that of 300-nm UVB. Furthermore, its longer wavelengths allow more of it to reach the dermis than does UVB. These considerations make it important to examine the effects of UVA and to compare it with those of UVB. In one study (17), hairless mice were irradiated with a solar simulator, filtered to provide UVA with a spectral distribution similar to that of solar UVA. Animals irradiated with this source for 34 weeks, with a total UVA dose of 3000 J/cm^2 developed a significant degree of elastosis. Although the deposition of elastic fibers was less dense than that produced by a total UVB dose of 5 J/cm^2, the elastosis extended more deeply into the dermis (Fig. 7). Ultraviolet-A from a black light source (340–400 nm), with peak emission at approximately 365 nm, produced only mild elastic fiber hyperplasia, despite a

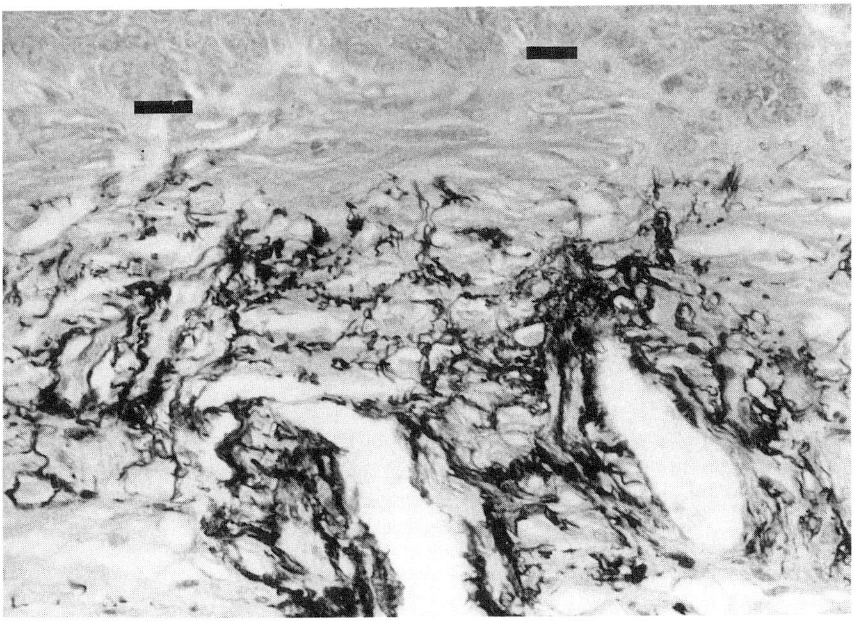

Figure 7 Mouse skin. Delicate, deeply deposited UVA-induced elastosis. Luna's stain (×380). Dermal–epidermal junction (bars).

total UVA dose of 13,000 J/cm^2. This strongly suggests that the UVA wavelengths of 320–340 nm (UVA II), abundant in solar UVA, but lacking in the black light spectrum, may be most responsible for UVA-induced photodamage. Despite this, the longer wavelengths of solar-simulating UVA (340–400 nm: UVAI) cannot be ignored. Studies have shown elastic fiber hyperplasia and increased GAGs with high, but not unrealistic doses (47). In contrast with UVB, neither full spectrum UVA nor UVAI had a histological effect on collagen. Supporting the hypothesis that it is the proteolytic enzymes of the inflammatory infiltrate that degrade collagen, prolonged UVA, at the doses used, did not induce inflammation. Biochemically, however, UVA has a profound effect on collagen. Both full-spectrum and UVAI rendered collagen highly resistant to digestion by pepsin in a dose-related manner (48,49). Only 20–40% is digestable, whereas in normal and UVB-irradiated skin 85–100% can be digested (48,51). It has been suggested that the pepsin resistance may be due to increased cross-linking of the collagen by UVA. Other evidence suggests complexities that further research will need to clarify. For example, Yamauchi et al. (50) found a decrease in a specific collagen cross-link in exposed, compared with nonexposed human skin. These authors postulated that UV radiation may hinder maturation of this particular cross-link.

Similar to UVB, UVA radiation produces an increase in GAGs and PGs, as visualized with Mowry's stain. The distribution, however, was different. With UVB, the granular blue-staining material localized in the upper dermis appeared to replace collagen, whereas UVA-induced GAGS and PGs were deposited throughout the dermis, imparting a bluish hue to the collagen bundles. Additionally, a densely staining material was deposited at the dermal–epidermal junction.

Recent ultrastructural studies in the hairless mouse (52) have shown that chronic exposure to UVA is exceedingly damaging to skin and in ways that differ from UVB. Notably, vascular endothelial cells are severely damaged, and the surrounding basement membrane is highly reduplicated (Fig. 8), whereas that of the dermal–epidermal junction is not. With prolonged UVB radiation, endothelial cells show little or no damage and no membrane duplication, but

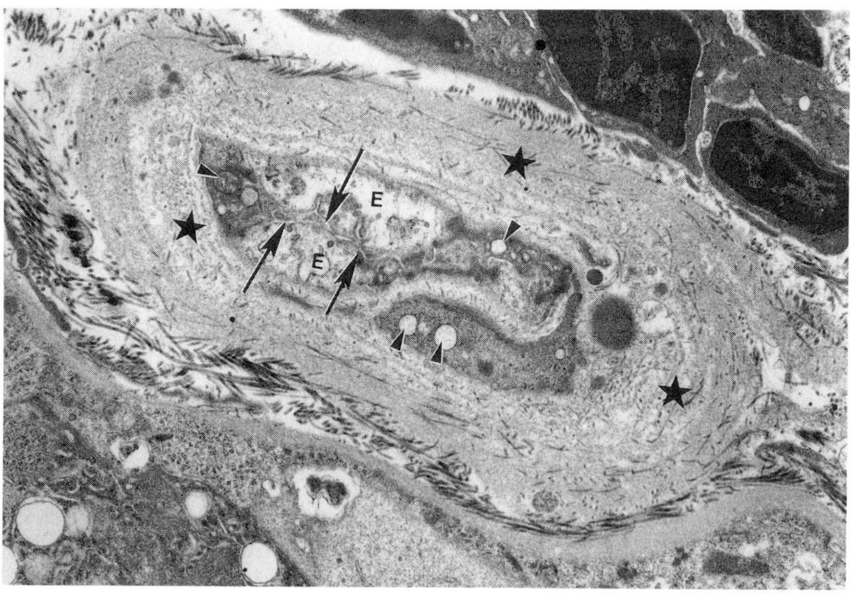

Figure 8 Damaged vessel from UVA-irradiated hairless mouse skin. Basement membrane is greatly reduplicated (★). Endothelial cells are severely vacuolated (E), mitochondria are swollen with disrupted cristae (▶), and the vessel lumen is collapsed (→). (×11,200).

the basal lamina of the dermal–epidermal junction is reduplicated (52). The UVA doses used in these various experiments were realistic in terms of human exposure. For example, a 1-h exposure at midday in June at 40° north latitude will provide approximately 10 J/cm^2 UVA. Within 4 years, summertime exposure alone could result in the accumulation of 3000 J/cm^2. Of concern also is the greatly increased absorption of UVA by persons protected with a sunscreen that filters out only UVB. Furthermore, even with most sunscreens that contain UVA absorbers, little or no protection is provided against UVAI. Many more hours in the sun are now possible without the warnings of a sunburn. In view of these considerations, the experimental findings provide compelling evidence that UVA radiation is capable of inducing profound photodamage to dermal connective tissue and other skin structures. Interestingly, UVA appears to be more effective in inducing elastosis than erythema. A recently constructed action spectrum for elastosis, although resembling the erythema action spectrum, has shown that a 50% increase in elastic fiber hyperplasia requires only 20 times more UVA than UVB, whereas for erythema, 500–1000 times more is required (53). Despite a similarity of the two spectra, it is clear that long-term repeated exposure has different consequences than the brief exposures that result in sunburn.

C. Solar-Simulating Radiation and Photoaging

The effects of UVA alone are interesting from a scientific perspective. However, because UVA and UVB are inseparably linked in sunlight, it is the cumulative effect of the entire UV spectrum that is of prime concern. In one of the studies described in the foregoing (17), a group of animals were exposed to full-spectrum, solar-simulating radiation (SSR). The total UVB dose, delivered during the 34-week experiment, was approximately 5 J/cm^2, with a 100-fold more solar-simulating UVA. In some aspects, the photodamage reflected the influence of both wavebands, whereas in others it was like that of UVB alone. Elastosis was severe, but was not confined to the upper dermis, as it is with UVB. More dense than UVA elastosis, it extended more deeply into the dermis than with UVB alone. Collagen was damaged to an extent comparable with UVB alone. The GAGs were equal to or slightly increased compared with UVB-irradiated tissue, with deeply stained material at the dermal–epidermal junction, providing evidence for the additional effect of UVA. There is a scarcity of other reports on the effects of SSR on the hairless mouse. Those published (34,54) mainly support these histological findings and add observations on some physical properties. Solar-simulated radiation appears to induce UVB type wrinkles, with a decrease in the microtopography of skin markings. Skin-fold thickness increases as does transepidermal water loss. Clearly, more research is needed in this vital area.

D. Prevention of Photoaging: Sunscreens

In the hairless mouse model, broad-spectrum sunscreens, with a sun protection factor (SPF) of 15, provided effective protection against ultraviolet-induced connective tissue damage. This was especially true for animals irradiated with UVB at approximately 6 MED per exposure (37). Histological specimens were virtually indistinguishable from normal age-matched controls. Animals irradiated with solar-simulating radiation (about 2 MED UVB plus UVA) and protected by the same sunscreens showed some mild dermal damage (17). Elastosis was prevented, but elastic fibers were mildly hyperplastic. Collagen appeared undamaged and, although dermal cellularity was increased, massive inflammation did not occur. The GAGs were slightly, but not remarkably, increased compared with unirradiated controls.

In both experiments, the sunscreen applications began at the time of the first exposure. This does not conform to human practice in which sunscreens are generally not used until some degree of photodamage (such as the first wrinkle or actinic keratosis) is clinically visible. By that time, the connective tissue has already sustained severe photodamage. Previously, it was believed that this damage was irreversible. Early studies with hairless mice demonstrated that considerable repair could occur within a few months, even after extensive photodamage, if the UV exposures were stopped (37). Broad regions of new collagen were deposited in the subepidermal dermis, pushing downward the old elastotic material (Fig. 9a). The new collagen was histochemically and ultrastructurally normal; new elastic fibers in this region were sparse and delicate, as in unirradiated tissue. Interestingly, evidence for similar repair is seen in human skin when former sunworshippers avoid further sun exposure (see Fig. 9b).

Because of these findings, it was of interest to determine whether the skin would repair itself in a like manner under conditions that simulated human experience (14). Groups of animals were irradiated for a total of 30 weeks with UVB (3 MED per exposure). Sunscreens of SPF 6 and 15 were applied to separate groups after 10 or 20 weeks of irradiation. Control animals, irradiated without sunscreen protection for the full 30 weeks, sustained severe connective tissue damage that included extensive elastosis, loss of mature collagen, and greatly increased GAGs. Protected animals had only the degree of damage that reflected the weeks of irradiation before sunscreens were applied. As expected, animals irradiated for 20 weeks before sunscreens were applied had considerable photodamage. Nevertheless, during the final 10 weeks of irradiation, with sunscreen protection, a repair zone of new subepidermal collagen was deposited (Fig. 10).

These experiments demonstrated that further damage could be prevented, and repair could occur, even in the face of continuing UVB irradiation, once sunscreens were applied. In general earlier application and higher SPF provided the

(a)

(b)

Figure 9 Post-UV dermal repair is characterized by a large region of new collagen (★) that displaces, downward the elastosis (→) formed during chronic irradiation. (a) Mouse skin 15 weeks after cessation of irradiation. Note fine new elastic fibers in the repair zone. Luna's stain (×380). (b) Human skin (bald scalp biopsy) after 10 years abstinence from sunbathing. Luna's stain (×380).

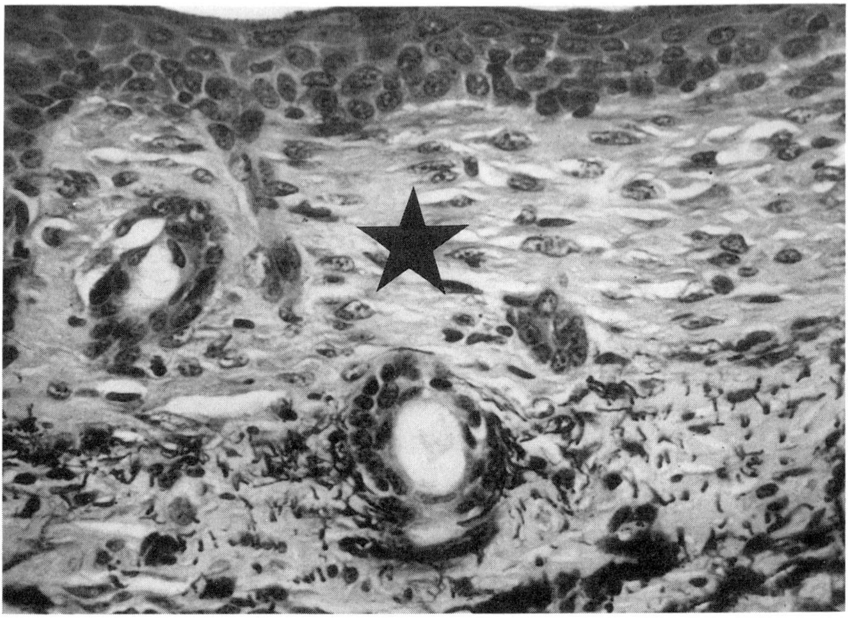

Figure 10 Repair zone (★) in mouse skin irradiated for 30 weeks, but protected with an SPF-15 sunscreen for the final 10 weeks. The new collagen has the same staining qualities as the mature collagen in the lower dermis. Luna's stain (×380).

greatest degree of protection against photoaging. However, even with an SPF 15 sunscreen, at any given time, radiation equal to 1/15th of an MED is reaching the skin. This "breakthrough" radiation, similar to any UV exposure is cumulative and has the potential to damage skin. Experiments with hairless mice have shown that 1/2 MED of UVB, followed by <1 MED of UVA additively enhanced connective tissue damage and synergistically enhanced tumorigenesis, compared to 1/2 MED of UVB alone. The UVA alone produced no tumors (55). Thus, an argument could be made for the use of SPFs higher than 15 in areas of high insolation and in persons with severe sensitivity to sunlight.

Until recently, most broad-spectrum sunscreens contained sun filters that absorbed effectively only up to approximately 330 nm, seemingly providing little protection against a substantial portion of UVA and none against infrared radiation. Recent experiments with hairless mice have shown some oxybenzone-containing SPF 15 sunscreens to be almost as protective against prolonged exposure to UVAI as an avobenzone (Parsol 1789)-containing sunscreen (56). Unexpectedly, a few oxybenzone-containing sunscreens not only failed to protect, but actually resulted in greater damage to UVA-irradiated hairless mouse skin than did UVA alone (56,57). We have termed this new phenomenon photoirritation (57) and ascribe it not to the oxybenzone, but to a still to be determined component of the vehicle.

For infrared radiation, this broad waveband (700–1 million nm) cannot be ignored in connection with photoaging. Inseparably linked to sunlight, infrared radiation is perceived as heat. The harmful effect of heat is ancient knowledge (58). Human skin persistently exposed to infrared radiation often develops the condition known as erythema ab igne, with its characteristic mottled pigmentation, resulting from damaged, leaking blood vessels. Severe elastic fiber hyperplasia, extending deeply into the dermis, is often present, along with thermal keratoses that are almost identical with those produced by UV radiation. Experimentally, in a guinea pig model, infrared radiation alone can produce elastosis and can exacerbate the elastosis produced by UVB, adding to photodamage (59). Although not tested in the guinea pig model, it might be expected that the new titanium dioxide sunscreens would afford some protection against this portion of the electromagnetic spectrum.

IV. SUMMARY

In summary, in recent years, there has been a growing awareness that many of the so-called attributes of aging skin are, instead, a reflection of environmental assault on exposed areas of the body. Of special import are the deleterious effects of solar radiation on dermal connective tissue, leading to the visible manifestations of photoaging. Often termed *premature aging*, the salient features of the process are distinctly different from those found in normal intrinsic aging.

Experimental studies with animal models have confirmed the notion that the shorter, more energistic portion of the ultraviolet spectrum (UVB) is responsible for much of the dermal connective tissue destruction observed in photoaged skin. Ultraviolet A and infrared radiation contribute significantly to photoaging, producing, among other changes, severe elastosis. Because the three broad wavebands are inseparably linked in terrestrial sunlight, all are of concern in the photoaging of human skin.

Photoaging changes have been thought to be irreversible. However, our findings indicate that destruction and repair go on simultaneously under continued assault by solar radiation. The balance is shifted toward repair when the radiation stress is relieved. Both epidermis and dermis are capable of moderate self-restoration when exogenous injury ceases, either by avoidance of sunlight or by the use of broad-spectrum, high-SPF sunscreens. Although protection from sunlight early in life, before severe photodamage occurs, is most desirable, it is deemed advisable to counsel even older persons with photoaged skin to adopt protective measures, thereby allowing repair processes to occur.

REFERENCES

1. Unna, P. Histopathologie der Hautkrankeiten, A. Herschwald, Berlin, 1894.
2. Dubreuilh, W. Des hyperkeratoses circonscrites, *Ann. Dermatol. Syphiligr.* (Ser. 3), *7*:1158–1204 (1896).
3. Montagna, W. and Carlisle, K. Structural changes in aging human skin, *J. Invest. Dermatol.*, *73*:47–53 (1979).
4. Plewig, G. and Kligman, A. M. Proliferative activity of the sebaceous glands of the aged, *J. Invest. Dermatol.*, *70*:314–317 (1978).
5. Cooper, Z. K. Aging of skin, *Cowdry's Problems of Aging* (A. E. Lansing, ed.), Williams & Wilkins, Baltimore, 1952, pp. 764–790.
6. Kligman, L. H. and Kligman, A. M. *Photoaging, Dermatology in General Medicine*, 4th ed. (T. Fitzpatrick, A. Z. Eisen, K. Wolff, I. M. Freedberg, and K. F. Austen, eds.), McGraw Hill, New York, 1993, pp. 2972–2979.
7. Balin, A. and Kligman, A. M. *Aging and the Skin*, Raven Press, New York, 1989.
8. Gilchrest, B. A. *Skin and Aging Process*, CRC Press, Boca Raton, FL, 1984.
9. Kligman, A. M. Early destructive effects of sunlight on human skin, *JAMA*, *210*: 2377–2380 (1969).
10. Braverman, I. M. and Fonkerko, E. Studies in cutaneous aging. I. The elastic fiber network, *J. Invest. Dermatol.*, *78*:434–444 (1982).
11. Lavker, R. M. Structural alterations in exposed and unexposed aged skin, *J. Invest. Dermatol.*, *73*:59–66 (1979).
12. Knox, J. M., Cockerell, E. G., and Freeman, R. G. Etiological factors and premature aging, *JAMA*, *179*:136–142 (1962).
13. Chen, V. L., Fleischmajer, R., Schwartz, E., Palaia, M., and Timpl, R. Immunochemistry of elastotic material in sun-damaged skin, *J. Invest. Dermatol.*, *87*:334–337 (1986).
14. Kligman, L. H., Akin, F. J., and Kligman, A. M. Sunscreens promote repair of ultraviolet radiation-induced dermal damage, *J. Invest. Dermatol.*, *81*:98–102 (1983).
15. Lovell, W. W. Ultraviolet irradiation of dermal collagen in vivo, *Trans. St. Johns Hosp. Dermatol. Soc.* (*Lond.*), *59*:166–174 (1973).
16. Shuster, S. and Bottoms, E. Effect of ultraviolet light on skin collagen, *Nature*, *199*:192–193 (1963).
17. Kligman, L. H., Akin F. J., and Kligman, A. M. The contributions of UVA and UVB to connective tissue damage in hairless mice, *J. Invest. Dermatol.*, *84*:272–276 (1985).
18. Lavker, R. M. and Kligman, A. M. Chronic heliodermatitis. A morphologic evaluation of chronic actinic damage with emphasis on the role of mast cells, *J. Invest. Dermatol.*, *90*:325–330 (1988).
19. Werb, Z., Banda, M. J., and Jones, P. A. Degradation of connective tissue matrices by macrophages. I. Proteolysis of elastin, glycoproteins and collagen by proteinases isolated from macrophages, *J. Exp. Med.*, *152*:1340–1357 (1980).
20. Schwartz, T. and Luger, T. Effect of UV irradiation on epidermal cell cytokine production, *J. Photochem. Photobiol. B*: *Biol.*, *4*:1–13 (1989).

21. Wlaschek, M., Bolsen, K., Hermann, G., Schwarz, A., Wilmroth, F., Heinrich, P. C., Goerz, G., and Scharffetter-Kochanek, K. UVA induced autocrine stimulation of fibroblast-derived collagenase by IL-6: A possible mechanism in dermal photodamage? *J. Invest. Dermatol.*, *101*:164–168, (1993).
22. Smith, J. G., Davidson, E. A., Sams, W. M., and Clark, R. D. Alterations in human dermal connective tissue with age and chronic sun damage, *J. Invest. Dermatol.*, *39*:347–350 (1962).
23. Oikarinen, A. and Kallioinen, M. A biochemical and immunohistochemical study of collagen in sun exposed and protected skin, *Photodermatology*, *6*:24–31 (1989).
24. Bentley, J. P. Aging of collagen, *J. Invest. Dermatol.*, *73*:80–83 (1979).
25. Yamauchi, M., Woodley, D. T., and Mechanic, G. L. Aging and cross-linking of skin collagen, *Biochem. Biophys. Res. Commun.*, *152*:898–903 (1988).
26. Lavker, R. M., Zheng, P., and Dong, G. Aged skin: A study by light, transmission electron and scanning electron microscopy, *J. Invest. Dermatol.*, *88*:44–51s (1987).
27. Sams, W. M. and Smith, J. G. The histochemistry of chronically sundamaged skin, *J. Invest. Dermatol.*, *37*:447–452 (1961).
28. Andrew, W., Behnke, R. H., and Sato, T. Changes with advancing age in the cell population of human dermis, *Gerontologica*, *10*:1–19 (1964–1965).
29. Kligman, A. M. Perspectives and problems in cutaneous gerontology, *J. Invest. Dermatol.*, *73*:39–46 (1979).
30. Braverman, I. M. and Fonferko, E. Studies in cutaneous aging II. The microvasculature, *J. Invest. Dermatol.*, *78*:444–448, (1982).
31. Gilchrest, B. A., Stoff, J. S., and Soter, N. A. Chronologic aging alters the response to UV-induced inflammation in human skin, *J. Invest. Dermatol.*, *79*:11–16 (1982).
32. Urbach, F. Geographic pathology of skin cancers, *The Biologic Effects of Ultraviolet Radiation* (F. Urbach, ed.), Pergamon Press, Oxford, 1969, p. 635.
33. Lesnik, R. H., Kligman, L. H., and Kligman, A. M. Agents that cause enlargement of sebaceous glands in hairless mice. II. Ultraviolet radiation, *Arch. Dermatol. Res.*, *284*:106–108 (1992).
34. Kligman, L. H. The hairless mouse and photoaging, *Photochem. Photobiol.*, *54*: 1109–1118 (1991).
35. Cole, C. A., Davies, R. E., Forbes, P. D., and D'Aloisio, L. C. Comparison of action spectra for acute cutaneous responses to ultraviolet radiation: Man and albino hairless mouse, *Photochem. Photobiol.*, *37*:623–631 (1983).
36. Kligman, L. H. and Kligman, A. M. Histogenesis and progression of ultraviolet light-induced tumors in hairless mice, *JNCI*, *67*:1289–1297 (1981).
37. Kligman, L. H., Akin, F. J., and Kligman, A. M. Prevention of ultraviolet damage to the dermis of hairless mice by sunscreens, *J. Invest. Dermatol.*, *78*:181–189 (1982).
38. Bissett, D. L., Hannon, D. P., and Orr, T. V. An animal model of solar-aged skin: Histological, physical and visible changes in UV-irradiated hairless mouse skin, *Photochem. Photobiol.*, *46*:367–378 (1987).
39. Sams, W. M., Smith, J. G., and Burk, P. G. The experimental production of elastosis with ultraviolet light, *J. Invest. Dermatol.*, *43*:467–471 (1964).
40. Hirose, R. and Kligman, L. H. An ultrastructural study of ultraviolet-induced elastic fiber damage in hairless mouse skin, *J. Invest. Dermatol.*, *90*:697–702 (1988).

41. Plastow, S. R., Lovell, C. R., and Young, A. R. UVB-induced collagen changes in the skin of the hairless albino mouse, *J. Invest. Dermatol.*, *88*:145–148 (1987).
42. Kligman, L. H., Gebre, M., Alper, R., and Kefalides, N. A. Collagen metabolism in ultraviolet irradiated hairless mouse skin and its correlation to histologic observations, *J. Invest. Dermatol.*, *93*:210–214 (1989).
43. Schwartz, E., Cruickshank, F. A., Perlish, J. S., and Fleischmajer, R. Alterations in dermal collagen in ultraviolet irradiated hairless mice, *J. Invest. Dermatol.*, *93*: 142–146 (1989).
44. Margelin D., Fourtanier, A., Thevenin, T., Medaisko, C., Breton, M., and Picard, J. Alterations of proteoglycans in ultraviolet irradiated skin, *Photochem. Photobiol.*, *58*:211–218 (1993).
45. Kaidbey, K. H. and Kligman, A. M. The acute effects of longwave ultraviolet radiation on human skin, *J. Invest. Dermatol.*, *72*:253–256 (1979).
46. Gilchrest, B. A., Soter, N. A., Hawk, J. L. M. Barr, R. M., Block, A. K., Hensby, C. N., Mallet, A. I., Greaves, M. W., and Parrish, J. A. Histologic changes associated with ultraviolet-A induced erythema in normal human skin, *J. Am. Acad. Dermatol.*, *9*:213–219 (1983).
47. Kligman, L. H., Kaidbey, K. H., Hitchins, V. M., and Miller, S. A. Long wavelength (>340 nm) ultraviolet-A induced skin damage in hairless mice is dose dependent, *Human Exposure to Ultraviolet Radiation*: *Risks and Regulation* (W. F. Passchier and B. F. M. Bosnajkovic, eds.), Elsevier Science, New York, 1987, pp. 77–81.
48. Kligman, L. H. and Gebre, M. Biochemical changes in hairless mouse skin collagen after chronic exposure to UVA radiation, *Photochem. Photobiol.*, *54*:233–237 (1991).
49. Trautinger, F., Trenz, A., Raff, M., and Kokoschka, E. M. The influence of UV radiation on dermal collagen content in hairless mice, *Arch. Dermatol. Res.*, *281*: 144a (1989).
50. Yamauchi, M., Prisayanh, P., Haque, Z., and Woodley, D. T. Collagen cross-linking in sun-exposed and nonexposed sites of aged humans skin, *J. Invest. Dermatol.*, *97*:938–941 (1991).
51. Kligman, L. H. UVA-induced biochemical changes in hairless mouse skin collagen: A contrast to UVB effects, *Biologic Responses to UVA Radiation* (F. Urbach, ed.)., Valdenmar Press, Overland Park KS, 1992, pp. 209–215.
52. Zheng, P. and Kligman, L. H. UVA-induced ultrastructural changes in hairless mouse skin: A comparison to UVB-induced damage, *J. Invest. Dermatol.*, *199*: 194–199 (1993).
53. Kligman, L. H. and Sayre, R. M. An action spectrum for ultraviolet induced elastosis in hairless mice: Quantification of elastosis by image analysis, *Photochem. Photobiol.*, *53*:237–242 (1991).
54. Fourtanier, A., Berrebi, C., Charles, N., Medaisko, C., and Noblet, J. P. Influence of UV light on natural skin aging in hairless mice, *Cosmetic Dermatology*, Vol. 1 (P. Morganti and W. Montagna, eds.). International Ediemme, Rome, 1986, pp. 289–300.
55. Kligman, L. H. and Crosby, M. J. The synergistic effects of UVA and UVB in tumorigenesis, *Photochem. Photobiol.*, *55*:108s (1992).

56. Kligman, L. H. and Zheng, P. The protective effect of a broad spectrum sunscreen against chronic UVA radiation in hairless mice: A histologic and ultrastructural assessment, *J. Soc. Cosmet. Chem.*, Jan/Feb 1994; *45*:21–33.
57. Kligman, L. H., Zheng, P., and Kligman, A. M. Photoirritation, a new photobiologic phenomenon induced by long wavelength UVA radiation in hairless mice treated with broad spectrum sunscreens, *Acta Dermatol. Venereol (Stockh)*, 1995; *75*: 205–211.
58. Kligman, L. H. and Kligman, A. M. Reflections on heat, *Br. J. Dermatol., 110*: 369–375 (1984).
59. Kligman, L. H. Intensification of ultraviolet-induced dermal damage by infrared radiation, *Arch. Dermatol. Res.*, *272*:229–238 (1982).

7

Antioxidants and Carotenoids as Potential Photoprotectants

Homer S. Black
Veterans Affairs Medical Center and Baylor College of Medicine, Houston, Texas

Betty Rajan
Texas Tech University Health Science Center, Lubbock, Texas

I. INTRODUCTION

It is becoming more apparent that many of the maladies that afflict humans are the direct result of environmental interactions. Certainly ample evidence exists to affirm that most cutaneous cancers, and many of the associated symptoms attributed to cutaneous aging, are the result of exposure to solar radiation. In dealing with this etiological agent of environmental origin, it is almost certain that preventative modalities will continue to be the most reasonable approach to abating the symptoms resulting from actinic insult. Intuitively, it would seem that photoprotectants will continue to be the major weapon in this preventive campaign.

Conventionally, the protective index, or sun protection factor (SPF), to erythema has been the standard by which the effectiveness of a photoprotective agent is evaluated. Actinically induced erythema represents an easily observable, immediate response. One should be mindful, however, that this end-point represents only one of the myriad of events triggered in skin by ultraviolet (UV) insult. Erythema itself is a manifestation of a very complex physiological response, compounded by a host of variables that influence it. Indeed, it has been demonstrated that some agents that are capable of inhibiting the inflammatory response, exhibited as erythema, do so with little or no effect on some of the more serious, prolonged manifestations of actinic damage. For example, sunscreens that were effective against a range of physical and histological UV alterations were ineffective against the UV-enhanced growth of melanoma cells injected into murine recipients (1). Melanoma cell growth is controlled, in part, by UV-modulated immunological responses. This example vividly demonstrates that different UV effects on the skin, mediated by different mechanisms, manifest different sensitivities relative to the photoprotective properties of sunscreens. Thus, in our quest to find more efficacious photoprotectants to erythema, one must be cognizant that some of the more important aspects of photoprotection are not overlooked. This is not to say that photoprotectants to erythema do not protect against more serious long-term responses, for many do (2,3), but rather, as we have already alluded, some may not. Ideally, then, an effective photoprotectant should not only prevent erythema, but carcinogenesis and actinic dermal damage, as well. To find such agents, we must first understand the manner by which photoprotectants act.

II. HOW PHOTOPROTECTANTS WORK

Wiskemann (4) has outlined the means by which photoprotectants might work. Although these potential mechanisms were specifically discussed relative to systemic protection, they are equally applicable to topical agents:

1. Agents that attenuate the level of actinic insult (e.g., by absorbing or otherwise acting as a barrier to UV rays): This mechanism of action represents that typically exhibited by topical photoprotectants, and several excellent overviews of such agents can be found (5–8).
2. Agents that compete with the target molecule for the damaging products of the offending agent: For example, UV might induce the formation of a variety of reactive species, or free radicals, in the skin. Agents possessing antioxidant or free radical-scavenging properties would compete with potential target molecules and divert the damaging effects. Many such effects, such as lipid peroxidation, occur at the membrane level, but could also affect nucleic acids (9,10).
3. Agents that provide restorative action (i.e., repair or aid the repair of the damaged target molecule): Some agents, such as nucleotides, delay UV-mediated edema (4) and have been used in the treatment of photosensitive disease (11). Although certain proprietary cosmetic products imply restorative action of actinic damage, for the most part, this line of investigation has either not been vigorously pursued or widely reported in the literature and remains anecdotal. However, a recent study has demonstrated that delivery of a dimer-specific repair enzyme after UV radiation not only decreased the number of cyclobutane dimers in epidermis, but

prevented UV suppression of delayed and contact hypersensitivity (12). This approach shows promise not only in restoring immune function, but may be applicable to the prevention of other UV-induced alterations resulting from DNA damage. Studies with tretinoin also provide promising prospects for restorative action of photoprotectants (13–15).

4. Agents that suppress various stages of the inflammatory response and, thus, mask certain manifestations of UV damage: We have already alluded to the fact that some anti-inflammatories appear to inhibit normal protective mechanisms of the skin that are mobilized on UV insult, while providing no true increase in resistance to the insult itself. Jarvinen (16) first reported that high doses of cortisone suppressed reactivity of patients to UV radiation. Aspirin, phenylbutazone, and indomethacin are also able to suppress the inflammatory reaction to UV (17–19). More recently, Lim et al. (20) demonstrated that indomethacin, although inhibiting the erythema response to UV, did not prevent other manifestations of UV-mediated damage. Whereas the anti-inflammatory agents may be useful in treatment of the symptoms of UV insult, they cannot be considered a universal means of photoprotection.

With the qualifying caveats accompanying mechanisms 3 and 4, it is apparent that the most promising category of agents demonstrating photoprotective properties are those chemicals with potential for acting by mechanisms 1 and 2.

Interest in antioxidants and carotenoids as photoprotectants can be traced to the earlier studies with radioprotectors. In vivo radioprotective effects were reported in 1949 (21). Many of the agents tested in biological systems for effectiveness possessed antioxidant properties (i.e., capable of inhibiting free radical reactions). Subsequent research in this area led to several proposed mechanisms to explain radioprotection (22,23). These included

1. Formation of mixed disulfides.
2. Lowering oxygen tension of target tissues. This resulted in reduction of free radical formation occurring from radiolysis of tissue H_2O.
3. Scavenging harmful free radicals.
4. Donation of hydrogen atoms to permit rapid repair of target molecules.
5. Chelation of metal ions essential for enzyme function.
6. Energy traps that spare more vital molecular functions.

It is readily apparent that most of these potential mechanisms involved some aspect of free radical-initiated damage. Thus, in 1962, when Norins (24) provided evidence for free radical formation in UV-radiated skin (later confirmed by Pathak and Stratton; 25), the basis for free radical involvement in UV-mediated cutaneous damage was provided. As a corollary, if antioxidants were effective radioprotectors, they should also affect photoprotection. In vitro assays were developed to test antioxidants in this role. Epstein et al. (26) tested some 98 compounds in an assay based on the inhibition of oxygen-dependent, photodynamic toxicity of benzo[*a*]pyrene to *Tetrahymena pyriformis*. The compounds tested were, at that time, conventionally grouped into:

1. Primary antioxidants (i.e., those considered to function as free radical acceptors)
2. Secondary antioxidants (i.e., those agents that act synergistically with primary antioxidants)
3. Ultraviolet absorbers (i.e., those that block the catalytic effect of UV and visible light)
4. A miscellaneous group that comprises largely reducing agents with free —SH groups, including some radioprotectors that are not generally considered as antioxidants

Among the notable results obtained from this study, the "hindered phenols" (e.g., butylated hydroxytoluene; BHT) were without effect when compared with the inhibitory action of DL-α-tocopherol, whereas "polyenes" (e.g., β-carotene) were about 11 times more potent. Ultraviolet absorption was not a factor in either category of compounds. The fact that some agents with known antioxidant properties were ineffective in inhibiting photodynamic toxicity, suggested that, most probably, both the pathomechanism of the process, as well as the action of the antioxidants, reflected a reasonable degree of specificity. The implication was that individual agents would have to be tested against specific UV responses. That this was indeed so was shown by Kahn and Curry (27) who, after testing 30 antioxidant agents in both in vitro (protophorphyrin-sensitized photohemolysis) and in vivo tests, found no correlation between the two. In the in vitro tests, β-carotene actually enhanced hemolysis while providing a slight degree of protection to erythema. In general, most free radical scavenging drugs did not protect well (to erythema) when applied topically. These authors concluded that even after enhancing penetration and increasing effectiveness, the agents tested would not be useful clinically.

Since those earlier studies, the role of free radicals in inflammatory process, carcinogenesis, and aging has become more apparent (28–30). If one embraces a more global view of photoprotection, one that encompasses all of these major responses to UV insult, then antioxidants and carotenoids could play an important auxiliary role in future photoprotective strategies. Some of the photoprotective properties of these respective agents will now be considered.

III. ANTIOXIDANTS

There are many low molecular weight agents that serve as free radical quenchers. Compounds capable of inhibiting free radical reactions are routinely defined as antioxidants, and lists of such common agents have been previously published (26,27,31). Likewise, there are numerous mechanisms by which a cell may deal with electronically activated species and some mention is warranted of natural, endogenous free radical defense mechanisms. Examples include several enzyme systems. Among the latter, superoxide dismutase (SOD) plays a prominent role in regulating cellular superoxide anion levels, and does so by dismutation to hydrogen peroxide (H_2O_2). As an hyaluronidase-derived protein (32,33), SOD is effective in the treatment of a wide range of inflammatory responses (34). Acute sunlight-induced insult manifests most of the common inflammatory responses (i.e., erythema, edema, and neutrophil infiltration; 35), and thus SOD represents an obvious avenue by which to investigate enzymatic antioxidant mechanisms. In fact, recent studies have indicated that endogenous SOD activity is important in the severity and treatment of a wide range of inflammatory skin diseases. In a study in which lipid peroxides and SOD levels were measured in skin biopsies of patients suffering from various inflammatory lesions, lipid peroxide levels were markedly elevated (36). Lipid peroxidation has been suspected of being a source of sunlight-induced cellular damage and initiated by free radical attack (37,38). The SOD activity in those patients with severe disease who were highly recalcitrant to therapy was only slightly elevated. In contrast, patients with mild disease who responded well to therapy demonstrated marked elevations in SOD activity. It was concluded that the lipid peroxide/SOD ratio was an accurate prognostic indicator, and that the severity of the disease and response to treatment was partly governed by the degree to which the patient's SOD activity was up-regulated. Others report similar findings for inflammatory dermatitis (39). Importantly, several phototoxic and photoallergic photosensitivity reactions can be prevented by treatment with SOD, occasionally by topical application to the skin (40–42). Several nonenzymatic superoxide anion-scavenging antioxidants also have a photoprotective effect on repeated UV-induced cutaneous damage (43).

In addition to SOD, a membrane-associated thioredoxin reductase (TR) has been described that, in the presence of NADPH, acts as a free radical trap at the external surface of the epidermis (44). Both SOD and TR are involved in the dismutation of superoxide anion to H_2O_2, the reduction of which is catalyzed by either catalase or glutathione peroxidase.

Other, nonenzymatic antioxidants also prevent inflammatory responses. For example, anthraline is often used as a therapeutic agent in the treatment of psoriasis. This agent auto-oxidizes readily through a free radical mechanism and induces lipid peroxidation in microsomes isolated from the skin. Whereas anthraline produces clearing on many psoriatic patients, it can sometimes cause severe inflammation of the perilesional skin. Finnen et al. (45) have topically applied several free radical scavengers to anthralin-treated areas and found that α-tocopherol, BHT, and the acetate and palmitate esters of retinol were effective in preventing the erythema and edema normally resulting from anthraline treatment. α-Tocopherol and the retinol esters were also effective inhibitors of anthraline-induced lipid peroxidation.

Free radical-mediated lipid peroxidation, involving reactive oxygen species, provides the rationale on which a preventive role for antioxidants rests. It seemed reasonable that these chain reactions should be terminated by antioxidant intervention. The UV-induced formation of one such photooxidative product, cholesterol epoxide (a putative carcinogen), was inversely related to the cutaneous antioxidant level after feeding a broad spectrum of water- and lipid-soluble antioxidants (including ascorbic acid [vitamin C], reduced glutathione, α-tocopherol [vitamin E], and BHT; 46). These findings suggested possible prophylactic effects of this antioxidant mixture, not only on the formation of this putative carcinogen, but on the subsequent pathological conditions that might result from, or concurrent with, its formation. Indeed, this antioxidant mixture provides significant protection against UV carcinogenesis (47). Pauling et al. (48) corroborated these observations with the antioxidant mixture, but found no protective effect with either ascorbic acid or α-tocopherol alone, at concentrations present in the mixture. A subsequent study from Pauling's group (49) found that higher levels of ascorbic acid could significantly inhibit UV carcinogenesis. Recent dietary studies with α-tocopherol have demonstrated its photoprotective properties to UV carcinogenesis, albeit at doses that were toxic (50). Topical application of α-tocopherol also inhibits UV carcinogenesis (51). The evidence for the photoprotective effects of vitamin E has been reviewed (52).

One of the aforementioned antioxidants, BHT, acts as an antagonist to the toxicity and carcinogenicity of a wide range of chemical and physical agents, including UV (53–55). In 1978, DeRios et al. (56) demonstrated that BHT provided an SPF of 2.0 when systemically administered to mice. Topical application of the agent provided only marginal protection to erythema. Subsequent studies showed that BHT accruing in the skin would absorb no more than 1% of the incident UV, thus making it doubtful that the screening of erythemic wavelengths by BHT per se could account for the agents photoprotective action (57). However, orally ingested BHT increased the spectral absorption of mouse epidermis and inhibited UV induction of epidermal ornithine decarboxylase (ODC) activity (57,58). The induction of ODC has been used as a marker of inflammatory response (59) and is thought by some to be an obligatory step in tumor promotion (60).

The effects of BHT could be explained by the promotion of epidermal hyperplasia, with resultant diminution of UV penetration to potential target sites. Nevertheless, BHT failed to induce changes in epidermal protein, DNA synthesis, or the cell-labeling index, all of which would be expected to be elevated if enhanced epidermal proliferation had occurred (61). Koone and Black (62) reported that BHT decreased the spectral transmission of mouse stratum corneum

(SC), while leaving the physical dimensions (thickness) of this tissue unchanged. Also, BHT-inhibited ODC activity could be restored to near normal levels if the SC were stripped away before irradiation. Induction of ODC is a useful measure of topically applied sunscreen efficacy (63). Thus, it was concluded that most of the protective effect of BHT resided in the SC, and it was suggested that BHT photoprotection was related to changes in the chemical properties of this tissue. These investigators theorized that BHT, acting in its antioxidant capacity, retarded the normal maturation (oxidation) of keratin and, thereby, reduced the degree of disulfide cross-linking which, in turn, would be reflected in the optical characteristics of the SC. Results from crossover feeding studies supported the contention that BHT exerts its major anticarcinogenic effects during the period in which the initiating incitant (UV) was administered and were compatible with a mode of action mediated by UV dose reduction (64).

Several structurally related phenolic antioxidants, including butylated hydroxyanisole (BHA), vanillin, and propyl gallate, have been compared with BHT for protective properties against UV carcinogenesis (65). Only BHT provided a significant increase in tumor latency. However, BHT and vanillin significantly reduced tumor multiplicity, whereas BHT and BHA both significantly inhibited UV induction of ornithine decarboxylase. It would appear that structural characteristics of BHT, when compared with compounds of like lipophilicity, such as BHA, determine the former's anticarcinogenic activity. The additional *t*-butyl group of BHT may hinder systemic metabolism of the phenol and allow low levels of the unaltered compound to reach cutaneous sites at which its antioxidant properties are exerted. The polyphenolic fraction of green tea also provides significant protection against UV carcinogenesis (66).

We have noted from the earlier studies with radioprotectors the potential role of metal chelators in modulating free radical reactions. We have also observed that H_2O_2 is an intermediate in the four-electron reduction sequence of molecular oxygen to H_2O, a sequence that would normally be completed by catalase or glutathione peroxidase. However, when heavy ions (e.g., Fe^{3+}) are present, superoxide anion ($O_2^{\cdot-}$) may reduce them by the Haber–Weiss cycle [Eq. (1)]. The reduced metal ions may then reduce H_2O_2 by a Fenton reaction [Eq. (2)] to produce a hydroxyl radical ($^{\cdot}OH$).

$$Fe^{3+} + O_2^{\cdot-} \rightarrow Fe^{2+} + O_2 \tag{1}$$

$$Fe^{2+} + H_2O_2 \rightarrow Fe^{3+} + OH^- + {}^{\cdot}OH \tag{2}$$

The hydroxyl radical is the most electrophilic of the oxy radicals and, as such, is extremely reactive. Increased levels of nonheme iron in skin results from prolonged UV exposure (67). Furthermore, Fe^{2+} ions are released from ferritin on exposure to UVA (68). These could drive Fenton reactions to produce $^{\cdot}OH$. In addition, there is evidence that UV radiation of H_2O_2 can directly result in $^{\cdot}OH$ formation (69,70).

Aside from the potential damage that could be wreaked from ˙OH formation, iron may act as a catalyst in UV-induced lipid peroxidative reactions as in Eq. (3), where iron acts on lipid hydroperoxides (LOOH) to produce lipid alkoxyl radicals (˙LO).

$$Fe^{2+} + LOOH \rightarrow Fe^{3+} + {}^{\cdot}LO + HO^{-} \quad (3)$$

Evidence that such may occur has recently been reported with the observation that a lipid-derived free radical is formed in UV-radiated skin (71). Topical application of an iron chelator before UV exposure significantly reduced the lipid radical electron paramagnetic resonance signal. Topical application of certain iron chelators delayed the onset of UVB-induced skin damage (i.e., skin wrinkling and tumorigenesis; 67). The role of iron chelation should be given serious consideration in any photoprotective strategy.

IV. CAROTENOIDS

Carotenoids are lipid-soluble, yellow to orange-red pigments that are universally found in photosynthetic tissues of plants, algae, and some bacteria. Chemically, these pigments are tetraterpenoids, consisting of eight isoprenoid residues. Carotenoids serve a protective role in green plants and certain photosynthetic bacteria to photosensitization by endogenous photosensitizers, such as the porphyrin-containing photosynthetic pigments (72). In 1951, Kesten (73) reported that oral β-carotene delayed erythema onset in a patient suffering severe photosensitivity. Armed with the knowledge of the carotenoid's protective role in porphyrin-related photosensitization and the presumptive evidence of Kesten, Mathews-Roth theorized that carotenoid pigments might serve a protective role in humans, similar to that in plants (74,75). Support for this thesis was obtained when it was demonstrated that β-carotene was effective in preventing lethal hematoporphyrin photosensitization in mice (76). Subsequently, she and her colleagues (77) then demonstrated that β-carotene was effective in the treatment of erythropoietic protoporphyria (EPP), a disease characterized by abnormally elevated levels of protoporphyrin in the blood and skin, which is manifest by increased sensitivity to visible light (380–560 nm). Of 133 patients treated, over 75% reported at least a four-fold increase in tolerance to sunlight. Only marginal improvement was seen in patients suffering other photosensitive diseases, including polymorphic light eruption, porphyria cutanea tarda, solar urticaria, and actinic reticuloid. Other investigators have reported a similar lack of effect of β-carotene therapy for these diseases (78). Furthermore, β-carotene did not alter the degree of erythema in controlled clinical trials, and it only slightly (although statistically significant) elevated the threshold dose of sunlight required to produce a minimal erythema (79). These observations, along with the fact that porphyrin levels changed only slightly during β-carotene treatment of EPP patients, suggested a rather specific photoprotective action. Although some carotenoid pigments absorb strongly in the UVB region (80), β-carotene does not function in the skin as an optical filter, and its protective action is probably related to its ability to quench specific reactive species and their reactions (81).

Singlet oxygen is produced on illumination of protoporphyrin (82), and a similar mechanism might be involved in photosensitivity. β-carotene is a very efficient quencher of singlet oxygen (83). Moreover, the pigment's ability to protect against photosensitization is related to the length of the conjugated double-bond chain, supporting an antireactive species action (84). Mathews-Roth (85), using a singlet oxygen probe, has demonstrated that β-carotene and other carotenoid pigments (canthaxanthin and phytoene) can quench singlet oxygen-driven photochemical reactions that occur in mouse epidermis. However, β-carotene is also capable of quenching other free radical reactions, including those involved in lipid peroxidation (86).

Aside from the ameliorating effect on EPP, β-carotene, when administered intraperitoneally, provides protection against UV-induced carcinogenesis in hairless mice (87). Photoprotection is manifested in both delayed tumor appearance and reduced tumor growth rates. These observations have been confirmed in dietary studies and extended to include other carotenoid pigments displaying no vitamin A activity (88,89).

Peto et al. (90) have reviewed the experimental and epidemiological data on the potential effects of β-carotene for prevention of various types of cancers. This study served to stimulate interest in carotenoids as cancer-preventive agents. A study by Kune et al. (91) reported a significant inverse relation between the risk of skin cancer and the level of serum β-carotene. They also observed a significant inverse relation between the risk of skin cancer and a diet rich in β-carotene. However, Greenberg et al. (92) found no significant reduction in the occurrence of new skin cancers in patients who received daily β-carotene supplements. In this study, the daily 50-mg supplements did not result in carotenemia, but did elevate plasma carotenoid levels to approximately 170 μg/100 ml. This level is only about 40% of that found to be therapeutically effective in the treatment of EPP. Mathews-Roth et al. (77) found β-carotene effective in the treatment of EPP only after patients developed carotenemia and had plasma β-carotene levels of at least 400 μg/100 ml. It is possible that the earlier study did not achieve adequate levels of β-carotene to be effective in preventing carcinogenesis. Thus, evidence for the involvement of β-carotene in the prevention of human skin cancer remains equivocal and merits further study.

V. CONCLUSION

From the foregoing discussion, it is clear that neither the carotenoids nor any single antioxidant now satisfactorily fulfills the requirements of a universal photoprotectant (93,94). This is especially true if photoprotection is viewed from the more global prospectus suggested. At the same time, selected antioxidants and carotenoids exhibit photoprotective properties against specific photoresponses. The experimental evidence suggests that each category of agent is capable of intercepting reactive species that are specific to the pathomechanism of the respective response. This, coupled with the observation that neither category of agent acts as an effective barrier to UV penetration, implies that such "mopping-up" action is consistent with a supportive role in a preventive campaign. Thus, some combination of agents exhibiting both UV-screening and free radical-scavenging action would be of value. To this end, both carotenoids and selected antioxidants, when applied topically to skin, inhibit some UV-initiated, radical-mediated photoresponses (95). Certain of these agents have recently found their way into sunscreen formulations. Their value and effectiveness in diminishing the detrimental effects of solar radiation on skin are only now being investigated, and it is probable that their role, as adjuncts to current photoprotective strategies will be expanded in the future.

REFERENCES

1. Wolf, P., Donawho, C. K., and Kripke, M. L. Effect of sunscreens on UV radiation-induced enhancement of melanoma growth in mice, *JNCI*, *86*:99–105 (1994).
2. Synder, D. S. and May, M. Ability of PABA to protect mammalian skin from ultraviolet light-induced skin tumors and actinic damage, *J. Invest. Dermatol.*, *65*: 543–546 (1975).
3. Kligman, L. H., Akin, F. J., and Kligman, A. M. Sunscreens prevent ultraviolet photocarcinogenesis, *J. Am. Acad. Dermatol.*, *3*:30–35 (1980).
4. Wiskemann, A. Restoration protection by nucleotides and nucleosides, *Sunlight and Man* (T. B. Fitzpatrick, M. A. Pathak, L. C. Harber, M. Seiji, A. Kukita, eds.), University of Tokyo Press, Tokyo, 1974, pp. 771–778.
5. Groves, G. A. Sunburn and its prevention, *Aust. J. Dermatol.*, *21*:115–141 (1980).
6. Pathak, M. A. Sunscreens: Topical and systemic approaches for protection of human skin against harmful effects of solar radiation. *J. Am. Acad. Dermatol.*, *7*:285–312 (1982).
7. Murphy, G. M. and Hawk, J. L. M. Sunscreens, *J. R. Soc. Med.*, *79*:254–256 (1986).
8. Klein, K. Formulating sunscreen products, *Sunscreens: Development, Evaluation, and Regulatory Aspects* (N. J. Lowe and N. A. Shaath, eds.), Marcel Dekker, New York, 1990, pp. 235–266.
9. Schaich, K. M. and Black, H. S. Free radicals, antioxidants, skin cancer and related diseases: An introduction to the symposium, *Lipids*, *23*:568–569 (1988).
10. Borg, D. C. and Schaich, K. M. Iron and hydroxyl radicals in lipid oxidation: Fenton reactions in lipid and nucleic acids co-oxidized with lipid, *Oxy-Radicals in Molecular Biology and Pathology* (P. A. Cerutti, I. Fridovich, and J. M. McCord, eds.), Alan R. Liss, New York, 1988, pp. 427–441.

11. Gajdos, M., Gajdos-Torok, M., Mantz, I.-M., and Schiradin, H. Protoporphyrine erythropoietique. Traitement par l'inosine, *Presse Med.*, *73*:119–122 (1965).
12. Kripke, M. L., Cox, P. A., Alas, L. G., and Yarosh, D. B. Pyrimidine dimers in DNA initiate systemic immunosuppression in UV-irradiated mice, *Proc. Natl. Acad. Sci. USA*, *89*:7516–7520 (1992).
13. Kligman, A. M., Grove, G. L., Hirose, R., and Leyden, J. J. Topical tretinoin for photoaged skin, *J. Am. Acad. Dermatol.*, *15*:836–859 (1986).
14. Weiss, J. S., Ellis, C. N., Headington, J. T., Tincoff, T., Hamilton, T. A., and Voorhees, J. J. Topical tretinoin improves photoaged skin. A double-blind vehicle-controlled study, *JAMA*, *259*:527–532 (1988).
15. Lever, L., Kumar, P., and Marks, R. Topical retinoic acid for treatment of solar damage, *Br. J. Dermatol.*, *122*:91–98 (1990).
16. Jarvinen, K. A. J. Effect of cortisone on reaction of skin to ultra-violet light, *Br. J. Med.*, *2*:1377–1378 (1951).
17. Winder, C. V., Wax, J., Burr, V., Been, M., and Rosiere, C. E. A study of pharmacological influences on ultraviolet erythema in guinea pigs, *Arch. Int. Pharmacodyn. Ther.*, *116*:261–292 (1958).
18. Gupta, N. and Levy, L. Delayed manifestation of ultraviolet reaction in the guinea-pig caused by anti-inflammatory drugs, *Br. J. Pharmacol.*, *47*:240–248 (1973).
19. Law, E. and Lewis, A. J. The effect of systemically and topically applied drugs on ultraviolet-induced erythema in the rat, *Br. J. Pharmacol.*, *59*:591–597 (1977).
20. Lim, H. W., Thorbecke, G. J., Baer, R. L., and Gigli, I. Effect of indomethacin on alteration of ATPase-positive Langerhans cell density and cutaneous sunburn reaction induced by ultraviolet-B radiation, *J. Invest. Dermatol.*, *81*:455–458 (1983).
21. Patt, H. M., Tyree, E. B., Straube, R. L., and Smith, D. E. Cysteine protection against X irradiation, *Science*, *110*:213–214 (1949).
22. Foye, W. O. Radiation-protective agents in mammals. *J. Pharm. Sci.*, *58*:283–300 (1969).
23. Carr, C. J., Huff, J. E., Fisher, K. D., and Huber, T. E. Protective agents modifying biological effects of radiation. A selective review, *Arch. Environ. Health*, *21*:88–98 (1970).
24. Norins, A. L. Free radical formation in the skin following exposure to ultraviolet light, *J. Invest. Dermatol.*, *39*:445–448 (1962).
25. Pathak, M. A. and Stratton, K. Free radicals in human skin before and after exposure to light, *Arch. Biochem. Biophys.*, *123*:468–476 (1968).
26. Epstein, S. S., Saporoschetz, I. B., Small, M., Park, W., and Mantel, N. A simple bioassay for antioxidants based on protection of *Tetrahymena pyriformis* from the photodynamic toxicity of benzo[*a*]pyrene, *Nature*, *208*:655–658 (1965).
27. Kahn, G. and Curry, M. C. Ultraviolet light protection by several new compounds. *Arch. Dermatol.*, *109*:510–517 (1974).
28. Black, H. S. Role of reactive oxygen species in inflammatory process. *Pharmacology and the Skin*, Vol. 2, *Nonsteroidal Anti-inflammatory Drugs* (N. J. Lowe and C. N. Hensby, eds.), Karger, Basel, 1989, pp. 1–20.
29. Pryor, W. A. Free radical biology: Xenobiotics, cancer, and aging, *Vitamin E: Biochemical, Hematological, and Clinical Aspects* (B. Lubin and L. J. Machlin, eds.), New York Academy of Science, New York, 1982, pp. 1–22.

30. Black, H. S. Potential involvement of free radical reactions in ultraviolet light-mediated cutaneous damage, *Photochem. Photobiol.*, *46*:213–221 (1987).
31. Proctor, P. H. and Reynolds, E. S. Free radicals and disease in man, *Physiol. Chem. Phys. Med. NMR*, *16*:175–195 (1984).
32. Huber, W., Schulte, T., Carson, S., Goldhamer, E., and Vogin, E. Some chemical and pharmacologic properties of a novel anti-inflammatory protein, *Toxicol. Appl. Pharmacol.*, *12*:308 (1968).
33. McCord, J. M. and Fridovich, I. Superoxide dismutase. An enzymatic function for erythrocuprein (hemocuprein), *J. Biol. Chem.*, *224*:6049–6055 (1969).
34. Wilsmann, K. Ten years of clinical experience with SOD treatment of inflammatory disorders, *Superoxide and Superoxide Dismutase in Chemistry, Biology and Medicine* (G. Rotilio, ed.), Elsevier, Amsterdam, 1986, pp. 500–507.
35. Epstein, J. H. The pathological effects of light on the skin, *Free Radicals in Biology III* (W. A. Pryor, ed.), Academic Press, New York, 1977, pp. 219–249.
36. Niwa, Y., Kanoh, T., Sakane, T., Soh, H., Kawai, S., and Miyachi, Y. The ratio of lipid-peroxides to superoxide dismutase activity in the skin lesions of patients with severe skin diseases: An accurate prognostic indicator, *Life Sci.*, *40*:921–927 (1987).
37. Meffert, H., Diezel, W., and Sonnichsen, N. Stable lipid peroxidation products in human skin: Detection: ultraviolet light-induced increase, pathogenic importance, *Experientia*, *32*:1397–1398 (1976).
38. Ogura, R. and Sugiyama, M. Cellular membrane damage and lipid peroxide following exposure of the skin to ultraviolet rays, *Cutaneous Aging* (A. M. Kligman and Y. Takase, eds.), University of Tokyo Press, Tokyo, 1988, pp. 459–471.
39. Miyachi, Y., Uchida, K., Komura, J., Asada, Y., and Niwa, Y. Auto-oxidative damage in cement dermatitis, *Arch. Dermatol. Res.*, *277*:288–292 (1985).
40. Pathak, M. A. and Joshi, P. C. Production of active oxygen species ($^1O_2$ and $O_2^{\cdot -}$) by psoralens and ultraviolet radiation (320–400 nm), *Biochim. Biophys. Acta*, *798*: 115–126 (1984).
41. Miyachi, Y. and Niwa, Y. Effects of psoriatic sera on the generation of oxygen intermediates by normal polymorphonuclear leucocytes, *Arch. Dermatol. Res.*, *275*: 23–26 (1983).
42. Miyachi, Y., Horio, T., and Imamura, S. Sunburn cell formation is prevented by scavenging oxygen intermediates, *Clin. Exp. Dermatol.*, *8*:305–310 (1983).
43. Bissett, D. L., Chatterjee, R., and Hannon, D. P. Photoprotective effect of superoxide-scavenging antioxidants against ultraviolet radiation-induced chronic skin damage in the hairless mouse, *Photodermatol. Photoimmunol. Photomed.*, *7*:56–62 (1990).
44. Schallreuter, K. U. and Wood, J. M. Free radical reduction in the human epidermis. *Free Radic. Biol. Med.*, *6*:519–532 (1989).
45. Finnen, M. J., Lawrence, C. M., and Shuster, S. Inhibition of dithranol inflammation by free-radical scavengers, *Lancet*, *2*:1129–1130 (1984).
46. Lo, W. B. and Black, H. S. Inhibition of carcinogen formation in skin irradiated with ultraviolet light, *Nature*, *246*:489–491 (1973).
47. Black, H. S. and Chan, J. T. Suppression of ultraviolet light induced tumor formation by dietary antioxidants, *J. Invest. Dermatol.*, *65*:412–414 (1975).

48. Pauling, L., Willoughby, R., Reynolds, R., Blaisdell, B. E., and Lawson, S. Incidence of squamous cell carcinoma in hairless mice irradiated with ultraviolet light in relation to intake of ascorbic acid (vitamin C) and D,L-α-tocopheryl acetate (vitamin E), *Int. J. Nutr. Res.*, *23*:53–82 (1982).
49. Dunham, W. B., Zuckerkandl, E., Reynolds, R., Willoughby, R., Marcuson, R., Barth, R., and Pauling, L. Effects of intake of L-ascorbic acid on the incidence of dermal neoplasms induced in mice by ultraviolet light, *Proc. Natl. Acad. Sci. USA*, *79*:7532–7536 (1982).
50. Gerrish, K. E. and Gensler, H. L. Prevention of photocarcinogenesis by dietary vitamin E, *Nutr. Cancer*, *19*:125–133 (1993).
51. Gensler, H. L. and Magdaleno, M. Topical vitamin E inhibition of immunosuppression and tumorigenesis induced by ultraviolet irradiation, *Nutr. Cancer*, *15*: 97–106 (1991).
52. Fryer, M. J. Evidence for the photoprotective effects of vitamin E, *Photochem. Photobiol.*, *58*:304–312 (1993).
53. Babich, H. Butylated hydroxytoluene (BHT): A review, *Environ. Res.*, *29*:1–29 (1982).
54. Burlakova, E. B., Gaintsevz, V. D., Slepukhina, L. V., Khrapova, N. G., and Emanuel, N. M. Connection between radiation-protective and antitumor action of antioxidants, *Dokl. Akad. Nauk SSSR*, *164*:934–936 (1965); *Chem Abstr.*, *64*:2394 (1966).
55. Black, H. S., Chan, J. T., and Brown, G. E. Effects of dietary constituents on ultraviolet light-mediated carcinogenesis, *Cancer Res.*, *38*:1384–1387 (1978).
56. DeRios, G., Chan, J. T., Black, H. S., Rudolph, A. H., and Knox, J. M. Systemic protection by antioxidants against UVL-induced erythema, *J. Invest. Dermatol.*, *70*: 123–125 (1978).
57. Black, H. S., Kleinhans, C. M., Hudson, H. T., Sayre, R. M., and Agin, P. P. Studies on the mechanism of protection by butylated hydroxytoluene to UV-radiation damage, *Photobiochem. Photobiophys.*, *1*:119–123 (1980).
58. Peterson, A. O., McCann, V., and Black, H. S. Dietary modification of UV-induced epidermal ornithine decarboxylase, *J. Invest. Dermatol.*, *75*:408–410 (1980).
59. Lowe, N. J. and Breeding, J. Antiinflammatory drug effects on ultraviolet light-induced epidermal ornithine decarboxylase and DNA synthesis, *J. Invest. Dermatol.*, *74*:418–420 (1980).
60. O'Brien, T. G., Simsiman, R. C., and Boutwell, R. K. Induction of the polyamine-biosynthetic enzymes in mouse epidermis by tumor-promoting agents, *Cancer Res.*, *35*:1662–1670 (1975).
61. Black, H. S., Lenger, W., Gerguis, J., and McCann, V. Studies on the photoprotective mechanism of butylated hydroxytoluene, *Photochem. Photobiol.*, *40*:69–75 (1984).
62. Koone, M. D. and Black, H. S. A mode of action for butylated hydroxytoluene-mediated photoprotection, *J. Invest. Dermatol.*, *87*:343–347 (1986).
63. Gange, R. W. and Mendelson, R. Sunscreens block the induction of epidermal ornithine decarboxylase by ultraviolet-B radiation: A new way of evaluating sunscreen efficacy in vivo, *Br. J. Dermatol.*, *107*:215–220 (1982).

64. Black, H. S. and Mathews-Roth, M. M. Protective role of butylated hydroxytoluene and certain carotenoids in photocarcinogenesis, *Photochem. Photobiol.*, *53*:707–716 (1991).
65. Black, H. S. and Tigges, J. Evaluation of structurally-related phenols for anti-photocarcinogenic and photoprotective properties, *Photochem. Photobiol.*, *43*:403–408 (1986).
66. Wang, Z. Y., Agarwal, R., Bickers, D. R., and Muktar, H. Protection against ultraviolet B radiation-induced photocarcinogenesis in hairless mice by green tea polyphenols, *Carcinogenesis*, *12*:1527–1530 (1991).
67. Bissett, D. L., Chatterjee, R., and Hannon, D. P. Chronic ultraviolet radiation-induced increase in skin iron and the photoprotective effect of topically applied iron chelators, *Photochem. Photobiol.*, *54*:215–223 (1991).
68. Aubailly, M. A., Santus, R., and Salmon S. Ferrous ion release from ferritin by ultraviolet-A radiations, *Photochem. Photobiol.*, *54*:769–773 (1991).
69. Floyd, R. A., West, M. S., Eneff, K. L., Hogsett, W. E., and Tingey, D. T. Hydroxyl free radical mediated formation of 8-hydroxyguanine in isolated DNA, *Arch. Biochem. Biophys.*, *262*:266–272 (1988).
70. Roza, L., van der Schans, G. P., and Lohman, P. H. M. The induction and repair of DNA damage and its influence on cell death in primary human fibroblasts exposed to UV-A or UV-C irradiation, *Mutat. Res.*, *146*:89–98 (1985).
71. Jurkiewicz, B. A. and Buettner, G. R. Ultraviolet light-induced free radical formation in skin: An electron paramagnetic resonance study, *Photochem. Photobiol.*, *59*:1–4 (1994).
72. Goodwin, T. W. Carotenoids, *The Encyclopedia of Biochemistry* (R. J. Williams and E. M. Lansford, eds.), Reinhold Publishing, New York, 1967, pp. 200–205.
73. Kesten, B. M. Urticaria solare, *Arch. Dermatol. Syphilol.*, *64*:221–223 (1951).
74. Mathews-Roth, M. M. Carotenoids in medical applications, *Carotenoids as Colorants and Vitamin A Precursors* (J. C. Bauernfeind, ed.), Academic Press, New York, 1981, pp. 755–785.
75. Mathews-Roth, M. M. Beta-carotene therapy for erythropoietic protoporphyria and other photosensitivity diseases, *The Science of Photobiology* (J. D. Regan and J. A. Parrish, eds.), Plenum Press, New York, 1982, pp. 409–440.
76. Mathews, M. M. Protective effect of β-carotene against lethal photosensitization by haematoporphyrin, *Nature*, *203*:1092 (1964).
77. Mathews-Roth, M. M., Pathak, M. A. Fitzpatrick, T. B., Harber, L. H., and Kass, E. H. Beta-carotene therapy for erythropoietic protoporphyria and other photosensitivity diseases, *Arch. Dermatol.*, *113*:1229–1232 (1977).
78. Kobza, A., Ramsay, C. A., and Magnus, I. A. Oral β-carotene therapy in actinic reticuloid and solar urticaria. Failure to demonstrate a photoprotective effect against long wave ultraviolet and visible radiation, *Br. J. Dermatol.*, *88*:157–166 (1973).
79. Mathews-Roth, M. M., Pathak, M. A., Parrish, J., Fitzpatrick, T. B., Kass, E. H., Toda, K., and Clemens, W. A. A clinical trial of the effects of oral beta-carotene on the responses of human skin to solar radiation, *J. Invest. Dermatol.*, *59*:349–353 (1972).

80. Mathews-Roth, M. M. and Pathak, M. A. Phytoene as a protective agent against sunburn (>280 nm) radiation in guinea pigs, *Photochem. Photobiol.*, *21*:261–263 (1975).
81. Sayre, R. M. and Black, H. S. Beta-carotene does not act as an optical filter in skin, *J. Photochem Photobiol.*, *12*:83–90 (1992).
82. Dalton, J., McAuliffe, C. A., and Slater, D. H. Reaction between molecular oxygen and photo-excited protoporphyrin IX, *Nature*, *235*:388 (1972).
83. Foote, C. S. and Denny, R. W. Chemistry of singlet oxygen. VII. Quenching by beta-carotene, *J. Am. Chem. Soc.*, *90*:6233–6235 (1968).
84. Mathews-Roth, M. M., Wilson, T., Fujimori, E., and Krinsky, N. I. Carotenoid chromophore length and protection against photosensitization, *Photochem. Photobiol.*, *19*:217–222 (1974).
85. Mathews-Roth, M. M. Carotenoids quench evolution of excited species in epidermis exposed to UV-B (290–320 nm) light, *Photochem. Photobiol.*, *43*:91–93 (1986).
86. Krinsky, N. I. and Deneke, S. M. Interaction of oxygen and oxy-radicals with carotenoids, *JNCI*, *69*:205–210 (1982).
87. Epstein, J. H. Effects of beta-carotene on ultraviolet induced cancer formation in the hairless mouse skin, *Photochem. Photobiol.*, *25*:211–213 (1977).
88. Mathews-Roth, M. M. Antitumor activity of β-carotene, canthaxanthin, and phytoene, *Oncology*, *39*:33–37 (1982).
89. Mathews-Roth, M. M. and Krinsky, N. I. Carotenoid dose level and protection against UV-B induced skin tumors, *Photochem. Photobiol.*, *42*:35–38 (1985).
90. Peto, R., Doll, R., Buckley, D., and Sporn, M. B. Can dietary β-carotene materially reduce human cancer rates? *Nature*, *290*:201–208 (1981).
91. Kune, G. A., Bannerman, B. F., Watson, L. F., Cleland, H., Merenstein, D., and Vitetta, L. Diet, alcohol, smoking, serum β-carotene, and vitamin A in male nonmelanocytic skin cancer patients and controls, *Nutr. Cancer*, *18*:237–244 (1992).
92. Greenberg, E. R., Baron, J. A., Stukel, T. A., Stevens, M. M., Mandel, J. S., Spencer, S. K., Elias, P. M., Lowe, N., Nirenberg, D. W., Bayrd, G., Vance, J. C., Freeman, D. H., Clendenning, W. E., and Kwan, T. A clinical trial of beta-carotene to prevent basal-cell and squamous-cell cancers of the skin, *N. Engl. J. Med.*, *323*: 789–795 (1990).
93. Mathews-Roth, M. M. Systemic photoprotection, *Dermatol. Clin.*, *4*:335–339 (1986).
94. Black, H. S. Systemic photoprotective agents, *Photodermatology*, *4*:187–195 (1987).
95. Pugliese, P. T. and Lampley, C. B. Biochemical assessment of the anti-aging effects of cosmetic products, *J. Appl. Cosmetol.*, *3*:129–138 (1985).

8

Reactive Oxygen Species and Antioxidant Protection in Photodermatology

Douglas Darr
North Carolina Biotechnology Center, Research Triangle Park, North Carolina

Sheldon R. Pinnell
Duke University Medical Center, Durham, North Carolina

I. INTRODUCTION

Skin, the outermost defense against noxious stimuli, is constantly assaulted by environmental insults. These consist of, but are not limited to, pollutants, pesticides or herbicides, heat or cold, and most prominently, sun exposure. Common to all is the ability to generate or propagate the formation of reaction oxygen species (ROS).

Reactive oxygen species are a limited, but continually growing, family of small, oxygen-based molecules that either contain an unpaired electron [free radicals, e.g., superoxide (O_2^-), hydroxyl radical ($HO^{\cdot}$)] or are capable of forming one [hydrogen peroxide (H_2O_2)]. This unpairing of electrons makes for very reactive chemical compounds. Additionally, other species are potent oxidants that can wreak havoc without the need for radical intermediacy. Examples include single oxygen ($^1\Delta O_2$) and the hypohalous acids (e.g., HOCl), formed by activated phagocytic cells at the expense of H_2O_2. Finally, only recently appreciated, nitric oxide ($NO^{\cdot}$), a free radical in its own right, can react with other reactive oxygen species; most notably O_2^- to form a particularly potent oxidant, peroxynitrite ($ONOO^{\cdot}$).

In the not-so-distant past, free radicals and reactive oxygen species were considered an academic curiosity. It is now accepted that these are relevant at all levels of biological science. In this chapter we will briefly review the formation, chemistry, and protective mechanisms against ROS damage. We will then focus on photodamage, the role of ROS in this damage, and the potential for antioxidant protection.

II. FORMATION AND REACTIVITY OF REACTIVE OXYGEN SPECIES

This section is not meant to be a treatise on the subject. Those interested in greatly detailed texts on the formation and chemistry of free radicals and ROS are referred to several of the numerous excellent references on the subject (1–8).

A. Superoxide

Physical dictates require that oxygen be reduced one electron at a time with the intermediate formation of various radical species including superoxide (Fig. 1). In cellular respiration, this is typically done in the cytochrome-*c* oxidase complex that has evolved to be very efficient at preventing the release of the partially reduced oxygen before its complete reduction to water. Other segments of the electron transport chain, however, are somewhat "leaky" and electrons are occasionally lost to oxygen, generating the superoxide radical. Additionally, certain cellular enzymes, such as xanthine oxidase and aldehyde oxidase, release significant amounts of O_2^- during their catalytic cycle. Membrane-associated NADPH oxidases [abundant in macrophages and neutrophils, but also seen in nonphagocytic cells such as fibroblasts (9)] quantitatively convert oxygen to superoxide, typically in response to some extracellular stimuli. In total, it has

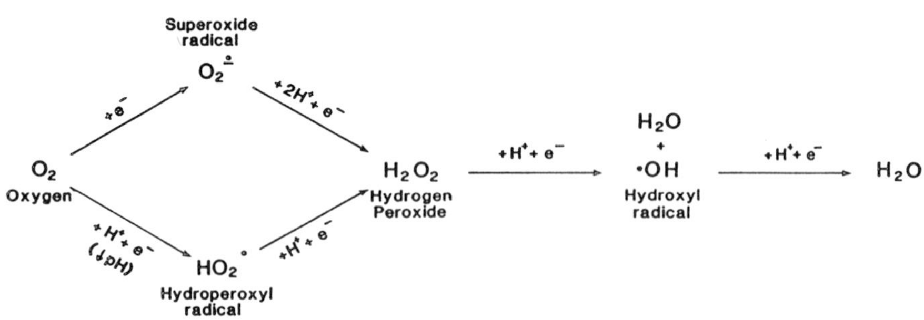

Figure 1 Ultraviolet reduction pathway for molecular oxygen: In several cellular locales, localized pH may be quite acidic, leading to formation of the protonated form of superoxide in spite of the low p*K*a (≈4.8).

been estimated that upward of 1% of all oxygen consumed in aerobic organisms is converted to O_2^-. Considering the enormous amounts of oxygen used per day, this is certainly a significant amount of O_2^-. In addition, there are numerous compounds (xenobiotics) that can increase the net O_2^- production by subverting normal electron flow using various cellular "reductases" and then passing those electrons one at a time to molecular oxygen. These compounds includes quinones and viologens among others (10). They exert their toxicity not only because of the formation of O_2^- itself, but also because they divert electron flow from normal productive pathways (Fig. 2).

Superoxide can act as an oxidant as well as a reductant. This is evident during its dismutation reaction (either spontaneous or enzyme-catalyzed). Thus, one O_2^- acts as a reductant, whereas the other acts as an oxidant (i.e., $O_2^- + O_2^- + 2H^+ \rightarrow H_2O_2 + O_2$).

Superoxide can oxidize many biologically important molecules, such as catecholeamines, polyphenols, sulfite, and ascorbate, to name a few. On the other hand, it can reduce compounds, such as cytochrome-*c*, tetranitromethane, and nitroblue tetrazolium; these latter reactions are used as methods for the assay of superoxide dismutase (SOD) activity (11, and references therein).

Superoxide can exert its toxicity through either pathway; however, the former may be more relevant in vivo. Thus, although O_2^- can serve as an oxidant, its association with cationic centers vastly increases this property. Protonation of O_2^- (to $HO_2^{\cdot}$), which occurs with a pK_a of roughly 4.9, allows the oxidation of polyunsaturated fatty acids, an unfacile reaction of O_2^- alone. Another example is provided by vanadate (V^{+5}). Superoxide does not oxidize NADPH appreciably. In the presence of vanadate, however, the peroxovanadyl species rapidly does; this reaction initiates a free radical chain reaction during which one O_2^- can lead to the oxidation of many molecules of NADPH, depleting cellular stores of these important reducing equivalents (12,13). Finally, superoxide can oxidize a family of cellular enzymes (dehydratases) that contain [4Fe-4S] clusters at their active site presumably causing a degradative release of Fe^{2+} from the enzyme, inactivating it (14,15). Besides inactivation of the enzymes, the release of the iron poses a tremendous threat as we shall see.

B. Hydrogen Peroxide and the Hydroxyl Radical

Hydrogen peroxide is produced directly inside the cell by a variety of oxidases. It is also produced, as shown previously, by the dismutation of superoxide. Because of its small size and lack of a charge, it must be considered potentially very dangerous because of its ability to readily diffuse across cell membranes considerable distances from its site of production; dangerous because it can be univalently reduced to OH^- and $HO^{\cdot}$, the hydroxyl radical. This lat-

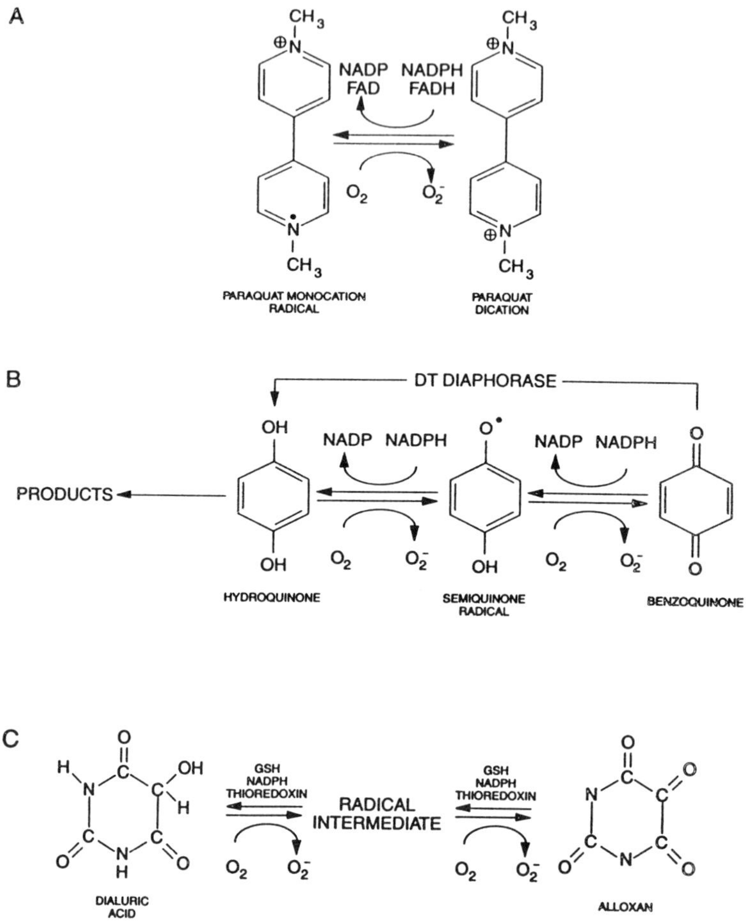

Figure 2 Examples of redox-cycling, superoxide-generating compounds: Nonspecific diaphorases use cofactors (NADPH, FADH, thioredoxin, and others) to subvert normal electron flow to these drugs. In the presence of oxygen, the electrons are transferred back, with the resulting formation of O_2^-. In B, a DT diaphorase directly induces a quinone by two electrons; this compound is then rapidly metabolized, circumventing radical production.

ter species is a highly reactive and a mostly indiscriminant oxidant. The reduction of hydrogen peroxide is largely carried out by redox-active metal cations, such as Cu^{2+} and Fe^{2+}. If these metals are found bound to critical targets (e.g., enzyme active sites, DNA, or cell membranes), then the $OH^{\cdot}$ will be generated adjacent to these critical cellular targets.

Superoxide can collaborate with H_2O_2 in the generation of the hydroxyl radical. This can happen two ways; in vitro, O_2^- can reduce Fe^{3+} or Cu^{2+} to Fe^{2+} and Cu^+, respectively. These, in turn, can than reduce H_2O_2 to OH^- and $HO^\cdot$. The relevant reactions are:

$$Fe(III)/Cu(II) + O_2^- \rightarrow Fe(II)/Cu(I) + O_2$$

$$Fe(II)/Cu(I) + H_2O_2 \rightarrow Fe(III)/Cu(II) + HO^- + HO^\cdot$$

$$\text{Summation:} \quad O_2^- + H_2O_2 \xrightarrow[Cu(II)]{Fe(III)} HO^- + HO^\cdot + O_2$$

Ordinarily, this is an in vitro-relevant reaction; in vivo there is practically no free iron or copper (16). However, O_2^- is capable of oxidatively releasing Fe^{2+} from the [4Fe-4S] centers as well as causing the release of Fe^{2+} from ferritin, thus indirectly leading to hydroxyl radical formation (15).

C. Singlet Oxygen

There is more than enough energy in a photon of near ultraviolet (UV) radiation to elevate molecular oxygen to an excited ''singlet'' state, the biologically relevant species being $^1\Delta g$. Oxygen does not absorb radiation in this region; many other compounds do, however, and their resultant excited states can transfer that excitation energy to O_2. This is the basis for photosensitized production of $^1\Delta g$ in many adverse, light-driven biological processes (17,18). By absorbing energy, singlet oxygen overcomes the physical constraints of reactivity imposed on ground state oxygen and is thus extremely reactive (it will rapidly attack the double bonds in unsaturated lipids, initiating a chain reaction leading to lipid peroxidation).

D. Nitric Oxide

It has been known for years that endothelial cells secrete a short-lived substance that is capable of causing vasorelaxation of vascular smooth muscle (endothelium-derived relaxing factor; EDRF). The fleeting nature of this substance formed a barrier to its identification. Finally, however, EDRF was found to be nitric oxide (NO), which is now also involved in biological responses such as neurotransmission and the cytocidal properties of activated macrophages. Even before it was identified as NO per se, EDRF was seen to be stabilized by superoxide dismutase. As O_2^- and NO are both free radicals, they would be expected to react very rapidly, and they do, at nearly diffusion-limited rates. Thus, O_2^- is capable of eliminating NO (and its vasorelaxation properties) and thus acts as a hypertensive agent. The SOD then would serve as a vasorelaxant itself, and this has been seen (19).

The product of the reaction between O_2^- and NO is potentially a very potent toxin in its own right, peroxynitrite ($ONOO^-$). It is biocidal, capable of nitrating proteins, as well as causing other biological damage in vivo (19,20).

E. Products of the Respiratory Burst

Activation of neutrophils is accompanied by a 15- to 20-fold increase in the rate of oxygen consumption by these cells. Virtually all of this can be accounted for by the formation of O_2^- by the membrane-bound NADPH oxidase (21). Through the dismutation reaction, copious amounts of H_2O_2 are formed, which is then used by neutrophil myeloperoxidase (MPO), together with chloride or bromide, to form the corresponding hypohalous acid.

$$H_2O_2 + Cl^- \underset{MPO}{\rightarrow} OH^- + HOCl.$$

The hypochlorous acid can chlorinate amines to *N*-chloramines; both species being potent oxidants that have the added property of high diffusability. The HOCl can also react with yet another O_2^- providing another pathway for the formation of the hydroxyl radical:

$$O_2^- + HOCl \rightarrow O_2 + Cl^- + HO^{\cdot}$$

III. CONSEQUENCES OF ROS AND FREE RADICAL FORMATION

Suffice to say that the early groundwork proving biological relevance for free radicals showed conclusively that ROS could react with, and compromise, every compartment in a biological system. Perhaps the premier target is the lipid bilayer making up the cell membrane. It is made of significant amounts of unsaturated lipids that react very well with many of the aforementioned ROS. Typically, a hydrogen atom is stripped away from a conjugated double bond in a fatty acid chain of a phospholipid by a free radical, a chemically permissive reaction. This satisfies the free radical's thirst for an electron, but creates another radical in the fatty acid chain. After molecular rearrangements, this lipid free radical removes another hydrogen atom from a neighboring lipid molecule propagating the reaction; that is, a chain reaction begins where one free radical can lead to the disruption of many lipid molecules, compromising the integrity of the entire cell membrane.

Nucleic acids are an obvious choice for potential ROS damage. DNA and RNA both can be efficiently degraded by free radicals in time through the binding of metal ions or through the intercalation of redox-active drugs into the helices. This explains the mutability of many free radical species. The excised, oxidized products of DNA reactions with ROS have been identified and detected in the urine of several mammalian species and in cells exposed to UV light (22).

Finally, proteins, although not perhaps receiving as much attention as the foregoing biomolecules, are certainly adversely affected by reaction with ROS. All the components of connective tissue are altered or degraded by free radicals. Enzymes can certainly be inactivated, as has been alluded to, but perhaps more importantly, certain enzyme inhibitors can be rendered nonfunctional; for example, α1-antiproteinase, which keeps enzymes such as elastase in check, has a critical methionine residue that, when oxidized, causes the loss of function of this molecule (23). This deregulates proteolytic balances, causing unwanted tissue destruction. Oftentimes, proteins subjected to these oxidative results become refractory to normal turnover and begin accumulating both inside and out of cells. Collagen exposed to free radicals or UV radiation becomes increasingly insoluble (24,25).

IV. DEFENSES

We are continually exposed to sources or generators of free radicals, either directly or indirectly derived from molecular oxygen. This would soon spell death for the organism if it were not for a multilayered defensive system that has evolved over the eons.

The superoxide radical is kept at very low concentrations (perhaps 10^{-10} *M*) intracellularly by two superoxide dismutases: a copper and zinc-containing enzyme and a manganese-containing one. Humans also contain an extracellular copper-based SOD, which binds to cell surfaces and collagen by virtue of a heparin-binding domain on the enzyme (26). This helps maintain O_2^- levels low in areas where extracellular O_2^- production would be significant, as they are, for example, at sites of inflammation and presumably areas exposed to UV radiation.

Hydrogen peroxide, a potent oxidant, must be eliminated. There are two enzymatic mechanisms for this. First, catalase, a predominantly peroxisomal enzyme, dismutates two H_2O_2 into H_2O and O_2. Because of its extremely high K_m for H_2O_2 it is often assumed that catalase, although slow-acting at the reasonably low cellular levels of H_2O_2, provides a great reservoir of peroxide-removing potential if or when the cell finds itself facing a high H_2O_2 flux. The second hydrogen peroxide-scavenging system uses glutathione to reduce H_2O_2 to H_2O and O_2 under the catalytic influence of a selenium-containing glutathione peroxidase (GSHPx). Thus:

$$2\ \mathrm{GSH} + H_2O_2 \xrightarrow{\mathrm{GSHPx}} \mathrm{G\text{-}S\text{-}S\text{-}G} + 2H_2O$$

There are actually several locales where GSHPx isoenzymes work. A soluble enzyme removes H_2O_2 or simple alkyl hydroperoxides from the cytoplasm. A second is membrane-associated and is responsible for reducing phospholipid hydroperoxides (PLOOH) to the corresponding alcohol. Similar to SOD, there is also an extracellular form. Again, all these enzymes use a selenocysteine at the active site. The peroxides are destroyed at the expense of glutathione, so there must be a mechanism in place to regenerate this important compound. This is accomplished by a glutathione reductase (GR) that catalyzes the reaction:

$$\text{G-S-S-G} + H^+ + \text{NADPH} \xrightarrow{\text{GR}} + \text{GSH} + \text{NADP}^+$$

Finally, two systems not completely characterized in skin, but that certainly may play a role in enzymatic protection, are thioredoxin reductase and DT diaphorase. Thioredoxin reductase has been reported to remove O_2^- (29) and, interestingly, appears to act at significant levels at the skin surface (not only at the site of highest concentration of oxygen and lipid, but also the recipient of the highest flux of UV radiation). It may well serve as an important antioxidant defense against environmentally generated active oxygen species.

Quinones are reduced in cells by various flavoproteins to the semiquinone radical and then to the hydroquinone (see Fig. 2). Whereas either of these reduced species can react with oxygen to generate superoxide, the hydroquinone is rapidly metabolized. DT diaphorases, in reducing the quinone by two electrons directly to the metabolized hydroquinone, bypasses the certain superoxide generating semiquinone radical (20). As skin is frequently exposed to pollutants such as quinones, this may have substantial benefit to the tissue.

Low molecular weight antioxidants, which serve to interrupt free radical chain reactions, serve as another level of defense. As we have seen before, lipid peroxidation (leading to the destruction of all kinds of membranes) is a free radical-initiated chain reaction. Vitamin E (α-tocopherol) serves such a chain-breaking role in the cell membranes by intercepting reactive, lipid-derived radical intermediates before they can propagate the damage (29). Thus, a lipid radical once formed rapidly reacts with oxygen, generating a lipid hydroperoxyl radical:

$$\underset{\text{(lipid radical)}}{\text{LH}^{\cdot}} + O_2 \rightarrow \underset{\text{(lipid hydroperoxyl radical)}}{\text{LHOO}^{\cdot}}$$

This species can abstract another hydrogen atom from a neighboring phospholipid, propagating the chain.

$$\underset{\text{(lipid peroxyl radical)}}{\text{LHOO}^{\cdot}} + \underset{\text{(phospholipid)}}{\text{LH}_2} \rightarrow \underset{\text{(lipid hydroperoxide)}}{\text{LHOOH}} + \underset{\text{(lipid radical)}}{\text{LH}^{\cdot}}$$

And on it goes. Vitamin E, however, can react with the $\text{LHOO}^{\cdot}$ before it can react with another lipid molecule, breaking the chain:

$$\text{LHOO}^{\cdot} + \underset{(\alpha\text{-tocopherol})}{\text{TOC-H}_2} \rightarrow \text{LHOOH} + \underset{\text{(tocopherol radical)}}{\text{TOC-H}^{\cdot}}$$

The tocopherol radical is not sufficiently reactive to propagate the chain reaction on its own. Vitamin E is very good at what it does; so much so that, relatively speaking, there are only a few vitamin E molecules for every thousand lipid molecules in membranes. To act catalytically (which is must), there must be a mechanism to regenerate the parent vitamin E molecule. The cells have several:

Vitamin C (L-ascorbic acid) is a prominent water-soluble antioxidant in its own right, capable of quenching reactive oxygen species such as singlet oxygen and superoxide (30). In plasma, it is the predominant inhibitor of low-density lipoprotein (LDL) oxidation (31), so that, on a whole, vitamin C is valuable in preventing free radical damage. Its other main function is to reduce the tocopherol radical back to α-tocopherol, thereby sparing the vitamin E and allowing it to inhibit yet another chain reaction in cell membranes (29). Vitamin C is much more plentiful than vitamin E and so can be "sacrificed" here. In some organisms, the vitamin C can, in turn, be spared by glutathione, either directly or at the expense of NADPH, through a semidehydroascorbate reductase. Several other cellular compounds are effective at regenerating vitamin E in lipid membranes. Ubiquinone (which also resides in cell membranes) and lipoic acid have the capability (32,33); it remains to be seen how prevalent the reactions are in vivo.

Finally, a defense against free radical damage with certain relevance to skin and UV exposure is heme oxygenase. This enzyme is responsible for catabolism of heme to biliverdin, freeing iron, which presumably acts as an inducer for ferritin (a redox-inactive iron-transport protein). Ultraviolet A-irradiated fibroblasts markedly up-regulate heme oxygenase (both protein and mRNA), which subsequently increases levels of ferritin and leads to protection against a subsequent UV exposure (34,35). This provides clear evidence of a UV-induced oxidative stress (mediated through "free" iron) and a new understanding of the cell's adaptive response.

V. ULTRAVIOLET LIGHT-GENERATED FREE RADICALS IN SKIN

Harkening back to work done 30 years ago, irradiation of intact skin generates a free radical signal measurable by electron spin resonance (ESR) spectroscopy (36). This first observation is due to the generation of a stabilized melanin radical. Subsequent work has confirmed these results. The use of spin traps has allowed a somewhat more definitive characterization of the UV-elicited free radical species. Thus, UV irradiation of skin extracts or whole skin leads to the generation of the ascorbate radical (consistent with its antioxidant function) and carbon-centered radicals, thought to be derived from lipid alkoxyl radicals (37). The UV-induced increase in the ESR signal from epidermis was oxygen-dependent and SOD inhibitable (38). It thus appears quite irrefutable that UV exposure of skin leads to the generation of a host of free radical species by as yet uncharacterized mechanisms.

A. Effect of Ultraviolet Radiation on Skin Antioxidant Defenses

From the preceding section, it is seen that UV irradiation of the skin imparts an oxidative stress to this tissue. A typical organismal response to this kind of insult is the up-regulation of defensive systems. Paradoxically, and unfortunately for the skin, a reproducible finding is that there is a marked inhibition of antioxidant defenses in response to brief UV radiation. Thus, in vitro irradiation of murine skin leads to a significant inhibition of superoxide dismutase and catalase activities (39–42). A less reproducible loss of glutathione peroxidase has also been seen, however. More importantly, activity of these two antioxidant enzymes does not return to normal for hours and even days after irradiation. Nonenzyme antioxidants are not spared. Vitamin E and ubiquinols (lipophilic antioxidants) are dramatically reduced after UVB irradiation (39,41). The hydrophilic antioxidant ascorbic acid is also depleted in both murine (41) and porcine skin (43) after UV exposure. Glutathione (reduced form) levels are diminished. Studies on isolated cells show that the antioxidant capacity of both keratinocytes and fibroblasts are adversely affected following UV irradiation; hence, both epidermis and dermis are at risk. A recent study reconfirms the initial UV effects on antioxidant enzymes; a slight induction was noted early after prolonged exposure, but this was lost with time (44). Thus, skin can respond positively to UV, but it is easily overwhelmed.

Although certain of the foregoing experiments were run with solar-simulated UV light, the preponderance of observed changes was brought about by UVB. However, several investigators have assessed the effects of UVA on the antioxidant defenses in cultured skin cells. Again, catalase is considerably inhibited after UVA irradiation (45–47). SOD and $GSHP_x$ were not inhibited by UVA alone; however, UVA significantly augmented UVB-induced inhibition of SOD, catalase, and $GSHP_x$, suggesting an important biological role for these wavelengths (46).

In total, there is considerable evidence that UV irradiation of skin leads to a significantly compromised antioxidant defense system. Catalase, by far, seems to be the most sensitive (to both UVB and UVA wavelengths). Presumably, some of this loss can be attributed to a direct photoinactivation (48). Detailed studies on this phenomenon as it pertains to SOD and $GSHP_x$ are unavailable. Another possibility is the inactivation of certain antioxidant defensive enzymes by substrates for the others. Thus, H_2O_2 inactivates both forms of cellular SOD (Cu/Zn and Mn-containing), whereas O_2^- reciprocates and inactivates both catalase and glutathione peroxidase (49 and references therein). It will be important to ascertain mechanisms of this UV-mediated enzyme loss and its consequences to be able to therapeutically intervene.

The depletion of nonenzymatic antioxidants bespeaks an ongoing attempt on the part of the skin to combat the UV-mediated oxidative stress. Indeed this is a reasonable marker for the existence of just such a stress. As lipid membranes are a primary target for radical damage and vitamin E is the most important inhibitor of this damage, it would seem reasonable that levels of this vitamin would be maintained preferentially at the expense of more sacrificial antioxidants. This, indeed, has been shown. Thus, irradiation of skin homogenates generates a clear ascorbate radical ESR signal, which is to be expected as, again, one of ascorbate's known roles is the reduction of the tocopherol radical to regenerate vitamin E. The ascorbyl radical signal persists until depletion of the vitamin C is complete, at which time the ESR signal characteristic of the vitamin E radical appears (50). Besides vitamin C, several other molecules thought to be important in regenerating vitamin E are UV-depleted, sparing vitamin E. Ubiquinols completely disappear during UV irradiation, helping recycle vitamin E. Again, glutathione, although presumably not directly reacting with vitamin E radical, can recycle vitamin C, extending the series of sacrificial antioxidants. Glutathione depletion has been noted in UV-irradiated skin (39).

B. Effects of Antioxidants on Ultraviolet Damage to Skin

From the foregoing, it would appear natural that antioxidants should be photoprotective; this is true. Over 10 years ago, Pauling showed that diets high in ascorbic acid were beneficial in reducing UVB-induced neoplasms in mice (51), confirming and extending earlier work showing diets high in a mixture of antioxidants reduced UV-induced carcinogenesis (52). Feeding mice high levels of vitamin E alone increases skin levels of the vitamin and inhibits a solar-simulated increase on lipid hydroperoxides (53).

Since that time, effort has been expended studying whether antioxidants applied topically would significantly enhance the aforenoted skin protection. By far the most extensively studied is topical vitamin E (natural or as the acetate derivative). It is clear that this vitamin is capable of protecting skin from UV damage. Thus, more than a decade ago, it was shown that topical application of vitamin E (but not the acetate derivative) modestly prevented UVB erythema, even if applied several minutes after the UV exposure (54). With rabbits, the same group noted that vitamin E (again not the acetate) significantly protected against a phototoxic (PUVA) insult. In this instance, the antioxidant had to be in place during irradiation to be effective (55). These results have been confirmed. There are reports of the biological usefulness of vitamin E acetate in photobiological experiments (56). This suggests that, to be effective, the acetate form of vitamin E must be metabolically converted; recent work confirms that this conversion does occur in skin (57), but that it is limited and may take days to occur (56,58). Tocopherols have an absorption spectrum (Abs_{max} ≈297 nm) partially in the biologically relevant UVB range; thus, they are capable of acting as sunscreens. In experiments, in which the acetate form is inactive, this is controlled for, and one may ascribe protection to the antioxidant properties of vitamin E.

Vitamin C (L-ascorbic acid) is one of the most effective antioxidants in vivo, and should be a logical candidate for topical photoprotection. It is, however, an extremely unstable molecule (known to formulation chemists for decades); at or near neutral pH, it begins to degrade in minutes and must be used fresh. Acidic pH helps stabilize vitamin C significantly and increases percutaneous absorption without any undue skin irritation (59). Few studies, therefore, have been carried out assessing vitamin C's effectiveness as a topical photoprotectant. Two that have, have shown that ascorbic acid (5–10% w/v) does moderately protect against UVB damage, measured as erythema, sunburn cell formation, and tumor formation (43,60). One study assessed vitamin C protection against UVA (wrinkling in mouse skin) and found no significant effect (60). The other looked at UVA-mediated phototoxic response (PUVA) in pig skin and found a significant level of protection (43). This may point to differences between the UVA and PUVA models or, because the vitamin C formulations were at different pHs (≈7 for UVA experiment vs. ≈2.5 for the PUVA experiments), differences in the percutaneous absorption of the antioxidant. Thus, at neutral pH, ascorbic acid would be virtually 100% ionized and hindered from entering the skin.

Similar to vitamin E, stabilized derivatives of ascorbic acid have been incorporated into topical preparations, and as with vitamin E, these must be converted metabolically to vitamin C proper to be effective. This conversion apparently can occur at least with the derivatives, ascorbic acid-2-phosphate (61), ester-C (62), and with ascorbic acid-2-glucoside (63). Although an early report of UV photoprotection exists for ascorbic acid phosphate, detailed studies of these compounds must still be carried out.

Ascorbic acid palmitate is derivatized on the number 6 carbon (numbers 2 and 3 being important for redox chemistry and the molecule's attendant instability) imparting great lipophilicity to the molecule. This compound, too, has been reported to be moderately protective against UVB sunburn, even if applied after the insult, illustrating an anti-inflammatory property (64).

Vitamin C interacts with vitamin E in vitro and in vivo as discussed before. It seems natural then to assess what effectiveness a combination formulation would have on preventing ultraviolet-induced skin damage. In preliminary experiments, although vitamin E was superior to vitamin C in preventing UVB damage, the opposite was true in a PUVA model. In both cases, the combination of the two antioxidants fared better than either alone (65; and unpublished observations). To date, these are the only experiments directly assessing the combination of vitamin C and vitamin E in photobiology experiments and suggests that combinations of these antioxidants may provide broader coverage (i.e., against both UVB and UVA-mediated phototoxic reactions).

Besides vitamins C and E, another natural antioxidant that has been tried as a photoprotectant is β-carotene. Chemically, this compound is an efficient quencher of singlet oxygen (generated in many photochemical reactions), and indeed this vitamin has proved capable of inhibiting UV-induced tumor formation (66) and photosensitization reactions, even if given orally. Topically, it inhibits UVB-mediated lipid damage (67) and PUVA-induced erythema (68) in animal models. Other antioxidant compounds have been investigated as topical photoprotectants, but certainly less vigorously. Quite interesting are several reports of the biological effectiveness of the application of the enzyme SOD in preventing UV damage. This dimeric protein has a molecular weight of 32,000, making it highly unlikely to penetrate even the most superficial layers of the stratum corneum, yet it was capable of inhibiting PUVA-induced erythema (68), strongly suggesting that the signaling pathway for this endpoint is very much at the skin surface. SOD has also been encapsulated into liposomes, where it inhibits UV-induced lipid peroxidation (69) and prevents the UV-mediated loss in intrinsic SOD activity (70). Penetration studies would be of interest to assess if or where the SOD went into the skin. Dihydrolipoate, a dithiol compound involved in the reduction of thioredoxin, has antioxidant properties, presumably by its ability also to regenerate vitamin E, vitamin C, and glutathione. It provides local photoprotection to the hairless mouse when applied before or immediately after UVB exposure (71). Given the selenium requirement of the hydroperoxide-eliminating glutathione peroxidase, selenomethionine has also been used and is also topically effective at inhibiting UV damage to mouse skin (72). Even more esoteric, extracts of green tea (known to be naturally high in polyphenoic catechol-like compounds with antioxidant properties) has been tested on animals and prevents UV tumorigenesis in mice (73, and references therein). Finally, all the antioxidants heretofore mentioned are "natural." Synthetic antioxidants have also been tested for UV-protective effects. Butylated hydroxytoluene (BHT) proved more efficacious than vitamin E, C, or glutathione when given orally in preventing short-wavelength UV damage in rats; however, this may have had more to do with the compound's ability to alter the physical properties of the skin, thereby inhibiting penetration of UV wavelengths (74).

From the foregoing, it can be appreciated that the overwhelming evidence suggests that antioxidants are effective at inhibiting UV damage to skin. Why, then, are there not more antioxidant skin care products on the market? Formulating difficulties have been alluded to. Cost is a factor, as the concentrations of these compounds necessary for biological effectiveness is high (anywhere from 1 to 10% w/v). Finally, quite simply, antioxidants have not proved to be as good (particularly against UVB radiation) as a decent sunscreen in preventing skin damage. In the UVB-wavelength range, this is to be expected. The major chromophore for these wavelengths in cells is DNA; it absorbs the energy from the UV and can be damaged, without the need for free radical intermediacy. As a result, it could be suggested that most of damage seen in UVB-irradiated skin is ROS-independent; thus, antioxidants could not be effective. UVA, no longer thought innocuous, appears to have a larger oxygen dependence for its effects on skin. However, besides PUVA-induced phototoxicity, model systems for quantitatively measuring said damage are either lacking or are not particularly user-friendly. There are reports showing a combination antioxidant formulation can reduce UVA-mediated erythema and pigment changes (75). We have observed inhibition of UVA-induced pigment changes in human skin with topical vitamin C (76). The biological importance is not altogether clear, but further research is clearly warranted.

Do antioxidants have a place in photoprotection products? It seems clear they do, but presumably in combination with available sunscreens. It is unclear what the biological consequences of life-time use of "chemical" sunscreens will be. Use of safe, natural antioxidants will permit products using less sunscreen to have similar photoprotective properties compared with high-level sunscreen products alone. In the UVB range antioxidants provide additive protection with sunscreens, bolstering this concept. In the UVA range, antioxidants may provide even greater protection in a combination product. Thus, in swine treated in a UVA-phototoxic model, sunscreen plus antioxidants (in this case vitamins C and

Table 1 Increased Protection from PUVA-Induced Phototoxicity by a Combination of Antioxidant Vitamins and Sunscreen

Experimental condition	No. of "sunburn cells" per 4-mm biopsy
Vehicle-treated (4)	TMTC
Vitamin C/vitamin E-treated (4)	59.5 ± 16.7
Oxybenzone-treated (4)	$31.2 \pm 11.9^{++}$
VC/VE ± oxybenzone-treated (4)	$4.3 \pm 1.1^{*,+}$

Experimental sites (10 cm^2) were treated for 3 days with ≈ 0.1 ml of the antioxidant, antioxidant/sunscreen, or vehicle in a cream base. Sixty minutes before UVA exposure, 0.1 ml of a 0.1% (w/v in ethanol) 8-MOP solution was applied to all sites; 30 min later the last treatment application was made. Triplicate 4-mm biopsies were taken from all sites 48 h after UVA exposure, processed for histology, H&E stained and sunburn cells enumerated, averaged, and normalized to the 4-mm biopsy diameter. Vitamin C concentration: 10% (w/v); vitamin E: 2% (w/v); oxybenzone: 0.25% (w/v). TMTC: Too many to count (> 100 SBCs, epidermal necrosis); $^{++}$: not significantly different from VC/VE treated ($p > 0.1$); *; $p < 0.05$ compared with oxybenzone treated; + $p < 0.01$ compared with VC/VE treated. Numbers in parenthesis indicate animal number. Values given are the mean ± standard error of averaged values for the four animals.

E) provided more than additive protection when combined with a UVA/B sunscreen, oxybenzone (Table 1). There is a growing perception that non-UVB radiation may play a role in processes ranging from photoaging of the skin to melanoma formation, and that strong UVB filters are either unable to prevent the changes or may be part of the problem themselves (77–79). Antioxidants may provide important protection here.

In conclusion, free radicals are clearly generated by exposure of the skin to UV radiation, and this is not good. Antioxidants are, and presumably will continue to be, effective adjuncts to skin photoprotective products. Insofar as photodamage to skin can be used as a model for inflammation, in general, antioxidant-containing anti-inflammatory products are a natural extension and would be predicted to have great therapeutic potential.

REFERENCES

1. Rice-Evans, C. A. and Burdon, R. H. (eds.), *Free Radical Damage and Its Control*. Elsevier Press, New York, 1994.
2. Packer, L. and Fuchs, J. (eds.). *Oxidative Stress in Dermatology*, Marcel Dekker, New York, 1993.
3. Fuchs, J. (ed.). *Oxidative Injury in Dermatopathology*, Springer-Verlag, Berlin, 1992.
4. Scandalios, J. G. (ed.). *Molecular Biology of Free Radical Scavenging, Systems*, Cold Spring Harbor Laboratory Press, Cold Spring Harbor, NY, 1992.
5. Sies, H. (ed.). *Oxidative Stress: Oxidants and Anti-oxidants*, Academic Press, London, 1991.
6. Halliwell, B. and Gutteridge J. M. C. (eds.). *Free Radicals in Biology and Medicine*, 2nd ed., Clarendon Press, London, 1989.
7. Hayaishi, O., Imamura, S., and Miyachi, Y. (eds.). *The Biological Role of Reactive Oxygen Species in Skin*, Elsevier Press, New York, 1987.
8. Darr, D. The biology of oxygen free radicals and their relevance to skin. In *Cutaneous Aging*, A. Kligman and Y. Takase, eds., University of Tokyo Press, 1988, pp. 415–433.
9. Meier, B., Cross, A. R., Hancock, J. T., Kaup, F. J., and Jones O. T. Identification of a superoxide-generating NADPH oxidase system in human fibroblasts, *Biochem. J.*, *275* (pt. 1):241–245 (1991).
10. Hassan, H. and Fridovich, I. Intracellular production of superoxide radical and of hydrogen peroxide by redox active compounds, *Arch. Biochem. Biophys.*, *196*: 385–395 (1979).
11. Fridovich, I. Superoxide radical: An endogenous toxicant, *Annu. Rev. Pharmacol. Toxicol.*, *23*:239–257 (1983).
12. Darr, D. and Fridovic, I. Vanadate and molybdate stimulate the oxidation of NADH by superoxide radical, *Arch. Biochem. Biophys.*, *232*:562–565 (1984).
13. Liochev, S. and Fridovich, I. Vanadate stimulated oxidation of NAD(P)H in the presence of biological membranes and other sources of O_2^-, *Arch. Biochem. Biophys.*, *279*:1–7 (1990).

14. Gardner, P. and Fridovich, I. Superoxide sensitivity of the *E. coli* 6-phosphogluconate dehydratase, *J. Biol. Chem.*, *266*:1478–1483 (1991).
15. Liochev, S. I. and Fridovich, I. The role of O_2^- in the production of HO: In vitro and in vivo, *Free Radic. Biol. Med.*, *16*:29–33 (1994).
16. Gutteridge, J. M. C. and Halliwell, B. Radical promoting loosely-bound iron in biological fluids and the bleomycin assay, *Life Chem. Rep.*, *4*:113–119 (1987).
17. Morliere, P. Drug-induced photosensitivity: Phototoxic and photoallergic reaction —a few molecular aspects, *Biochemie*, *68*:849–855 (1986).
18. Laustriat, G. Molecular mechanisms of photosensitization, *Biochemie*, *68*:771–778 (1986).
19. Beckman, J. S., Beckman, T. W., Chen, J., Marshall, P. A., and Freeman, B. Apparent hydroxyl radical production by peroxynitrite: Implications for endothelial injury from nitric oxide and superoxide, *Proc. Natl. Acad. Sci. USA*, *87*:1620–1624 (1990).
20. Nakazano, K., Watanabe, N., Matsuno, K., Sasaki, J., Sato, T., and Inoue M. Does superoxide underlie the pathogenesis of hypertension? *Proc. Natl. Acad. Sci. USA*, *88*:10045–10048 (1991).
21. Loschen, G. Superoxide formation accounts for almost all extra oxygen consumption upon stimulation of leukocytes by PMA. In *Superoxide and Superoxide Dismutase in Chemistry, Biology, and Medicine*, G. Rotilio, ed., Elsevier Press, New York, 1986, pp. 373–375.
22. Beehler, B. C., Przybyszewski, J., Box, H. B., and Kulesz-Martin, M. F. Formation of 8-hydroxydeoxyguanosine within DNA of mouse keratinocytes exposed in culture to UVB and H_2O_2, *Carcinogenesis*, *13*:2003–2007 (1992).
23. Carp, H., Miller, F., Hoidal, J. R., and Janoff, A. Potential mechanism of emphysema: α_1 Proteinase inhibitor recovered from lungs of cigarette smokers contains oxidized methionine and has decreased elastase inhibitory activity, *Proc. Natl. Acad. Sci. USA*, *79*:2041–2045 (1987).
24. Carbonate, M. D. and Pathak, M. Skin photosensitizing agents and the role of reactive oxygen species in photoaging, *J. Photochem. Photobiol.*, *14*:105–124 (1992).
25. Kligman, L. H. and Gebre, M. Biochemical changes in hairless mouse skin collagen after chronic exposure to ultraviolet-A radiation, *Photochem. Photobiol.*, *54*:233–238 (1991).
26. Hjalmarsson, K., Marklund, S., Engstrom, A., and Edlund, T. Isolation and sequence of cDNA coding human extracellular superoxide dismutase, *Proc. Natl. Acad. Sci. USA*, *84*:6340–6344 (1987).
27. Schallreuter, K. U. and Wood, J. M. Free radical reduction in the human epidermis, *Free Radic. Biol. Med.*, *6*:519–532 (1989).
28. Lind, C., Hochstem, P., and Ernster, L. DT-diaphorase as a quinone reductase: A cellular control device against semiquinone and superoxide formation, *Arch. Biochem. Biophys.*, *216*:33–40 (1982).
29. McCay, P. Vitamin E: Interaction with free radicals and ascorbate. *Annu. Rev. Nutr.*, *5*:323–340 (1985).
30. Niki, E. Action of ascorbic acid as a scavenger of active and stable oxygen radicals. *Am. J. Clin. Nutr.*, *54*:1119S–1124S (1991).

31. Frei, B., England, L., and Ames, B. Ascorbate in an outstanding antioxidant in human blood plasma, *Proc. Natl. Acad. Sci. USA*, *86*:6377–6381 (1989).
32. Kagen, V. E., Serbinova, E. A., and Packer, L. Anti-oxidant effects of ubiquinones in microsomes and mitochondria are mediated by tocopherol recycling, *Biochem. Biophys. Res. Commun.*, *169*:851–857 (1990).
33. Kagen, V. E., Khans, S., Swanson, C., Shvedova, A., Serbinova, E., and Packer, L. Anti-oxidant action of thioctic and dihydrolipoic acid (abstr.), *Free Radic. Biol. Med.*, *9*(suppl):15 (1990).
34. Vile, G. F. and Tyrrell, R. M. Oxidative stress resulting from ultraviolet A irradiation of human skin fibroblasts leads to a heme oxygenase-dependent increase in ferritin, *J. Biol. Chem.*, *268*:14678–14681 (1993).
35. Vile, G. F., Basu-Modak, S., Waltner, C., and Tyrell, R. M. Heme oxygenase 1 mediates an adaptive response to oxidative stress in human skin fibroblasts, *Proc. Natl. Acad. Sci. USA*, *91*:2607–2610 (1994).
36. Pathak, M. A. and Stratton, K. Free radicals in human skin before and after exposure to light, *Arch. Biochem. Biophys.*, *123*:468–476 (1968).
37. Jurkiewicz, B. A. and Buettner, G. R. UV light induced free radical formation in skin: An EPR study (abstr.), *Free Radic. Biol. Med.*, *15*:528 (1994).
38. Ogura, R., Sugiyama, M., Nishi, J., and Haramaki, N. Mechanisms of lipid radical formation following exposure of epidermal homogenate to ultraviolet light. *J. Invest. Dermatol.*, *94*:1044–1047 (1991).
39. Fuchs, J., Huflejt, M. E., Rothfuss, L. M., Wilson, D. S., Carcamo, G., and Packer, L. Impairment of enzymic and nonenzymic antioxidants in skin by UVB irradiation, *J. Invest. Dermatol.*, *93*:769–773 (1989).
40. Pence, B. C. and Naylor, M. E. Effects of single dose ultraviolet radiation in skin superoxide dismutase, catalase and xanthine oxidase in hairless mice, *J. Invest. Dermatol.*, *95*:213–216 (1990).
41. Shindo, Y., Witt, E., and Packer, L. Anti-oxidant defense mechanisms on murine epidermis and dermis and their responses to ultraviolet light, *J. Invest. Dermatol.* *100*:260–265 (1993).
42. Miyachi, Y., Imamura, S., and Niwa, Y. Decreased skin superoxide dismutase activity by a single exposure of ultraviolet radiation is reduced by liposomal superoxide dismutase pretreatment, *J. Invest. Dermatol.*, *89*:111–112 (1987).
43. Darr, D., Combs, S., Dunston, S., Manning, T., and Pinnell, S. Topical vitamin C protects swine skin from ultra-violet radiation-induced damage, *Br. J. Dermatol.*, *127*:247–253 (1992).
44. Iizawa, O., Kato, T., Tagami, H., Akamatsu, J., and Niwa, Y. Long term follow-up study of changes in lipid peroxide levels and the activity of superoxide dismutase catalase and glutathione peroxidase in mouse skin after acute and chronic UV irradiation, *Arch. Dermatol. Res.*, *286*:47–52 (1994).
45. Fuchs, J., Huflejt, M. E., Rothfuss, L. M., Wilson, D. S., Carcamo, G., and Packer, L. Acute effects of near ultraviolet and visible light on the cutaneous antioxidant defense system, *Photochem. Photobiol.*, *50*:729–744 (1989).
46. Punnonen, K., Puntala, A., and Ahotupa, M. Effects of ultraviolet A and B irradiation on lipid peroxidation and activity of the anti-oxidant enzymes in keratinocytes in culture, *Photodermatol. Photoimmunol. Photomed.*, *8*:3–6 (1991).

47. Moysan, A., Marquis, I., Gaboriau, F., Santus, R., Dubertret, L., and Morliene, P. Ultraviolet A-induced lipid peroxidation and anti-oxidant defense systems in cultured human skin fibroblasts, *J. Invest. Dermatol.*, *100*:692–698 (1993).
48. Aronoff, S. Catalase: Kinetics of photooxidation. *Science*, *150*:72–73 (1965).
49. Fridovich, I. Biologic effects of the superoxide radical, *Arch. Biochem. Biophys.*, *246*:245–256 (1986).
50. Kagan, V., Witt, E., Goldman, R., Scita, G., and Packer, L. Ultraviolet radiation induced generation of vitamin E radicals and their recycling. A possible photosensitizing effect of vitamin E in skin, *Free Radic. Res. Commun.*, *16*:51–64 (1992).
51. Dunham, W. B., Zuckerhandl, W., Reynolds, R., Willoughby, R., Marcuson, R., Barth, R., and Pauling, L. Effect of intake of L-ascorbic acid on the incidence of dermal neoplasms induced in mice by ultraviolet light, *Proc. Natl. Acad. Sci. USA*, *79*:7532–7536 (1982).
52. Black, H. S. and Chan, J. T. Suppression of ultraviolet light-induced tumor formation by dietary anti-oxidants. *J. Invest. Dermatol.*, *65*:412–414 (1975).
53. Packer, L. Ultraviolet radiation (UVA, UVB) and skin antioxidants. *Free Radical Damage and Its Control*, C. A. Rice-Evans and R. H. Burdon, eds., Elsevier, New York, 1994, pp. 239–255.
54. Roshchupkin, D. I., Pistov, M. Y., and Potapenko, A. Y. Inhibition of ultraviolet light-induced erythema by antioxidants, *Arch. Dermatol. Res.*, *266*:91–94 (1979).
55. Potapenko, A. Y., Abiev, G. A., and Piquett, F. Inhibition of erythema of skin photosensitized with 8-methoxypsoralen by α tocopherol, *Bull. Exp. Biol. Med.*, *89*: 611–615 (1980).
56. Trevithick, J. R., Xiong, H., Lee, S., Shum, D. T., Sanford, S. E., Karlick, S. J., Norley, C., and Dilworth, G. R. Topical tocopherol acetate reduces post-UVB sunburn associated erythema, edema and skin sensitivity in hairless mice, *Arch. Biochem. Biophys.*, *296*:575–582 (1992).
57. Norkus, E. P., Bryce, G. F., and Bhagavan, H. N. Uptake and bioconversion of α-tocopherol acetate to α-tocopherol in skin of hairless mice, *Photochem. Photobiol.*, *57*:913–915 (1993).
58. Beijersbergen van Henegouwen, G. M. J., de Vries, H., van den Brocke, L. T., and Tunginger, H. E. *R,R,R*-tocopherols and their acetates as a possible scavenger of free radicals produced in the skin upon UVA exposure—an in vivo screening method, *Fat Sci. Techol.*, *94*:24–27 (1992).
59. Darr, D. and Pinnell, S. R. Stabilized ascorbic acid compositions. US Patent 5,140,043 (1992).
60. Bissett, D., Chaterjee, R., and Hannon, D. Photoprotective effect of superoxide scavenging antioxidants against ultraviolet radiation-induced chronic skin damage in the hairless mouse, *Photoderm. Photoimmunol. Photomed.*, *7*:58–62 (1990).
61. Imai, Y., Usui, T., Matsuzaki, T., Yokotani, H., Mima, H., and Aramaki, Y, *Jpn. J. Pharmacol.*, 17:317–324 (1967).
62. Verlangieri, A. J., Fay, M. J., and Bannon, A. W. Comparison of the anti-scorbutic activity of L-ascorbic acid and ester-C in the non-ascorbic synthesizing osteogenic disorder Shinogi (ODS) rat, *Life Sci.*, *48*:9–12 (1991).

63. Wakamiya, H., Suzuki, E., Yamamoto, I., Akiba, M., Otsuka, M., and Arakawa, N., Vitamin C activity of 2-*O*-alpha-D-glucopyranosyl-L-ascorbic acid in guinea pigs. *J. Nutr. Sci. Vitaminol.* (Tokyo), *38*:235–345 (1992).
64. Perricone, N. V. The photoprotective and anti-inflammatory effects of topical ascorbyl palmitate, *J. Geriatr. Dermatol.*, *1*:5–10 (1993).
65. Darr, D., Dunston, S., Kamino, H., Foust, H., and Pinnell, S. R. Effectiveness of a combination of vitamin C and E in inhibiting UV damage to porcine skin (abstr.), *J. Invest. Dermatol.*, *100*:597 (1993).
66. Epstein, J. H. Effects of β-carotene on ultraviolet induced cancer formation in the hairless mouse skin, *Photochem. Photobiol.*, *25*:211–2136 (1977).
67. Pugliese, P. T. and Lampley, C. B. Biochemical assessment of the anti-aging effects of cosmetic products, *J. Appl. Cosmet.*, *3*:129–138 (1985).
68. Carroro, C. and Pathak, M. A. Studies on the nature of in vitro and in vivo photosensitization reaction by psoralen and porphyrins, *J. Invest. Dermatol.*, *90*:267–275 (1988).
69. Ogura, R., Sugiyama, M., Sakanashi, T., and Hidaka, T. Role of oxygen in lipid peroxide of skin exposed to ultraviolet light. *The Biological Role of Reactive Oxygen Species in Skin*, O. Hayaishi, S. Imamura, and Y. Miyachi, eds., Elsevier, New York, 1987, pp. 55–63.
70. Hamanaka, H., Miyachi, Y., and Imamura, S. Photoprotective effect of typically applied superoxide dismutase on sunburn reaction in comparison with sunscreen, *J. Dermatol.*, *17*:595–598 (1990).
71. Fuchs, J., Milbrandt, R., and Zimmer, G. Dihydrolipoate inhibits reactive oxygen species mediated skin inflammation (abstr), *J. Invest. Dermatol.*, *94*:526 (1990).
72. Burke, K. E., Combs, G. F., Gross, E. G., Bhuyan, K. C., and Abu-Libdeh, H. The effects of topical and oral L-selenomethionine on pigmentation and skin cancer induced by ultraviolet irradiation, *Nutr. Cancer*, *17*:123–137 (1992).
73. Mukhtar, H., Katiyar, S. K., and Agarwal, R. Green tea and skin-anticarcinogenesis effects, *J. Invest. Dermatol.*, *102*:3–7 (1994).
74. Black, H. S., Lenger, W., Gerguis, J., and McCann, V. Studies on the photoprotective mechanism of butylated hydroxytoluene, *Photochem. Photobiol.*, *40*:69–75 (1984).
75. Muizzuddin, N., Marenus, K., and Maes, D. Assessment of UV-A protection provided by sunscreen ingredients and products. Poster presented at American Academy of Dermatology, Dallas, Texas, 1992.
76. Murray, J., Darr, D., Reich, J., and Pinnell, S. P. Photoprotection of human skin by topical vitamin C (abstr.), *Clin. Res.*, *40*: 143 (1992).
77. Wolf, P., Donawho, C. K., and Kripke, M. L. Effects of sunscreens on UV radiation-induced enhancement of melanoma growth in mice, *JNCI*, *86*:99–105 (1994).
78. Setlow, R., Grist, E., Thompson, K., and Woodhead, A. D. Wavelengths effective in induction of malignant melanoma, *Proc. Natl. Acad. Sci. USA*, *90*:666–6670 (1993).
79. Knowland, J., McKenzie, E. A., McHugh, P. J., and Cridland, N. A. Sunlight induced mutagenicity of a common sunscreen ingredient, *FEBS Lett.*, *324*:309–313 (1993).

9

Dosimetry of Ultraviolet Radiation

Brian L. Diffey
Dryburn Hospital, Durham, England

I. INTRODUCTION

Dosimetry is the science of radiation measurement. There are two principal reasons why ultraviolet radiation (UVR) should be measured (1):

1. To allow consistent radiation exposure of subjects over many months and years within a local laboratory
2. To allow the results of irradiations made in different laboratories to be published and compared

It is important to distinguish between these two objectives. The first requires *precision*, or reproducibility. The radiometer is used as a monitor to give a reference measurement, and so it needs to be stable. *Accuracy*, that is, absolute calibration against some accepted standard, is not essential. The second objective requires both precision and accuracy. Here, the radiometer must not only be stable from one day to the next, but also the display (in, say, milliwatts per square centimeter) must be traceable to absolute standards. Although electro-optical technology has improved over the years, resulting in the availability of versatile and precise ultraviolet radiometric equipment, these improvements have not always been accompanied by improved accuracy.

II. RADIOMETRIC TERMS AND UNITS

In photobiological UVR dosimetry it is customary to use the terminology of radiometry, rather than that of photometry, because photometry is based on visible light measurements that simulate the human eye's photopic response curve and, strictly speaking, a source that emits only UVR has a zero intensity in photometric terms.

The common radiometric terminology is listed in Table 1. Terms relating to beam of radiation passing through space are the *radiant energy* and *radiant flux*. Terms relating to a source of radiation are the *radiation intensity* and the *radiance*. The term *irradiance*, which is the most commonly used term in photobiology, relates to the object (e.g., UV sensor) struck by the radiation. The radiometric quantities in Table 1 may also be expressed in terms of wavelength by adding the prefix *spectral*.

The time integral of the irradiance is strictly termed the *radiant exposure*, but is sometimes expressed as *exposure dose*, or even more loosely as *dose*. The term "dose" in photobiology is analogous to the term "exposure" in radiobiology and not to "absorbed dose." As yet the problems of estimating the energy absorbed by critical targets in the skin remain unsolved.

A. Radiometric Calculations

The most frequent radiometric calculation is to determine the time for which a subject, who has been prescribed a certain dose (in J/m^2), should be exposed when the radiometer indicates an irradiance in watts per square meter (W/m^2). The relation between these three quantities (time, dose, and irradiance) is simply:

$$\text{Exposure time (min)} = \frac{\text{prescribed dose (J/m}^2\text{)}}{60 \times \text{measured irradiance (W/m}^2\text{)}} \tag{1}$$

Table 1 Radiometric Terms and Units

Term	Unit	Symbol
Wavelength	nm	λ
Radiant energy	J	Q
Radiant flux	W	ϕ
Radiant intensity	$W\ sr^{-1}$	I
Radiance	$W\ cm^{-2}\ sr^{-1}$	L
Irradiance	$W\ cm^{-2}$	E
Radiant exposure	$J\ cm^{-2}$	H

B. Units of Biologically Effective Ultraviolet Radiation

In addition to the radiometric quantities given in the foregoing, derived quantities of effective irradiance and dose related to a specific photobiological action spectrum are often used in photobiology. The effective irradiance is obtained by weighting the spectral irradiance of the radiation at wavelength λ nm by the effectiveness of radiation of the wavelength to cause a particular photobiological effect (e.g., minimal erythema) and summing over all wavelengths present in the source spectrum. This can be expressed mathematically as:

$$\Sigma E(\lambda)\, S(\lambda)\, \Delta\lambda \ \mathrm{W/m^2} \tag{2}$$

$E(\lambda)$ is the spectral irradiance in watts per square meter per nanometer ($\mathrm{W/m^2/nm}$) at wavelength λ nm, and $\Delta\lambda$ is the wavelength interval used in the summation. $S(\lambda)$ is a measure of the effectiveness of radiation of wavelength λ nm relative to some reference wavelength in producing a particular biological endpoint. As it is a ratio, $S(\lambda)$ has no units. The effective irradiance is equivalent to a hypothetical irradiance of monochromatic radiation having a wavelength at which $S(\lambda)$ is equal to unity. The time integral of effective irradiance is the effective radiant exposure (also called the *effective dose*).

A unit of effective dose commonly used in cutaneous photobiology is the *minimal erythema dose* (MED). One MED has been defined as the lowest radiant exposure of UVR that is sufficient to produce a visible erythema reaction with sharp margins 8–24 h after exposure. Another endpoint often used is a just perceptible reddening of exposed skin. The dose of UVR necessary to produce this *minimal perceptible erythema* is sometimes also referred to as an MED. Furthermore, in unacclimatized white skin there is a four- to fivefold range in MED for exposure to UVB radiation. When the term MED is used as a unit of exposure dose, however, a representative value is chosen for sun-sensitive individuals. If, in Eq. (2), $S(\lambda)$ is chosen to be the reference action spectrum for UV erythema in human skin (2) and a value of 200 $\mathrm{J/m^2}$ at wavelengths for which $S(\lambda)$ is equal to unity (i.e., $\lambda \leq 298$ nm) is assumed for the MED (3), the dose (expressed in MEDs) received after an exposure period of t seconds is

$$\frac{t\Sigma E(\lambda) S(\lambda) \Delta\lambda}{200} \tag{3}$$

Notwithstanding the difficulties of accurately interpreting the magnitude of such an imprecise unit as the MED, it has the advantage over radiometric units of its relation to the biological consequences of the exposure.

C. Exposure Time in Sunlight Necessary to Cause a Minimal Erythema

The spectral irradiances of midday summer terrestrial UVR on an unshaded, horizontal surface on a clear day in Durham, UK (latitude 55°N; solar altitude 58°) is given in Table 2, together with the relative effectiveness of different wavelengths of UVR in causing a minimal erythema in unacclimatized, white skin (the action spectrum). If the minimal erythema dose of monochromatic 290-nm radiation is taken as 200 J/m^2, the exposure time necessary to receive sufficient radiation to result in a minimal erythema in a subject who is sunbathing is 200/0.168 s, which is approximately 20 min.

III. DETECTION OF ULTRAVIOLET RADIATION

Techniques for the measurement of UVR fall into three classes: physical, chemical, and biological. In general, physical devices measure power, whereas chemical and biological systems measure energy.

The use of chemical methods, which measure the chemical change produced by the radiation, is called *actinometry*. These techniques usually form the basis of personal UV dosimeters (see Sec. VIII).

Table 2 The Spectral Irradiance [$E(\lambda)$] of Midday, Clear Sky, Summer Sunlight in Durham (Latitude 55°N) and the CIE Reference Action Spectrum for Erythema [$\varepsilon(\lambda)$] Measured in Steps [$\Delta(\lambda)$] of 10 nm

λ, nm	Spectral irradiance $E(\lambda)$, W m^{-2} nm^{-1}	Erythema action spectrum, $\varepsilon(\lambda)$	$E(\lambda)\varepsilon(\lambda)\Delta\lambda$, W m^{-2}
290	0	1.00	0.0000
300	0.004	0.65	0.026
310	0.11	0.074	0.081
320	0.37	0.0086	0.032
330	0.64	0.0014	0.009
340	0.66	0.00097	0.006
350	0.65	0.00068	0.004
360	0.75	0.00048	0.003
370	0.88	0.00034	0.003
380	0.80	0.00024	0.002
390	0.88	0.00017	0.001
400	1.2	0.00012	0.001
	$\sum E(\lambda)\epsilon(\lambda)\Delta(\lambda)$		0.168

Biological techniques of measurement are generally limited to the use of viruses and microorganisms.

A. Physical Ultraviolet Radiation Detectors

A physical UVR detector consists of an element that absorbs the radiation and a means of measuring the resulting change in some property of the element. There are two basic physical types: thermal and photon.

Thermal Detectors

Thermal detectors respond to heat or power and have a broad spectral response with near-uniform sensitivity from the UV to the infrared (IR). The absorption of radiation increases the temperature of the element, and this change can be detected in a variety of ways.

In the thermopile, the temperature rise of the element causes a small voltage to be generated at the junction of two dissimilar metals. The heat-sensitive element in a pyroelectric detector consists of a slab of ferroelectric material that produces a change in current proportional to the rate of change of temperature of the surface of the slab which, in turn, is proportional to the rate of change of irradiance. A bolometer works by sensing the change in resistance of the absorbing element. In the Golay cell, the temperature rise is sensed by the expansion of an enclosed gas that presses against a flexible mirror, altering its focal length. Finally, photoacoustic detectors use a microphone to detect the small fluctuations in pressure that occur when the heat generated by the radiation absorption is coupled into the gas contained in an acoustic resonator.

Photon Detectors

Photon detectors operate on the principle of the liberation of electrons by the absorption of a single quantum of radiation and, consequently, have a nonlinear spectral response. Examples of photon detectors include photoemissive types (vacuum phototube, gas-filled phototube, and photomultiplier tube); junction photodetectors (e.g., Si, GaAsP, GaP photodiodes) that may be operated with a *zero bias* (sometimes called *photovoltaic mode*) or a *reverse bias* (sometimes called *photoconductive mode*); and photoconductors (e.g., CdS, CdSe, PbS, and InAs).

More detailed descriptions of physical optical radiation detectors can be found elsewhere (4–6).

IV. SPECTRORADIOMETRY

It is common practice to talk loosely of *UVA lamps* or *UVB lamps*. However, such a label does not adequately characterize UV lamps, because nearly all UV lamps will emit both UVA and UVB, and even UVC, visible light, and IR radiation. The only correct way to specify the nature of the emitted radiation is by reference to the spectral power distribution. This is a graph (or table) that indicates the radiated power as a function of wavelength. The data are obtained by a technique known as spectroradiometry. Figure 1 shows the spectral power distribution of ultraviolet and visible radiation from a xenon arc lamp after it has passed through a 2-mm-thick color glass filter (Schott WG320). This combination of lamp and filter produces a spectrum of radiation that is similar to that of terrestrial sunlight. Consequently, it is often referred to as a *solar simulator* and is used as a laboratory source for determining photoprotection achieved by sunscreens. In Figure 1a, the relative power emission is plotted on a linear scale. Although lamp spectra are commonly plotted using a linear scale, this representation may not be the most appropriate. Because the erythemal sensitivity of normal skin varies by four orders of magnitude over the spectral region of 200–400 nm, it is helpful to be able to discern components of the spectrum that may be of small amplitude in physical terms, but nonetheless are important in photobiological terms. In Figure 1b, the spectrum from the same lamp is shown with the relative power plotted on a logarithmic scale.

A. Components of a Spectroradiometer

The three basic requirements of a spectrometer system are (1) the input optics, designed to conduct the radiation from the source into (2) the monochromator, which usually incorporates a diffraction grating as the wavelength dispersion element, and (c) an optical radiation detector.

Input Optics

The spectral transmission characteristics of monochromators depend on the angular distribution and polarization of the incident radiation as well as the position of the beam on the entrance slit. For measurement of spectral irradiance, particularly from extended sources, such as linear arrays of fluorescent lamps or daylight, direct irradiation of the entrance slit should be avoided. There are two types of input optics available to ensure that the radiation from different source configurations is depolarized and follows the same optical path through the system: the integrating sphere, or diffusers made of either ground quartz or polytetrafluorethylene (PTFE). Both these types of input optics produce a cosine-weighted response, because the radiance of the source, as measured through the entrance aperture, varies as the cosine of the angle of incidence.

Monochromator

A blazed, ruled diffraction grating is normally preferred to a prism as the dispersion element in the monochromator used in a spectroradiometer, mainly because of better stray radiation characteristics.

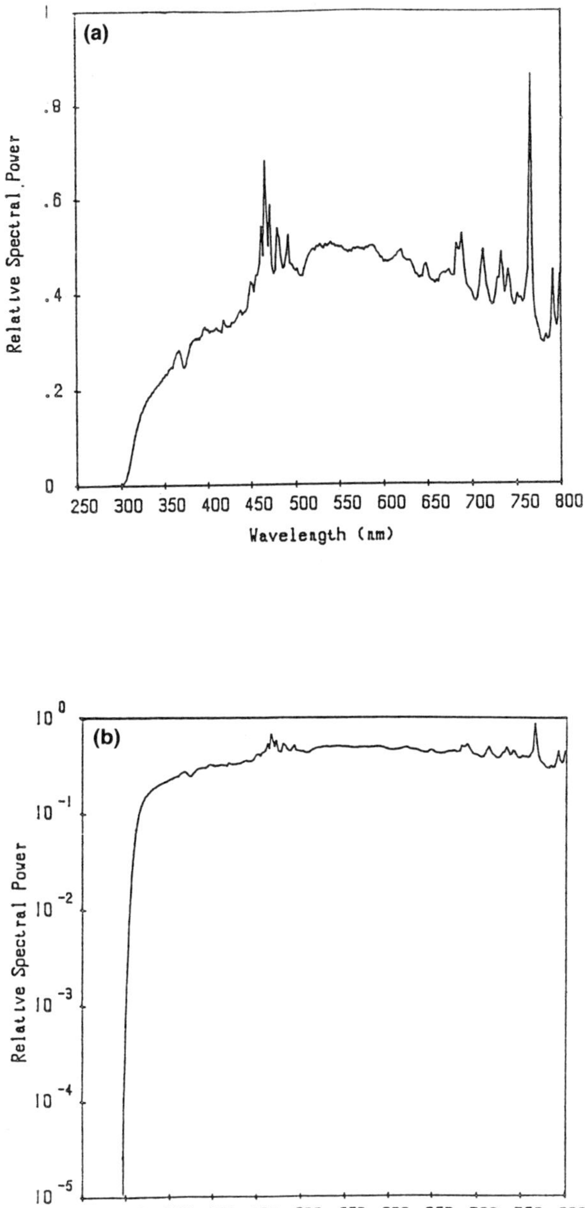

Figure 1 The spectral power distribution of ultraviolet and visible radiation from a xenon arc lamp used in conjunction with Schott WG320 filter (2 mm thick): (a) linear scale on ordinate; (b) logarithmic scale on ordinate.

High-performance spectroradiometers, used for determining low UV spectral irradiances in the presence of high irradiances at longer wavelengths, demand extremely low stray radiation levels. Such systems may incorporate a double monochromator; that is, two single-ruled grating monochromators in tandem, or laser holographically produced concave diffraction gratings can be used in a single monochromator.

Detector

Photomultiplier tubes, incorporating a photocathode with an appropriate spectral response, are normally the detectors of choice in spectroradiometers. However, if radiation intensity is not a problem, solid-state photodiodes may be used, as they require simpler and cheaper electronic circuitry.

B. Calibration

It is important that spectroradiometers are calibrated over the wavelength range of interest using standard lamps. A tungsten filament lamp operating at a color temperature of about 3000 K can be used as a standard lamp for the spectral interval 250–2500 nm, although workers concerned solely with the UV region (200–400 nm) may prefer to use a deuterium lamp.

C. Source of Error in Spectroradiometry

Accurate spectroradiometry, even when only relative spectral power distributions are used, requires careful attention to detail. Factors that can affect accuracy include wavelength calibration, bandwidth, stray radiation, polarization, angular dependence, linearity, and calibration sources.

V. BROADBAND RADIOMETRY

Although spectroradiometry is the fundamental way to characterize the radiant emission from a light source, radiation output is normally measured by techniques of broadband radiometry. Broadband radiometers generally combine a detector (such as vacuum phototube or a solid-state photodiode) with a wavelength-selective device (such as a color glass filter or interference filter) and suitable input optics (such as quartz hemispheric diffuser or PTFE window).

A. The Problem of Spectral Sensitivity

Naive users of broadband radiometers often gain the impression from commercial literature that instruments are readily available to measure UVA, UVB, or UVC. To meet the criterion for a UVB radiometer, say, the sensor should have a uniform spectral response from 290 to 320 nm (the UVB waveband) with zero response outside this interval. In other words, the electrical output from the sensor should depend on only the total power within the UVB waveband received by the sensor and not on how the power is distributed relative to wavelength. In practice no such sensor exists with this ideal spectral response (neither does one exist that measures UVA or UVC correctly, for that matter). All radiometers that combine a photodector with an optical filter have a nonuniform spectral sensitivity within their normal spectral band.

B. A Common Pitfall in Ultraviolet Radiometry

The detector illustrated in Figure 2 is marketed as a notional UVB radiometer. The spectral sensitivity of the sensor peaks at 310 nm and falls to 1% of its peak response at approximately 250 and 370 nm (manufacturer's data on typical spectral response of the UVX-31 sensor). The manufacturer claims that the radiometer is calibrated to indicate the total irradiance from a fluorescent UVB

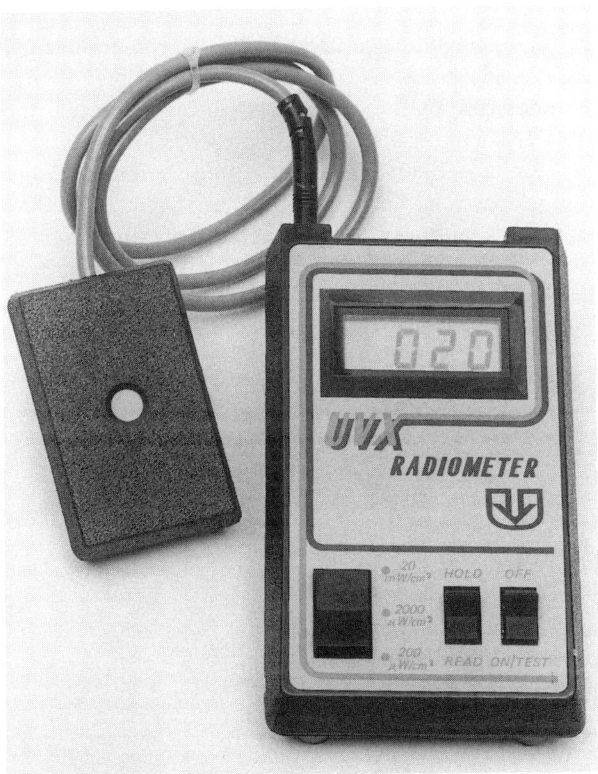

Figure 2 A UVX-31 radiometer intended for use as a notional UVB radiometer.

sunlamp (e.g., Westinghouse FS40 or Philips TL12). The irradiances measured by this radiometer from a variety of lamp and filter combinations are compared in Table 3, with the true UVB irradiances determined by the author using a high-quality, double holographic-grating spectroradiometer. It can be seen from Table 3 that the UVB radiometer overestimates the true UVB irradiance in every case, but by differing amounts. The reason that the sensor overestimates the UVB irradiance from the fluorescent UVB sunlamp (the lamp against which it is calibrated) is that only 80% of the radiation output from this type of lamp is in the UVB (290–320 nm), yet the detector is calibrated to record the total irradiance. The unsuspecting user who wanted to measure the UVB irradiance from a xenon lamp filtered by a Schott WG320 filter (a typical combination used as a source of simulated sunlight) would record a value that was 2.83 times the true value.

C. Radiometers that Simulate Biological Action Spectra

In experimental techniques for evaluating sunscreens, the delayed erythemal response of the skin is often considered the biological endpoint. It may be desirable, therefore, to use a radiometer with a spectral sensitivity to match the action spectrum for cutaneous erythema. This can be achieved reasonably well by appropriate choice of detector and filter combination, and the most widely used instrument that has this property is the Robertson-Berger meter (7). The result is that the radiometer can be calibrated to read the erythemally weighted irradiance directly. The reading is then correct for any spectral power distribution, with an accuracy limited by the degree to which its spectral response differs from the erythema action spectrum.

Table 3 Ratio of Irradiance Measured with a UVB Radiometer[a] to True UVB (290–320 nm) Irradiance

Lamp	UVB radiometer/true UVB
Fluorescent UVB sunlamp (Philips TL12)	1.25
Fluorescent UVA lamp (Philips TL09)	9.50
Medium-pressure mercury arc lamp	1.34
Unfiltered xenon arc lamp	1.76
Xenon lamp with 2-mm Schott WG305	1.82
Xenon lamp with 2-mm Schott WG320	2.83

[a]The UVB radiometer is a model UVX-31 (Ultra-Violet Products, Inc., San Gabriel, CA 91778). The radiometer is calibrated by the manufacturer to indicate the total irradiance from a fluorescent UVB sunlamp.

D. Calibration of Radiometers

Broadband radiometers need to be calibrated separately for each type of light source for which they are to be used. The method of choice is as follows:

1. Measure the spectral irradiance from the desired light source using a spectroradiometer.
2. From the spectral irradiance data, determine the irradiance within the waveband of interest (either unweighted or spectrally weighted according to some predefined action spectrum).
3. Place the entrance aperture of the sensor at the same point in space as the entrance optics of the spectroradiometer and note the reading.
4. Either adjust the gain control of the radiometer electronics to read the correct irradiance, or determine a multiplicative correction factor by diving the correct irradiance by the displayed irradiance. If the radiometer is intended to be used with more than one type of lamp the latter method is preferred.

VI. WAVELENGTH-INDEPENDENT RADIOMETRY

In this branch of radiometry a detector is used that responds equally to all wavelengths of optical radiation. The most common detector of this type is the thermopile, and this is especially useful in measuring the irradiance from an irradiation monochromator used for determining erythema action spectra. Until a few years ago, commercial thermopiles were handmade, expensive, and fragile devices. A major advance came with the production of multiple junction thermopiles based on thin-film technology. These devices are rugged, are much less expensive, and are typified by the Dexter range of thermopiles (Dexter Research Center, Michigan, USA).

A. Calibration

Thermopiles measure absolute radiant power and calibration can be satisfactorily achieved only by national standards laboratories (8), such as the National Physical Laboratory in the United Kingdom and the National Bureau of Standards in the United States.

VII. RADIOMETER STABILITY

It should be remembered that the sensitivity of all radiometers will change with time; frequent exposure to high-intensity sources of light will accelerate this change. Therefore, it is always a sound policy to acquire two radiometers, preferably of the same type, one of which has a calibration traceable to a national standards laboratory. This radiometer should be reserved solely for intercomparisons with the other radiometer(s) used for routine purposes. A measurement of the same source is made with each radiometer and a ratio calculated. It is the stability of this ratio over a period of months and years that indicates long-term stability and good precision.

VIII. PERSONAL ULTRAVIOLET RADIATION DOSIMETRY

We have seen that ultraviolet radiation is generally measured with thermal or photon detectors, often used in conjunction with optical filters. A different yet complementary approach is the use of various photosensitive films as UVR dosimeters. The principle is to relate the degree of deterioration of the films, usually in terms of changes in their optical properties, to the incident UVR dose. The principal advantages of the film dosimeter are that it provides a simple means of integrating UVR exposure continuously and that it allows numerous sites, inaccessible to bulky and expensive instrumentation, to be compared simultaneously. Personal ultraviolet dosimetry can be helpful in establishing the effect of photoprotective agents in the treatment of photosensitivity (9).

A. Requirements of Personal Ultraviolet Dosimeters

Ultraviolet dosimeters designed for personal use should have the following characteristics:

1. The physical or chemical change produced in the dosimeter (e.g., increase in optical absorbance) should, ideally, increase linearly with UV dose. If not, the dose–response curve should at least be monotonic, that is, any given dosimeter response is effected by only one radiation dose.
2. The dosimeter should exhibit photoaddition; each wavelength acts independently, and the effect of polychromatic radiation is the sum of the effects of all wavelengths involved.
3. The dosimeter response should depend only on dose and be independent of dose rate.
4. The spectral sensitivity of the dosimeter should, ideally, match the action spectrum of the photobiological effect being monitored.
5. The dosimeter response should be independent of temperature and humidity; it should exhibit no *dark effect* (continuing response when radiation exposure terminated); and it should be stable on long-term storage.
6. The dosimeter should be easy to handle and not impose restrictions on the activities of the wearers.

7. The dosimeter should not require laborious processing, and it should be easy to convert the physical or chemical response to a measure of UV exposure dose.
8. The cost per dosimeter should be low so that large-scale monitoring is feasible.

B. Types of Personal Ultraviolet Dosimeters

Polysulfone Film

Perhaps the most commonly used material for studies of personal UV dosimetry has been the thermoplastic, polysulfone, which was first suggested as a possible dosimeter for UVR by Davis et al. (10). Since then, the use of polysulfone film as a personal UV dosimeter has been exploited for monitoring both environmental and artificial UVR (11), including studies in which the efficacy of a sunscreen in preventing the common photodermatosis, polymorphic light eruption, was assessed (12). The basis of the method is that when film is exposed to UVR at wavelengths less than 330 nm, its UV absorption increases. The increase in absorbance, measured at a wavelength of 330 nm, increases with UV dose. In practice the film (40 μm thick) is mounted in cardboard photographic holders and normally worn on the lapel site.

Photosensitive Papers

One drawback of the film dosimeters described in the foregoing is that they require laboratory equipment to facilitate readout. An alternative approach is to use a system whereby the photochemical process initiates a color change so that visual comparisons with stable printed color standards enable the user to obtain a reasonably accurate and continuously readable integrated measure of his or her exposure to UVR. An example of such a dosimeter is based on the reversible color change of photochromic aziridine formulations (13). The colorless aziridine undergoes isomerization following UV exposure to form the blue-colored azomethine ylide.

Photosensitive papers form the basis of dosimeters designed for consumer use while sunbathing (14).

Polycarbonate Plastic

On exposure to UVR, the transparent plastic, CR-39 (allyl diglycol carbonate), alters its optical properties. These changes are the basis of its use as UVR dosimeter (15). After exposure the plastic is etched in 6 *N* KOH at 80°C for 3 h, rinsed, and allowed to dry. The degree of UV dose-dependent front surface damage, visible as opacification, is quantified by measuring the transmission at 700 nm.

REFERENCES

1. Mackenzie, L. A. UV radiometry in dermatology, *Photodermatology*, *2*:86–94 (1985).
2. McKinlay, A. F. and Diffey, B. L. A reference action spectrum for ultraviolet induced erythema in human skin, *Chem. Ind. Eng.*, *6*:17–22 (1987).
3. Urbach, F. Man and ultraviolet radiation, *Human Exposure to Ultraviolet Radiation: Risks and Regulations* (W. F. Passchier and B. F. M. Bosnjakovic, eds.), Excerpta Medica, Amsterdam, 1987, pp. 3–17.
4. Rabek, J. F. *Experimental Methods in Photochemistry and Photophysics*, John Wiley & Sons, New York, 1982.
5. Dereniak, E. L. and Crowe, D. G. *Optical Radiation Detectors*, John Wiley & Sons, New York, 1984.
6. Wilson, A. D. Optical radiation detectors, *Radiation Measurement in Photobiology* (B. L. Diffey, ed.), Academic Press, London, 1989, pp. 23–45.
7. Berger, D. The sunburning ultraviolet meter: Design and performance, *Photochem. Photobiol.*, *24*:587–593 (1976).
8. Goodman, T. M. Calibration of light sources and detectors. *Radiation Measurement in Photobiology* (B. L. Diffey, ed.), Academic Press, London, 1989, pp. 47–70.
9. Corbett, M. F., Hawk, J. L. M., Herxheimer, A., and Magnus, I. A. Controlled therapeutic trials in polymorphic light eruption, *Br. J. Dermatol.*, *107*:571–581 (1982).
10. Davis, A., Deane, G. H. W., and Diffey, B. L. Possible dosimeter for ultraviolet radiation, *Nature*, *261*:169–170 (1976).
11. Diffey, B. L. Ultraviolet radiation dosimetry with polysulphone film. *Radiation Measurement in Photobiology* (B. L. Diffey, ed.), Academic Press, London, 1989, pp. 135–159.
12. Proby, C. M., Baker, C. S., Morton, O., and Hawk, J. L. M. New broad-spectrum sunscreen for polymorphic light eruption, *Lancet*, *341*:1347–1348 (1993).
13. Fanselow, D. L., Pathak, M. A., Crone, M. A., Ersfield, D. A., Raber, P. B., Trancik, R. J., and Dahl, M. V. Reusable ultraviolet monitors: Design, characteristics and efficacy, *J. Am. Acad. Dermatol.*, *9*:714–723 (1983).
14. Moseley, H., Mackie, R. M., and Ferguson, J. The suitability of suncheck patches and Tanscan cards for monitoring the sunburning effectiveness of sunlight, *Br. J. Dermatol.*, *128*:714–718 (1993).
15. Wong, C. F., Fleming, R., and Carter, S. J. A new dosimeter for ultraviolet B radiation, *Photochem. Photobiol.*, *50*:611–615 (1989).

10

Reduction of Actinic Keratoses by Sunscreens

Robin Marks
University of Melbourne and St. Vincent's Hospital, Melbourne, Australia

I. INTRODUCTION

For over 20 years, sunscreens have been recommended as a useful way to prevent skin cancer, both nonmelanoma skin cancer and melanoma. These recommendations are based on the knowledge that skin cancer is clearly related to sunlight exposure, particularly the ultraviolet radiation component of sunlight (1); sunscreens absorb ultraviolet radiation (2); sunscreens reduce the number of nonmelanoma skin cancers produced by ultraviolet radiation in animals (3); sunscreens reduce erythema (sunburn) induced by ultraviolet radiation in humans (4); sunburn is a risk factor for both melanoma and nonmelanoma skin cancer (5,6).

Despite this knowledge, until recently there have been no studies in humans that demonstrated that regular use of sunscreens can actually prevent skin cancer. In fact, there have been several reports by persons who doubt whether they are capable of this effect, and some that suggest that they may even enhance skin carcinogenesis (7,8).

This chapter discusses recent work demonstrating that regular use of sunscreens can prevent the development of actinic keratoses and aid in their remission. This is the first time that it has been demonstrated that regular sunscreen use can prevent precancerous lesions of the skin, and by implication, reduce the risk of skin cancer in the long-term.

II. WHAT ARE ACTINIC KERATOSES?

Actinic keratoses are dysplasia of keratinocytes. The dysplasia occurs predominantly at the basal layer of the epidermis and is associated with abnormalities of maturation and keratinization of the cells involved in the dysplastic process.

They occur on areas of maximum light exposure, such as the head and neck, dorsum of the hands, forearms, upper back, and lower legs in some persons. Histologically, the dysplasia always occurs in association with the changes of solar elastosis in the dermis. The dysplasias are highly antigenic, inducing an inflammatory response that is apparent histologically. Clinically, this is manifest as erythema associated with the adherent scale of the keratosis. The scale varies in color from pale white, through yellow, to brown, on occasion. There may be associated telangiectasiae. The lesions often have an indefinite edge.

Actinic keratoses are generally asymptomatic, but may sting on occasion, particularly if they are picked at. They burn following sunlight exposure. They may also become more apparent clinically after periods of sunlight exposure.

III. ACTINIC KERATOSES AS A MODEL FOR INTERVENTION STUDIES DESIGNED TO REDUCE THE RISK OF SKIN CANCER IN HUMANS

Actinic keratoses are in situ dysplasias. They are not true invasive cancer. Similar to other in situ dysplasias, such as cervical dysplasia or bronchial mucosal dysplasia, they do not necessarily inevitably progress to invasive carcinoma. They are believed to be precursors of squamous cell carcinoma of the skin, but the rate of malignant transformation over time appears to be very low (9). They may remit spontaneously, particularly in those persons who have the ability to reduce their sunlight exposure (10).

Actinic keratoses are epidemiologically a major risk factor for nonmelanoma skin cancer (11). In fact, they are a necessary accompaniment in those persons who develop squamous cell carcinoma, although not necessarily always present in those who develop basal cell carcinoma. They, along with nonmelanoma skin cancer, are also a risk factor for melanoma (12).

The actinic keratoses are very much more common than true invasive nonmelanoma skin cancers. In some Australian populations, they have occurred with a prevalence of up to 50% of the normal population aged 40 years and older (13). They occur with increasing frequency with increasing age, as does skin cancer. They occur in a ratio of about 15 persons with actinic keratoses for every 1 person with nonmelanoma skin cancer in the general population. The average number of actinic keratoses in persons with these lesions is about six actinic keratoses per person. In other words, they appear to be a more sensitive indicator of sunlight exposure than invasive nonmelanoma skin cancer.

Finally, spontaneous remission of these lesions occurs in a substantial proportion of individuals within 12 months (10). Therefore, they seem to come and go relatively quickly, making them ideal to study the effects of a given intervention.

On the basis of all the foregoing data, actinic keratoses appear suitable as surrogates for risk of skin cancer. They also appear sensitive to sunlight so that they can be used as a rapidly responsive model tumor to measure the effect of interventions designed to reduce sunlight exposure in the short-term, and skin cancer in the long-term.

IV. METHODS FOR MEASURING THE EFFECT OF REGULAR SUNSCREEN USE ON ACTINIC KERATOSES

A study was undertaken in Victoria, Australia, in 1991–1992 using the following method (14). Persons 40 years of age or older, with at least one actinic keratosis were involved in a double-blind, randomly controlled trial of a sun protection factor (SPF) 17 sunscreen compared with a base cream without the sunscreen active agents. The sunscreen cream blocked out at least 90% of the UVA in the wavelengths 320–360 nm.

The number of actinic keratoses in the participants ranged from 1 to 30 lesions. The 588 subjects who were enrolled in the study were stratified according to sex and self-rated skin type (burn only and never tan, burn first and then tan, or tan only and never burn). They were all examined by one member of a medical team, consisting of a dermatologist, dermatology trainees, and a local medical officer with an interest in skin cancer.

The actinic keratoses were present either on the head, neck, hands, or forearms, and were recorded on grid maps at the initial presentation. The diagnosis was based on clinical appearance of the skin lesions, with a randomized subsample of subjects selected for biopsy of a lesion from the hands or forearms to ensure that the lesions being studied were actually actinic keratoses.

The sunscreen was a cream containing 8% (wt/wt) 2-ethylhexyl *p*-methoxycinnamate and 2% (wt/wt) 4-*tert*-butylmethoxy-4-dibenzoylmethane, rated according to the Australian Standard 2604 (1986) as having a sun-protection factor (SPF) of 17 for ultraviolet B (resulting in a 94% reduction of radiation in the wavelengths 290–320 nm) (15). The base cream, minus the two active ingredients, was made up to the same consistency as the sunscreen by adding 10% (wt/wt) mineral oil.

Subjects kept daily diaries in which they recorded the time of application of the cream. They were instructed to apply approximately 1.5 ml to the head and neck and the same amount to each hand and forearm once every morning, and to reapply if necessary during the day. They were warned that 50% of them might be using a placebo and, therefore, to use other methods of sun protection apart from the cream they were given.

The participants were followed over the 7-month maximum sunlight period in southern Australia (from spring until autumn) and were examined on three occasions. At each examination, the total number of keratoses, remissions, and new lesions were recorded. The amount of cream used was determined by weighing the bottles, and the diaries were examined. When possible, the same examiner saw the participants on the first and final occasions, when the lesions were recorded for the statistical analysis.

V. THE EFFECT OF REGULAR SUNSCREEN USE ON ACTINIC KERATOSES

A total of 431 subjects completed the study successfully: 157 persons withdrew from the study or failed to attend the final examination. Eighty-seven of the total of 211 reasons given for withdrawal from the study were related to problems using the creams: 114 (18.9%) subjects developed an adverse reaction to the cream that they were applying (16). Almost all of these were inflammatory reactions consistent with an irritant contact dermatitis. None of the persons tested by patch testing, photopatch testing, or scratch testing were allergic to any of the sunscreen active ingredients. A higher than expected proportion of the subjects who developed the inflammatory reaction had a personal history consistent with atopy. Other reactions included acneiform eruptions and contact urticaria. Overall, about 15% had acute contact dermatitis related to sunscreen use. The frequency of reactions was the same in both the sunscreen-treated group and those using the placebo.

The subjects in the sunscreen group initially had a total of 1856 actinic keratoses (mean, 9 ± 7 lesions per subject), and the subjects in the base cream group had a total of 1642 actinic keratoses (mean, 8 ± 6 lesions per subject). At the end of the study there was a mean (± SE) increase of 1 (± 0.3) in the total number of actinic keratoses per subject in the base cream group, compared with a mean decrease of 0.6 (± 0.3) lesions per subject in the sunscreen group (difference 1.53; 95% confidence interval, 0.8 to 2.25).

Figure 1 shows the estimated relative changes in the sunscreen and base cream groups according to sex. The change, although smaller in women than men, was not modified by age or skin type. Table 1 summarizes the results according to development of new lesions and the number of remissions.

The sunscreen-treated group had a total of 333 new lesions, whereas the base cream group had a total of 508 (1.6 versus 2.3 mean lesions per subject). The difference in the mean numbers of new lesions per subject between the two groups was 0.72 (95% confidence interval, 0.15 to 1.28; $p = 0.014$).

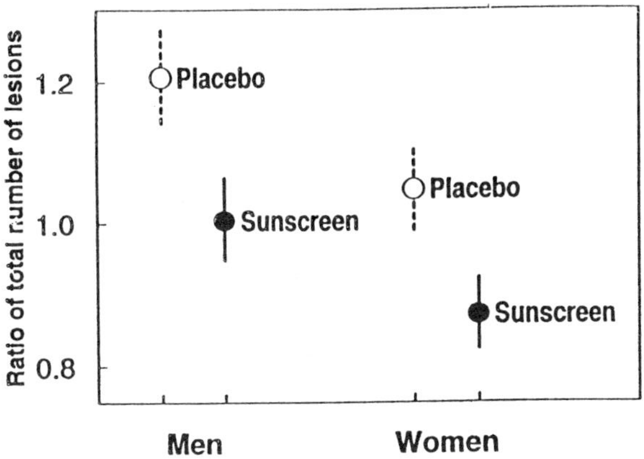

Figure 1 Ratio of the total number of solar keratoses at the beginning and end of the study, according to treatment group and sex. The vertical bars are the 95% confidence intervals for the ratio. (From Ref. 14.)

At the completion of the study, 25% of the lesions present at the baseline had remitted in the sunscreen group, compared with 18% in the base cream group (odds ratio 1.45; 95% confidence interval, 1.10 to 1.88). The mean percentage of lesions remitting throughout the study was 28% in the sunscreen group, compared with 20% in the base cream group (difference, 8%; 95% con-

Table 1 Number of Base-Line Lesions, Remissions of Lesions, and New Lesions, According to Site, in the Base-Cream and Sunscreen Groups

Variable	Base-cream group (n = 221)	Sunscreen group (n = 210)
No. of base-line lesions	1642	1856
Hands and forearms	796	943
Face and neck	846	913
No. of remissions	301	457
Percent of base-line lesions remitting	18	25
Hands and forearms	20	28
Face and neck	17	21
No. of new lesions	508	333
Mean no. of new lesions per subject	2.3	1.6
Hands and forearms	1.1	0.7
Face and neck	1.2	0.9

Source: Ref. 14.

fidence interval, 2 to 13%). The likelihood of remission was greater in the sunscreen group than in the base cream group (odds ratio, 1.53; 95% confidence interval, 1.29 to 1.80).

There was a dose–response relation for both the number of new lesions and the probability of remission according to the amount of cream used. Figure 2 shows a significantly greater number of remissions and significantly fewer new lesions developing in the sunscreen group compared with the base cream group, with increasing amounts of cream used over the 7-month study period.

VI. WHAT DO THESE RESULTS MEAN?

These results show quite clearly that regular use of sunscreens can prevent the development of actinic keratoses and hasten the remission of existing ones. It is the first time in humans that these products have been shown to be useful in prevention of skin carcinogenesis.

There are some important implications from this work that are related to the role of sunlight in the development of skin cancer and, thus, the position of sunscreens in preventing these tumors. The enhancement of remission of actinic keratoses by regular use of sunscreen confirms that, although keratinocytes may undergo dysplasia, this need not inevitably be followed by the development of

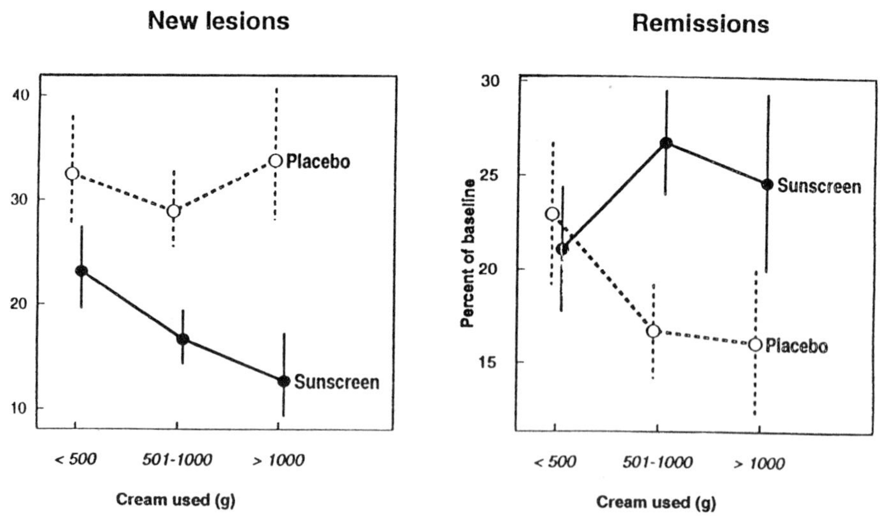

Figure 2 The relations between the quantity of cream used and the number of new lesions and number of remissions, according to treatment group. The values are expressed as the percentages of the number of lesions present at base line. The vertical bars are the 95% confidence inervals. (From Ref. 14).

true invasive carcinoma. Reduction of sunlight by the use of sunscreens was sufficient to reverse the dysplasia that was initiated by sunlight exposure, presumably many years before entering into the study, and that had been promoted for a sufficient period to create clinically apparent lesions. This finding supports the recommendation that regular use of sunscreens, even in adults who have had sufficient sunlight exposure in the past to induce cutaneous dysplasias, is worthwhile in preventing or reducing problems in the future. So often one hears the response by adults to public education programs recommending sunlight reduction, "It is too late, I have already done the damage." The data presented refute this statement.

The finding that the development of new actinic keratoses can be prevented by regular sunscreen use in adults is another important outcome. It has been presumed that the amount of sunlight necessary to develop skin cancer, particularly nonmelanoma skin cancer, is considerable, and that the time required to accumulate a sufficient amount should be many years. The average age of onset for squamous cell carcinoma in fair-skinned people in hot sunny climates, such as Australia, is in the order of 60 years. Once again, the finding that the regular use of sunscreens can prevent the development of actinic keratoses, even in adults who have previously had considerable sunlight exposure, is an indication that continuing exposure as an adult is important in the development of these tumors. This is also further evidence of the potential for sunscreen use in preventing development of skin cancer in the future, even when used for the first time as adults.

In the long-term, it is skin cancer prevention that is the prime aim of the regular use of sunscreens. Nevertheless, actinic keratoses are a major risk factor for nonmelanoma skin cancer, as well as being a risk factor for melanoma. They are widely accepted as precursors of squamous cell carcinoma, although the risk of malignant transformation is low. The use of similar precursors as risk factors for epithelial cancer is a well-accepted method for assessing the effect of an intervention on the prevention of cancer. For example, adenomatous polyps of the colon have been used to predict the effect of dietary interventions on the risk of carcinoma of the colon (17). The use of actinic keratoses, in this case as a surrogate for skin cancer, enables a relatively rapid answer to be obtained on whether or not these products are likely to be valuable in preventing true skin cancer. Use of skin cancer as the measure of outcome of such an intervention would require a very long trial, with a very large number of subjects. There would also be ethical problems involved in recommending long-term use of a placebo compared with a sunscreen when the predicted outcome in the placebo-treated group is an increase in cancer. Consequently, studies on actinic keratoses have an important place in the assessment of the efficacy of sunscreens in skin carcinogenesis.

The participants in this study were advised to use other methods of sun protection, as well as the cream they were given. This means that the effect of the sunscreen may well have been muted. The use of sunscreen alone, in the absence of any other forms of solar protection, may well have shown a greater difference between the treated and untreated groups. Nevertheless, sunscreens are not normally recommended for use alone. Their role is as an adjunct to other protection, such as clothing, hats, and avoidance of the sun near the midday, rather than being a substitute for these more natural methods of sun protection.

Interestingly, even though these subjects were highly motivated, with two-thirds having taken part in previous studies of skin cancer and more than a quarter having a personal history of skin cancer, there was still an enormous variation in the total amount of sunscreen use among individuals. It is unlikely in a normal education program that rates of compliance and response would be as high in the whole population after the institution of a public education program as they were in the study population. However, one of the other outcomes of the study was the clear measurement of the prevalence of reactions to the regular use of these products, which had not been demonstrated on a longitudinal basis before. A frequency of inflammatory reactions to these products of over 15% when they are used on a regular basis over a summer period is likely to be a major barrier to compliance with public education messages. There is still much work to be done on the formulation of these products to manufacture them in a way that makes them acceptable for regular and long-term use by the public. The finding that there was an equal frequency of reactions between the sunscreen-treated and the placebo groups suggests that the problems may well be related as much to the vehicle as to the type and concentration of the sunscreen active ingredients.

The sunscreen used in this study was a broad-spectrum product blocking out not only a high proportion of UVB, but also a large proportion of UVA in the range of 320–360 nm, when tested for the latter by an in vitro thin-film method. Therefore, it is not possible to say blocking of which component of UVR was most efficacious in affecting the development of actinic keratoses. These days most sunscreens, particularly those with a high SPF, do incorporate broad-spectrum activity in the formulation necessary to achieve such high numbers. Nevertheless, the sunscreen in this case was only an SPF 17. Whether or not a substantially increased effect could have been achieved with the use of some of the very high SPF number sunscreens available on the market today is debatable. The finding of a dose-related response in relation to remissions and new lesions by weight of cream does not necessarily mean that the more cream individuals put on in a given day will make a great deal of difference. Those people using over 1000 g of the cream were merely complying with the recommendation that approximately 4.5 g of cream be used every day. The dose–response applies more to the regular application every day than to putting on a very large amount each time the creams were used.

VII. CONCLUSION

In summary, the data presented in this chapter indicate for the first time that regular sunscreen use can prevent the development of cutaneous dysplasias; here, actinic keratoses. Actinic keratoses and nonmelanoma skin cancers are the most common dysplasias and cancers in white persons in many countries throughout the world. The incidence has been increasing in recent decades and concern about ozone depletion has led to predictions of even further increases in the future. These preliminary data confirm the appropriateness of the health education messages promoted by an increasing number of public health organizations throughout the world. As predicted for many years, sunscreens do have an important role, along with other methods of protection from sunlight, in reducing the risk of skin cancer in the long-term.

ACKNOWLEDGMENTS

We are indebted to Drs. P. Foley, C. Hadwen, D. McColl, F. Bruce, F. Watkins, S. Gilmore, and J. Horton, who examined the participants; to Dr. R. O'Keefe, Department of Anatomical Pathology, St. Vincent's Hospital, who examined the biopsies; to members of the Maryborough Rotary, Mr. R. Were, the City of Maryborough, and the Maryborough Community Health Centre for assistance and use of the facilities; and to Jenny Greed, our nurse, for her dedicated effort throughout the study.

REFERENCES

1. Urbach, F. Environmental risk factors for skin cancer. *Skin Carcinogenesis in Man and in Experimental Models*, E. Hecker, E. G. Jung, F. Marks, and W. Tilgen, eds. Recent Results in Cancer Research Series 128. Springer Verlag, Berlin, 1993, pp. 243–262.
2. Pathak, M.A. Sunscreens: Topical and systemic approaches for protection of human skin against harmful effects of solar radiation. *J. Am. Acad. Dermatol.*, *7*:285–312 (1982).
3. Gurish, M. F., Roberts, L. K., Krueger, G. G., and Daynes, R. A. The effect of various sunscreen agents on skin damage and the induction of tumour susceptibility in mice subjected to ultraviolet irradiation. *J. Invest. Dermatol.*, *76*:246–251 (1981).
4. Pathak, M. A., Fitzpatrick, T. B., Greiter, F. J., and Kraus, E. W. Principles of photoprotection in sunburn and suntanning, and topical and systemic photoprotection in health and diseases. *J. Dermatol. Surg. Oncol.*, *11*:575–579 (1985).

5. Green, A. and Battistutta, D. Incidence and determinants of skin cancer in a high risk Australian population. *Int. J. Cancer, 46*:356–361 (1990).
6. Marks, R. and Whiteman, D. Sunburn and melanoma: How strong is the evidence? *Br. Med. J., 308*:75–76 (1994).
7. Garland, C. F., Garland, F. C., and Gorham, E. D. Rising trends in melanoma: An hypothesis concerning sunscreen effectiveness. *Am. J. Epidemiol., 3*:103–110 (1993).
8. Skolnick, A. A. Sunscreen protection controversy heats up. *JAMA, 265*:3218–3220 (1991).
9. Marks, R., Rennie, G., and Selwood, T. S. Malignant transformation of solar keratoses to squamous cell carcinoma. *Lancet, 1*:795–797 (1988).
10. Marks, R., Foley, P., Goodman, G., et al. Spontaneous remission of solar keratoses: The case for conservative management. *Br. J. Dermatol., 115*:649–655 (1986).
11. Marks, R., Rennie, G., and Selwood, T. S. The relationship of basal cell carcinomas and squamous cell carcinomas to solar keratoses. *Arch. Dermatol., 124*:1039–1042 (1988).
12. Green, A. C. and O'Rourke, M. G. Cutaneous malignant melanoma in association with other skin cancers. *JNCI, 74*:977–980 (1985).
13. Marks, R. and Selwood, T. S. Solar keratoses: The association with erythemal ultraviolet radiation in Australia. *Cancer, 56*:2332–2336 (1985).
14. Thompson, S. C., Jolley, D., and Marks, R. Reduction of solar keratoses by regular sunscreen use. *N. Engl. J. Med., 329*:1147–1151 (1993).
15. Australian standard 2604—1986. Sunscreen products—evaluation and classification. Sydney: Standards Association Australia, 1986.
16. Foley, P., Nixon, R., Marks, R., et al. The frequency of reactions to sunscreens: Results of a longitudinal population-based study on the regular use of sunscreens in Australia. *Br. J. Dermatol., 128*:512–518 (1993).
17. Bruce, W. R., Eyssen, G. M., Ciampi, A., Dixon, P. W., and Boyd, N. Strategies for dietary intervention studies in colon cancer. *Cancer, 47*(Suppl):1121–1125 (1981).

III

REGULATORY ASPECTS

11

Regulatory Aspects of Sunscreens in the United States

Emalee G. Murphy
Bryan Cave LLP, Washington, D.C.

I. STATUTORY BACKGROUND TO THE REGULATION OF SUNSCREEN DRUG PRODUCTS IN THE UNITED STATES

For many years, the U.S. Food and Drug Administration (FDA) recognized that products intended to prevent sunburn were subject to regulation as drugs, whereas products represented as "suntan" preparations could be sold as cosmetics. As early as 1940, only two years after passage of the Federal Food, Drug, and Cosmetic Act (the act) (1) FDA clarified its position in a trade correspondence to the industry (T.C. No. 61, Feb. 15, 1940), in which it stated:

> Articles which refer to sunburn or any other disease condition are drugs under Section 201(g), but articles which are represented exclusively for the production of an even tan will be regarded as cosmetics under Section 201(i).

The distinction was based on the statutory definitions for drugs and cosmetics. Section 201(g)(1) of the act defines *drugs* as:

> (B) articles intended for use in the diagnosis, cure, mitigation, treatment or prevention of . . . disease . . . ; and (C) articles . . . intended to affect the structure of any function of the body . . . ; and (D) articles intended for use as a component [of these articles] . . . (2).

Cosmetics, on the other hand, are defined in Section 201(i)(3) as:

> (1) articles intended to be rubbed, poured, sprinkled, or sprayed on, introduced into, or otherwise applied to the human body or any part thereof for cleansing, beautifying, promoting attractiveness, or altering the appearance, (2) articles intended for use as a component of any such articles; except that such term shall not include soap (3).

Therefore, according to these definitions, a product's regulatory status as either a drug or cosmetic largely depends on its intended use as expressed in the product labeling and advertising.

Classification as drugs, rather than as cosmetics, subjects sunburn prevention products to the more stringent FDA drug regulations.

Before 1938, drugs could be marketed in the United States without prior FDA review or approval. Following passage of the 1983 act, FDA was authorized to conduct premarket review of the safety of drugs not generally recognized as safe for their intended uses and subsequent 1962 amendments extended FDA's premarket approval authority to include the safety and efficacy of all "new" drugs (4). *New drugs* are defined as those not generally recognized, among experts, as safe and effective for their intended uses, or that although shown to be safe and effective, have not been marketed for a material time or to a material extent. (5).

II. FDA REVIEW OF OTC DRUG SUNSCREEN PRODUCTS

To facilitate the review of nonprescription, or over-the-counter (OTC) drug products on the U.S. market (6), FDA established a procedure (7) by which drug categories and ingredients, including sunburn prevention products, could be classified as "generally recognized by qualified experts as safe and effective and not misbranded under prescribed, recommended or suggested conditions of use" and thus as drugs that could be marketed in the U.S. without an approved new drug application. Accordingly, FDA appointed advisory review panels composed of experts in the technical, scientific, and medical fields applicable to each category of OTC drugs to evaluate the safety and efficacy of OTC active ingredients.

The FDA Advisory Review Panel on Topical Analgesic, Antirheumatic, Otic, Burn, and Sunburn Treatment and Prevention OTC Drug Products (the panel) included representatives of the medical and academic communities, as well as nonvoting representatives of consumer organizations, The Proprietary Association (now the Nonprescription Drug Manufacturers Association), and The Cosmetic, Toiletry, and Fragrance Association. Food and Drug Administration representatives served as panel liaisons and administrative staff (8).

Regulations governing OTC drug review procedures (9) require FDA to publish an initial Advance Notice of Proposed Rulemaking (ANPR), followed by a Notice of Proposed Rulemaking, and ultimately, the Final Rule. Public comments must be considered by FDA following publication of each proposal.

The FDA published its Advance Notice of Proposed Rulemaking for OTC drug sunscreen products for human use in the August 25, 1978 *Federal Register* (10). The ANPR included the unaltered recommendations of the FDA Advisory Panel concerning the composition, testing, and labeling of sunscreen products.

On May 12, 1993, FDA issued its Notice of Proposed Rulemaking for OTC drug sunscreen products which dramatically changed the current guidelines for sunscreen regulation (11). In the Proposal, FDA announced its intention to revoke T.C. No. 61, and redefined cosmetics that (1) contain sunscreens and use the term *sunscreen* or similar terms in labeling (12); (2) bear an SPF value; or (3) make any claims to therapeutic attributes of sunscreen ingredients, as drugs. In proposed rule §700.35(a), the FDA stated that

> [A] product that includes a sunscreen active ingredient and the term "sunscreen" in its labeling or in any other way represents or suggests that it is intended to prevent, cure, treat, or mitigate disease to affect a structure or function of the body comes within the definition of a drug in section (201)(g)(1). . . (12).

The proposal also made changes in the regulation of (1) cosmetics containing sunscreens, (2) the maximum SPF allowed, (3) sunscreen efficacy testing, (4) UVA and other claims in labeling, (5) permitted active sunscreen ingredients, (6) required warning statements, and (7) "waterproof" claims (13).

Throughout the OTC drug review procedure, FDA has emphasized that the review "is an active, not an inactive, ingredient review." Nevertheless, the OTC panels occasionally made recommendations relative to inactive ingredients; these recommendations were made for public awareness and were not intended to be included in the OTC drug monographs (14). Thus, the proposals apply only to the active drug ingredients in these products.

Although the proposed rule is not final, FDA has stated that unless a product is the subject of an approved New Drug Application (NDA), marketing a sunscreen with a formulation or labeling that is not in accordance with an ANPR or proposed rule can result in regulatory action against the product, the marketer, or both (15). Therefore, most companies in the United States used the ANPR as a guide for drug sunscreen composition, testing, and labeling, and are now having to adjust to the changes made in the proposed rule.

For example, the proposed rule adopts the common or usual names for several sunscreen active ingredients and drops padimate A from category I to category III (16). Sunscreen ingredients are reviewed as safe and effective and not misbranded (category I) if used at the recommended concentration levels, provided they exhibit a sun protection factor (SPF) value of at least "2." The sunscreen active ingredients considered category I include:

Aminobenzoic acid	Up to 15%
Cinoxate	Up to 3
Diethanolamine methoxycinnamate (previously named diethanolamine *p*-methoxycinnamate)	Up to 10
Digalloyl trioleate	Up to 5
Dioxybenzone	Up to 3
Ethyl 4-[bis(hydroxypropyl)] aminobenzoate	Up to 5
Octyl methoxycinnamate (previously named ethylhexyl *p*-methoxycinnamate)	Up to 7.5
Octyl salicylate (previously named 2-ethylhexyl salicylate)	Up to 5
Glyceryl aminobenzoate	Up to 3
Homosalate	Up to 15
Lawsone with dihydroxyacetone	Up to 0.25
Menthyl anthranilate	Up to 5
Octocrylene (previously named 2-ethyhexyl 2-cyano-3,3-diphenylacrylate)	Up to 10
Oxybenzone	Up to 6
Padimate O	Up to 8
Phenylbenzimidazole sulfonic acid (previously named 2-phenylbenzimidazole-*S*-sulfonic acid)	Up to 4
Red petrolatum	Up to 100
Sulisobenzone	Up to 10
Titanium dioxide	Up to 25
Trolamine salicylate (previously named triethanolamine salicylate)	Up to 12

Four ingredients were reviewed and classified as category II (not generally recognized as safe and effective or recognized as misbranded), and five as category III (insufficient data to permit final classification), including zinc oxide.

The Agency eliminated the lower concentration limits on active ingredients and placed more emphasis on final drug testing and SPF determination. The FDA stated that "[b]ecause the formulation in a sunscreen drug product influences the effectiveness of the active ingredient in the product . . . the presence of a required minimum amount of sunscreen ingredient in a drug product does not guarantee effectiveness'' (17).

Furthermore, the FDA proposes to allow a combination of two category I sunscreen ingredients in a sunscreen formulation, as long as the finished product provides a minimum SPF value of ''2.'' Lower limits on concentrations of category I ingredients originally proposed in the ANPR were kept in combination formulations to prevent manufacturers from misleading or deceiving consumers by including ingredients for promotional purposes, yet in ineffective quantities. Combinations of sunscreens and category I skin protectant ingredients would likewise be permitted.

III. DETERMINATION OF SUNSCREEN EFFICACY

In determining the protective power of sunscreen products, the OTC drug sunscreen panel and FDA recognized the need for a standard method to quantify sunscreen effectiveness. The proposed rule, therefore, includes clinical procedures to measure the SPF value that specify, among other provisions, (1) a standard reference formula of 8% homosalate with a known SPF mean value of 4.47 for testing sunscreens with SPF values up to 15 (there is at this time no standard formula for testing sunscreens with SPF above 15 although FDA has proposed an upper limit of SPF 30) (18); (2) a preferred source of artificial light (the xenon-arc solar simulator providing 290–400 nms; (3) test panel size (minimum of 20 human test subjects and a maximum of 25) and composition (only skin types I, II, or III) (19); (4) the test area (back); (5) radiation exposure increments; (6) test material application density (2 mg/cm^2); and (7) data evaluation procedures.

Additional procedures were recommended for determining product SPF values using natural sunlight and for determination of sweat and water-resistant properties using an artificial light source.

The 1993 proposed rule set a cap on the SPF allowable in sunscreens at SPF 30 (20). FDA concluded that the additional benefit from sunscreens with SPF above 30 is outweighed by the increased exposure to sunscreen ingredients. The increase in protection from products with SPFs higher than 30 is negligible, especially since SPF 30 sunscreens block 96.7% of the UV energy from the sun. In addition, a sunscreen's effectiveness is highly dependent on variable factors, such as individual's skin and the latitude of the location of sun exposure.

The question of appropriate methodology to determine the effectiveness of sunscreen products as expressed by the SPF number is of considerable international industry concern. European, American, and Japanese sunscreen products compete in world markets. Although the need for a comparable, if not uniform, method for SPF determination is recognized, national marketing concerns, tradition, and less than unanimous scientific opinion have hampered agreement on several significant testing details.

In the proposal, FDA noted that it is considering four petitions and one comment requesting reconsideration of four active sunscreen ingredients marketed in various countries, but not approved in the United States as category I OTC drugs (21). FDA concluded that is has not reached a decision on comparative testing methods for drugs approved in foreign countries or on the use of foreign marketing data as the sole basis for inclusion of an ingredient in OTC drug review. Therefore, except for butyl methoxydibenzoylmethane, FDA chose to defer on its decision on whether to include these foreign ingredients in OTC drug review.

In the proposed rule, the FDA eliminated the term *waterproof* from sunscreen drug product labeling, stating that waterproof is an absolute claim and misleads consumers about the limits of sunscreen protection. A product that retains the same product category description after 80 min of water immersion may make the claim of being "very water-resistant" rather than waterproof. Products that retain the same product category description after 40 min may make the claim of being "water-resistant" (22). Products that carry claims such as water-resistant or very water-resistant may also claim to be "sweat-resistant" because FDA concluded that the "washoff" effects of sweat emerging from underneath the product are less than the washoff effects from water running over the product.

In addition, the proposed rule acknowledged the need for protection against ultraviolet (UV)A radiation (320–400 nm) which, according to scientific data, penetrates the skin more efficiently than UVB radiation (23). However, FDA also acknowledged that there are no appropriate, established testing methods to measure UVA protection. A sunscreen drug product may claim it protects against UVA radiation if (1) it contains active ingredients that absorb UVA radiation extending to 360 nm or above, and (2) passes the testing procedures the FDA will be proposing for UVA protection (24).

To help consumers choose the appropriate sunscreen products, the proposal includes descriptive categories of common sunburn and tanning histories by which consumers can classify their skin types and recommends five product category designations (PCD) based on SPF values obtained by the proposed test method. Product category designation (PCD) labels are optional on sunscreen drug products. (Table 1).

The FDA proposed a new Recommended Sunscreen Product Guide that would be required on all sunscreen drug products. (Table 2). Additional labeling

Table 1 Product Category Designations Based on SPF Values

Category description	SPF
Minimal	SPF 2 to < 4
Moderate	SPF 4 to < 8
High	SPF 8 to < 12
Very High	SPF 12 to < 20
Ultrahigh	SPF 20–30

indications describe directions for use, general and specific warnings, and product performance claims.

The FDA also proposed that a warning statement or a "sun alert" appear on all sunscreen drug products. The sun alert warning states:

> SUN ALERT: The sun causes skin damage. Regular use of sunscreens over the years may reduce the chance of skin damage, some types of skin cancer, and other harmful effects due to the sun.

IV. USE OF SUNSCREENS IN COSMETIC PRODUCTS

A. History

The cosmetic industry was quick to recognize the protective value of sunscreens and, from 1978, began to add sunscreens to various skin care cosmetics, although generally without accompanying claims that would place the products clearly in the drug category. In general, before 1985 FDA took little notice of

Table 2 Recommended Sunscreen Product Guide

Sunburn and tanning history	Recommended sun protection product
Always burns easily; rarely tans	SPF 20–30
Always burns easily; tans minimally	SPF 12 to < 20
Burns moderately; tans gradually	SPF 8 to < 12
Burns minimally; always tans well	SPF 4 to < 8
Rarely burns; tans profusely	SPF 2 to < 4

Source: Ref. 23.

products containing sunscreens, as long as they were correctly labeled as cosmetics.

Nevertheless, FDA's internal opinion concerning the legal classification of products containing sunscreens has shifted over the years. In 1976, prior to publication of the ANPR, FDA stated in a correspondence to industry counsel:

> [T]he Agency has moved away from the position stated over the past several years. We have concluded that a product containing a sunscreen ingredient . . . even when labeled solely as a tanning aid, is both intended and understood to be a preventative of sunburn, and is therefore a drug. . . . Because we have not yet publicly revoked the old Trade Correspondence, . . . we will, for the present adopt an enforcement posture that is consistent with the Trade Correspondence (26).

FDA reaffirmed the policy expressed in its 1976 letter in a memorandum of meeting with industry in 1985 (27). In the same year, FDA also stated:

> Whether an ingredient is regulated as a drug, cosmetic, or both, depends on its intended use in the product in which it is used. In the case of (ingredient), the intended use, as expressed in the labeling of each product in question, is that of a sunscreen active ingredient in a sunscreen product which is not exclusively represented for the production of an even tan. Thus, we regard the product to be a drug (28).

Although FDA expressed a preference for classifying all products containing sunscreens as drugs, the agency was not prepared to challenge the presence of approved sunscreens in products labeled solely for cosmetic purposes.

However, in 1987, FDA launched a major regulatory offensive against skin care products labeled as "antiaging" cosmetics, the claims for which suggested that the products were intended to affect the structure or function of the body. In regulatory letters to more than 23 companies, FDA cited the products as drugs within the meaning of Section 201(g) of the act (the drug definition). Several of the products also contained sunscreens with accompanying antiaging or sun protection claims. In these instances, the agency cited the presence of the sunscreens, plus reference to their presence in product labeling, as evidence of the product's drug status (29).

Furthermore, where the cosmetics contained a sunscreen as yet unapproved by FDA either through the OTC Drug Review or New Drug Application process, FDA stated that the products were "new" drugs within the meaning of Section 201(p) of the act (the new drug definition) (30).

In the case of the particular new sunscreen agent thus classified, FDA had in an earlier regulatory action taken the position that its

> intended use . . . as a UVA agent has been well established through labeling, promotion and advertising by the manufacturer of the ingredient, as well as through the

> expression of its intended use in the labeling of other recently marketed sunscreen drug products. . . . [E]ven if the ingredient were regarded as an inactive ingredient, its presence in the products causes them to be new drugs (31).

In other words, FDA took the view that addition of this particular sunscreen to products indicated an objective intent that the products were intended to be used for therapeutic purposes. As Heinz Eiermann, director of FDA's Division of Color and Cosmetics stated:

> The view that a sunscreen product should be regulated as a drug, regardless of its label claims, may run counter to the definition of the term ''intended use'' as expressed in the 1934 Senate Report . . . and the various court decisions. . . . However, it must also be noted that in *National Nutritional Foods v. Mathews*, the circuit court expressed the opinion that proof in the record supporting the agency's conclusion that therapeutic usage far outweighs nontherapeutic usage may show ''an objective intent'' that a product is used ''in the mitigation and cure of disease'' and may appropriately be classified as a drug. Accordingly, a factual showing that sunscreens are used in suntan products to prevent sunburn may be sufficient to uphold a decision classifying such products as drugs if the decision were challenged in court (32).

B. Currently

The proposed sunscreen rule further illuminates FDA's policy distinguishing drug and cosmetic use of sunscreens by stating that cosmetic products that use the term ''sunscreen'' in product labeling will be considered drugs, unless qualified to describe a cosmetic benefit provided by the sunscreen, for example, ''this product contains a sunscreen to protect hair from damage by the sun.'' Furthermore, suntanning products and sunless tanning products that do not contain sunscreens must carry warnings that they provide no protection from the sun (33).

In the proposed rule, FDA recognized that some cosmetic products contain sunscreen for ''nontherapeutic'' uses and strictly for cosmetic benefit, such as protecting hair or the color of nail polish from sun damage. However, FDA concluded in the proposed rule that when consumers are aware a product contains sunscreen, consumers expect the product to protect them from the effects of the sun, regardless of what the label states about the product's purpose (34).

A cosmetic containing a sunscreen is not considered a drug by FDA if it meets one of three requirements. A product is strictly a cosmetic if the sunscreen ingredient is added solely because it (1) is a color additive, (2) is added as a protector of the color of the product, or (3) is not intended to function as a drug and no claims are made about the ingredient on the product.

The FDA did make the distinction between ''beach'' and ''nonbeach'' products, such as lipsticks and cosmetics. FDA did not require separate statements of identity or definitions to distinguish nonbeach from beach products, but stated it ''is aware that some of the labeling recommended by the Panel for sunscreen drug products may not be appropriate for [products solely as cosmetics] (35).

The concentration ranges, limitations, warnings, and directions accompanying sunscreen active ingredients do not apply to the active ingredients in products intended solely as cosmetics. In the case of nonbeach OTC drug sunscreens, FDA proposed modified indications more suitable to the product's intended use as both cosmetic and drug (36). In addition, both beach and nonbeach products containing sunscreen must contain on the label of the product the following warning: ''Discontinue use if signs of irritation or rash appear. If irritation or rash persists, consult a doctor'' (38). On products intended as cosmetics and drugs, cosmetic claims may not appear in the boxed area designated ''Approved Uses'' (37) to avoid confusing consumers.

Because FDA is unaware of any substantial scientific evidence about the safety or effectiveness of cosmetics that claim to reverse the aging process, prevent or reduce aging, or prevent wrinkling, the FDA labeled such cosmetics ''new drugs'' and incorporated them into category II (39). The only acceptable labeling proposed regarding premature aging of the skin include:

1. Sunscreen may reduce the chance of skin aging caused by exposure to the sun.
2. While biological aging is inevitable, sunscreen may help protect skin from aging caused by exposure to ultraviolet radiation from the sun (40).

V. CONCLUSION

In summary, the regulation of sunscreen ingredients and products continues to evoke debate and interest within and beyond the United States.

In the meantime, FDA has made it clear that sunscreen agents not eligible under the OTC Drug Review process must be approved for use in the United States through submission and approval of a New Drug Application. Therefore, the final regulation may not appreciably increase the list of monograph sunscreen ingredients found to be safe and effective, although there may be some modification of permitted concentration levels for some ingredients, and possible inclusion of ingredients approved and marketed outside the United States.

REFERENCES AND NOTES

1. 21 U.S.C. §301 *et seq.*
2. 21 U.S.C. §321(g).

3. 21 U.S.C. §321(i).
4. Harris-Kefauver Amendments to the Federal Food, Drug, and Cosmetic Act, Public Law 87-781, 76 Stat. 780 (1962).
5. 21 U.S.C. §201(p).
6. 21 C.F.R. §330.12 (regulations governing final date and conditions for inclusion in the review).
7. The OTC drug review was first announced by FDA in 37 *Fed. Reg.* 85 (1972).
8. Panel members included Thomas G. Kantor, M.D., Chairman, John Adriani, M.D., Col. William A. Akers, M.D., Maxine Bennett, M.D., Minerva S. Buerk, M.D., Walter L. Dickison, Ph.D., and Jerry H. Shuck, M.D.
9. 21 C.F.R. pt. 330.
10. 43 *Fed. Reg.* 38206 (1978).
11. 58 *Fed Reg.* 28194 (1993).
12. Possible terms that will classify a product as a drug are (1) *prevents freckling*, (2) *shields from the sun*, (3) *blocks out the rays of the sun*, (4) *protects from the sun*, (5) *prevents redness*, and (6) *prevents uneven coloration*. 53 *Fed. Reg.* 28204.
13. See 14 Food, Drug, and Cosmetics Report No. 20, FDA's OTC Sunscreen Tentative Final Monograph 1, 1–13 (1993); Ken Klein, "The Sunscreen TFM: Questions, Inconsistencies, Issues and Implications," 109 *Cosmet. Toiletries* 39(11):109 (1994); 16 *Chain Drug Rev.* No. 9, Groups Dispute FDA Sunscreens Study RX34 (1994).
14. 53 *Fed. Reg.* 7078 (1988).
15. 21 C.F.R. §330.13(b)(2).
16. Category I: Safe and effective and not misbranded.
    Category II: Not generally recognized as safe and effective or recognized as misbranded.
    Category III: Insufficient data to permit final classification.
17. 53 *Fed. Reg.* 28214.
18. The standard reference formula published in the ANPR was subsequently modified in the Proposed Rule. 53 *Fed. Reg.* 28250.
19. Selection of fair-skin subjects
    Skin type and sunburn and tanning history (*based on first 30–45 min of sun exposure after a winter season of no sun exposure*).

    I Always burns easily; never tans (sensitive).
    II Always burns easily; tans minimally (sensitive).
    III Burns moderately; tans gradually (light brown) (normal).
    IV Burns minimally; always tans well (moderate brown) (normal).
    V Rarely burns; tans profusely (dark brown) (insensitive).
    VI Never burns; deeply pigmented (insensitive).

20. Cavanaugh, T. The numbers game; sunscreen standards. *Chem. Marketing Rep.* SR 18(6):244 (1993).
21. Eusolex 6300 is currently a category III drug. FDA is awaiting new data on its international marketing history. FDA is deferring its consideration of PEG-25

PABA, and isoamyl-*p*-methoxycinnamate. In addition, FDA has classified and has reviewed and approved butyl methyloxydibenzoylmethane in a new drug-product.
22. Proposed rule §352.76.
23. Groves G. A. UVA: Toward more effective protection. *Drug Cosmet. Ind.* 37(2): 155 (1994).
24. 53 *Fed. Reg.* 28233.
25. 53 *Fed. Reg.* 28217.
26. Letter from S. Fine to R. F. Kingham of Covington & Burling, June 17, 1976. Note that the Proposed Rule revokes T.C. No. 61.
27. Memorandum of meeting between E. I. Lambert of Covington & Burling and R. Apodaca, R. Heller, and G. P. Clarke, April 16, 1985.
28. Letter from R. Apodaca to E. I. Lambert of Covington & Burling, March 29, 1985.
29. In the April 24, 1987 letter to J. Correze, Chairman of the Board, Cosmair, Inc., FDA stated, "Additionally the labeling declares the presence of 'octyl dimethyl PABA' and 'benzophenone-3,' two sunscreen ingredients, and makes the claim that it 'contains sunscreen,' therefore, representing and suggesting that the product is useful as a sunscreen."

    In an April 17, 1987 letter to H. Waldron, Chairman of the Borad of Directors, Avon Products, Inc., FDA cited the presence of approved sunscreens listed as inactive ingredients in the cosmetic ingredient statement. "The labeling declares the presence of octydimethyl PABA and benzophenone-4, two sunscreen ingredients." The product was additionally labeled to "prevent lines and wrinkles by guarding against UV damage . . ."
30. FDA cited the presence of an unapproved sunscreen ingredient in letters to several companies. The language used in the May 6, 1987 letter to R. Perelman, Chairman of the Board, Germaine Monteil Cosmetiques Corp., is typical. It states:

    > [The product] is labeled to contain the sunscreen ingredient 'butyl methoxydibenzoylmethane" in addition to other sunscreen ingredients. We are not aware of any over-the-counter (OTC) sunscreen drug product that has been marketed in the United States on or before December 4, 1975, containing the sunscreen ingredient. . . . , nor are we aware of any substantial scientific evidence that establishes that this ingredient is generally recognized as safe and effective for its intended use. . . . Therefore the product is a new drug within the meaning of section 201(p) of the Act.

31. Memorandum of meeting between W. R. Pendergast, J. Gubernick, R. Dereshinsky, and D. L. Michels with R. Apodaca, R. Heller, and G. P. Clarke, April 22, 1986.
32. Eiermann H. Presented at the meeting of the Southeast Chapter of the Society of Cosmetic Chemists at Jekyll Island, Georgia, March 19, 1982.
33. Proposed rule §740.19.
34. In addition, SPF 0 or 1 will not be allowed on product labels, since such labels would mislead consumers into believing the product contains sunscreen. *Cf.* 154 Drug & Cosmetic Industry No. 6, Regulatory Rumblings 58 (1994) (since FDA's dramatic change in regulation of sunscreen-augmented make-up, there is an increased possibility that the FDA may classify moisturizers as drugs).

35. 58 *Fed. Reg.* 28203.
36. The following indications are proposed:

> (''Filters'' or ''screens'' out the (''sun's rays,'' ''sun's harsh rays,'' or ''sun's harmful rays'') to prevent (''lip damage,'' ''skin damage,'' ''freckling,'' or ''uneven coloration''). (''Protects from,'' or ''Shields from'') (''the harmful rays of the sun'' or ''the sun'') to help prevent (''lip damage,'' ''skin damage,'' ''freckling,'' or ''uneven coloration'').

37. 53 *Fed Reg.* 28206.
38. 53 *Fed. Reg.* 28231.
39. 53 *Fed. Reg.* 28287.
40. *Id.*

12

Regulatory Aspects of Sunscreens in Europe

Angela Janousek
Quest International, Ashford, Kent, England

I. DEFINITION OF ULTRAVIOLET FILTERS

In Europe sunscreen products are considered to be cosmetics, as their function is to protect the skin from sunburn. The European Economic Community (EEC) has issued a directive to its member states relating to cosmetic products (1). Article 1 says:

> A "cosmetic product" means any substance or preparation intended for placing in contact with the various external parts of the human body or with the teeth and mucous membranes of the oral cavity with a view exclusively or principally to clean them, perfuming them or protecting them in order to keep them in good condition, change their appearance or correct body odours.

Because of this definition there is a need to differentiate between the legal status of sunscreens in the United States and in European countries.

II. THE EEC DIRECTIVE PUBLISHED NOVEMBER 28, 1983

The Third Amendment (2) gives a definition and list of ultraviolet (UV) filters that cosmetic products may contain. This list is divided into two parts: UV filters that are fully permitted, and those that are provisionally permitted. This provisional list has since been updated by transferring some of the substances to the fully permitted list.

The chemical nomenclature used is IUPAC or INN and requires interpretation into trade names, which are commonly understood by a cosmetic chemist. This is tabulated in Table 1. The maximum concentration, other limitations, and requirements and warnings that must appear on the label are also listed in the table. The numbers prefaced with "S" indicate COLIPA numbers (COLIPA: Committee De Liaison Des Associations Europeans de L'Industries de la Parfumerie, De Produits Cosmetiques et de Toilette, which is the trade association representing the cosmetic industry in those European countries that are members of the Common Market).

A definition of UV filters is given in the preamble of the Annex VII EEC Directive:

> UV filters are substances which, contained in cosmetic sunscreen products, are specifically intended to filter certain UV rays in order to protect the skin from certain harmful effects of these rays. These UV filters may be added to other cosmetic products within the limits and under the conditions laid down in this Annex.

Other UV filters not listed in the table can be used for product protection (stabilizing colors, and so forth).

The status of provisionally permitted UV filters is being reviewed by the EEC working party dealing with the safety of cosmetic products, called the Ad Hoc Working Party on Cosmetic Directive (AHWP). This group consists of member states government departments, as well as representatives from industry and consumer groups. As this working party contains a mix of people, not necessarily scientifically trained, they are being advised by the Scientific Committee on Cosmetology (SCC) set up by the EEC Commission. Committee members are eminent scientists, mainly dermatologists and toxicologists from different countries. They study and review the data provided to them from various sources, including different government health boards, industry, hospitals, from databases, or others. Their remit is to form an opinion on the safety of chemical substances used in cosmetics under normal conditions of use and then publish their findings in a report.

Meanwhile, the cosmetic industry, through the EEC Countries Trade Association, COLIPA, plays an important role in securing a reasonable list of permitted UV filters and continues to collect, review, and present data and information on usage. Many of the provisionally listed UV filters are not defended because they are of no interest to industry. Those for which COLIPA is still trying to obtain full listing are denoted in Table 1 (f) (3).*

---

*Since this chapter was written, a revised list of UV filters has been published incorporating the proposed changes mentioned above (11).

The individual countries' trade associations deal with the regulatory bodies of their own governments. COLIPA has not only contributed to the preparation of a sensible definition and list of UV filters but has also been collecting safety data on the chemicals and presented them to the EEC commission in the required format.

COLIPA has a subcommittee on sun products that deals with UV filters and a working party on sun protection factor (SPF) measurement. There is a standard procedure for adding substances to the list of UV filters. This procedure is laid down in Article 8 (2) of the EEC Cosmetic Directive which reads: "the amendments necessary for adapting Annexes II to VII to technical progress shall be adapted in accordance with the same procedure, after consultation of the Scientific Committee for Cosmetology at the initiative of the Commission or of a member State."

This, in practice, means that if a new cosmetic ingredient, for example, UV filter, is discovered the following action needs to be taken.

COLIPA sends a written request to the EEC commission, supported by a file with safety data on the ingredient.

EEC commission also submits the summary of the file to the working party or Cosmetic Directive (AHWP).

SCC presents a written opinion to the EEC commission.

This opinion is then circulated to all members of AHWP.

The AHWP discusses the proposed addition to the list, usually in the presence of an industry expert. This procedure takes time, usually two or more meetings.

In case of a favorable attitude by EEC member states' delegates, the EEC commission then prepares an official proposal for an amending Commission Directive.

The proposal is sent to the Committee for Adaptation to Technical Progress (CATP), which is then convened by the EEC commission.

CATP discusses the proposed amendment and approves or rejects it by qualified majority voting.

In case of approval, the EEC commission then prepares the amending directive, this is signed by an EEC commissioner and officially distributed to all EEC member states and subsequently published in the *Official Journal of the European Communities.*

It should be clear from the foregoing that the procedure is laborious and time-consuming and some quicker action on behalf of the authorities is being looked for. It is hoped that with a more rational approach innovation will not be inhibited, but on the contrary, scientific research into new more effective compounds will increase.

Table 1 List of UV Filters That Cosmetic Products May Contain

COLIPA reference number	EEC reference number	Substances	Maximum authorized concentration (%)	Other limitations and requirements	Conditions of use and warnings which must be printed on the label	Trade names, other chemical or trivial names
Permitted UV filters:						
S1	1	4-Aminobenzoic acid	5			PABA, Paraminol, Pabacidum, Amben, Pabanol, Mexoryl SK
S57	2	*N,N,N*-trimethyl-4-(2-oxoborn-3-ylidenemethyl) anilinium methyl sulphate	6			(3-(4′-Trimethylammonium-benzylidene)-lbornan-2-one-methylsulfate
S12	3	Homosalate (INN)	10			Homomenthyl salicylate, 3,5,5 trimethyl cyclohexyl salicylate, Heliophan, Filtrosol A, benzoic acid 2-hydroxy-3,3,5 trimethyl cyclohexyl ester, HMS
S38	4	Oxybenzone (INN)	10		Contains oxybenzone[a]	2-Hydroxy-4-methoxy-benzophenone, Uvistat 24, Uvinul M40, Eusolex 4360, Cyasorb UV-9, Spectrasorb UV-9, benzophenone-3 (CTFA), Neo-Heliopan BB, Syntase 62, Uvasorb Met
S46	5	3-(Imidazol-4-yl) acrylic acid and its ethyl ester	2 (expressed as acid)			Urocanic acid

S45	6	2-Phenylbenzimidazole-5-sulfonic acid and its potassium, sodium, and triethanolamine salts	8 (expressed as acid)			Eusolex 232, Novantisol, Neo-Heliopan Hydro
Provisionally approved UV filters:						
S2 (f)[b]	1	2-Ethyl-4-bis (hydroxypropyl) aminobenzoate, mixed isomers	5			Ethyl dihvdroxy propyl PABA, Amerscreen P, (*N*-propoxylated-ethyl-4-aminobenzoate) (mixed isomers)
S3 (f)	2	Ethoxylated 4-aminobenzoic acid	10			Ethoxylated PABA ethyl ester Lusantan 25 (Ethoxylated-ethyl-4-aminobenzoate)
S5	3	Padimate (INN)	5		Contains Padimate	Amyl 4-dimethylaminobenzoate mixed isomers, mixture of pentyl, isopentyl, and 2-methylbuthyl 4-dimethylamino benzoates
S6 (f)	4	Glycerol 1-(4-aminobenzoate)	5	Free from benzocaine (INN)		Glyceryl-*p*-aminobenzoate Glyceryl PABA, GMPA, PABA monoglyceride, Escalol 106 (part), 4-aminobenzoic acid glyceryl esters
S8 (f)	5	2-Ethylhexyl 4-dimethylaminobenzoate	8			Sunarome PLUS-Arlatone UVB, Octyldimethyl PABA, Escalol 507, Padimate O, Eusolex 6007, 2-ethylhexyl *p*-dimethyl amino benzoate
S13 (f)	6	2-Ethylhexyl salicylate	5			Octyl salicylate (CTFA), Sonnelan UV, Sunarome WMO, Benzoic acid, 2 hydroxy-2-ethyl hexyl ester

Table 1 (Continued)

COLIPA reference number	EEC reference number	Substances	Maximum authorized concentration (%)	Other limitations and requirements	Conditions of use and warnings which must be printed on the label	Trade names, other chemical or trivial names
S18	7	3,3,5-Trimethylcyclohexyl-2-acetamidobenzoate	2			
S19	8	Potassium cinnamate	2			
S24	9	4-Methoxycinnamic acid salts (potassium, sodium, and diethanolamine salts)	8 (expressed as acid)			
S25	10	Propyl 4-methoxycinnamate	3			
S9	11	Salicylic acid salts (potassium, sodium, and triethanolamine salts)	2 (expressed as acid)	The pH of the finished product shall be such that the acid is not liberated	Not to be used for children under 3 years of age	Triethanolamine salicylate, Sunarome W, Sunarome G
S27 (f)	12	Amyl 4-methoxycinnamate, mixed isomers	10			Isopentyl-4-methoxycinnamate, Neoheliopan E 1000
S28 (f)	13	2-Ethylhexyl 4-methoxycinnamate	10			Parsol MCX, Neoheliopan AV, *p*-methoxycinnamic acid 2-ethylhexyl ester, Sunarome OMC
S29	14	Cinoxate (INN)	5			2-Ethoxyethyl 4-methoxycinnamate

S55	15	3,4-Dihydroxy-5-(3,4,5-trihydroxybenzoyloxy) benzoic acid, trioleate	4		Digalloyl trioleate
S39 (f)	16	Mexenone (INN)	4	Contains mexenone[c]	2-Hydroxy-4-methoxy-4′ methyl benzophenone, Uvistat 2211, Benzophenone-10 (CTFA), Aduvex 2211
S40 (f)	17	Sulisobenzone (INN) and sulisobenzone sodium (INN)	5 (expressed as acid)		2-Hydroxy-4 methoxy-benzophenone-5-sulfonic acid and salt Uvinul MS-40; benzophenone-4 (CTFA), Uval, Aduvex 112, C yasorb UV-284, Uvistat 112, Uvasorb S-5, Syntase 230
S42	18	2-Ethylhexyl 2-(4-phenylbenzoyl)-benzoate	10		2-Ethylhexyl 4′-phenyl-benzophenone-2-carboxylate
S47	19	5-Methyl-2-phenylbenzoxazole	4		
S50	20	Sodium 3,4-dimethoxyphenylglyoxylate	5		
S52	21	1,3-bis (4-methoxyphenyl) propane-1,3-dione	6		
S56	22	5-(3,3-Dimethyl-8,9,10-trinorborn-2-ylidene) pent-3-en-2-one	3		
S58	23	Alpha-(2-oxoborn-3-ylidene)-*p*-xylene-2-sulfonic acid	6		
S59 (f)	24	Alpha-(2-oxoborn-3-ylidene) toluene-4-sulfonic acid and its salts	6 (expressed as acid)		3-(4′-Sulfobenxylidene) camphor, Mexoryl SL

Table 1 (Continued)

COLIPA reference number	EEC reference number	Substances	Maximum authorized concentration (%)	Other limitations and requirements	Conditions of use and warnings which must be printed on the label	Trade names, other chemical or trivial names
S60 (f)	25	3-(4-Methylbenzylidene) bornan-2-one	6			3-(4′-Methylbenzylidene)-d-l-camphor, Eusolex 6300, Eusolex 8021 (part)
S61 (f)	26	3-Benzylidenebornan-2-one	6			3-Benzylidene camphor, Mexoryl SD, Ultren BK, Ultraoyd
S62	27	Alpha-cyano-4-methoxycinnamic acid and its hexyl ester	5			
S64 (f)	28	1-*p*-Cumenyl-3-phenylpropane-1,3-dione	5			4-Isopropyl dibenzoylmethane 1-(4′isopropylphenyl)-3-phenylpropane-1,3-dione, Eusolex 8020, Eusolex 8021 (part)
S16 (f)	29	4-Isopropylbenzyl salicylate	4			Megasol
S30	30	Cyclohexyl 4-methoxycinnamate	1			
S66 (f)	31	1-(4 *tert*-Butylphenyl)-3-(4-methoxyphenyl) propane-1,3-dione	5			Parsol 1789, 4-*t*-Butyl-4′-methoxy dibenzoyl methane
S69		2,4,6-Trianilino-(p-carbo-2-ethylhexyl-1-oxi)-1,3,5 triazine	5			

[a]Not required if concentration is 0.5% or less and when it is used for product protection.
[b]The symbol (f) represents substances in which the industry has an interest, and these are actively used in Europe. The substances in the probisional list not marked with an (f) are not used, and therefore may be delisted in the future. In these cases trade names are not given.
[c]Not required if concentration is 0.5% or less and when it is used for product protection.
Source: Adapted from Annex VII of the EEC Cosmetic Directive (1983).

The future of the provisionally listed UV filters is still uncertain. According to the present timetable, decisions have to be made by December 31, 1988. Precise analytical specification of each substance has to be presented. The amount of safety data required to defend these chemicals is unlikely to decrease. Possibly even more data will be required, such as results of investigations predicting the interactions between the chemicals, UV light, and skin.

It is very likely, however, that the timing for the provisional list of UV filters will be extended for at least 1 year. It is anticipated that in some cases the maximum concentration allowed for use in finished products will be reduced.

Additionally, certain EEC countries, including France and Italy, require labeling of the UV filters and concentration present in the finished product, if a sunscreen claim is made on the packaging. Analytical evidence of presence of the declared chemical is needed. However, no published standard methods of analysis are yet available.

III. PHYSICAL ULTRAVIOLET FILTERS

Substances that have an inherent physical capacity and have the ability to screen out light and act as a light block or barrier are not regulated according to the EEC directive. These materials include zinc oxide, titanium dioxide, mica, and pigments, to name a few. However, they favorably contribute to the increase in sun protection factor and they are often used to boost the efficiency of a sun product mainly in total block-type formulae.

IV. NON-EEC COUNTRIES: REGULATORY STATUS

Most non-EEC countries (e.g., Sweden, Turkey, Switzerland) have not laid down legislation dealing with sunscreens but largely follow the EEC directive as a code.

Austria has a separate list of pharmacologically active substances including sunscreens (4). Finland has Decree 974 of December 1985 (5). The Ministry of Trade and Industry issued on January 15, 1986 a definition of "Sunscreen Agents" and a list of permitted substances subject to restrictions. The warning statement must be in the Finnish language. Eastern European countries do not have a restricted list; however, they follow the United States and European trends.

V. SUN PROTECTION FACTOR

In determining the protective power of sunscreen products a variety of different methods have been used throughout the world. There are procedural discrepancies in the methodology recognized by the U.S. Food and Drug Administration (FDA) and the European methods (6).

The only European country that has actually published an official method for measuring SPF is Germany, and this method is generally recognized throughout Europe. The Deutsches Institut für Normung published the method as DIN 67.501 (7). The procedure is done clinically under medical supervision, and it standardizes many parameters:

Standard ref. formula with SPF value 3.7 ± 0.3 containing 2-ethylhexyl 4-methoxycinnamate concentration 2.7%
Light source: Osram-Ultravitalux 300 W. Maximum use of the lamps is 500 hr
Panel of minimum of 20 healthy volunteers, male and female of average UV sensitivity.
Test area: back 0.3 m × 0.3 m; test surface separated by 1-cm-wide plaster; 0.4-cm^2 test area.
Radiation exposure: 20 min after the first preparation is applied, observing the burning time t_{th}; increments of $\sqrt{2}$.
Test material application: Evenly 150 ± 15 mg/100 cm^2.
Data evaluation procedures 24 ± 4 h after the beginning of irradiation. The sun protection factor Q_i of the individual preparations is calculated.
Q (log): The mean sun protection factor is to be determined.

The foregoing parameters are only a rough outline of the methods and a detailed study of the standard is recommended before any further comparisons with the FDA method are made or SPF measurements conducted (8).

The method concerns itself neither with using natural sunlight nor evaluation of water-resistant or waterproof properties of UV filters. No descriptive categories of skin type or degree of sunburn are given.

The sun protection factor Q is a measurement of the effectiveness of the sunscreen product. The sun protection factor Q_i is the ratio of the time required to reach the erythema for protected skin t_i and for unprotected skin t_o.

$$Q_i = \frac{t_i}{t_o}$$

There are other methods known in Europe for determining SPF. The supporters for in vitro methods (9,10) argue that one can eliminate the limitations with biological experiments (i.e., variation from one test person to the other). Spectroscopic measurements are commonly known.

Representatives of the cosmetic industry in France, Germany, the United Kingdom, and the United States have been discussing an international harmonization of the methodology for the measurement of the SPF for sunscreen products since 1984.

In January 1988 an FDA public meeting took place to discuss the appropriate testing procedures for sunscreen products. COLIPA also submitted a position paper, and it is likely that a unform method will be published in the near future.

There is, after all, no regulatory requirement to substantiate the SPF numbers on cosmetic packaging.

The lack of legislation has advantages as well as disadvantages for both the manufacturer and the consumer. Is the cosmetic industry going to help the consumer to understand the "number game"? Perhaps harmonized guidelines throughout the world may benefit everyone.

VI. FUTURE

In recent years epidemiological evidence has accumulated data that indicate that skin cancer and degenerative skin changes (e.g., aging) are partly related to excessive exposure to UV rays. Cosmetic manufacturers tend to use UVA and UVB filters in many products, not only those that are used to prevent sunburn.

This gives the cosmetic scientist reason to believe that UV filters and other substances that have the ability to filter out UV light will become even more important in reducing the risk of premature skin aging and skin cancer. Further regulatory restriction would only harness flexibility and innovation. Guidelines, rather than regulations, would be well respected. The consumer's safety is the main obligation of a manufacturer and, therefore, the scientific and ethical approach to sunscreens should be left with the experts.

REFERENCES

1. OJ of European Communities Number L262/170 (September 27, 1976).
2. OJ of European Communities Number L332/38 (November 11, 1983).
3. Janousek, A. Current status of UV filters in the EEC. *Cosmet. Toilet.*, *102*:55–60 (July 1987).
4. BG BI Number 12/83. Pharmacologically active substances permitted in cosmetics. Austrian Federal Ministry of Health and Environment.
5. Ministry of Trade and Industry Decision Number 21 on Cosmetic Preparations, Helsinki, January 15, 1981.
6. Martini, M. C. Comparison des Méthodes de determination des SPF. *Int. J. Cosmet. Sci.*, *8*(5):215–224 (1986).
7. DIN 67.501 Beuth Verlag GmbH, Berlin (December 1985).
8. Department of Health, Education and Welfare, U.S., F.D.A. Sunscreen drug products for over-the-counter human drugs. Proposed safety, effective and labelling conditions. *Federal Register*, *43*:38206–38269 (1978).
9. Stockdale, M. Sun protection factors. *Int. J. Cosmet. Sci.*, *7*: (1985).
10. Stockdale, M. A novel proposal for the assessment of sunscreen product efficacy against UVA. *Int. J. Cosmet. Sci.*, *9*(5):85–98 (1987).
11. OJ of European Communities Number L64/10 (March 8, 1989).

13

The Sunscreen Industry in Australia: Past, Present, and Future

Gordon A. Groves
Unique Laboratories Pty., Ltd., Kenmore, Queensland, Australia

I. INTRODUCTION

With its prolonged periods of intense solar radiation the environment in Australia is such that its predominantly white population should take measures to provide adequate protection against the damaging effects of excessive exposure to sunlight. Only since the early 1970s, however, have the interests of industry and government been sufficient to generate an awareness of the significant role that sunscreens can play in providing this protection. As a consequence of publications by investigators, publicity releases by health authorities, and sales tax incentives provided by the government to manufacturers and suppliers to encourage development, Australia now has a wide range of highly effective sunscreens available. This chapter discusses the development of the sunscreen industry in Australia and the attempts that are being made to educate the public of the necessity for their use.

Medical authorities have long recognized that the incidence of nonmelanoma skin cancer in Australia is among the highest in the world. Both squamous and basal cell carcinomas, as well as other related skin lesions, are prevalent in all areas of Australia. This very high incidence of skin damage can primarily be attributed to the presence of a largely white population, mainly of Celtic origin, in a tropical environment. In addition, the Australian's desire for an outdoor life and the ready availability of pleasant beaches, excellent golf courses, and other venues that encourage long periods of exposure to sunlight is a major contributing factor, resulting in the early appearance of sun-damaged skin. Publicity releases by tourist agencies and related organizations have, for many years, exemplified the nature of life in Australia by using as a symbol of the country the "Bronzed Aussie" in reference to the well-tanned skin that a large majority of the residents possess. This healthy tanned appearance is deceiving, however, and the long periods of exposure to high-intensity ultraviolet (UV) radiation has also left many inhabitants with a coarse and leathery skin texture. Most residents of the country visibly appear to be much older than their age-matched counterparts in Europe or England because of the premature aging of the skin brought about by this outdoor existence.

Australians, for many years, have been aware of the dangers associated with excessive exposure to solar radiation; however, like so many people around the world, they have, in the past, chosen to ignore the possible consequences to their health in their pursuit of an outdoor life. Only in the last decade, as a result of intensive efforts by concerned authorities to educate the public, has there been an effort made by ever-increasing numbers of Australians to take protective measures during outdoor activities. Recently, concerned investigators conducted a series of clinics at popular beaches—only to discover that approximately 20% of the people examined had either active carcinomas of the skin or evidence that indicated the possibility of a carcinoma in the process of development. These studies indicate that the educational program must not only be continued, but must also be intensified.

II. THE ORIGIN OF SUNSCREENS IN AUSTRALIA

Although no official records have been maintained, discussions with interested parties indicate that one of the first sunscreens available in Australia was developed by a dermatologist associated with the Queensland Radium Institute in the early 1930s. The Institute, which is a division of the Royal Brisbane Hospital, is the main cancer treatment center in the state of Queensland. This original product was rather primitive by today's standards. The product was simply a 10% dispersion of phenyl salicylate (salol) in soft paraffin. The formulation was intended for patients of the Radium Institute and was not available commercially. Shortly after this product was developed, however, ICI Proprietary Limited introduced their version of a sunscreen product that was intended for the commercial market. This product was the first commercial sunscreen to be sold to the general public in Australia. This sunscreen also contained 10% salol as the ultraviolet absorber. The vehicle, however, marketed as a cream based on a beeswax–borax emulsion was more cosmetically acceptable than the original hospital formulation. The popularity of this product, UV Filter Cream, can be recognized by the fact that, even as late as 1975, many Australians were still calling all sunscreens, irrespective of their origin, UV Filter Cream.

A report published in 1959 (1), by Mowatt and Robertson of the Queensland Radium Institute, stated that salol had been used for many years by the Institute in barrier creams, zinc oxide cream, lanolin, and alcohol, B.P. The statement made in this report was that, although these preparations provided excellent protection, they were defective in the sense of vehicle "instability." The nature of this instability was not explained. Further objections to some of these formulations were that they were greasy and often lacked adhesive power, as they were easily rubbed off.

The authors of this article presented a formula for "Brown Protection Cream" which had been formulated in the early 1950s by a pharmacist named Maroney, who was in charge of the dispensary at the Townsville Base Hospital. The name was derived from the brown color of the cream caused by the tannic acid contained in the product in addition to salol. The product was reported to have had a high level of skin substantivity. The tannic acid component, however, stained clothing and linens. To overcome this disadvantage a "White Protection Cream" from which the tannic acid had been eliminated was formulated. Although the vehicle used in both creams was almost identical, it was claimed that this latter version was not as substantive to the skin as the cream containing tannic acid and was inclined to be removed by water and perspiration. The tannic acid-containing cream was preferred by the Radium Institute because of its greater resistance to removal during swimming or as a result of perspiration.

There was some indication in this publication that the period of protection provided by either of these products was not of great duration. From our present knowledge of the sun protection factor (SPF) concept in relation to the protective ability of sunscreens, we are aware that a sunscreen containing 10% salol would be rated only approximately SPF 4. The protection provided, thus, would be for only approximately 1 h with either of the two sunscreens described by Mowatt and Robertson (1). When it was necessasry to use a product that would provide hours of protection, the use of salol in concentration ranging from 10 to 30% in soft paraffin remained the ultraviolet absorber of choice of medical practitioners in the Radium Institute.

Mowatt and Robertson claimed that only one patient had ever exhibited sensitivity to salol. This sensitivity had arisen after several years of using the salol-containing sunscreen. To overcome this untoward effect, a cream based on homomenthyl salicylate was developed. The patient did not exhibit any evidence of sensitivity to this sunscreen. In later years numerous cases of sensitivity to the salol cream were reported. In many instances the irritation produced was the result of using improperly stored creams. In these creams hydrolysis of the phenyl salicylate into phenol and salicylic acid resulted in the production of a highly irritating product. As a consequence and because of its low protective power, creams based on phenyl salicylate are no longer available on the commercial market in Australia.

The *Australian Pharmaceutical Formulary* (APF), a publication of the Australian Pharmaceutical Society, contains formulas for various pharmaceutical mixtures. In the 12th version, which appeared in 1978, this publication was still providing formulations for a cream intended to act as a sunscreen. The formula shown was based on triethanolamine stearate and contained 10% salol as did the previous versions of this sunscreen.

III. SUNSCREEN EVALUATION

In 1972, Groves and Robertson (3) published a paper in which the significance of sunscreens and their use in the protection of the skin from the damaging effects of ultraviolet radiation was discussed. The protective ability of sunscreens in relation to their absorptive power and the significance of film thickness in relation to this absorptive power was discussed in detail. Reference was also made to the desirability of using sunscreens that showed adhesion to the skin. These factors were also discussed in relation to the different purposes for which the sunscreens might be used. The ultraviolet transmissions of approximately 80 sunscreens were tabulated at three wavelengths. The transmission figure recorded at 360 nm was intended to represent the range from 320 to 400 nm. The transmission figure listed for 305 nm was supposed to be indicative of erythema-producing wavelengths between 280 and 320 nm, whereas the figure given for the transmission at 260 nm was considered to be illustrative of the transmission below 260 nm. The table containing these values was intended to provide a guide from which consumers and their medical and cosmetic advisers might select a sunscreen appropriate for the specific needs of the individual concerned. This list was not complete and included only those preparations that were readily available to the authors. In 1974 (4) a further publication included practically all sunscreens that were on the Australian market at that time. The list at that time contained 150 sunscreens and could be considered to be an indication of the Australian public's increasing interest in sunscreens. This second list classified the sunscreens into six groups, based on the percentage transmission at 305 nm shown by an 0.01-mm film of the product in its original form. Group 1 consisted of all those products for which the transmission level was less than 1%, whereas group 2 consisted of products with a range of transmissions between 1 and 3%. Groups 3 and 4 comprised those products with transmission values at 305 nm between 3 and 9% and 9 and 20%, respectively. Group 5 was subdivided into products with transmission values of over 20%, but less than 50%, and those for which the percentage transmission was over 50%, but less than 70%. All products with transmission values for the 0.01-mm film greater than 70% were consigned to group 6.

In 1977 Robertson and Groves (5) listed all the products available on the Australian market in October of that year. The list by this time had grown to 195 products. As a further service to the consumer, this list not only classified sunscreens into the same six groups previously employed, but also categorized the products in each major group into subgroups indicative of the nature of the sunscreen. These designations included the terms aerosol, solution, creams, gels, lip, nose and face products, milks, and oils.

In the initial two lists (3,4) the classification of the sunscreens was based entirely on the percentage transmission shown through a 0.01-mm film of the product in the form in which it would be applied to the skin. In the third (5), the concept of the percentage effective transmission was employed, rather than the single transmission value at 305 nm previously used. This procedure required the determination of the percentage transmission at 5-nm intervals through the erythemal range 295–330 nm. Each of these values was rated for sunburning effectiveness and the combined total of the eight values was used to express the percentage effective transmission over this erythemal range. The effective transmission values were considered by the authors to more clearly represent the protective ability of the sunscreen, rather than the single value at 305 nm previously employed, because the value obtained represented the complete absorption spectrum throughout the erythemal range. Quantitative determinations using a UVB meter with sunlight and a fluorescent sunlamp confirmed the correctness of this modification.

In all three studies (3–5) the percentage transmission of the sunscreens was determined using the so-called thin-film procedure. This procedure employed a modified Hilger-Watts spectrophotometer in which the cell compartment was replaced with a unit designed to hold the plates containing a thin film of the sunscreen product. The sample unit consisted of a brass plate containing a quartz window. A second brass plate of similar dimensions contained a rectanglar opening into which a quartz prism could be inserted. This prism contained a precisely ground depression on its flat surface into which the sample was placed. The two plates were tightly clamped together with the flat surface of the prism against the quartz window. The entire unit was then fastened securely to the flat end-face of a photomultiplier tube. This unit was then placed in an upright position with the prism in the cell holder compartment of the spectrophotometer. This procedure for evaluating the protective effectiveness of sunscreens only saw use in Australia. The evaluation of sunscreens by this technique has been discontinued for a number of years because of the many anomalies encountered.

Although these publications showing the thin-film–transmission values were instrumental in generating considerable interest in sunscreens in Australia, our present knowledge of these products indicates that many of the preparations listed would not now even rate classification as sunscreens.

IV. SUNSCREENS AND THE CONSUMER

In 1974, three visiting dermatologists from the United States carried out a limited sunscreen study in Australia on behalf of an international sunscreen manufacturer. The study was conducted using what is now described as the sunscreen protective factor (SPF) procedure. These investigators caused considerable concern at both state and federal government levels when they were reported in the press as saying that only two of the approximately 180 sunscreens on the Australian market at that time were capable of preventing erythema. When it is realized that the study was conducted on only three products, the reports by the media would have to be seen as being grossly exaggerated. Investigators at the University of Queensland, including the author, quickly stated that their studies had shown that sunscreens available to the Australian consumer provided varying levels of protection and did include a large number of products providing an extremely high level of protection. These comments were based on the results that had been obtained spectrophotometrically by Groves and Robertson (3,4).

The press statements by the U.S. investigators concerning the general lack of effectiveness of sunscreens available to the consumer aroused government interest. The Federal Department of Science and Consumer Affairs was requested to undertake an evaluation of all sunscreens on the Australian market. The sole purpose of this evaluation was to provide a guide from which consumers could select the product that most met their needs. The federal government analyst made arrangements for the evaluation to be conducted by Dr. Groves and Dr. Robertson of the University of Queensland. The initial method of evaluation was the thin-film procedure previously described. Subsequently, however, a solution method was adopted as being more representative of the level of transmission allowed by an 0.008-mm layer of the sunscreen on the skin.

The lists prepared by these investigators were published on an annual basis by the Australian Cancer Society and were distributed free to any interested person. As a consequence of the publication of these lists, considerable public interest was aroused and greater awareness of the dangers of excessive exposure to ultraviolet radiation became evident. Publication of these lists ceased in 1983 with the introduction of the SPF method of evaluation as a standard procedure in Australia.

V. STANDARDS AUSTRALIA AND SUNSCREENS

Standards Australia is, as the title indicates, a government-regulated body charged with developing standards for a wide variety of materials and operations. As the result of an approach to the government by an international manufacturer of sunscreens, concerned about the supposedly poor quality of some of the sunscreens available, this association was asked by the Federal Department of Health to undertake the development of a procedure that could be used for determining the effectiveness of sunscreens. As a consequence of this request Subcommittee 42 of the Standards Association was formed consisting of representatives of government, cosmetic and pharmaceutical manufacturers, professional associations, and universities.

This committee met at regular intervals over several years. After considerable discussion Australian Standard 2604–1983 was finalized. The procedure finally accepted by the committee was a standardized version of the technique originally proposed by Schulze (6). In addition to providing a method of evaluation, this publication also gave directions for the labeling of sunscreens.

The test procedure proposed by Australian Standard 2604–1983 was essentially identical with that proposed for adoption by the U.S. Food and Drug Administration (FDA) (7). In both procedures the xenon arc lamp was recommended as the preferred source of radiation. Unlike the FDA-proposed protocol, however, its Australian counterpart makes no provisions for testing sunscreens using natural sunlight. However, in the critical area of film thickness, which can significantly influence the results obtained, both procedures require the application of 2 mg/cm^{2}m. Table 1 compares the major steps in the two procedures.

In 1986 a revised standard was published as Australian Standard 2604–1986. The title of the standard remained the same as that of the 1983 version, that is, "Sunscreen Products—Evaluation and Classification." This version of the standard differed from the original by introducing the concept of the "secondary sunscreen." This descriptive term was introduced, supposedly, to clarify the requirements for multipurpose sunscreen products relative to labeling. This revision of the standard also contained guidelines for the testing of both broad-spectrum and water-resistant sunscreen products as did the 1983 edition.

Table 1 A comparison of the U.S. and Australian Procedures for the Determination of the SPF

Parameter	United States	Australia
No. of panelists	20	10
Preferred radiation source	Xenon arc	Xenon arc
Product dose (mg/cm^2)	2	2
Exposure increment (n)	1.25	1.25
Water resistance (min)	40	40
Waterproof (min)	80	Not used
Standard test formula[a]	Yes	Yes

[a]The same standard formula is used in both procedures.

The latest edition of Australian Standard 2604 was published in 1993 as a joint publication between Standards Australia and Standards New Zealand. New Zealand sunscreen suppliers must now provide only products that comply with this new publication, known as Australian/New Zealand Standard 2604–1993, if they wish their product to be included in a list of sunscreens published by the New Zealand government. This marks the initiation of regulations for the sunscreen industry in New Zealand.

VI. SUNSCREEN PRODUCTS: EVALUATION AND CLASSIFICATION

Australian Standard—1993 describes procedures for determining the performance of sunscreens in terms of their sun protection factor as did the two previous editions of the standard. As indicated previously, it also provides guidance on the testing of both broad-spectrum and water-resistant sunscreen products. The standard also specifies labeling requirements in some detail. The following sections describe more fully the precise requirements applying to sunscreens as indicated in the standard.

A. Classification of Sunscreens

Sunscreens in Australia are classified either as primary or as secondary sunscreen products. This is a somewhat arbitrary division of these preparations into those that have as their main or primary function the protection of the skin against the effects of ultraviolet radiation and those products that are primarily intended to serve some other purpose, but contain an ultraviolet absorber as an additive action to their main functions. The description of a product as a "sunscreen with added moisturizer" would designate the preparation as a primary sunscreen. Alternatively, if the product is described as a "moisturizer with added sunscreen," the classification would be as a secondary sunscreen. The parameters that differentiate products into either of these two categories have never been defined. The principal reason for establishing the two categories initially was to overcome the concern of cosmetic manufacturers that some cosmetics, such as lipsticks, would not have room on their container to include all the labeling requirements that a primary sunscreen must meet.

Whether a product is used for the purpose of sunbathing at the beach or as a daytime moisturizer, it is, for all intents and purposes, a sunscreen and should be labeled as such in my opinion.

B. Sunscreen Categories

Both primary and secondary sunscreen products are classified into four categories, based on the label SPF value. These category descriptions are shown in

Table 2. Products with a protection factor of less than 2 are not considered to be sunscreens.

In the first two editions of Australian Standard 2604 it was mandatory that the sunscreen label must display the sunscreen category into which the product fell. The Australia–New Zealand Standard 2604–1993 has removed the requirement that the sunscreen category must be shown on the label. In the opinion of Standards Australia the public is now sufficiently educated on the meaning of the SPF value that the category designation had become superfluous. The SPF value is now the only mandatory requirement of a sunscreen label relative to the level of effectiveness provided.

The 1993 standard also saw the category description of "maximum" changed to the designation "very high protection." It was felt that the term maximum implied "absolute" protection and, therefore, was not appropriate. A further change in categories was the elimination in the 1993 edition of the table relating the category description to skin type. This change was made because it was considered that this advice was no longer appropriate, as it could be interpreted that people with category IV skins and above would not benefit from a higher SPF product.

C. Broad-Spectrum Sunscreens

The term *broad-spectrum* is used in Australia to describe those products that, in addition to providing protection in the UVB region between 280 and 320 nm, also provide protection against UVA radiation higher than 320 nm. The original standard published in 1983 restricted the use of this term solely to sunscreens that were described at that time as providing maximum protection. In the 1986 standard, however, this terminology was extended so that it could be applied to sunscreens categorized as moderate and high protection also. Sunscreens with SPF values of less than 4, which are considered to be minimum protection products, cannot be described as having broad-spectrum capabilities.

Table 2 Categorization of Sunscreens in Australia

Category	SPF level
Minimum protection	At least 2, but < 4
Moderate protection	At least 4, but < 8
High protection	At least 8, but < 15
Very high protection	15 or more[a]

[a]Australian standards state that products with SPF values greater than 15 may only be labeled SPF 15^{+}.

The procedures described for the determination of the UVA protection capabilities of a sunscreen are based on spectrophotometric determinations. These techniques were first described in Australian Standard 2604–1983 and have been revised for inclusion in the 1993 edition. The procedures may be basically divided into two types: those using an appropriate strength solution of the sunscreen in a spectroscopically clear solvent; and those in which the transmission of a thin film of the sunscreen in its originally form are employed. These latter procedures are intended primarily for the evaluation of products containing insoluble pigments.

The 1993 standard states that for a sunscreen to be considered capable of providing adequate protection against UVA radiation, an 0.008-mm film must not show more than 10% transmission of radiation at any wavelength between 320 and 360 nm.

The procedures described in Australian Standard 2604–1993 cannot be considered to be representative of the true function of the product on the skin in its ability to protect the skin from the effects of UVA radiation. With the possible exception of the thin-film procedures, spectrophotometric procedures do not consider the role of the vehicle in contributing to the level of protection provided. Another criticism of the standard is the definition of what constitutes a broad-spectrum product. There have been no studies to determine whether a product that shows 10% transmission as an 0.008-mm film is providing adequate protection against skin damage by UVA radiation.

Until such time as a reliable and accurate biological method is developed for the determination of the protective ability of a sunscreen in terms of UVA radiation, the Australian standard will continue to include the inappropriate spectrophotometric methods.

D. Water-Resistant Sunscreens

The Australian standard outlines procedures that can be used for determining the water-resistance of a sunscreen. The use of either a swimming pool or a spa is permitted. The period of immersion required is 40 min. This 40-min study requires an initial immersion of 20 min in either the spa or pool, followed by a 5-min rest period before a second immersion period of 20 min is undertaken.

In the first two editions of the Australian Standard 2604 it was stated that for a sunscreen to meet the water-resistant specifications it should not show an SPF value after immersion that is less than 50% of the SPF value determined before immersion. This proviso has been deleted in favor of a requirement that the SPF value stated on the label must be that value determined after the period of immersion. One criticism of this requirement is that some manufacturers, to obtain the postimmersion value desired, are developing products that possess a much higher SPF value for the preimmersion determination. This implies that a customer using one of these sunscreens is actually getting a much higher level of protection than the label indicates. This could be contrary to the expectations of the user. Another criticism of this new requirement is that there is no limitation to the percentage that the SPF value is lowered following immersion in water. As an example, the static SPF value might be 50 and the postimmersion value 10, which is hardly indicative of a water-resistant sunscreen. As long as this product is labeled SPF 10, however, it satisfies the requirements stated in Australian Standard 2604–1993.

There has been a trend in Australia toward the development of sunscreens that will provide resistance to removal by water for immersion periods of 2, 4, or even 8 h. Although this level of substantivity is easily obtained through the use of resins and polymers in the formulation, Standards Australia has placed a restriction on the claims that may be made for water resistant products. There was concern that some of the claims being made for water resistance could give the impression that the product was effective for a longer period than the SPF value would allow. As an example, there is a definite anomaly in stating on the label of an SPF 4 product that it is water resistant for up to 4 or even 8 h. Because the times claimed for water resistance should not exceed the times for which the sunscreen would be expected to give adequate protection, Australian Standard has placed limits on the length of time immersed that may be claimed for a water-resistant sunscreen. Table 3 shows these limits.

The term *waterproof* is not permitted under current regulations. The product may be described only as being water-resistant, regardless of the period of immersion to which it might have been subjected.

VII. GOVERNMENT REGULATIONS

Before 1992, sunscreens in Australia were considered to be in the same category as cosmetics, and regulations governing the marketing of such products were the concern of the various state governments. In 1992, however, sunscreens were declared to be drugs and came under the control of the Therapeutic Goods Administration (TGA), a division of the Federal Department of Health.

Table 3 Water-Resistance Claims

SPF value after immersion	Maximum water-resistance claimed
At least 2, but < 4	Shall not be claimed
At least 4, but < 8	1 h
At least 8, but < 15	2 h
15 or more	4 h

As a consequence of this development, sunscreens are under more rigid control than they previously were. In addition to ensuring that the product complies with the current Australian standard the manufacturer must comply with the Code of Good Manufacturing Practice. The active ingredients must be stated on the label, and only those ingredients that have been recognized in various government regulations are permitted. In addition, sunscreens must now carry an expiration date, similar to other drug products.

VIII. ANTICANCER COUNCILS AND SUNSCREENS

The knowledge that the incidence of skin cancer has almost reached epidemic proportions in Australia has led to concerted efforts by government health departments and anticancer organizations to increase public awareness of the dangers associated with unwise exposure to sunlight. The incidence of nonmelanotic skin cancer on an Australia-wide basis is 938:100,000, ranging from a high of 1,348:100,000 in Queensland, which is the northern subtropical region to a low of 311:100,000 in Tasmania (R. Marks, private communication, 1988), where high solar intensity levels are present for much shorter periods of the year than they are in the northern regions.

To encourage manufacturers to market sunscreens, and also to keep the cost of these within the reach of lower-income families, the Australian government exempts from sales tax all primary sunscreens showing an effectiveness level of SPF 4 or higher. Secondary sunscreens are not given this exemption. The exemption for primary sunscreens is also dependent on the nature of the claims made on the label. The label must stress the protective nature of the product and not its cosmetic virtues.

Anticancer councils in each of the states are very active in their fight against skin cancer. Central to most of the states' attempts to educate the public was a "Slip, Slap, Slop" campaign. This was an animated video that was introduced about 10 years ago and had monumental success. The video was widely broadcast by television stations and did more to contribute to the fight against skin cancer than any other approach. Unfortunately, this approach to advertising has been discarded in favor of several different advertising efforts.

Several of the states make use of education officers to take the message about the dangers of skin cancer to schools and other community groups. In Queensland, volunteers are trained to deliver the education message about sunscreens. Education about the danger of skin cancer is started in most schools in the first grade, with follow-up programs in later years. In Queensland, special kits have been developed to assist teachers in preparing lessons for their pupils on the use of sunscreens. In addition, each child on starting school is provided with a sample of sunscreens and literature to supplement the information obtained from the teacher or the nurse educator.

In some areas of Australia, schools are actually providing sunscreens for their students, and no student is allowed out of the class room and into the playground without applying sunscreen.

Pharmacies, doctors' offices, and community health centers all display signs containing information about sunscreens, sunlight, and the skin, along with other forms of health literature.

These efforts over the last two decades appear to have succeeded in making the public reasonably well informed on the selection and use of sunscreens and other protective measures required to prevent skin damage by ultraviolet radiation. Despite these attempts, the incidence of skin cancer is still very high and, for melanoma, the incidence is actually increasing.

IX. SUNSCREENS IN THE FUTURE

Since 1970, there has been a tremendous increase in our knowledge of the relation between exposure to ultraviolet radiation and skin damage. Concurrent with this has been an increase in our knowledge of the level of protection a sunscreen must provide to adequately protect the skin. The statements of a decade or so ago that, if a sunscreen reduced the percentage transmission to 10%, the level of protection would be adequate, is now known to be without foundation. SPF 15 sunscreens often reduce the transmission of ultraviolet radiation to 0.001% when applied as a 20-μm film to the skin.

Unfortunately, occasionally, the desire to increase the protective level of a sunscreen has led to the incorporation of what might be considered to be unacceptable high levels of ultraviolet absorbers. Despite the knowledge that most of these agents have been thoroughly tested for absence of deleterious effects on the skin, there is still some concern about the possible long-term effects that might result from the use of these higher concentrations. This concern has led to attempts to improve the vehicle used for the ultraviolet absorber to allow the incorporation of these agents at much lower concentrations.

In the last decade the development of daytime moisturizers and other products of a cosmetic nature containing ultraviolet absorbers has increased significantly. Most of these products possess SPF values of between only 4 and 8 and are not intended for use while sunbathing. There are indications that, in the future, it might be difficult to find any cosmetic product in Australia intended for use on areas that will be exposed to sunlight that does not contain an ultraviolet absorber. Products intended for use on the hair that contain ultraviolet absorbers are also being made available.

The standard of sunscreens provided to the Australian consumer has increased dramatically since 1972. There is no reason to doubt that this advancement will not continue as new agents and new formulations become available. In the near future there is a strong possibility that sunscreens, which are now restricted to being labeled SPF 15+ when the SPF value is greater than 15 will be able to show on the label the actual SPF value that has been determined for the product.

REFERENCES

1. Mowatt, K. S. and Robertson, D. F. Solar protection cream. *Med. J. Aust.*, *2*:1306 (1959).
2. Mowatt, K.S. and Robertson, D. F. Solar lipstick: A protective device. *Med. J. Aust.*, *2*:222–223 (1958).
3. Robertson, D. F. and Groves, G. A. The selection and use of topical sunscreens. *Med. J. Aust.*, *2*:1445–1451 (1972).
4. Groves, G. A. and Robertson, D. F. Further studies on topical sunscreens. *Med. J. Aust.*, *2*:898–901 (1974).
5. Robertson, D. F. and Groves, G. A. Classification of sunscreen preparations. *Aust. J. Dermatol.*, *18*:109–118 (1977).
6. Schulze, R. Einige Versuche und Bermerkungen zum Problem der Handelsublichen lichtschutzmittel. *Parf. Kosmetik.*, *37*:300–316 (1956).
7. Sunscreen products for over the counter use. *Federal Register*, *43*:38206–38269 (1978).

14

The Sunscreen Industry in Japan: Past, Present, and Future

Minoru Fukuda and Masako Naganuma
Shiseido Research Center, Yokohama, Japan

I. INTRODUCTION

The chronic and acute damage to human skin caused by the ultraviolet rays in sunlight has attracted attention both in Western society and in Japan, although to different extents. The necessity of sunscreens to prevent such damage is also recognized in both cultures. However, Japanese and Caucasians have different skin colors and sensitivities to ultraviolet rays. Therefore, their attitudes toward suntanning and their expectations of what sunscreen products should do differ likewise. Furthermore, the legal regulations covering the manufacture of sunscreen products differ from one country to another.

In this report, we introduce the Japanese attitude toward sunburn and suntan, and describe the sunscreen products developed to meet demands arising from it. Current information on ultraviolet (UV) absorbers permitted for sale in Japan is presented. Finally, we discuss current and future market trends in sunscreen products, current problems related to evaluating sunscreen efficacy, and subjects for future research.

II. WHAT FUNCTION DO SUNSCREENS SERVE?

Billions of people live on the earth, and each has his or her own color of skin. In humans, skin color depends on the quantities of melanin, carotene, and oxygenated or reduced hemoglobin contained in the skin, as well as in the skin's thickness, water content, and diaphanousness. Among other factors, the quantity of melanin that is distributed in the skin determines its fairness or darkness, and greatly influences the human "complexion." At the same time, melanin plays a part in minimizing the damage that UV rays cause in the skin. This means that a person's color determines his or her degree of resistance to UV, which is one measure of the adaptability of people to earth's environment. Moreover, skin colors determine not only attitudes toward UV, but also attitudes toward prevention of UV damage.

The so-called sun belt zone circling the equator, where people receive strong UV rays, is part of the torrid or subtropical zone; the people who live there generally are dark skinned. In the middle latitudes away from the equator, where sunlight contains moderate UV, the majority of inhabitants have yellowish skin tones. People with white skin live in high latitudes, far from the equator.

The Japanese live in an island country that extends from latitude 25 N to latitude 45 N, and ethnically belong to the Mongolian peoples. They have yellowish skin, with moderately developed melanin productivity. The gradations of skin color of the Japanese vary widely from fair to dark. Most of those with fair skin have the skin brightness and hue common to Caucasians; those with dark skin have aspects in common with Negroes (1). In our classification of dorsal skin colors of the Japanese by the sensuous parameter of visual sensation in addition to the colorimetric parameters of value, chroma, and hue, the colors observed were divided basically by the parameters of value (V) and chroma (C), and were arranged within the spectrum from fair skin through dark (2; Fig. 1). Naturally there are differences in sensitivity to erythema and tanning between Japanese and Caucasians, and even among Japanese individuals with various skin tones. Figure 2 compares the process of suntanning between the Japanese and Caucasians by means of measurement of the MED/MMD ratio using sunlight as the light source (2–5). The result showed that the Caucasians tanned hardly at all when exposed to single UV radiation unless erythema (sunburn) had been induced to some degree; the Japanese were more prone to tan than the Caucasians.

It is well known that chloasma, one of the main pigmentation disorders, is caused or aggravated by UV (4–6). Because Japanese women have higher levels of melanin productivity than Caucasians and they have median-colored skin, chloasma is prominent on their skin and can be a source of trouble. For Caucasians, or Negroes whose bare skin is black, chloasma does not pose so serious a problem. A comparison of conceptions and preferences about suntanning between the Japanese and Caucasian women revealed quite characteristic differences (Tables 1 and 2).

Japanese women's concepts about suntans are different for the face and the body (see Table 1). Where the face is concerned, 61% of the Japanese female

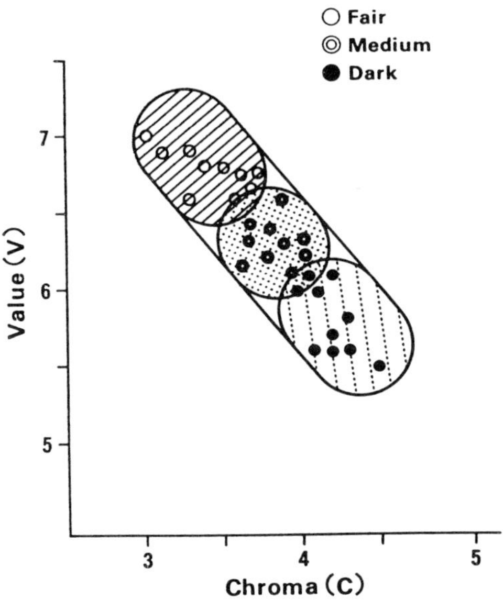

Figure 1 Distribution of Japanese skin color.

consumers dislike tanning, and only 14% are "suntan enthusiasts" who want to be bronze-colored or lightly tanned. For the body, the proportion who want to tan reached 37%. However, 35% of the Japanese female consumers were basically "sunscreen lovers."

Among the American female consumers, 82% of them prefer to tan on both the face and body (see Table 2). If we add the 18% who prefer to tan only on the body, but not on the face, 100% of the American female consumers desire a tanned skin; there was no one in our sample who disliked suntanning. In this respect, there was a great difference of opinion between Japanese and American women.

Furthermore, among Japanese women, most of those who gave affirmative answers, wish to recover their natural skin tone when the summer is over. In contrast, about 90% of American and European women said they like to remain tanned into the autumn or winter. These differences mean that Japanese and American women also differ in their expectations of sunscreen products. American women with white skins expect sunscreens to allow them to develop a deep, beautiful tan that lasts for a long time, without the adverse effects of burning, skin cancer, or photoaging (89%). Only a small fraction of American women require an effective protecting function that permits neither sunburn nor suntan (11%).

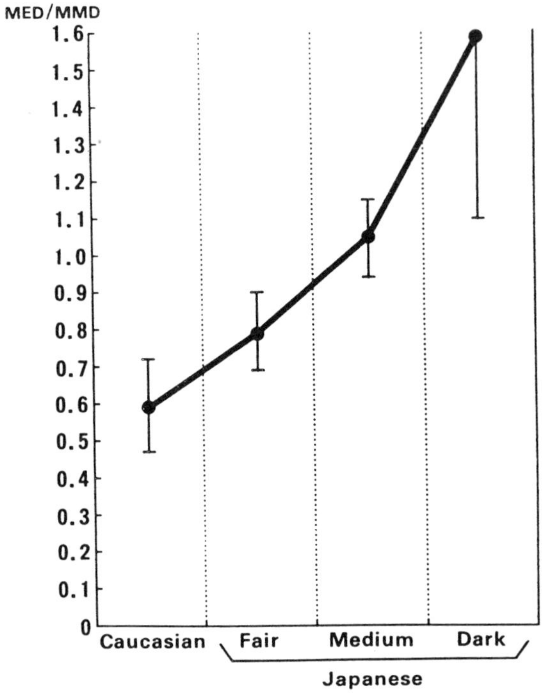

Figure 2 Tanning capacity by Japanese with various skin colors and Caucasian.

On the other hand, in Japan, the word "suntan" means melanogenesis, and sunscreens are used as preventives against tanning. Although Japanese women expect sunscreens to prevent photoaging or sunburn, as do American women, they further require that the product retard or prevent tanning. They will not accept sunscreens that lack this antitanning function.

Table 1 Japanese Women's Attitudes Toward Suntans

	Face (%)	Body (%)
Would like a quick, deep bronze tan	3	13
Would like a slow, mild, and beautiful tan	11	24
Suntan acceptable	25	28
Would rather not tan	24	20
Suntan unacceptable	37	15

n = 261.

Table 2 Preferences of American Consumers

What type of suntan do American consumers prefer?	
Dark tan on body and face	62%
Mild tan on body and face	20
Dark tan on body and no tan on face	15
Mild tan on body and no tan on face	3
No tan on either body or face	0
What American consumers expect from sun care products	
Deep beautiful tan without dryness or aging	43%
Deep and beautiful tan in safety	26
Mild and deep tan without sunburn	11
Complete protection with no tan and no sunburn	11
Quick and long-lasting tan	9

In Japan, fair skin has been considered from ancient times to be an essential factor in beauty: the Japanese aesthetic sense is represented by the old saying, "A fair complexion hides seven defects." Following this traditional aesthetic sense, skin colors and UV reactivities, which differ from those of Caucasians have determined Japanese women's thinking about suntanning and have led to different expectations of sunscreen performance between Japanese and American women. In addition, Japanese skin is more sensitive to stimulation by cosmetics and other products than that of Caucasians. This is an important point that should be taken into account in the design, development, and sale of sunscreens.

III. THE DEVELOPMENT OF ULTRAVIOLET ABSORBERS

The Japanese skin has greater skin reactivity to UV and skin sensitivity to stimulations than either American or European Caucasians. Furthermore, there is a difference even in the function they expect sunscreens to perform: Japanese expect total UV preventive functions, including a defense against tanning. In Japan, therefore, sunscreens are required to have proved safety and efficacy for skin protection.

Sunscreens are classified as cosmetics in Japan. Although cosmetics are composed of several kinds of materials that are approved for use as cosmetics, notification of the purpose of the preparation is not required, as it is for drugs. Therefore, cosmetics manufacturers produce sunscreens using the materials that have the best UV preventive effects among the approved ones, which are listed in Table 3. The UV spectra of typical absorbers are shown in Figure 3.

Of these absorbers, *p*-aminobenzoic acid (PABA; 7–9), glyceryl *p*-aminobenzoate (10–12), amyl *p-N,N*-dimethylaminobenzoate (13), and 2-ethoxyethyl

Table 3 UV Absorbers Used in Japan

UV absorber	COLIPA no.	Category I of FDA	Trade name
Derivatives of PABA			
p-Aminobenzoic acid (PABA)	1	+	
Glyceryl *p*-aminobenzoate (glyceryl PABA)	6	+	Escalol 106
Amyl *p-N,N*-dimethylaminobenzoate(amyldimethyl PABA, padimate A)	5	+	Escalol 506
2-Ethylhexyl *p-N,N*-dimethylaminobenzoate(octyldimethyl PABA, padimate O)	8	+	Escalol 507
Ethyl *p*-aminobenzoate		−	
Derivatives of cinnamic acid			
Methyl 2,4-diisopropylcinnamate	23	−	
Ethyl 2,4-diisopropylcinnamate	22	−	
Potassium *p*-methoxycinnamate			
Sodium *p*-methoxycinnamate	24	−	
Isopropyl *p*-methoxycinnamate		−	
2-Ethylhexyl *p*-methoxycinnamate (octyl methoxycinnamate)	28	+	Parsol MCX, Neo Heliopan AV
2-Ethoxyethyl *p*-methoxycinnamate (Cinoxate)	29	±	Giv Tan F
Ethyl *p*-methoxycinnamate			
Sunguard B		−	
Derivatives of benzophenone			
2,4-Dihydroxybenzophenone (benzophenone 1)	33	−	ASL-23, Uvinul 400
2,2′,4,4′-Tetrahydroxybenzophenone (benzophenone 2)	34	−	Uvinul D-50
Sodium 2-hydroxy-4-methoxy-5-sulfobenzophenone (benzophenone 5)	40	−	ASL-24S
2-Hydroxy-4-methoxybenzophenone-5-sulfonic acid (benzophenone 4)	40	+	Uvinul MS-40
2-Hydroxy-4-methoxybenzophenone (Oxybenzone, benzophenone 3)	38	+	ASL 24
2,2′-Dihydroxy-4,4′-dimethoxybenzophenone (benzophenone 6)	35	−	Uvinul D49
Sodium 2,2′-dihydroxy-4,4′-dimethoxy-5-sulfobenzophenone (benzophenone 9)	36	−	Uvinul DS-49
Derivatives of salicylic acid			
2-Ethylhexyl salicylate (octyl salicylate)	13	+	Sunarome WMO
Phenyl salicylate	14	−	
3,3,5-Trimethylcyclohexyl salicylate (homosalate)	12	+	Filtrosol A
Others			
2-(2′-Hydroxy-5′-methoxyphenyl)benzotriazole	47	−	Tinuvin P
4-*t*-Butyl-4′-methoxybenzoylmethane	66	−	Parsol 1789

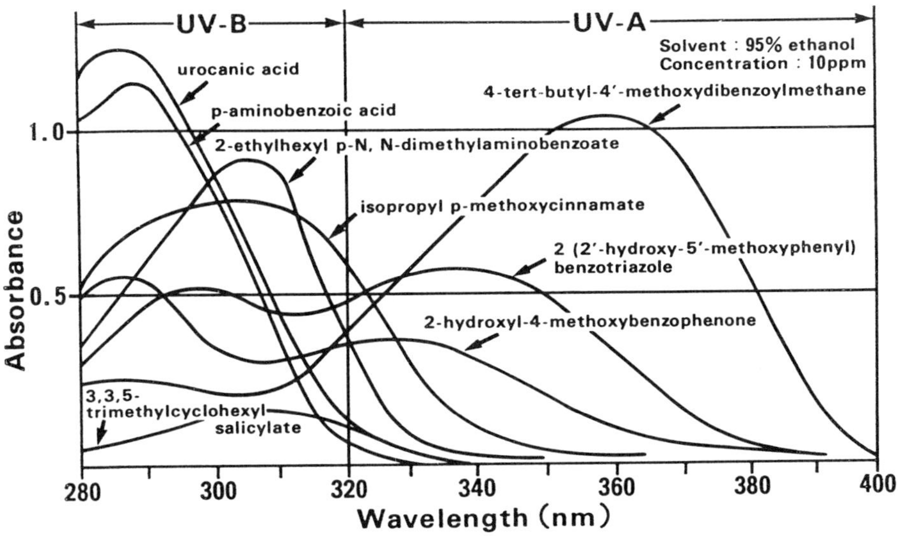

Figure 3 UV absorption spectrum.

p-methoxycinnamate (14,15) are considered to be of questionable safety, although they have superior UV-absorbing effects. 3,3,5-Trimethylcyclohexyl salicylate has inferior UV absorbance and thus must be added in large amounts. Urocanic acid, which is a biogenic constitutent, is safe and has high UV absorbance, but its absorbance peaks at near 287 nm, with weak absorbance near 310 nm, so that no practical effect can be expected from it. In the 1970s, it was found that 2(2′-hydroxy-5′-methylphenyl)benzotriazole induces allergic response and Riehl's melanosis (16,17), and this gave rise to public controversy in Japan. Since then, this substance has not been used.

The most popular material with Japanese manufacturers on the basis of its safety and effectiveness is octyl *p*-dimethylaminobenzoate, followed by ethylhexyl *p*-methoxycinnamate and 2-hydroxy-4-methoxybenzophenone. Since the harmful effects of UV rays have become widely known in recent years, the demand for safer UV absorbers has increased, and manufacturers are working to develop products that can intercept even the UV rays we are exposed to in daily life. Sungard B was the first such screening agent to be developed, and some sunscreens that contain it have appeared on the market (18). We have so far described mainly UVB protective agents. The Japanese have also shown an interest in the harmful effects of UVA, especially in its tanning effect. UVA rays penetrate the skin more deeply than UVB (19), and reach the earth in high doses (20). Normally our skin is exposed to greater amounts of UVA than to UVB rays. Several experiments have proved that prolonged exposure of skin to UVA rays causes abnormalities in the connective tissue of the dermis (21,22) and accelerates the toxic effects of UVB, such as carcinogenesis (23). Furthermore, it has become clear that the UV rays that induce chloasma, a source of concern for Japanese women, range from UVA to UVB, and that the main causative wavelength lies in UVA (6). Of the UV absorbers shown in Table 3, benzophenone derivatives and 2(2′-hydroxy-5′-methylphenyl)benzotriazole seem to provide the best UVA coverage. However, these substances have low absorbancy of UVA rays; moreover, the safety of the latter is questionable, as described earlier (16,17). Thus, neither of them is satisfactory as a UVA absorber. Some attempts were made to develop effective UVA absorbers, and the use of 4-*tert*-butyl-4′-methoxydibenzoylmethane (Parsol 1789) and its derivatives for cosmetics were approved (18). This UVA absorber, however, has not yet gained approval in the United States. One of the reasons may be that UVA absorber cannot effectively prevent sunburn caused mainly by UVB. In the United States, UV absorbers are evaluated on the basis of their preventive effect against sunburn (24). Now that the harmful effect of UVA has been proved, this UVA absorber should be accepted even in the United States.

The foregoing discussion has dealt with UV absorbers. Next we take up UV-scattering agents, in the form of inorganic powder, as UV preventive cosmetic materials. The scattering effects of several kinds of inorganic powders are shown in Figure 4. Superior UV preventive effect is observed in titanium oxide, ferrous oxide, and zinc oxide.

To improve the texture of cosmetics using these inorganic powders, while taking advantage of their UV preventive effect, some attempts have been made to change the forms of the powders or to prepare complexes of several inorganic powders or combinations of inorganic powder and organic substances. The UV protective effect of inorganic powder varies with the diameter of the particle. Titanium oxide, with its fine grain 0.03 μm in diameter, has high UV-scattering effect and is permeable to visible rays (25). It does not have a conspicuous color when applied to the skin, which is an advantage. Similar superfine-grained forms of zinc oxide are now being developed. Furthermore, for the purpose of improving the affinity for the cosmetics base, the surfaces of particles are often processed with oil to make them lipophilic. Because titanium oxide has poor spreading quality, some procedures, such as adding a close covering of the surface of spherical nylon particles with titanium oxide, were devised to overcome this drawback (26). The oxide complex of silicon and titanium was also developed; this complex has superior UV intercepting effect and permeability to visible rays. Moreover, it is possible to process inorganic powder with UV absorbers to give the powder a UV protective effect. Thus, many kinds of powders have been developed, and are being used in the newer sunscreens on the market.

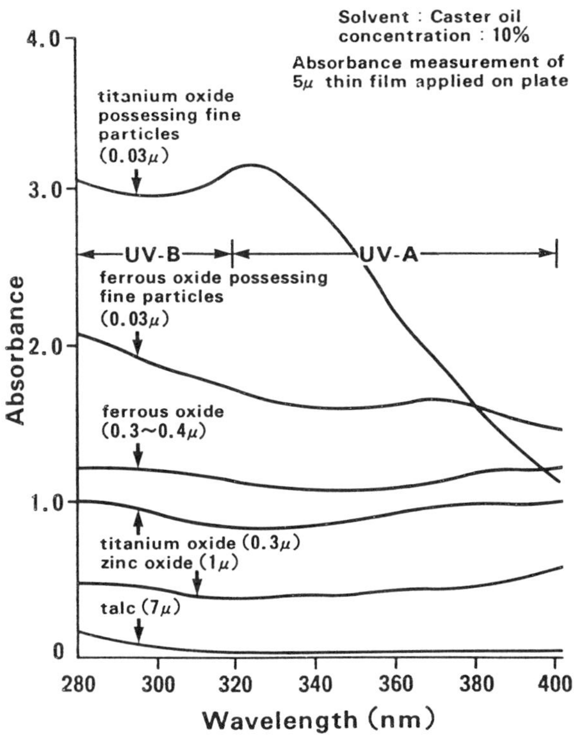

Figure 4 UV protective effect of UV-scattering agent.

Safety and complete UV protective effect are essential factors for sunscreens sold in Japan. Therefore, UV absorbers and UV-scattering agents are usually added in sunscreens in various combinations, rather than as a single ingredient in large amounts.

IV. THE PAST, PRESENT, AND FUTURE OF SUNCARE PRODUCTS IN JAPAN

Cosmetic products with the added attraction of UV preventive effect first appeared in Japan in 1978. Before then, partly as a result of advertising posters showing beautifully tanned women in swimsuits, it had been fashionable to get a tan by sunbathing at the beach. However, young women often became excessively sunburned or pigmented in their eagerness to become tanned in a short time. At the same time, many people disliked tanning on the face. In this situation, product labeling enabled consumers to use sunscreens to the best effect for their own purposes and according to their skin sensitivities to UV.

As the dangers of UV rays, especially the potential of aggravating stains and freckles, became known to the public, people came to desire sunscreens with greater UV preventive effect. Nevertheless, sunscreens were still considered seasonal products, used only for leisure, on the beach or in the mountains in summer, and on the ski slopes in winter.

In Japan seasonal changes in climate are very distinct. UV radiation, temperature, and humidity are all lower in winter (Fig. 5). After the vernal equinox, the temperature rises and the weather becomes suitable for outdoor sports. UV radiation exposure increases, and the skin needs some care. Around the summer solstice, the rainy season begins, and most days are humid and wet. With the passing of the rainy season, the hot summer weather begins, sun rains down its full strength from the sky. In autumn, the weather becomes comfortable again, although it gets dry toward the beginning of winter.

It has become popular to alter the basic cosmetics and the style of makeup with the changing seasons. In winter, mositurizing agents or oily cosmetics are often used to protect the skin from drying, whereas in summer, people prefer more astringent cosmetics. Emulsive or comparatively oily foundations are used in winter, whereas foundation cakes for wet or dry use are preferred in summer. In recent years, it has become common practice to add UV protective substances to the makeup foundations intended for use in summer when the skin is exposed to higher doses of UV rays.

Because most foundations normally contain titanium oxide and ferrous oxide as coloring materials, they have some degree of UV protective effect. This effect

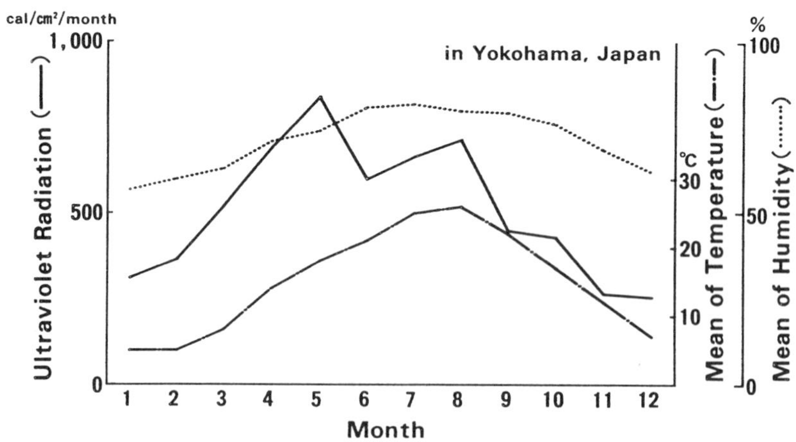

Figure 5 Seasonal variation in UV radiation, temperature, and humidity.

is enhanced by increasing the amount of these materials used in foundations by adding UV absorbers. Elaborate procedures of ingredient arrangement and combination have been developed so that a change in prescription will not spoil the original effect of cosmetics.

Thus, with the popularization of UV protection, UV care has become part of the daily routine, at least in the summer months. In addition, people now realize the necessity of UV protection throughout the year owing to warnings by dermatologists that daily exposure to UV rays can accelerate aging of the skin. Accordingly, some milky lotions and other basic cosmetics that contain UV protective agents have appeared on the market. This development of effective UVA absorbers, and the combination of UVA absorbers and the UVB absorbers has spurred the development of cosmetics with the ability to intercept UV rays over the whole spectra, without containing any powder. In the future, more products of this sort will be developed, and daily UV care will become even more routine.

V. WHAT IS THE EFFECT OF SUNSCREENS?

A. Sun Protection Factor

Since the idea of sun protection factor (SPF) was popularized and numerical data on the sunscreening products became available, one problem has become apparent. Because the standard value of SPF varies, depending on the manufacturer or country, it is difficult to judge the UV protective effect of sunscreens on the basis of a comparison of numerical data.

If SPF is to be classified solely on the basis of the UV protective effect of sunscreening products, a standardization of measuring conditions, whether in vitro or in vivo, could solve the problem. However, when SPF is considered a

Table 4 SPF Determined with Four Types of Light Source

	Light source			
			Xenon arc solar simulator	
Sample	Natural sunlight	Fluorescent sunlamp	290–400 nm	290–2457 nm
Sunscreen 2	2.3 ± 0.9[a]	2.8 ± 0.6	1.8 ± 0.3	1.6 ± 0.3
Sunscreen 4	2.4 ± 0.8	7.1 ± 2.4	N.D.	N.D.
Sunscreen 6	5.1 ± 1.2	22.0 ± 4.9	13.1 ± 1.8	7.1 ± 2.0
Sunscreen 10	6.9 ± 1.2	26.8 ± 5.0	N.D.	8.6 ± 3.0

[a]Mean ± SD.
Source: Ref. 27.

criterion that enables consumers to judge the effectiveness of a sunscreen in preventing sunburn (telling how long the product can provide safe sunbathing, compared with or without any sunscreens), some problems have to be solved in measurement of effect.

The first is the problem of the light source used for the SPF measurement. According to our examination, the SPF value of sunscreens varies depending on whether natural sunlight, with all its rays from UV to IR, is employed or artificial light sources (including solar simulators, which can radiate the substantially faded spectra of IR) are employed. The numerical values are obtained from the latter (27,28).

The second problem is the individual differences among the subjects selected for SPF measurement. The higher a subject's sensitivity to UV, the lower the minimal erythema dose (MED) shown, and the larger the SPF value becomes (27).

The third problem is the difference in the amount of product applied to the skin between actual use and experimental use for SPF measurement. We asked Japanese women to use on their faces several kinds of sunscreening products sold on the market, and calculated the consumption per unit of surface area of the face on the basis of the weight of the product used. This experiment revealed that there is a great difference in the amount of the various products applied to the skin (Table 5). The smallest amounts consumed were observed in the solid powder product containing less oil, which gave the value of 0.5 mg/cm^2, and in the solid powder product containing large amounts of oil, for which consumption was about 1.3 mg/cm^2. They were followed by the values for the oil-type product, about 2.0 mg/cm^2, and the oil stick-type product, about 2.1 mg/cm^2. The largest consumption was observed in the lotion product, which gave a value of about 2.7 mg/cm^2. In some products, the application amount at the time of SPF measurement (2 mg/cm^2) reached about two to four times the practical consumption. This indicates a great discrepancy in conditions between SPF measurement and practical use.

Table 5 Usual Dose of Various Types of Sunscreens

Sunscreen type	Dose (mg/cm^2)
Solid powder type I	0.46 ± 0.29
Solid powder type II	1.28 ± 0.21
Oil type	1.97 ± 0.59
Lotion type	2.72 ± 0.89
Stick type	2.09 ± 1.29

Actual efficacy of the sunscreening product is determined by three factors: (1) UV sensitivity of the subject; (2) amount of exposure and quality of UV at the time of use of the product; and (3) UV protective effect of the product (SPF). Furthermore, SPF value varies depending on the quality of the light source, sensitivity of the subject, and amount applied. The factors that influence the SPF value and actual efficacy of sunscreens include those—such as differences among subjects, types of products, and differences in application amounts between subjects—that cannot be standardized internationally.

We proposed that sunscreens should be evaluated under the conditions in which the product would be used by employing sunlight as the light source, if the products are intended for domestic use by Japanese consumers (29).

It is as important to compare UV protective effects of products under standardized conditions as it is to grasp the efficacy of products at the time of actual use. However, in the present situation, the meanings of comparison and of actual efficacy are occasionally confused.

B. Tanning Protection

Japanese are more prone to tanning than are Caucasians. Although Americans and Europeans expect sunscreens to allow them to tan safely, Japanese expect them to prevent tanning; exactly the opposite expectation. When the reactivity to UV and attitudes about tanning of Japanese are considered, it may be necessary to introduce the idea of controlled tanning as a parameter for judging the effect of sunscreens, in addition to SPF.

At present, SPF is the only universal parameter that is used for judging the effect of sunscreens. This SPF, however, indicates the protective effect against erythema caused by UV, but not that against tanning, which is the most serious problem for Japanese women.

We have evaluated the effect of five kinds of marketed sunscreening products on Japanese subjects under sunlight using two parameters (Table 6; 28): $SPF_{erythema}$ and $SPF_{tanning}$, the coefficient of prevention against tanning.

Table 6 $SPF_{erythema}$ and $SPF_{tanning}$

Sunscreen	Subjects	$SPF_{erythema}$	$SPF_{tanning}$
Sunscreen oil	18	1.39 ± 0.35	1.54 ± 0.30
Sunscreen lotion	35	2.07 ± 0.61	1.94 ± 0.59
Sunscreen cream	35	3.03 ± 1.01	2.90 ± 0.79
Sunscreen stick I[a]	18	8.09 ± 2.27	6.72 ± 1.71
Sunscreen stick II[b]	8	9.17 ± 2.58	6.51 ± 1.36

[a]Stick I: $SPF_{erythema}$ vs. $SPF_{tanning}$ $p < 0.05$.
[b]Stick II: $SPF_{erythema}$ vs. $SPF_{tanning}$ $p < 0.05$.

In the evaluation of the oils, lotions, and creams that have comparatively lower UV protective effect, there appeared to be no significant difference between $SPF_{erythema}$ and $SPF_{tanning}$. However, in stick I and stick II, which have high UV preventive effect, $SPF_{tanning}$ was significantly lower than $SPF_{erythema}$. This result can be interpreted as follows: (1) It is more difficult to block UV rays involved in tanning than to block those involved in erythema; (2) the sunscreening products tested had not been sufficiently well designed from the viewpoint of prevention of tanning; (3) there is a possibility that SPF should be taken as an extravagant indication by Japanese consumers, who consider the prevention of tanning as priority; and (4) the efficacy of sunscreening products should properly be evaluated by using two parameters, $SPF_{erythema}$ and $SPF_{tanning}$.

C. Evaluation of the Ultraviolet A Protective Effect

Ultraviolet A protection has so far been studied mainly for the purpose of clinical treatment, using patients sensitive to UVA (30) or by artificially producing a state of sensitivity to UVA by oral or topical application of 8-methoxypsoralen (8-MOP) (31), trimethylpsoralen (32), or tetracycline (33) to patients with psoriasis vulgaris and vitiligo.

However, although those methods are superior in clinical practice, it is not possible to compare the UVA protective effect of sunscreens quantitatively with each other when experimental facilities are different, owing to discrepancies in sensitivities of the subjects and action spectra.

Johnson (34), Barth (35), Akin (36), and Groves (37) and their co-workers reported in experiments with rats, mice, and guinea pigs that the UVA protection effect of a sunscreen can be determined by inducing hypersensitivity in the animals to UVA by oral, subcutaneous, intraperitoneal, and topical applications of photosensitizers such as 8-MOP, 5-MOP, chlorpromazine, and anthracene. However, the measuring conditions yielding differential effects taking into account species of animal, type of photosensitizer, dose, route of administration, and light source, have not yet been established.

We recommend two methods: (1) oral administration of 8-MOP to guinea pigs and (2) onset of immediate pigment darkening on the skin of healthy humans, as indicators.

The first experimental system is a combination of the methods of Akin et al. (36) and Johnson et al. (34). Guinea pigs show the same reaction as seen on the human skin after administration of 8-MOP; they have a larger skin area for testing than mice and rats, and are easier to handle. Furthermore, because topical application necessitates mixing the test material with photosensitizers on the same patch of skin, which involves a risk of determining the UVA protective effect of the test material incorrectly, oral administration was employed.

In an experiment with guinea pigs photosensitized by 8-MOP, Hartley albino guinea pigs were orally given 5 mg of 8-MOP, and the minimum phototoxic dose (MPD) of the applied and nonapplied areas was obtained using a Toshiba FL-BLB lamp equipped with a glass filter as the UVA light source to calculate PF_A (protection factor to UVA) by the formula:

$$PF_A = \frac{\text{MPD of applied area}}{\text{MPD of nonapplied area}}$$

The immediate pigment darkening in the second experimental system is said to be a phenomenon wherein the colorless melanin precursor existing in epidermal keratinocyte is photooxidized to darken; the action spectrum of this reaction is reported to range from the UVA to the visible region (38).

The effect of UVA absorbers on the prevention of immediate pigment darkening in human skin caused by UVA was measured by determining minimum immediate pigment darkening dose (MIPD) as the index; the effect of UV absorbers in protecting the skin from UVA exposure (PF_A) was calculated by dividing MIPD in the sample-applied area by MIPD in the nonapplied area.

The experimental system based on immediate pigment darkening is a practical method using healthy men as subjects; it causes no serious skin damage, and allows determination of UVA protective effects within a short time. In evaluating the effect of UV absorbers on the prevention of UVA exposure using these two methods (Table 7), there was a little difference in PF_A between the action spectrum for 8-MOP (maximum effective wavelength: 330 nm) and that

Table 7 PF_A of UV Absorbers Determined with Two Test Methods

UV absorbers	Protection factor to UVA: 8-MOP method[a]	Protection factor to UVA: IPD method[b]
Octyl dimethylaminobenzoate (Escalol 507)	1.2 ± 0.3	1.0 ± 0.2
Titanium dioxide (P-25)	1.6	1.9 ± 0.8
Di-(*p*-methoxycinnamoyloxy)mono-(2′-ethylhexanoyloxy)propane (Sunguard B)	1.6	1.0
2-Hydroxy-4-methoxybenzophenone (ASL-24)	2.6 ± 0.9	2.1 ± 0.4
2-(2′-Hydroxy-5′-methylphenyl)benzotriazole (Tinuvin P)	3.2 ± 0.7	2.4 ± 0.6
4-*tert*-Butyl-4-methoxydibenzoylmethane (Parsol 1789)	5.8 ± 3.0	3.8 ± 1.3

[a]8-MOP method—Concentration: 2%; solvent: castor oil or olive oil; subject: 4 ~ 8 8-MOP-sensitized guinea pigs.
[b]IPD method—Concentration: 3%; solvent: emulsion; subjects: 10 male Japanese.

for immediate pigment darkening (maximum effective wavelength: 380–550 nm). The effect of UV absorbers on the prevention of UVA absorption should be evaluated using at least two test systems.

VI. PROBLEMS AND PROSPECTS FOR THE FUTURE

One goal of cosmetics scientists is to be able to prevent the skin from aging. Recovery and maintenance of "fresh skin" is a common desire of all people. Aging phenomena in the skin include stains, wrinkles, yellow skin, flabby skin, and tumors. These skin changes are most commonly observed in the sites usually exposed to the light, especially the face, and they are said to be caused by UV accumulation and external stimulation.

According to Kligman et al. (21,39), skin aging caused by UV rays is characteristically accompanied by accelerated degeneration in dermal elastic fibers and in collagen fibers. They say that these symptoms disappear and skin can easily recover its previous condition if proper sunscreens are applied to the skin to intercept UV.

Therefore, the most important countermeasure to skin aging is defense of the skin against UV rays, and it is not too far off to say that this can be achieved by the use of effective and safe sunscreens. UV absorbers are roughly classified by the range on the UV spectrum that they absorb into UVB absorbers, UVA absorbers, and UVAB absorbers. Recent study has indicated that UVB and UVA accelerate skin aging, and that both must be intercepted to prevent this aging.

When the reactivity of Japanese people's skin to tanning and the effect of UVA on the dermis are considered, the emphasis on research on sunscreens is placed on development of UV absorbers that have a practical protective effect against both UVA and UVB. Japanese are more prone to tanning than Americans or Europeans, although most dislike it, and so Japanese expect sunscreens to provide superior blocking of UV rays. Obtaining the best effect requires using a high concentration of UV absorbers. However, the Japanese government considers safety the most important point in the approval of drugs in cosmetics. This is true of UV absorbers, and additive UV absorbers in most products have been restricted to less than 4–5%.

Therefore, to design sunscreens that give large values of $SPF_{erythema}$ and $SPF_{tanning}$, manufacturers often use inorganic powders of titanium dioxide, zinc oxide, and ferrous oxide, which have the physical function of reflecting and scattering of UV, in addition to UV absorbers. Various combinations of powders that have specific functions have been developed (25,26).

However, the types of cosmetics in which these kinds of inorganic powders can be used are limited, and products in which large amounts are used sometimes have unappealing colors or textures. Inorganic powder cannot be used in lotion or oil sunscreens with the ''light texture'' preferred by Japanese consumers.

In the future, skin care cosmetics for daily use as well as those for leisure use, will also be required to have sunscreen function. Accordingly, there is a continuing need for new UV absorbers that can be used in sunscreens in sufficient and necessary amounts and, at the same time, that are safe enough for use on sensitive Japanese skin (40).

The dosages of sunscreens are determined not only by the safety levels of UV absorbers, but also by their solubility in the base or solvent. High solubility in cosmetics bases and solvents, such as oil, alcohol, and water, is an essential factor for cosmetics materials.

Furthermore, UV absorbers must be water resistant. Because most sunscreens are used on the beach or in the mountains in summer, UV absorbers would lose their practical value if they were easily dissolved by sea water or perspiration. UV absorbers have little practical value without water-resistant qualities, even if they have high UV absorbance. Therefore, development of UV absorbers having superior water resistance as well as UV protective effect, safety, and high solubility is highly desirable.

In addition, development of a low-priced UV dosimeter, which can be used by the average consumer to measure UV exposure on the spot, is also essential for controlling UV rays and for promoting proper use of sunscreens by consumers. The paper sensor developed by Suetugu et al. is made of 0.3-mm-thick paper, coated with photoactivator, discolorator, and UV absorbers (41). This paper sensor reddens as it absorbs UV rays; it is easy to use and gives good quantitative results.

VII. CONCLUSION

A variety of UV absorbers will be developed in the future. At present it is difficult to compare available products because of the vast differences in evaluation methods among manufacturers and research institutes. Only recently has an attempt been made to standardize the procedure for SPF measurement in Japan, the United States, West Germany, and Australia. International standardization should be encouraged. The necessity of UVA prevention is evident from medical proof of the dermal toxicity of UV. Therefore, measures should immediately be taken to develop a method of evaluating UVA blocking and to identify UVA absorbers on an international scale.

Quite a few data have indicated that cosmetics are useful for protecting the skin from cumulative skin damage caused by external environmental factors. Interception of UV rays by sunscreens is a significant tool for the prevention of acute or chronic skin damage. In the future, it is supposed in Japan that not only sun care products, but also other cosmetics for daily use, including foundations, will be required to have sunscreen functions.

REFERENCES

1. Morikawa, F., Nakayama, Y., Iikura, T., Nakajima, K., Ohta, O., and Ishihara, M. The application of photographic techniques for the differentiation of the location of melanin pigment in the skin, *Biology and Diseases of Dermal Pigmentation* (T. B. Fitzpatrick, et al., eds.), University of Tokyo Press, Tokyo, 1981, p. 231.
2. Fukuda, M., Nagashima, M., Munakata, A., Nakajima, K., and Ohta, S. Effect of biological and physical factors on ultraviolet erythemal and pigmentary response. *J. Soc. Cosmet. Chem. Jpn.*, *13*:20 (1979).
3. Fukuda, M. and Naganuma, M. Sunscreen, *Cutaneous Aging* (Y. Takase, et al., eds.), Seiji Shoin, Tokyo, 1978, p. 369.
4. Fukuda, M., Naganuma, M., and Maeda, K. Protection against cutaneous aging induced by repeating sun exposure, *Cutaneous Aging* (A. M. Kligman, et al., eds.), University of Tokyo Press, Tokyo, 1988, p. 589.
5. Fukuda, M. and Nakajima, K. Sunscreen. *J. Jpn. Cosmet. Sci. Soc.*, *5*:73 (1981).
6. Toda, K. and Ohta, M. Female facial melanosis, *Biology and Diseases of Dermal Pigmentation* (T. B. Fitzpatrick, et al., eds.), University of Tokyo Press, Tokyo, 1981, p. 225.
7. Horio, T. Photocontact dermatitis from *p*-aminobenzoic acid. *Dermatologica*, *156*: 124 (1978).
8. Marmelzat, J. and Rapaport, M. J. Photodermatitis with PABA. *Contact Dermatitis*, *6*:230 (1980).
9. Thune, P. Photosensitivity and allergy to cosmetics. *Contact Dermatitis*, *7*:54 (1981).
10. Thune, P. Contact and photocontact allergy to sunscreens. *Photodermatology*, *1*:5 (1984).
11. Meltzer, L. Sensitization to monoglyceryl *p*-aminobenzoate. *J. Invest. Dermatol.*, *12*:21 (1949).
12. Fisher, A. A. Dermatitis due to benzocaine present in sunscreen containing glyceryl PABA. *Contact Dermatitis*, *3*:170 (1977).
13. Kaidbey, K. H. and Kligman, A. M. Phototoxicity to a sunscreen ingredient (padimate A). *Arch. Dermatol.*, *114*:547 (1978).
14. Frosh, P. J. and Kligman, A. M. A method for appraising the stinging capacity of topically applied substances. *J. Soc. Cosmet. Cham.*, *28*:197 (1977).
15. Davies, M. G., Hawk, J. L., and Rycroft, R. J. Acute photosensitivity from the sunscreen 2-ethoxyethyl-*p*-methoxycinnamate (Giv-Tan). *Contact Dermatitis*, *8*:190 (1982).
16. Cronin, E. Photosensitizers, *Contact Dermatitis*, Churchill–Livingstone, Edinburgh, 1980, p. 102.
17. Hayakawa, R., Kobayashi, M., and Matunaga, K. Leucomelanosis from Tinuvin P. *Aller. Pract.*, *21*:38 (1983).

18. Fukuda, M., Naganuma, M., and Nakajima, K. Laboratories studies on UVA protection with Parsol A. *Nishinihon J. Dermatol.*, *49*:88 (1987).
19. Everett, M. A., Yeargers, E., Sayre, R. M., and Olson, R. L. Penetration of epidermis by ultraviolet rays. *Photochem. Photobiol.*, *5*:533 (1966).
20. Fukuda, M., Naganuma, M., and Nakajima, K. Ultraviolet radiation of sunlight in Japan. *Acta Dermatol. (Kyoto)*, *82*:551 (1987).
21. Kligman, L. H., Akin, F. J., and Kligman, A. M. The contributions of UVA and UVB to connective tissue damage in hairless mice. *J. Invest. Dermatol.*, *84*:272 (1985).
22. Poulsen, J. T., Staberg, B., Wulf, H. C., and Brodthagen, H. Dermal elastosis in hairless mices after UV-B and UV-A applied simultaneously, separately or sequentially. *Br. J. Dermatol.*, *110*:531 (1984).
23. Staberg, B., Wulf, H. C., Klemp, P., Poulsen, T., and Brodthagen, H. The carcinogenic effect of UVA irradiation. *J. Invest. Dermatol.*, *81*:517 (1983).
24. Sunscreen drug products for over-the-counter-human use. *Federal Register*, *43*: 38206 (1978).
25. Saitoh, T., Murui, Y., and Tate, K. Function and role of foundation. *Fragrance J.*, *73*:10 (1985).
26. Tokubo, K. Application of powders to cosmetics—their properties and reform of their function. *Fragrance J.*, *80*:60 (1986).
27. Fukuda, M., Naganuma, M., Nakajima, K., and Mitui, T. Studies on several factors affecting SPF. *J. Soc. Cosmet. Chem. Jpn.*, *19*:38 (1985).
28. Fukuda, M. Sunscreen and sun protection factor in Japan. *J. Jpn. Cosmet. Sci. Soc.*, *8*:315 (1984).
29. Fukuda, M. Current state and problems concerning SPF measurement in Japan. Proceedings of the 6th International Information Center of Cosmetic Industries Conference, Geneva, Switzerland, Session IIIB (1984).
30. Rice, E. G. Dihydroxyacetone naphthoquinone protection against photosensitivity. *Dermatologica*, *153*:38 (1976).
31. Jarratt, M., Hill, M., and Smiles, K. Topical protection against long-wave ultraviolet A. *J. Am. Acad. Dermatol.*, *9*:354 (1983).
32. Parrish, J. A., Pathak, M. A., and Fitzpatrick, T. B. Prevention of unintentional overexposure in topical psoralen treatment of vitiligo. *Arch. Dermatol.*, *104*:281 (1971).
33. Dhalen, R. F., Shapiro, S. I., Berry, C. Z., and Schreiber, M. M. A method for evaluating sunscreen protection from longwave ultraviolet. *J. Invest. Dermatol.*, *55*: 164 (1970).
34. Johnson, J. A., Czaplewski, R. R., and Fusaro, R. M. Protection against long ultraviolet light with dihydroxyaceton/naphtoquinone. *Dermatologica*, *147*:104 (1973).
35. Barth, J. Mouse screening test for evaluating protection to longwave ultraviolet radiation. *Br. J. Dermatol.*, *99*:357 (1978).
36. Akin, F. J., Rose III, A. P., Chamness, T. W., and Marlow, E. Sunscreen protection against drug-induced phototoxicity in animal models. *Toxicol. Appl. Pharmacol.*, *49*:219 (1979).

37. Groves, G. A. and Forbes, P. D. A method for evaluating the photoprotective action of sunscreens against UV-A radiation. *Int. J. Cosmet. Sci.*, *4*:15 (1982).
38. Pathak, M. A., Riley, F. C., and Fitzpatrick, T. B. Melanogenesis in human skin following exposure to longwave ultraviolet and visible light. *J. Invest. Dermatol.*, *39*:435 (1962).
39. Kligman, A. M. 3. Intrinsic aging vs photoaging. *Cutaneous Aging* (Y. Takase, et al., eds.). Seiji Shoin, Tokyo, 1985, p. 33.
40. Rapaport, M. J. Patch testing in Japanese subjects. *Contact Dermatitis*, *11*:93 (1984).
41. Suetugu, M., Tomita, K., Higashi, T., and Kawamura, F. Development of ultraviolet sensor. *J. Soc. Cosmet. Chem. Jpn.*, *23*:207 (1989).

IV

SUNSCREEN INGREDIENTS AND FORMULATIONS

15

The Chemistry of Sunscreens

Nadim A. Shaath
KATO Worldwide, Ltd./Nickstadt-Moeller, Mt. Vernon, New York

I. INTRODUCTION

The number of new cases of skin cancer reported each year worldwide has reached epidemic proportions (1). This rise in cases may be attributed to several factors, including changing lifestyles and pursuit of leisure activities that lead to excessive exposure to the sun's damaging rays, demographic changes, earlier disease detection techniques, the depletion of the earth's protective ozone layer (2), and the recent evidence linking the ultraviolet A (UVA) rays to more serious damage, beyond the fashionable and sought-after tan for which UVA rays were originally responsible (3).

These important findings prompted the cosmetic industry to create new sunscreen products that would afford the consumers more efficient protection. Because the cosmetic industry began formulating many sunscreen active agents into an array of functional products, it has thus become necessary for the cosmetic chemist to know more about the chemical structure and reactivity of sunscreening chemicals and their potential interaction with, and effect on, cosmetic formulations.

This chapter, which is based in part on a series of articles published by this author in 1986 and 1987, reviews the chemistry of UV filters, the theory behind UV absorption, and the effect of vehicle on the efficacy of sunscreen chemicals (4).

II. THE ELECTROMAGNETIC SPECTRUM

Human skin and hair are constantly subjected to solar radiation (5). The radiation emitted by the sun is of an electromagnetic character and differs from other forms of electromagnetic radiant energy in its spectrum, as described by its energy (E), wavelength (λ), or the frequency (ν). Thus, electromagnetic radiation is energy governed by the following relation:

$$E = h\nu \tag{1}$$

where E = energy (ergs), h = Planck's constant = 6.62×10^{-27} erg/s, and ν = frequency (cycles per second or Hertz [Hz] units).

An important physical relation governing the properties of electromagnetic waves is described by Eq. (2)

$$\nu = \frac{c}{\lambda} \tag{2}$$

where c = speed of light = 3.0×10^{10} cm/s and λ = wavelength (cm or m). Equation (3) is derived by substituting Eq. (2) into Eq. (1), which explains the relation of the mechanism of action of sunlight on humans:

$$E = \frac{hc}{\lambda} \tag{3}$$

Energy (E) and the wavelength (λ) have a reciprocal relation seen in Eq. (3); that is, as the wavelength increases, the energy associated with it decreases and vice versa. Thus, for example, we see from Figure 1 that x-rays ($\lambda = 10^{-9}$ m) have higher energies associated with them than visible light ($\lambda = 10^{-6}$ m). The significance of this relation between energy and wavelength will become evident when the effects of UV radiation on the skin are discussed.

Solar radiation that is visible to the eye is only a very small segment of the total range of the electromagnetic waves, which can roughly be divided into three regions (see Fig. 1): electrical or radiowaves; optical rays; x-ray, gamma, and cosmic rays. The optical region of the solar spectrum is subdivided into three regions: ultraviolet, visible, and infrared (Fig. 2). The UV region has the shortest wavelengths and the higher energies associated with it. These rays are sufficiently energetic to cause photochemical reactions, resulting in both immediate and delayed damage to the skin and hair, termed *photochemical effect*. The visible region or the *light effect* is where the color of the spectrum is exhibited (violet to red). The longest wavelength (hence, lower energy) is the infrared (IR) region, which is responsible for the *heat effect*. The UV rays, which have been demonstrated to be the most damaging to humans, can be further subdivided into three regions (Fig. 3).

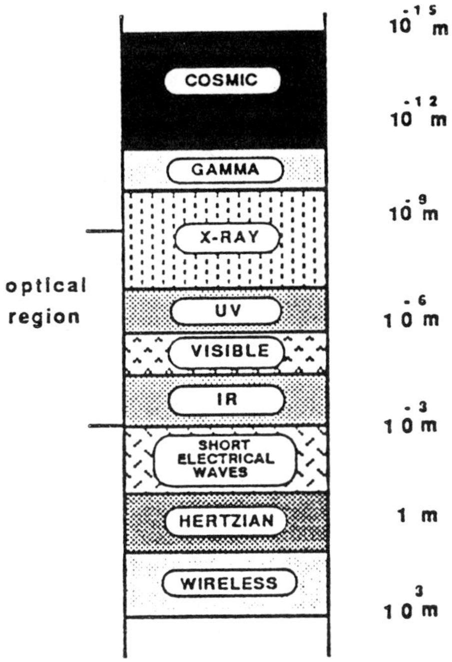

Figure 1 The total spectrum of radiant energy.

The most damaging of the UV radiations is the UVC, also called the *germicidal region*, because it has the highest energy associated with it (the lowest wavelengths, 200–290 nm). Fortunately, the harmful rays are filtered by the stratospheric ozone layer; thus, none of the rays reach the earth. The depletion of this layer through the continued use of chlorofluorohydrocarbons (CFCs) poses a major threat to mankind if left unabated. Nevertheless, artificial light sources (tanning salons, mercury arc, or welding arcs) do contain UVC radiation and should be avoided or used only with adequate protection. It is the UVA and UVB regions, which are not filtered out by the ozone layer, that are sufficiently energetic to cause damage to the skin and hair. The UVB region, with wavelengths ranging from 280 to 320 nm, also called the *burning* or *erythemal region*, is responsible for most of the immediate damage to the skin, resulting in erythema or sunburn.

Parrish et al. (3) list ten reasons why UVA rays are extremely important and should be dealt with:

1. The amount of solar UVA reaching the earth's surface is enormously greater than that of UVB.

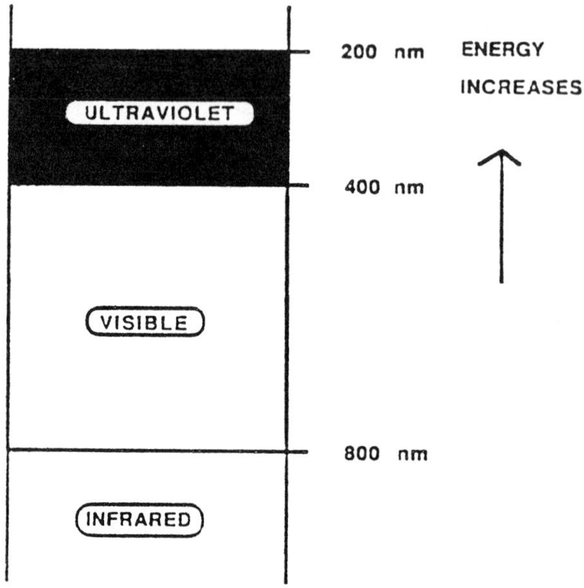

Figure 2 The solar spectrum.

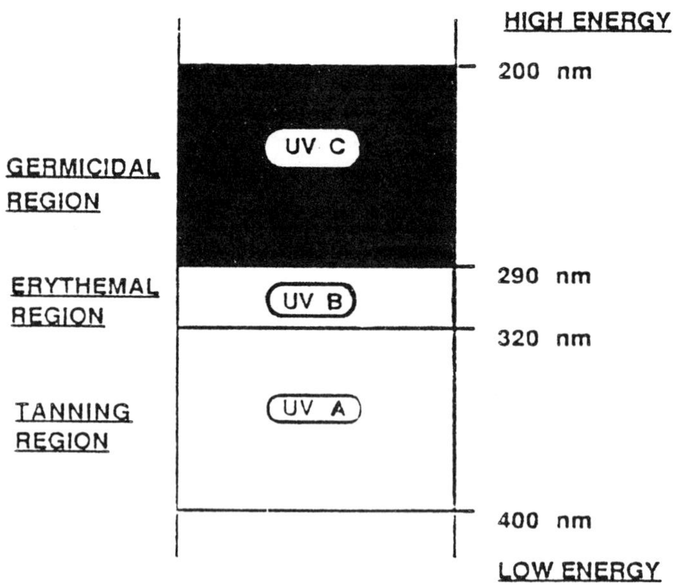

Figure 3 The ultraviolet spectrum.

2. Photosensitivity reactions (phototoxicity and photoallergy) are mostly mediated by UVA.
3. High doses of UVA can cause redness of human skin; moreover, UVA may potentiate or add to the biological effects of UVB.
4. High-intensity sources of UVA are now available for basic research and clinical studies.
5. The development of sunscreens that effectively block or diminish the highly erythemogenic UVB permits prolonged sun exposures; however, many of these sunscreens do not significantly alter the amount of UVA reaching the skin. Furthermore, UVA is transmitted by most window glass and many plastics that do not transmit UVB.
6. Recent studies suggest that UVA can affect cells and microorganisms.
7. UVA-induced photopolymerization and photochemical reactions are being used in industry to alter rubber, plastic, glass, metal, paper, and photographs.
8. Photopolymerization of certain chemicals provides convenient ways to apply dental and orthopedic appliances and "photocure" them in place.
9. The use of UVA in conjunction with photosensitizing drugs has opened up new therapeutic possibilities in chronic skin disorders, such as psoriasis, mycosis fungoides, and eczema.
10. There is experimental and epidemiological evidence to suggest that solar UVA is one of the possible etiological agents for certain kinds of cataracts in humans.

III. EFFECT OF ULTRAVIOLET RADIATION ON THE SKIN

The skin, which is the largest organ in the body, has several functions, including the regulation of body temperature, protection from the environment, partial regulation of water loss and retention, and serves as a temporary storage site for glucose when blood glucose is elevated. Other biochemical properties associated with the skin are melanin formation, epinephrine stimulation of the sweat glands, and the regeneration of viable epidermal cells (8).

The skin is composed of three layers: the stratum corneum, the epidermis, and the dermis. The dermis contains melanocytes, which generate the melanin pigment responsible for the color of the skin. Exposure to rays with wavelengths in the UVA region will stimulate the formation of melanin, which acts as a protective layer on the skin (Fig. 4). The skin is shown along with the amount of UV radiation that penetrates each layer. UV radiation near 300 nm (UVB) penetrates both the stratum corneum and the epidermis and is sufficiently energetic to cause severe burning (erythema)) of the skin, especially in fair-skinned individuals. Radiation higher than 350 nm starts penetrating the third layer, the

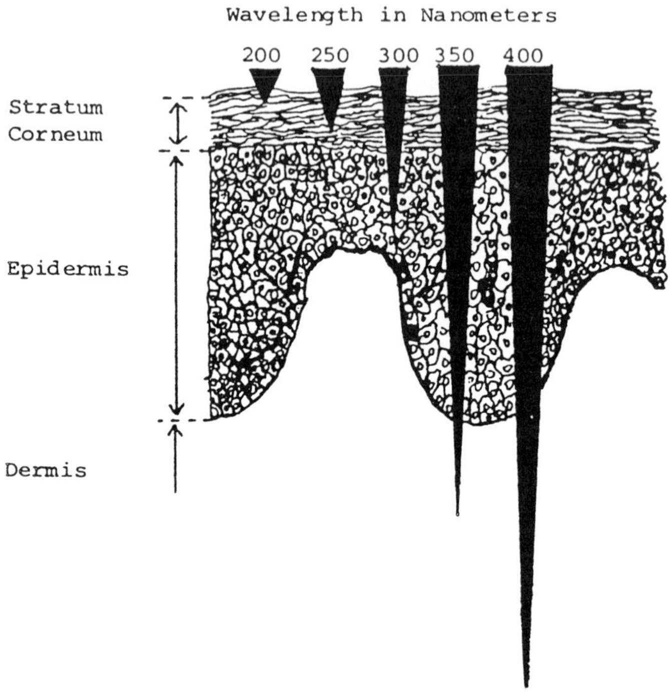

Figure 4 Schematic representation of light penetration into the skin. (Adapted from Ref. 8.)

dermis, thereby stimulating the formation of melanin, producing a tan that protects the skin from an immediate sunburn. Unfortunately, although UVA rays are of lower energy than the UVB rays, UVA rays can penetrate into the dermis, causing elastosis (loss of structural support and elasticity of the skin).

IV. CLASSIFICATION OF SUNSCREEN CHEMICALS

Sunscreening chemicals may be classified according to the type of protection they offer as either physical blockers or chemical absorbers.

A. Physical Blockers

The physical blockers are chemicals that reflect or scatter the ultraviolet radiation. Examples of physical blockers include zinc oxide, titanium dioxide, and red petrolatum. Physical blockers, if present in sufficient quantities, will reflect all the ultraviolet, visible, and infrared rays. They are currently being used in conjunction with chemical absorbers to achieve high sun protection factors (SPF). New forms of the metal oxides are currently being introduced that claim to enhance the sun protection without imparting the traditional opaqueness that is aesthetically unappealing in cosmetic formulations. A Japanese patent discloses the use of transparent metallic oxides, with an average diameter smaller than 300 Å. They include titanium dioxide, zinc oxide, chromium oxides, cobalt oxides, and tin oxides (9). Another Japanese patent reveals the use of siloxane-treated titanium dioxide with an average diameter of 40–70 μm (10). Other attempts have been made to change the physical form of the inorganic powders or to complex them with organic substances.

Of the physical blockers available, titanium oxides, iron oxides, and zinc oxides have been used in a multitude of particle sizes and suspensions and are widely used in cosmetic formulations. For a review of physical blockers consult the monograph by Nakada and Konishi (12).

B. Chemical Absorbers

The chemical absorbers absorb the harmful ultraviolet radiation. Table 1 lists the sunscreens approved for use in the United States (13) and Table 2 those UV filters approved in Europe (14).

Chemical absorbers are classified into either UVA or UVB blockers, depending upon the type of radiation they protect:

UVA absorbers are chemicals that tend to absorb radiation in the 320 to 360-nm region of the ultraviolet spectrum (benzophenones, the anthranilates, and the dibenzoyl methanes).

UVB absorbers are chemicals that absorb radiation in the 290 to 320-nm region of the ultraviolet spectrum (*p*-aminobenzoic acid [PABA] derivatives, salicylates, cinnamates, and camphor derivatives).

The best classification of chemical UV absorbers is one based on the chemical properties of the sunscreens. Hence, all sunscreens available on the market can be classified as follows:

1. Cinnamate derivatives are all UVB absorbers and have the following general chemical structure:

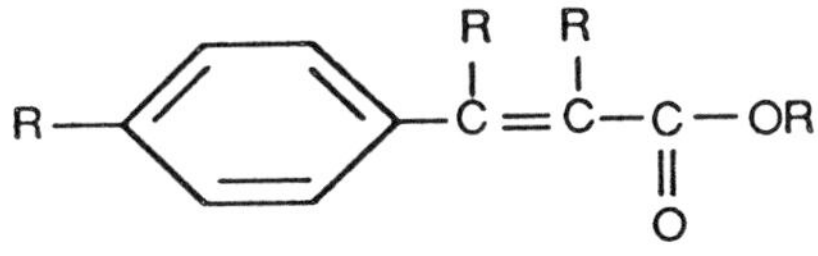

1

There are nine cinnamate derivatives approved for use worldwide: 2-ethoxyethyl-*p*-methoxycinnamate (FDA, 2.14, S.29) (15), diethanolamine-*p*-methoxy-

Table 1 FDA–OTC Panel Category I Sunscreens

Chemical	Approved %
UVA absorbers	
Oxybenzone (1)[a]	2–6
Sulisobenzone (2)	5–10
Dioxybenzone (3)	3
Methyl anthranilate (8)	3.5–5
UVB absorbers	
Aminobenzoic acid (13)	5–15
Amyldimethyl PABA (NA)	1–5
2-Ethoxyethyl *p*-methoxycinnamate (NA)	1–3
Diethanolamine *p*-methoxycinnamate (4)	8–10
Digalloyl trioleate (NA)	2–5
Ethyl 4-bis(hydroxypropyl)aminobenzoate (5)	1–5
2-Ethylhexyl-2-cyano-3,3-diphenylacrylate (9)	7–10
Ethylhexyl *p*-methoxycinnamate (11)	2–7.5
2-Ethylhexyl salicylate (12)	3–5
Glyceryl aminobenzoate (6)	2–3
Homomenthyl salicylate (7)	4–15
Lawsone with dihydroxyacetone (NA)	0.25 with 33
Octyldimethyl PABA (10)	1.4–8
2-Phenylbenzimidazole-5-sulfonic acid (14)	1–4
Triethanolamine salicylate (15)	5–12
Physical	
Red patrolatum	30–100
Titanium dioxide	2–25

[a]Encyclopedia listing number is used. The numbers refer to chemicals listed in Ref. 13a.

cinnamate (FDA, 2.9, S.24), octyl-*p*-methoxycinnamate (FDA, 2.13, S.28), 2-ethylhexyl-2-cyano-3,3-diphenylacrylate (FDA), potassium cinnamate (2.8, S.19), propyl-4-methoxycinnamate (2.10, S.25), amyl-4-methoxycinnamate (2.12, S.27), α-cyano-4-methoxycinnamic acid, hexyl ester (2.27, S.61), and cyclohexyl-4-methoxycinnamate (2.30, S.30).

2. *p*-Amino benzoate (PABA) derivatives are all UVB absorbers and have the following general chemical structure:

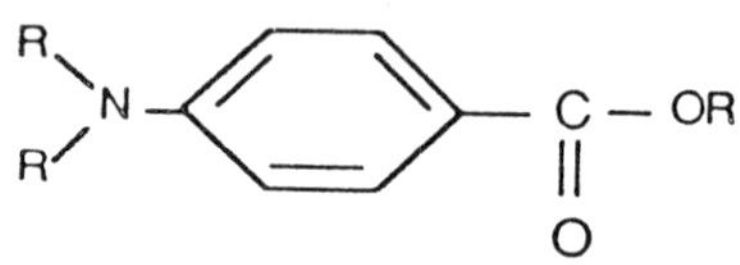

2

There are six PABA derivatives approved for use worldwide: PABA (FDA, 1.1,

Table 2 EEC and COLIPA Numbers for UV Filters Used in Europe

EEC no.	COLIPA no.	Chemical name	Encyclopedia listing no.	Max (%/conc)
1.1	S 1	4-Aminobenzoic acid	13	5
1.2	S 57	*N,N,N*-Trimethyl-4-(2-oxoborn-3-ylidene methyl)anilinium methyl sulfate	—	6
1.3	S 12	Homosalate	7	10
1.4	S 38	Oxybenzone	1	10
1.5	S 46	3-Imidazol-4-ylacrylic acid and ethyl ester	—	2
1.6	S 45	2-Phenylbenzimidazole-5-sulfonic acid and salts	14	8
2.1	S 2	Ethyl-4-bis(hydroxypropyl)aminobenzoate	5	5
2.2	S 3	Ethoxylated 4-aminobenzoic acid	—	10
2.3	S 5	Amyl 4-dimethylaminobenzoate	—	5
2.4	S 6	Glyceryl 1-(4-aminobenzoate)	6	5
2.5	S 8	2-Ethylhexyl 4-dimethylaminobenzoate	10	8
2.6	S 13	2-Ethylhexylsalicylate	12	5
2.7	S 18	3,3,5-Trimethylcyclohexyl-2-acetamidobenzoate	—	2
2.8	S 19	Potassium cinnamate	—	2
2.9	S 24	4-Methoxycinnamic acid salts	4	8
2.10	S 25	Propyl 4-methoxycinnamate	—	3
2.11	S 9	Salicylic acid salts	15	2
2.12	S 27	Amyl 4-methoxycinnamate	—	10
2.13	S 28	2-Ethylhexyl 4-methoxycinnamate	11	10
2.14	S 29	Cinoxate	—	5
2.15	S 55	Digalloyl trioleate	—	4
2.16	S 39	Mexenone	—	4
2.17	S 40	Sulisobenzone	2	5
2.18	S 42	2-Ethylhexyl 2-(4-phenylbenzoyl)benzoate	—	10
2.19	S 47	5-Methyl-2-phenylbenzoxazole	—	4
2.20	S 50	Sodium 3,4-dimethoxyphenylglyoxylate	—	5
2.21	S 52	1,3-Bis(4-methoxyphenyl)propane-1,3-dione	—	6
2.22	S 56	5-(3,3-Dimethyl-2-norbonylidene)-3-penten-2-one	—	3
2.23	S 58	α′-(2-Oxoborn-3-ylidene)-*p*-xylene-2-sulfonic acid	—	6
2.24	S 59	α-(2-Oxoborn-3-ylidene)toluene-4-sulfonic acid and its salts	—	6
2.25	S 60	3-(4-Methylbenzylidene)bornan-2-one	16	6
2.26	S 61	3-Benzylidenebornan-2-one	—	6
2.27	S 62	α-Cyano-4-methoxycinnamic acid and its hexyl ester	—	5
2.28	S 64	1-*p*-Cumenyl-3-phenylpropane-1,3-dione	21	5
2.29	S 16	4-Isopropylbenzyl salicylate	—	4
2.30	S 30	Cyclohexyl 4-methoxycinnamate	—	1
2.31	S 66	1-(4-*tert*-Butylphenyl-3-(4-methoxyphenyl)propane-1,3-dione	22	5

S.1), glyceryl PABA (FDA, 2.4, S.6), amyldimethyl PABA (FDA, 2.3, S.5), ethyl-4-bis(hydroxypropyl)aminobenzoate (FDA, 2.1, S.2), ethoxylated-4-aminobenzoic acid (2.2, S.3), and octyldimethyl PABA (FDA, 2.2, S.3)

3. Salicylate derivatives are all UVB absorbers and have the following general structure:

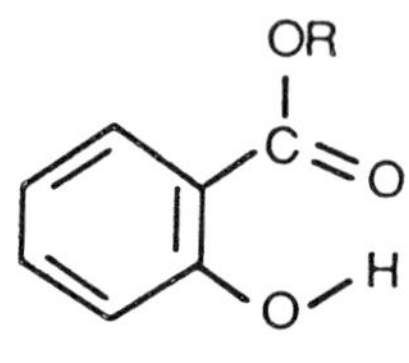

3

There are five salicylate derivatives approved for use worldwide: octyl salicylate (FDA, 2.6, S.13), homomenthyl salicylate (FDA, 1.3, S.12), triethanolamine salicylate (FDA, 2.11, S.9), salicylic acid salts (2.11, S.9), and 4-isopropylbenzyl salicylate (2.29, S.16).

4. Benzophenone derivatives are all UVA absorbers and have the following general chemical structure:

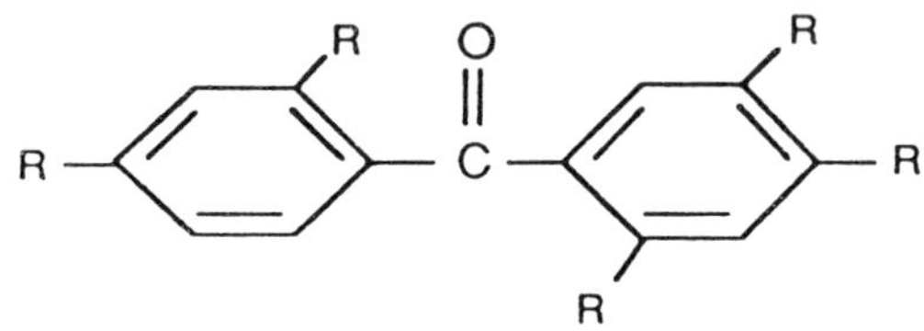

4

There are five benzophenone derivatives approved for use worldwide: oxybenzone (FDA, 1.4, S.38), dioxybenzone (FDA), sulisobenzone (FDA, 2.17, S.40), mexenone (2.16, S.39), and 2-ethylhexyl-2-(4-phenylbenzoyl)benzoate (2.18, S.32).

5. Camphor derivatives are all UVB absorbers and have the following general chemical structure:

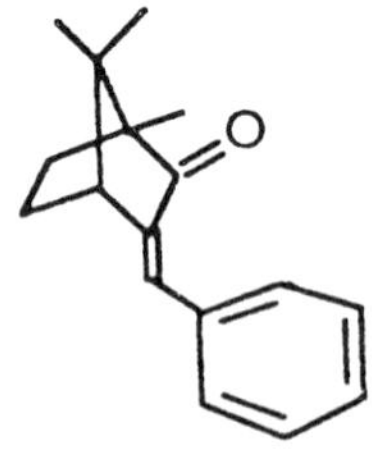

5

There are six camphor derivatives approved in Europe (none approved in the United States: *N,N,N*-trimethyl-4-(2-oxoborn-3-ylidenemethyl)anilinium methyl sulfate (1.2, S 57), 5-(3,3-dimethyl-2-norbonylidene)-3-pentene-2-one (2.22, S 56), α′-(2-oxoborn-3-ylidene)-*p*-xylene-2-sulfonic acid (2.23, S 58), α′-

(2-oxoborn-3-ylidene)toluene-4-sulfonic acid (2.24, S 59), 3-(4-methylbenzylidene)bornan-2-one (2.25, S 60), and 3-benzylidenebornan-2-one (2.26, S61).

6. Dibenzoylmethane derivatives are all UVA absorbers and have the following chemical structure:

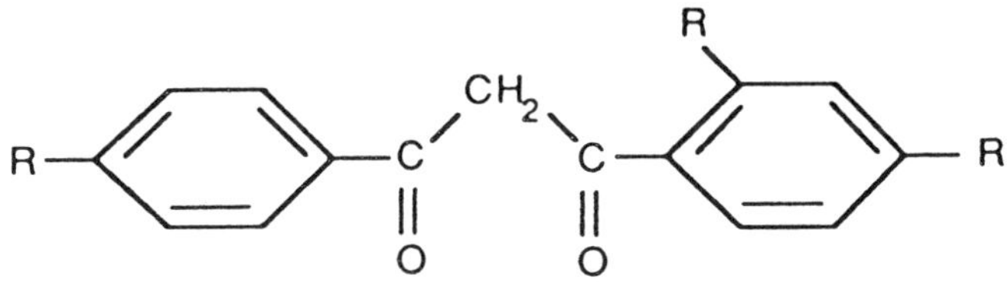

6

There are three dibenzoylmethane derivatives approved in Europe, but not approved by the FDA for use in the United States: 1-(4-*tert*-butylphenyl-3-(4-methoxyphenyl)propane-1,3-dione (2.31, S.66),* 1-*p*-cumenyl-3-phenylpropane-1,3-dione (2.28, S.64), and 1,3-bis(4-methoxyphenyl)propane-1,3-dione (2.21, S.52).

7. Anthranilate derivatives are UVA absorbers and have the following chemical structure:

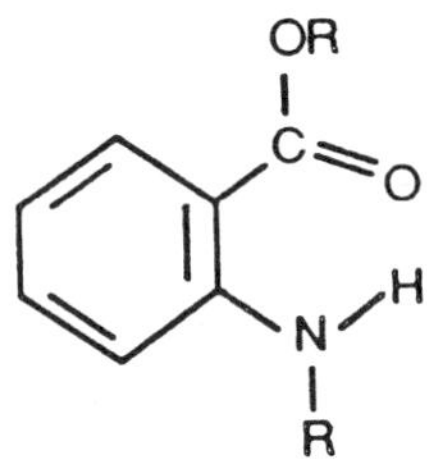

7

There are two anthranilate derivatives approved worldwide: menthylanthranilate (FDA), and homomenthyl-*N*-acetylanthranilate (2.7, S.18).

8. Miscellaneous compounds: The remaining six compounds are all UVB sunscreens and are approved for use in various countries: Digalloyl trioleate (FDA, 2.15, S.55), Lawsone with dihydroxyacetone (FDA), 2-phenylbenzimidazole-5-sulfonic acid (FDA, 1.6, S.45), 3-imidazol-4-ylacrylic acid (1.5, S.46), 5-methyl-2-phenylbenzoxazole (2.19, S.47), and sodium 3,4-dimethoxyphenyl glyoxylate (2.20, S.50).

V. MECHANISM OF SUNSCREEN ACTION

Sunscreen chemicals are generally aromatic compounds conjugated with a carbonyl group (16). In many examples, an electron-releasing group (an amine or

*Has received FDA approval for use in a specific formulation at a level of 3%.

Figure 5 General chemical structure of most sunscreen chemicals approved for use in the United States, where Y = OH, OCH_3, NH_2, $N(CH_3)_2$; and X = no substituent or —CH═CH—, and R = C_6H_4Y, OH, OR′ (R′ = methyl, amyl, octyl, menthyl, or homomenthyl).

a methoxyl) is substituted in the *ortho*- or *para*-position of the aromatic ring as shown in Figure 5.

Chemicals of this configuration absorb the harmful short-wave (high-energy) UV rays (250–340 nm) and convert the remaining energy into innocuous longer wave (lower-energy) radiation (usually above 380 nm). Quantum mechanical calculations have shown that the energy of the radiation quanta present in the UVB and UVA region lies in the same order of magnitude as that of the resonance energy of electron delocalization in aromatic compounds (Fig. 6). Thus, the energy absorbed from the UV radiation corresponds to the energy required to cause a "photochemical excitation" in the sunscreen molecule (17). In other words, the sunscreen chemical is excited to a higher energy state (π^*) from its ground state (n) by absorbing this UV radiation. As the excited molecule returns to the ground state, energy is emitted that is lower in magnitude than the energy initially absorbed to cause the excitation. Thus, the energy is emitted in the form of longer wavelengths because the energy is lower than the shorter wavelengths originally absorbed.

The longer wavelength radiation is emitted in one of several ways (Fig. 7). If the loss in energy is quite large, that is, the wavelength of the emitted radiation is sufficient length that it lies in the infrared region, it may be perceived as a mild heat radiation on the skin. This miniscule heat effect is undetected because the skin receives a much larger heat effect by being directly exposed to the sun's heat.

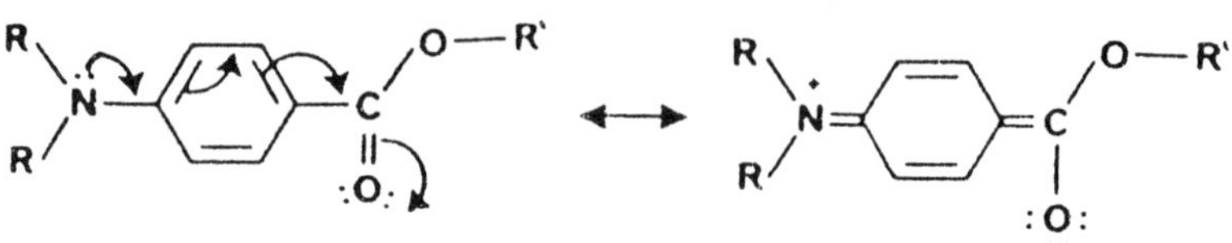

Figure 6 Resonance delocalization in a *p*-aminobenzoate molecule.

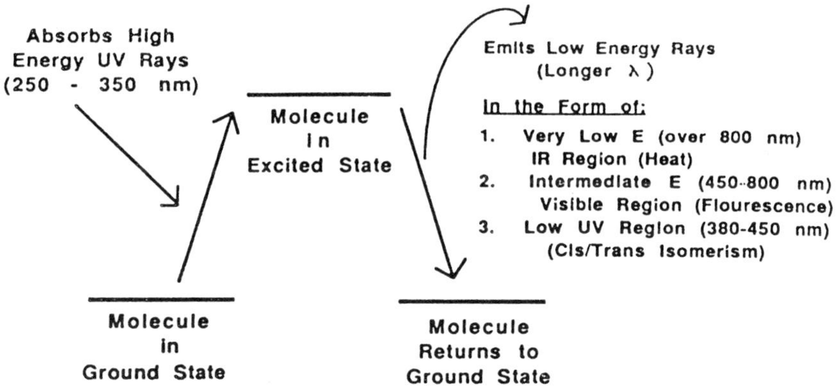

Figure 7 Schematic representation of the process in which a sunscreen chemical absorbs the harmful high-energy rays and renders them relatively harmless low energy rays. (Data from Ref. 4b.)

If the emitted energy lies in the visible region, then it may be perceived as either a fluorescent or a phosphorescent effect. This is common in the imidazoline-type sunscreens for which a slight bluish haze may be seen on the skin or in formulations.

In the extreme case, the emitted radiation is sufficiently energetic (lower wavelength) that it may cause a fraction of the sunscreen molecule to react photochemically. Cis–trans or keto–enol photochemical isomerization has been observed in some organic molecules, causing a mild shift in the λ_{max} of the chemical (18).

VI. EFFECT OF VEHICLE ON THE ULTRAVIOLET ABSORBANCE OF SUNSCREENS

A. Effect of pH

The UV absorption spectra of acidic and basic compounds may be affected by pH. In acidic compounds, the use of alkaline conditions (pH over 9) will assist in the formation of anions that tend to increase delocalization of electrons (19). This electron delocalization would decrease the energy required for the electronic transition in the UV spectrum; hence, a bathochromic shift is observed (longer wavelength or λ_{max}). For example, phenol in an alkaline environment will experience this anticipated bathochromic shift owing to the formation of the phenolate anion (Fig. 8). This phenolate anion will participate in resonance

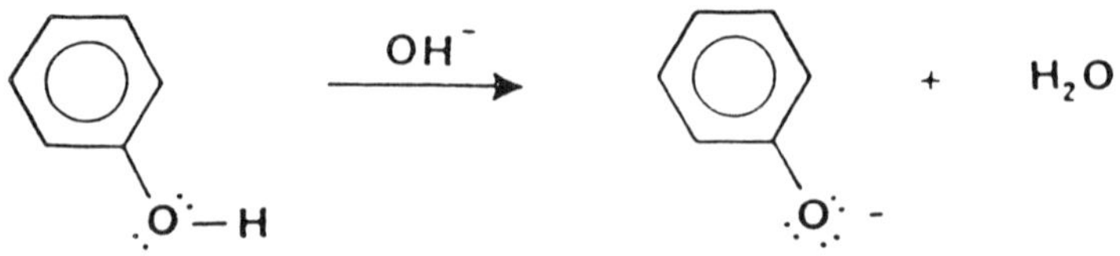

Figure 8 Phenolate anions formed by the action of alkali in phenol.

delocalization of electrons (Fig. 9). Acidic conditions (pH below 4) will assist in the formation of cations with aromatic amines. A hypsochromic shift toward lower wavelength occurs because the protonation of the unbounded lone pair of electrons with acid would prevent any resonance delocalization of the electrons. Thus aniline, for example, would form the anilinium cation at low pH (Fig. 10) and a considerable hypsochromic shift occurs.

B. Effect of Solvent

Solvent shifts in sunscreen chemicals have been observed and their findings published (4c, 20–23). The use of different solvents in cosmetic formulations may profoundly influence the effectiveness of a sunscreen chemical. The shifts in the ultraviolet spectrum are due to the relative degrees of solvation by the solvent of the ground state and the excited state of the chemical. Thus, to predict the effect the solvent has on a particular chemical, the interaction (mostly hydrogen bonding) between the solvent and the sunscreen chemical must be understood.

The solvation of polar sunscreens (e.g., PABA; see Table 3), with polar solvents, such as water and ethanol, will be quite extensive. This extensive solvation stabilizes the ground state, thereby inhibiting electron delocalization, leading to the excited state illustrated in Figure 11, where Y = NH_2 and R = OH. The net result would be a hypsochromic shift to lower wavelengths. This extensive interaction (hydrogen bonding) between PABA and the solvent would

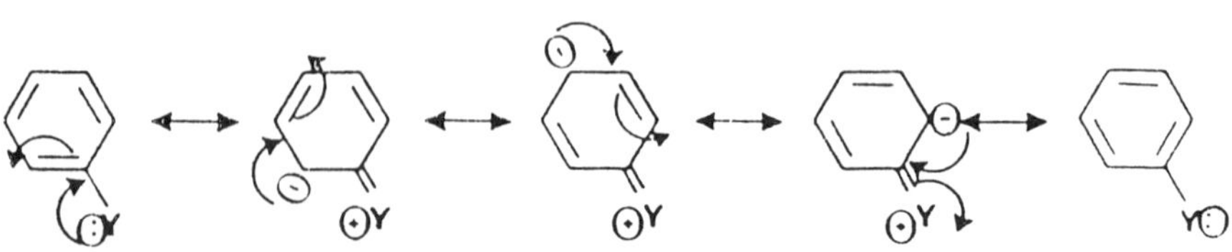

Figure 9 Resonance delocalization for monosubstituted (Y) derivatives with an unshared pair of electrons, where Y = O^-, NH_2, NR_2, or other such.

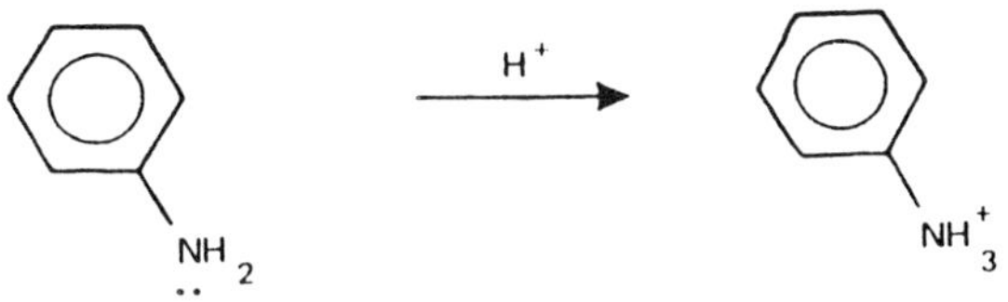

Figure 10 Anilinium cation formed by the action of acid on aniline.

hinder electron delocalization, thereby the formation of the excited state. Figure 11 pictorially represents this solvent–solute interaction.

Table 3 (4c) summarizes the results of a solvent study on sunscreen chemicals. Most commercially available sunscreen chemicals were analyzed by UV spectroscopy in several polar and nonpolar solvents. As predicted, polar compounds, such as PABA, dioxybenzone, sulisobenzone, and oxybenzone, all experienced a hypsochromic shift of −27 nm, −26 nm, −10 nm, and −8 nm, respectively.

For less polar sunscreen compounds, such as octyldimethyl PABA, the solvent–solute interaction (hydrogen bonding) is different because the excited state is more polar than the ground state. The net result is stabilization of the excited state by polar solvents. This then lowers the energy requirements for the electronic transition; hence, a higher λ_{max} would be expected, and a bathochromic shift occurs. Molecules such as ethylhexyl *p*-methoxycinnamate, octyldimethyl PABA, and butylmethoxydibenzoyl methane experienced bathochromic shifts of +23 nm, +16 nm, and +9 nm, respectively (see Table 3).

Hypsochromic and bathochromic shifts in sunscreen chemicals with different solvents is pictorially represented by the energy diagram in Figure 12. Note the difference in the solvent shifts in PABA (shifts of −27 nm) as compared with the shift in its derivative, octyl dimethyl PABA (shifts of +16 nm). The reac-

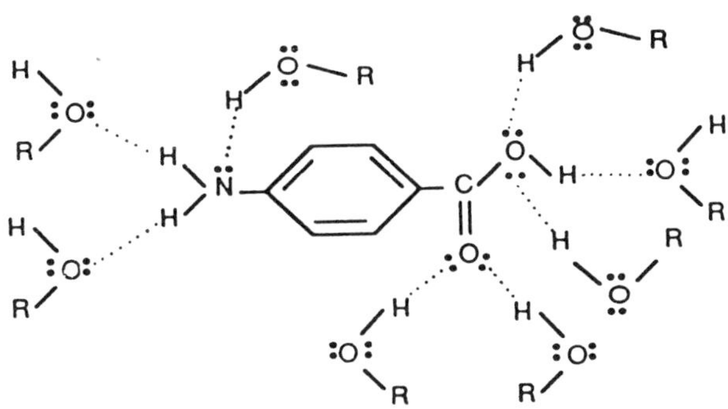

Figure 11 Solute–solvent interaction in PABA.

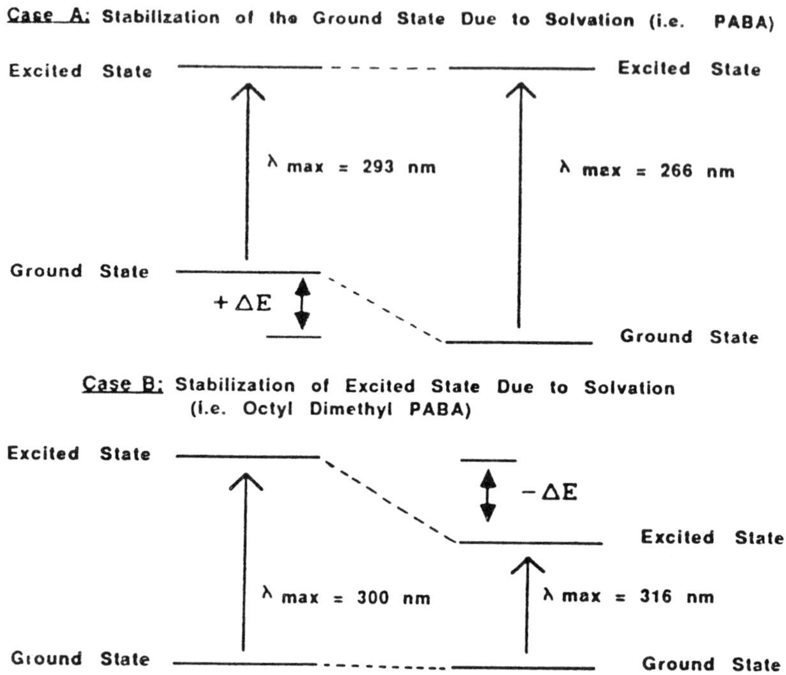

Figure 12 Energy diagram depicting the stabilization of the ground state and the excited state.

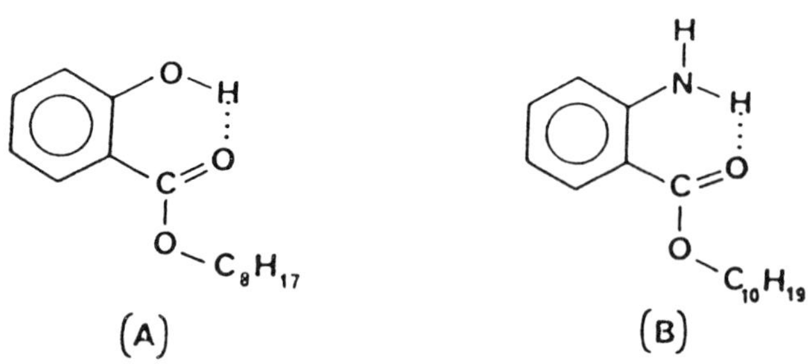

Figure 13 "Through space" hydrogen-bonding interaction in (A) octylsalicylate and (B) menthyl anthranilate.

Table 3 Summary of UV Absorption Data of Sunscreens in Combination with Polar and Nonpolar Solvents

No.	Sunscreen	$\Delta\lambda$ $(\lambda_2-\lambda_1)$	λ_1 max nonpolar solvent[a]	λ_2 max polar solvent[b]	Extinction coefficient (ε)[b]
1	PABA	−27	293	266	13,600
2	Dioxybenzone	−26	352	326	9,400
3	Sulisobenzone	−10	334	324	8,600
4	Oxybenzone	−8	329	321	9,300
5	Octylsalicylate	−2	308	306	4.900
6	Homomenthyl salicylate	−2	310	308	4,800
7	Menthylanthranilate	+2	334	336	5,600
8	Butylmethoxydibenzoyl methane	+9	351	360	31,000
9	Octyldimethyl PABA	+16	300	316	28,400
10	Ethylhexyl *p*-methoxycinnamate	+23	289	312	24,200

[a]Nonpolar solvent is hexane or mineral oil.
[b]Polar solvent is a mixture of 70% ethyl alcohol and 30% water.
Source: Ref. 22.

tivity of *ortho*-disubstituted compounds, such as salicylates and anthranilates, are subject to the "ortho" effect which supersedes other resonance delocalization effects for the observed ultraviolet transitions. The six-membered ring formation (Fig. 13) reduces the energy requirements for the electronic transition in the molecule by loosening the electrons in the carbonyl group that is conjugated to the aromatic ring. This lower-energy transition is thus reflected in a higher than usual λ_{max}. Most of the available electrons are involved in the six-membered cyclic arrangement and are not available for interaction with the solvent molecules. Thus, salicylates and anthranilates do not exhibit any significant solvent shift. The results depicted in Table 3 confirm a negligible λ_{max} shift observed in the following chemicals: homomenthyl salicylate shifts by −2 nm, octyl salicylate by −2 nm, and menthylanthranilate by +2 nm.

C. Effects on the Extinction Coefficient

The value of the extinction coefficient (ε) is the basis on which the effectiveness of a sunscreen chemical is assessed. Therefore, chemicals with a high extinction coefficient are more efficient in absorbing the energy of the harmful UV radiation than chemicals with a lower extinction coefficient.

All the electronic transitions for any compound may be characterized as *symmetry allowed* or *symmetry forbidden* (23,24). Symmetry-allowed transitions generally have high extinction coefficients, and symmetry-forbidden transitions have lower extinction coefficients. Nevertheless, trends in extinction coefficients for sunscreen chemicals can be arrived at qualitatively by studying both the spatial requirements and the electronic transition responsible for the observed UV spectrum. The degree of resonance delocalization in a molecule can predict the relative λ_{max}, and a similar qualitative prediction for its extinction coefficient is possible.

The more efficient the electron delocalization in a molecule, the higher its extinction coefficient. Compare, for example, octyldimethyl PABA and homomenthyl salicylate. In PABA, the two substituents on the benzene ring are in a *para*-relation, whereas the two substituents in the homomenthyl salicylate are in a sterically hindered *ortho*-relation. In *ortho*-disubstituted aromatic compounds, the two groups are close to one another, causing a deviation from planarity. The slightest deviation from coplanarity will significantly reduce resonance delocalization (17); hence, a lower extinction coefficient is observed in homomenthyl salicylate compared with octyldimethyl PABA. For the same reason, octyl salicylate and homomenthyl salicylate (both *ortho*-disubstituted) have lower extinction coefficients than the *para*-disubstituted compounds (see Table 3 for comparative results). Increased conjugation, allowing for increased resonance delocalization, will also result in higher extinction coefficients. For example, the extinction coefficient of ethylene is 15,000, that of 1,3-butadiene is 21,000, that of 1,3,5-hexatriene is 35,000, and for the highly conjugated molecule, β-carotene, it is 152,000 (25).

VII. THE FUTURE OF ULTRAVIOLET FILTERS

The ultimate sunscreen chemical should have the following characteristics:

1. The sunscreen chemical should absorb the harmful UV radiation in the region 280–360 nm. If broad-spectrum protection is not possible using one sunscreen chemical, then the use of two or more ingredients that filter the 280–320 (UVB region) and the 320–360 nm (UVA region), may be necessary.
2. The sunscreen chemical should possess a large molar extinction coefficient (ε) at the wavelength (λ_{max}) at which it absorbs maximum UV radiation. Values exceeding 20,000 would be extremely desirable. This would afford the maximum possible protection with the least amount of sunscreen added in the cosmetic formulations.
3. The λ_{max} and the molar extinction coefficient (ε) should not be affected by solvents. Excessively polar sunscreen chemicals are stabilized by polar solvents, thereby lowering the energy requirements of the ground state

of the sunscreen. This, in turn, will cause a hypsochromic shift (to shorter wavelengths) in polar solvents. On the other hand, sunscreens that are not too polar in their ground state, but more polar in their photochemical excited states, will experience a bathochromic shift (to longer wavelengths) in polar solvents (4c). The ideal sunscreen would be one in which the polarity of the ground state and that of the photochemical excited state are similar in nature. Hence, the hypsochromic shift (owing to the solvent stabilization of the ground state) will be counterbalanced by the bathochromic shift (owing to the solvent stabilization of the photochemically excited state).

4. The sunscreen should have excellent photostability and be photochemically inert. If isomerization such as cis–trans or keto–enol, is possible in the molecule, then the degradation quantum yields should be low, indicating that the isomerization is reversible.
5. For waterproof formulations, the sunscreen should be totally insoluble in water. Water-soluble sunscreens will still have a role to play in sunscreen formulations, such as in hair preparations or when boosting the SPF is required.
6. The sunscreen chemical should not be toxic, sensitizing, nor phototoxic.
7. The sunscreen should be compatible with cosmetic vehicles and ingredients and should be easy to use and handle.
8. Because UV filters constitute a significant portion of the cosmetic formulation, occasionally exceeding 15% of the formula, then it may also be desirable to have the sunscreen impart additional characteristics. Examples of such properties include emolliency, solubilizing or emulsifying properties, moisturizing, or possibly imparting a mild pleasant aroma that can cover base notes in formulations that are fragrance-free.
9. The sunscreen should not discolor skin, stain clothes, cause a stinging sensation, deposit crystals, cause drying of the skin, or produce off-odors when applied to the skin or hair.
10. The UV filter should be available isomerically pure, be chemically stable for prolonged storage, and be chemically inert to other cosmetic ingredients.
11. The ideal sunscreen should be inexpensive to use.

The foregoing conditions are obviously a wish list of the theoretically ideal sunscreen candidate. Unfortunately, no sunscreen chemical on the market today can claim to possess all of these properties. Nevertheless, the sunscreen chemicals available, whether through deliberate design or through serendipity, arm the cosmetic chemists with a reasonable arsenal of UV filters that are effective, possess a number of the "ideal" properties and have only a few undesirable effects.

REFERENCES

1. Robins, P. *The Skin Cancer Foundation Journal*, Skin Cancer Foundation, New York, 1985.
2. Scientists warn 1 in 7 Americans will develop skin cancer. *The New Journal*, March 10, 1987, p. 3A.
3. Parrish, J. A., Anderson, R. R., Urbach, F., and Pitts, D. The spectrum of electromagnetic radiation: UV-A in perspective. *UV-A*. Plenum Press, New York, 1978, pp. 5–6.

4a. Shaath, N. A. The chemistry of sunscreens. *Cosmet. Toilet.*, *101*(3):55–65 (1986).

4b. Shaath, N. A. On the theory of ultraviolet absorption by sunscreen chemicals. *J. Soc. Cosmet. Chem.*, *82*:193–207 (1987).

4c. Agrapidis-Paloympis, L. E., Nash, R. A., and Shaath, N. A. The effect of solvents on the ultraviolet absorbance of sunscreens. *J. Soc. Cosmet. Chem.*, *38*:209–221 (1987).

5. Semat, H. *Fundamentals of Physics*, Holt, Rinehart & Winston, New York, 1966.
6. Daniels, F. and Johnson, B. Normal physiological and pathological effects of solar radiation on the skin, *Sunlight and Man* (M. Pathak, L. Harber, M. Seije, and A. Kukita, eds.). University of Tokyo Press, Tokyo, 1974, p. 117.
7. Urbach, F. Photocarcinogenesis, *Dermatoxicity*, 2nd ed. (M. Marzulli and H. Maibach, eds.). Hemisphere Publishing , New York, 1983, p. 415.
8. Marzulli, F. and Maibach, H. *Dermatoxicity*, 2nd ed., McGraw-Hill, New York, 1983, p. 327.
9. Shirogane, K., Kida, Y., and Ito, S. Cosmetics containing transparent metal oxides as absorbers of ultraviolet light. Okamura Oil Mill, Ltd., Jpn. Kokai Tokkyo Koho. Japanese patent 62 84017 (1987).
10. Kao Corp. Japanese patent 62,145011 (June 29, 1987).
11. Saitoh, T., Murui, Y., and Take, K. *Fragrance J.*, *73*:10 (1985).
12. Nakada, S. and Konishi, H. Sun protection effect of nonorganic materials. *Fragrance J.*, *15*:64–70 (1987).
13. *Federal Register*, *43*;266:38206–38269 (August 25, 1978).

13a. Shaath, N. A. The encyclopedia of UV absorbers for sunscreen products, *Cosmet. Toilet.*, *3*:21 (1987).

14. Janousek, A. *Cosmet. Toilet.*, *98*(3):79 (1983).
15. The chemical names of the sunscreens are followed by three items of reference; approval by the FDA, European Economic Council Reference no. and the COLIPA no., respectively.
16. Morrison, R. and Boyd, R. *Organic Chemistry*, Allyn & Bacon, Boston, 1973.
17. Jaffe, H. H. and Orchin, M. *Theory and Applications of Ultraviolet Spectroscopy*. John Wiley & Sons, New York, 1970.
18. Rao, C. N. R. *Ultraviolet and Visible Spectroscopy*, 3rd ed. Butterworths, London, 1975, p. 100.
19. Morrison, R. and Boyd, R. *Organic Chemistry*. Allyn & Bacon, Boston, 1973, pp. 787–814.
20. Riegelman, S. and Penna, R. Effect of vehicle components in the absorption characteristics of sunscreen components. *J. Soc. Cosmet. Chem.*, *11*:280–291 (1960).

21. Groves, G. Evaluation of solar protective preparations. *Cosmet. Perf.*, *90*:36–50 (1975).
22. Cumpelik, B. Spectral shift of lambda max of PABA in hydroalcoholic systems. *Van Dyke Tech. Bull.*, Belleville, NJ (1977).
23. Jaffe, H. H. and Orchin, M. *Theory and Applications of Ultraviolet Spectroscopy*. John Wiley & Sons, New York, 1970, pp. 111–146.
24. Streitwiezer, A., Jr. *Molecular Orbital Theory for Organic Chemists*. John Wiley & Sons, New York, 1961.
25. Scott, A. I. *Interpretation of the Ultraviolet Spectra of Natural Products*, Pergamon Press, New York, 1964, p. 271.

16

Sunscreen Products: Formulation and Regulatory Considerations

Kenneth Klein
Cosmetech Laboratories, Inc., Fairfield, New Jersey

I. INTRODUCTION

With the publication of the Tentative Final Sunscreen Monograph (TFM) on May 12, 1993, the "playing field" for marketers of finished goods became more complex and now offered some new opportunities for "marketing handles." Although some claims could no longer be made, such as those relating to aging and wrinkling, we now see the appearance of strong claims dealing with ultraviolet A (UVA) radiation. In fact, the U.S. Food and Drug Administration (FDA) has even published a list of those sunscreens for which usage permits UVA reference. Some additional claims that have begun to appear deal with protection against infrared (IR) radiation. The FDA TFM does not deal with IR in any way. Thus, IR claims are not drug claims. They must be substantiated for the Federal Trade Commission (FTC), but not the FDA. Marketers also have made claims for "all-day" waterproof protection. The FDA no longer allows a waterproof claim. They feel that this term suggests an absolute—waterproof—which is not borne out by the currently available test methods. The sweatproof claim can now be made, if the product meets the requirements for a very water-resistant claim.

The last several years has seen the appearance of products on the shelf that make the claim "chemical-free" sunscreens. This claim is silly and has no basis in reality. Products that make this claim generally are formulated with particulate sunscreens (zinc oxide or titanium dioxide). In the minds of some consumers, "chemicals" are bad things to put on their skin, so marketers try to take advantage of this. This is a questionable approach, at best, and quite misleading. The FDA has not looked kindly at this "claim."

The cosmetic chemist has at his or her disposal many vehicles from which to choose; these include, but are not limited to, emulsions, oils, gels, sticks, mousses, aerosols, and ointments. There are many factors, some technical and some not, that determine which will be the vehicle of choice. Each of the vehicles has strengths and weaknesses that can have a significant influence on the cosmetic chemist's choice. To make a wise choice relevant factors and questions must be addressed:

What is the target sun protection factor (SPF)? This is a market and marketing-driven question. The trend toward higher SPFs seen in the mid-1980s to the early 1990s has appeared to level off. Although there have been products introduced with SPFs as high as 50, almost all sunscreen products have SPFs of 30 or less. One should question the real need for a product with this degree of protection. If the average minimum erythemal dose (MED) is taken as 20 min (for a person with type II skin in New Jersey exposed to noonday sun), then a SPF 50 product would permit exposure to noonday sun for 1000 min (16.67 h). Where can one get this type of exposure? Certainly not in New Jersey. Although it is true that other latitudes do indeed yield much shorter MED times, an SPF of 50 would seldom, if ever, be required by most of the population at large.

Who is the target group? If it is persons with very sensitive skin then maybe *p*-aminobenzoate (PABA)-"free" should be the approach of choice. The FDA feels that a PABA-free statement should not appear on the label. First, the correct name for PABA is aminobenzoic acid. Second, by making the statement that this product is PABA-free you are implying that PABA is either unsafe, or is ineffective. Because it has been placed into category I (safe and effective) by the FDA, they feel that such a statement should not be made. Although it is true that more persons are sensitive to PABA than some of the other approved materials, the PABA-free statement implies a degree of safety that may not be present. Certainly, there have been ample questions raised concerning the safety profile of oxybenzone, yet in this country, there has been no major push to market products that are "oxybenzone-free." From this situation one could conclude that marketers and raw material suppliers at least occasionally put economic considerations before technical and safety ones.

What are the cost of materials limitations? The sunscreen business, as with all mass merchandise efforts, is extremely sensitive to pricing. The cost of sunscreen active agents is quite high compared with other typically employed cosmetic raw materials. This is reflected in the high selling price of finished goods in this category.

How important are water-resistant or very water-resistant claims? This area is receiving more attention from consumers and thus must be addressed by the manufacturers. How will the formula be packaged? Because sunscreens are very good solvents, particular care must be paid to this factor.

How important are aesthetic considerations? Although sunscreens are drugs, the consumers who purchase them have become accustomed to elegant vehicles, such as those that deliver their cosmetics. If the sunscreen gives a greasy, occlusive feel, or leaves an opaque residue, then the purchaser will probably look elsewhere.

II. FORMULA TYPES: GENERAL CONSIDERATIONS (STRENGTHS AND WEAKNESSES)

A. Emulsions

By far the most popular of all vehicles used for sunscreens, the emulsion, offers almost unlimited versatility. Lotions are more popular than creams owing to their easier spreadability on the skin and dispensability from bottles. An emulsion is termed a cream or lotion on the basis of its viscosity (resistance to flow). It is difficult to determine the exact point at which a lotion becomes a cream and, sometimes, the designation may be quite arbitrary. Lotions typically have viscosities below 50,000 centipoise (cps) and will flow when the container is tilted or squeezed. The viscosity of creams can run into the millions of centipoise, but usually is in the 150,000–500,000 cps range. Although emulsions are the most popular vehicle, they are also the most difficult to stabilize. We will discuss methods of stabilizing emulsions later in this chapter. If the highest SPF possible, at the lowest possible cost, is the goal, then the emulsion vehicle must be strongly considered. To appreciate why emulsions are so effective one must understand the factors (1) that lead to consistently high SPFs:

Uniform sunscreen film
Thick sunscreen film
Nontransparent sunscreen film
Minimum ingredient interaction with the sunscreen's active agent

In each of these important areas, emulsions exhibit good performance. Emulsions allow easy incorporation of sunscreen active agents, which are typically oils that can be readily emulsified. Emulsions can be prepared that contain a large percentage of water; thus, they can be, and are, a most cost-effective vehicle. From an aesthetic viewpoint, emulsions are an elegant medium that can give the skin a smooth silky feel without being greasy. They can accommodate an almost infinite variety of raw materials; thus, you can tailor the formulation to suit your needs. On the negative side, emulsions are quite difficult to stabilize, particularly at elevated temperatures. Additionally, emulsions present a perfect medium for microbial contamination and eventual product breakdown. As more and more consumers seek the protection of very high SPF products (30 and higher) that are water-resistant or very water-resistant, the cosmetic chemist finds it increasingly difficult to achieve these goals with an emulsion vehicle. As can be readily seen, emulsions present great opportunities for the creative, experienced formulators and great difficulties for the novice. Without doubt they will remain the vehicle of choice for the foreseeable future.

B. Oils

Oils are one of the oldest and most easily formulated sunscreen vehicles. There is only one phase, so excellent product stability is readily achievable. Because most sunscreen active agents are lipophilic, they are soluble in sunscreen oils; thus, manufacturing processes are greatly simplified (many can even be prepared at room temperature); certainly when compared with the manufacture of emulsions. Oils are easily applied to the skin, spreading rapidly to cover a wide surface area. Unfortunately, there are numerous negatives associated with sunscreen oils. Their excellent spreadability results in a very thin, transparent sunscreen film, and thus a lowered SPF (2). Sunscreen oils exhibit the poorest SPF performance of any vehicle. This poor performance is further explained by the interaction between the most popular sunscreens (nonpolar esters) and the very nonpolar oil vehicles. Nonpolar oils, such as mineral oil, cause the position of the ultraviolet (UV) curves to shift to shorter wavelengths (3). Thus, part of the curve is moved into the area below 290 nm where it is wasted. The result is a poorer-performing sunscreen; that is, a lowered SPF rating (4). This shift to the shorter wavelengths is due to a stabilization of the ground state by the nonpolar vehicle. An additional negative for oils relates to packaging. Sunscreen esters are excellent solvents. When they are combined with other esters in sunscreen oils, the resulting product can attack the plastic packaging typically used. Certainly, low-density polyethylene is a poor choice, unless expensive internal coatings are employed. Another obstacle to their use relates to cost of goods. Because this is an anhydrous system, there is no water to lower the cost of expensive raw materials. Sunscreen oils are one of the most expensive systems found.

C. Gels

The crystal clarity of cosmetic sunscreen gels projects an aura of purity and elegance. On initial consideration, they would seem to present, from an aesthetic perspective, the ideal sunscreen vehicle. Unfortunately, this perception is quite premature. There are four major formula categories of sunscreen gels: (1) aqueous, (2) hydroalcoholic, (3) microemulsion, and (4) geled oleaginous (oily anhydrous). All of these present the cosmetic formulator with an abundant supply of formulating difficulties.

The aqueous gel must use water-soluble sunscreens or employ solubilizers (usually nonionic surfactants, which may be ethoxylated fatty moieties or phosphate esters) at sufficiently high levels to ensure that the gel will remain clear at all temperature extremes. Neither of these approaches provide good answers, for the resultant product is very prone to wash off when exposed to water or perspiration. The high level of surfactant can often make the finished formula both expensive, and more importantly, quite irritating. Add to all of this the great batch-to-batch variation these gels exhibit, owing to the very delicate raw material balance required to achieve clarity. Manufacturing these gels presents rather unique problems in itself. Aeration must be avoided, because deaeration can be time-consuming, expensive, and sometimes impossible. The gel formation is often accomplished through the incorporation of an appropriate "gum." Although cellulose-based materials can be used, they must be incorporated at very high levels (2–3%) to obtain the desired gel structure. At these use levels, they can be quite sticky on application. The carbomers, synthetic carboxyvinyl polymers, first introduced by the B. F. Goodrich Company, are by far the most popular thickener used in this category, allowing the cosmetic formulator to achieve crystal clarity at reasonable use levels (0.5–1.5%). Unfortunately, the carbomers have compatibility, aesthetic, and performance problems. Because they are anionic when neutralized, care must be taken not to include any cationic ingredients, as gel breakdown may occur. The carbomers can impart a transient tackiness that is most evident during rubout. This is very apparent in gels for which use of emollient esters must be minimized or eliminated to ensure optimum clarity. In addition to these difficulties, the carbomers are quite sensitive to the electrolytes found on the skin. Thus, when applied, the carbomer gel "breaks," and the product exhibits very poor application characteristics. This effect is exacerbated by swimming in salty water or by heavy sweating. Lastly, it is very difficult to obtain high SPFs in this vehicle because the clarity issue rules out the use of high levels of the best sunscreen active molecules, such as octyldimethyl PABA or octyl *p*-methoxycinnamate. Many of the comments concerning the aqueous gel also apply to the hydroalcoholic gel. A real benefit that is afforded by this vehicle is the desirable cooling effect as it is applied to the skin. This is particularly refreshing on a hot summer day. As its name implies, this gel employs the use of alcohol (usually ethanol) in conjunction with water, as the main carrier components. The use of alcohol greatly reduces the need for additional solubilizers because most of the lipophilic sunscreens are readily miscible in ethanol. Some of the problems associated with the hydroalcoholic gel are the following:

1. Water washability: By incorporating resins and other film formers, water resistance can be improved, but the inherent nature of this vehicle limits its effectiveness.
2. Facial or eye sting: The high levels of alcohol tend to cause facial and eye stinging on certain individuals. Use of humectants can ameliorate this but not eliminate it entirely.
3. Packaging: The high volatility of alcohol gives it a great tendency to evaporate. This is most evident at high temperatures, such as those encountered at the beach in the summer. Packaging must use barrier coatings to retard this, which further increases costs. Closures should be tight-fitting and easily resealable.
4. Efficacy: It is in this key area that hydroalcoholic gels exhibit their poorest performance. As mentioned earlier, to realize high SPFs, one must lay down a uniform sunscreen film on the skin. On exposure to heat and sunlight, the alcohol flashes off rapidly, leaving a porous or discontinuous film. Thus, there are areas where there is no coverage, and a lowered SPF is the result. Use of film formers can help, but in the end the level of sunscreen must be increased to overcome this effect; in reality, it is never completely conquered.

In microemulsion gels, the particle size is so small (less than 0.5 μ) that light seems to pass right through the emulsion and a clear gel is the result. These gels can have an elegant skinfeel and can lay down a smooth, thick, and uniform film on the skin, thus optimizing the SPF. Unfortunately, as with the previous gels, there are numerous negative factors to contend with:

1. Cost: To achieve clear microemulsions, it is necessary to employ very high levels of emulsifiers. Levels as high as 15–25% are not uncommon. This results in a very costly product.
2. Safety: In many emulsion systems the emulsifier is often one of the most irritating ingredients in the formula. At these use levels one must indeed be wary of this factor.
3. Water resistancy: The high emulsifier levels makes the sunscreen easily removed by swimming or perspiring.
4. Manufacturing problems: Slight variations in raw materials can, and often do, lead to hazy or cloudy products. A slight change in the ethoxylation distribution may be all that is needed for disaster.

The microemulsion route is certainly a precarious one at best.

The oleaginous gels share many attributes with the ointments (addressed later), but they are clear, whereas the ointments are translucent. There are no popular products of this type on the market, as they are difficult to manufacture and are quite expensive. They are generally produced by geling a combination of mineral oil and sunscreen with special silicas.

D. Sticks

A relatively new form of sunscreen vehicle, the stick, brings with it a real convenience of usage. Sticks are mostly used on the lips or nose. Because they cover a very small surface area on each application it is not practical to use them on large areas of the body. The vast majority of sticks are composed of oils and oil-soluble sunscreens which in turn are thickened through the incorporation of waxes and petrolatum. They tend to have an oily or greasy skinfeel and are probably the most water-resistant of any product form. With the addition of an opacifying agent, such as titanium dioxide (an approved sunscreen) or zinc oxide (a category III sunscreen), the resultant product is often seen on the noses of lifeguards. Not very aesthetic, but quite functional. Clear sticks can be made (using organic sunscreens), with the use of alkanolamides as gelants, but stability and cost considerations have not allowed marketing of these products.

E. Mousses

The sunscreen mousse was first introduced several years ago. It has not attained much consumer acceptance and, thus, remains a niche product at best. A mousse is typically an emulsion lotion to which has been added some propellant. When the product is exposed to the lower pressure of the atmosphere (as compared with the inside of the container), the propellant flashes off and a foam is formed. It must be emphasized that, in the strict sense, mousses are not aerosols, an aerosol being a colloidal dispersion of liquid in the atmosphere. These foams have more in common with shave creams than with other product types. Most use anonic (triethanolamine stearate) emulsifier systems, often coupled with a lesser amount of nonionic (ethoxylates). The sunscreen active agents can be readily incorporated into the internal phase of these oil-in-water emulsions. They mimic most of the functional and aesthetic characteristics of the nonaerosol emulsions. Their primary functional reason for being lies in their ease of use. Great care must be taken to ensure that the cans are able to withstand the temperature extremes encountered at the beach that give rise to a significant increase in internal pressure and could potentially present a safety hazard.

F. Aerosols

The sunscreen aerosols have not enjoyed the same popularity as have the other sunscreen vehicles for several reasons. Typically, they are oil-based, which makes them rather expensive and often reduces their effectiveness (see foregoing discussion). Additionally, it is difficult to see where you have applied the sunscreen, and unless you are careful the sunscreen could be sprayed accidentally into the eyes. Aerosols also put down a discontinuous film onto the skin; this, too, reduces their effectiveness. They can be either mechanical (pump spray) or propellant-based, at this time neither has made any significant market penetration.

G. Ointments

These oily products closely mirror the composition and function of the stick products. They are generally based on mineral oil and esters that have been thickened by the addition of petrolatum and some waxes. Their main benefit is that they are very difficult to remove with water; thus, they are used by persons who must have a sunscreen that stays on the skin no matter how much swimming or physical activity is done. They are, however, not very aesthetic to use because they are oily and greasy.

III. FORMULATING BASICS

Because almost all sunscreen products are creams and lotions, it is important to understand the intricacies of formulating this most challenging product category.

A. Principles of Emulsification

Although it is clear that it is extraordinarily difficult to produce emulsions that meet the primary market objectives of efficacy, aesthetics, and cost parameters, an ever-present, but underlying, requirement is product integrity; or stated another way: emulsion stability. At times it may seem that obtaining emulsion stability is quite an elusive goal, in fact, perhaps unattainable. But by understanding the underlying principles of emulsification, coupled with a healthy mix of experience gained through many failures, we can make real inroads in this quest.

Even though emulsions come in two basic varieties, oil-in-water and water-in-oil, and two basic "styles," creams and lotions, the ideas presented in this chapter are quite universal; thus, they can be successfully applied to all.

It is good to start with a healthy appreciation for the problem at hand; improving emulsion stability. That which you are trying to do (make a stable emulsion) is quite impossible (from a thermodynamic viewpoint) (5). An emulsion is a dispersion (the internal phase) of one immiscible liquid in another (the external phase) in the form of tiny droplets. These droplets are constantly trying to come back together, coalesce, to form a single large droplet (phase). This is the continuing battle of the cosmetic formulator against the laws of thermodynamics. Can the battle ever be won, or is the only hope to put off the inevitable for a long enough period to successfully market the product? Sometimes this emulsion breakdown can be put off for several years. Some very viscous creams have remained "stable" for more than 10 years through rather harsh temperature fluctuations. The basic goal of all cosmetic chemists who formulate emulsions must be to prevent, or more realistically, minimize particle interaction. If this can be accomplished, the formulator is well along the road to success. There is an equation that describes this interaction: Stokes law (6)

$$V = \frac{d^2(p1 - p2)g}{18\eta}$$

where:

V = velocity of sedimentation (interaction) of spherical particles
d = diameter of the particles of the dispersed phase
$p1$ = specific gravity of the dispersed phase
$p2$ = specific gravity of the external phase
g = gravitational constant
η = viscosity of the external phase

After studying this simple equation it becomes apparent that our goal is to minimize the value of *V*, the velocity of sedimentation (coalescence). This can be done in only two ways.

1. Reduce the value of the numerator
2. Increase the value of the denominator

Let us first deal with the numerator, *d*: Reducing the size of the dispersed droplets is indeed a very good approach that may be accomplished in several ways. Choice and level of emulsifier (see discussion later) is certainly of primary importance. Choice of optimum manufacturing procedure is also key. Generally, it is advisable to add the external phase to the internal phase. In this way the emulsion can go through a definite phase inversion. This occurs because of the predominance of the internal phase relative to the external phase. The phase volume ratio does not permit the formation of the expected emulsion type. As more of the external phase is added, the phase volume ratio permits the formation of the predicted emulsion type and, thus, a phase inversion is seen. The result will be a reduced particle size and, just as important, a narrow distribution of particle sizes. Unfortunately, this type of phasing is seldom available during production, as usually the phase volume ratio between the internal and external phases is such that there is not enough internal phase to reach the mixing blade; thus, one cannot use this method. The particle size can also be reduced by employing the method of Lin (7). An emulsion concentrate is formed, composed of all of the internal phase (oil, emulsifier, sunscreen, and other components) and a part (approximately 30%) of the external phase (water, emulsifier, or such). The water phase can be added to the oil phase to go through a phase inversion. The percentage of emulsifier is higher in the concentrate than it will be after all of the water is added; thus, you have effectively increased the level of emulsifier, without adding more emulsifier. Very often a smaller particle size will be the result. The remainder of the water (at room temperature) can then be added to the emulsion concentrate. It will cool the emulsion and dilute the external phase. This addition of water will not have a detrimental effect on the particle size. Lin's technique can dramatically reduce production costs and produce better emulsions.

The particle size can also be reduced through mechanical means. If the finished emulsion is subjected to high shearing forces, the large particles will be reduced in size. Numerous devices have been designed to accomplish this feat, which are categorized broadly as homogenizers or (colloid) mills. Although they often do an excellent job in reducing the particle size, great care must be taken with their use, as they can degrade certain emulsion components, such as gums, with their high shear. Another benefit of using homogenizers is that they give a more uniform particle size distribution. Very often this is even more important than reducing the average particle size. If the range of particle size is quite narrow, there will be lower probability of coalescence when two particles of similar masses collide, as the collisions tend to be more elastic (less transfer of energy) under these conditions. Thus, the emulsion will exhibit better stability. Another way to decrease the numerator is to make the factor ($p1 - p2$) as small as possible. In other words, make the specific gravities of the internal and external phases as close as possible to each other. Since the external phase is primarily composed of water (oil-in-water emulsion) *the specific gravity* (SG) (ratio of the density of a substance relative to the density of water) (8) is approximately 1. The internal phase is mostly composed of oil, which has a specific gravity of 0.8–0.95. Therefore, to make these two phases close to each other in density we must either lower the SG of the water phase or raise the SG of the oil phase. Because there is no easy way to raise the SG of oil, we will lower the SG of the water phase. The obvious way to accomplish this is by adding alcohol to the water phase. At first glance this seems to work quite well. But the emulsion rapidly breaks down because of the alcohol's excellent solvency with various emulsifiers; hence, because of partition coefficient considerations, it extracts the emulsifier from the oil–water interface and the emulsion breaks. Adding alcohol to the water phase is thus not a worthwhile approach.

The last factor in the numerator is g, the gravitational constant. This cannot be altered by even the most creative cosmetic chemist (at least not in New Jersey).

The denominator has only one factor that we can work with, η (the viscosity of the external phase). If we could increase this, it would decrease V (velocity of sedimentation coalescence), and a more stable emulsion would be the result. Here, we have the most often used technique (successfully) by cosmetic chemists. There are many ways to increase the viscosity of the external phase.

1. Add more internal phase: This can effectively be used in both oil-in-water (o/w) and water-in-oil (w/o) emulsions. It will certainly increase the vis-

cosity of the emulsion at room temperature, but at elevated temperatures, the emulsion will thin out and instability may very well result.

2. Reduce the particle size, so that the same amount of internal phase will seem to occupy a greater volume, and the viscosity will increase.
3. Add a fatty moiety to form liquid crystals in the external phase. The following phenomenon has been observed by many cosmetic chemists: When a small amount of fatty alcohol is added to an oil-in-water emulsion the viscosity is seen to dramatically increase. According to Suzuki et al. (9) "The self-bodying action of fatty alcohols was caused by the formation of a network structure of liquid crystalline phase in the emulsion system." In addition, the fatty alcohol will complex with emulsifiers at the o/w interface to strengthen it and improve emulsion stability.
4. Add a thickener ("gum") to the external phase. Indeed this is a simple and very effective approach. There are a bewildering number of gums available to serve this purpose. Cellulosics are often used and can be very successful. Unfortunately, they do not improve the high-temperature stability to any great extent. There are several gums that can do this: the carbomers and xanthan are the most popular of them. Both of these allow the emulsion to retain a real measure of viscosity at elevated temperatures. Consequently, the emulsion is much less likely to show signs of instability. One must, however, be careful to only use these last two thickeners in nonionic (the emulsifiers are uncharged) or anionic (the fatty part of the emulsifiers are negatively charged) systems because they are anionic in character themselves.

B. Selecting Key Ingredients

To formulate successfully you must have a thorough understanding of the principles and rules by which we choose the appropriate ingredients for each formula. Nowhere is this more true than in the formulation of sunscreen emulsions, which are drug vehicles. One must appreciate the potential for ingredient interaction and keep in mind the formula requirements laid down by the various interested internal groups: marketing, R&D, packaging, claim substantiation, and safety. Certainly, all ingredients selected must be cosmetically acceptable (color, odor, skinfeel), have good safety parameters, be stable to chemical or UV attack, and must not interfere with the efficacy of the sunscreen active agent in any way. The following are some comments on how to select ingredients with these thoughts in mind.

Emulsifiers

The oldest and most widely used method of selecting emulsifiers is the hydrophilic–lipophilic balance (HLB) system. This system, invented by Griffin (10) in 1949, was the first attempt to put a scientific basis behind the choosing of emulsifiers. The HLB was determined for each emulsifier: that is, the relative percentage by weight of the molecule that is water-loving versus the oil-loving part. If a greater weight percentage was water-loving, the emulsifier would be assigned a high HLB number and this would be judged to be an oil-in-water emulsifier. The oil phase components are assigned a required HLB; thus, by choosing emulsifiers that match the HLB requirements of our oil phase, a fairly good emulsion can be formed (11). Experience has shown that this system works better for nonionic than for ionic emulsifiers. Although the HLB system has great usefulness for choosing a starting point for emulsions, there are several major concerns about the use of this system that must be addressed.

Temperature Effect on Nonionics (Ethoxylates) As the temperature increases, the apparent HLB of emulsifiers decreases. Because ethoxylates (emulsifiers that employ ethylene oxide to increase their polarity and, hence, water solubility) are soluble as a consequence of hydrogen bonding to water, they become less soluble as increased kinetic energy (heat) causes a breakage in the hydrogen bonds. This is often referred to as the inverse solubility of nonionics. As the temperature increases they become less soluble. This is not what we typically see dealing with chemicals. Thus, an emulsifier that orients toward o/w emulsions at one temperature may indeed orient toward w/o at some higher temperature. The temperature at manufacture will play a significant role in the quality (type and particle size) of the emulsion. This is one of the reasons that many nonionic emulsions are not so easily scaled-up from the benchtop to the 5000 gal kettle.

Ingredient Interactions Anionic and cationic emulsifiers will interact to produce a precipitate and break the emulsion, but the HLB system does not take this into account.

Fatty Alcohols According to the HLB system fatty alcohols are oils that have a required HLB. However, experience has shown us that indeed fatty alcohols are effective emulsion stabilizers and can be considered to be secondary emulsifiers.

Phase Volume Relations The HLB system ignores the relative sizes of the internal and external phases. As we know, this factor can and does have a profound effect on the emulsion. Yet, while all of these drawbacks to the HLB system exist, it should not be abandoned by cosmetic chemists. Rather, it should be used in conjunction with sound experience and other systems to choose emulsifiers.

The phase inversion method of choosing emulsifiers was proposed by Shinoda (12). This system takes into account many of the drawbacks of the HLB system, yet its underlying principles are closely linked to it. The inverse solubility phenomenon can be used to choose emulsifiers. After the oil and water phases have been combined, the cooling process begins. During this cooling, the effective HLB of the emulsifiers becomes higher. There is a temperature, the phase inversion temperature (PIT), at which the hydrophilic and lipophilic tendencies of the emulsifiers balance. At this temperature, a phase inversion can be observed as a dramatic change in viscosity. Additionally, one can notice a stable pH below the PIT. If the PIT is at least 20°C above the highest storage temperature, you can be fairly confident that the emulsion will be a stable one. You thus choose the level and type of emulsifier to achieve this PIT temperature.

Lin (13) has proposed a rather unique method of selecting emulsifiers. All of the oil phase components are combined (esters, waxes, emulsifiers, sunscreens) and heated until liquid. Then water is slowly added (titrated in) while stirring. As the first drop of water hits the oil mixture, a turbidity is formed that slowly disappears. According to Lin the more water that can be solubilized into the oil phase, the smaller the emulsion droplets; hence, the better the emulsion will be. You can modify the emulsifier types and concentrations to increase the water solubilization capacity of the oil phase. There is ample experience to show that this method does indeed work. There is an inverse correlation between the amount of water soluble in the oil phase and the particle size of the final emulsion.

Up to this point I have addressed mainly nonionic emulsifiers. Perhaps the most widely used emulsifier is, however, anionic: soap; to be specific, triethanolamine (TEA) stearate. This emulsifier, prepared in situ, is one of the most powerful and inexpensive of the oil-in-water emulsifiers known. This accounts for its unmatched popularity. Other monovalent soaps can be used (sodium and potassium), but the interfacial film they form is not as flexible as that of TEA stearate. Emulsion stability will be further enhanced if combinations of emulsifiers are employed. Thus, use of several types of emulsifiers (anionic and nonionic) is recommended. The variety of emulsifiers is almost unlimited, by using one or more of the foregoing systems you can minimize the trial-and-error approach, and maximize your chances for success.

Emollients

Emollients represent one of the most important classes of emulsion components. Although there are both oil-soluble and water-soluble emollients, by far the oily materials predominate. They provide a silky skinfeel on application, while acting as the vehicle in which the oil-soluble sunscreen is delivered. There are several categories of fatty emollients: esters (liquid and solid), waxes, fatty alcohols, mineral oils, and silicone materials.

Esters are a very widely used class of compounds in sunscreen emulsions. Quite simply, they are the condensation product of an alcohol and an acid. By varying the fatty moiety of each, an almost endless variety of esters can be produced, each with slightly different properties from the next. Some general rules can be made that should provide guidance in choosing the best ester for your requirements.

1. Chain length increase: As the chain length increases, the ester becomes more viscous, eventually becoming solid; it becomes more difficult to emulsify; it becomes less irritating to the skin; it acquires an oilier skinfeel and becomes less polar (this has an effect on the position of the UV curve that may influence the effectiveness of the sunscreen actives).
2. Branching: As the ester becomes more branched it produces a drier skinfeel; it becomes less polar.
3. Unsaturation: As the ester becomes more unsaturated it becomes less polar; it becomes harder to emulsify; and it seems to disappear into the skin during rubout.

Waxes are not used to any great extent in sunscreen emulsions. Certainly, carnuba, ceresin, beeswax, and many others are available to us, but they tend to give a rather draggy feel to the skin and, thus, their use is held to a minimum.

Fatty alcohols provide a skin-smoothing during rubout, a matte, nongreasy feel that is quite desirable. They can, however, if used at too high a level (2% maximum), become draggy on the skin, certainly not a desirable feature. As would be expected, as the chain length increases the melting point increases. Thus, stearyl alcohol (18 carbon atoms) has a higher melting point than cetyl alcohol (16 carbon atoms) and has a slightly heavier feel on the skin.

Mineral oil is one of the main constituents of many emulsions. Its widespread use can be easily understood when one considers its low cost, broad chemical compatibility, and safety. There are several grades of viscosity to consider, depending on the skinfeel and emulsion viscosity desired (minor effect). One must remember that mineral oil can have a rather oily skinfeel, which can be reduced by combining it with esters. Because it is so nonpolar, it usually has a detrimental effect on the sunscreen efficacy.

Silicone-based "oils" have enjoyed an almost explosive increase in popularity in recent years. Dimethicone (polydimethylsiloxane) is the most widely used material of this type. It provides a smooth initial skinfeel, reduces skin whitening (soaping), and minimizes tacky afterfeel. The cyclomethicones, also called volatile silicone, greatly improve initial skinfeel without contributing to final greasiness.

Film Formers

In recent years, there has been a noticeable trend toward increased efficacy in sunscreen products. This could be manifested in a higher SPF or a product that maintains its SPF after exposure to water. Film formers play a key role in both of these endeavors. One of the primary factors that influence the ability of a sunscreen product to achieve a high SPF is the uniformity of the sunscreen film on the skin. It is absolutely crucial that this film be thick and uniform. Several materials have been used to develop this type of film. Cellulosic gums lay down a rather even film and can be readily incorporated into almost any emulsion. Polyvinylpyrilidone (PVP) also serves this purpose. Unfortunately, none of these ingredients imparts any measure of water resistancy to the formulation. There are, fortunately, materials that can function as film formers that impart washoff resistance. The two that are used most are based on acrylic–acrylate copolymer or PVP–hexadecene copolymer. Recently, polyethylene has also been used. This material can act to increase the film thickness and additionally can improve both the SPF and the water resistancy.

Stabilizers/Protectants

Emulsions are very delicate and are subject to attack from many directions. If any of these attacks are successful, the emulsion will exhibit instability and will not be sellable. We have previously discussed the thickeners (gums) as stabilizers and will not do so again. Antioxidants represent a most useful and functional class of materials. When oil-phase components are used that are unsaturated (contain one or more double bonds) it is advisable to employ an antioxidant to prevent oxidation of the double bond, with its associated rancidity and discoloration. BHA, BHT, propyl gallate and tocopheryl (acetate) are frequently used. Antioxidants can also act as free radical scavengers and, thereby, reduce chances of skin damage by these highly reactive moieties. It has been suggested that use of antioxidants can actually increase the SPF. It is likely that this effect is due to a reduction in erythema, rather than any increase in UV absorbance. Chelating agents find use in many cosmetic formulations, ranging from emulsions to hair products (shampoos). They can improve the preservative efficacy of conventional preservatives and can tie up unwanted metal ions that can cause discoloration and viscosity of the emulsion. Preservatives must be considered one of the most important parts of the emulsion. No emulsion is complete or safe without them. As with emulsifiers, they seem to work best when several are combined at the same time. The subject of preservation is so important and complex that the reader is urged to study it in some depth using the many excellent texts available. This is no area for the novice. Some suggestions that should be kept in mind when formulating emulsions include:

- When using gums or proteins, enhance the preservative system.
- When using ethoxylated nonionic emulsifiers, paraben use levels should be increased (the *para*-hydroxyl group of the parabens can hydrogen bond to the ethylene oxide chains and thus effectively inhibit it from acting as a preservative).
- Always place the preservative in the water phase (this is the phase that must be protected)

Sunscreens

We are now ready to choose the primary functional ingredient in the sunscreen formulation—the sunscreen. At the present time (in the United States) our choice is limited to 20 sunscreen chemicals (14). Although the list of chemicals may seem to be extensive, in reality only very few are seen in finished products (15). Most of the 20 materials have serious drawbacks associated with them:

Aminobenzoic acid: One of the oldest sunscreens, *p*-aminobenzoic acid, commonly known as PABA has many drawbacks: Its UV curve peaks at the low end of the UVB spectrum; it is too water-soluble (owing to high polarity at both ends of the molecule); a segment of the population exhibits some sensitivity to it; and it can stain clothing and skin.

Padimate A: Amylidimethyl PABA is no longer used in the United States because a significant segment of the population exhibits some facial stinging (16) to this chemical.

Diethanolamine p-methoxycinnamate: This screen exhibits very good UVB absorbance, but is quite expensive and is water-soluble.

Ethyl 4-bis(hydroxypropyl) PABA: An excellent sunscreen, this material has not had good acceptance among cosmetic formulators owing to its poor solubility in most cosmetic solvents. It is no longer being manufactured.

Octocrylene (2-Ethylhexyl-2-cyano-3,3 diphenylacrylate): This has not seen wide acceptance owing to its relatively low extinction coefficient. In recent years, however, its usage has dramatically increased. This is probably due to its ability to boost the SPF and also improve the water-resistancy of many formulations.

Octyl salicylate and homomenthyl salicylate: Both of these compounds have a long history of use. Both were supplanted by the more efficient PABA and cinnamate derivatives. With the trend toward high SPFs, a significant increase in the use of octyl salicylate has been observed. This has been accelerated by another marketing trend: PABA-free. Octyl salicylate, like all salicylates, is *ortho*-disubstituted and, thus, exhibits a rather low extinction coefficient. It has the ability to solubilize oxybenzone.

Glyceryl aminobenzoate: A close relative of PABA, this screen has a very high extinction coefficient; but its peak occurs at 297 nm, slightly to the left (shorter wavelength) of optimum. Additionally, its water solubility limits its applicability in waterproof formulations.

Trolalamine salicylate: A weak absorber, it has poor oil solubility (good water solubility).

Oxybenzone and dioxybenzone: The best known of the UVA absorbers, they are seldom if ever used alone, but are combined with UVB screens to give high SPFs. They have poor solubility and relatively low extinction coefficients.

Sulisobenzone: This sulfonated version of oxybenzone is water-soluble and, thus, is not used in water-resistant formulations. As with oxybenzone, it is a rather reactive molecule; thus, care must be taken when formulating with these materials from a stability and safety viewpoint.

Menthyl anthranilate: This UVA screen is not widely used because of its low absorptivity (*ortho*-disubstituted). This screen may see increased acceptance as the trend toward higher SPFs and UVA claims continues. It should be combined with other screens to achieve this end.

Titanium dioxide: This physical blocker had not gained acceptance for many years owing to its poor aesthetics (skin whitening). Usage has skyrocketed with the appearance of microfine grades of this material that substantially address this problem.

Padimate O (octyldimethyl PABA) and *octyl p-methoxycinnamate*: By far the two most popular sunscreens. Both of these chemicals exhibit very strong absorbance in the middle of the UVB range (310 nm) and have the attributes most sought after in a sunscreen chemical. They are not water-soluble (will not wash off easily); will not stain the skin or clothing; are very safe, chemically inert, and UV stable; stay on the skin (minimal percutaneous absorption); have minimal odor; will not color the emulsion; and are relatively inexpensive. They seem like ideal sunscreens and, in fact, they come very close to the ideal. Practically all formulations in the world use one of these two as the primary sunscreen. As the cosmetic chemist reaches for higher and higher SPF values, many formulations are found that combine both of these sunscreens.

Zinc oxide: This material has been used for many years in both cosmetic and toiletry products. It was not on the original list of 21 sunscreens approved by the FDA. With the publication of the TFM it was placed into category III (needs more data). I, as well as most others involved in the sunscreen business, expect zinc oxide to achieve category I status (meets monograph conditions) when the final sunscreen monograph issues.

Phenylbenzimidalole sulfonic acid: This material becomes water-soluble when neutralized by an appropriate base (triethanolamine, NaOH, or others). It has found use in clear gels owing to its water solubility. It also exhibits a very high efficiency. This is probably due to its ability to partition into the upper layers of the stratum corneum; thereby providing a very uniform film on the skin. Additionally, its high polarity does not allow it to be adversely affected by the nonpolar skin lipids.

Fragrances

The choice of a proper fragrance can enhance the aesthetics of any sunscreen product. Because these materials can be quite irritating, their use level must be kept as low as possible. Additionally their breakdown products can be photosensitizers; hence, great care must be taken in choosing them.

C. Achieving Formula Goals

1. To achieve high SPFs
   a. Avoid ''bad'' ingredients. Use of ethanol at high levels can reduce the SPF of the finished formulation. It is likely that ethanol (very volatile) flashes off and leaves a porous film on the skin. This discontinuous (17) film is not conducive to high SPFs. The use of very nonpolar materials, such as mineral oil, petrolatum, and highly branched esters, results in a shift of the λ_{max} (wavelength of maximum absorption), of nonpolar sunscreens, to a shorter wavelength. This is very detrimental to the efficiency of these sunscreens.
   b. Use the correct vehicle. Emulsions in general, and creams in particular, can achieve the highest SPFs attainable. They put down a very uniform film on the skin.
   c. Combine sunscreens: Use sunscreens that absorb into the short UVB area (320–340 nm), as well as sunscreens that show good absorbance in the erythemal range (290–320 nm). Although these wavelengths are much less effective at producing erythema than the 290- to 320-nm wavelengths, they must be blocked if very high SPFs are to be realized.
   d. Obtain uniform film on skin: use of film formers (see foregoing) can optimize the SPF.
2. To achieve water resistance
   a. Use water-resistant sunscreens (18).
   b. Use high-level oil phases in oil-in-water emulsions.
   c. Use water-resistant resins.
   d. Use minimum levels of hydrophilic emulsifiers. This will minimize the reemulsification of films left on the skin when exposed to water.
   e. Have the oil-in-water emulsion break on the skin during rubout. This leaves an oily (water-resistant) film on the skin.
3. To obtain mild formulations
   a. Use minimum emulsifier levels, particularly soaps.
   b. Minimize use of fragrances.
   c. Minimize use of preservatives, making sure, however, to use adequate levels.
   d. Use the minimum level of sunscreen to achieve the target SPF.
   e. Use long-chain esters which are less irritating than their short chain analogs.

D. Stability Evaluation

The stability evaluation of sunscreen products is not unlike that required for other cosmetic or toiletry products. The finished formulation, in the production package, must be subjected to both high and low temperature extremes. A typical minimal program should include the following:

25°C for 6 months
37°C for 3 months
45°C for 2 months
50°C for 1 month
4°C for 6 months
−20°C for 6 months
Freeze–thaw (45°–25°C) for five cycles
UV exposure
High-humidity exposure

The following parameters must be monitored:

Particle size and distribution (for emulsions)
Color
Odor
Preservative efficacy
Weight loss
Viscosity
pH
UV profile (curve)
Closure torque

E. Organoleptic Considerations

Although consumers purchase sunscreen products for protection from the damaging rays of the sun, they will not buy them a second time if they do not possess the cosmetic attributes they have come to expect from all toiletries. Use of fragrance to provide a pleasant note on application and hide the unpleasant odor of some needed ingredient is a must. Use of silicones or other emollients will reduce the dragginess on rubout often found in high glyceryl monostearate (GMS), stearic acid, or fatty alcohol-containing emulsions. The formulation should exhibit easy spreadability on the skin, without any dragginess. When particulate materials (TiO_2 and zinc oxide) are used, great care must be taken to ensure that minimum skin whitening is seen. Additionally, these formulations have a tendency to rewhiten when one goes into the water or sweats profusely. The cosmetic chemist has not completed his assignment until he has optimized his formula for efficacy, safety, stability, and last, but not least, organoleptic considerations. A successful formula has all of these areas addressed.

III. FORMULATIONS

Following is a compilation of formulations that serves to illustrate the diversity of formula types currently available in the marketplace. By no means is this collection of formulations complete, but it is representative of the state of art and science (as of this moment).

Formula type: Cream
Comments: Emollient skinfeel
Expected SPF: 25+

Ingredients	% w/w
Phase A	
Octyl dimethyl PABA	8.0
Benzophenone-3	5.0
Octyl methoxycinnamate	6.0
Cyclomethicone	10.0
Glyceryl stearate SE	5.0
Phenyldimethicone	2.0
Cetearyl alcohol (and) ceteareth-20	2.0
Cetyl alcohol	1.0
Octyl palmitate	10.0
Phase B	
Water	qs
Preservative	qs
Glycerin	5.0
Diethanolamine *p*-methoxycinnamate	8.0
Titanium dioxide	3.0
Xanthan	0.2
Hydroxyethylcellulose	0.1
Phase C	
Fragrance	0.3

Formula type: Lotion
Comments: Inexpensive lotion with outstanding high temperature stability
Expected SPF: 15

Ingredient	% w/w
Phase A	
Octyl dimethyl PABA	7.0
Benzophenone-3	3.0
Stearic acid XXX	4.0

Ingredient	% w/w
Mineral oil	10.0
Myreth-3 myristate	4.0
Phase B	
Water	qs
Preservative	qs
Carbomer 1342	0.2
Propylene glycol	5.0
Phase C	
Triethanolamine 99%	0.7
Phase D	
Fragrance	0.3

Formula type: Cationic lotion
Comments: Cationic lotion offers excellent skinfeel
Expected SPF: 4–6

Ingredient	% w/w
Phase A	
Glycol stearate	5.0
C12~15 alcohol benzoate	3.5
Octyl methoxycinnamate	5.0
PEG-40 stearate	1.5
Phase B	
Water	qs
Preservative	qs
Stearamidopropyl PG-dimonium chloride	
Phosphate	3.5
Glycerin	4.0
Phase C	
Fragrance	0.3

Formula type: Water-resistant lotion
Comments: Minimum emulsifier level and film former makes this lotion water resistant
Expected SPF: 15

Ingredient	% w/w
Phase A	5.0
Mineral oil	1.0

Sorbitan sesquioleate	7.5
Octyl dimethyl PABA	7.0
Benzephenone-3	3.0
Octyl salicylate	4.0
Laureth-23	2.0
Stearic acid XXX	2.0
PVP/eicosene copolymer	2.5
Phase B	
Water	qs
Sorbitol 70%	5.0
Carbomer 1342	0.2
Preservative	qs
Phase C	
Triethanolamine 99%	0.5
Phase D	
Fragrance	0.5

Formula type: Lotion
Comments: Nongreasy lotion offering high protection
Expected SPF: 18

Ingredient	% w/w
Phase A	
Octyl methoxycinnamate	7.5
Benzephenone-3	5.0
Menthyl anthranilate	5.0
Glyceryl monostearate SE	2.0
PEG-40 stearate	1.5
Cetyl alcohol	0.5
Diisopropyl adipate	5.0
Phase B	
Water	qs
Carbomer 941	0.15
Preservative	qs
Phase C	
Triethanolamine 99%	0.15
Phase D	
Fragrance	0.25

Formula type: Lotion
Comments: Excellent stability and smooth application
Expected SPF: 6–8

Ingredient	% w/w
Phase A	
DEA-cetyl phosphate	2.0
Glyceryl monostearate SE	2.0
Cetearyl alcohol	1.0
Phenyldimethicone	1.0
Cyclomethicone	2.0
Octyl dimethyl PABA	7.0
Phase B	
Water	qs
Xanthan	0.3
Propylene glycol	5.0
Preservative	qs
Phase C	
Fragrance	0.25

Formula type: Sunscreen oil
Comments: Octyl palmitate reduces oiliness of mineral oil while minimizing UV curve shift
Expected SPF: 4

Ingredient	% w/w
Octyl methoxycinnamate	5.0
Octyl palmitate	qs
Lauryl lactate	15.0
Mineral oil	35.0
Isocetyl alcohol	10.0
Fragrance	1.0

Formula type: Sunscreen oil
Comments: Menthyl anthranilate provides protection into the UVA region
Expected SPF: 4–6

Ingredient	% w/w
Octyl salicylate	5.0
Menthyl anthranilate	3.5
Jojoba oil	2.0
Cocoa butter solid	2.0

Isocetyl alcohol	15.0
Fragrance	1.0
Octyl palmitate	qs
Vitamin E acetate	0.1
Mineral oil	40.0

Formula type: Sunscreen gel
Comments: Poloxamine aids in sunscreen solubilization
Expected SPF: 12

Ingredient	% w/w
Phase A	
Carbomer 940	1.5
Poloxamine 704	7.0
Ethanol	qs
Phase B	
Dioctyl malate	10.0
Octyl methoxycinnamate	7.5
Benzophenone-3	4.0
Phase C	
PEG-15 cocamine	3.5
Water	12.5

Formula type: Aftershave gel
Comments: Cellulose-based gel incorporating emollient ester that has good hydroalcoholic solubility
Expected SPF: 2

Ingredient	% w/w
Phase A	
Ethanol	qs
Hydroxypropyl cellulose	2.00
Phase B	
Fragrance	5.0
Dioctyl malate	5.0
Padimate O	2.0
Phase C	
Water	15.0
Preservative	qs

Formula type: Nose protectant
Comments: Contains sunscreen and opacifying agents to achieve a very high degree of protection
Expected SPF: 25+

Ingredient	% w/w
Phase A	
Octyl dimethyl PABA	7.0
Benzephenone-3	3.5
Mineral oil (and) lanolin alcohol	5.0
C12–15 Alkyl benzoate	15.0
Phase B	
Water	qs
Preservative	qs
Phase C	
Zinc oxide	25.0
Titanium dioxide	30.0

Formula type: Sunscreen cream
SPF: 27.6

Ingredients	% w/w
Phase A	
Water	27.65
Tetrasodium EDTA	0.10
Propylene glycol	2.00
Phase B	
Octyl methoxycinnamate	7.50
DEA-cetyl phosphate	1.50
Cyclomethicone	1.00
Tridecyl neopentanoate	7.50
Octyl salicylate	5.00
Cetearyl alcohol	2.00
Octocrylene	10.00
Tocopheryl acetate	0.25
Ceteareth-20	0.50
Tricontanyl PVP	3.00
Phase C	
Zinc oxide	6.00

Phase D	
Preservative	1.00

Sunscreen lotion
SPF: 18.25

Ingredients	% w/w
Phase A	
Octyl methoxycinnamate	7.50
Menthyl anthranilate	5.00
Isopropyl myristate	3.00
PEG-40 stearate	2.00
Glyceryl stearate	3.00
Cetearyl alcohol and ceteareth-20	2.00
PVP/eicosene copolymer	4.00
Dimethicone	2.00
Phase B	
Water	qs
Carbomer 940 (2% soln)	20.00
Propylene glycol	2.00
Tetrasodium EDTA	0.10
Phase C	
Phenylbenzimidazole Sulfonic acid (30% TEA salt)	6.67
Phase D	
Triethanolamine 99%	0.50
Phase E	
Preservative	qs
Fragrance	qs

REFERENCES

1. Stockdale, M. Sun protection factors. *Int. J. Cosmet. Sci.*, *7*:235–246 (1985).
2. Ibid.
3. Agrapidis-Paloympis, L., Nash, R. A., and Shaath, N. The effects of solvents on the ultraviolet absorbance of sunscreens. *J. Soc. Cosmet. Chem.*, *38*:209–221 (July/Aug.: 1987).
4. Klein, K. and Doshi, I. Sunscreen solvent interactions: An in-vitro evaluation. 14th IFSCC Congress, Barcelona, Spain, September 1986.
5. Klein, K. Improving emulsion stability. *J. Cosmet. Toilet.*, *99*:121–126 (Mar. 1984).

6. Stokes, G. G. *Philos. Mag.*, *1*:337 (1851).
7. Lin, T. J. Low energy emulsification-l-principles and applications. *J. Cosmet. Chem.*, *29*:117–125 (March 1978).
8. Martin, A., et al. *Phys. Pharm.*, *2*:4–5 (1969).
9. Suzuki, T., Tsutsumi, H., and Ishida, A. Secondary droplet emulsion: Mechanism and effects of liquid crystal formation in o/w emulsions. *Dispersion Sci. Technol.*, *5*(2):119–141 (1984).
10. Griffin, W. C. Classification of surface active agents by "HLB." *J. Soc. Cosmet. Chem.*, *1*:311–326 (1940).
11. The HLB System. ICI America, Inc. Wilmington, DE (1984), p. 19.
12. Shinoda, K. Comparison between phase-inversion-system and the hydrophile–lipophile balance-value system for emulsifier selection. *Nippon Kagaku Zasshi*, *89*: 1–8 (1968).
13. Lin, T. J., Kurhara, H., and Ohta, H. Prediction of optimum o/w emulsification via solubilization measurements. *J. Soc. Cosmet. Chem.*, *28*:457–479 (August 1977).
14. Docket No. 78N-0035, Sunscreen Drug Products for over-the-counter Human Use; 58 *Federal Register* 28194 (May 12, 1993); 58 *Federal Register* 53460 (October 15, 1993).
15. Shaath, N. The chemistry of sunscreens. *Cosmet. Toilet.*, *101*:55–70 (Mar. 1986).
16. Frosch, P. and Kligman, A. A method for appraising the stinging capacity of topically applied substances. *J. Soc. Cosmet. Chem.*, *28*:197–209 (May 1977).
17. Eierman, H. U.S. Patent 3,341,419 (1967).
18. Formulating vehicles for sunscreens. Norda Shimmel Briefs, 935: (Feb. 1966)

ADDITIONAL READING

Martin, A., Swarbrick, J., and Cammarata, A. (eds.). *Physical Pharmacy*. 2nd ed. Lea & Febiger, Philadelphia, 1969.

Balsam, M. S. and Sagarin, E. (eds.). *Cosmetics: Science and Technology*. Vol. 1. Wiley-Interscience, New York, 1972.

17

Particulate Sun Blocks: General Principles

David Fairhurst and Mark A. Mitchnick
sunSmart, Inc., Wainscott, New York

I. INTRODUCTION

The particulate sunscreens have gained wide acceptance over the past decade, going from relative obscurity to being considered mainstream ingredients. Zinc oxide, for instance, is not even in the index of the 1990 version of this text. In 1996 it, and its class, warrant their own chapter. The reasons for this increase in popularity are many. Chief among them are the low toxicity potential and the effectiveness of the particulates. Still, there remains a great deal of misunderstanding about these materials. How do they work? What can and what can they not accomplish? How do you formulate best with them? How do you evaluate them? This chapter will attempt to give an overview of the particulates, but it will not go into the specific "brands" available. There are several excellent particulate manufacturers, each with a very competent support staff that can answer questions about individual particulates. Because each particulate is unique, we strongly encourage formulating chemists to obtain detailed information and support from their supplier of choice.

Ironically, the particulates were among the first sunscreens used. The protective value of zinc oxide has been appreciated, although not well understood, for centuries. The first volume of the *Journal of the Society of Cosmetic Chemists*, published in 1947 (1), contained an article titled "Zinc Oxide in Face Powders." In that article, the author thoroughly discloses the absorption spectrum for several metal oxides, including zinc oxide and titanium dioxide. He even went as far as to predict the value of zinc oxide as an ultraviolet (UVA) block—a rather prophetic observation.

The particle size versus opacity versus ultraviolet radiation (UVR) attenuation relations have also been known for the better part of this century (2). In retrospect, the particulate sunblocks failed to advance because there was no commercial need for them. Now, with the increased awareness of UVR damage, especially UVA-related, and the general desire to minimize our exposure to organic chemicals, the proper conditions exist for nurturing this technology.

For this chapter, the word *particulate* will be used to describe any material that is insoluble in the chosen medium. The particulate may be organic, inorganic, or a combination of materials. Commonly recognized particulates include zinc oxide, titanium dioxide, talc, iron oxide, kaolin, bentonite, silica, and mica. Also included are the insoluble melanins, other metal oxides, pure metals, and a new variation that will be briefly discussed later, encapsulated organic sunscreens.

There is a natural tendency to treat all particulates as being alike. This, however, makes no more sense than treating all the soluble sunscreens as being the same. Each particulate will have a different attenuation spectrum, surface chemistry, and mode of action. Furthermore, there is probably greater manufacturer-to-manufacturer variation with the particulates than with the soluble sunscreens. Add to this the surface treatments that are often applied to particulates and one can see that these materials need to be treated as discrete entities. This is not to say that they are difficult to use, just to point out that they have their own chemical personality that must be recognized to obtain optimal performance.

Zinc oxide and titanium dioxide account for almost all of the particulate sunscreens used today, because they are the two most effective and widely available materials. Consequently, much of this chapter will be devoted to the discussion of these two particulates, but where appropriate, other materials will be discussed.

II. ADVANTAGES OF PARTICULATES

As we will see in the following, particulates can be stable, efficient, cost-effective, and benign sunscreens. Their use is attractive on many levels, including UVR attenuation efficiency. Today, however, the sun protection factor (SPF) is but one of several very important criteria on which to judge a sunscreen. Safety and attenuation spectrum are also of equal, and at times of greater, importance. The incorporation of sunscreens into daily-wear products and the increasing use of beach sunscreens will inevitably push the existing soluble screens to their limit. Issues of photo-induced and nonphoto-induced toxicity and allergy are becoming increasingly important (3). This, at a time when higher SPFs are being demanded by the consumer, leaves the formulator in a difficult position. In Japan, for instance, the total amount of soluble organic sunscreen that can be used (without special testing) is limited to 10%. Particulates offer a way to formulate high-SPF products without the addition of more soluble organic sunscreens. In addition, for lower SPF (less than 20), products that are based solely on particulates are today easily achieved. This can be especially important in children's products, daily-wear formulations, and those topicals aimed at individuals with sensitive skin. To date, though, most marketed products that use particulates do so in combination with the soluble sunscreens. As the particulates mature, we can expect to see more and more products based solely on them.

Attenuation of UVA is another emerging area of importance for today's sunscreens. Evidence is mounting that UVA is at least a contributing factor to both visible aging and carcinogenesis. Some even advocate UVA as the causative agent in melanoma formation. Whatever the exact role of UVA in human biology, it is certainly detrimental, and blocking it (UVA) is rapidly becoming mandatory in any product claiming skin protection. Some particulates (see later) offer very effective UVA attenuation. This is of even greater importance when one takes into account that, among the soluble sunscreens, discussion of the UVA agents tends to be focused on the problem of irritancy (3) and, thus, are potentially limited in the current atmosphere of increasing scrutiny.

III. MICROFINE

When discussing particulates as sunscreens, one is usually talking about relatively small particles and, more importantly, a well-defined particle size distribution. The term *microfine* is most commonly used to describe such particles.

Zinc oxide and titanium dioxide are commercially available in pigmentary and microfine grades. Microfine is also termed micronized, nanosize, or ultrafine. This loose terminology has led to a great deal of confusion. By definition, *micronized* refers to a pigment that has been physically reduced to its given particle size. To our knowledge, it is virtually impossible to "grind" or "mill" a pigment to smaller than about 1 μm. Practically speaking, this means that micronized pigments are of little value as sunscreens because 1 μm is not an efficient size (see Sec. V).

Microfine is a term without a universally agreed on meaning. It is used commercially to describe particles with claimed particle sizes anywhere from 4 nm (0.04 μm) to several microns. It is our hope that, in the future, the terminology will be standardized so that microfine will be used to describe a well-defined particle size distribution. One such definition might be particles larger than 80 nm, but smaller than 250 nm. The majority of particles would have to fall into this range and not just some small percentage. This definition, or a similar one, would serve to distinguish microfine particles from the high-quality pigmentary grades the particles of which have an average diameter of about 250 nm.

This size range (100–250 nm) is commercially and technically very meaningful, for it reflects a refinement in available particle size. It is not, however, meant to infer the creation of a new particle size. For decades, humans have been using materials with particles within this size range. There has always been a considerable number of particles in the 100- to 250-nm range in commercially used zinc and titanium dioxide powders (this is also true for carbon blacks and iron oxides). One only needs to look at the literature and the quoted particle sizes of these products. These are the ''fines'' with which industry has long been used to dealing. Their presence, however, is optically overwhelmed by the larger particles and their UV attenuation is poor, but nevertheless, they are present. The suggested concept of microfine is not so much that some particles are in the preferred size range, but that the particle size distribution is controlled so that most particles are in the desired range. In this way, the bulk behavior of the powder is that of the microfine particles and not of the larger ones. Practically speaking, controlled mass production of microfine particles is not easy, and it is only recently they have become available.

Regardless of the final parameters arrived at, it is absolutely critical that some definition be established so that we, the cosmetic industry, do not force a third party to establish a definition for us. The particulate sunscreens are too important. They are quite likely one of the major public health advancements in recent years, and it is incumbent on us to handle them as such. Microfine, as defined in the foregoing, simply describes a refinement of particle size distribution and not the creation of previously unavailable particle sizes.

Along with size, purity must be considered. Efficacy will be poor with a less than optimal size distribution. Therefore, industry will regulate this aspect (size) because incorrectly sized materials will simply be commercially nonviable. Chemical purity, on the other hand, is a matter for the various standards organizations (such as the *U.S. Pharmacopeia USP*). It is the authors' opinion that the particulate sunscreens should be held to the highest standards of purity. There seems to be no reason to accept less from a sunscreen than one would from another topically applied drug.

IV. TYPES OF PARTICULATES

Particulates can be conveniently divided into two groups: organic and inorganic. The inorganic include the metal oxides (zinc and titanium being the most common), kaolin, barium sulfate, mica, glass, plastics, metal powders such as gold, and many others. Organic particulates include materials such as melanin and encapsulated organic sunscreens. Virtually anything that can be made into particles will have some effect, albeit often insignificant, on UVR. Of the inorganics, only zinc oxide and titanium dioxide are currently viable as widely acceptable UVR-blocking agents. This is true for both physical (see Sec. V) and regulatory reasons. These are the only two that are globally allowable as sunscreen active ingredients.

A. Zinc Oxide

Zinc oxide, ZnO, is a metal oxide that has been used topically for centuries. It is the main constituent of calamine. It is manufactured either through the oxidation of zinc vapor or through ''wet'' methods (4). Most zinc oxide used in health-related applications is used in skin protection products. It is well known as an antimicrobial and wound-healing agent. Before the invention of antibiotics, it was the preferred dressing for burns and other wounds. Its mechanism of action is not completely understood, but is presumably related to zinc's presence in many of the body's metaloenzymes (5).

Manufacture

Zinc oxide, for health care uses, is usually manufactured using the ''French process''—the oxidation of zinc vapor. The key is that the zinc vapor is produced from a highly refined form of zinc that has little or no contaminants. Particles formed using the French process typically have sizes of about 1 μm or larger, but there is tremendous variability, and control is difficult.

Wet methods for producing zinc oxide usually consist of precipitation of a zinc intermediate, which is calcined to form the oxide. Although particle sizes produced using wet methods can range from 10 nm upward, most commercially available products have particle sizes in the 200 nm and larger range. Wet method technology offers the chance for greater variability in particle content and morphology, but suffers from surface contamination. There are inevitably some remains of the salts used to precipitate the intermediate. These contaminants, often sodium ions or sulfate ions, may or may not be significant in a given formulation.

Last, there are several proprietary methods for zinc oxide production that attempt to eliminate some of the problems just discussed.

B. Titanium Dioxide

Titanium dioxide, another metal oxide, is currently the most used of the particulates. It is best known as a whitening agent and is used for this purpose in virtually all industries. In its microfine form, the whitening is reduced, and it has gained wide acceptance as a UVR screen. As a sunscreening agent, it is usually sold as a surface-treated pigment. This is done to minimize any photoreactivity and to aid in the formulation process (see following paragraph).

Manufacture

Titanium dioxide can also be manufactured either through thermal or wet methods. Much of the titanium dioxide available is manufactured by the hydrolysis of titanium tetrachloride or titanyl sulfate (4). Commercially available titanium dioxide exists in either the rutile or anatase form. Although historically anatase was the first form of titanium dioxide pigment sold generally, rutile is preferred as a pigmenting agent for coatings that may be exposed to sunlight. Although more expensive to produce, rutile provides greater reflectivity for visible light and has reduced photoreactivity. Most titanium dioxides sold for sunscreen purposes are surface-modified with either an inorganic such as silica, or an organic layer, such as stearate.

Some titanium dioxides manufactured by vapor deposition are beginning to appear on the market. These offer the potential advantage of very tight particle size distribution control, but may suffer from increased cost.

C. Other Particulates

The other commonly available particulates are predominately UV and visible radiation scatters; therefore, that cannot be used as effective transparent UV blocks. They include, among others, talc, mica, kaolin, barium sulfate, and the iron oxides. The average particle sizes are typically in the micron range. Interestingly, the iron oxides do show some absorbency in the UV, but their cutoff is in the visible. This makes it difficult to exploit the UVR attenuation properties except in colored products. Even then, the UVR attenuation abilities of these materials is surprisingly low. For instance, a lipstick containing 10% total pigment (not microfine zinc oxide or titanium dioxide) can be expected to have an SPF of only 2–4. These agents can, however, provide useful "boosts" to products in which color is desired, such as in lip products and, when incorporated at reasonable levels (2–5%), they can be expected to boost the SPF 1 or 2 points especially in high-SPF products. The iron oxides are also used in many commercially available titanium dioxide-containing preparations to "mask" the whitening.

V. MECHANISM OF ACTION

The development of particulate-based sunscreen requires knowledge of both the optical properties and the surface chemistry of the particulate material relative to the dispersion medium. The surface chemical aspect will be dealt with in Section VI: Fundamentals of Particles in Liquid Dispersions.

In discussing the optical properties of particulate sunscreens one must bear in mind that there are two separate aspects to consider:

1. The color strength and opacity of the "film" (whether aqueous or nonaqueous) on the skin: We are concerned here with light in the visible part of the spectrum from about 400 to 720 nm. For most purposes, we want the sunscreen agent, the particulate, to have as little color and opacity as possible. In other words, we want it to be transparent.
2. The ability of this film to act as a sunscreen by stopping light in the UV region of the spectrum (200–400 nm) from reaching the skin: More specifically, we are concerned with radiation from 300 to 400 nm, the earthly UV solar spectrum. Certain occupations, such as welding, may require broader protection, but we will not consider this in our discussion.

Both aspects are uniquely related to the particle size of the material. The first is an aesthetic concern, the second is one of performance (SPF), but they are equally important, for, without proper aesthetics (minimal whitening), no level of sunscreen performance is high enough because most persons simply will not use it.

A. The Fundamentals of Light

Light of any wavelength will interact with, and be modified by, the material on which it falls. When a given wavelength of light interacts with a particle, the light will be absorbed or scattered (Fig. 1); in most instances, both processes contribute. For any inorganic particulate, the color strength (staining power) increases with light-absorbing capacity, whereas the opacity (hiding power) increases with the light-scattering power.

Before continuing, let us define the terms absorption and scattering. *Absorption* is the process in which light is "lost" when it falls on a material: The light is not actually lost, but is converted into some other energy, such as heat. Most color effects arise from absorption of different amounts of incident light at dif-

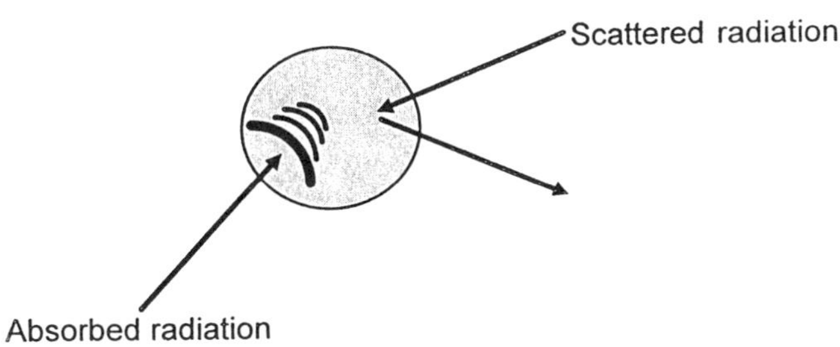

Figure 1 Scattered and absorbed light.

ferent wavelengths. A material that completely absorbs incident radiation of all wavelengths will be absolutely black. Such a material does not exist. Actual "black" pigments vary in the completeness with which they absorb incident light; in other words, they also scatter some light. Light absorption is a function of molecular and crystal structure and also of particle size.

Scattering is the process in which light is redirected so that some fractions of it are traveling in all possible directions, including forward. Scattering provides translucency and opacity to materials. *Transparent materials* are those in which there is no internal scattering: Examples are pure liquids. An ideal white pigment absorbs no radiation and, therefore, is colorless. However, when we compare different white objects, such as pieces of paper, we can often "see" a difference. How white, then, is white? The whiteness results from the scattering of incident radiation by multiple reflection, and again, refraction and is a function of the molecular and crystal structure as well as particle size.

Scattering and absorption are not restricted to the optical part of the spectrum—all radiation, whatever the wavelength, is either scattered or absorbed, or both.

The parameter that describes the scattering–absorption characteristic of a material is the *refractive index*, *n* or RI. This is an oversimplification. In reality, it is a complex function. For crystalline material, such as TiO_2 and ZnO, the different refractive indices (RI) for the three axes of each crystal must be accounted for. Having said this, the single number usually referred to as the RI allows reasonable real-world predictions.

The actual "whiteness" of a dispersion is a function of the ratio of the refractive indices of the material and its surrounding medium. Thus, if a particle and medium have identical refractive indices, then the particle will appear "invisible." The light will not be altered as it goes from medium to particle and back to medium. For TiO_2 at visible wavelengths, the RI is approximately 2.6. Most cosmetic vehicles are made of components with RI values less than 1.5. Thus, it is very difficult to "hide" TiO_2 in a topical product. Zinc oxide, because of its lower RI (1.9), is a somewhat easier case. This is graphically depicted in Figure 2, which shows that, at virtually all particle sizes, TiO_2 scatters more than zinc oxide. The observation that TiO_2 in sunscreens becomes whiter on rewetting is due not only to agglomeration, which increases the effective particle size, but also to the stripping away of the outer oil layer, thereby increasing the RI difference.

B. The Effect of Particle Size

The refractive index is an intrinsic property of a material itself; therefore, it is fixed, but the particle size (PS) can be deliberately chosen.

As a rule of thumb, the scattering power of a material is highest when its PS is in an optimum range, equal to about half the wavelength of the light to be

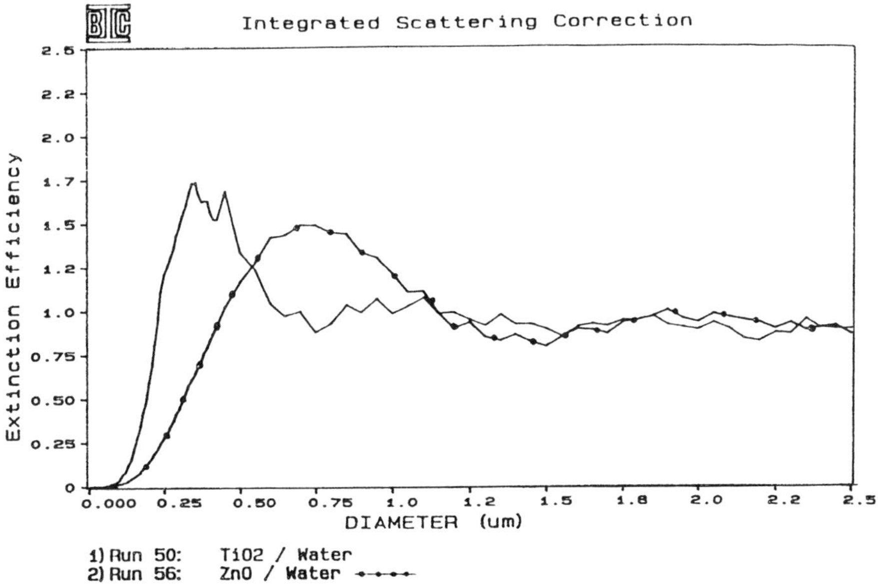

Figure 2 Comparison of scattering efficiencies of TiO_2 and ZnO.

scattered. For monochromatic light, this is straightforward. The problem is the visible spectrum (~400–720 nm), which constitutes "white" light. The human eye has a peak response at about 560 nm in the green portions of the spectrum. In addition, the greater the number of particles in a unit volume of film the more effective the scattering is.

There is, therefore, no single optimum PS for a white pigment; hence, the different whiteness for different particle size distributions. Indeed, the whole matter of color, opacity, and appearance is very subjective (6). Whatever the so-called primary size of the particles, it is the state of dispersion of the final product that determines the effective PS and thus the optical properties.

The dramatic effect of particle size on scattering properties of materials can be estimated using the Lorenz–Mie (LM) theory (7) of which there are four limiting cases, depending on particle size, refractive index, and wavelength (Table 1).

The scattering efficiency calculated using LM theory is shown in Figure 3 for three different materials in water: carbon black, quartz, and polymer latex. Functionally, they are all the same. They all show maxima and minima—"wiggles"—that become progressively damped as the PS gets larger. The effect of decreasing the wavelength is to produce a greater frequency of wiggles and

Table 1 Limiting Cases of Lorenz–Mie Theory[a]

$d \ll \lambda$	Rayleigh	$p \ll 1$
$d < \lambda$	Rayleigh-Gans-Debye	$p < 1$
$d > \lambda$	Anomalous diffraction	$p > 1$
$d \gg \lambda$	Fraunhoffer diffraction	$p \gg 1$

[a] d is diameter, λ is wavelength, $p = 2\pi d(m - 1)/\lambda$, $m = n_p/n_o$ = relative refractive index.

a larger amplitude. The more absorption of a given wavelength, the more smooth the scattering function will be.

Figure 2 shows the scattering efficiency of zinc oxide and titanium dioxide calculated using an integrated value for the wavelength of white light. For all PS in the colloidal size range (smaller than 1 μm) the theory predicts that titanium dioxide is a much more efficient light scatterer than is zinc oxide—TiO_2 is whiter than ZnO. The function for ZnO is smoother than for TiO_2, reflecting the fact that ZnO is a more efficient absorber. The theory also predicts that maximum scattering occurs for TiO_2 at a PS of approximately 250 nm. In practice, maximum scattering is usually achieved at about 220 nm.

The theoretical calculations are based on oversimplified assumptions, and it is not possible to derive a mathematical treatment in any rigorous form for the optical properties of an actual film, such as a sunscreen emulsion, although attempts have been made for paint films (8–10).

When we come to wavelengths shorter than the visible spectrum, the optical behavior of materials—especially particulates such as TiO_2 and ZnO—becomes more complex. In addition to PS, we now have to consider the crystal structure itself and the energy gap between valence and conduction bands. For all crystals there is some characteristic wavelength that will excite an electron from the valence band to the conduction band. This is the process through which these metal oxides owe their semiconductor properties. Wavelengths that are shorter than the characteristic wavelength and fall on the crystal will, therefore, be dissipated (i.e., absorbed by the crystal structure).

The zinc oxide crystal structure efficiently absorbs radiation over virtually the whole UV region (380 nm and shorter). Titanium dioxide is predominately a very efficient UVB absorber, but a UVA scatterer. At a wavelength of 360 nm, about 90% of TiO_2's attenuation is by scattering, whereas, for ZnO, about 90% is due to absorbance.

How does this relate to TiO_2 and ZnO as sunscreens for human application? Both are good UVB absorbers and, in fact, are used as such today. Zinc oxide, because it is an absorber in the UVA, and not a scatterer, remains an efficient effective sunblocker. TiO_2, which is dependent on scattering for attenuation in

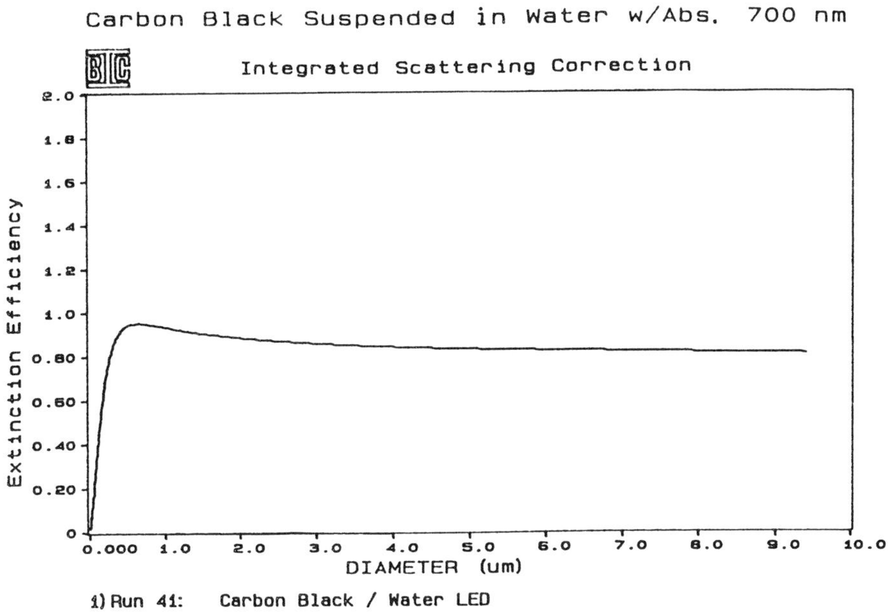

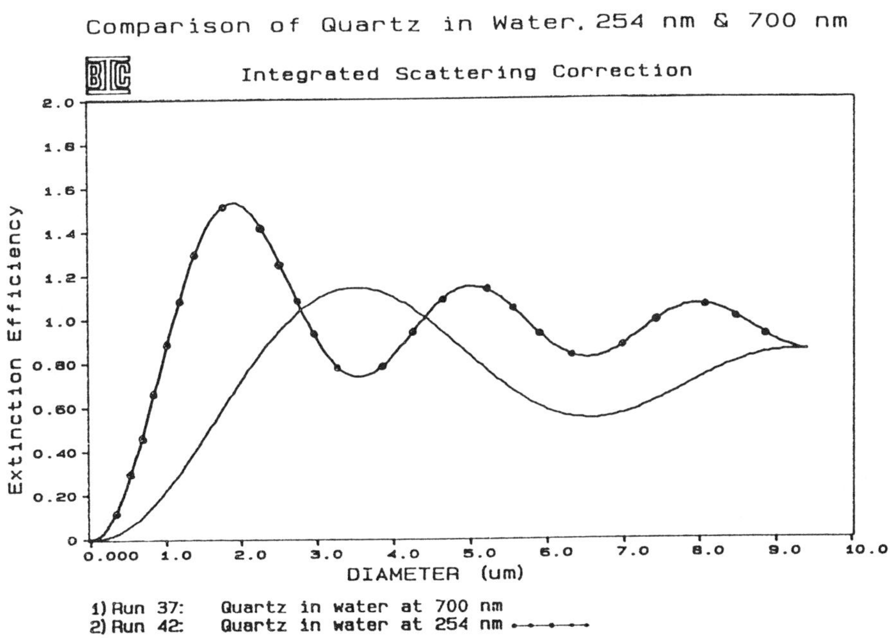

Figure 3 Scattering efficiencies calculated from Lorenz–Mie theory for various materials.

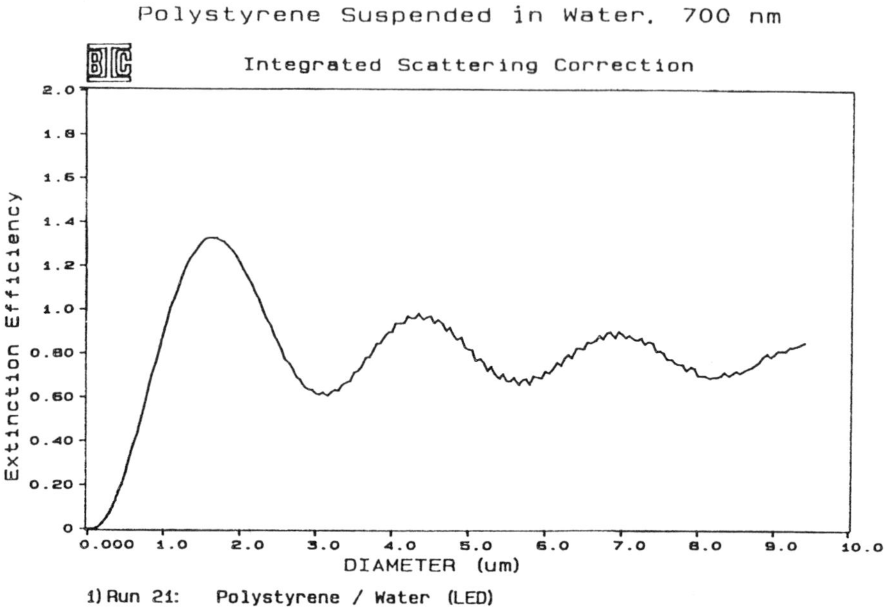

Figure 3 Continued

the UVA, is a relatively poor UVA screen as a microfine particle (even so, it is still better than most of the organics). Larger, whiter, TiO_2 particles are better in the UVA, but it is at the expense of the product's aesthetics. (Fig. 4).

We have seen that both scattering and absorption are a function of particle size. The important point is that the degree of scattering and absorption both vary *differently* with particle size. This is shown schematically in Figure 5 (11). For use as a sunscreen, we need maximum attenuation; that is, we need maximum absorption of UV wavelengths with minimum scattering of visible wavelengths. Empirically, for ZnO, a size range from about 80 to about 150 nm is optimal. This is in good agreement with the calculated optimal PS range.

VI. FUNDAMENTALS OF PARTICLES IN LIQUID DISPERSIONS

Dispersions of solids in liquids find application in a very wide variety of industrial formulations, from cosmetic sunscreens to automotive paints and pharmaceutical preparations. However, the preparation of wet suspensions having a defined dispersed state, with specific properties, is difficult to achieve because complete knowledge of the physicochemical conditions of the dispersed system

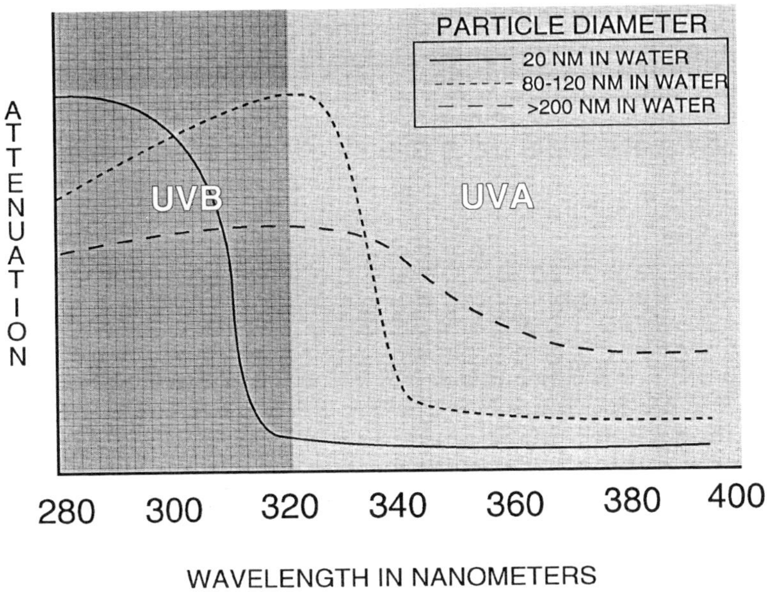

Figure 4 TiO_2 size-dependent UV properties.

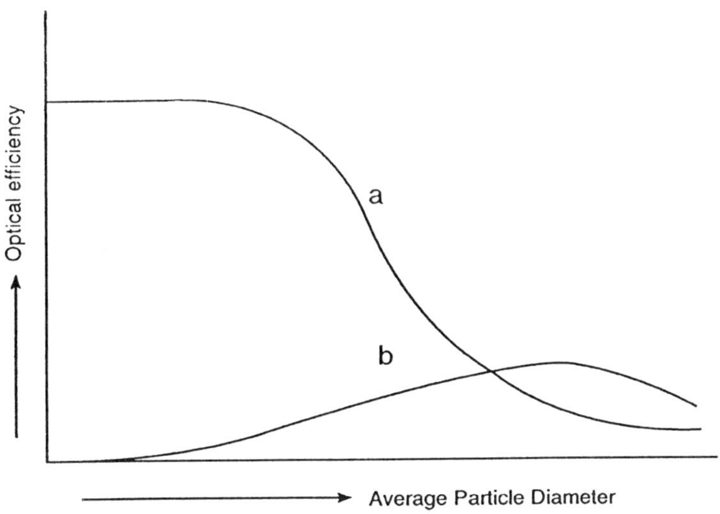

Figure 5 Schematic of scattering and absorption of typical pigments as a function of their particle size. Curve a shows absorption; curve b shows scattering power.

is usually lacking. It is probably fair to say that most industrial methods of formulation have been developed and optimized empirically, or on the basis of working hypotheses that provide adequate guidance to the industrial chemist, but are insufficient as a basis for understanding how and why suspensions of particles in liquids behave as they do and, many times, not as the formulator wants.

The formulation of particulate sunscreens requires that a stable (nonagglomerating) reproducible particle size distribution (PSD) be achieved. Just as the SPF of an organic–soluble system is based on solubilizing the organic sunscreen, the SPF of a particulate system is based on the PSD. So, just as we learned to correctly solubilize the organics, we must learn to properly disperse the particulates. Although new, in many ways, the particulates are simpler to deal with—it is easy to see when the formulation aggregates. The goal of what follows is to bring together some general recommendations and guidelines that, if followed, will result in suspensions having reproducible, consistent, dispersion characteristics. The reader should not assume that the following is complete. On the contrary, these few pages constitute nothing more than an outline of the fundamental parameters involved and a glimpse at their significance. A deeper, more detailed understanding is beyond the scope of this chapter. Information can be sought from the patent literature and other publications; however, much remains as "in-house expertise."

There are, unfortunately, no real textbooks on the preparation of solid–liquid dispersions. The closest is *Dispersion of Powders in Liquids* (12). There are also chapters in various physical chemistry-based textbooks (13,14), books on mineral materials, and review articles on particle size analysis (15,16).

A. The Size Nature of Particles

Aggregates and Agglomerates

Depending upon the material properties, conditions of manufacture, and storage, three distinct species of particles must always be expected, the existence of which results from a variety of different binding mechanisms. The three species (Fig. 6) are

1. *Crystals* or primary single particles are crystalline or organic structures held together by atomic or molecular bonding. They are the "fundamental" particles, often called the *primary particles.* They cannot be separated into smaller particles except by extreme physical comminution. In any given sample they are usually present at only a fraction of a percent.
2. *Aggregates* comprise two or more primary particles tightly bound together by rigid chemical bonding resulting from sintering or cementation during manufacture. Aggregates are typically formed when powders are heated, compressed, or dried from suspension. They have a large interfacial area

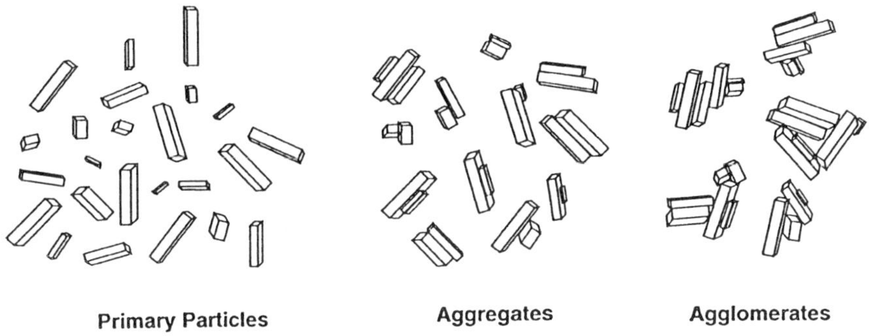

Figure 6 The size nature of dry powders.

of contact between each particle, and the force necessary to rupture these bonds is considerable. Aggregates constitute, for all practical purposes, the largest single fraction of any particle size distribution (PSD) that one can hope to achieve in formulation.

3. *Agglomerates* are aggregates of aggregates, loosely held together at point-to-point contact by weak electromagnetic forces, van der Waals forces, mechanical friction, and interlocking. Agglomerates are formed when fine powders are handled, shaken, rolled, or stored in a single position. They can be readily broken apart with proper dispersion techniques.

In addition, particle physical properties vary widely, and this must be taken into consideration when preparing the suspension. Some materials are brittle, prone to fracture even with gentle handling. Thus, the choice of mechanical processing will be a critical factor. For very strongly aggregated materials, excess dispersive energy may result in fracturing at defect crystal planes. This will produce a new ''primary'' particle that did not exist following the original manufacture. Some materials are partially soluble, such as $CaCO_3$ in water. Here, volume fraction, temperature, pH, and solution composition, all will affect equilibrium dissolution. The wet suspension will ''age'' on standing or during storage through a process known as *Ostwald ripening* (17,18), a phenomenon whereby large particles in suspension grow at the expense of smaller ones, owing to the greater solubility of the smaller particles. Ostwald ripening is a phenomenon that affects not only solid particulate materials, but also emulsions having a range of droplet sizes and foams with a range of bubble sizes. Some powders are prone to pick up an electrical charge and form granules when handled in the dry state. This effect can be eliminated by precoating with an antistatic agent. However, any such coating may actively intervene at some stage of the subsequent dispersion process.

The Importance of the Colloid Size Range

A typical wet suspension will have a PSD in which the majority of particles will be aggregates. The tails of the distribution will be primary particles and agglomerates (Fig. 7). This PSD will result in the suspension having, for example, a given viscosity. Any subsequent change in PSD will affect the flow properties of the suspension. Again, the total number of particles within the distribution will result in a given surface area, and any change in PSD will affect coating or adsorption characteristics of the suspension. Now, for a given system, as the PS becomes smaller, either by mechanical attrition or by improved dispersibility, the total surface area per gram of material increases. This becomes of extreme importance when particles are within what is called the colloid size range. A *colloid* is defined as a particle with at least one dimension less than 1 μm and all dimensions more than 10 nm. As can be seen from Figure 8, the surface area per given mass for a colloid is orders of magnitude larger than it is for particles of only a few micrometers in size. It is a simple matter to show that a particle of dimension of a cube of 1 cm on a side will have a total surface area of 6 m^2. If the whole particle (cube) is now broken down into 10 nm–sized particles the total surface area increases to a staggering 600 m^2. In addition, as material is broken down, the hitherto internal surface becomes exposed and with it a change in the number or type of surface chemical sites

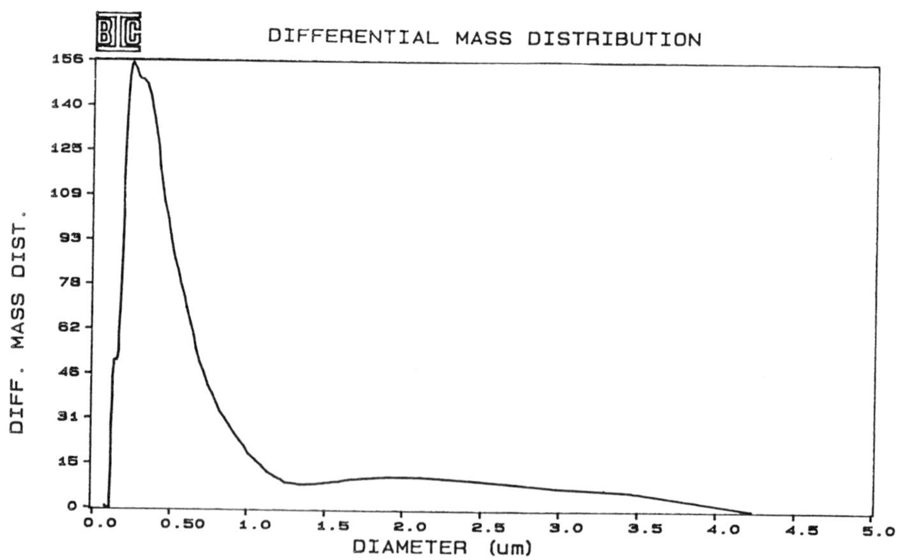

Figure 7 Typical PSD of a metal oxide dispersed in water.

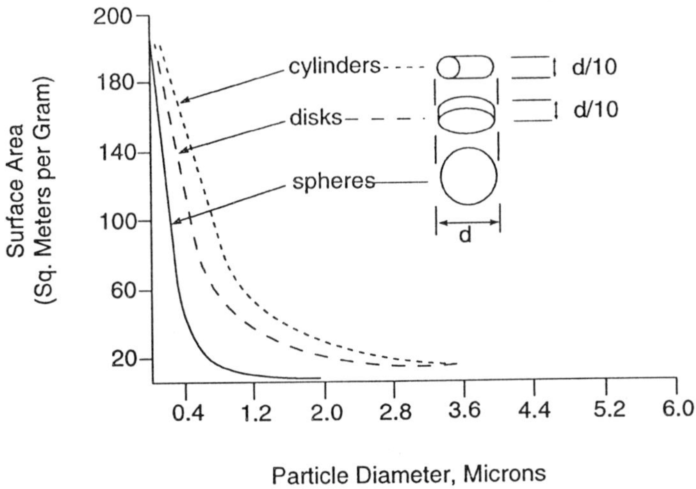

Figure 8 Variation of surface area with particle size.

and groups. Referring again to our particle (cube) of 1 cm only two or three molecules in 10 million are surface molecules and these have negligible influence on its properties. However, when divided into 10 nm particles, the ratio rises to nearly 1:4.

Thus, surface effects that are normally negligible become dominant in the description of colloidal behavior. The consequence of overgrinding or excess mechanical attrition during formulation, producing "fines," can be very significant to product performance.

B. The Dispersion Process

The act of dispersion entails overcoming the various binding forces between particles by use of both physicochemical and mechanical means. The final dispersed state is influenced by three fundamental processes that take place.

1. Wetting of the solid surface
2. Deagglomeration of particle aggregates ("dispersing")
3. Stabilization of the dispersed suspension

It is critical that these three steps be viewed as separate, and that they also be dealt with in the correct order. The physicochemical ways in which all three processes can be influenced are too numerous to cite here. They are covered in detail in most textbooks of colloid and surface chemistry (19–21). The following questions, however, must be addressed.

1. Selection of the suspension liquid
2. Selection of the type and concentration of the various dispersion aids (i.e., wetting agent, deagglomerating agent, and stabilizing agent)
3. Selection of the type and duration of the mechanical dispersing treatment

C. Wetting

Wetting is the act of getting the particle into the chosen liquid. Surfaces that are easily wetted are termed *lyophilic.* Those that are difficult to wet are termed *lyophobic.* When the suspension liquid is water it is more common to refer to the terms *hydrophilic* and *hydrophobic*, respectively. With lyophilic material there is an affinity between the material surface and the liquid (solvent). Thus, the material will disperse (or even dissolve) spontaneously in the liquid. Examples of lyophilic (with water) materials are proteins, polyelectrolytes, and metal hydroxides. With a lyophobic material, there is no affinity between the surface and the solvent. As a consequence, thermodynamic work is required to disperse the material. Examples of lyophobic materials are metals, inorganic crystals, and salts (such as AgI, TiO_2, ZnO, $CaCO_3$), polymer surfaces such as PTFE, waxes, and graphitic carbon. It is important to remember that the terms lyophilic and lyophobic are relative. For example, zinc oxide is lyophilic compared with ethanol, but lyophobic compared with benzene. Similarly, zinc oxide is less lyophilic (or more lyophobic) than titanium dioxide when water is the liquid. It is possible to make a lyophobic surface lyophilic and vice versa. A clean glass surface can be made hydrophobic to water by a coating of paraffin wax; the wax surface can be rendered hydrophilic by adsorbing a protein. Whatever the chemical nature of the surface there are three consecutive processes to be considered in the wetting of a solid (Fig. 9). They are

1. *Adhesion,* which refers to the solid and liquid surfaces being brought into face to face contact.
2. *Spreading,* in which a liquid spreads over the surface as a duplex layer i.e., a film with two independent interfaces (in contrast with a monolayer).
3. *Penetration,* in which the liquid permeates into the bed of particles or into the open pores of a particle.

In reality, most solid surfaces are rough and irregular on a microscope scale. This affects the wetting process, such that all three processes may contribute simultaneously.

D. Adhesion and Spreading

Consider now a drop of liquid sitting on a solid surface, both in equilibrium with the vapor (Fig. 10). The basic equations (known as the Young–Dupré

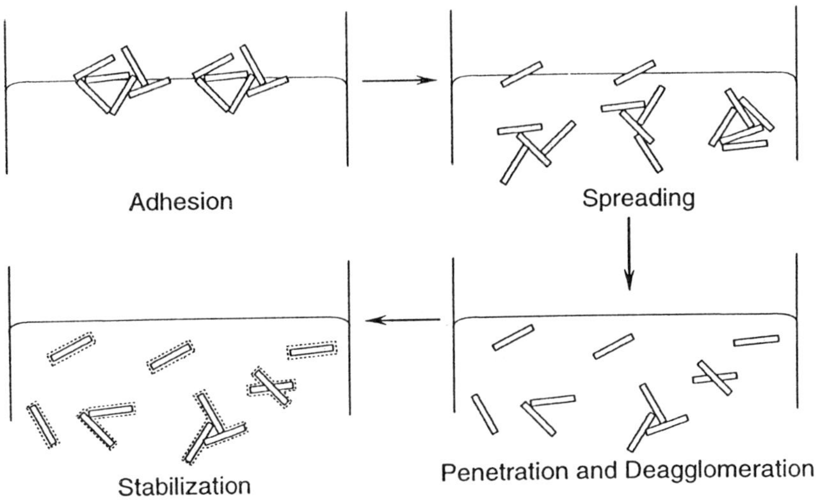

Figure 9 The dispersion process.

equations) governing adhesion and spreading are the following:

$$\gamma_{SV} = \gamma_{SL} + \gamma_{LV} \cos \theta \tag{1}$$

$$\omega_A = \gamma_{SV} + \gamma_{LV} - \gamma_{SL} \tag{2}$$

where
γ_{SV} is the interfacial tension between the solid and the vapor;
γ_{SL} is the interfacial tension between the solid and the liquid;
γ_{LV} is the interfacial tension between the liquid and the vapor;
θ is the contact angle at the liquid–solid interface; and
ω_A is the work of adhesion.

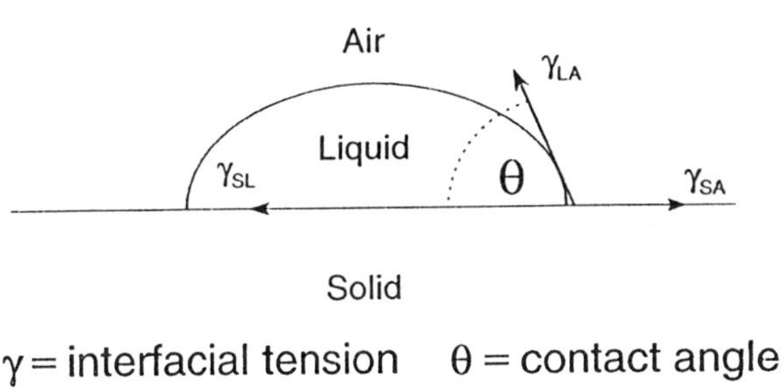

Figure 10 Forces at the solid–liquid–vapor interface.

A surface, therefore, will be fully wetted by a liquid as both γ_{LV} and cos θ tend to zero. To achieve this condition it is often necessary for lyophobic surfaces to add a wetting agent to the liquid. A wetting agent absorbs at the vapor (air)–liquid interface, reducing the interfacial tension γ_{LV} and also the contact angle of the liquid at the particle surface. A wetting agent, therefore, is "surface active." Hence, the term *surface-active agent*, or *surfactant.* Because water has such a large interfacial tension (72 dyne/cm) use of a surfactant is generally unavoidable in the preparation of aqueous suspensions of lyophobic surfaces.

Surfactants

Surface-active materials owe their strong affinity for interfaces to their unique molecular structure. They are amphipathic (i.e., the molecule is comprised of both lyophilic and lyophobic regions; 22, 23). Their properties are determined by the balance of the lyophilic and lyophobic components. This is often referred to, rather confusingly, as the hydrophilic–lypophilic balance (HLB). The lyophobic moiety may be derived from a variety of groups, including linear or branched aliphatic chains, alkyl-substituted phenols, alkyl naphthalenes, alkyl succinates, and fluorinated hydrocarbons.

There are four basic types of lyophilic moiety:

1. *Nonionic*: Here, the moiety is polar, but uncharged. Common examples are polyethylene oxides.
2. *Anionic*: The moiety contains a negatively charged group, such as a sulfonate, sulfate ester, or phosphate.
3. *Cationic*: The charged group is cationic. Examples are the quaternary ammonia compounds and fatty acid amines.
4. *Amphoteric*: The molecule contains groups that can be positive or negative, depending on the solution conditions (particularly pH). An example is alkyl betaine.

It follows, then, that there is a wide choice of individual surfactant, and an even greater number of surfactant combinations. The guidelines for use are, regrettably, diffuse, and many successful formulations have been developed on a largely empirical basis. The following are some points to bear in mind (24–26).

1. The degree of lowering γ_{LV} and contact angle are both a function of the concentration of the surfactant in solution. The maximum lowering of γ_{LV} occurs at the critical micelle concentration (CMC) of the surfactant.
2. Each surfactant will imbue the wet suspension with particular "performance" features. For example, you may find that oleates give the best feel, whereas stearates may give you the best viscosity control. At the same time, perhaps, laurates are more economical (i.e., they work at lower concentrations).

3. All nonionic surfactants have a cloud point. This is the temperature at which a surfactant solution separates into two phases: one that is rich in the surfactant, and one that is almost pure liquid (solvent). It is distinguished by the sudden onset of turbidity.
4. All ionic surfactants have a Krafft point. This is the temperature interval above which the solubility of a surfactant sharply increases. At this (Krafft) temperature the solubility of the surfactant is equal to the critical micelle concentration.
5. Cationic surfactants are among the most substantive and efficient adsorbates known; depletion from bulk solution is very rapid (27). It is this property that distinguishes cationics from anionics and nonionics.

E. Penetration

The rate of penetration of liquid into the pores and capillaries between particles is an important factor. It is desirable that this be as large as possible. An indication of the major variables is given by the Washburn equation:

$$l = \left[\frac{rt\ \gamma_{LV}\ \cos\theta}{2\eta}\right]^{1/2} \tag{3}$$

Where l is the depth of penetration into a pore of radius r in time t and η is the viscosity of the liquid. Thus, the maximum penetration is obtained for the largest interfacial tension γ_{LV}. However, we have seen in the foregoing that addition of a surfactant to a liquid reduces γ_{LV}.

In general, the overall wetting process (adhesion, spreading, penetration) is more spontaneous the lower the contact angle θ and the higher the liquid interfacial tension γ_{LV}. These two factors tend to operate in opposite senses and, because any surfactant affects both, the best agent for promoting such opposing effects may be difficult to find and will vary from one system to another.

In any event, it is necessary to use the *absolute minimum* of wetting agent. Many industrial formulations suffer from a severe case of overdosing when using surfactants. Another downside to the use of too much surfactant is to increase the potential for foaming (as a consequence of lowering γ_{LV}) and this may cause separation of the finer particles. Should foaming occur, and it is not feasible either to reduce the concentration of the surfactant or to replace the surfactant, then an antifoaming agent will be needed, which, itself, may have the potential for causing problems in subsequent stages of the formulation.

A simple test for wetting efficiency is to make up suspensions using the same concentration of powder, but different agents, and allow the suspension to settle out. Slow settling, a visible interface between clear upper liquid and turbid lower layer, and a small depth of sediment indicates the best agent. It is important that no vibration be imparted to the test containers during settlement, which may take several hours.

In preparing a suspension it is essential to avoid the formation of partially wetted agglomerates. The initial mix should be kept as stiff as possible by limiting the volume of liquid phase. The consequence of not achieving complete wetting is to profoundly affect the suspension properties. The region of any surface that is not wetted will

1. Not be deaerated, and so colloidal-sized particles will be subjected to flotation separation.
2. Not contribute to the particle surface charge and so the particles will be less stable and prone to reaggregation.
3. Provide completely different adsorption characteristics.

It is important to realize that if a solid wets on its own (i.e., it is lyophilic), then a wetting agent should not be used at all. In this situation a non–surface-active compound should be used as the deagglomerating agent (''dispersant'').

F. Deagglomeration

Deagglomeration is the process of separation of particles from each other once they are wetted. A deagglomerating agent adsorbs only at the solid–liquid interface. Such materials are not ''surface active.'' They do not significantly affect the interfacial tension γ_{LV}. Their role is to chemically aid separation of agglomerated particles by increasing the electrostatic forces of repulsion between the particles. This allows further liquid penetration into the interparticle spaces which, in turn, enhances the separation process.

For easily wetted material, this enhanced penetration of liquid into the voids between particles may provide sufficient force alone to bring about complete dispersion. More often than not, however, some mechanical energy is required. The choice of the correct agent requires consideration of

1. The fundamental chemical nature and charge on the particle surface
2. The solution conditions with which the particle will be challenged such as pH and electrolyte concentration.

The Solid–Liquid Interface

A brief, simplified description of the solid–liquid interface follows. More detailed information is available (19–21). All solids acquire a surface charge when placed in polar liquids. This is the rule, not the exception. The charge can arise in many ways. Among the more common are

Ionization of Surface Groups, Such as Carboxyl and Amino Groups The ionization of these groups, and so the net molecular charge, strongly depends

Table 2 Isoelectric Points of Some Ionic Solids

Ionic solid	IEP (pI units)
CaF_2	pCa3
$BaSo_4$	pBa7
AgCl	pCl8
Ag_2S	pAg10

on the pH of the solution. The pH at which the net charge is zero is called the isoelectric point (IEP).

Unequal Dissolution of the Ions That Make Up the Surface Molecules of Ionic Solid Materials, Such as Metal Halides, Sulfates, Carbonates, and Oxides Here, the surface charge can be changed by the addition of ions that make up the molecular crystal into solution. For example, the addition of chloride ions (Cl^-) to a dispersion of silver chloride (AgCl) will render the surface negative. The addition of as little as 10^{-3} *M* calcium ions (Ca^{2+}) to a dispersion of fluorite (CaF_2) can effectively reduce the charge to zero (Table 2).

That such small concentrations of ions can fundamentally alter the surface charge has implications in terms of the purity and hardness of water used in formulations.

Solution pH also plays an important role, especially for oxides. Because most oxide crystal surfaces are anisotropic, the surface charge can be either positive or negative. Hence, these materials will also have an isoelectric pH (Table 3). Oxide surfaces with an IEP below pH 7, such as silica (SiO_2), are acidic whereas those above pH 7, such as magnesia (MgO) are basic (Fig. 11).

Clays and other aluminosilicates exhibit even more complex surface chemical behavior in that crystal edges can be positively charged and, at the same time,

Table 3 Isoelectric Points of Some Oxides

Oxide	IEP (pH Units)
SiO_2	2
TiO_2 (rutile)	6
ZnO	10
MgO	12

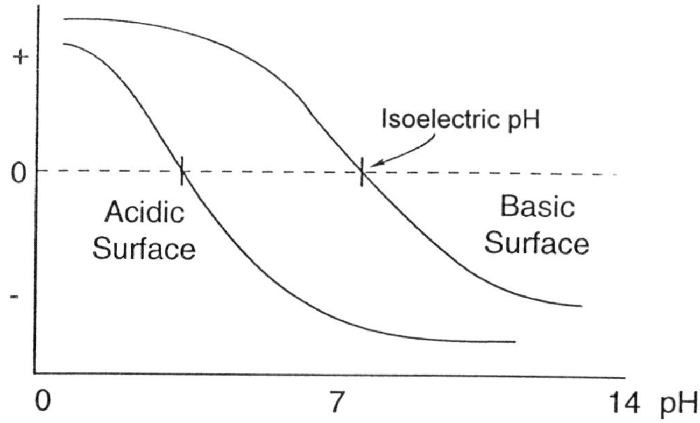

Figure 11 Effect of pH on particle charge.

crystal faces can be negative. Face–face, edge–edge, and face–edge particle associations result in many different structures being possible in solution—from extended ribbons to voluminous gels.

Although the extent of such associations can be a consideration, the difficulty in working with clay-like materials is that these structures and, hence, their viscoelastic solution properties, are very sensitive to changes in solution pH and electrolyte concentration. The IEP pH of these materials varies considerably with the ratio of oxide species at the solid–liquid interface. Typically, the net charge will be negative in neutral solutions and in the absence of any other ions (see below).

The values of IEP in Table 3 are meant as guides only and should be used with reservation. The actual value for a particular material can be influenced by its source or preparation method, pretreatment, and presence of trace impurities (4,28) and should be determined batch-to-batch and lot-to-lot for each material. For example, rutile is often surface-modified with silica and alumina. The IEP varies dramatically with different amounts of silica–alumina. However, the bulk percentages of each determined from elemental analysis are not a reliable indicator of how the surface will behave in solution (Table 4).

Adsorption from Solution of Simple Electrolytes, Multivalent Ions, Polyelectrolytes, and Ionic Surfactant Molecules Here, the surface charge is dependent on a variety of factors. The process is referred to as "specific adsorption" and the mechanisms involved can be quite complex (29).

The concentration of even simple electrolytes is an important factor; both coions (ions of the same sign of charge as the solid surface) and counterions (ions of opposite charge) are involved (Fig. 12). Starting with a negative surface

Table 4 Effect of Surface Modification on the IEP of TiO_2

Bulk %		IEP
SiO_2	Al_2O_3	(pH units)
—	—	6.8
—	4.5	8.4
6.5	3.5	5.8
8	8	4.6

in water at low electrolyte concentration, specific adsorption may initially make the surface more negative. Above some critical concentration, the charge will become less negative, to the point of even reducing the charge to near zero, thus promoting aggregation. The effect is magnified as the valence of the co-ion is increased. The important point to remember is that dry powders can already contain sufficient electrolytic debris, as impurity from the manufacturing process, to produce an unstable dispersion. This must either be removed, or its effect negated by use of additional surface charge modifiers or stabilizing agents (see following section).

A negative surface can be made positive by specific adsorption of a high-valence counterion such as Al^{3+} (see Fig. 12). The concentration of Al^{3+} ions

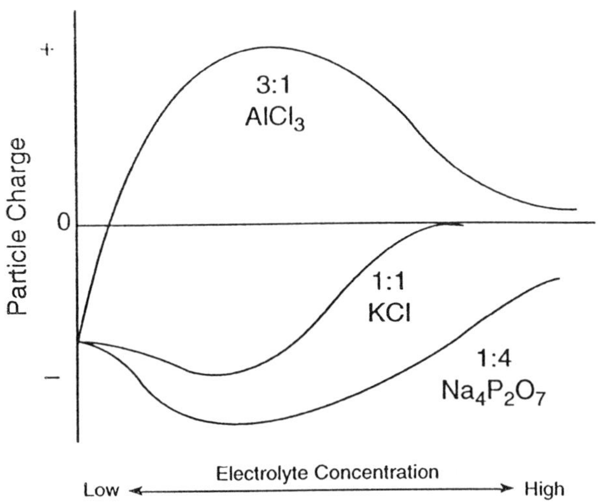

Figure 12 Effect of electrolyte concentration on particle charge.

is critical because there will be some lower concentration of Al^{3+} ions that is sufficient to bring the surface to its IEP. The particle will be unstable and will agglomerate. This low concentration is called the critical coagulation concentration (CCC). Indeed, alum (aluminum sulfate) is one of the most widely used, inexpensive coagulating agents available. Just sufficient Al^{3+} ions must be added to produce a restabilized suspension. Overdosing, however, results in destabilization. Calcium (Ca^{2+}) and magnesium (Mg^{2+}) ions present in "hard" water can also act as very effective coagulating agents.

The problem with using polyelectrolytes and ionic surfactants is that adsorption can also take place by lyophobic interactions as well as hydrogen bonding and chemisorption and is compounded by the heterogeneous nature of the solid surface, which is always crystalline. So each particle will present different faces, with differing surface energies. However, in aqueous systems, specific adsorption usually is nonspecific although the tenacity of the adsorbed species may not be strong (i.e., it can be readily desorbed through changes in temperature or by replacement with another molecule).

G. Dispersing Agents

As with wetting agents, it follows that there is a wide choice of dispersing agents. Dispersing agents can be broadly classified into

1. Inorganic compounds, such as (poly)silicates and polyphosphates
2. Polymeric materials derived from natural sources, such as starch, tannin, and lignin sulfonates
3. Synthetic polymeric materials, such as polyelectrolytes, phosphate esters, polyglycol ethers, and block copolymers having acid–base characteristics

Particulate oxides, silicates, carbonates, sulfates, and the like, can, as the preceding has shown, be readily dispersed in water in the absence of any additives by adjustment of the solution pH to some value above or below the IEP. Dispersion fluidity (and subsequent stability) can be enhanced by use of polyelectrolytes. Low concentrations of polyphosphates are particularly effective in that they can irreversibly chemisorb to surfaces. They are, however, susceptible to hydrolysis, with the result that the particulate suspension may not be stable over long periods. Organic polyelectrolytes are more hydrolytically stable.

For nonaqueous systems, phosphate esters are particularly effective dispersants for highly polar materials, such as pigments and fillers. Alkyd resins that contain acid phthalate ester terminal groups have been used as pigment dispersants for solvent-based paints. Ideally, the organic ester part of the molecule should be highly compatible with the nonaqueous medium.

Block copolymers having acid–base characteristics can also be effective, provided that one component of the copolymer is relatively insoluble in the solvent medium or has specific affinity for the surface. Block copolymers can also impart a high degree of stabilization to the resulting suspension.

Attempts have been made to classify pigments using an acid–base concept to provide a guide to formulators (30). Perversely, oxide surfaces that are acidic in water tend to be basic in nonpolar media because of the difference in electron acceptor or donor properties of the surface relative to the medium (31).

It is possible to make an acidic surface basic and vice versa. For example, many carbon blacks are acidic and are, thereto, often quite polar. It is difficult to disperse such blacks in nonpolar media. Preadsorption of a phthalocyanine with an amino substituent group provides a strongly basic surface that can react with an acidic dispersant such as those based on hydroxystearic acid. Finally, it is worthy of reemphasis that all dispersing agents can, and will, function as depressants (i.e., promote coagulation or flocculation) depending on, for example, their concentration in solution and pH regulation. Water-soluble polymerics have one other variable—molecular weight and molecular structure—which distinguishes their performance as dispersant or depressant. Typically, the former are of low molecular weight (in the range 10^4) and may be of branched structure, whereas the latter are mainly linear and are characterized by high molecular weight (about or above 10^6).

H. Stabilizing

The stability of a wetted, deagglomerated system requires the presence of a stabilizing agent to keep particles from reaggregating. Unfortunately, the term *stability* is an arbitrary concept because it depends on the time scale of observations. In simple terms, stability implies no change in the number of particles with time. The formulator must know the answer to the question, "how long?" and then choose the stabilizing agent(s) accordingly. To do this requires some knowledge of the stabilization process.

All dispersed systems, including emulsions, are thermodynamically unstable and will, over time, reaggregate because of random motion of the particles. On a macroscopic basis, it is predicted, from the second law of thermodynamics, that small particles will coalesce into big ones to reduce the surface area. The frequency of collision depends on particle size, particle concentration, liquid viscosity, and temperature. Whether two approaching particles will combine or not depends on the potential energy barrier between them. This potential energy barrier arises as a consequence of the difference in magnitude of repulsive forces in comparison with attractive forces.

The process of settling is not necessarily an indication of instability. Particulates, especially those such as TiO_2 and ZnO, that have a large density will simply settle over time under gravity. Obviously, if the particles are not well stabilized, they will then eventually aggregate. However, a well-stabilized, but settled, system will redisperse quite readily with agitation.

Forces Between Particles

There are three types of forces that act on a particle (19,20).

London–van der Waals Dispersion Forces These are forces of attraction. On a microscopic basis, attractive forces are always present, pulling the particles together. The attractive forces are present even for completely nonpolar particles. Such forces have their origin in the momentary, induced dipole effects that arise when, owing to diffusion, nonpolar particles approach each other at random. These forces, variously called London, van der Waals, or dispersion forces, are also responsible for the liquefaction of the noble gases, such as helium, neon, and argon.

The dispersion forces are often just as large as any permanent dipole forces of attraction. This may seem unusual until you look at some typical cases, as found in most physical chemistry textbooks. Another surprise is the long range over which these forces are important. For simple molecules, the dispersion forces (and, indeed, all the attractive forces), vary as r^{-7}, where r is the intermolecular distance. The attractive component of the total potential energy varies as r^{-6}; but for fine colloidal particles, this simple description is insufficient. For such particles, the attractive force is the sum of the individual molecular forces, summed over all pairs. The final result depends on particle shape, particle size dimensions, and distance of approach and is given by:

$$V_A = \frac{Aa}{12H^n} \tag{4}$$

where

V_a is the force of attraction.
A is called the Hamaker constant.
a is the particle size.
H is the shortest distance between particles.

The factor n varies between 1 and 6. Thus, fine particles attract each other over much longer ranges than individual molecules, and this will lead to aggregation, unless a repulsive force is present.

To maintain the highest degree of dispersion stability, the forces of attraction should be kept as low as possible. When a medium is present between like particles the effective Hamaker constant is given by

$$A_{eff} = [A_p^{1/2} - A_m^{1/2}]^2 \tag{5}$$

where

A_p is the Hamaker constant for the particle.
A_m is the Hamaker constant for the medium.

For quartz in water A_{eff} is approximately 3×10^{-21} J, whereas for quartz in a hydrocarbon, such as hexane, A_{eff} is about 15×10^{-21} J. The more similar the Hamaker constants are for the particle and the liquid medium, the smaller will be the energy of attraction. This is a very complex matter and beyond the scope of this chapter. Suffice it to say that materials such as oxides and metals have very unfavorable Hamaker constants. Surfactants, polyelectrolytes, polymers, and proteins all have Hamaker constants closer to that of a liquid, such as water. Thus, an adsorption layer can "mask" the Hamaker constant of the material surface. The forces of attraction increases as the polarity of the medium decreases. This is very important in nonaqueous systems in which the presence of small amounts of less polar impurities in a solvent will promote aggregation.

Coulombic Forces Associated with Electrostatically Charged Particles Coulombic forces are those of repulsion. For suspensions in water and other highly polar liquids electrostatic repulsion is common. The name "electrostatic" is misleading in the sense that the particles, solvent molecules, and ions are always in motion. It is used, however, because much of the theory of electrostatic stabilization is based on relatively simple electrostatic equations and because it is assumed that the time-averaged results can be calculated from an electrostatic theory. The force of electrostatic repulsion can be represented by Eq. (6).

$$V_R = Da\psi_0^2 \text{ [geometric term]} \qquad (6)$$

where

D is the (dimensionless) dielectric constant of the liquid medium.
a is the particle size.
ψ_0 is the surface potential, which depends directly on the surface charge.

The geometric team is complex, but is an exponential term containing the parameters of particle size, ionic strength (proportional to electrolyte concentration), and the distance of separation of the particles. For maximum stability the electrostatic force of repulsion should be as large as possible. From Eq. (6) it can be seen that

1. For a fixed medium, particle and surface potential (charge), the repulsive forces become smaller as the ionic strength increases. Thus, the lowest electrolyte concentration possible should be used.
2. For a fixed medium, surface potential (charge), and ionic strength, the repulsive force decreases as the particle size becomes smaller. Thus, the particle size should be as large as possible. Because there is always a particle size distribution, the PSD should be of lowest polydispersity and, certainly, without any tail of fines.

3. For a fixed medium, particle size, and ionic strength, the repulsive force increases with surface potential charge.

Earlier we dealt in detail with ways in which particles can acquire a charge. Any or all of the ionic ''dispersing'' agents will provide a layer of ionized surface groups and, hence, a degree of electrostatic repulsion. To make use of this mechanism necessitates some prior knowledge of the surface chemistry of the material to be dispersed. A very useful parameter is the IEP and this can be readily determined using what are called zeta potential measurements (29). A plot of zeta potential versus pH is exceptionally useful in that it shows

1. The solution pH at which the surface develops its maximum natural charge
2. The rate of change of potential around the IEP. This indicates how precisely the pH must be controlled to produce repeatable formulations.

Adsorption of polyphosphates can provide an increase in charge groups. Linear polyphosphates form anions of the general formula $[P_nO_{3n+1}]$ $(n + 2)^-$. They adsorb either through hydrogen bonding (e.g., kaolinite, rutile, and alumina) or by chemical bonding (e.g., on calcium carbonate and barium sulfate). The valence of the polyphosphate anion depends on the solution pH, as does its ability to hydrogen bond to oxide surfaces. The zeta potential measurements are, again, an ideal way to determine the optimum conditions.

Steric Forces Arising from Absorbed Layers and Solvation Steric forces, similar to coulombic forces, are forces of repulsion. Here particles are stabilized in either aqueous or nonaqueous media by solvated polymeric or oligomeric moieties adsorbed on the particle surface. The adsorbed layers form a structure–mechanical barrier (called a steric barrier) around interacting particles preventing them from approaching one another to such a distance that there is a significant attractive force. In the simplest terms, the force of repulsion is given by Eq. (7).

$$V_S = BC^2[\text{geometric term}] \tag{7}$$

Where B is a virial coefficient that can be positive or negative and C is the concentration of macromolecule in the adsorbed phase. The geometric term is again complex, but is linear containing the parameters of adsorbed layer thickness and distance of separation of the particles.

The use of macromolecules to provide stabilization is almost as much an art as a science and various conditions must be met if stable dispersions are to be prepared.

1. The medium must be a good solvent for the absorbed macromolecule. This means that the macromolecular (polymers) chains should extend as far as possible into solution.

2. The surface coverage must be complete and the adsorbed moiety must be firmly attached to the surface, but not so strongly that it will be collapsed.
3. The depth of the steric barrier must be large.

Figure 13 illustrates schematically just a few difficulties in achieving these three conditions simultaneously. The characteristic feature of a macromolecule in solution is its high degree of conformational freedom. The shape and dimensions of a macromolecule in solution are critically affected by the macromolecule concentration, the nature of the solvent, and the temperature. This, in turn, will affect how the molecule absorbs onto a surface.

In Figure 13a, the conditions are such that the macromolecule is well solvated and adopts a configuration of an extended chain. For complete surface coverage the molecule must absorb flat-on. Although there may be firm attachment, it produces only a thin layer and virtually no steric barrier.

In Figure 13b, the conditions are such that the macromolecule is a tightly bunched random coil. Although adsorption onto the surface produces a thick layer, there are only a few point contacts. Displacement can readily take place, particularly under shear conditions. Surface coverage is not optimum.

Figure 13c is the ideal situation. Good steric stabilization is achieved when the conditions for absorption are such that the macromolecule absorbs onto the surface by many contacts. The open structure of the "loop" segments provides

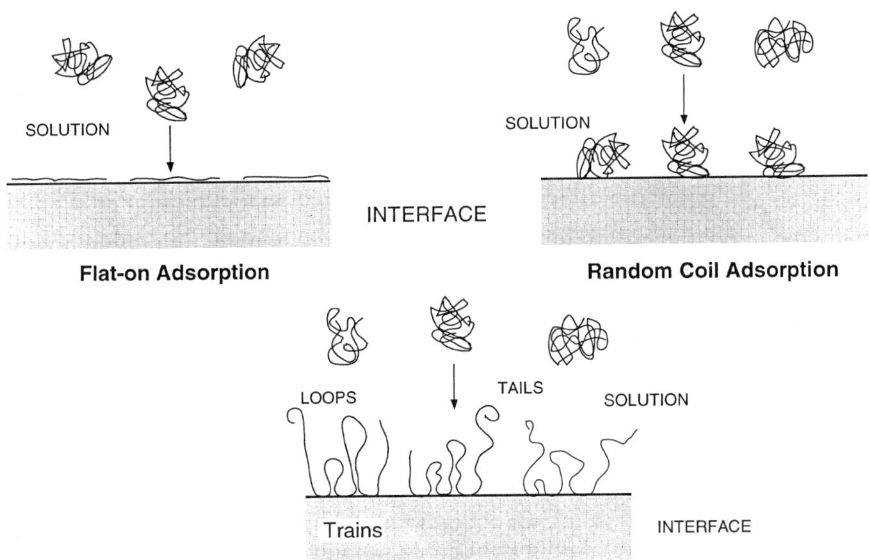

Figure 13 Conformations of a macromolecule absorbed at an interface.

an additional solvation layer and the "tails" provide additional depth to the barrier. An excellent textbook dealing with the whole subject is now available (32).

In practice, the most useful types of macromolecules for steric stabilization are block and graft copolymers. These have been synthesized for industrial use, chiefly in nonaqueous media. In aqueous media, block or graft copolymers are less easy to use except in their low molecular weight form as nonionic surfactants. Steric stabilizers can take form other than synthetic polymers. Naturally occurring polymers, such as gelatin, casein, gum arabic, and vegetable oils, have long been used in paints and inks. In modern solvent-based coatings alkyd and acrylic polymers are used. The advantages of steric stabilization over simple charge stabilization are as follows:

1. It provides stability in nonaqueous systems where coulombic repulsion is weak.
2. It provides tolerance against high electrolyte levels in aqueous media.
3. It allows much greater particle concentrations to be tolerated (i.e., high solids loading).
4. It provides shear stability over a wide range of particle sizes.

The disadvantage is that a suspension will be destabilized (flocculation) if

1. There is incomplete coverage of the particle surface.
2. There is excess polymer in solution.

A whole textbook has been devoted to this subject! (33).

Finally, solvation itself provides an additional repulsive force. Because, by definition, the absorbed "layer" is composed of segments that are lyophilic, these will be heavily solvated. There is a resistance to squeezing out the intervening layer because it entails removal (or displacement) of the solvent (34).

Theories of Stability

It was mentioned earlier that particles can be prevented from aggregating if a potential energy barrier exists between them.

The DLVO theory (35) deals with the interplay of dispersion and coulombic forces (i.e., it is concerned only with electrostatic stability). Named after its proposers—Derjaguin and Landau (Russia), and Verwey and Overbeek (The Netherlands), its success over the last 50 years is testimony to the reasonable character of the assumptions underlying the theory. Although few practical applications have been fully interpreted with DLVO theory, it is a good foundation for beginning to understand stability, dispersion, and aggregation processes in systems where simple chemical dispersants have been used to produce strong electrostatic repulsion between interacting particles. Essentially, the DLVO theory deals with the pairwise interaction of particles by the summation of the potential energy of electrostatic repulsion (V_R) and London–van der Waals attraction (V_A):

$$V_T = V_R + V_A \tag{8}$$

This is schematically shown in Figure 14. Because attraction decays as an inverse (hyperbolic) power of distance [see Eq. (5)] and repulsion decays as an exponential of distance [see Eq. (6)], then at both short and long distances of separation between particles, attraction outweighs repulsion. When particles aggregate into the secondary minimum this process is termed *flocculation*. It is reversible by application of mechanical or thermal energy, because the depth of this minimum is not great. The minimum is situated at some distance from a particle surface, so essentially two particles do not touch, but have a layer of solvent between them. The height of the potential energy barrier V_{max} is important. If V_{max} has an energy of more than 15 k_BT where k_B is Boltzmann's constant and T the absolute temperature, the system is stable. If V_{max} is reduced to zero, then particles will aggregate into the primary minimum, and this process is termed *coagulation*: It is irreversible.

The main weakness of the DLVO approach for practical formulations is that it assumes perikinetic conditions (i.e., those in which the liquid is stationary and

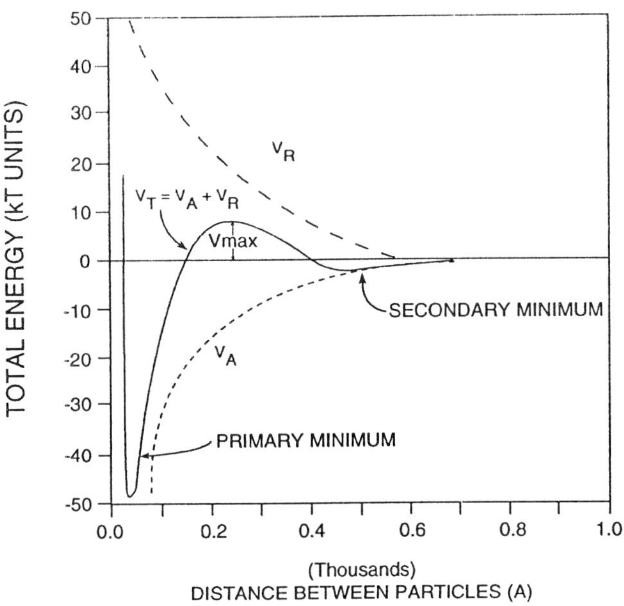

Figure 14 DLVO theory: electrostatic stabilization. The primary minimum yields permanent coagulation; the secondary minimum yields a reversible flocculation.

collisions occur from Brownian motion; and are thermal in origin and follow random statistics). In other words, it is adequate for studies of shelf storage stability, but will not predict stability under orthokinetic or shear conditions that are encountered in the preparation of industrial dispersions.

The theoretical aspects of the mechanism of steric stabilization are not well understood. They generally fall into two main groups (36):

1. Entropic repulsion, in which the layers of absorbed polymers do not penetrate when the particles collide, but the compressional effects cause a decrease in entropy. This results in an increase in (Gibbs) free energy.
2. Osmotic repulsion, in which the polymer chains can penetrate on collision causing an increase in (Gibbs) free energy because of the reduction of polymer chains in contact with the dispersion medium, which results in an increase in entropy. In addition, the configurational entropy of the absorbed polymer is reduced. In either event, the total interaction of Eq. (8) must be modified to include an additional repulsive term V_S to account for this repulsive component caused by overlapping absorbed polymer segments.

$$V_T = V_R + V_A + V_S \tag{9}$$

Schematically this is shown in Figure 15. The requirement for stability is that a particle must have some minimum film thickness of absorbed layer of $H_{min}/2$ and that $V_{min} > k_B T$.

The development of a theory for structural forces caused by hydration in aqueous dispersions is still in its infancy (37). Generally, hydration forces produce either repulsion in liquids that are able to wet a surface and attraction when the solid is poorly wetted.

I. A Few Final Points To Remember

1. Stabilization is the last step in the dispersion process. If done too early, one just stabilizes agglomerates. Hence, the order of addition of ingredients is always of concern. The three processes—wetting, deagglomeration, and stabilizing are fundamentally different and need to be considered separately. From what has been written, it is important to recognize that a single molecular species cannot satisfactorily perform all three tasks—at best two out of the three.
2. So-called dispersion aids are often ambiguous in their primary function and, as a consequence, are used interchangeably, often with deleterious results.
3. Adding agents together as a mixture creates a competitive absorption situation—a proverbial rat race, because all molecular species will, and do, adsorb at surfaces.

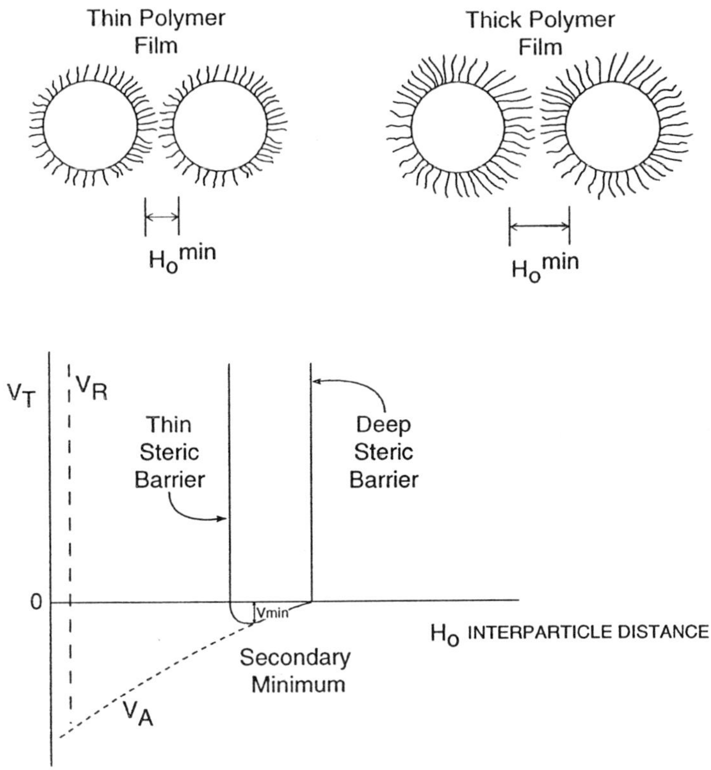

Figure 15 Steric stabilization.

4. "Thickeners" are often added to increase the fluid viscosity. Although useful, this should not be viewed as a substitute for true stabilization. Increasing the viscosity simply reduces the rate of settling of the larger, denser particles. In any event thickeners should be added after the stabilization step.
5. In setting out to prepare a suspension, it is essential to consider the mode of stabilization in relation to the end use. Sometimes more than one dispersing agent or stabilizing agent may be desirable.

J. Surface Coatings

We have seen in the foregoing how the surface of a particle dictates its behavior. For decades, if not centuries, it has been common practice to surface modify pigments, usually to alter their chemical reactivity while maintaining various physical characteristics. This is traditionally done by the pigment manufacturer before sale. A very common example is the surface modification of colored pigments to allow formulation in different applications without losing tint or hue. Surface treatments can be inorganic, such as oxides and silicones, or organic, such as fatty acids and organosilanes. Most microfine titanias and some microfine zinc oxides are surface treated. Surface treatments can affect all aspects of dispersion: wetting, deagglomerating, stabilization. Surface treatments are, however, not always beneficial in a given system. One must be aware of the surface treatments chemistry in the same way one needs to pay attention to an untreated particle's surface. No one surface treatment will be compatible with all mediums. A silicone-coated particle will behave like a silicone droplet. Naturally, it will be hydrophobic, but it will not disperse well (obtain optimal PSD) in a medium in which silicone is not miscible. In such a case, one must remember the basic three steps in dispersing as outlined previously. There are three general approaches to surface treatments. They are as follows.

Physically Adsorbed

A physical adsorbed coating is one that sits on the surface of the particle. The type and strength of the bond holding the adsorbate molecule on the surface can be electrostatic (charge), lyophobic interaction, or hydrogen bonding, depending on the surface chemistry of the material to be coated and the chemical structure of the adsorbate. The adsorbate stays there because there is a particular active site that makes adsorption of the molecule energetically favored over remaining in solution. The common dispersants are of this type. This type of surface modification will, and has, sufficed for many applications. The shortcoming of this sort of coating is that the coating can be displaced by a different molecule with an even stronger affinity for the particle surface. Also the coating can "choose" to leave if a more desirable surface presents itself. In addition, any physically adsorbed coating can, and often will, be displaced during formulation processes because of the severe thermal and mechanical agitation employed. This is the fundamental problem with predispersed systems. The particle can be left naked and, in that state, is very likely to agglomerate. Despite this, however, physical adsorption has been the foundation of particle dispersion for years and has been quite successful.

Chemically Bound or Grafted

Chemically bound coatings are much more durable. Inorganic oxide coatings on pigments are, for all intents, part of the surface and can be removed only by extreme chemical means. Organic coatings that are grafted are very strongly anchored. However, it is possible to rupture segments of the molecule by excessive mechanical agitation, especially for high molecular weight grafts. Chemical grafting is widely used in the preparation of polymer lattices to generate very effective steric stabilizers "in situ." Because grafting requires specific radical sites, the total number of grafts is limited. This means that the underlying surface can be accessible to other molecules during formulation of a sunscreen.

Encapsulated

An encapsulated system consists of a particle totally surrounded by a matrix of some sort. In this case the particle is theoretically totally isolated from its surroundings. Typical types of encapsulation matrices are waxes and synthetic and natural macromolecules. Some of the major reasons for encapsulation are

1. To protect the underlying particle from the surrounding environment or vice versa (where the core material is hazardous or toxic)
2. To separate components, allowing control of the incompatibility of components
3. Masking undesired properties of the active component (e.g., odor or chemical activity)

Although grafting and encapsulation offer many advantages, the coating will provide completely different performance characteristics and, hence, must be chosen with care.

K. Mechanical Dispersing Treatment

Following initial wetting of a powder (and sometimes deaeration)—often called the premix stage—comes penetration and deagglomeration. More often than not this necessitates mechanical agitation. Comminution of the agglomerates may be brought about by four basic processes (Table 5): impact; shear; extension; cavitation.

The degree of turbulence and shear vary considerably with the type of operation and design of equipment. If, however, the various steps outlined in the foregoing are followed then the minimum of mechanical agitation will suffice.

Table 5 Typical Milling Processes

Impact	Kinetic–dispersion mill	Ball mill Sand mill	
Shear	Roller mills	High-speed disk impeller	2-blade pug mill
Extension			Extruder
Cavitation	Ultrasonic dispersion		

Excess mechanical agitation results in (1) changes in particle size distribution, (2) changes in the total surface area, and (3) changes in the surface chemistry.

All of the foregoing impinge on the specific properties of the suspension. In addition, even with a high energy barrier (see Fig. 14), agglomeration can recur. An increase in mechanical or thermal energy increases the possibility of the collision of particles. Now the maximum collision energy attained using even an ultrasonic bath of low power (50 W) is well in excess of a typical V_{max} of 10–15 k_BT. High-power sonicators and homogenizers often generate 400 W of power, and they also generate considerable heat—enough to boil water. All electrostatically stabilized systems have a critical coagulation temperature, and all sterically stabilized systems have an upper critical flocculation temperature. Thus, the degree of useful mechanical agitation is limited by the energy maximum V_{max} and the prevailing thermal conditions.

VII. SUN PROTECTION FACTOR TESTING

The fondest hope of the sunscreen industry is that a meaningful in vitro test that is reliably predictive of SPF will be established. This would cut down on the need for human experimentation and should greatly decrease the cost of development. Several systems have been proposed and are used by different companies. In our experience, as well as some others (38), the SPF of particulate sunscreens has been notoriously difficult to predict using these methods. In over 50 in vitro or in vivo comparisons, we were unable to draw reliable correlation for particulate-based systems, especially when zinc oxide was used. Consequently, we strongly recommend that in vivo testing be used for the majority of evaluations and that in vitro systems be reserved for comparing, in house, known formulations with minor changes. Several groups have apparently had more success with in vitro methods than we have, and the foregoing is strictly our opinion based on our own results.

VIII. FUTURE DEVELOPMENTS

Undoubtedly, the major advancements in metal oxides will come in the form of surface treatments. The metal oxides are inherently efficient UVR attenuators and are limited only by their youth. Surface treatments will likely solve any formulation difficulties still present. Already on the market are surface-treated zinc oxides, titanium dioxides, and iron oxides. Through increased dispersion, the surface treatments provide increased aesthetic and SPF performance.

In addition, other oxides will be developed into sunscreens. These will, however, have to undergo regulatory review, which will delay, or perhaps even prevent, their widespread use.

Very recently, polymeric particulates, designed to increase the optical path length, have been introduced. This is a well-recognized tactic. Silica, for example, has been used in the past for this purpose. The usefulness of these new, specifically designed agents has yet to be established, but they certainly warrant evaluation.

Another development is the encapsulated organic sunscreens. Although still in development, they typically consist of 200- to 300-nm particles in an aqueous suspension. The particles are composed of a matrix and the sunscreen. Matrices include, among other things, waxes, polymers, lipids, proteins, and glass. Because they are particulate in nature, they can, if properly designed, offer the advantages of all particulates—isolation of the sunscreen from the skin and minimal ingredient interaction. They also allow the use of the oil-soluble sunscreens in oil-free systems. In addition, when properly produced, they provide increased performance relative to the unencapsulated screen. For instance a 5% octyl methoxycinnamate formulation using organoparticulates has an in vivo SPF of 17, compared with a control of SPF 8, using the same formulation that incorporated neat octyl methoxycinnamate. The encapsulated organic sunscreens may provide a solution to many of the problems inherent in using the soluble sunscreens. They could potentially decrease bioavailability, improve efficacy, and increase ease of formulation.

IX. CONCLUSION

The cost of creating new chemicals and of meeting regulatory requirements significantly limits new ingredient creation in an area such as sunscreens, for which the profit potential is minor in comparison with other drugs. It seems likely that most innovation will come from improvements in existing technology, rather than from totally new chemicals. The particulates are perfect for such exploitation. Their benefits are clear: lack of toxicity, broad-spectrum coverage, and stability, to name a few. They have already found a considerable audience and it appears inevitable that their use will continue to increase.

REFERENCES

1. Grady, L. D. Zinc oxide in face powders. *J. Soc. Cosmet. Chem. 1*(1):(1974).
2. Stutz, G. *Ind. Eng. Chem., 19*:52 (1927).
3. Szczurko, C., Dompmarttin, A., Michel, M., Moreau, A., and Leroy, D. Photocontact allergy to oxybenzone: Ten years experience. *Photodermatol. Photoimmunol. Photomed. 10*:144–147 (1994).
4. Solomon, D. H. and Hawthorne, D. G. *Chemistry of Pigments and Fillers.* John Wiley & Sons, New York, 1983.
5. Prasad, A. S. *Biochemistry of Zinc.* Plenum, New York, 1994.
6. Hunter, R. S. *The Measurement of Appearance.* Academic Press, New York, 1975.

7. Kerker, M. *The Scattering of Light and Other Electromagnetic Radiation.* Academic Press, New York, 1969.
8. Brockes, A. *Optiks; 21*:550 (1964).
9. Orchard, S. R. *Oil Color Chem. Assoc., 51*:44 (1968).
10. Kubelka, P. and Mouk, F. *Z. Tech. Phys., 12*:593 (1931).
11. Webber, T. G., ed. *Coloring of Plastics.* John Wiley & Sons, New York, 1979.
12. Parfitt, G. D., ed. *Dispersion of Powders in Liquids.* Applied Science Publishers, New York, 1981.
13. Tadros, T. F. *Solid–Liquid Dispersions.* Academic Press, New York, 1987.
14. Laskowski, J. S. and Ralston, J. D., eds. *Colloid Chemistry in Mineral Processing.* Elsevier, New York, 1992.
15. Allen, T. *Particle Size Measurement.* 4th ed. Chapman & Hall, New York, 1993.
16. Bernhardt, C. *Adv. Colloid Interfac. Sci., 29*:79 (1988).
17. Higuchi, H. and Misra, A. *J. Pharm. Sci., 51*:459 (1962).
18. Nancollas, G. H. *Adv. Colloid Interfac. Sci., 10*:215 (1979).
19. Everett, D. H. *Basic Principles of Colloid Science.* Royal Society of Chemistry Publications, 1988.
20. Hiemenz, P. C. *Principles of Colloid and Surface Chemistry.* Marcel Dekker, New York, 1986.
21. Kitahara, A. and Watanabe, A. *Electrical Phenomena at Interfaces.* Marcel Dekker, New York, 1984.
22. Shimoda, K. *Colloidal Surfactants.* Academic Press, New York, 1969.
23. Tadros, T. F., ed. *Surfactants.* Academic Press, New York, 1984.
24. Corkhill, J. M. and Goodman, J. F. *Adv. Colloid Interfac. Sci., 2*:297 (1969).
25. Rosen, M. F. *Surfactants and Interfacial Phenomena.* John Wiley & Sons, New York, 1978.
26. Lindeman, B. and Wennerstron, H. *Micelles—Amphiphile Aggregation in Aqueous Solution.* Springer-Verlag, Berlin, 1980.
27. Jungerman, E., ed. *Cationic Surfactants.* Marcel Dekker, New York, 1970.
28. Parfitt, G. D. and Sing, K. S. W., eds. *Characterisation of Powder Surfaces.* Academic Press, New York, 1976.
29. Hunter, R. J. *Zeta Potential in Colloid Science.* Academic Press, New York, 1981.
30. Sorenson, P. *J. Paint Technol., 47*:31 (1975).
31. Fowkes, F. M. and Mostafa, M. A. *Ind. Eng. Chem. Prod. Res. Dev., 64*:726 (1978).
32. Fleer, G. H. et al. *Polymers at Interfaces.* Chapman & Hall, New York, 1993.
33. Napper, D. H. *Polymeric Stabilization of Colloidal Dispersions.* Academic Press, New York, 1983.
34. Silberberg, A. *Prog. Colloid Poly. Sci., 59*:33 (1976).
35. Kruyt, H. R., ed. *Colloid Science.* Vol. 1. Elsevier, New York, 1960.
36. Sato, R. and Ruch, R. *Stabilization of Colloidal Dispersions by Polymer Adsorption.* Marcel Dekker, New York, 1980.
37. Derjaguin, B. V. *Theory of Stability of Colloids and Films.* Consultants Bureau, New York, 1989.
38. Kelly, K. A., Ewing, G. D., Dromgoole, S. H., Lichtin, J. L., and Sakr, A. A. In vitro sun protection factor evaluation of sunscreen products. *J. Soc. Cosmet. Chem., 44*:139–151 (May/June 1993).

18

Broad-Spectrum Physical Sunscreens: Titanium Dioxide and Zinc Oxide

Michael W. Anderson, Julian P. Hewitt, and Stephen R. Spruce
Tioxide Specialties, Ltd., Durham, England

I. INTRODUCTION

A. General Considerations

Titanium is the ninth most abundant element (1,2) and makes up 0.6% of the earth's crust. It exists in vast deposits of high-purity ores such as mineral rutile (93–97% titanium dioxide), or in conjunction with iron oxides in ores such as ilmenite (45–75% titanium dioxide). Titanium dioxide, TiO_2, is generally considered as being almost chemically inert, being insoluble in all but the strongest mineral acids. This inertness is clearly demonstrated by the existence in nature, over millennia, of the ore mineral, rutile.

A global industry, manufacturing nearly 3 million metric tons of TiO_2 per year, has developed with the aim of further processing the ores to provide a form of TiO_2 sufficiently pure and inert to be suitable for both pigmentary use and for cosmetic, food, and topical applications.

This industry has now developed grades of TiO_2 that are highly effective in attenuating the ultraviolet (UV)B portion of the electromagnetic spectrum and are thus sold for use in topical applications where sun protection is required.

Zinc is another widely dispersed element, although nowhere near as abundant as titanium. Its principal ores are blende (sulfide), smithsonite (carbonate), calamine (silicate) and franklinite (zinc, manganese, and iron oxide). Zinc oxide

(ZnO) is extensively used in industrial products, such as paints, plastics, and rubbers, and also in the pharmaceutical and cosmetics industries, where its use as a topical antiseptic is widely appreciated. Increasing awareness of the dangers of exposure to UVA damage has spurred the industry into producing a grade of zinc oxide that is designed to attenuate the UVA portion of the spectrum. This tends to be used to supplement existing UVB attenuators and also in products providing all-day broad-spectrum protection.

B. Pigmentary Materials: Background

Titanium dioxide, as a pigmentary product, has a long, well-documented history spanning over 70 years (3). Two main processes are used to make this material, based on either gas-phase, high-temperature oxidation (the ''chloride'' process) or liquid-phase precipitation–crystallization (the ''sulfate'' process) (4,5). Both these processes are designed to effect the purification of the naturally occurring mineral forms of TiO_2 to give a high purity end product. Once the pure material is obtained, then both processes are developed to achieve a product that has the correct physical properties for its end use as a pigment, specifically a closely defined mean particle size, with a defined particle size distribution. This is accomplished by the use of various techniques, such as crystal size control at precipitation, calcination (high-temperature heat treatment), or micronization (fluid energy milling). [The optimum particle size is calculated using the refractive index of TiO_2 and the wavelength of visible light (500–800 nm). Once the correct particle size is achieved, a coating is added to the TiO_2 conferring the property of ease of dispersability while further enhancing chemical stability.]

Zinc oxide is more limited in its applications as a pigmentary material, as it has a lower opacity as a result of its lower refractive index (ZnO is 2.01–2.03 whereas TiO_2 is 2.4–2.7). Pigmentary ZnO material is sometimes known as ''Chinese white'' and is produced by two main methods (6):

1. The oxidation of zinc vapor (the American process, in which ZnO residues are reduced to zinc vapor then oxidized, or by the French process based on the oxidation of pure, vaporized zinc metal)
2. A wet chemical route in which zinc salts (carbonates or hydrosulfites [from bleaching products]) are precipitated then calcined to the oxide.

Again particle size control is paramount to the efficacy of the final product, and this is achieved by control of the process parameters during manufacture (time, temperature, concentration, and so forth).

The prime purpose of both of these pigmentary materials is to give a high degree of whiteness (i.e., act as a colorant) and to give good opacity (i.e., act as a barrier to visible light) while remaining inert to the medium in which they are used. These materials have found not only widespread industrial applications in paints, paper, and plastics, but also in food, pharmaceutical, and cosmetic applications (7).

C. Transparent Fine Particle Materials: Background

The consumer has a need for a range of UV attenuators that provide protection from UV radiation, while in themselves being inert and safe to use. To fulfill these needs, the pigment industry has developed "transparent" grades of both TiO_2 and ZnO, manufactured using modifications to the standard processes currently used. These materials are specifically designed to attenuate the UVB and UVA portions of the electromagnetic spectrum (290–320 and 320–400 nm) while minimizing interactions with the visible portion (i.e., remaining transparent on the skin). This change in the optical properties of these materials is achieved purely by modification of their mean particle size, the chemical structure of the base materials remaining unchanged. The materials are marketed as fine-particle, subpigmentary, ultrafine, micronized, and other descriptions.

[The term *micronized* when applied to fine-particle materials is not a generic description for all such material. Micronization is a process that involves attrition milling with high gas velocities and generally follows a calcination step, which tends to give hard in-process agglomerates, requiring intensive milling. Some materials are produced without this milling step, using thermodynamic control to obtain the correct crystal size.]

The major requirements to ensure safe and effective UV-attenuating products using these inorganic materials are their purity and the control of their particle size.

Purity

This must be controlled to enable the products to meet the specifications of the relevant pharmacopeial bodies. This is achieved by use of quality-assured raw materials, the operation of an appropriate manufacturing route designed to minimize inclusion of impurities, and by the application of accurate and sensitive monitoring tests to assess purity and consistency of production.

Particle Size

Particle size is controlled both in the plant, by efficient process control of the thermodynamics of crystal growth, and in the product, by maintaining a stable effective dispersion of the particles within the formulation and, ultimately, when applied to the skin of the consumer.

II. RELATION BETWEEN PARTICLE SIZE AND ULTRAVIOLET PROTECTION

Pigmentary TiO_2 has a mean particle size of between 150 and 300 nm (8,9), with a geometric standard deviation (GSD) around this mean, giving sizes down to 50 nm (GSD 1.35 gives ~5% < 100 nm). Pigmentary TiO_2 interacts with (scatters) visible light; hence, appearing white and opaque (Fig. 1). Particle size is critical in determining its efficacy spectrum, and theoretical studies (10) carried out developed the graph shown in Figure 2, showing the relation between particle size and the degree of scattering. The data have been corroborated by other researchers (11).

Titanium dioxide specifically designed to optimally attenuate UVB has a mean crystal size that ranges between 20 and 150 nm (11,12). The efficacy of attenuation can be judged by reference to the UV–visible spectra of such material (Fig. 3). Comparison of pigmentary and UV-attenuating grades of TiO_2 (Fig. 4) illustrates this size difference. Maintenance of this particle size is critical to the efficacy of the products. Should agglomeration occur (Fig. 5), the material reverts to behaving as a pigmentary material, thereby losing efficacy in the UV and increasing interaction in the visible, hence appearing white. Fine-particle materials have a tendency to form agglomerates and the manufacturers have developed several technologies to minimize this effect: coatings, dispersion techniques, and formulation methods.

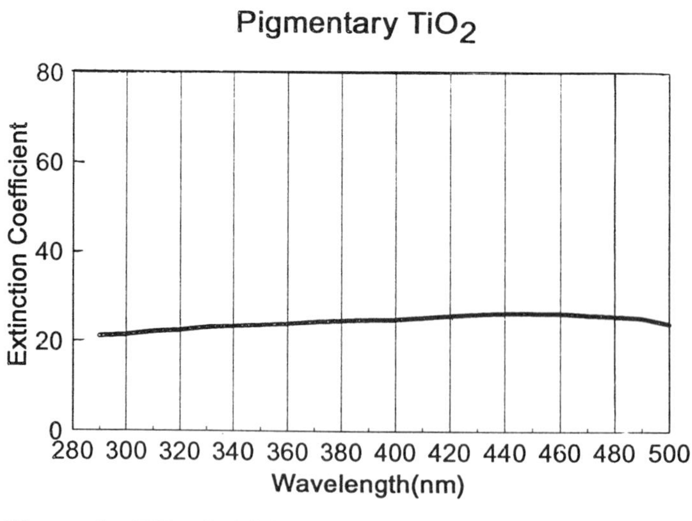

Figure 1 UV and visible spectrum of pigmentary TiO_2.

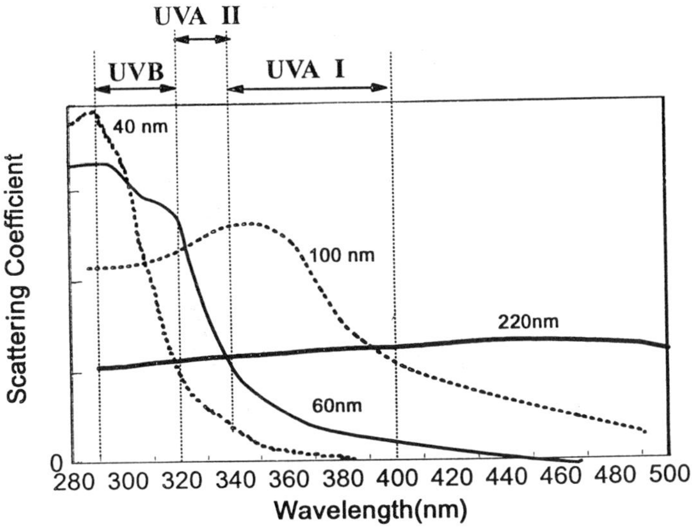

Figure 2 Scattering versus particle size for TiO_2.

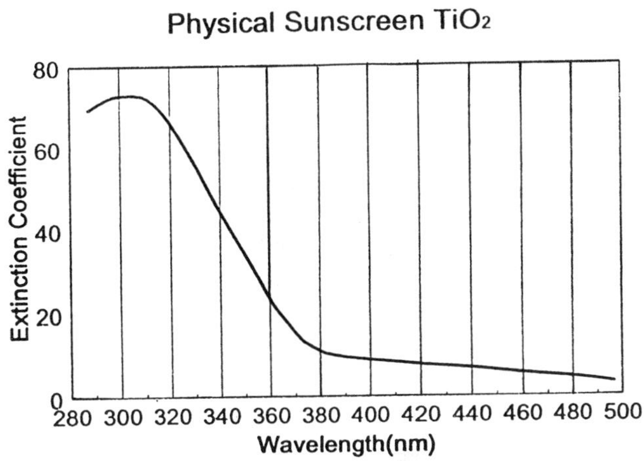

Figure 3 UV and visible spectrum of UV-attenuating TiO_2.

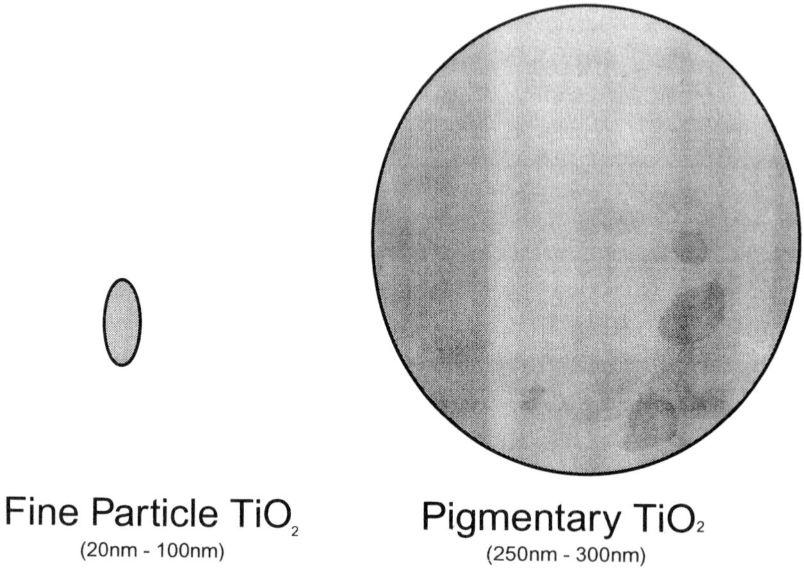

Figure 4 Relative sizes of UV-attenuating and pigmentary grades of TiO_2.

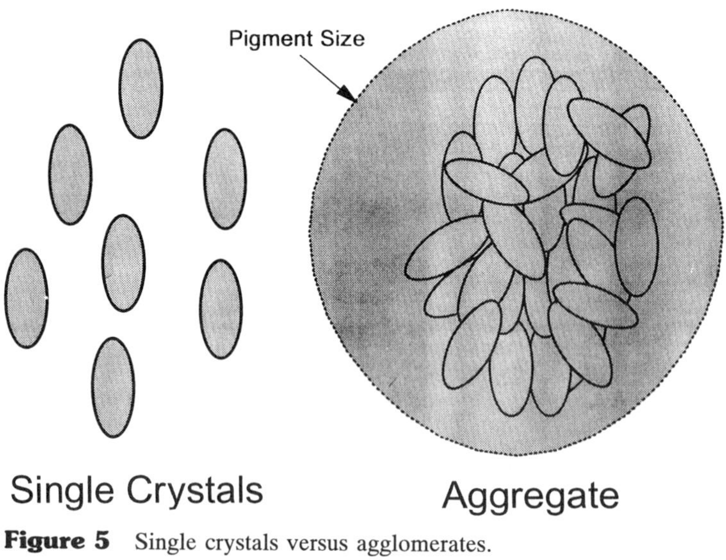

Figure 5 Single crystals versus agglomerates.

Along a similar vein, a fine-particle ZnO industry is developing using the same technology of particle size control. Figure 6 illustrates the effect of particle size on the degree of attenuation of UVA, based on a wavelength of 360 nm. Thus, as with TiO_2, maintenance of particle size and minimization of agglomeration is essential. Pigmentary ZnO has a particle size in the region 200–400 nm (6), whereas the UV-attenuating grades typically have particle sizes between 40 and 100 nm (13,14). Comparison of pigmentary and UV-attenuating grades of ZnO shows the significant effect of particle size control on the attenuation of UV and visible light (Figs. 7 and 8). The aim of the fine-particle ZnO is to provide a degree of attenuation in the UVA I and II parts of the electromagnetic spectrum to either supplement existing UVB attenuators or to provide a broad-spectrum protection product for everyday use (e.g., in cosmetics and decorative products).

III. FLEXIBILITY OF ULTRAVIOLET PROTECTION AFFORDED BY PHYSICAL SUNSCREEN AGENTS

A. Ultraviolet A and Ultraviolet B Radiation

Ultraviolet B Radiation

The initial raison d'etre of the sunscreen industry was to protect consumers from the harmful acute effects of UV radiation that resulted in painful dermal erythema and could ultimately lead to the formation of skin cancers. Sunscreen products were developed that filtered out a large proportion of the most erythmogenic UV rays and, thus, had the effect of reducing the most obvious physiological effect of skin damage (i.e., sunburn). This process influenced the artificial separation of the UV spectrum into UVB and UVA, when most people began to associate UVB with erythema and UVA with aging. UVB is obviously of high energy, it causes significant damage to the skin, and is a biological hazard. However, as pointed out by Diffey (15) in a recent article on broad-spectrum sunscreens, "... when we are outdoors we are not exposed to either UVB or UVA but to solar ultraviolet radiation which always contains both. Also biological effects that we may wish to protect against by using sunscreens are mediated by all wavelengths in the solar UV spectrum."

Ultraviolet A Radiation

It is only relatively recently that the physiological importance of UVA radiation has been widely recognized. One reason for this is that the effects of excess exposure are long-term, and it may be many years before any adverse effects become apparent. Contrast this with the well-known effects of overexposure to

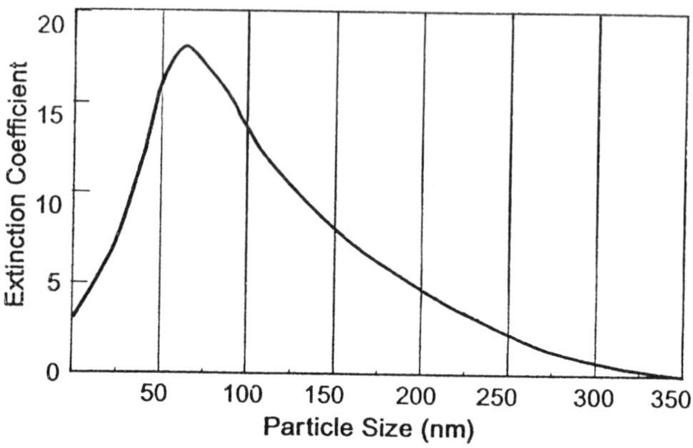

Figure 6 Effect of particle size of ZnO on extinction coefficient of ZnO at 360 nm.

UVB radiation: the painful erythema and peeling of skin will follow within 48 h of exposure.

Ultraviolet A radiation is of lesser energy than UVB radiation, but, at the surface of the earth, it is more intense (Fig. 9). Also, because it is of longer wavelength, it penetrates deeper into the skin than UVB radiation and reaches the dermis. Here, it is believed to interact with collagen fibers, and it is claimed that this is the primary mechanism behind premature aging. Furthermore, UVA radiation is thought to act synergistically with UVB radiation in the formation of skin cancers.

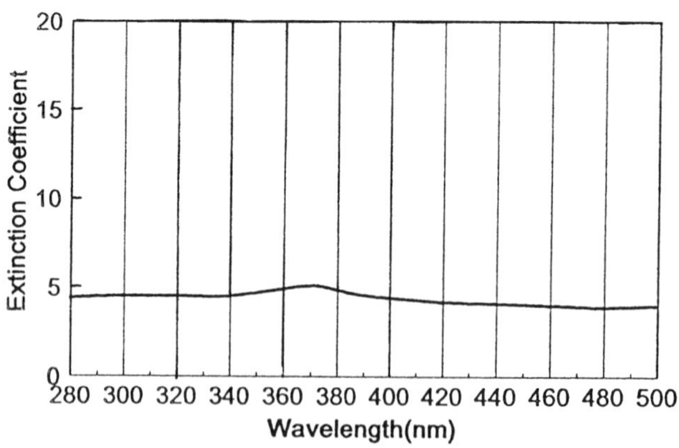

Figure 7 UV and visible spectrum for pigmentary zinc oxide.

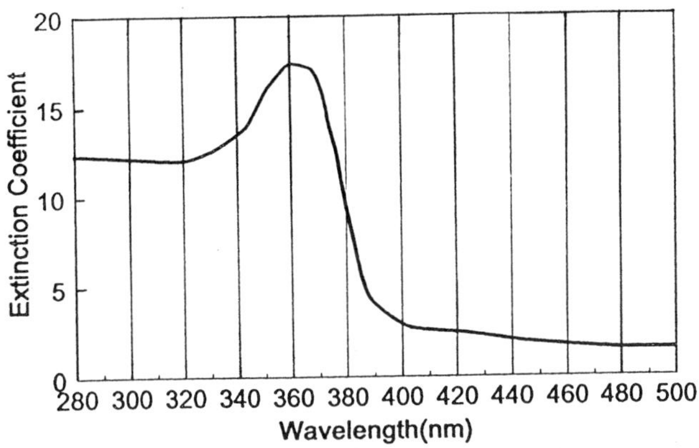

Figure 8 UV and visible spectrum for fine particle zinc oxide.

UVA screens are being incorporated into many daily-wear products, for use throughout the year.

For suncare products there is a danger that, if only UVB protection is provided, the consumer is encouraged to spend longer periods in the sun and receive an inordinate dose of UVA radiation. Manufacturers are now providing products with a balanced UVB–UVA protection to avoid this.

B. Custom Ultraviolet Protection

Suncare products are thus being produced with a variety of UV protection shapes; however, the prime driver for product choice remains the sun protection factor (SPF). The consumer is now faced with an impressive array of products varying in both SPF and UVA–UVB protection.

Sunscreens are not solely employed in suncare products, and there is growing consumer demand for year-round protection in products such as moisturizers, foundations, lipsticks, and eye makeup. It is reasonable to anticipate that the prime choice criteria for these products will not be SPF, and that they will be chosen on the basis of UVA–UVB protection profile. As with suncare products we are now beginning to see a widening choice of products that are primarily aimed at broad-spectrum protection and have a secondary benefit of sun protection. Consequently, the formulator requires materials that are flexible enough to give a wide variety of protection profiles and yet do not bring added complications of complex mixtures and increased inventory.

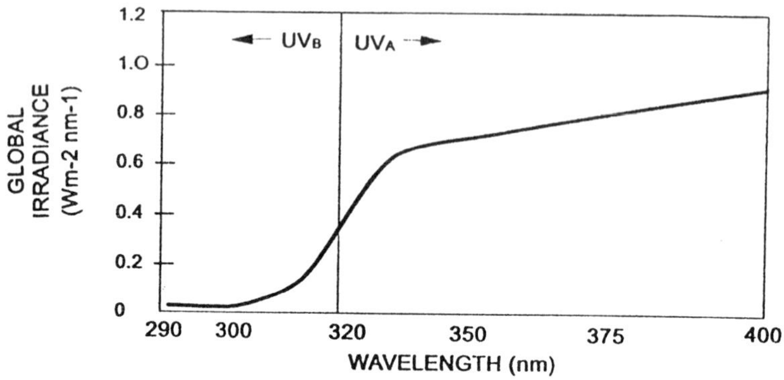

Figure 9 Global irradiance spectrum at midday, midsummer, latitude 40°N.

C. Physical Sunscreens as an Elegant Solution to Flexible Protection

Section II outlined UV protection profiles obtainable from fine-particle TiO_2 and fine-particle ZnO (see Figs. 3 and 8).

In these examples, TiO_2 is an excellent UVB blocker, whereas ZnO is extremely effective at attenuating UVA radiation: TiO_2 is an excellent candidate for a broad-spectrum high SPF product (typically SPF 20 and UVA/UVB ratio of 0.59–0.60 from 7.5% active); ZnO is an excellent choice as a completely balanced UVA–UVB block, with a moderate SPF (typically, SPF 6 and UVA/UVB ratio of 0.91 utilizing 6% ZnO). If, however, the protection profile of neither of these is suitable for the application in mind, then there are the infinite number of possibilities available by blending TiO_2 and ZnO together in a formulation.

Two examples of blends are shown in Figures 10 and 11. The formulation used for Figure 10 contained equal amounts (weight for weight) of TiO_2 and ZnO (2.5% each) and results in an SPF of 9 with a UVA/UVB ratio of 0.68 (i.e., significantly higher than the UVA/UVB ratio for TiO_2 alone; 0.59–0.60).

The formulation used for Figure 11 contained 7.5% ZnO and 2.5% TiO_2 and yielded an SPF of 18.5 and a UVA/UVB ratio of 0.81—perhaps one of the most comprehensive high-SPF broad-spectrum protection curves available.

This approach to blending physical sunscreens gives the formulator a very wide scope to produce a custom-made unique protection product to match the needs of his or her chosen consumer niche.

The SPF values and UVA/UVB ratios of the formulations discussed are tabulated in Table 1.

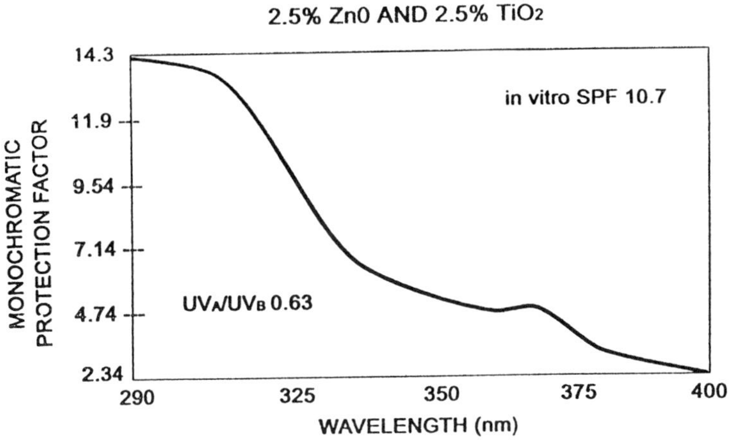

Figure 10 Sunscreen formulation with both ZnO and TiO_2 at 1:1 ratio.

IV. MECHANISM OF ACTION

Within the cosmetics industry, it was originally considered that the fine-particle grades of both TiO_2 and ZnO acted as pure scatterers of UV light, as do the pigmentary versions for visible light. This has now been shown to be a false premise, and the attenuation of UV takes place through a combination of scattering and absorption in both products (16,17). A degree of absorption of UV

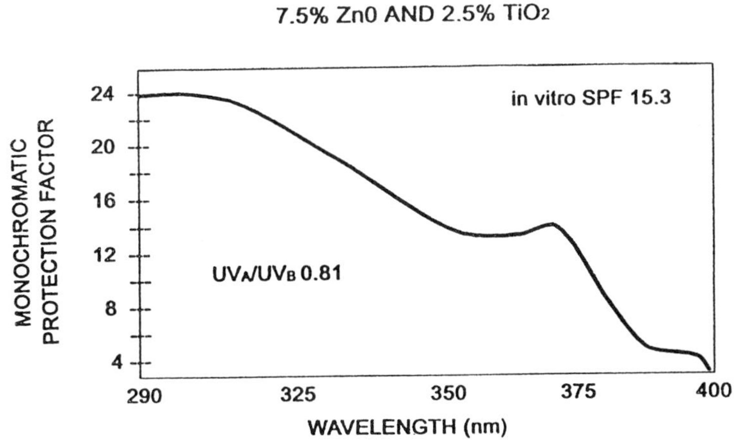

Figure 11 Sunscreen formulation with both ZnO and TiO_2 at 3:1 ratio.

Table 1 Efficacy Results for TiO_2 and ZnO Sunscreens

TiO_2 %	ZnO %	SPF[a]	UVA/UVB ratio[b]
7.5	0	20	0.59–0.60
0	6	6	0.90
2.5	2.5	12	0.63
2.5	7.5	20	0.81

[a]In vivo, FDA protocol.
[b]Ref. 22.

occurs with both pigmentary grades and transparent grades of TiO_2 and ZnO (see Sec. IV.C).

A. Overall View

Reference to Figure 12 shows that in a physical sunscreen system the incident radiation (I_o) undergoes several interactions during its path through the sunscreen to the skin. Initially, approximately 4% is lost by specular reflection (I_r) at the sunscreen surface before the incident ray being refracted into the sunscreen layer. Here, it is attenuated by interaction with the physical sunscreen particles by scattering (I_s) and by absorption (I_a), with the residue being transmitted into the skin layer (I_t). The simple equation (1) describes this effect, in that the residual transmitted ray *must* be equal to the incident ray minus all those processes designed to attenuate the light.

$$I_t = I_o - (I_r + I_s + I_a) \tag{1}$$

[A component of the scattered radiation will be forward-scattered and transmitted to the skin, and I_t is the total of all radiation reaching the skin. I_s in the equation relates to that component attenuated, thus not reaching the skin.]

The aim of manufacturing the physical sunscreen agents to a specific particle size and size distribution is to optimize the amount of scattering and absorption in the UV, thereby maximizing the total amount of attenuation and minimizing transmission to the skin.

B. Scattering

Several theories have been developed to describe the action of light scattering with small particles. The most widely quoted are the Mie (18) and the Rayleigh (19) theories. These theories predict that the scattering of a given wavelength of light can be optimized by using an appropriate particle size material (Rayleigh

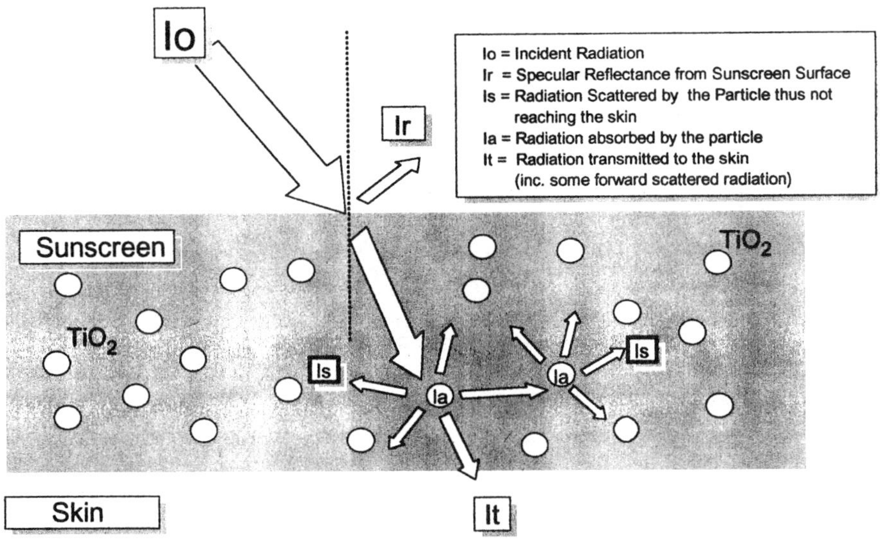

Figure 12 Schematic of attenuation processes.

states scattering α $1/\lambda^4$). The optimal particle size is also dependent on the refractive index of the material, which itself varies with wavelength. This latter fact explains why the optimum particle size for a given wavelength is different for both zinc oxide and titanium dioxide. [Both these theories are rather limited in interpretation, as they deal with isolated particles as the scattering model. Attempts have been made to increase the sophistication of these theories but, as yet, none exactly define the practical situation, which is highly complex.]

C. Absorption

Both ZnO and TiO_2 are semiconductors (i.e., under certain excitation conditions, they can mobilize electrons within their atomic structure). In both cases, the electrons are excited from the valence band up to the conduction band, thus absorbing a portion of the excitation energy, which for these materials is in the form of UV light.

The energy gap (i.e., the amount of energy required to promote the electron) is 3.06 eV for rutile TiO_2 and 3.23 eV for ZnO. By using quantum theory [Eq. (2)], these gaps are calculated to correspond to UV wavelengths of 405 and 385 nm, respectively.

$$\lambda = \frac{hC}{E} \tag{2}$$

where

λ = wavelength
h = Planck's constant
C = speed of light
E = energy gap

This means that all wavelengths of UV light shorter than 405 and 385 nm for TiO_2 and ZnO, respectively (higher energy, where energy is inversely proportional to wavelength), will be capable of being absorbed. Radiation at wavelengths longer than these figures will only be attenuated by scattering. Wavelengths at or near these values are said to be at the "absorption edge."

The energy absorbed is then released as longer-wavelength radiation as the electron relaxes back to a lower energy state (Fig. 13). This whole process is extremely rapid, being completed in less than 10^{-6} s.

Photoactivity Minimization

Pigmentary or physical sunscreen grades of TiO_2 are, as stated earlier, required to be photostable. Over the many years of production of pigmentary grades the industry has developed various technologies to increase this stability. This includes the use of the rutile rather than the anatase crystal form, and the design of coating systems specifically to minimize photoactivity. These technologies have been so successful that modern paint systems, which contain up to 50% by weight TiO_2 intimately dispersed in an organic resin, now have lifetimes measured in several years even when exposed to the most rigorous of environ-

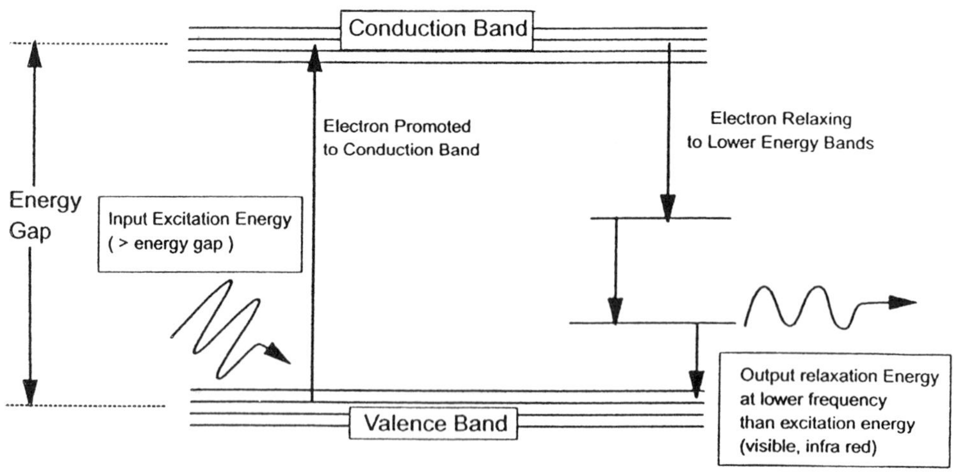

Figure 13 Absorption of UV radiation by electronic transition.

ments. The coatings used are generally based on inorganic oxides (e.g., SiO_2 or Al_2O_3) and serve a further purpose aiding the dispersion of TiO_2 in both organic and aqueous systems.

Physical sunscreens have now been developed using these technologies, such that they have a photostability comparable with or better than that of existing grades of TiO_2, used for many years in both food and cosmetic applications (20).

V. FORMULATING WITH PHYSICAL SUNSCREEN AGENTS

A. Factors Influencing Efficacy in Formulations

Formula Types

In principle, physical sunscreens can be incorporated into any of the vehicles commonly used for organic sunscreens; for example, emulsions, oils, gels, and sticks. As with the more traditional sunscreens, the choice of vehicle is influenced by various factors, both technical and commercial. However, with the current trend toward ever-increasing SPF values, often one of the most important considerations is how to obtain the maximum possible efficacy from the sunscreen active ingredient, to give a high SPF while maintaining a safe and aesthetically elegant product.

Emulsions As with the organic sunscreens, emulsions are by far the most popular vehicle for the inorganic materials. Emulsions offer great flexibility to the formulator in terms of product viscosity (lotion or cream), SPF, placement of active agents, aesthetics, and incorporation of other auxiliaries, such as emollients, humectants, and moisturizers. With titanium dioxide as the sole sunscreen active agent, SPF values ranging from 2 to more than 30 can be achieved with an emulsion system. Because TiO_2 is available in both hydrophilic and lipophilic forms, the active agent can be placed in either phase of an emulsion, or even in both phases. Either oil-in-water or water-in-oil systems can be used, thereby allowing any type of skin feel to be achieved. Any other desired ingredients, including organic sunscreens, can be readily incorporated. Also, because water usually constitutes at least 50% of the composition of an emulsion, this vehicle is often the most cost-effective way of achieving the desired results. The major disadvantage of emulsion systems is the difficulty of stabilizing them, and the introduction of significant quantities of particulate material can often make this more complex. Stabilization of the emulsion, while maintaining the efficacy of the inorganic sunscreen, is the major challenge facing the formulator.

Oils Oils have traditionally been the easiest sunscreen vehicle to formulate, because there is only one phase, and this is also true with physical sunscreens.

If the physical sunscreen, usually titanium dioxide, is adequately dispersed in the oil, product stability is relatively easy to achieve. Adequate dispersion usually requires either a lipophilic coating on the TiO_2, or use of a suitable dispersing agent to ensure that the TiO_2 is wetted out by the oil and the particles are prevented from flocculating.

The principal disadvantage of oil systems, with organic or physical sunscreens, is low efficacy. The rheological properties of the oils mean that, although it is possible to obtain excellent spreading properties on the skin, the oil tends to ''pool'' in the wrinkles of the skin, thereby giving uneven coverage and a lower SPF. This situation can be remedied, to some extent, by the introduction of rheological additives (e.g., waxes) to add structure to the system, thus creating an ointment-type product. Such a product, however, tends to have poor aesthetic properties, being rather oily and greasy. Another potential barrier to consumer acceptance of this type of product with a physical sunscreen as the active agent is that the products are opaque or, at best, translucent in the bottle, whereas consumers expect sunscreen oils to be transparent.

Gels The major appeal of gel-type sunscreen products is their crystal clarity; unfortunately, this is extremely difficult to achieve if any significant amount of titanium dioxide or zinc oxide is incorporated into such a product. The optical path length means that such a product will always be translucent or opaque in the bottle, although it may well be transparent in a thin film on the skin. It may be possible to formulate a transparent product by using TiO_2 with an extremely small particle size (< 15 nm), but such a small particle size results in decreased attenuation in the UVB region of the spectrum and virtually no attenuation in the UVA (17); hence, the degree of protection is dramatically reduced. Nevertheless, there is considerable interest in gel products containing the physical sunscreens, so they are worthy of some further discussion.

The formulating difficulties inherent in the various types of gel systems are well known (21), and the introduction of physical sunscreens adds further complications. The most popular means of creating an aqueous gel is by use of carbomers. However, at the use levels required to give a rigid gel structure (> 0.5%), the carbomers are often incompatible with physical sunscreens. Even if this can be overcome, SPF values tend to be low, because a carbomer gel does not provide a very good structure for support of the physical sunscreen after spreading on the skin; particles tend to aggregate, resulting in lower effectiveness and poor aesthetics (whitening on the skin).

In hydroalcoholic systems, there is the additional problem that hydrophilic grades of TiO_2 often aggregate in the presence of ethanol. Incorporation of an appropriate thickener can overcome this by suspending the TiO_2 in the gel structure before introduction of the ethanol. As with the aqueous gels, however, efficacy is often low.

Little work has been carried out on the use of physical sunscreens in microemulsion gels. Again, it is unlikely that clear products could be achieved, but it is possible that this type of system could give elegant and highly effective products. The major disadvantages are related to the high levels of emulsifiers required, leading to problems of cost, safety, and water-resistancy.

Sticks Physical sunscreens are readily incorporated into stick formulations, often giving effective and water-resistant products. Unfortunately, fine-particle materials can inhibit the recrystallization of some (but certainly not all) waxes, and many waxes vary considerably in their properties according to their source, so careful selection of waxes is required to obtain a suitable solid product.

As will be apparent from the foregoing, emulsions are often the vehicle of choice for physical sunscreens, and it is for this reason that the bulk of this section is devoted to the formulation of emulsion systems containing fine-particle titanium dioxide or zinc oxide, or both.

Rheology

The effectiveness of any sunscreen product is critically dependent on the formation of an even, continuous product film on the skin, and a homogeneous distribution of physical sunscreen within that film. Therefore, the rheological behavior of the formulation has a large effect on the SPF.

To fully understand the factors influencing efficacy of TiO_2 (or ZnO) as a sunscreen, it is necessary to consider what actually happens when the emulsion is applied to the skin. For a water-in-oil (w/o) emulsion, we can discuss this in terms of the properties of the emulsion as such, because the evaporation of water here is relatively slow. However, with an oil-in-water (o/w) system, much of the external water phase evaporates quite rapidly, and we must then consider the properties of whatever is left—a mixture of oils and waxes, emulsifiers, humectants, and active ingredients.

The process of rubbing a cosmetic cream or lotion onto the skin involves high shear forces (shear rates of the order of $10^3-10^4\ s^{-1}$), and it is the behavior of the product both during and after this high shear process that defines its film-forming properties. There are three principal parameters that affect the efficacy of a product based on a physical sunscreen.

- High shear viscosity
- Recovery time (i.e., the time taken for a product to recover its structure after this has been broken down under high shear)
- Thixotropy

To spread well over the uneven skin surface, a product should have a low viscosity at high shear. This is reflected in the data shown in Figure 14. This shows in vitro SPF plotted against high-shear viscosity for a series of samples

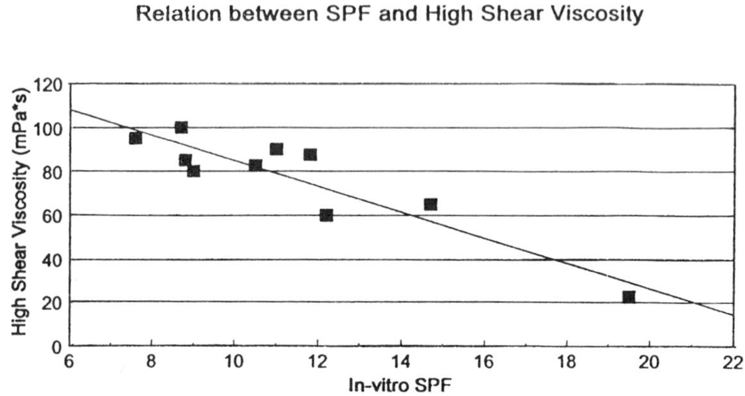

Figure 14 Influence of high shear viscosity on SPF.

based on the same o/w formulation; only the manufacturing procedure was changed, resulting in different product rheologies, as indicated by the viscosity data. The product contained 5% of an aqueous dispersion of TiO_2, and 5% of an oil dispersion; each dispersion contained 40% w/w TiO_2, thus giving a total of 4% TiO_2, in the final formulation. In vitro SPF values were measured by the method of Diffey and Robson (22), using Transpore tape as substrate. It can be clearly seen that a decrease in high-shear viscosity leads to an increase in SPF value, owing to the better spreading properties of the formulation.

However, after the product has been spread over the skin, it must recover its structure quickly to maintain an even film. If the viscosity remains low, the product will flow into the wrinkles of the skin, resulting in uneven coverage. This can be seen from the data in Figure 15, which shows in vitro SPFs for w/o formulations (each containing 4% TiO_2) containing different consistency-enhancing waxes. The SPF data are plotted against "tan δ," where

$$\tan \delta = G''/G' \tag{3}$$

G' is the storage modulus, and indicates the extent of elastic or solid-like behavior in a viscoelastic material. G'' is the loss modulus, and indicates the extent of viscous or liquid-like behavior. In the experiments represented in Fig. 2, the products were subjected to high shear to break them down, and G' and G'' were measured over time. If we look at tan δ at a certain time after the high shear stress has been removed, we can assess the extent to which the product has recovered its structure. The lower tan δ, the more solid-like is the behavior, indicating that the product has recovered to a greater extent. It can be seen from

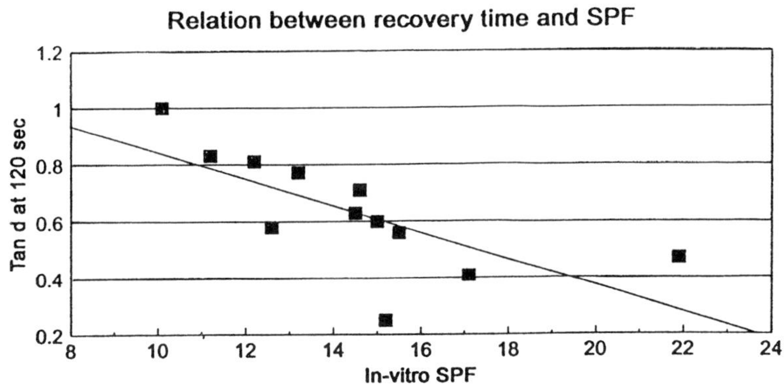

Figure 15 Influence of recovery time on SPF.

the data in Figure 15 that shorter recovery times lead to higher in vitro SPF values.

The relation between SPF and thixotropy is more complex, because thixotropy, in turn, depends on other parameters, including the high-shear viscosity and the recovery time. It must be remembered that many aspects of rheological behavior are interlinked, and that by modifying a formulation to obtain the optimum value of one parameter, we may inadvertently change another, and the expected increase in SPF may not occur.

Distribution of Physical Sunscreen: Phase Considerations

The second major factor influencing the efficacy of a physical sunscreen is the location of the sunscreen particles after application of the product. For optimal efficacy of a physical sunscreen, the maximum amount of active agent must be incorporated into the product film that remains on the skin after dry-down. Figures 16 and 17 show schematically good and poor product distribution on skin. With a good distribution, there is an even oil film, with sunscreen particles evenly distributed within this film. A discontinuous oil film, with some of the sunscreen excluded from it and, therefore, likely to aggregate and be less effective, gives a lower SPF. This can be demonstrated by a simple experiment: an aqueous dispersion of titanium dioxide is applied to Transpore tape and an in vitro SPF test is carried out; the result is little more than 1.0 (i.e., virtually no protection at all). This is because, once the carrier medium (water) has evaporated, the TiO_2 particles simply aggregate in small areas, leaving most of the surface with no coverage. If a similar experiment is carried out with an oil dispersion, a high SPF is recorded, because the carrier does not evaporate, and the particles remain dispersed.

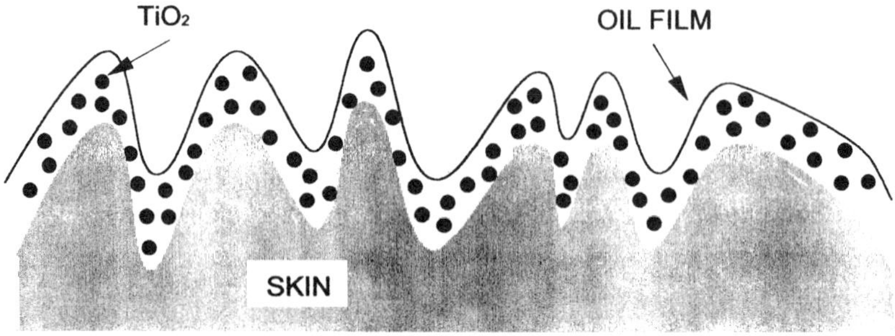

Figure 16 Even (satisfactory) distribution of sunscreen on skin surface.

Therefore, we might conclude that, for optimum efficacy of a physical sunscreen, the active agent should always be placed in the oil phase of the emulsion. However, several factors conspire to make the situation rather more complex.

First, the film that is deposited on the skin does not consist only of oils; also present will be emulsifiers, humectants, and active ingredients. Emulsifiers, in particular, can affect the distribution of physical sunscreen particles in the film by formation of liquid crystalline structures (23).

Second, titanium dioxide and zinc oxide are both inherently hydrophilic; zinc oxide, in fact, shows significant solubility in water at pH values below 6.0. Therefore, when placed in the oil phase of an emulsion, both will tend to migrate from the oil phase to the aqueous phase; therefore, they are likely to aggregate on application of the product. This tendency can be reduced to varying degrees by applying a hydrophobic coating to the particles or by incorporating a suitable dispersant to keep the sunscreen dispersed in the oil phase. Even so, the possibility of migration is a factor the formulator must be aware of to design effective formulations.

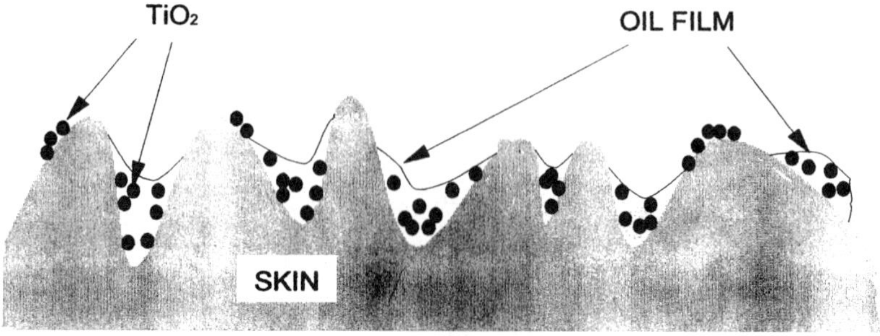

Figure 17 Uneven (poor) distribution of sunscreen on skin surface.

B. Emulsion Systems with a Single Physical Sunscreen

Because titanium dioxide is the most widely used physical sunscreen, this section will initially concentrate on the use of this ingredient. The particular issues that must be considered when formulating with zinc oxide will then be discussed.

Interactions in a "Suspoemulsion" System

The term *suspoemulsion* was originally coined to describe agricultural products that consisted of an emulsion in which were suspended particles of solid pesticide. The term could be equally well applied to a cosmetic emulsion containing a physical sunscreen. The various interactions present in such a system are represented in the following scheme (24):

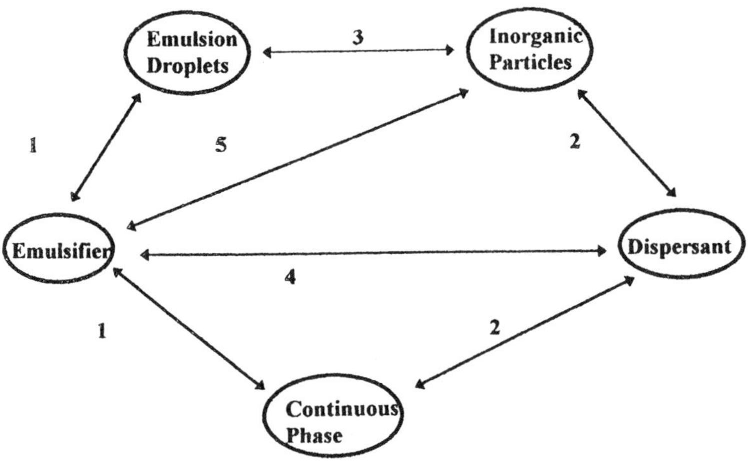

Positive Interactions

1. Emulsification: Action of the emulsifier to stabilize the emulsion droplets within the continuous phase.
2. Dispersion: Action of the dispersing agent to stabilize the inorganic particles within the continuous phase [in this context, *continuous phase* means the phase in which the inorganic particles are dispersed, which could be either the internal or external phase of the emulsion].

Negative Interactions

3. Flocculation: If the particles of physical sunscreen flocculate around the emulsion droplets, this can result in reduced SPF, because the physical sunscreen may aggregate. In some cases, however, this interaction has a

positive effect on emulsion stability, because the solid particles increase the barrier to coalescence of the emulsion droplets.

4. Desorption: Interaction between emulsifier and dispersant can result in either or both being partially removed from the interfaces they are intended to stabilize.

Other

5. Networking: Interaction of the inorganic particles with liquid crystalline structures formed by the emulsifier(s) can be positive or negative, depending on the nature of the structures formed.

Clearly, the aims of the formulator must be to maximize the positive interactions by suitable choice of emulsifier(s) and dispersant, while minimizing the negative interactions.

Form of the Physical Sunscreen: Powder Versus Dispersion

Both titanium dioxide and zinc oxide sunscreens are available as powders or predispersions. Which should the formulator choose? As with most such questions, there are advantages and disadvantages to both types, and the final choice will depend on which factors are the most important in the particular circumstances.

Predispersions have the advantage in terms of handling and ease of use because they are liquid ingredients that can be readily mixed with the appropriate phase of the emulsion. Any aggregates should have already been broken down, so that the dispersion contains active material with the optimum particle size. Powders are prone to aggregation, and high-energy processes are required to break down the aggregates. An optimized dispersion can also help avoid one of the negative interactions mentioned in the foregoing; namely, desorption. Ideally, only polymer-stabilized systems should be used, with the concentration of dispersant optimized such that there is no free surfactant that could affect the emulsifier system (23). Use of a powder usually necessitates addition of a dispersant to the formula; the type and concentration of this dispersant must be carefully selected to maintain efficacy of the physical sunscreen while avoiding negative interactions with the emulsifier(s).

The principal advantage of powders is in terms of formulation flexibility; when a dispersion is used, incorporating a certain level of active agent automatically fixes a minimum amount of the carrier medium (oil or water) that also must be included in the formulation. This is not usually a problem in aqueous dispersions, because most cosmetic emulsions contain at least 50% water anyway, but it can be a problem with oil-based dispersions.

Types of Emulsion Systems

Leaving aside multiple emulsions or multiphase systems, there are two types of emulsion used for personal care products: oil-in-water and water-in-oil. Titanium dioxide sunscreens are available in both hydrophilic and lipophilic forms. Therefore, we might envisage four types of emulsion systems that make use of TiO_2 sunscreens:

1. Oil-in-water emulsions containing TiO_2 in the aqueous phase
2. Oil-in-water emulsions containing TiO_2 in the oil phase
3. Water-in-oil emulsions containing TiO_2 in the oil phase
4. Water-in-oil emulsions containing TiO_2 in the aqueous phase

Stable, effective formulations of all four types have been made, but types 1 and 3 (i.e., in which the TiO_2 is in the *external* phase of the emulsion) tend to be the easiest to formulate, to provide the most versatile systems, and give the highest SPFs. It is also possible to incorporate titanium dioxide in both phases of the emulsion, and there is evidence that this occasionally gives improved SPF values (23,25,26). This synergistic effect is believed to arise from a more uniform coverage of the skin with the sunscreen; similar synergistic effects have been reported for combinations of oil- and water-soluble organic sunscreens (27).

Oil-in-Water Emulsions Containing Titanium Dioxide in the Aqueous Phase

This is the most versatile type of system and tends to be the easiest to work with. Because o/w cosmetic emulsions typically contain more than 70% water, there is plenty of scope for inclusion of high concentrations of a hydrophilic titanium dioxide, either as powder or dispersion, for high-SPF products.

A wide range of different emulsifier systems can be used. However, observance of certain guidelines will help to give the optimum products:

1. Avoid inclusion of electrolytes and ethanol.
2. If hydrocolloids are used (e.g., carbomers, veegum, xanthan gum), these should be fully dispersed before the TiO_2 is added. Carbomers should not be used at concentrations of greater than 0.2% and should be neutralized before addition of the TiO_2.
3. Use vigorous mixing when adding the TiO_2.
4. Ensure that polyols (e.g., glycerol, propylene glycol) are well mixed with the water phase before the TiO_2 is added.

Also, the point of zero charge (PZC) of the titanium dioxide should be avoided. All inorganic solids bear positive and negative charges at their surface; the net charge depends on the pH of the environment. At low pH the overall charge is positive; at high pH it is negative. Each material has a particular pH at which there is no net charge. This pH is the point of zero charge, and is characteristic of the nature of the surface—hence, different grades of TiO_2 have different PZCs because they have different coatings. If the pH of the aqueous phase is in the region of the PZC, there is no net charge on the TiO_2 particles, so there is no electrostatic repulsion between them and there is a danger of agglomeration.

That this type of system, with the active dispersed in the aqueous phase of the emulsion, gives very good efficacy, apparently contradicts the earlier statement that effective dispersion in the *oil* phase, after dry-down, is required for high SPF. Clearly, there must be some mechanism by which the TiO_2 is transferred from the aqueous phase to the oil phase after application of the product to the skin. The most likely mechanism is by *liquid crystalline structures.*

Liquid crystals are anisotropic structures that have one- or two-dimensional order; they are intermediate between solids, which have three-dimensional order, and liquids, which are disordered. Liquid crystals are present in many cosmetic emulsions, particularly o/w systems (28–30), but only relatively recently have they been studied in any depth and used as the basis for improving skin care formulations. These liquid crystals can form various different structural phases; of most interest is the *lamellar* phase, in which layers of water are trapped between layers of lipid emulsifier. These lamellae, in turn, can take different forms; perhaps the most well-known are *oleosomes*, in which the lamellae form concentric spheres around the oil droplets of an emulsion (29–31). A more recent innovation is the use of low-HLB emulsifiers to form *lamellar gel networks* (30,32) in which lamellar structures form a network throughout an o/w emulsion, thereby stabilizing it.

Electron microscopy of emulsions containing hydrophilic grades of fine-particle titanium dioxide has indicated that the TiO_2 has a tendency to locate itself in and around liquid crystalline structures of this type, suggesting that this is the means by which the TiO_2 is transferred into the oil film after dry-down of the emulsion on skin. Further studies (23,26) have shown that polymer-stabilized aqueous dispersions of TiO_2 are particularly suitable for incorporating into lamellar gel networks; in fact, such dispersions appear to reinforce such structures, rendering them more stable. Attempts to incorporate powder TiO_2 into the lamellae were less successful, but this would depend on the surface treatment of the TiO_2 and on any dispersant present.

Oil-in-Water Emulsions Containing Titanium Dioxide in the Oil Phase

This type of system is a little more limiting in terms of the amount of active agent that can be incorporated, because high concentrations of TiO_2 in the internal phase of an o/w emulsion can lead to crowding effects that will cause aggregation, thus reducing efficacy and transparency of the product. However, the wide range of oil dispersions and hydrophobically modified grades of TiO_2 that are now available mean that stable, effective formulations can be designed.

As indicated earlier, one factor governing the effectiveness of this type of product is the potential for the hydrophilic TiO_2 to migrate out of the oil phase and into the external water phase, in which it is likely to be significantly less effective. The degree to which this occurs depends on the surface treatment of the TiO_2 and the effectiveness of the dispersant in keeping the active within the oil phase. The spreading properties of the oil components themselves also exert a significant influence; varying the type of ester incorporated in a formulation with TiO_2 can result in wide variations in SPF (33). Dahms (34) has proposed a relation between the measured SPF and the surface and interfacial tensions of the used emollient system.

Water-in-Oil Emulsions Containing Titanium Dioxide in the Oil Phase

In theory this should be the most efficient system of all: the TiO_2 is present in the oil phase, and because this is the external phase, a homogeneous, continuous covering of the skin should be readily achieved. Practical experience thus far does seem to bear this out; experimental systems have given SPF values of higher than 15 from 4% TiO_2 (26), and it is likely that further progress will be made in the future.

In this case in particular, the rheology of the formulation plays a major role in determining the SPF of the product. Figure 18 shows in vitro SPF data for simple w/o emulsions, each containing 4% titanium dioxide, with various amounts of consistency-enhancing waxes. It can be seen that SPF can be increased from relatively low values (7–8) up to 15 or even 20 or higher, and this can be directly related to the modification of the rheological properties of the emulsion. Of course, the spreading properties of the emollients will also have an effect, as outlined in the previous section.

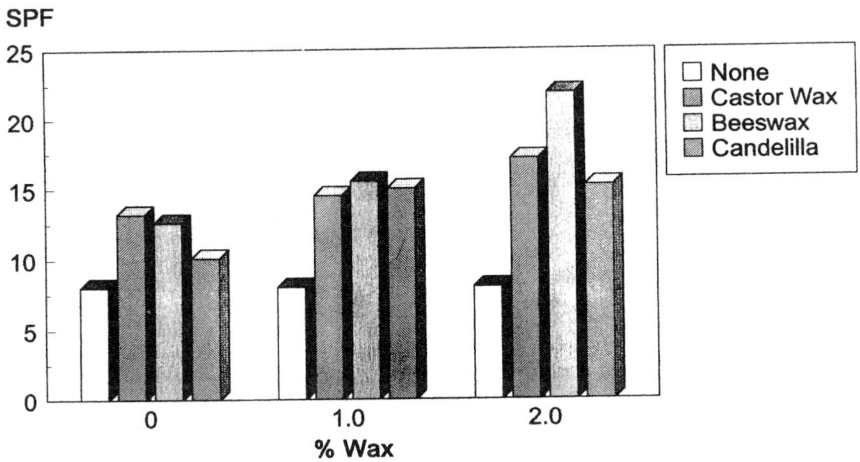

Figure 18 Enhancement of SPF by inclusion of waxes.

The stability of a w/o emulsion depends primarily on the following parameters:

Phase/volume ratio
Emulsifier concentration
Electrolyte concentration

The phase/volume ratio is the ratio of internal (i.e., water) to external (oil) phases. The optimum phase/volume ratio depends on emulsifier structure, oil phase polarity, and wax concentration. It should be remembered that if TiO_2 is present, the amount of TiO_2 will also affect the phase/volume ratio.

Emulsifier concentration is critical. Instability will result from too much or too little emulsifier. Results thus far indicate that products containing fine-particle TiO_2 often require lower emulsifier concentrations than more conventional w/o emulsions. In fact, frequently inclusion of fine-particle TiO_2 has a positive effect on the stability of a w/o emulsion (23). Reducing the level of emulsifier has several advantages.

Reduced cost
Improved cosmetic elegance
Reduced irritancy potential

Water-in-Oil Emulsions Containing Titanium Dioxide in the Aqueous Phase

Although it is possible to make stable, effective formulations of this type, it is generally not recommended. An electrolyte is normally required in a w/o formulation, and this is likely to cause agglomeration of TiO_2, thus reducing SPF.

Formulating with Zinc Oxide

Some of the principles of formulating with fine-particle TiO_2 also apply to formulating with fine-particle grades of zinc oxide (35,36), particularly in w/o emulsions. However, there are some significant differences which the formulator needs to be aware of.

Zinc oxide is much more hydrophilic than titanium dioxide, so keeping it in the oil phase to obtain maximum efficacy is rather more difficult. Also, the solubility of ZnO below pH 6 (Fig. 19) can affect emulsion stability. The ZnO needs to be dispersed in the oil phase and kept there, either by lipophilic coating or use of a suitable dispersant. If there is significant migration of ZnO, this tends to manifest itself as an increase in pH, owing to the formation of alkaline zinc complexes in the aqueous phase.

Use of a predispersed form of ZnO is one way to avoid migration, but even here, some limited migration may be observed. This may be reduced or eliminated by the following measures:

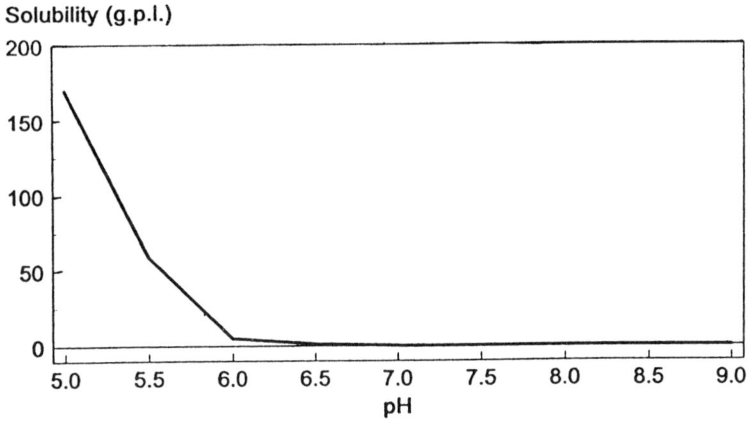

Figure 19 Solubility of ZnO against pH.

Use less polar oils, thereby increasing the interfacial tension between oil and water phases.

Incorporate a small amount (0.5–1.5%) of a hydrophobic w/o emulsifier (e.g., sorbitan oleate). This has the effect of increasing the interfacial viscosity, increasing the barrier to migration.

In any event, it is wise to maintain a pH of higher than 6.0, to avoid significant dissolution of ZnO. Other hints concerning the use of ZnO are

Avoid the use of stearic acid in o/w systems, because this will react to some extent to form zinc stearate, thereby reducing both stability and efficacy of the product.

Xanthan gum appears to be the best thickener to use in zinc oxide formulations. Carbomers should be used only in small concentrations ($< 0.2\%$) and should be neutralized before addition of the zinc oxide.

C. Combinations of Physical Sunscreens

Titanium dioxide and zinc oxide are complementary in terms of their sun protective effects. Titanium dioxide shows broad-spectrum protection across the UVA and UVB regions of the spectrum, but its primary function is for UVB protection. Zinc oxide provides much less protection than TiO_2 in the UVB, but has a ''flatter'' spectrum and is intended primarily for UVA protection. It follows, therefore, that it is often desirable to use the two together to obtain true broad-spectrum coverage (35).

Thus far, the most effective way to do this, as with a single physical sunscreen, is with w/o emulsions. Both TiO_2 and ZnO can be dispersed in the oil phase of a w/o system and, provided the correct rheology can be developed, high SPFs combined with good UVA protection are readily achievable. However, although w/o emulsions are increasing in popularity because of their functional benefits and improving aesthetics as a result of better emulsification technology, o/w emulsions remain by far the most popular vehicle for suncare and skincare products.

In an o/w system, it is also possible to disperse both TiO_2 and ZnO in the oil phase, and for low levels of active agents this is probably the best course of action. However, for higher concentrations of physical sunscreens it leads to problems caused by "overcrowding" of the internal phase of the emulsion. Migration of the active agents, particularly zinc oxide, into the water phase is encouraged, leading to reduced effectiveness and poor stability. With a suitable liquid crystalline structure, it may be possible to disperse both active agents in the aqueous phase, but the solubility of ZnO at low pH and the surface properties of the two materials make this a difficult system to stabilize. The problem arises from the different points of zero charge of the two materials. The PZC for an inorganic powder depends on its surface properties and, hence, on any coating present. But if the two inorganic sunscreens have significantly different PZCs, then any pH in between the two PZCs must be avoided, because at such a pH the two actives will bear opposite charges. This leads to an electrostatic attraction, resulting in aggregation. For example, the PZC for uncoated ZnO is about 9.5; a typical sunscreen grade of TiO_2 might have a PZC of about 7. For such a system, pH must be below 7 for the two actives to both bear a net positive charge, but at the same time must be kept above 6 to avoid dissolution of ZnO. This creates a very narrow pH range within which the formulator can work.

The system of choice, then, is often to disperse the TiO_2 in the aqueous phase, and the ZnO in the oil phase. This avoids crowding problems and should give good overall coverage of the skin. However, if there is significant migration of ZnO, then the difficulties outlined in the foregoing may still arise, so an awareness of the PZCs of the particular grades being used is always helpful.

D. Combinations of Physical Sunscreens with Organic Ultraviolet Filters

There is no reason why physical sunscreens should not be combined with the more traditional organic UV filters in the same formulation, provided that there is no reaction between the two that could lead to production of new species, the safety of which would have to be assessed. This approach has a number of advantages:

1. Lower levels of organic sunscreens can be used than in the traditional sunscreen compositions, reducing the potential for irritancy.
2. Simpler mixtures can be used to obtain high SPFs, again improving the product's safety profile. This should also make the formulation less complex and less expensive.
3. If whitening by the physical sunscreen is perceived to be a problem, use of an organic to complement it allows the concentration of physical sunscreen to be reduced, giving a more elegant product.
4. Use of a combination is a less radical change for the formulator than switching from organic UV-absorbers to an inorganics-only product. This will tend to reduce development times, and make the products more acceptable to consumers and marketers.

The major advantage of combined systems, however, is that significant synergistic effects have been observed between physical sunscreens, particularly titanium dioxide, and the organic sunscreens. The SPF values obtained from formulations containing both types of sunscreen have been significantly higher than might be expected from the SPFs of products containing the individual sunscreens alone. For example, a nonoptimized formulation containing 4% TiO_2 (from an aqueous dispersion) gave an in vitro SPF of 8. The same formula containing only 1% octyl methoxycinnamate gave a result of 3. With both active agents present, the SPF was 14. Other examples are given in Table 2.

One possible explanation of these synergistic effects is that scattering of light by the physical sunscreen increases the optical path length through the film. This increases absorption of light by the organic sunscreen. However, further work

Table 2 Efficiency of Mixed Organic/Inorganic Sunscreens

Physical sunscreen	Organic sunscreen	In vitro SPF
4% TiO_2		6.8
—	3% Octyl triazone	1.5
4% TiO_2	3% Octyl triazone	14.8
—	3% Octyl salicylate	2.5
4% TiO_2	3% Salicylate	10.1
—	3% Octyl dimethyl PABA	4.7
4% TiO_2	3% Octyl dimethyl PABA	16.2
4% TiO_2	—	7.2
—	3% Benzophenone-4	5.4
4% TiO_2	3% Benzophenone-4	20.1
4% TiO_2	—	5.2
—	3% Phenylbenzimidazole sulfonic acid	8.1
4% TiO_2	3% Phenylbenzimidazole sulfonic acid	21.7

is required to fully elucidate the mechanism and to optimize these effects. The results just given indicate considerable potential for the development of high-SPF products with relatively small amounts of sunscreen actives.

VI. EFFECTIVENESS MEASUREMENTS AND QUALITY CONTROL

A. In Vitro Sun Protection Factor

Sun protection factor is the definitive measure of the effectiveness of any sunscreen product. In vivo SPF testing using human subjects is time-consuming and expensive, so although this should always be carried out on finalized products before they are marketed, it is not feasible to measure every trial formulation during development of a new sunscreen product.

In recent years considerable effort has been expended in attempts to provide an in vitro SPF test that is rapid, inexpensive, and most importantly, provides a reliable prediction of in vivo results. Thus far, the most widely used method is that of Diffey and Robson (22), which is now the basis of a commercial instrument. It is this method that was used to generate the in vitro SPF data quoted in this chapter.

The method involves a spectroradiometric measurement, in which a tape substrate is used. The tape, Transpore (3M Industries), was chosen because it is UV-transparent and has a textured surface that distributes topically applied sunscreen product in a manner similar to human skin. The quantity of UV radiation transmitted through the substrate, with and without sunscreen applied, is determined automatically by recording photocurrent in 5-nm steps from 290 to 400 nm.

The instrument used (Fig. 20) comprises a continuous UV-VIS source, color-compensating filters, diffusion plates, a grating monochromator, and a photomultiplier detector. UVA and UVB radiation is provided by a 75-W xenon arc lamp. Both the operation of the scanning monochromator and the collection of UV light transmission readings are controlled by a computer. The computer also performs all the mathematical calculations necessary to determine the SPF and the UVA/UVB ratio.

Product is applied at 2 $\mu L/cm^2$, by applying a series of small dots over the surface of the tape and then spreading these evenly using a gloved finger.

Measurements of monochromatic protection factor (MPF) are made at each 5-nm increment between 290 and 400 nm. MPF is defined as the ratio of the detector signal intensity without sunscreen applied to the Transpore tape, to that

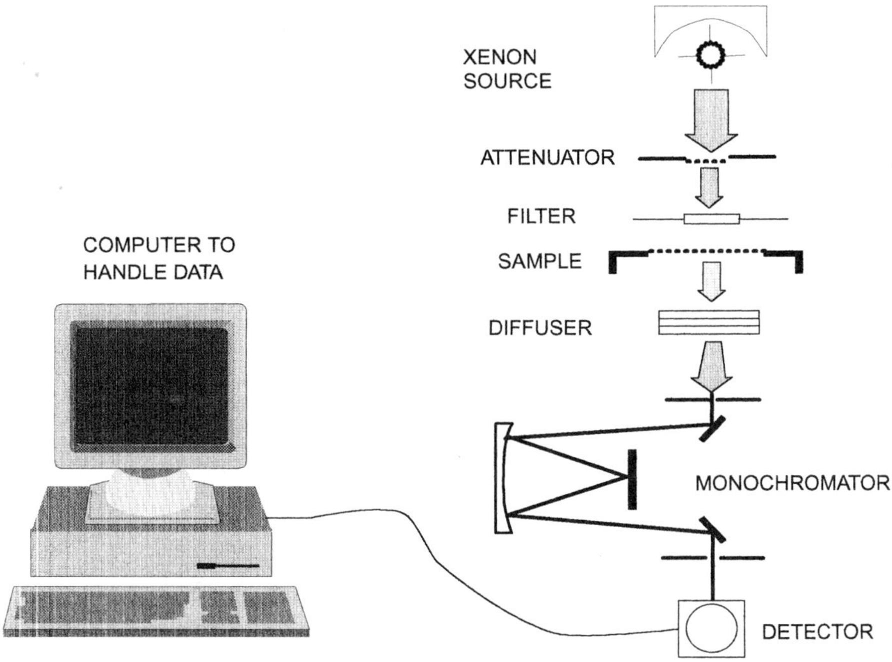

Figure 20 Optometrics in vitro measurement equipment.

with sunscreen applied to the tape:

$$\text{MPF}(\lambda) = \frac{\text{mean photocurrent at wavelength }(\lambda)\text{ without sunscreen applied}}{\text{mean photocurrent at wavelength }(\lambda)\text{ with sunscreen applied}} \quad (4)$$

A graph of MPF versus wavelength is plotted to illustrate the protection profile of the sunscreen product across the UV spectrum. For sunscreen products containing TiO_2 and ZnO, these graphs typically have the same shape as the UV spectra characteristic of these materials.

The SPF is calculated using the following formula:

$$\text{SPF} = \sum_{290}^{400} E(\lambda)\varepsilon(\lambda) \Bigg/ \sum_{290}^{400} \frac{E(\lambda)\varepsilon(\lambda)}{\text{MPF}(\lambda)} \quad (5)$$

where $E(\lambda)$ is the spectral irradiance of terrestrial sunlight under defined conditions, (e.g., midday midsummer sunlight for southern Europe (latitude 40 N;

solar zenith angle 20°; ozone layer thickness 0.305 cm]; and $\varepsilon(\lambda)$ is the relative effectiveness of UV radiation, at wavelength λ nanometers, in producing delayed erythema in human skin (the erythema action spectrum), and $PF(\lambda)$ is the monochromatic protection factor at wavelength λ nanometers.

The UVA/UVB ratio is also determined using this method. This is defined as the ratio between the mean absorbance in the UVA region (320–400 nm) and the mean absorbance in the UVB region (290–320 nm). This is one way of assessing the extent of UVA protection provided by a product, and is the basis of a rating system in the United Kingdom to indicate the level of UVA protection provided by a product.

With good product application technique, this method, as shown in Table 3, is capable of giving good agreement with in vivo SPF data, for example (37,38).

However, the so-called Diffey method suffers from several shortcomings. The Transpore tape is of variable quality, with pore size varying from batch to batch and even between different parts of the same roll. This can lead to wide variations in results, because the pore size affects the spreading of the sample. Also, some cosmetic ingredients seem to react with either the tape itself or the adhesive on it, leading to inflated SPF values. The tape cannot mimic any chemical or biochemical effect that skin may exert on the efficacy of the product. Nevertheless, this in vitro method is a valuable screening tool for the formulator. Development samples can be measured rapidly (a single measurement takes less than 10 min), allowing the formulator to judge which formula modifications increase SPF.

B. In Vivo Sun Protection Factor

It is not proposed to discuss in vivo methods in detail here, because they are described elsewhere in this book. However, it is appropriate to comment on in vivo testing of products containing physical sunscreens.

Those familiar with SPF testing will be aware that product application technique exerts a major influence on the results obtained. Product must be applied evenly and rubbed in well, without beginning to remove product from the test site. Effective rub-in seems to be particularly important when physical sunscreens are concerned. Insufficient rub-in can lead to a patchy distribution of product, with the required even film not being formed. This significantly reduces the measured SPF. Variations in results have been noted between laboratories in different parts of the world, while applying the same protocol to the same products.

C. Optical Density

Theoretical calculations for TiO_2 indicate that the shape of the UV–visible attenuation curve is unique for a given particle size with a given particle size distribution. Thus, by taking optical density readings at three defined points on

Table 3 Comparison of In Vivo and In Vitro Data

Emulsion type	TiO_2 %	ZnO %	In Vitro SPF	In Vivo SPF
o/w	2.5		6.0	5.5[a]
w/o	3.6	—	12.0	12.9[b]
o/w	5.0	—	19.3	17.3[a]
w/o	5.0	—	14.3	15.3[b]
w/o	6.0	—	18.1	17.7[b]
o/w	7.5	—	19.9	17.3[a]
w/o	7.5	—	21.6	24.2[a]
w/o	—	2.0	2.0	2.99[a]
o/w	—	6.0	4.4	6.16[a]
w/o	2.5	2.5	9.2	11.77[a]
w/o	2.5	7.5	18.0	20.52[a]

[a]FDA protocol (20 subjects).
[b]FDA protocol (5 subjects).

the curve and calculating a set of ratios from these figures it is possible to give figures for particle size and particle size distribution. This method is already in use in the pigment industry (39) and is now being applied to UV-attenuating grades of TiO_2. Because of the shape of the absorption curve for ZnO, this method cannot be used. The changes seen with particle size and size distribution give rise to areas of commonality and, thus, ambiguity in using ratios (no uniqueness exists).

As with all particle size methods the main failing is that of sample preparation. To present the sample to the UV visible spectrometer (photomultiplier on monochromator) the sample must be altered by dilution, mixing, and so on. Thus, the size measured may not bear any similarity with the particle size seen in actual use, which is the critical parameter in maintaining efficacy. Future work is now targeted on measuring the optical density ratios of dried films, thus giving a more realistic particle size measurement.

D. Analytical Methods

Both TiO_2 and ZnO are used in cosmetic or drug type products and, as such, it is a requisite of the manufacturing process to entail analytical methods that monitor product consistency. These methods are based on assuring efficacy and purity.

Efficacy

The efficacy of both TiO_2 and ZnO is due to a combination of both the amount added and their degree of dispersion. TiO_2 assay is well documented, and being nonvolatile, TiO_2 can be assayed after ashing the formulation. Several methods are available (40) with the U.S. Pharmacopeia (*USP*) monograph version being commonly used (41). The ZnO assay is also well documented with many variations being used, again the *USP* monograph method (42) is extensively used.

The best method for assessing efficacy is based on measuring the UV visible spectrum of the base material. Here, a known weight of the active material is made up into a stable dilute dispersion and placed in a quartz cuvette in a UV–visible spectrometer. The UV–visible spectrum is measured and an extinction value calculated at several wavelengths (Fig. 21)

$$\text{Extinction value} = \frac{\text{absorbance value}}{\text{TiO}_2 \text{ concentration}} \tag{6}$$

The example for TiO_2 given in Figure 11, shows extinction values measured at 308 nm (the point of maximum erythemal effectiveness), 360 nm (UVA region), and 500 nm (blue end visible region). By using these values it is possible to calculate ratios that tend to give more consistent results than absolute absorbence. In this example, the extinction value at 308 nm is 71, at 360 nm is 24, and at 524 is 4. This give the following ratios;

$$\frac{308}{500} = 17.75 \quad \text{(indicating high absorption in UVB and low interaction with visible)} \tag{7}$$

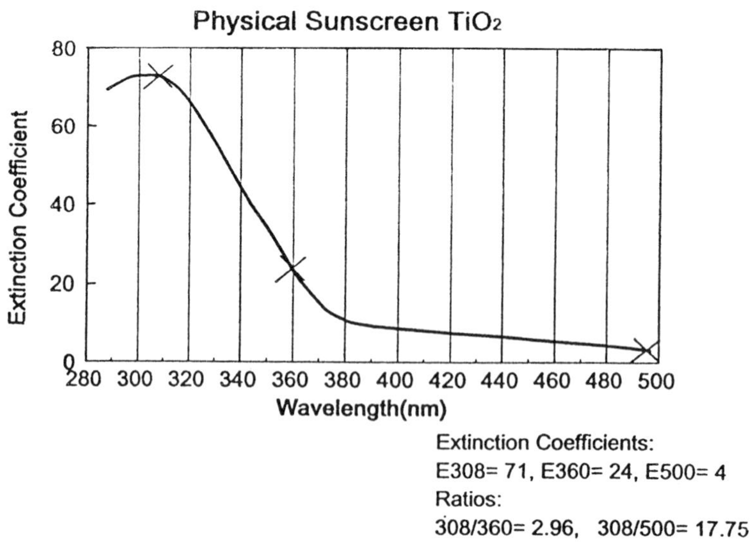

Figure 21 UV visible spectrum for efficacy measurement.

$$\frac{308}{360} = 2.96 \quad \text{(indicating efficacy of UVB versus UVA absorption; a lower value would show a broader spectrum coverage)} \qquad (8)$$

These ratios give an overview of the potential of the product. Pigmentary material would, using the same ratios, show typical values of 1–1.3.

Purity

Purity specifications are based on assessing the amount of contaminants present alongside the base active material. Most of the global pharmacopeial agencies include methods for the heavy metals, water-soluble salts, acid-soluble salts and other ancillary tests (color, alkalinity and such). Tables 4 and 5 set these data out in a collated format. It can be seen that there is considerable variation in specifications. Closer examination of the quoted monographs will also highlight

Table 4 Global TiO_2 Purity Specifications[a]

Specification	BP 1993	EP 2 ed	*USP* XXII	JP 1986	FDA
Identification	√	√	√	√	
Solubility	√	√		√	
Clarity/color	√	√			
Acid/alkalinity	√	√		√	
TiO_2 (%)	>98.0	>98.0	>99.0	>98.5	>99.0
Loss at 105 (%)			<0.5	<0.5	<0.5
Loss at 800 (%)			<0.5		<0.5
Water sol. (hot) (%)	<0.5	<0.5			
Water sol. (cold) (%)			<0.25	<0.25	<0.3
Acid sol. (%)			<0.5	<0.35	<0.5
Total AS (ppm)	<5	<5	<1	<10	
Total Sb (ppm)	<100	<100			
Total Hg (ppm)			<1		<1
Total Fe (ppm)	<200	<200			
Total Pb (ppm)				<60	
Acid sol. As (ppm)					<1
Acid sol. Sb (ppm)			<2		<2
Acid sol. Ba (ppm)	<5	<5			
Acid sol. Pb (ppm)[b]	<20	<20	<10		<10

[a]The specification for acid-soluble heavy metals in BP/EP is actually acid-soluble Pb.
BP, British pharmacopeia; EP, European pharmacopeia; USP XXII, United States Pharmacopeia, 22 ed.; JP, Japanese pharmacopeia; FDA, U.S. Food and Drug Administration; sol., solubility.

Table 5 Global ZnO Purity Specifications

Specification	BP 1993	EP 2 Edn	*USP* XXII	JP XI 1986	JSCI-II
Identification	√	√	√	√	√
Solubility	√	√		√	
Clarity/color	√	√		√	√
Acid/alkalinity	√	√	√	√	√
ZnO (%) (on ignited)	>99.0	>99.0	>99.0	>98.0	>99.5
Loss at 500 (%)	<1.0	<1.0	<1.0	<1.0[a]	<1.0
Water sol. (hot) (%)					<0.1
Acid insol/carbonates	√	√	√	√	√
Total AS (ppm)	<5	<5	<6	<4	<2
Total sulfate (%)				<0.096	<0.096
Total Fe (ppm)	<200	<200	√[b]	√	<200[c]
Total Pb (ppm)	<50	<50	√	√	<40
Total Cd (ppm)	<10	<10			

[a]Ignition at 850°C.
[b]Iron and other heavy metals.
[c]Iron or other metals.

the many different methods used to analyze these materials. Presently, there is a global harmonization exercise being carried out for TiO_2 (43).

GLOSSARY

Agglomeration The formation of large stable particles by the bonding together of smaller particle/crystal units (c.f. flocculation). Agglomerates can be very difficult to disperse.

Anatase Naturally occurring crystal form of TiO_2. Less dense packing than rutile form, hence, lower refractive index (2.55 vs. 2.70). More photoactive than rutile.

Attenuation The lessening in the effect of . . . (i.e., the reduction in the intensity of ultraviolet radiation).

Flocculation The formation of large less stable particles by the bonding together of smaller particle or crystal units. Flocculates can be dispersed with the input of a moderate amount of energy.

Photoactivity The degree of effect that a substance has on other chemicals as a result of the action of exposure to electromagnetic radiation.

Photocatalysis The changing of the rate of a chemical reaction as a result of the presence of a photoactive species.

Photostability The immunity of a substance to the action of incident electromagnetic radiation.

Rutile Naturally occurring crystal form of TiO_2. More dense packing than anatase form, hence, higher refractive index (2.70 vs. 2.55). Less photoactive than anatase.

REFERENCES

1. Heslop, R. B. and Robinson, P. L. *Inorganic Chemistry.* 3rd ed. New York: Elsevier, 1982, p. 634.
2. Weast. *Handbook of Chemistry and Physics.* 68th ed. Boca Raton, FL: CRC Press, 1987, p. B-39
3. Lynd, E. L. *Titanium, Mineral Facts and Problems.* 1985, pp. 859–866.
4. Whitehead, J. *Kirk-Othmer Encyclopedia of Chemical Technology* (23), 3rd ed. New York: John Wiley & Sons, 1983, pp. 143–149.
5. Tioxide Publicity Publication, *Manufacture and General Properties of Titanium Dioxide.* February 1992.
6. Brown, H. E. *Zinc Oxide, Properties and Applications.* International Lead Zinc Research Organization, New York, 1976. LCCN 76-9932.
7. OTC Vol 0-60001.
8. Craft, R. W. *Modern Paint Coatings*, October: 156–160 (1989).
9. Bohren, C. F. and Huffman, D. R. *Absorption and Scattering of Light by Small Particles.* John Wiley & Sons, New York, 1983.
10. Tioxide Technical Data: Personal communication.
11. Judin, V. P. S. *Chem. Bri.*, June: 503–505 (1993).
12. Kerker, M. *The Scattering of Light and Other Electromagnetic Radiation.* Academic Press, New York, 1969.
13. Mitchnik, M. A. *Cosmet. Toiletries 107*:111–116 (October 1992).
14. Catlow, B. *S.P.C. Mag.* March 29–30 (1993).
15. Diffey, B. L. *Int. J. Cosmet. Sci. 16*:47–52 (1994).
16. Sayre, R. *J. Cosmet. Chem., 41*(2):103–109 (1990).
17. Robb, J. L., et al. *D.C.I. Mag* (March 1994).
18. Mie, G. *Phys. Lpz., 25*:377 (1908).
19. Lord Rayleigh. *Philos Mag., 41*:447 (1871).
20. Bickley, R. I. and Hogg, L. T. University of Bradford, Internal Publication (January 1994).
21. Klein, K. Formulating sunscreen products. *Sunscreens* (edited by N. J. Lowe and N. A. Shaath). New York: Marcel Dekker, 1990.
22. Diffey, B. L. and Robson, S. *J. Soc. Cosmet. Chem. 40*:127–133 (1989).
23. Dahms, G. H. *Cosmet. Toiletries, 107*:87–92 (Oct. 1992).
24. Taelman, M.-C. (ICI Surfactants, Everberg, Belgium); private communication.
25. UK patent 2 243 781 A.
26. Dahms, G. H. (IFAC GmbH, Duisburg, Germany); private communication.

27. Johncock, W. *Euro Cosmet., 7*:22–24 (1994).
28. Junginger, H. E. *Skin Care Forum (Henkel KGaA)*, No. 5, (Feb. 1993).
29. Dahms, G. H. *SÖFW, 10*:388–392 (1990).
30. Loll, P. *Cosmetics and Toiletries Manufacture Worldwide.* 1994, 108–120.
31. Dahms, G. H. *Cosmet. Toiletries, 101*:113–115 (Nov. 1986).
32. Dahms, G. H. *In-Cosmetics*. Conference, Paris, March 1993.
33. Woodruff, J. *Cosmetics and Toiletries Manufacture Worldwide.* 1994, pp. 179–185.
34. Dahms, G. H. *Cosmet. Toiletries, 109*:45–52 (Nov. 1994).
35. Spruce, S. R. *In-Cosmetics*. Conference, Barcelona, March 1994.
36. Mitchnick, M. A. *Cosmetics and Toiletries Manufacture Worldwide.* 1994, pp. 192–197.
37. Hewitt, J. P. Third International Symposium on Cosmetic Sciences, Taichung, Republic of China; May 1994.
38. Tioxide Specialties Limited. Technical data sheet.
39. Tunstall, D. F. US Patent 4279512, 21st July, (1981).
40. Norris, J. D. *Analyst 109*:1475–1482 (Nov 1984).
41. *USP Monograph*, XXII, pp. 1380.
42. *USP Monograph*, XXII, pp. 1464.
43. Artiges, A. and Schorn, P. J. *Pharmeuropa, 6*(1):29–32 (March 1994).

APPENDIX: SAMPLE FORMULATIONS CONTAINING PHYSICAL SUNSCREENS

The following formulations all contain dispersions of titanium dioxide and/or zinc oxide. In each case, the titanium dioxide dispersions contain 40% solids, while the zinc oxide dispersions contain 60% solids.

O/W Sunscreen Lotion (SPF 12)

Phase A	*% W/W*
Glyceryl stearate (and) PEG-100 stearate	5.00
Stearyl alcohol	0.50
Sorbitan stearate	0.50
Polysorbate 60	0.90
Petrolatum	3.00
Dimethicone	1.00
Sweet almond oil	5.00
Mineral oil	8.00
Evening primrose oil	0.50

Phase B	*% W/W*
Deionized water	56.76
Sorbic acid	0.20
Glycerine	5.00
Water (and) titanium dioxide	**12.50**
Xanthan gum	0.10

Phase C	*% W/W*
Deionized water	1.00
2-Bromo-2-Nitropropane-1,3-diol	0.04

Manufacture

Heat the deionized water, sorbic acid, and glycerine to 75°C, stirring occasionally until all the sorbic acid has dissolved.

Sprinkle in the xanthan gum and homogenize until this is fully dispersed. Maintain temperature at 75°C.

Add the TiO_2 dispersion and homogenize briefly.

Heat phase A to 75°C.

Add A to B and homogenize for appropriate time. Paddle stir to cool.

Add phase C at 40°C. Continue cooling with stirring to 25°C.

O/W Sunscreen Lotion (SPF 10)

Phase A	*% W/W*
Isopropyl myristate	4.00
Mineral oil	6.50
Grape seed oil	2.50
Stearyl alcohol	1.00
Petrolatum	2.00
Titanium dioxide [and] mineral oil [and] caprylic-capric triglyceride	**5.00**

Phase B	*% W/W*
Deionized water	62.90
Sucrose cocoate	0.10
Sucrose stearate	6.00
Glycerine	4.00
Allantoin	0.20
d-Panthenol	0.80

Phase C	*% W/W*
Titanium dioxide [and] water	**5.00**

Phase D	
Preservative	qs

Manufacture

Heat A and B separately to 80°C.
Add A to B, stirring intensively. Add C with intensive stirring.
Homogenize.
Allow to cool for 25°C with slow agitation. Add D.

O/W Baby Sunscreen Lotion (SPF 15)

Phase A	*% W/W*
Dimethicone [and] trimethylsiloxysilcate	7.00
Mineral oil	6.00
Myristyl lactate	2.00
Cetyl alcohol	2.00
Glyceryl stearate SE	1.20
Ceteareth-20	0.40
Stearic acid	1.00
Propylparaben	0.05

Phase B	*% W/W*
Deionized water	58.45
Glycerine	1.00
Triethanolamine	1.00
Aloe vera gel 10:1	0.50
Methylparaben	0.15
Titanium dioxide [and] water	**18.75**

Phase C	*% W/W*
Perfume	0.20
Imidazolidinyl urea	0.30

Manufacture

Heat phases A and B separately to 70–75°C.
Add phase A to phase B and homogenize for appropriate time.
Cool down to 50°C with slow stirring, then add perfume.
Continue cooling with stirring. Add preservative at 35°C. Stop stirring at 30°C.

W/O Sun Protection Cream (SPF 15+)

Phase A	*% W/W*
Dimethicone	2.00
Hydroxyoctacosanyl hydroxystearate	5.00

Methoxy PEG-22/dodecyl glycol copolymer	2.00
PEG-45/dodecyl glycol copolymer	4.00
Cocoa butter	3.00
Propyl paraben	0.08
Mineral oil	3.125
Caprylic/capric triglyceride	3.125
Zinc oxide [and] mineral oil [and] caprylic/capric triglyceride	**12.50**
Titanium dioxide [and] mineral oil [and] caprylic/capric triglyceride	**6.25**

Phase B	*% W/W*
Propylene glycol	5.00
Methyl paraben	0.10
Demineralized water	53.82

Manufacture

Heat phase A to 75–80°C with occasional stirring, until fully molten and homogeneous.

Premix the propylene glycol and methyl paraben. Add to the water, mix, and heat to 75°C.

When both phases are completely dissolved and at 75–80°C, add phase B to phase A while mixing with a high shear mixer. Mix for 5 min and then commence cooling.

Cool to approximately 25°C with slow speed mixing.

Mix briefly at high speed until smooth.

O/W Organic/Inorganic Sunscreen Lotion (SPF 30)

Phase A	*% W/W*
Isopropyl myristate	4.00
Mineral oil	6.50
Grape seed oil	2.50
Stearyl alcohol	2.00
Petrolatum	2.00
Octyl methoxycinnamate	5.00

Phase B	*% W/W*
Deionized water	51.80
Sorbitan stearate	6.00
Disodium ricinoleamido MEA-sulfosuccinate	0.20
Glycerine	4.00

Allantoin	0.20
d-Panthenol	0.80
Titanium dioxide [and] water	**15.00**

Phase C	*% W/W*
Preservative	qs

Manufacture

Heat A and B separately to 80°C.
Add A to B, stirring intensively.
Homogenize.
Allow to cool for 25°C with slow agitation. Add C.

O/W Sunscreen Lotion (SPF 20)

Phase A	*% W/W*
Glyceryl stearate	2.00
Dilauryl trimethylolpropane siloxy silicate	3.00
Potassium cetyl phosphate	0.50
Sorbitan palmitate	3.50
Dimethicone	1.50
Beeswax; white pellets	3.50
Titanium dioxide [and] C12–15 alkyl benzoate	**18.75**

Phase B	*% W/W*
Xanthan gum	0.20
Magnesium aluminum silicate	0.80
Water; pure	61.45
Diazolidinyl urea [and] methylparaben [and] propylparaben [and] propylene glycol	1.00
Polysorbate 20	3.50
Sodium lactate 70%	0.30

Manufacture

Disperse xanthan and magnesium aluminum silicate in water.
Add remaining phase B ingredients and heat to 75°C.
Heat phase A to 75°C.
Add A to B with high-shear mixing. Mix for appropriate time and then commence cooling.
Cool to room temperature with slow stirring.

Day Cream with UV Protection (SPF 6)

Phase A	*% W/W*
Glyceryl stearate	3.00
Steareth-21	2.50

Cetyl alcohol	1.00
PPG-15 stearyl ether [and] cyclomethicone	6.00
Light mineral oil	7.00
Sweet almond oil	2.00
Zinc oxide [and] mineral oil [and] caprylic/capric triglyceride	**10.00**

Phase B	*% W/W*
Sorbeth-30	4.00
Water; pure	63.50

Phase C	*% W/W*
Diazolidinyl urea [and] methylparaben [and] propylparaben [and] propylene glycol	1.00

Manufacture

Heat phase A (except zinc oxide dispersion) and phase B to 75°C.
Add ZnO dispersion to phase A with stirring, and readjust temperature to 75°C.
Add A to B with vigorous stirring.
Homogenize.
Begin cooling with stirring. Add phase C at 45°C.
Continue cooling with stirring to 25–30°C.

Day Cream with UV Protection (SPF 15)

Phase A	*% W/W*
Glyceryl stearate	3.00
Steareth-21	2.50
Cetyl alcohol	1.00
PPG-15 stearyl ether [and] cyclomethicone	6.00
Light mineral oil	7.00
Sweet almond oil	2.00
Zinc oxide [and] mineral oil [and] caprylic/capric triglyceride	**10.00**

Phase B	*% W/W*
Sorbeth-30	4.00
Water; pure	63.50
Titanium dioxide [and] water	**10.00**

Phase C	*% W/W*
Diazolidinyl urea [and] methylparaben [and] propylparaben [and] propylene glycol	1.00

Phase D	*% W/W*
Citric acid; 30% solution	2.00

Manufacture

Heat phase A (except zinc oxide dispersion) and phase B to 75°C.
Add ZnO dispersion to phase A with stirring, and readjust temperature to 75°C.
Add A to B with vigorous stirring.
Homogenize.
Begin cooling with stirring. Add phase C, then D, at 45°C.
Continue cooling with stirring to 25–30°C.

W/O Sunscreen Lotion (SPF 25)

Phase A	*% W/W*
Dimethicone	1.50
Wheatgerm oil	3.00
Octyl palmitate	4.50
Isohexadecane	6.00
Cetyl dimethicone copolyol	2.00
Cyclomethicone [and] aluminum/magnesium hydroxide stearate	10.00
Candelilla wax	1.00

Phase B	*% W/W*
Sodium chloride	0.50
Water; pure	51.30

Phase C	*% W/W*
Titanium dioxide [and] C12–15 alkyl benzoate	**20.00**

Manufacture

Heat phases A to B separately to 75°C.
Heat phase C to >60°C.
Slowly add B to A with intensive stirring.
Add C with intensive stirring.
Cool to 45°C, maintaining the same stirrer speed. Add D.
Continue cooling to 25–30°C.
Homogenize.

Skin Care Lotion (SPF 5)

Phase A	*% W/W*
Glyceryl stearate SE	2.20
Avocado oil	1.00
Sweet almond oil	6.00

Cetyl dimethicone	2.00
Stearyl alcohol	0.60
Propylene glycol	4.00
Ceteareth-20	0.40
Myristyl lactate	1.00
Light mineral oil	4.00
Cetyl alcohol	1.00

Phase B	*% W/W*
Glycerine	3.00
Water; pure	66.06
Titanium dioxide [and] water	**7.50**
Aloe vera powder 200:1	0.20

Phase C	*% W/W*
Water; pure	1.00
2-Bromo-2-nitropropane-1,3-diol	0.04

Manufacture

Heat water and glycerine to 70°C. Add TiO_2 dispersion and homogenize.
Add aloe vera and homogenize. Maintain temperature at 70°C.
Mix oil phase ingredients and heat to 70°C.
When oil phase is fully homogeneous, add to water phase at 70°C.
Homogenize for appropriate time.
Cool with stirring to 40°C. Add preservative.
Continue cooling with stirring to 35°C.

19

Photoprotection of Hair: A Review

Arthur C. W. Georgalas
Tri-K Industries, Inc., Emerson, New Jersey

I. INTRODUCTION

Results of using ultraviolet (UV) radiation absorbers or sunscreens in products and on the skin may seem miraculous from their striking efficacy at preventing color fading and sunburn, respectively, but effecting this magical feat is simply a matter of adding an alternate target to absorb the UV energy in accordance with physicochemical principles. This can be accomplished on any substrate that sunlight ravages. Sunscreens can protect hair from structural damage to the fibrillar protein matrix or network, and they can also shield hair color from fading and distortion.

The UV absorbers dissipate the energy from sunlight's UV rays as well as that emanating from artificial sources, such as indoor fluorescent spectroradiometry. They may have one or more absorption maxima (λ_{max}) in the UV region, with broad or narrow absorption bands around them and with a wide range of molar absorptivities. These sunscreens absorb energy that would otherwise cause the breakdown of structural components or colors in the hair. The absorbed energy is converted in a number of ways. It can degrade the sunscreen itself, cause reemission of lower wavelength light (fluorescence), or cause vibrational heat to be generated. This will usually protect the protein or dye we wish to save.

II. EARLY WORK ON FIBERS AND HAIR

Researchers have sought to protect the hair and other fibers from ultraviolet radiation (UVR) for both structure and hue integrity for many years. Early work investigating benzophenone derivatives for treatment of fabrics was conducted by workers at GAF Co. and American Cyanamid in the late 1950s, soon after patents for synthesis of these chemicals were issued to both of these companies. They found that 2-hydroxy-benzophenone was photostabile, thus making it useful for prolonged light exposure (1) and that a dihydroxy-methoxybenzophenone protected numerous dyestuffs as well as prevented tensile strength loss in nylon fiber from UVR (2). Much later, Milligan and Holt in Australia relate their findings on the usefulness of almost two dozen sulfonated benzophenones for protecting wool by measuring breaking loads, yellowing indices, and tear strengths. Alkyl-substituted 2-hydroxy derivatives gave the best results in retarding the phototendering and fading of dyed wool (3,4). Other studies at the University of Nebraska seem to contradict this in their evaluation of mordant-dyed textiles for museum storage (5). Solvent-soluble benzophenones protected only linen fabric from strength loss, not silk or wool, but they did prevent dye fading. Surprisingly, the water-soluble benzophenone actually accelerated both fading of dyed wool and phototendering of all fabrics, but the sunscreen, a sulfonated benzophenone, may not have been buffered up to a near neutral pH, and its high acidity could have contributed to the destructive effects.

Researchers in Yugoslavia (6) and the former USSR (7) confirm the destructive effects of UVR on both wool and the hair cover of sealskins. More recently, further work reported in the textile research literature (8,9) again show that exposure to sunlight causes wool fabric to become more yellow, less elastic, stiffer, and rougher. This continues to be a subject for the fiber and fabric community in assessing the nature and degree of damage as well as means of prevention.

For human hair, as far back as 1969, specialty chemical companies were synthesizing chemicals for just such purposes, and formulators and marketers were putting sunscreens into a variety of hair care products. In one case, National Starch patented a hair spray containing a UV-absorbing film-forming copolymer (10,11). Although not now commercially available, this resin 28-3307, was an ingredient that protected the hair with an absorbing monomer in the hair spray film-former itself. At that time, the market need never materialized. Later, in the mid-1980s, Van Dyk took their expertise and reputation for making and marketing personal care sunscreens and tried to sell a cationic sunscreen, Escalol 537Q, based on *p*-aminobenzoic acid (PABA) chemistry for use in hair care. This quaternary ammonium compound, Cosmetic, Toiletries, and Fragrance Association (CTFA) name—dimethyl PABA ethyl cetearyldimonium tosylate—had limited solubility in commonly used solvents and had low molar absorptivity in primarily the UVB range (12). It also met a relatively unreceptive market and is no longer available among the commercially active agents.

III. MARKET INTEREST AND SUBSTANTIATION

Although the retail market for sun protection hair care has been variable, it now appears to be gaining momentum, based on consumer awareness of sun damage of skin as well as education and promotion by marketers. Publicity about loss of protective stratospheric ozone fuels the interest. Medical and scientific journals, as well as fashion magazine and newspaper writers, have been preaching an antisolar campaign. Consumers have their reasons to seek skin protection from yearly bouts of sunburn, rising rates of skin cancer, reports of UVR damage to the systemic immune system, and implication as the major cause of premature aging of the skin (13,14). Acceptance of products for hair protection depends on demonstrating both the damage to hair structure and color by UVR and subsequent photoprotection afforded by use of these products.

Some recent additions to the market suggest that believability is growing. Colorvive® remains the foremost example of a full line devoted to hair photoprotection—in this case, integrity of hair dyes. The L'Oreal patent proscribing the use of their specialty quat sunscreen, camphor benzalkonium methosulfate, for skin, hair, and product color protection was issued December 6, 1977 (15). Only with the Colorvive® line introduction in the 1990s has it come into widespread commercial use in the United States. In other mass-merchandised lines domestically we see the inclusion of sunscreens to afford protection against sun damage. Usually, however, their inclusion seems to be just an afterthought for promotional purposes, with little efficacy expected or demonstrated. Seasonally, the opposite tactic from sun protection is sometimes employed in products to enhance the lightening effect that many people experience with lengthy summertime sun exposure. In these cases a small percentage of hydrogen peroxide is employed to augment the oxidative effect of UVR on wet hair. This adds credence to the findings that irradiation causes damage by secondary oxidative reactions.

There seem to be two sides to the debate on this need for photoprotection of the hair. In one view, all such formulations and claims for hair care sunscreens are mere puffery. In the other view, there are distinct needs for photoprotection that can be substantiated. Advocates are defining the needs and communicating them to both technical readers and consumer markets. In the first edition of these monographs, Nacht reported on the validity of using sunscreens on hair (16). He discussed the weathering of hair as consisting of the selfsame damage seen with wool—discoloration, reduction of tensile strength, and free radical damage to structural integrity. His experiments demonstrated the efficacy of octyl dimethyl-PABA from both a solution soak and shampoo formulations that were both rinsed. Ratios of yield slopes from Instron tensile testing were used to define hair protection factors (HPF), analogous to the accepted measure of acute skin damage—sun protection factor (SPF). Dubief of L'Oreal used both natural sunlight and two different solar simulators to assess the effects on natural brown hair (17). Photooxidation of hair elements, such as melanin and cuticle proteins, occurred as seen under scanning electron microscopy (SEM). Photobleaching and diminution of stretching strength further demonstrated photooxidation that was exacerbated by water presence during irradiation. Dubief cites the work of Deflandre (18) also of L'Oreal as corroborating his findings. This earlier work looked at infrared (IR) spectra that identify oxidation products of the cystine —S—S—bonds.

Formulators should not, however, plunge into product development and haphazardly include sunscreens in any number of hair treatments, making vague unsubstantiated claims. Researchers can decide what true benefit they want to deliver and work from there. From an investigation of prior reported research and a modicum of preformulation laboratory work, a product development chemist should be able to imagine a variety of concepts and formulate prototypes to fit them.

We need to consider three things before going to the laboratory bench. If our goal is to provide adequate protection, we need to identify the specific target and understand our substrate, the hair. Second, we need to identify the active ingredients at our disposal—what they can and cannot do, properties, and availability. Last, we need to consider what product types or vehicles we can formulate to meet consumer needs. Do we have to invent and promote a new vehicle in a new package? Can we make use of multiple or multifunctional active ingredients?

IV. HAIR COMPOSITION AS SUBSTRATE

Many detailed descriptions of hair structure, chemistry, and life cycle exist in the literature. One such review occurs in C. R. Robbins' volume *Chemical and Physical Properties of Human Hair*, now in its second edition (19), which can serve along with the Charles Zviak collection, *The Science of Hair Care* (20), as reference. The hair is a complex multicomponent fiber consisting of protofibrils, giving rise to microfibrils that associate to form macrofibrils, encased in the cortex and sheathed in an overlapping array of flattened cuticles, sometimes with an amorphous or hollow medulla in its center. In this fiber, there are lipids, both free and protein bound in the epicuticle, belying that this is indeed a filament of cellular origins, not a man-made fiber. The keratin of hair gives rise to its final physical form and structural properties (21). This primary component consists of a hard, keratinized protein, with a high level of sulfydryl amino acid and virtually none of the marker amino acid, hydroxyproline, found in fibrillar connective tissue, such as collagen. In some places in the hair, for example the "A" layer of the cuticle, the cystine content can be as high as 35%. Other UV-absorbing amino acids present in the hair include the aromatics, tyrosine, and to a lesser extent tryptophan (19). All three of these protein monomers in the pure state and in situ are photolabile and give rise to active oxygen species, chain reaction-propagating free radicals, such as singlet oxygen ($O^{\bullet}$) when irradiated (22–24). These reactive chemicals can also cause chain scission by breaking amide linkages, or can reverse cross-linking by oxidizing disulfides to cysteic acid, as demonstrated by detection of sulfonate and sulfinite IR bands in wool (25). More recently, a similar tactic of comparing chemically oxidized hair with actual weathered hair (distal or tip ends of virgin hair) convinced Robbins and Bahl (26) that the fission occurs at the —C—S— bond in UV-catalyzed cleavage versus the —S—S— bond in peroxide oxidation.

Active oxygen species can also react with dye molecules as well as depolymerize and decolorize melanin granules (19,20). Proposed structures of melanin are complex polyaromatic aggregates that absorb light radiation over a wide range of wavelengths and, as they are polyphenolic, may act as a free radical trap or sink for active oxygen species. Some of melanin's precursors, however, may also generate free radicals (27–30). These natural pigments may be protective, yet may be susceptible to light-induced oxidation and bleaching (19,31). Dyestuffs used for permanent hair coloring are formed in situ from various aromatic phenolic and anhilinic compounds that couple to form complex resonance structures that absorb in the visible and strongly in the UV wavelengths. Absorbed UVR in the presence of water and oxygen can cause degradation. Likewise, aromatic cationic dyes, which are used for temporary hair coloring, can be photobleached. Choice of the right sunscreen application could be photoprotective (15,19,20).

V. HAIR SUNSCREEN AFFINITY

The hair is an inert, but complex, structure, with both hydrophilic and lipophilic properties. Although without surfactants, it may be hard to wet with bulk water owing to its hydrophobicity, it does pick up atmospheric water at higher relative humidities, acting as a humectant, and it has a tendency to hold onto that moisture. Suggestions are that hydrophilic amino acid side chains contribute to these properties. Concurrently, it has lipophilicity based on its free lipid content and hydrophobic alkyl side chain amino acids. This amphiphilic character suggests that it falls within the midrange of solubility parameter values, because pig skin stratum corneum is reported in 9.8 units. These values relate the relative polarities of materials and, hence, to affinities or miscibilities with other materials, especially solvents. It is instructive to realize that many of the so-called oil-soluble sunscreens are in this solubility parameter range (32). Many sunscreens' partition coefficients between oil and water promote a favorable affinity for untreated hair (33). Treated hair has already been structurally oxidized and made more hydrophilic, but yet seemingly drier to the touch. Materials such as fatty quaternary ammonium conditioning agents may be needed to treat the hair before or concurrently with sunscreen application, to provide a more receptive surface. Alternatively, substantivity to the hair can occur by ionic effects, such as anion–cation attraction.

VI. PHOTOPROTECTIVE ACTIVE AGENTS

Sunscreen active agents that are available and useful include a limited number of monomeric aromatic compounds capable of strongly absorbing UVR and safe for use in contact with the skin. Since the establishment of an over-the-counter drug (OTC) monograph for sunscreens for skin protection in 1978, several sunscreen active agents have departed from the marketplace, either because of inadequate safety data, or new questions about safety have been raised. A recent issuance of a new, tentative final monograph on May 12, 1993 does not delist any more actives from skincare use and only speaks to the issue of hair photoprotection in a late paragraph in Part 700—General section 700.35 "Cosmetics containing sunscreens," because hair products are considered cosmetics (34). To avoid consumer misunderstanding, any mention of the term "sunscreen" needs to have "in direct conjunction with the term 'sunscreen' . . . qualifi(cation) by describing the cosmetic benefit provided by the sunscreen. For example: 'This product contains a sunscreen that assists in protecting the hair from damage by the sun' " (34). The market for these UV absorbers revolves around their use in sunscreens and sunblocks for the skin or for protection of inert materials such as plastics, textiles, and outdoor finishes—paints and lacquers. Notably, a popular car vinyl interior protective finish now includes and promotes sunscreen. The availability of sunscreen actives, as with all specialty chemicals, should be kept in mind when formulating a new product. Will the raw ingredient I am incorporating outlive my own product so I will always have the raw materials I need for manufacturing during my product's life cycle? Will I need to find an alternative when that essential ingredient is phased out? With this in mind, the formulator should consult current vendors to determine availability of active agents as well as inert excipients, to ensure that the formulation can be manufactured in the future. This should be considered, especially when looking at old formularies, patents, or the OTC monographs.

Active ingredients for sun protection of hair are more extensive than those for skin, simply because the inert nature of hair means that its treatment is categorized as cosmetic, with the qualifications mentioned earlier. Product development chemists as well as marketers should carefully screen their claims to ensure they are not crossing the gulf into OTC drugs. Hair protection factors and the like would have to be carefully worded to avoid misleading the consumer into thinking he or she is getting a conventional skin-protecting and, therefore, drug regulated, sunscreen. The OTC monograph-sanctioned, category 1 sunscreens are to date limited to about 12 or 13 commercially available molecules. They fall into chemical classes of benzophenones. *p*-aminobenzoic acid (PABA) and its derivatives, salicylates, methoxycinnamates, phenylbenzimidazole sulfonic acid, a phenylacrylate, and the pigment titanium dioxide. When picking an active agent for hair, we expand our palette by quite a few more ingredients, because all active agents available have not had to go through the OTC review process for skin sunscreens, but do need some extensive toxicity testing to ensure safety. Besides the foregoing listed group, this would include the dibenzylmethanes, one of which, avobenzone, has been used in two new skin protection sunblocks under new drug applications (NDAs). Others are triazines, camphor and borneol derivatives, and benzotriazoles (Table 1). Among the nonmonograph sunscreens are some readily available benzophenones used to protect plastics, maintain the color of consumer products, and the like. Other absorbers, such as PEG-25-PABA, as well as the once available triPABA panthenol, seem to have been designed to service just this market for hair photoprotection (35–40). Sunscreen active agents' summary information should be reviewed to initially screen for choice of agent by both substrate target and product form (see Table 1).

In general, the benzophenones are moderately effective for absorbing both the short wavelength (UVB) and long wavelength (UVA). Table 1 is not a comprehensive list, but those cited here encompass the primary group used in personal care. Work at Henkel Dusseldorf originally reported at the 16th IFSCC demonstrates the superiority of benzophenone in protecting both hair color (melanin) and structure (keratin) in either a water-based setting lotion or an alcohol/silicone-based shine spray (41). Although the shine spray was actually dipped to coat the hair, both 2-hydroxy-4-methoxybenzophenone and its sulfonated water-soluble analogue were more effective than either the aminobenzoate ester, the phenylbenzimidazole sulfonate salt, or the dibenzoylmethane derivative. This efficacy was partially attributed to the broader-spectrum absorption, different types of binding to the hair, and photodegradation of active ingredients. The substituted benzophenones have varying, but mostly polar solvent, solubility unless they are sulfonated, in which case they become very water-soluble and strong acids. The sulfonates are usually a light yellow, but as they are neutralized by raising the pH with alkali or alkanolamines, they become progressively deeper yellow, but not significantly until they approach or go over about pH 6. The oil-soluble versions prefer more polar oils and solvents and, although not cited in Vaughan's review (32), probably have a higher solubility parameter (near 11) than many of the typical emollient esters employed in formulations

Table 1 Sunscreens

CTFA name	Alternative: OTC monograph or trivial name	Chemical name	Absorption range	Solvents
Benzophenone-1		2,4-Dihydroxybenzophenone	UVB, UVA	Alcohols, polar oils, glycols
Benzophenone-2		2,2′,4,4′-Tetrahydroxybenzophenone	UVA, UVB	Alcohols, glycols
Benzophenone-3	Oxybenzone	2-Hydroxy-4-methoxy benzophenone	UVB, UVA	Alcohols, polar oils
Benzophenone-4	Sulisobenzone	2-Hydroxy-4-methoxy benzophenone-5-sulfonic acid	UVB, UVA	Water, alcohols, glycols
Benzophenone-6		2,2′-Dihydroxy-4,4′-dimethoxybenzophenone	UVA, UVB	Slight in alcohol, glycols, polar oils
Benzophenone-8	Dioxybenzone	2,2′-Dihydroxy-4-methoxy benzophenone	UVB, UVA	Alcohols, polar oils, glycols
Benzophenone-9		Benzophenone-5,5′disodium sulfonate	UVB, UVA	Water, slight in glycols
Benzophenone-11		Mixture of 2,2′,4,4′di and trihydroxy-mono and dimethoxybenzophenones	UVA, UVB	Moderate in alcohol, glycols, polar oils

Benzophenone-12		2-hydroxy-4-*n*-octoxy benzophenone	UVB, UVA	Slight in alcohols, polar oils
PABA	PABA	*p*-Aminobenzoic acid	UVB	Alcohols
Octyl dimethyl PABA	Padimate-O	*N,N*-Dimethyl 2-ethylhexyl-*p*-aminobenzoic acid	UVB	Oils, alcohols
PEG-25 PABA		4-Bis(polyethoxy)-*p*-aminobenzoic acid polyethoxyethyl ester	UVB-weak	Water, alcohols, glycols
DEA methoxycinnamate	DEA methoxycinnamate		UVB	Water > pH 7, glycols, alcohols
Octyl methoxycinnamate		2-Ethylhexyl methoxycinnamate	UVB	Oils, alcohols
Etocrylene		2-Ethyl-2-cyano-3,3′-diphenyl	UVB—weak	Oils, alcohols
Octocrylene	Octyl cyanodiphenylacrylate	2-Cyano-3,3′-diphenylacrylic acid 2′-ethylhexyl ester	UVB—weak	Oils, alcohols
Butylmethoxydibenzoylmethane	Avobenzone	Butylmethoxydibenzoyl methane	UVA broad	Oils, alcohols
Isopropyl dibenzoylmethane		4-Isopropyldibenzoyl methane	UVA broad	Oils, alcohols
TEA salicylate		Triethanolamine salicylate	UVB—weak	Water
Octyl salicylate		2-Ethylhexyl salicylate	UVB—weak	Oils, alcohols
Homosalate	Homomenthyl salicylate		UVB—weak	Oils, alcohols

Table 1 Continued

CTFA name	Alternative: OTC monograph or trivial name	Chemical name	Absorption range	Solvents
Menthyl anthranilate		Menthyl-*o*-aminobenzoate	UVA—weak	Oils, alcohols
Octyl triazone		2,4,6-Trianilino-*p*-(carbo-2-ethylhexyl-1′-oxy)-1,3,5,-triazine	UVB	Slight to moderate in alcohols, polar oils
Phenylbenzimidazole sulfonic acid		2-Phenylbenzimidazole-5-sulfonic acid	UVB	Neutralized in water, alcohols, glycols
4-Methylbenzylidene camphor		3-(4-Methylbenzylidene)-camphor	UVB	Alcohols, oils, glycols
Camphor benzalkonium methosulfate	Mexoryl SO		UVB	Water
Drometrazole		2-(2′-Hydroxy-5′-methylphenyl)benzotriazole		Oils, alcohols
Octrizole		2-(2′-Hydroxy-5′-*t*-octylphenyl)benzotriazole		Oils, alcohols

Source: Refs. 36, 75, 76.

(near 8), and distinctly far from that of mineral oil (7.09). They occur as crystalline solids, so that rather than just being immiscible in formulations, sparingly soluble benzophenones slowly form needle-like crystals that are distinctly gritty and sharper in some aged samples of high-SPF sunscreens. This slow crystallization out of solution needs to be closely monitored at different storage temperatures and times. Effective levels in hair treatment may be lower than the monograph dosages for skin products (2–6%), but the foregoing researchers used 5% in their tests. When the solvent evaporates, the film remaining must be capable of keeping the active agent, dissolved, or some efficacy is lost, as well as product aesthetics. Ethanol, at 12.55 on the solubility parameter scale, is a good solvent, but other inert ingredients, such as emollients, humectants, nonvolatile and polymeric film formers, must provide a residue for the active chemical to remain on the hair shaft.

Further investigations that help confirm the efficacy of benzophenones include work done on hair as well as on some more remote, but related, substrates. National Starch's patent, previously mentioned, was based on polymerization of a benzophenone derivative methacrylate monomer, Permasorb MA, [2-hydroxy-4-(2-hydroxy-3-methacrylyloxy)propoxy benzophenone] into a film-forming hair spray resin effective in protecting artificial hair colorants (10,11). In actual use, however, film formation on the entire hair shaft from an aerosol is neither possible, without saturating the head, nor desirable; thus, use of this resin was limited, although it was commercialized. Saettone and associates (42,43) continue to evaluate alkyl benzophenone quaternary ammonium salts for irritation and antimicrobial properties to provide presumably safer and more keratin substantive agents. This necessary involvement of water for photodegradation was also recently reported in the experiments of Dubief (17).

The other class of broad-spectrum UVA sunscreens are dibenzoylmethane derivatives. Although promoted for more complete coverage from this longer-wavelength radiation, which might be expected to trigger free radical chain reactions when absorbed by chromophores in the hair, only Bernhardt et al. (41) have reported their evaluation on hair. In this test, it afforded limited protection against color fading, but was better than the PABA derivative that was tested. As with all actives, an assessment of the photodegradation characteristic of these sunscreens should be made. A review of active agents subject to these losses has been reported by Shaath et al. (44). Sunscreen degradation was solvent-dependent, but demonstrated the very good photostability of two benzophenones, poor photostability of a PABA and a cinnamate ester in some solvents, and variable moderate stability of six other active compounds.

The UVB sunscreens are the mainstay of skin protection products. They are more effective in preventing burns from low dosages of sun, less expensive, and more like cosmetic emollients in properties (solubility parameters for octyl dimethyl PABA = 9.01 and octyl methoxycinnamate = 9.10) (32). In emulsions, they also act similar to cosmetic esters, with required HLBs of 10 and 8, respectively (45). They are also readily dissolved in solvents and solubilized into aqueous systems. Some useful solubilizers include higher ethoxylates of oleyl alcohol, castor oil, and nonylphenol. Solvents include ethanol and volatile silicone. One currently marketed product is simply a solution of a low level of UVB sunscreen in cyclomethicone and alcohol, with silicone gum and fragrance. Besides these two popular active agents, other oil or solvent-soluble UVB actives include the diphenylacrylates, octocrylene—a viscous fluid—and etocrylene, a white powder. Both are lower in molar absorptivity than either the PABA derivatives or the methoxycinnamates by about half. To put them in perspective, though, they are more than twice as powerful as the salicylate esters. Other formulation considerations and cost may affect choice of active ingredient. Other PABA products have high absorptivity, depending on the dilution of the basic chromophore—PABA itself has high K value (absorbance in a 1-cm cell of a 1-mg/100 ml solution $\times$ 100) and PEG-25 PABA consequently has less. Octyl triazone, a solvent and oil-soluble PABA-type active is in between. As with the benzophenones, consideration of the finished product also needs to guide your choice of active agents. For example, is it better to solubilize octyl dimethyl PABA together with the fragrance and a high level of high-HLB nonionic, or can PEG-25 PABA with less nonionic be used to solubilize the fragrance? Stability, cost, ease of manufacture, and product performance all will guide the choice.

Noteworthy among efficacy studies is Nacht's work (16) with octyl dimethyl-PABA, in which he introduces the concept of hair protection factor (HPF). Analogous to skin SPF he uses hair tensile strength, from the yield slope of the stress–strain curve, for unprotected versus protected hair with nonirradiated values taken to normalize the data. His experimental findings included a confirmation of the photodamage to the hair, presumably by disulfide (—S—S—) bond cleavage, and effective protection afforded by minimal amounts of octyl dimethyl-PABA delivered in a shampoo (16). This may have promoted two major shampoo manufacturers to include this sunscreen in some of their recent introductions.

The other two available UVB classes of active agents, the camphor types and phenylbenzimidazole sulfonic acid, have high molar absorptivity. The oil-soluble Merck material is available as a crytalline solid. L'Oreal's patented quat is their own proprietary ingredient, listed as manufactured by Chimex, although it may be available in Europe. The patent limiting its use expires in 1994, so it may now be available for use. The phenylbenzimidazole sulfonic acid is available from a few vendors and needs to be neutralized with a base to provide solubility. This varies with the base used.

Other active components, both monomeric and polymeric, have been devised and synthesized, but have not been extensively commercialized. These include tetraalkyl piperidinyl derivatives (46) and triazine derivatives (47) as new monomers. Derivatives of current actives: benzenesulfonate quats of organic sunscreen acids (48) and PABA esters of benzophenones (49); and polymeric UV light absorbers (50–55).

VII. DELIVERY VEHICLES: FORMULATION GUIDELINES

Before the choice of active components, the formulator would choose the base for delivery. In most cases, photoprotection is an ancillary benefit, so that a functional hair care product prototype needs to be prepared first for sunscreen inclusion. General guidelines are available in a wide variety of publications, including documentaries of this book, specialty chemical supplier formularies, and various texts. Specific recommendations or suggestions can be made for the product development chemist to be aware of in formulating, but each formula and combination of ingredients, as well as order of addition can be important. Klein (56) gives some general formulation instruction for skin care sunscreens, in the first edition of this volume, that can also assist product development chemists with hair products. It goes without saying that specialty chemical vendors of both active and excipient ingredients also have extensive information for the formulator. As a starting point, in this review, generic formulas are presented based on conventional types and levels of ingredients. Since there are so many choices of raw materials in the armamentarium of chemicals available, with new ones introduced every week, the use of basic formulas with functionality may be a simpler way to inform.

A. Shampoos

Oil-soluble actives can be treated like fragrances (i.e., they may have to be solubilized and added as a final step). Alternatively, especially in larger amounts, they may need to be physically suspended by thixotropic gums or other suspending agents with high yield values, as in the current 2-in-1 and antidandruff shampoos. A representative so-called 3-in-1 shampoo preblends a hot mix of surfactants to add 2.0% of a PABA ester to a shampoo prototype. At low levels, premixing with liquid amides may help incorporation (57).

Water-soluble active substances are also possible additives, but being either salts or high ethoxylate nonionics, they will have an effect on associative, micellar, or electrolyte thickening (and thinning). Sulisobenzone has a wide pH range for solubility and stability, but with its very low p*K* it has a dramatic effect of viscosity, either in the acid or salt form. The other water-soluble sunscreen anions have limited solubility at an acidic pH and would probably find little use in these products because of high cost and low substantivity, and salts would have the same general effects. A recent European patent granted to Unilever (58), teaches the use of cationic guar gum to enhance the substantivity of sunscreens from shampoos. This specification as well as some of the basic work done by researchers at Clairol (59) should provide a good introduction to delivery of sunscreen from shampoos and conditioners (Table 2).

B. Conditioners

The thixotropic nature of fatty alcohol dispersions present in most conventional conditioners should be sufficient to suspend oil-soluble sunscreens, although, as in any emulsion, consideration must be given the total internal phase vis-à-vis required HLB, solubilities, and emulsifier package. Water soluble actives are possible here but observe possible incompatibility with cationic ingredients. Again recall the attempt by Van Dyk to market a cationic PABA derivative that had limited solubility, but good substantivity, for keratin fiber—a direct comparison with their octyl dimethyl-PABA was not conducted. Leave-in conditioners will have a better chance of delivering adequate sunscreen just as do the styling aids discussed in the next section. The same precautions need to be taken in looking for cationic incompatibilities and in observing if order of addition and pH adjustments can facilitate incorporation and enhance stability (Tables 3 and 4).

C. Styling Aids

Oil-solubles can be solubilized in Carbomer gels, but require large quantities of surfactant; alternatively, water-solubles can be used, but have a dramatic salt thinning effect on these gels. Other gelants, such as cellulosics and natural gums,

Table 2 Photoprotective Shampoo

Ingredients	Weight % range	
Water	qs to 100%	
Anionic surfactant	2.5–1.5% active	10–50% as supplied
Amphoteric surfactant	0–4% active	0–10% as supplied
Sunscreen—water- or oil-soluble	0.20–5%	
Alkanolamide	0–5%	
Polymeric/associative thickener	0–5%	
Preservative	qs	
Fragrance	qs	
pH adjuster	qs	
Electrolyte	qs	

Table 3 Photoprotective Conditioner

Ingredients	Weight % range	
Water	qs to 100%	
Cationic surfactant	0.5–10% active	0.5–40% as supplied
Quaternary ammonium salt		
Amphoteric surfactant	0–4% active	0–10% as supplied
Sunscreen—water- or oil-soluble	0.20–5%	
Emollient oil/ester/polyether/ silicone	0–5%	
Cetyl/stearyl fatty alcohol	0–10%	
Polymeric/cellulosic thickener	0–2%	
Nonionic surfactant	0–2%	
Preservative	qs	
Fragrance	qs	

may present better opportunities. A balance must be struck among levels of sunscreen, thickness, and clarity. Alternative low-viscosity cationic or cellulosic-thickened styling liquids are an appropriate vehicle. Cationic incompatibilities could be a concern with the anionics but, for one, sulisobenzone appears not to affect some water soluble or colloidal dispersions of cationic surfactants and polymers over a wide pH range. Note, however, a l'Oréal patent that may limit the use of this active (60). In the mousse category a Revlon patent describes the use of sunscreens with cationic surfactants and nonionic film formers (61) (Tables 5–7).

Table 4 Photoprotective Leave-In Conditioner

Ingredients	Weight % range	
Water	qs to 100%	
Cationic surfactant	0.25%–2% active	0.25–8% as supplied
Quaternary ammonium salt		
Cationic/nonionic film former	0–1% active	0–4% as supplied
Sunscreen—water-soluble	0.20–3%	
Nonionic/surfactant		
Wetting agent	0–2%	
Preservative	qs	
Fragrance	qs	

Table 5 Photoprotective Styling Gel

Ingredients	Weight % range	
Water	qs to 100%	
Acrylate or cellulosic thickener	0.2–2.0% active	0.2–8.0% as supplied
Film forming resin	0.5–4% active	0.5–20% as supplied
Sunscreen—water or solvent-soluble	0.10–2.0%	
Ethanol	0–70%	
Alkanolamine or alternative neutralizing agent	0–4%	Dependent on thickener and final pH necessary
Preservative	qs	
Fragrance	qs	

D. Hair Sprays, Spritzes, and Spray-On Shine Products

This whole gamut of finishing-touch products seems to be a likely candidate for sunscreen inclusion because of the solvent solubility and application being the last step in a hair care regimen. The experience of National Starch with discontinuing their sunscreen resin may be instructive. Possibly, the small droplets of resin that bind the hair to provide hold did not form a continuous film and, therefore, did not provide continuous protection or the perception of it. These product types are also poor candidates for application of sunscreen in that significant levels of oil-soluble actives would provide excess, unwanted plasticizing of resin. Solvent-soluble crystalline actives might flake off the hair

Table 6 Photoprotective Styling Liquid

Ingredients	Weight % range	
Water	qs to 100%	
Cellulosic thickener	0.2–2.0% active	
Film forming resin	0.2–4% active	0.2–10% as supplied
Sunscreen—water-soluble	0.10–2.0%	
Water-soluble cationic conditioner	0–2% active	0–8.0% as supplied
Alkanolamine or alternative in neutralizing agent	0–4%	Dependent on resin solution pH necessary
Preservative	qs	
Fragrance	qs	

Table 7 Photoprotective Mousse

Ingredients	Weight % range	
Water	qs to 100%	
Resin plasticizer	0–0.5%	
Film forming resin	Up to 2% active	Up to 10% as supplied
Cationic polymer/polyquaternuium	Up to 2% active	
Sunscreen—water- or solvent-soluble	0.10–2.0%	
Nonionic emulsifier/thickener/wetting agent	0.25–5%	
Ethanol	0–30%	
Alkanolamine or alternative neutralizing agent	0–4%	Dependent on resin used and solubility
Preservative	qs	
Fragrance	qs	
Hydrocarbon or alternative propellant	3–6%	Base/propellant ratio dependent on spray characteristics

or clogg valves and actuators. A potential skeleton formula in which solvent could be substituted for propellant in pump spray is provided (Table 8).

Looking again at our substrate, the hair, for possible targets to shield, we have the hair itself, as a keratinous fiber, either virgin or chemically treated with permanent waves or bleaches, the eumelanin-based natural coloration and the multitude of permanent and temporary dyes. All these dyes have light absorption of varying degrees in the visible spectrum and also the short and long UV—with many having a maximum near 300 nm—the same as many of the UVB actives. Melanin itself absorbs across many wavelengths and generates free radicals that may help explain the contribution of both water and oxygen in degradative processes—as in the rusting of iron. Lastly, keratin, with some amount of UV absorptivity, can break down at the disulfide bond linkages and cause the phototendering seen in natural fibers. Free radical generation from keratin amino acid irradiation has been substantiated. The degradative process caused by UVR has conclusively been proved in many keratinous materials, including wool and furs, as well as hair, mediated by oxygen, with water accelerating the reaction. Overall, benefits of particular treatments for photoprotection can be identified, and the mechanisms of photodamage and protection at least rudimentarily understood from the literature.

Table 8 Photoprotective Hairspray

Ingredients	Weight % range	
Water	qs to 100%	
Resin plasticizer	0–2.0%	
Film forming resin	2–8% active	2%–40% as supplied
Sunscreen—water- or solvent-soluble	0.10–2.0%	
Ethanol	0–70%	
Alkanolamine or alternative neutralizing agent	0–4%	Dependent on resin used and solubility
Preservative	qs	
Fragrance	qs	
Hydrocarbon or alternative propellant	10–40%	Base/propellant ratio dependent on spray characteristics

VIII. TEST METHODS

Test methods based on the research already reported can be revised to use natural sunlight or a standardized method of solar simulation (xenon arc solar simulators, mercury vapor tungsten bulbs, fluorescent lamps; 62–67) with measurement of hair structure by tensile testing (68,69), or color change by colorimetry (70,71). This two-pronged approach focuses on the two primary sites of damage by UVR—fiber elasticity and color. These and some further unusual methods using dyes, ESCA, IR, and extraction analyses provide the basis for formula modifications in the iterative process of product development and documentation for claim substantiation (72–74).

IX. REGULATORY AND SUMMARY

The most recent issue of the new Tentative Final Monograph, Proposed Rule for OTC Sunscreens does state the requirement that marketers of sunscreen for hair need to distinctly mark their packages as such, especially when making claims for protection and in proximity to the word "sunscreen." Since these are cosmetics, there is a need to substantiate the safety of final products, but the use of nonmonograph active substances is not proscribed. In the area of colors, cosmetic dyes need to be certified, and their use is also regulated by the Food and Drug Administration (FDA). However, hair dyes are specifically exempt from this certification. With this in mind how would someone classify optical brighteners when used for hair? Are they leuco dyes that do not need certification if used as hair dyes, or are they nonmonograph sunscreens if used to protect hair color? In any event, assurance of their use as safe would have to be determined.

In summary, the formulator should become familiar with available active agents and vehicles, choose a target based on a product idea, gather information about the chemistry involved in both damage and protection, from surveying the literature and experimentation. He or she can proceed to develop, concurrently evaluate performance and efficacy, test a battery of formulas, and eventually fine tune a finished product.

REFERENCES

1. Gantz, G. M. and Sumner, W. G. Stable ultraviolet light absorbers. *Textile Res. J.*, *27*:244–251 (1957).
2. Coleman, R. A. and Peacock, W. H. Ultraviolet absorbers. *Textile Res. J.*, *28*:784–791 (1958).
3. Milligan, B. and Holt, L. A. Ultraviolet absorbers for retarding wool photodegradation. *Polym. Degradat. Stabil.*, *10*:335–352 (1985).
4. Holt, L. A. and Milligan, B. Evaluation of the effects of temperature and UV absorber treatments on the photodegradation of wool. *Textile Res. J.*, *54*:521–526 (1984).
5. Crews, P. C. and Reagan, B. M. Evaluating UV absorbers for museum textiles. *Textile Chem. Color.*, *19*(11):21–26 (1987).
6. Klaic, B. and Friscio, V. Photodegradation of wool. *Tekstil*, *39*:717–720 (1990).
7. Goryacheva, M. M. Effect of ultraviolet irradiation on the chemical properties of the hair cover of sealskin. *Kozh.-Obuvn. Prom.-St.* [Russian] *18*(8):49–51 (1976): *CA*, *85*(24):178996x.
8. Baird, M. and Carr, C. M. Effect of photodegradation on the mechanical properties of wool and wool/polyester fabrics. *Textile Res. J.*, *61*:210–213 (1991).
9. Dunn, L. A. and Weatherall, I. L. Stress relaxation properties of wool fibers exposed to sunlight. *Proceedings of the 8th. International Wool Textile Research Conference.* Vol. 1 Christchurch, New Zealand (edited by G. H. Crawshaw), February 7–14, 1990.
10. Skoultchi, M. and Koehler, F. T. Hair spray compositions containing an ultraviolet absorbing film forming copolymer. U.S. patent 3,445,566. May 20, 1969.
11. National Starch & Chemical Corporation. Bridgewater, N. J. 08807. Datasheets: Permasorb MA, Resin 78-6121, Resin 28-3307.
12. Klein, K. Substantive cationic sunscreen—Escalol 537Q, Van Dyk Co., Belleville, NJ 07109, 1986.
13. Frederick, J. E., et al. Solar ultraviolet radiation at the earth's surface. *Photochem. Photobiol.*, *50*:443–450 (1989).
14. Frederick, J. E. Ultraviolet sunlight reaching the earth's surface: A review of recent research. *Photochem. Photobiol.*, *57*:175–178 (1993).
15. Kalopississ, G., et al. Anti-solar agent and compositions containing the same. U.S. patent 4,061,730. December 6, 1997.

16. Nacht, S. Sunscreens and hair. *Cosmet. Toilet.*, *105*(12):55–60 (1990) [originally published in *Sunscreens* (edited by N. Lowe and N. A. Shaath). Marcel Dekker, New York 1990:341–355.]
17. Dubief, C. Experiments with hair photodegradation. *Cosmet Toilet.*, *107*(10):95–102 (1992).
18. Deflandre, A., et al. Photoaging and photoprotection of natural hair. *16th IFSCC Congress, Cosmetic Science in the 1990s and Beyond*, New York City, Vol. 2. Oct 8–11, 1990, pp. 197–208.
19. Robbins, C. R. *Chemical and Physical Behavior of Human Hair*, 2nd ed. Springer-Verlag, New York, 1988.
20. Zviak, C., ed. *The Science of Hair Care.* Marcel Dekker, New York, 1986.
21. Feughelman, M. Physical properties of alpha-keratin fibers. *J. Soc. Cosmet. Chem.*, *33*:385–406 (1982).
22. Creed, D. Photophysics and photochemistry of near UV absorbing amino acids. I. Tryptophan and its simple derivatives: II. Tyrosine and its simple derivatives: III. Cystine and its simple derivatives. *Photochem. Photobiol.*, *39*:537–562; 563–575; 577–583 (1984).
23. Smith, G. Singlet oxygen produced by UV irradiation of wool keratin. *J. Photochem. Photobiol. B: Biol.*, *12*:173–178 (1992).
24. Pigault, C. and Gerard, D. Selective degradation of amino acids photosensitized by tryptophan in polypeptidic structures. *Photochem. Photobiol.*, *50*:23–28 (1989).
25. Strasheim, A. and Buijs, K. Infrared study of the oxidation of the disulfide bond in wool. *Biochim. Biophys. Acta*, *47*:538–541 (1961).
26. Robbins, C. R. and Bahl, M. K. Analysis of hair by electron spectroscopy for chemical analysis. *J. Soc. Cosmet. Chem.*, *35*:379–390 (1984).
27. Kollias, N. et al. Photoprotection by melanin. *J. Photochem. Photobiol. B: Biol.*, *9*:135–160 (1990).
28. Kaidbey, K. H. et al. Photoprotection by melanin—a comparison of black and Caucasian skin. *J. Am. Acad. Dermatol.*, *1*:249–260 (1979).
29. Rieger, M. M. Merits of melanin. *Cosmet. Toilet.*, *107*(5):41–42 (1992).
30. Johnson, B. E. Changes in sunburn and mechanism of protection. *J. Soc. Cosmet. Chem.*, *29*:31–44 (1978).
31. Wolfram, L. J. and Albrecht L. Chemical and photobleaching of brown and red hair. *J. Soc. Cosmet. Chem.*, *38*:179–191 (1987).
32. Vaughan, C. D. Solubility effects in product, package, penetration and preservation. *Cosmet. Toilet.*, *103*:47–69 (1988); Using solubility parameters in cosmetic formulation. *J. Soc. Cosmet. Chem.*, *36*:319–333 (1985).
33. Agrapidis-Paloympis, L. E. and Nash R. A. Effect of solvents on the ultraviolet absorbance of sunscreens. *J. Soc. Cosmet. Chem.*, *38*:209–221 (1987).
34. 21 CFR Parts 352, 700, 740. Sunscreen drug products for over-the-counter human use: Tentative final monograph: Proposed rule. *Fed. Reg.*, 58:28194–28302 (1993).
35. Uvinul. grades, UV absorbers for cosmetic products. Technical Information ME 266e, BASF Fine Chemicals (March 1993).
36. Shaath, N. A. Encyclopedia of UV absorbers for sunscreen products. *Cosmet Toilet.*, *102*(3):21–37, (1987).

37. Eusolex. UV filters for cosmetics. EM Industries, New Jersey, USA—E Merck, Darmstadt, Germany.
38. Neo Heliopan sunscreen filters. Haarmen and Reimer, Springfield NJ.
39. Katoscreens—technical bulletins. KATO Worldwide Ltd., Mount Vernon, NY, 10553.
40. Escalol technical information. ISP, Wayne NJ.
41. Bernhardt, P., Giesen, M., et al. UV filters for hair protection. *Int. J. Cosmet. Sci.*, *15*:181–199 (1993).
42. Saettone, M. F., et al. Substantivity of sunscreens: An appraisal of some quaternary ammonium sunscreens. *Int. J. Cosmet. Sci.*, *8*:99–109 (1986).
43. Saettone, M. F., et al. Substantivity of sunscreens: Preparation and evaluation of some quaternary ammonium benzophenone derivatives sunscreens. *Int. J. Cosmet. Sci.*, *10*:99–109 (1988).
44. Shaath, N. A., et al. Photodegradation of sunscreen chemicals: Solvent considerations. *Cosmet Toilet.*, *105*(12):41–44, (1990).
45. The HLB system: A time saving guide to emulsifier selection. ICI surfactants, Wilmington DE, May 1992.
46. Raynor, R. J. Selected 2,2,6,6,-tetraalkyl-4-piperidinyl derivatives and their use as light stabilizers. U.S. patent 4,751,073. June 14, 1988.
47. Richard, H., et al. *S*-Triazine derivatives carrying benzalmalonate substituents, and cosmetic UV screening compositions. U.S. patent 5,236,698. August 17, 1993.
48. Conner, D. E. Benzene sulfonate quaternary ammonium salts of organic sunscreen carboxylic acids. U.S. patent 4,680,144. July 14, 1987.
49. Haffey, J. G., et al. Photoprotection compositions having reduced dermal irritation. U.S. patent 5,087,445. February 11, 1992.
50. Akelah, A. Biological applications of functionalized polymers—a review. *J. Chem. Technol. Biotechnol.* *34A*:263–286 (1984).
51. Karg, G. Sunscreen method and composition containing film forming polyamide material. U.S. patent 3,895,104. July 15, 1975.
52. Ciaudelli, J. P. Substantive polymeric sunscreen composition. U.S. patent 3,864,473. February 4, 1975.
53. Jacquet, B., et al. Anti-solar polymers, method of making the same and cosmetic compositions containing the same. U.S. patent 3,980,617. September 14, 1976.
54. Gerecht, J. F., et al. Highly substantive sunscreening agents. U.S. patent 4,004,074. January 18, 1977.
55. Forestier S., et al. Polyaminoamides which filter ultraviolet radiation, process for their preparation and their use in the protection of the skin and hair. U.S. patent 4,004,074. September 12, 1989.
56. Klein, K. Formulating sunscreen products. *Sunscreens* (edited by N. Lowe and N. A. Shaath). Marcel Dekker, New York, 1990, pp. 235–266.
57. Beerling, J. Towards multifunctional shampoos. *Drug Cosmet. Ind.* April: 30–36; 92 (1994).
58. Birtwhistle, D. H. Shampoo composition. EP 0 386 898 B1. September 1, 1993.
59. Berthiaume, M. D. and Jachowicz, J. The effect of emulsifiers and oil viscosity on deposition of nonionic silicone oils from oil-in-water emulsions onto keratin fibers. *J. Colloid Interface. Sci.*, *141*(2): (1991).

60. U.S. 5,240,695. Dubief, C, et al. use of 2-hydroxy-4-methoxybenzophenone-5-sulfonic acid or its salts for protecting the hair against environmental attacking agents, and especially against light and process for protecting the hair using the compound. U.S. patent 5,240,695. August 31, 1993.
61. Ciaudelli, J., et al. Sunscreen composition for hair protection. U.S. patent 4,567,038. January 28, 1986.
62. Sayre, R. S., et al. The correlation of indoor solar simulator and natural sunlight. *Arch. Dermatol.*, *114*:1649–1651 (1978).
63. Solar Light Co., Philadelphia PA—solar simulators, radiometers.
64. Oriel Corp., Stratford CT—solar simulators, radiometers, optical components.
65. Heraeus DSET Laboratories, Inc. Phoenix AZ—solar simulators, test equipment.
66. Q-UV unit, Q-Panel Co., Cleveland OH—fluorescent light boxes, test equipment.
67. International Light Inc., Newburyport, MA—radiometers, solar simulators.
68. Diastron. Andover Hampshire, U.K. and Broomall, PA—tensile test equipment.
69. Instron Corp., Canton MA—tensile testing equipment.
70. MacBeth, division of Kollmorgen Instruments Corp., Newburgh NY—colorimeters, lighting equipment.
71. Minolta Instrument Systems, Minolta Corp., Osaka, Japan; Ramsey NJ—colorimeters, radiometers.
72. Togyesi, E. Weathering of hair. *Cosmet Toilet.*, *98*(10):29–33 (1983).
73. Sandhu, S. S., et al. A sensitive technique using dansyl chloride to assess hair damage. *J. Soc. Cosmet. Chem.*, *40*:286–296 (1989).
74. Pande, C. M., et al. Hair photodamage—measurement and prevention. *J. Soc. Cosmet. Chem.*, *44*:109–122 (1993).
75. *CFTA Cosmetic Ingredients Handbook.*
76. *CFTA International Cosmetic Ingredient Directory.*

20

Melanins

Miles R. Chedekel and Lisa Zeise
Mel-Co®, Orland, California

I. INTRODUCTION

There is no doubt that melanin in human skin attenuates the penetration of ultraviolet (UV) and visible light. The higher the concentration of melanin, the more the attenuation. This is a simple photochemical and photophysical fact. However, the important question is whether or not this attenuation results in human photoprotection. There are overwhelming data demonstrating that melanin in the skin confers protection from sunlight-induced skin cancer, and the greater the level of melanin, the greater the level of protection. However, this same level of melanin, which protects so well against skin cancer, is less effective against protection from sunlight-induced erythema and skin aging. It is the goal of this chapter to examine the photochemical and photophysical phenomena that are operative in human skin as a result of the presence of melanin, and to formulate potential explanations for the type and level of protection observed.

A. Melanin

Before evaluating the photochemistry and photophysics of melanin and its precursors, it is necessary to define and characterize the biopolymer melanin. As it is currently used, melanin is not a single unique polymer. Rather, melanin is a generic term that refers to a group of biopolymers. The *Merck Index*, refers to "melanins" as a group of polymeric "pigments" subdivided into classes of allomelanins, eumelanins, and phaeomelanins, according to the monomer units incorporated during biosynthesis. The situation is further complicated by the fact that the given classes of melanins are not homopolymers of a single monomer unit, but rather, are composed of multiple monomer units. In some cases, in situ chemical modification takes place after polymerization has occurred and, in other cases, it is a matter of the existence of multiple monomer pools during polymerization. In either event, the chemical composition and physical properties of a given class of melanin are highly dependent on the milieu from which it was formed. For example, one would not expect that eumelanin formed in the controlled environment of a melanosome would have the same composition and properties as eumelanin prepared in a buffered solution containing eumelanin precursors and the enzyme tyrosinase. Neither is right or wrong, they are simply different members of the eumelanin class.

An analogy to plastic is appropriate. Plastic is not a unique term, a plastic has many structures and many monomers; polyethylene, polypropylene, polymethyl methacrylate, and the like, all are plastics. Each class of plastic exhibits quite different properties. Even within a given class of plastic (e.g., polyethylene), such factors as level of cross-linking, average polymer chain length, cross-linking agents, and initiators give rise to forms of polyethylene exhibiting quite different properties (i.e., high-density polyethylene, low-density polyethylene, linear polyethylene, polyethylene microspheres, and low molecular weight soluble polyethylenes, to name a few). So it is with melanin.

Although we have drawn the analogy between melanins and plastics, it is not the intent to include Teflon in this analogy. Teflon is stable to treatment with heat, strong alkalis, and strong acids. This is not true of melanins. We now know that the use of such treatments in the ''isolation'' and ''purification'' of melanins results in significant structural modification of the polymer. Once again, this is not a case of right or wrong, simply different. The use of well-defined melanin preparations in scientific studies is always justified. However, the use of poorly or undefined melanin preparations is never justified; results and conclusions from such studies are of no use, and must be repeated using defined melanin preparations. Authors, researchers, and readers are cautioned to be specific in their choices and descriptions of melanins, and to limit their conclusions to the specific melanin polymer used and not, a priori, to an entire class of melanin.

Eumelanins and phaeomelanins are the two classes of melanins present in human skin. The relative concentrations in the epidermis of these two classes of melanin may or may not vary according to skin type. It is also possible that a mixed eumelanin—phaeomelanin copolymer may exist in some, all, or none of the skin types. For didactic purposes, the discussion of the photochemical and photophysical properties of epidermal melanins in this chapter will be limited to the eumelanin class.

B. Eumelanin

The highly cross-linked polymeric, dark brown to black, insoluble, nitrogenous pigments produced enzymatically from the amino acid tyrosine are known as eumelanins; the early stages of eumelanin biosynthesis are depicted in Figure 1. Eumelanins are composed of two distinct fractions that include

1. A peptide coating: The composition and profile of this peptide coating is unique for each eumelanin, and may be used as a fingerprint for identification purposes.
2. A chromophore backbone: A highly cross-linked, irregular polymer composed of a mixture of four major classes of monomer units; the monomers of the eumelanin backbone are depicted in Figure 2.

II. MELANIN AS A SUNSCREEN

The theme of this chapter is actually in vivo, in vitro, in sunscreens. For the purposes of this chapter, in vivo and in vitro are expendable. If it does not work in a sunscreen, then the topic of this chapter is worthless.

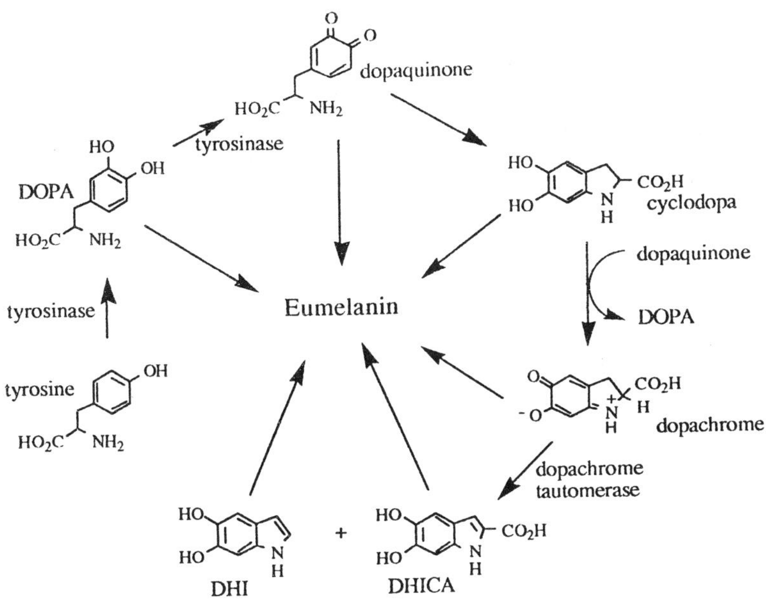

Figure 1 Early stages of eumelanin biosynthesis.

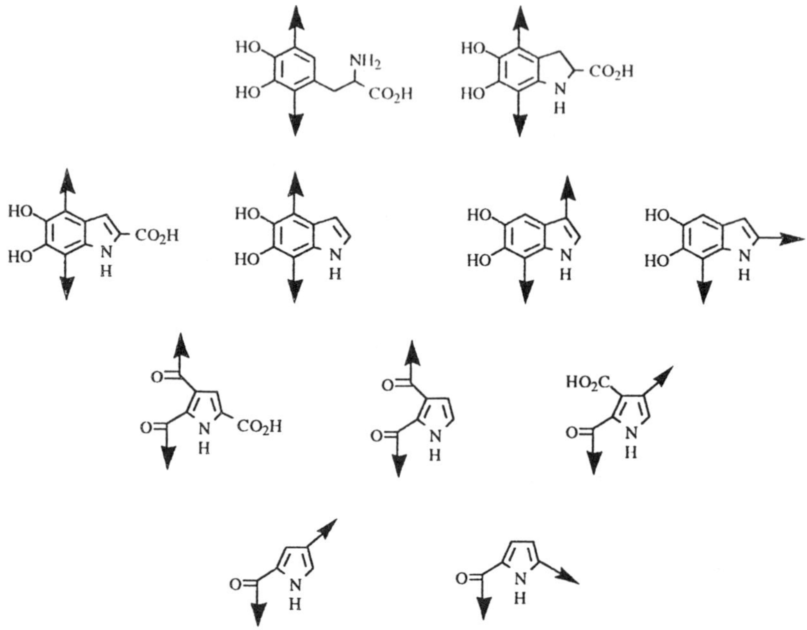

Figure 2 Eumelanin monomer units.

In dealing with the possibility of melanin in sunscreens, one has to first determine that commercial quantities of melanin are available. And one has to ask: "What are commercial quantities?" One should be aware that, even with a very minor success of melanin in sunscreens, we are talking about somewhere on the order of 25–50 tons of melanin per year. So, we had better have a reasonable source.

The three commercial sources of eumelanin are given in Table 1. There is a natural source, isolated and purified from the ink sac of cephalopods. There is a biosynthetic melanin, prepared from a genetically engineered microbe. And there is a synthetic melanin, produced from chemical precursors. This chapter will focus on these sources.

Table 1 Commercially Available Eumelanins

Natural: Isolated and purified from cephalopod ink sacs (e.g., *Sepia*)
Biosynthesis: Produced and isolated from genetically engineered organisms
Synthetic: Produced by chemical polymerization of one or more putative precursors

Now, to discuss potential modes of photoprotection as given in Table 2; again in sunscreens, not in vivo or in vitro. Light absorption is the simplest, followed by light scattering and free radical scavenging. Light absorption is due to the molar extinction coefficient. Extinction coefficients in the UV spectral region for soluble melanins are low. Concentrations of 3–10% are required to achieve significant absorption of UV light. Melanins have a high color index, and concentration levels of 2–7% are too dark to be of commercial significance.

When considering scattering, one must consider particle size and particle concentration. One also has to consider refractive index, which, unfortunately, is not known for melanin. For free radical scavenging, one has to consider the particle size, an aspect that will be covered toward the end of the chapter.

Most of the time, in commercial products, we are talking about a particle. But, when dealing with a particle, we are not really dealing with the same type of Lambert–Beer law as in solution. One cannot simply expect to increase the particle concentration and obtain higher and higher absorbances. In particle terms, it is the concentration of particles and, as a certain concentration is reached, where there are many particles per unit volume, you begin to develop multiple scattering events. This tends to be very beneficial. We can measure extinction of particles by a simple particle spectrophotometer in which we exclude all of the scattered light and look only at the light that penetrates straight through. Figure 3 is a schematic for measuring the extinction of particles. And the idea here is simply the total extinction of light, realizing that we have scattering and absorption occurring simultaneously.

Figure 4 is an extinction spectrum of melanin obtained on this type of a particle spectrophotometer. This is the extinction spectrum for highly dispersed *Sepia* melanin. These *Sepia* melanin particles were sized at 161 ± 10 nm, a fairly tight distribution of particles and at a very low concentration, but a very

Table 2 Potential Modes of Photoprotection

Light absorption
Molar extinction coefficient
Concentration
Light scattering
Particulate versus soluble melanin
Particle size
Particle concentration
Free radical scavenging
Concentration of soluble melanin
Particle concentration
Surface area of particulate melanin

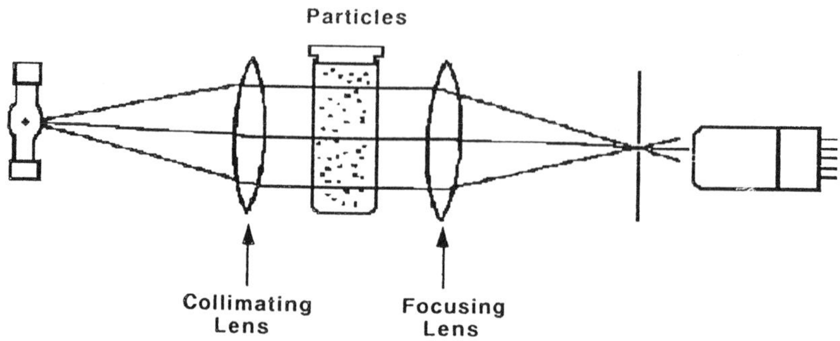

Figure 3 Measurement of extinction; the light source is on the left and the detector is on the right. The beam is focused through a pinhole before entering the detector.

good dispersion. These particles do not fall out of solution, and we observed a significant extinction on these; in fact, quite a bit of extinction for a 0.001% suspension. However, when discussing scattering, there are a lot of terms used (i.e., Mie scattering, Rayleigh scattering, forward scattering, back scattering, and multiple scattering, to name a few). These terms must be examined and understood relative to use with a product containing particles. Forward scattering and back scattering seem simple; however, multiple scattering is probably the key. The more scattering events that occur, the better a substance will become at

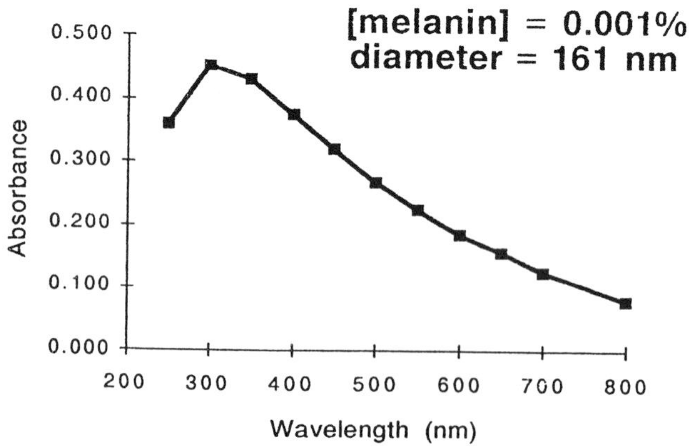

Figure 4 *Sepia* melanin extinction spectrum. The concentration of melanin is 0.001%. The diameter of the melanin particles is 161 nm.

causing extinction of light. When looking at scattering, we have to look at particle size.

Particle size versus the angle of light scattering is depicted in Figure 5. A refractive index of 1.5 was used to calculate the scattering one would expect from a spherical particle with an incident beam of 300 nm. One can see that if the particle diameter is ≥ 300 nm, predominantly forward scattering is observed. The particles have to be very small (i.e., on the order of one-tenth the wavelength of the light that is being used) before symmetric scattering occurs. However, both forward scattering and back scattering are still occurring. And we can see that melanin particles are clearly going to be forward scattering a lot of light. How can this possibly be of any use?

Even if a particle has a very small probability of absorbing light, given enough chances, the particle will absorb the light. If you can achieve a multiple scattering situation, as depicted in Figure 6, you can see that if you can achieve the light bouncing around between enough particles, eventually you have a very good chance of having the light absorbed after 10, 20, or 100 events. The other possibility is that, after multiple scattering events, the light will simply come right back out or off to the side in a direction that will also result in extinction of the light and be very useful. This type of multiple scattering and absorption or multiple scattering and back scattering is depicted in Figure 7 and is a very efficient way for very tiny particles, say 30-nm particles, to attenuate ultraviolet light.

What about forward scattering? This would at first seem to be counterproductive; and, in one sense, it is. Forward scattering in and of itself is probably

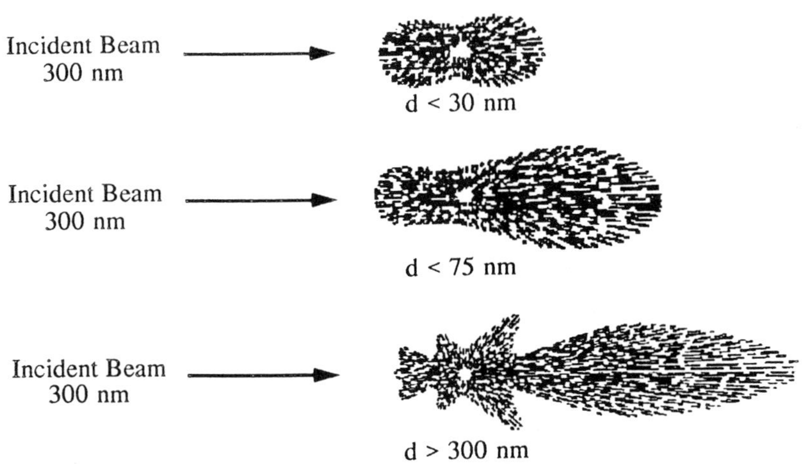

Figure 5 Particle size versus the angle of light scattering. The refractive index is 1.5.

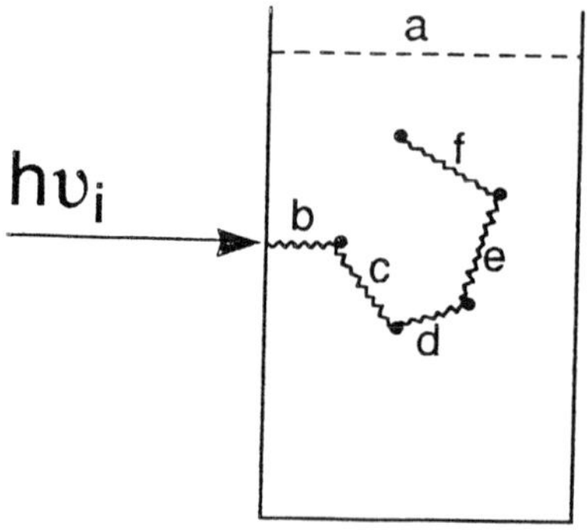

Figure 6 Multiple scattering/absorption. If the particles have a finite absorption cross section, multiple scattering/absorption will lead to efficient extinction.

useless. However, in conjunction with existing sunscreens, chromophores, or absorbers, multiple scattering has a very interesting effect. If one looks at Figure 8 as a depiction of a thin film of a sunscreen product and one looks at it without any particles or without any scattering, the light goes straight through, with a pathlength *a*. There is a certain concentration of absorbers and, therefore, you can calculate the absorption based on a traditional Beer–Lambert type calculation. Let us say that in this same suspension we have particles. And now the light bounces in a forward-scattering manner off one or more particles, but still goes straight through. How does this help? Clearly, we have extended the ef-

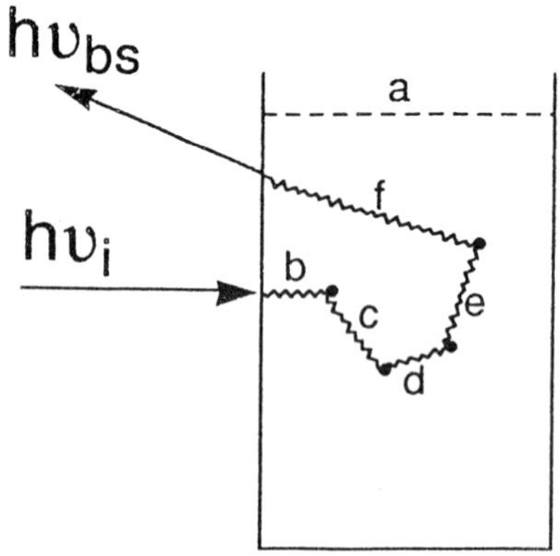

Figure 7 Multiple backward scattering. Backward scattering by a high concentration of nonabsorbing particles leads to efficient extinction of turbid suspensions.

absorbance = extinction • pathlength • concentration

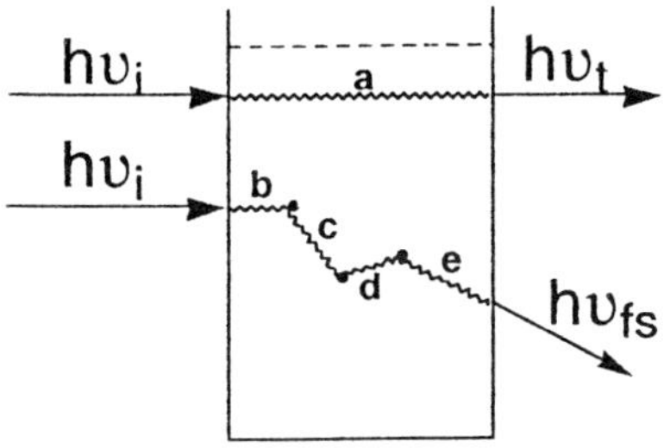

Figure 8 Multiple forward scattering. Forward scattering increases the optical path length and thus increases the absorption of a turbid suspension containing soluble chromophores.

fective pathlength; so, if there are one or more scattering events, the pathlength can be increased by 1.5, 2, or 3 times *a*. Therefore, the same concentration of chromophore in this solution will be much more effective. So, forward scattering can be effective. This is also a very good technique to try to exploit in the situation of a commercial sunblock. If you could put particles into sunblocks, you could make chemical sunblocks that are known to work much more effective at the same concentration.

Does this work in practice? Figure 9 is the absorption measured for a sun protection factor (SPF)-4 product straight off of the shelf. This product contained padimate O. Then, a small percentage of highly dispersed melanin particles was added. The result, the entire absorption curve was boosted evenly. This is an example of extension of the optical pathlength. All we have really done is make the same thickness of film more effective and, in this example, the measured SPF went from 4 to 8. This boost was observed by just putting in a small percentage of particles, but they have to be dispersed, dispersion is the key. Clearly, one big particle is not nearly as effective as hundreds of tiny particles. Figure 10 depicts this concept with a particle concentration of 1% (wt/wt). Here, the 0.15-μm particles are very well dispersed and are very effective at attenuating UV light. However, if these same 0.15-μm particles are formulated in such a way that they clump together and aggregate, so that the effective particle size is 10 μm, the same weight of melanin does absolutely nothing. At a particle size of 0.03 μm, the attenuation of UV light may be even more effective.

A. Conclusions on Absorption and Scattering

Nonparticulate melanin does not scatter UV light, although it may be efficient at absorbing light. However, the extinction coefficient is not very high. Up to

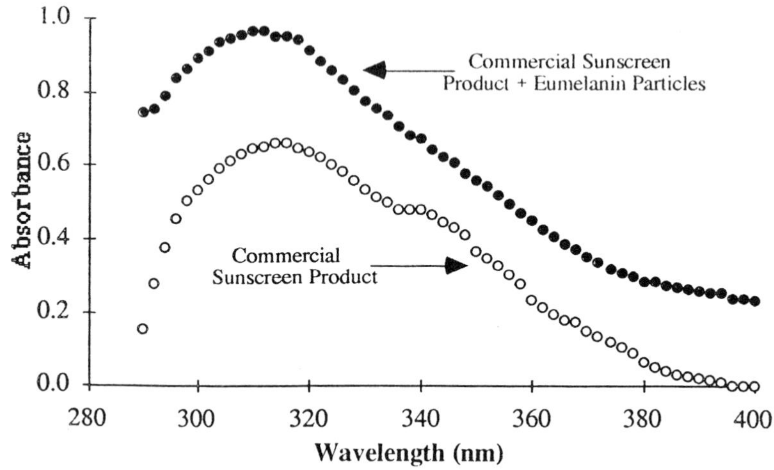

Figure 9 Particles and optical path length. Sunscreen formulations with and without particulate melanin. The commercial sunscreen formulation was an SPF 4 preparation containing padimate O.

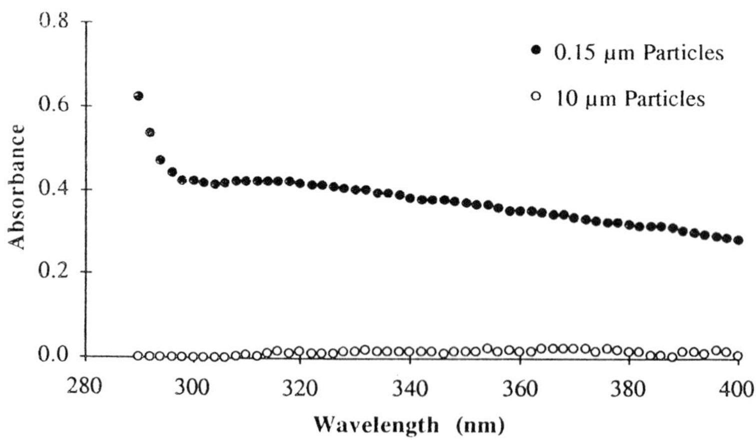

Figure 10 Particle size and UV absorption. The concentration of melanin is 1% (wt/wt).

5% melanin by weight is required to obtain appreciable absorption, and that type of a product would be very, very dark, so dark that it may be unappealing. If we use particles, a much smaller weight percentage is required. The particle melanin attenuates UV light tremendously, primarily by multiple scattering, so a little bit of melanin in the product does not do much good, a lot of particles are required to exert an effect. Obviously, the smaller the particles, and the better they are dispersed, the better they can potentially be as a multiscattering entity. Again, the important idea is dispersion. This concept is often overlooked, and people have put melanin in a formulation and said that it does not work. Most of the time if you take a look, you can see that they have clumps of melanin at the bottom of their formulation, which are absolutely useless. Nature knows this and formulators have since learned it.

The following are our conclusions on absorption and scattering:

Nonparticulate melanin is not efficient at absorption or scattering of UV light.
Particulate melanin attenuates UV light predominantly by a multiple scattering mechanism.
Particulate melanin can significantly increase the effective optical pathlength of UV light and thus can increase the efficacy of traditional soluble sunscreens.
Particle size, concentration, and dispersion must be controlled to optimize extinction by melanins.

III. MELANINS AND FREE RADICAL SCAVENGING

When we are dealing with melanin and free radical scavenging, there are many studies. Melanin pigments have been reported to exhibit superoxide dismutase (SOD) activity, inhibit lipid peroxidation, and react with active oxygen species. However, we must examine three distinct cases of melanins' ability to neutralize free radicals (i.e., in vivo, in vitro, and in sunscreens). Most of the studies were performed in vitro and in vivo, and there are some reports that melanin does trap free radicals in sunscreens. Of that, there is no doubt. The question is, "Can it be used in a product?" The answer is, "yes"; but we have to be careful how we use it and how we say it is being useful. Clearly, there are some questions that have to be taken into account. These questions include the following:

What wavelengths are most effective at producing free radicals?
In which layer of the skin are free radicals formed?
Is oxygen concentration important in the formation of free radicals?
What are the migratory habits and lifetimes of free radicals in the skin?
Can topically applied melanins neutralize or inhibit free radical damage?

Although intuitively, we may all know the answers to these questions, if we go back and take a hard look at the literature, many of these questions have to be reexamined. What wavelengths are most effective at producing free radicals? Clearly, the more energetic the wavelength, the more likely that free radicals are being produced. We have to look at the quantum yields.

A piece of information that we need to know if we are going to assume that melanin is going to be protective in a product is, "Where are the free radicals produced?" Melanin in a product has a marvelous chance of neutralizing free radicals produced in the stratum corneum, perhaps even in the upper layers of the epidermis. If free radicals are produced predominantly in the dermis, then a melanin product may not have a very high chance of protecting against them.

However, one also has to ask the question, "What does oxygen concentration do?" And as you start to shine light on skin, what are the equilibrium changes in oxygen concentration? Also, can a melanin product applied to the top of the skin alter this? Clearly, we have seen that, in many cases, if you can limit or deplete the oxygen concentration in the area of a potential photochemical event for free radical formation, you can significantly attenuate that event. If a melanin product can lower the oxygen flux entering the upper layers of the skin, you might be able to significantly decrease free radical production. Another question is, "What are the migratory habits and lifetimes of free radicals in the skin?" Are radicals produced on the skin going to go up or down or in both directions? And, how far down in the skin do they have to be produced before there is not the proverbial snowball's chance in hell that they are going to get to the surface? Clearly, melanin must work directly by reacting with the radical if it is going to neutralize it directly. Therefore, the radical has to reach the melanin. In products, the idea is not to force the melanin down into the skin, but rather the goal is to keep the melanin on top of the skin. Maybe this should be reevaluated.

A final question is, "Can topically applied melanins neutralize or inhibit free radical damage?" There is excellent evidence from several companies that have studied this; the answer is yes, it can. The exact mechanisms are unknown and need to be studied. If we are looking at a particulate melanin, which we usually are, what are the parameters of a particle interacting with a free radical? There are two: the particle can directly neutralize the free radical on contact; and diffusion-controlled free radical neutralization. However, a particle has an advantage over a solubilized melanin in that the particle can first adsorb the free radical and subsequently neutralize it. As we all know, melanin adsorbs many chemicals, many free radicals, many toxins, and many drugs. Melanin particles can adsorb free radicals without actually neutralizing them, and then the free radical can migrate around, until it comes to the correct site on the particle for surface neutralization. Thus, a very tiny particle with a free radical-neutralizing capability, and an adsorbing capability can be a most effective free radical sponge. However, we also have to ask ourselves, "What kind of particle?" Once again, the smaller the better. It is unlikely that a free radical is going to wiggle into the inside of a big particle. For the weight of melanin used, we want the highest surface area possible, and clearly the highest surface area/volume ratio is a trait of very small particles. Studies on particles and neutralization have shown that the smaller the particle, the better the neutralization.

A. Conclusions About Melanins and Free Radical Photoprotection

The detailed kinetics and mechanisms for neutralization of free radicals by melanins are not known. These unknown issues will, we hope, foster a reasonable amount of research on this topic. But we do know that submicron melanin particles are ideally suited for adsorption and neutralization of free radicals. Melanin particles can, if given a chance, act as free radical sponges. Many sunscreen ingredients and many of the fragrances when irradiated, will produce free radicals. Clearly, melanin in a formulation can protect against these free radicals by scavenging and neutralization. Free radicals produced in the stratum corneum and epidermis by sunlight can be neutralized by topically applied melanins. If the transport of these radicals from the top of the skin is facilitated by the formulation, then melanins can be most efficient at also pulling these out of the skin. We believe that products continue to be coming out that are making these claims. Some of the data in support of the free radical-trapping claims are compelling. However, we need to understand the mechanism and understand exactly how good this type of a mechanism is.

IV. PRODUCT APPROACHES

Several approaches have been taken to develop products. One unique approach, patented by Advanced Polymer Systems in the United States, is to make very tiny melanin-scattering centers suspended in an inert matrix. The melanin-containing particle is spherical, with a diameter reported to be 6 μm. What this approach has done, is to create inside this particle, little scattering centers of melanin, thousands of precipitated, nanometer-sized melanin particles, ideally suited for multiple scattering. Thus, when light enters this matrix, it is not likely to escape. This approach might be termed the light-trapping approach.

Another approach is to simply use melanin nanoparticles. Melanin particles of about 30 nm would give the ideal scattering profile for obtaining symmetric scattering. If suitably suspended in a lotion, these melanin particles could also be very effective at multiple scattering, absorption, and free radical neutralization. This approach is being taken.

A third approach is the use of soluble melanins that are dissolved just as another ingredient in the sunscreen. These are being used mainly for protection by free radical neutralization.

Finally, particles coated with melanin or melanic pigments have been cited in both the open and patent literature. This is understandable because a melanin coating on titanium dioxide (TiO_2) changes the color from white to tan and greatly improves the dispersibility of the TiO_2. This means that less TiO_2 may be used to obtain the same effect. An example of the effect of melanin as a coating on TiO_2 is given in Figure 11 and is described in Table 3.

At these levels, the melanin content is too low to be effective as either an UVA or UVB sunblock. Yet the data clearly depict an increase in the calculated SPF from 4.1 to 12.1, and an increase in the calculated UVA erythemal protection factor from 3.3 to 7.9 when TiO_2 and melanin are used together. Under suitable formulation conditions, melanin specifically coats the TiO_2 in situ and acts as a dispersive coating, thereby resulting in higher SPF and UVA protection values. Melanin also reduces the whitening problem associated with high concentrations of TiO_2.

V. CONCLUSIONS

Melanins are present in the human epidermis as submicron particles, and they are capable of conferring photoprotection. This photoprotection is due to both light scattering and free radical neutralization. Scattering seems to be the most prevalent mode of action. Topically applied melanins are being used in vehicles, and they have been demonstrated to confer photoprotection by both scattering and free radical neutralization. However, the key is to have a high dispersion of melanin particles and to give them a chance to work. If they aggregate or adsorb all of the other ingredients in the sunscreen and fall down to the bottom in a big clump, then the entire sunscreen is inefficient. Clearly, it is a matter of delivery, just as it is genetically in our skin, the persons who have melanin evenly disbursed throughout more layers of their skin have the best protection against sunlight, compared with those who just have melanin as one, flat little layer of a thin, thin film. So, high dispersion of the melanin particles and ease of free radical migration into the topically applied vehicle are key factors in the success of melanin particles in sunprotective products.

PATENT LITERATURE

Sources of Melanin

1. Method for enhanced melanin production
   Assignee: Biosource Genetics Corporation
   Patent numbers: WO 9,200,373; AU 9,182,954; EP 547,065; JP 6,500,011
   The method described comprises manipulating the growth medium, the

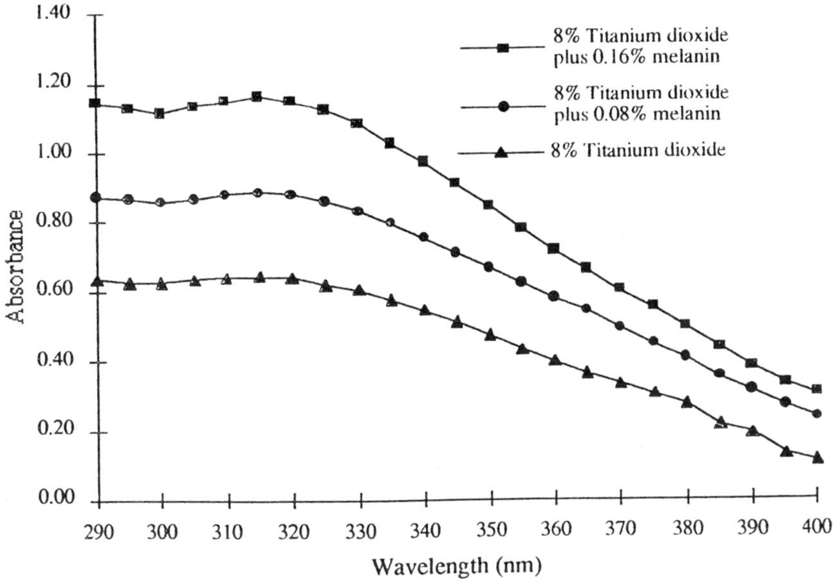

Figure 11 Titanium dioxide oil-in-water sunscreen formulations with and without melanin. TAYCA Micro Titanium Dioxide MT-500B was used in all formulations. Percentage composition (%) refers to weight of active melanin in the formulation.

Table 3 Titanium Dioxide[a] and Melanin Sunscreen Formulations In Vitro SPF and Erythemal UVA Protection Factors[b]

Formulation[c]	SPF	Erythemal UVA protection factor
8% TiO_2	4.1	3.3
8% TiO_2 plus 0.08% melanin	6.9	4.6
8% TiO_2 plus 0.16% melanin	12.1	7.9

[a]TAYCA Micro Titanium Dioxide MT-500B was used in all formulations.
[b]Forward scattering spectral measurements, and calculations of the SPF and erythemal UVA protection factors were performed using an Optometrics SPF 290S.
[c]Percentage composition (%) refers to weight of active melanin in the formulation.

fermentation conditions, or host microorganisms to achieve enhanced melanin production.

2. Melanin purification for manufacturing dyes, coatings, hair dyes, etc.
   Assignee: Inada, Y; Tainaka Kogyo KK
   Patent number: JP 4,175,377

This patent describes a method for isolating melanin from natural substances. The precipitated melanin can be used as a raw material for manufacturing dyes, coatings, hair dyes, and body lotions.

3. Enomelanine production
   Assignee: Gozhenko A. I.
   Patent number: RU 2,000,299
   This patent describes a method based on extraction of grape pressings with a solution of ammonia. The resultant material can be used in food production of grape pigment enomelanine which is used in dietary therapy for correction of various biochemical disorders.
4. Water-soluble melanin improved extraction from beet
   Assignee: Odess Food Ind. Tech.
   Patent number: SU 1,778,114
   Water-soluble melanin is extracted from table beets by extracting with a solution of ammonium hydroxide. This process is used in the food industry.
5. Extraction of melanin from black sesame
   Assignee: University of Qingdao
   Patent number: CN 1,059,540
   No abstract available.

Synthetic Melanins

The preparation of synthetic melanin is described in several patents. Many of these preparations use what little is known about melanin biosynthesis (see Fig. 1), that is, combining an oxidative enzyme with a putative melanin precursor, to form what is termed a melanin.

1. Preparation of pure synthetic melanin solution dispersible in plastic or monomer
   Assignee: Intercast Euro Spa
   Patent number: WO 9,323,480
   This preparation yields a solution of a pure synthetic melanin by polymerizing a melanin precursor in an aqueous solution. The resultant melanin is described as dispersible in plastic or plastic monomers and is for use as a pigment incorporated into sunglasses, contact lenses, and aircraft and automobile windows.
2. Soluble melanin(s) preparation
   Assignee: Yale University
   Patent numbers: US 5,227,459; WO 9,321,252
   This preparation yields soluble melanins by reaction of dopachrome with one or more dopachrome tautomerase enzymes in the presence of an indole. The resultant melanin is claimed to be soluble over a wide pH

range and useful for numerous applications requiring protection against UV damage.

3. Soluble melanin obtained from dihydroxyindole carboxylic acid
   Assignee: Yale University
   Patent number: WO 9,216,189; US 5,218,079; EP 548,110; US 5,225,435
   This patent claims a melanin that is soluble in an aqueous solution. Claims are made that the synthetic melanin can be added to compositions for use in sunscreens, for tinting glass, and for coloring coffee, tea, and barbecued potato chips.
4. Melanin-type pigments preparation
   Assignee: L'Oréal SA
   Patent numbers: EP 441,689; CA 2,035,516; LU 87,672; US 5,244,497
   This patent describes a method to prepare melanic pigments by enzyme oxidative polymerization of 5,6-dihydroxyindole derivatives in an aqueous oxidizing medium. The melanic pigment is used as a cosmetic powder.
5. Color modification of melanin for cosmetics
   Assignee: Advanced Polymer Systems, Inc.
   Patent numbers: WO 9,117,738; AU 9,179,597; US 5,188,844; AU 538,912; US 5,240,715; JP 5,507,091; EP 574,383
   This patent describes a method for modifying the color of melanin. The melanin may be employed in a variety of cosmetic formulations.

Melanin in Conjunction with Other Agents

1. Powder comprising fine colored particles carrying synthetic melanin pigment
   Assignee: L'Oréal SA
   Patent numbers: WO 9,313,743; FR 2,686,345; EP 575,603
   This patent describes a powder comprising fine colored particles carrying synthetic melanin pigment, which is formed in situ by oxidation of an indole compound. It is claimed that the product may be used in protection of human skin or in makeup for skin, hair, nails, eyelashes, and eyebrows.
2. Cosmetic powder
   Assignee: L'Oréal SA
   Patent numbers: EP 467,767; FR 2,664,498; CA 2,047,137; JP 4,230,305; US 5,205,837
   This patent describes a powder product that comprises inorganic or organic particles that have a lamellar structure with synthetic melanic pigments formed on the surface by in situ oxidation of an indole compound. The lamellar particles are claimed to uniformly disperse in cosmetic compositions.

3. Fine polymeric particles containing melanin-type pigment.
   Assignee: L'Oréal SA
   Patent numbers: EP 379,409; LU 87,429; CA 2,007,957; JP 2,232,264; DE 69,003,258
   This patent describes a product consisting of polymer particles with a melanin type pigment either in the polymer lattice or on the surface. The compositions are used as pigments.
4. Melanin-coated pigment for high safety and intense high saturation color tone
   Assignee: Kyowa Hakko Kogyo Co., Ltd.
   Patent number: WO 9,320,150
   This patent describes an inorganic or organic nylon carrier coated with alkali-soluble natural melanin isolated from *Aureobasidium*, *Aspergillus*, or *Streptomyces* genus. The pigment is claimed to be safe and to have an intense and high-saturation color tone.
5. Modified melanin
   Assignee: Taenaka Mining KK; Taenaka Mining Co., Ltd.
   Patent numbers: EP 455,175; US 5,157,075
   This patent describes the reaction of melanin and a polyethylene glycol derivative to yield a water-soluble melanin. This modified melanin can be used in cosmetics, medicines, hair dyes, and dyes.
6. Sunscreen compositions
   Assignee: Advanced Polymer Systems, Inc.
   Patent numbers: EP 313,380; AU 8,824,076; US 4,855,144; CA 1,303,913; EP 313,380; DE 3,886,675
   This patent describes new compositions that comprise melanin pigment entrapped in polymer particles. The melanin-containing particles used to absorb ultraviolet radiation in cosmetic and sunscreen compositions.
7. Sunscreen compositions
   Assignee: Advanced Polymer Systems, Inc.
   Patent numbers: US 4,806,360; ZA 8,807,834; JP 2,111,715; EP 313,380; DE 3,886,675
   This patent describes a composition that comprises porous, noncollapsible particles, with melanin pigment entrapped within the pores. The composition is useful as a component of odor-free, nontoxic topical sunscreen formulations.
8. Ultrafine inorganic powder of particle size below 20 μm
   Assignee: L'Oréal SA
   Patent numbers: GB 2,207,153; DE 3,824,094; FR 2,618,069; NL 8,801,810; LU 86,946; JP 1,092,273; BE 1,002,133; US 4,961,754; CH 676,243; IT 1,223,702; CA 1,301,069

This patent describes a powder that comprises inert particles with a pigment on their surface. The pigment is produced by oxidation of at least one indole dye. These compositions are used for dyeing of the hair or skin.

Melanin for Skin or Hair Treatment to Protect Against UV Radiation or Free Radicals

1. Melanin-containing UV protection formulation
   Assignee: Lancaster Group AG
   Patent numbers: DE 4,221,269; WO 9,400,097
   This patent describes a formulation for topical use that is characterized by melanin dissolved or dispersed in one or more fluorocarbons present in an aqueous system with asymmetrically lamellar phospholipid aggregates. The formulation is used to introduce melanin into the skin region above the epidermis.
2. Cosmetic composition containing melanin pigment and tocopherol
   Assignee: L'Oréal SA
   Patent numbers: WO 9,317,659; FR 2,688,136; AU 9,336,372
   This patent describes a cosmetic composition containing tocopherol, a melanin pigment, and a cosmetic carrier. The compositions are used for protecting human epidermis, mucous membranes, and hair against free radicals.
3. Dispersion for use in cosmetics
   Assignee: L'Oréal SA
   Patent numbers: WO 9,305,754; FR 2,681,329
   This patent describes a dispersion that comprises melanin pigment particles suspended in organic liquid. The dispersion is claimed to be stable and can be used in cosmetic compositions without isolation.
4. Cosmetic- or pharmaceutical-dispensing devices
   Assignee: L'Oréal SA
   Patent numbers: WO 9,212,911; FR 2,672,037
   This patent describes a device that comprises of a dispenser containing two immiscible viscous phases. The internal phase contains at least one active ingredient (e.g., melanin) and is disposed within an external phase. The devices can be designed so that the amount of active delivered decreases or increases over time.
5. Smooth texture cosmetic compositions for application to skin
   Assignee: Estée Lauder Inc.
   Patent numbers: EP 497,144; CA 2,059,379; JP 4,295,417
   These cosmetic compositions comprise a particulate styrene–ethylene–propylene copolymer, an emollient, and a colorant or a sunblock agent (e.g., melanin). The compositions have a smooth texture when applied to

the skin, are easily spreadable, and do not tend to flake-off the skin after application.

6. Safe and clean hair composition for dyeing gray hair
   Assignee: R. Saruya
   Patent number: JP 3,077,813
   This new hair composition contains a melanin pigment(s) with the pigment(s) being a melanin or polyphenol pigment(s) obtained from a mono- or diphenol(s) by the enzymatic action of a tyrosinase or a laccase. The composition dyes gray hair a natural black.
7. Composition for protecting skin against UV rays
   Assignee: F. C. Gaskin
   Patent numbers: CA 2,001,573; US 5,006,331; US 5,256,403
   This patent claims a composition for protecting skin against UV rays and comprises an effective amount of melanin, a substance to solubilize melanin, a substance to oxidize the melanin, and a vehicle for topical application. The composition is claimed to provide photoprotection of human epidermis, to accelerate healing of wounds, and to strengthen hair.

Unusual Applications of Melanin

1. Absorbing UV radiation in the atmosphere
   Assignee: D. L. Berliner; H. Leong
   Patent numbers: WO 9,322,773; US 5,286,979
   This patent describes introducing melanin into the atmosphere. The melanins are to absorb a quantity of UV radiation and lessen its overall effect on the earth's inhabitants.
2. New paramagnetic melanin products and metal complexes
   Assignee: University of Texas System
   Patent numbers: WO 9,218,166; AU 9,217,549; EP 580,734
   Paramagnetic melanins are produced by polymerizing mono- and dihydroxyphenyl derivatives in the presence of a radical initiator or polyphenol oxidase, with and without metals present. The paramagnetic melanins and complexes are useful as NMR contrast agents for magnetic resonance imaging.
3. Bandages with colored backing layer
   Assignee: Fleshtones Product Company, Inc.
   Patent number: US 5,120,325
   This patent describes sterile bandages for application to wounds on human skin with a backing layer colored with a pigmented composition containing one or more melanin pigments. The backing layer may be colored to match the color of skin having high melanin levels.
4. Treating melanin-deficient degenerative diseases
   Assignee: Biosource Genetics Corporation; D. L. Berliner

Patent numbers: WO 9,207,580; AU 9,190,288; EP 510,186; US 5,210,076; JP 5,503,308

This patent describes the administration of melanin derivatives for treating Alzheimer's and Parkinson's disease, schizophrenia, and dementia. The melanin derivatives would be used for the prophylaxis and treatment of degnerative diseases of the nervous system.

Melanin in Lenses or Plastic Sheets for UV Protection

1. Photoprotective hydrophilic polymers or lenses manufacturing
   Assignee: Lions Eye Institute of Western Australia, Ltd.
   Patent number: US 5,252,628

   This patent describes the oxidative polymerization of a melanin precursor to take place in situ within a polymer network. The pigmented polymer compositions are used for manufacture of hydrogel contact lenses which absorb 250 to 700-nm radiation to the same extent as the natural lens.
2. Soluble melanin production
   Assignee: Yale University
   Patent number: US 5,216,116

   This patent describes the production of a soluble melanin by polymerization of a dihydroxyindole carboxylic acid in an aqueous reaction medium in the presence of a sulfhydryl compound. This material may be used to provide a naturally appearing tan to mammalian skin and hair or as a sunscreen.
3. Apparatus for absorbing radiation
   Assignee: Photoprotective Technologies
   Patent number: US 5,047,447

   This patent describes an apparatus for radiation emitted from natural or synthetic sources that comprises a solid transparent material containing melanin. The apparatus is useful as an optical lens.
4. Optical filters using melanin pigment
   Assignee: Photoprotective Technologies
   Patent number: US 5,036,115

   This patent describes an optical filter with melanin present as a coating, a dispersion in a plastic layer, or as a copolymer with a liquid plastic monomer. The following forms of transparent filters are specifically claimed: solid lenses, sunglasses, aircraft windows, automobile windows, and welder's glass.
5. Radiation-absorbing devices
   Assignee: Photoprotective Technologies
   Patent numbers: WO 9,003,954; AU 8,946,226; US 5,112,883

This patent describes devices for absorbing radiation emitted by natural or artificial sources that comprise melanin connected with a solid material. The melanin solid substrate is claimed to be used as a sunglass, hydrophobic contact lens, intraocular device, packaging device, plastic film, window, umbrella, or canopy.

6. Hydrophilic contact lens containing melanin
   Assignee: Photoprotective Technologies; University of Texas System
   Patent numbers: WO 8,901,639; EP 311,290; AU 8,823,284; EP 380,518; JP 3,501,297; US 5,116,884

   This patent describes a hydrophilic contact lens that comprises a transparent filter having the same absorption spectrum as melanin. The lens is claimed to protect the eye from harmful radiation and superoxide.
7. Novel UV light-absorbing material used as sunscreen
   Assignee: Scopelead Ltd.
   Patent number: GB 2,203,437

   This material comprises neutralized extract of human or animal hair. The material may be incorporated as a sunscreen into cosmetic preparations

BIBLIOGRAPHY

1. Borovansky, J. and Duchon, J. Comparative study of the amino acid composition of some tumor and normal melanosomes. *Neoplasm, 22*:195–199 (1975).
2. Boyd, N. J. Melanin pigment: A redox free-radical scavenger. *Dissert. Abstr. Int., 33*/11-B:5185 (1972).
3. Brunauer, S., Emmett, P. H., and Teller, E. Adsorption of gases in multimolecular layers. *J. Am. Chem. Soc., 60*:309–319 (1938).
4. Bustamante, J., Bredeston, L., Malanga, G., and Mordoh, J. Role of melanin as a scavenger of active oxygen species. *Pigment Cell Res., 6*:348–353 (1993).
5. Chedekel, M. R., Murr, R. B., and Zeise, L. Melanin standard method: Empirical formula. *Pigment Cell Res., 5*:143–147, (1992).
6. Duchon, J., Borovansky, J., and Hach, P. Chemical composition of ten kinds of various melanosomes. *Pigment Cell, 1*:165–170 (1973).
7. Fitzpatrick, T. B., Quevedo, W. C., Jr., Levene, A. L., McGovern, V. J., Mishima, Y., and Oettle, A. G. Terminology of vertebrate melanin-containing cells: 1965. *Science, 152*:88–89 (1966).
8. Gan, E. V., Lam, K. M., Haberman, H. F., and Menon, I. A. Oxidizing and reducing properties of melanins. *Br. J. Dermatol., 96*:25–28 (1977).
9. Geremia, E., Corsaro, C. Bonomo, R., Giardinelli, R., Pappalardo, P., Vanella, A., and Sichel, G. Eumelanins as free radical traps and superoxide dismutase activities in amphibia. *Comp. Biochem. Physiol. 79*:67–69 (1984).
10. Ito, S. Reexamination of the structure of eumelanin. *Biochim. Biophys. Acta, 883*: 155–161 (1986).
11. Ito, S. and Fukita, K. Microanalysis of eumelanin and pheomelanin in hair and

melanomas by chemical degradation and liquid chromatography. *Anal. Biochem.*, *144*:527–536 (1985).

12. Ito, S., Kato, T., Shinpo, K., and Fujita, K. Oxidation of tyrosine residues in proteins by tyrosinase. *Biochem. J.*, *222*:402–411 (1984).
13. Ito, S., Novellino, E., Chioccara, F., Misuraca, G., and Prota, G. Co-polymerization of dopa and cysteinyldopa in melanogenesis in vitro. *Experentia*, *36*:822–823 (1980).
14. Kollias, N. Melanin and non-melanin photoprotection. *Melanin*: *Its Role in Human Photoprotection* (edited by L. Zeise, M. R. Chedekel, and T. B. Fitzpatrick), Valdenmar Publishing, Overland Park, 1994.
15. Kollias, N., Sayre, R. M., Zeise, L., and Chedekel, M. R. Photoprotection by melanin. *J. Photobiol. B*, *9*:135–160 (1991).
16. Lang, G. Melanin in protection from sunlight. *Proceedings of the 17th IFSCC International Congress*, *3*:1072–1094 (1992).
17. Lowell, S. and Shields, J. E. Density measurement. *Powder Surface Area and Porosity* (edited by B. Scarlett). Chapman & Hall, New York, 1984, pp. 217–220.
18. Menon, I. A., Leu, S. L., and Haberman, H. F. Electron transfer properties of melanin. Optimum conditions and the effects of various chemical treatments. *Can. J. Biochem.*, *55*:783–787 (1977).
19. *Merck Index*. 11th ed. Merck & Company, Rahway NJ, 1989, p. 911, entry 5692.
20. Nicolaus, R. A. Melanins. *Chemistry of Natural Products* (edited by L. Lendered), Hermann, Paris, 1968, pp. 68–91.
21. Piatteli, M., Fattorusso, E., Magnon, S., and Nicolaus, R. A. The structure of melanins and melanogenesis. III. The structure of sepiomelanin. *Tetrahedron*, *19*: 2061–2072 (1963).
22. Prota, G. *Melanins and Melanogenesis*. Academic Press, San Diego, 1993, pp. 88–118.
23. Quevedo, W. C., Jr., Fitzpatrick, T. B., Pathak, M. A., and Jimbow, K. Light and skin color. *Sunlight and Man* (edited by T. B. Fitzpatrick, et al.), University of Tokyo Press, Tokyo, 1974, pp. 165–194.
24. Rorsman, H., Agrup, G., Hansson, C., Rosengren, A.-M., and Rosengren, E. Detection of phaecomelanins. *Pigment Cell*: *Biologic Basis of Pigmentation*, Vol. 4 (edited by S. Klaus). Karger, Basel, 1979, pp. 244–252.
25. Sealy, R. C., Felix, C. C., Hyde, J. S., and Swartz, H. M. Structure and reactivity of melanins: Influence of free radicals and metals ions. *Free Radicals in Biology*, Vol. 4 (edited by W. A. Pryor). Academic Press, New York, 1980, pp. 209–259.
26. White, L. P. Melanin: A naturally occurring cation exchange material. *Nature*, *182*: 1427 (1958).
27. Yasunobu, K. T., Peterson, E. W., and Mason, H. S. The oxidation of tyrosine-containing peptides by tyrosinase. *J. Biol. Chem.*, *234*:3291 (1959).
28. Zeise, L. Bioanalytical Studies of Melanins. Doctoral Dissertation. The Johns Hopkins University, Baltimore, 1991, pp. 40–59.
29. Zeise, L., Murr, B. L., and Chedekel, M. R. Melanin standard method: Particle description. *Pigment Cell Res.*, *5*:132–142 (1992).
30. Zeise, L. and Chedekel, M. R. Melanin standard method: Titrimetric analysis. *Pigment Cell Res.*, *5*:230–239 (1992).

21

Sunscreens and Hair

Sergio Nacht
Advanced Polymer Systems, Inc., Redwood City, California

I. INTRODUCTION

Throughout the history of the development of sun protection, little attention has been given to the use of sunscreens on hair. This lack of interest is most likely because hair does not develop cancer or wrinkle with age as does its parent tissue, the skin. Nor does this skin appendage burn or become very uncomfortable from too much sun exposure.

Yet men and women through the centuries have experienced noticeable changes in the properties of their hair after exposure to ultraviolet (UV) radiation. Today, one can still find field workers wearing hats as they toil in the sun, and hair-conscious sunbathers will frequently don a scarf before seeking the sun's rays. These people have discovered that sunlight does produce undesirable changes in hair quality. Dryness, reduced strength, rough surface texture, loss of color and luster, stiffness, and brittleness have been reported by both the lay population and the research community (1–7), although only a sparse collection of quantitative data on the effects of ultraviolet light to human hair is available.

When absorbed, ultraviolet radiation is responsible for the photochemical degradation of many substances within the skin (8–11). Sunscreens reduce the chemical effects of ultraviolet light to the skin (12–15) and have become popular with consumers during the past decade. To determine the protective efficacy of topical sunscreen products, a sun protection factor (SPF) scale was proposed by the Over-the-Counter Food and Drug Administration (OTC–FDA) Advisory Review Panel in their proposed monograph addressing sunscreens as OTC drugs in 1978 (6). However, no mention is made in this monograph concerning sunscreen efficacy on hair protein.

In this chapter, we discuss some of the photochemical changes occurring in hair, the effects that sunscreens have on reducing these changes, and we propose a method and a numerical scale for the determination of sunscreen efficacy for hair protection, which we labeled the hair protection factor (HPF).

II. MECHANISMS OF HAIR WEATHERING

Hair is the recipient of a constant series of environmental assaults we term "weathering." Such potentially damaging influences as rain, air pollutants, wind, seawater, or chemicals in swimming pools contribute to the environmental processes known to cause structural and chemical degradation to hair (2).

Most damaging of all environmental factors is ultraviolet radiation (1–3). These effects are compounded when combined with other environmentally and chemically induced damage (as with hair bleaches, colorants, and permanent wave products) (2).

An example of this damage is the condition of a cuticle on a normal unexposed hair compared with a cuticle covering hair that has been exposed to ultraviolet radiation, either singularly or in association with other elements. The cuticle is considered to play a key role in determining the overall physical properties in keratin fibers (3), whereas the interior of the fiber, the cortex, is believed to have less importance for hair condition (1). The cuticle dictates the frictional properties of the hair fibers and is also largely responsible for the integrity of the interior. In *normal hair*, defined as hair that has not been exposed to the environment or chemical processing, the edges of the cuticle cells are smooth, rounded, and tightly packed onto the cortex. Long-term exposure of hair to sunlight and the atmosphere results in the breakdown of cuticle cells on the hair surface (1). Thus, cleavage of the cuticle cells that coat exposed hair has been observed within the endocuticle layer of these hairs in many instances (3). This results in the alteration of the mechanical properties of human hair.

III. ULTRAVIOLET RADIATION AND CHEMICAL DEGRADATION OF HAIR

Several photochemical reactions occur in hair after exposure to ultraviolet radiation. Discoloration, often called "highlights" observed by sunbathers the world over, is the most obvious chemical alteration caused by the sun. But other, subtler chemical changes occur that are evident to the human eye only after they have accumulated over time or by the intensity of exposure.

A. Discoloration

Fading of brown hair and yellowing of blond and red hair have been observed by several researchers (6,17). The fading of brown hair is attributed to the photooxidative bleaching of melanin hair pigment (2), whereas discoloration of blond hair has been explained in terms of the photodegradation of cystine, tyrosine, and tryptophan residues, producing yellowing and brown reaction products (18–21). Degradation of histidine, lysine, and proline have also been discovered (2) and are thought to be responsible for some of the yellow discolorations of blond hair (6).

Phaeomelanin, predominant in red and blond shades, is also present in brown hair, which appears to be a mixture of this lighter melanin pigment and the darker eumelanin. Eumelanin is chemically different from phaeomelanin, the most obvious difference being the higher sulfur content of the lighter pigment compared with that of eumelanin (22). It has been reported that superoxides are formed when phaeomelanin is degraded by irradiation with long-wavelength ultraviolet and visible light (23–26). This sensitivity of phaeomelanin can explain why the most obvious discolorations occur in lighter hair shades, whereas it is less apparent in darker tones of hair containing higher levels of eumelanin.

B. Sorption Characteristics

Changes in the sorption characteristics and the urea and bisulfite solubility of hair have been detected after extended weathering exposure (2,6,27). The increase in urea and bisulfite solubility suggests that a network of new types of cross-links is formed during the weathering process, especially as it relates to ultraviolet radiation (27). Although not much is known about the exact chemical structure of these cross-links, several isopeptides that are not present in intact hair were found in enzymic hydrolysates of weathered fibers (2,27,28).

C. The Free Radical Mechanism

A free radical mechanism has been proposed that may account for many of the photochemical changes occurring in weathered human hair (2,27,29). Ultraviolet radiation is absorbed by cystine, tyrosine, phenylalanine, and tryptophan residues, and possibly some peptide bonds, which results in the formation of free radicals. Homolytic scission of disulfide bonds then occurs.

In the presence of such free radical carriers as water, present as humidity in the atmosphere, a sequence of radical transfer reactions involving hydroxyl radicals can take place. This causes random displacements along the protein backbone that lead to an assortment of degradation products (2).

Free radical formation appears to be affected by melanin, which is believed to act as an energy sink (29), thus protecting keratin against radiation in the initial stages of weathering. Heavily pigmented hair has a high absorption capacity that is explained in terms of its comprehensive system of double bonds and conjugated carbonyl groups (2). Large fractions of ultraviolet light may be captured by the melanin with this absorption process, which also immobilizes free radicals as they are formed. This activity prevents the propagation of free radicals into underlying keratin structures.

D. Tensile Strength

Tensile strength of hair fibers decreases in correlation with ultraviolet radiation exposure time (2,6,17,30). Reduced cystine levels accompanied by the generation of cysteic acid are found in hair exposed to the environment, resulting in the destruction of stabilizing disulfide bonds (2,27–29,31). Pigmented fibers exhibit less cystine reduction than those that are nonpigmented (2). This reduction appears to be unrelated to the level of humidity present during exposure.

Although pronounced tensile alteration is due primarily to photochemical effects on disulfide linkages (30), deamination or decarboxylation of amino acids (32), disorientation of hydrogen bonds, and the chemical alteration of the aromatic nuclei in tyrosine and phenylalanine (33) also contribute to a chemical reduction in tensile strength.

Another factor in influencing the correlation of hair tensile strength to ultraviolet exposure is seasonal radiation; summer solar radiation is five times greater than that received by hair during the winter months (31).

IV. SUNSCREENS AND HAIR PROTECTION

Because of the obvious photochemical degradation of hair, several researchers have suggested the use of some form of chemical protection from the sun (2,8,17). Although many sunscreens are available for use in skin protection, only a few lend themselves to use in hair care products.

The ideal sunscreen for hair must be substantive to the hair cuticle, even in the presence of surfactants, without causing loss of sheen or adding a "gummy" quality to hair. Yet this sunscreen must also be easily incorporated into the largely water-based vehicles used on the hair by consumers. Few sunscreens possess these qualities while maintaining the water-resistance required for long-term hair protection. In a variety of tests (17), octyl dimethyl PABA was the most suitable of all sunscreens recognized in the OTC–FDA sunscreen monograph (16), whereas *p*-aminobenzoic acid (PABA) is the least desirable owing to its high water solubility and low substantivity to hair.

V. ASSESSING PROTECTIVE EFFICACY OF SUNSCREENS FOR HAIR

As sunscreen protection becomes an accepted practice in hair care, a simple, yet accurate, technique is needed to assess protective efficacy of sunscreen ingredients and formulations designed for hair use. A parametric scale for the comparison of the performance of these sunscreen products on the hair is also required.

A. Sun Protection Factor

To determine the protective efficacy of sunscreen products applied to the skin, a sun protection factor (SPF) scale was proposed by the OTC–FDA Advisory Review Panel in their monograph addressing sunscreens (16). This scale provides consumers with a simple, numerical identification of the different levels of protection offered by various commercial products.

The SPF is defined as the ratio of the ultraviolet energy required to produce minimal erythema on protected skin to that required to produce the same minimal erythema on unprotected skin in the same individual.

To ensure reproducibility and provide a meaningful comparison between products, the OTC–FDA panel recommended specific testing procedures that are now followed by most manufacturers of sunscreen products to determine the SPF ratings of their products.

B. Hair Protection Factor

The SPF scale is directly related to a perceivable change in the skin, notably sunburn. Hair damage caused by ultraviolet radiation does not result in a change so easily perceived by either the consumer or the researcher. Therefore, a scale that accurately measures hair damage and determines the protective effects of hair treatments containing sunscreens must use a different endpoint than that used for skin. This scale could be termed the *hair protection factor* (HPF).

A practical HPF can be determined by measuring the ultraviolet damage to hair, as detected by yield slope analysis of tensile strength assessments (17,30). When hair is extended with a tensile strength tester, the changes in the slope of the stress–strain curve in the yield region correlate well with the amount of ultraviolet radiation to which the hair has been exposed. Then, when the negative, inverse logarithm of the yield slope ($-1/\ln$) is plotted versus exposure time, a straight line is obtained between 0 and 3 h.

Because consumers have been educated, primarily by industry, to understand an SPF scale, which basically ranges from 2 through 15, we have chosen to arbitrarily construct a similar scale to facilitate the analogy. Therefore, to construct the HPF scale, the line obtained with a sample of virgin, untreated hair is arbitrarily divided into 15 divisions, with ''0'' set at the 3-h point for maximum damage, and ''15'' located at the no exposure value for no ultraviolet damage.

A perfect analogy with the SPF scale used for skin protection would have come from a definition such as:

$$\text{HPF} = \frac{\text{(yield slope, nonirradiated} - \text{yield slope, irradiated) unprotected hair}}{\text{(yield slope, nonirradiated} - \text{yield slope, irradiated) protected hair}}$$

However, simple calculations of protection factors with such an equation using the yield slope values obtained in our experiments result in an ''unnatural'' scale that provides exceedingly small numbers at the lower end of the scale, whereas it results in very large numbers (that tend to infinity as the denominator gets smaller) when the treatment provides good protection. These distortions make such a scale impractical and probably difficult for the consumer to understand because it could easily yield numbers as large as 200.

Consequently, we have chosen to use the seemingly more convoluted, but more practical, foregoing logarithmic scale.

VI. METHODOLOGY FOR DETERMINING HAIR PROTECTION FACTOR

A. Hair Tensile Strength as a Measurement

Hair passes through three phases while being extended under increasing force (Fig. 1). The first phase, the elastic region, is characterized by reversible extension. Hydrogen bonds are probably the only ones disturbed in this region. The second phase is characterized by a partially reversible transformation in which covalent and salt bonds are likely to be disturbed. This phase is known as the yield region. Leading to the breaking point is the postyield region, or third phase. This is where complete fiber breakage occurs.

Other research (34) has shown that the breaking point of hair is related to hair diameter, and may not necessarily be an indication of overall hair damage. Therefore, this parameter seems not be accurate or sensitive enough to measure subtle degrees of hair damage.

Because the yield region is the one most likely to correlate in general with covalent and possibly with disulfide bond breakage in hair, and because these are the types of bonds most probably affected by ultraviolet radiation, which results in intrinsic structural hair damage, evaluation of the yield slope can provide a practical and objective measurement of hair damage caused by ultraviolet exposure.

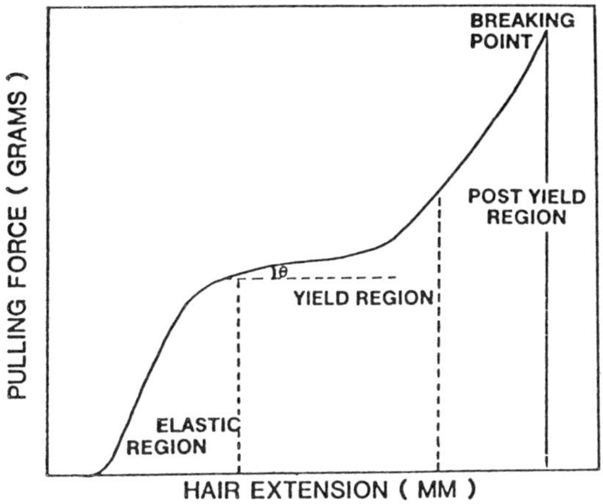

Figure 1 Typical stress–strain curve of a single hair fiber as obtained with an Instrom tensile strength tester.

B. Determination of Hair Tensile Strength

Hair tensile strength is determined by positioning a single strand of hair, about 30 cm long, between the two clamps of an Instron tensile strength tester (35) set at a standardized distance of 23 cm. The hair is then pulled with a 20-g load setting at a crosshead speed of 100 mm/min and at a chart speed of 200 mm/min. All measurements are made in an environmental chamber set at 22°C and 20% relative humidity (RH). At least 20 single-hair strands from each group are extended. From the stress–strain graphs obtained for each hair, the slope of the yield region (see Fig. 1) is measured and averaged for the group.

It is apparent that the differences in cross-sectional area of the hair can affect the accuracy of the yield–slope determination (Table 1). However, with a trained technician to pick hair of average size and similar cross-sectional area, it is possible to reproducibly select hair of average diameter.

C. Determination of Sun-Induced Damage to Hair

Virgin hair at least 30 cm long is spread apart in a single layer and uniformly exposed for different lengths of time to a Kratos solar simulator (36), adjusted to 20 V and 30 amp at a fixed distance of 64 cm. The solar simulator suggested for use in these tests provides a highly intensified spectrum of light which closely resembles solar radiation and, thus, is capable of accelerating the damage to hair produced by sun exposure. Although a small increase in temperature may

Table 1 Assessment of Different Hair Diameter by Yield Slope Analysis

		Average yield slope + SD[a]
Virgin untreated 30–50 units DIA	n = 20	0.159 ± 0.053
Virgin untreated 60–80 units DIA	n = 20	0.287 ± 0.103
Virgin untreated random sample[b] 30–80 units DIA	n = 40	0.230 ± 0.075

[a]Standard deviation.
[b]Average size hair randomly selected by the trained technician.

be observed after extensive exposure with the solar simulator, this temperature increase has no effect in the yield slope of hair (17).

After various times of exposure, the stress–strain curve of hair is determined as previously described. The mean yield slope for the 20 single-hair strands in each group of ultraviolet exposed hair is then calculated as indicated on the foregoing. These results can be compared with the mean yield slope of a similar group of virgin (untreated, unexposed) hair that represent the value for totally undamaged hair.

D. Evaluation of Hair Protection Factor

Swatches of hair in excess of 20 strands are prepared from the same sample of virgin hair as that used for standardization. These swatches are soaked for 30 min in the test solution and rinsed for 30 s with warm tap water (37°C). This procedure is repeated four times. Although this soaking cycle may seem exaggerated, results are produced that are similar to those obtained from 20 cycles of 2-min soakings in a shampoo, followed by rinsing (17). This regimen reflects accelerated 1-month in vivo usage conditions (Table 2). Consequently, the four-cycle regimen is recommended for simplicity. For leave-on products, such as hair sprays and mousses, the hair swatches are treated four times with the products as directed for use. After treatment by either shampoo or leave-on products, the hair swatches are air-dried overnight at room temperature.

These treated swatches are then exposed to the solar simulator for 3 h, as previously described. After exposure, hair strands are tested with the tensile strength tester, and the mean yield slope is calculated for each treatment group.

The HPF provided by the product undergoing testing can now be determined by calculating the mean yield slope obtained with the exposed, treated hair and interpolating that value in the standard curve determined with untreated samples of the same hair (Fig. 2).

Table 2 Effect of Treatment Regimen on HPF Value

Treatment[a]	Regimen	Yield slope + SD	HPF (0–15)
0.2% octyl dimethyl PABA	4 cycles 30′ per cycle	0.309 ± 0.073	6.6
0.2% octyl dimethyl PABA	20 cycles 2′ per cycle	0.313 ± 0.066	7.1
0.5% octyl dimethyl PABA	4 cycles 30′ per cycle	0.346 ± 0.057	11.6
0.5% octyl dimethyl PABA	20 cycles 2′ per cycle	0.348 ± 0.055	11.9

[a]All treatments consisted of a commercial shampoo (Pantene Shampoo for Normal Hair) containing various concentration of octyldimethyl-*p*-aminobenzoic acid.

VII. FACTORS INFLUENCING HAIR PROTECTION FACTOR

Various factors can influence the results obtained using the suggested method to determine HPF.

A. Sunscreen Concentration

When tested in a simple shampoo (37), the level of protection provided by the shampoo treatment rises with the increase in sunscreen concentration (Table 3) (17). Thus, using a scale where 15 represents total hair protection, 1% octyl dimethyl PABA results in an HPF of 9.8, whereas a 0.2% concentration of this same sunscreen in the same shampoo provides an HPF of 2.3. This value can be no different from that offered by a commercial shampoo without sunscreen (Table 4). Therefore, 0.2% of octyl dimethyl PABA may not be sufficient to provide measurable sun protection benefits to hair.

B. Shampoo Composition

Other ingredients frequently used in hair care products may enhance or reduce the spreadability and substantivity of the sunscreen and, therefore, may influence the sun protection ability of the total formulation. To determine these effects, the HPF of a commercial shampoo containing different levels of sunscreen were evaluated. The data (Tables 4 and 5) show that, at various levels of sunscreen concentration, a higher HPF is obtained with a commercial shampoo containing conditioners than those previously attained with a simple shampoo formula (17).

These results suggest that the conditioning ingredients found in many hair care formulations can enhance sunscreen protection of hair, perhaps by enhanc-

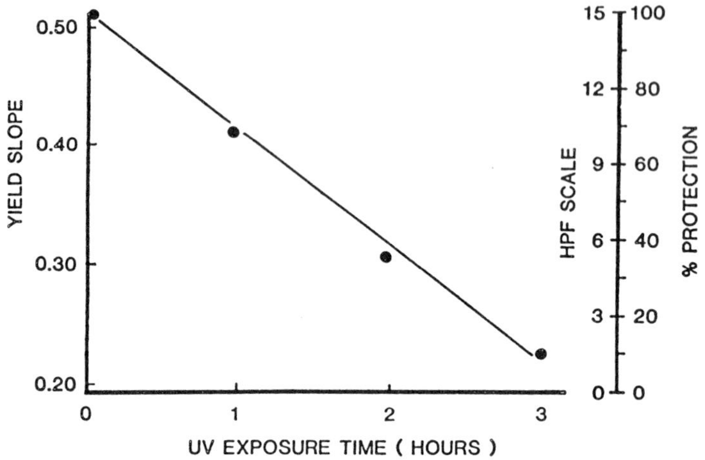

Figure 2 Decrease in yield slope with increased exposure of hair to UV radiation. Definition of the hair protection factor (HPF) scale.

Table 3 Effect of Sunscreen Concentration on HPF

Treatment	Yield slope + SD	HPF
Simple shampoo[a]	0.296 ± 0.061	0.4
0.2% octyl dimethyl PABA in simple shampoo	0.319 ± 0.033	2.3
0.5% octyl dimethyl PABA in simple shampoo	0.380 ± 0.059	7.1
1.0% octyl dimethyl PABA in simple shampoo	0.410 ± 0.041	9.8

[a]Shampoo consisting of sodium lauryl ether sulfate and citric acid.

Table 4 Hair Protection Factor (HPF) of Various Concentrations of Octyldimethyl-PABA in a Commercial Shampoo

Treatment	Yield slope ± SD	HPF
Commercial shampoo[a]	0.321 ± 0.032	2.7
0.1% sunscreen in commercial shampoo	0.354 ± 0.036	4.7
0.2% sunscreen in commercial shampoo	0.391 ± 0.045	8.7
0.5% sunscreen in commercial shampoo	0.438 ± 0.043	12.8
0.7% sunscreen in commercial shampoo	0.417 ± 0.037	11.0

[a]As in Table 2.

Table 5 Hair Protection Factor (HPF) of Various Commercial Shampoos

Treatment	HPF
Johnson's baby shampoo	1.5
Mill Creek henna shampoo (PABA)[a]	2.7
Silkience shampoo	3.0
Pioneer brand jojoba shampoo (PABA)[a]	3.9

[a]Type of sunscreen as identified in ingredients listing.

ing sunscreen substantivity or by promoting even distribution of the sunscreen over the entire hair strand.

C. Hair Type and Cosmetic History

Ultraviolet exposure of chemically processed hair increases the damage already present, frequently resulting in the maximum damage which can be produced to hair. Maximum damage for both virgin and bleached hair is achieved after 3 h of ultraviolet exposure to the solar simulator (Fig. 3). Interestingly, HPF values obtained using treatments with increasing percentages of sunscreen are similar between virgin and bleached hair (Table 6). This similarity in the correlation of ultraviolet exposure and hair damage observed with both groups suggests that

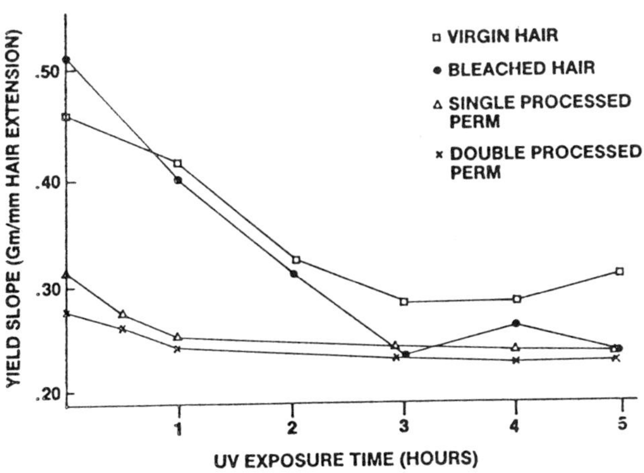

Figure 3 Comparison of the damage induced on different types of hair by increasing exposure to UV radiation.

Table 6 Effects of Sunscreen Concentration on HPF[a]

Treatment	Yield Slope			HPF		
	Virgin	Bleached	Permed	Virgin	Bleached	Permed
Base shampoo	0.189 ± 0.057	0.195 ± 0.053	0.178 ± 0.041	0	0	0
0.2% Octyl-dimethyl PABA in base shampoo	0.200 ± 0.052	0.200 ± 0.057	0.176 ± 0.042	4.0	4.0	0
0.5% Octyl-dimethyl PABA in base shampoo	0.204 ± 0.056	0.204 ± 0.046	0.180 ± 0.046	5.4	5.4	2.0
1.0% Octyl-dimethyl PABA in base shampoo	0.216 ± 0.060	0.210 ± 0.058	0.184 ± 0.051	9.6	10.7	6.6

[a]Different hair samples were used.

ultraviolet radiation disrupts covalent bonds at a similar rate in both of these hair types.

On the other hand, maximum damage induced by ultraviolet radiation in permed hair occurs within less than 1 h of exposure (Fig. 3). The waving solution of a permanent wave product reduces the disulfide bonds, but not all of those bonds are reformed during the neutralization phase. Therefore, permed hair is already severely damaged prior to ultraviolet exposure and suffers maximal damage much more quickly than other types of hair.

HPF can therefore be determined with virgin or bleached hair with good precision and reproducibility, but the use of severely damaged hair is not recommended.

VII. HPF AND THE FUTURE OF SUNSCREENS FOR HAIR PROTECTION

As outlined in this chapter, the hair protection factor has several major attributes in determining sunscreen protection in a broad spectrum of products.

1. HPF, as it is defined here, can be easily determined for a variety of hair types.
2. A variety of formulations and sunscreen ingredients may be tested accurately using this scale.
3. Unlike SPF, which is based on minimal erythemal dose and therefore responds only to UVB exposure, HPF reflects the effects of all segments of the ultraviolet spectrum.
4. Since the proposed HPF scale is based on the assessment of changes in the mechanical properties of hair which are a reflection of the breakage of various keratin bonds, this scale can be of value in determining also the effects of other nonsunscreen components that may help repair or reform these bonds (i.e., free radical scavengers, disulfide bond repairers, etc.)

Finally, as the effects of sunlight on hair are better understood and appreciated by both the consumer and research communities, sunscreens are becoming an accepted ingredient in many hair care formulations. Sunscreens are especially appropriate for the protection of hair that is already perceived by the consumer as being damaged, as in weathered or chemically processed hair. Although shampoos and conditioners are the obvious vehicles for such products, hair sprays, styling gels, mousses and other styling aids can also be considered for sunscreen additions.

REFERENCES

1. Breuer, M. M., Gikas, G. X., and Smith, I. T. Physical chemistry of hair condition. *Cosmet. Toilet.*, *94*:29–34 (1979).
2. Tolgyesi, E. Weathering of hair. *Cosmet. Toilet.*, *98*:29–33 (1983).
3. Hunter, L. D., Garcia, M. L., Newman, W., and Cohen, G. L. Observation of the internal structure of the human hair cuticle cell by SEM. *Text. Res. J.*, 136–140 (1974).
4. Berth, P. and Reese, G. Alteration of hair keratin by cosmetic processing and natural environmental influences. *J. Soc. Cosmet. Chem.*, *15*:659–666 (1964).
5. Rook, A. The clinical importance of ''weathering'' in human hair. *Br. J. Dermatol.*, *95*:111 (1976).
6. Tolgyesi, E., Peloquin, L., and Arnold, G. B. Unpublished results of an investigation conducted at Gilette Research Institute, Rockville, MD (1980).
7. Robbins, C. Weathering in human hair. *Text. Res. J.*, *37*:337 (1967).
8. Strobel, A. and Inserra, J. J. The use of UV absorbers in cosmetic products. *Am. Perf. Cosmet.*, *83*:25–30 (1968).
9. Kligman, L. H. and Kligman, A. M. The nature of photoaging: Its prevention and repair. *Photodermatology*, *3*:215–227 (1986).
10. Hersey, P., MacDonald, M., Henderson, C., Schibeci, S., D'Alessandro, G., Pryor, M., and Wilkinson, F. Suppression of natural killer cell activity in humans by radiation from solarium lamps depleted of UVB. *J. Invest. Dermatol.*, *90*:305–310 (1988).
11. Cleaver, J. DNA damage and repair in light-sensitive human skin. *J. Invest. Dermatol.*, *54*:181–195 (1970).
12. Pathak, M. A. Sunscreens: Topical and systemic approaches for protection of human skin against harmful effects of solar radiation. *J. Am. Acad. Dermatol.*, *7*:285–312 (1982).
13. Fitzpatrick, T. B., Pathak, M. A., Harber, L. C., Seiji, M., and Kukita, A. *Sunlight and Man*, University of Tokyo Press, Tokyo, 1974.
14. Grove, G. L. and Kaidbey, K. H. Sunscreens prevent sunburn cell formation in human skin. *J. Invest. Dermatol.*, *75*:363–364 (1980).
15. Hersey, P., MacDonald, M., Burns, C., Schibeci, S., Matthews, H., and Wilkinson, F. Analysis of the effect of a sunscreen agent on the suppression of natural killer cell activity induced in human subjects by radiation from solarium lamps. *J. Invest. Dermatol.*, *88*:271–275 (19xx).
16. FDA: Sunscreen drug product for over-the-counter human drugs; Proposed Rules, part II. *Federal Register*, 38206 (1978).
17. Nacht, S., Yeung, D., and Dunn, L. Unpublished results of an investigation conducted at Richardson-Vicks Inc., Shelton, CT (1982).
18. Speakman, J. B. and MacMahan, P. R. The action of light on wool and related fibers. *N. Z. J. Sci. Technol.*, *20*:2488 (1939).
19. Lennox, F. G. A spectrophotometric study of yellowing in wool fabric. *J. Text. Inst.*, *51*:1193 (1960).
20. Milligan, B. and Tucker, D. J. Studies on wool yellowing, Part III: Sunlight yellowing. *Text Res. J.*, *33*:773 (1963).

21. Holt, L. A. and Milligan, B. The involvement of tryptophan in the photoyellowing of wool. *J. Text. Inst.*, *67*:269 (1976).
22. Menon, I. A., Surujdeen, P., Haberman, H. F., and Kurian, C. J. A comparative study of the physical and chemical properties of melanins isolated from human black and red hair. *J. Invest. Dermatol.*, *80*:202–206 (1983)
23. Chedekel, M. R., Post, P. W., Deibel, R. M., and Kalus, M. Photodestruction of phaeomelanin. *Photochem. Photobiol.*, *26*:651–653 (1977).
24. Chedekel, M. R., Smith, S. K., Post, P. W., Pokora, A., and Vessell, D. L. Photodestruction of phaeomelanin: Role of oxygen. *Proc. Natl. Acad. Sci. (USA)*, *75*: 5395–5399 (1978).
25. Agin, P. R., Sayre, R. M., and Chedekel, M. R. Photodegradation of phaeomelanin: An in vitro model. *Photochem. Photobiol.*, *31*:359–362 (1980).
26. Chedekel, M. R., Agin, P. P., and Sayre, R. M. Photochemistry of phaeomelanin: Action spectrum for superoxide production. *Photochem. Photobiol.*, *31*:553–555 (1980).
27. Wolfram, L. J. The reactivity of human hair. *Hair Research Status and Future Aspects* (edited by C. E. Orfanos, W. Montagna, and G. Stuttgen). Springer-Verlag, Berlin, 1981, p. 479.
28. Wolfram, L. J. Unpublished results of an investigation carried out at CSIRO, Melbourne, Australia (1973).
29. Scott, E. and Wolfram, L. J. The weathering of pigmented hair. Proc. 4th Int. Wool Text. Res. Conf., Berkeley, 1970.
30. Beyak, R., Kass, G. S., and Meyer, C. F. Elasticity and tensile properties of human hair, II. Light radiation effects. *J. Soc. Cosmet. Chem.*, *22*:667–687 (1971).
31. Robbins, C. R. and Kelly, C. H. Amino acid composition of human hair. *Text. Res. J.*, *40*:89 (1970).
32. Luse, R. A. and McLaren, A. D. Mechanism of enzyme inactivation by ultraviolet light and the photochemistry of amino acids. *Photochem. Photobiol.*, *2*:343–360 (1963).
33. Alexander, P. and Lett, J. T. Effects of ionizing radiations on biological macromolecules. *Comp. Biochem.*, *27*:157–209 (1967).
34. Collins, J. D. and Chaikin, M. The stress–strain behavior of dimensionally and structurally, non-uniform wool fibers in water. *Text. Res. J.*, *35*:777–787 (1965).
35. Instron, Model #1122.
36. Kratos Solar Simulator, Model #LH153.

22

5-Methoxypsoralen-Containing Sunscreens

Antony R. Young
United Medical and Dental Schools of Guy's and St. Thomas's Hospitals, University of London, London, England

I. INTRODUCTION

A. Conventional Sunscreens

The purpose of most sunscreen preparations is to absorb solar ultraviolet (UV) radiation and prevent skin damage. In general, the active molecules absorb primarily in the UVB (290–315 nm) part of the solar spectrum (1). Efficacy of these preparations is assessed by their sun protection factors (SPFs) (2). More recently there has been an interest in UVA-absorbing (315–400 nm) sunscreens, especially in UVA-sensitive individuals (3,4), but the importance of such preparations on normal skin has yet to be fully determined.

It is now widely recognized that sunscreen use, especially in the more sun-sensitive skin types, I and II, is important in preventing the acute effects of solar exposure—namely, erythema (2)—and animal data suggest an important role in preventing or delaying established long-term effects of solar radiation, such as skin cancer (5) and photoaging (6,7). Increasingly, it is being advocated that sunscreens be used not just on the beach or during outdoor leisure activities, but also while outdoors generally (8,9), especially in sunny climates.

One of the, probably many, areas of resistance of sunscreen use is the reduction of tanning that is associated with good sunscreen photoprotection, especially in people who do not tan readily. It is obvious that a tan is still a highly valued cosmetic and social asset, despite much publicized dermatological opinion against excessive solar exposure and UVA sunbed use.

The application of an effective sunscreen will enable the user to (1) spend a longer time in the sun to obtain a given multiple of a minimal erythema dose (MED), or (2) to prevent or reduce erythema if the amount of time spent in the sun is similar to that spent without the sunscreen. The former option may actually increase the long-term risks of solar exposure (8), especially as the use of UVB sunscreens (to prolong solar exposure time) will result in increased exposure to solar UVA. It has been demonstrated experimentally that, on an MED equivalent basis, UVA and UVB show similar carcinogenic potential (10).

B. 5-Methoxypsoralen-Containing Sunscreens

About 15 years ago, UVB sunscreen preparations that contained 5-methoxypsoralen (5-MOP) appeared on the European market. 5-MOP is a natural furocoumarin present at about 3000–3600 ppm in the fragrant oil extracted from the rind of the bergamot fruit (*Citrus bergamia*) (11,12). The rationale behind the addition of 5-MOP to conventional sunscreens is that UVB erythema (and tanning) are inhibited by the sunscreens, but 5-MOP, at a final concentration of 10–50 ppm, in bergamot oil enhances melanogenesis in the presence of solar UVA (13). 5-MOP has long been recognized as the pigmentary agent in berloque dermatitis (14). The characteristics of this disorder were "pendant"-like areas of hyperpigmentation on sun-exposed sites of the neck after application of eau de cologne or other bergamot oil-containing perfumes. Berloque dermatitis is rarely seen nowadays, probably because of a reduction in the concentration of 5-MOP in perfume products. Current recommended maximum level of 5-MOP in final fragrance products is 15 ppm (International Fragrance Association, 1978).

II. PHOTOMUTAGENICITY AND PHOTOCARCINOGENICITY OF PSORALENS

Much of the photochemical and photobiological research on the psoralens has been concentrated on an isomer of 5-MOP, 8-methoxypsoralen (8-MOP), which is widely used in conjunction with UVA radiation (PUVA) in the treatment of psoriasis (15). 8-MOP and 5-MOP form unstable complexes with DNA in the dark (16,17), but in the presence of UVA, both psoralens photobind to DNA (18,19). Such photobinding, which is chemically stable, is associated with the many of the photobiological effects of these compounds. For example, 8-MOP and 5-MOP show potent mutagenicity in bacterial, yeast, and mammalian systems in the presence of UVA radiation (20,21). In general, 8-MOP and 5-MOP show similar photobiological properties, but often with quantitative differences. In the absence of UVA, the psoralens generally show no biological activity, but frameshift mutations have been observed in bacteria with 5-MOP and other psoralens (21).

Methoxypsoralen is a photocarcinogen in mouse skin (22,23), but the significance of psoralens in human skin carcinogenesis has not been unequivocally established. An extensive epidemiological study of psoriatic patients on 8-MOP photochemotherapy in the United States has provided convincing evidence that PUVA is a carcinogen (24). An important aspect of this study was the relation between PUVA dose and skin cancer. However, a long-term study published by a group of European investigators did not support the American observations that PUVA was a complete carcinogen, but indicated that it was a promoting agent (25). The reasons for the different conclusions are not established, but possible explanations have been discussed by Gibbs et al. (26).

Few data on human skin 5-MOP photocarcinogenicity are available. 5-MOP has been proposed as a better agent than 8-MOP in PUVA (27), but there have been no long-term follow-up studies in patients. An epidemiological study on workers in the bergamot oil production industry (28) did not find any evidence of an increased incidence of skin cancer in comparison with a control group. However, an International Agency for Research on Cancer (IARC) working group noted some limitations in this study (29).

Studies by Zajdela and Bisagni (30) and Young et al. (31) have clearly shown 5-MOP to be a skin photocarcinogen in mice after prolonged exposure. This has been recognized by an IARC working group (29). Comparative data from one study (30) indicated that 8-MOP was more potent than 5-MOP, but data from the study of Young et al. (31) indicated similar photocarcinogenic potential. The experimental photocarcinogenicity of 5-MOP is not surprising, given the potent mutagenicity of this molecule in a variety of systems in the presence of UVA (20,21,32). In addition, 5-MOP, as with 8-MOP, induces epidermal ornithine decarboxylase (ODC) activity (33). Induction of ODC activity has been associated with tumor promotion in chemical carcinogenesis of the skin (34,35).

The photocarcinogenicity data described in the foregoing studies were derived from 5-MOP (100–300 ppm) in oily or acetone vehicles without UVB sunscreens in albino mice. Zajdela and Bisagni (30) used a UVA source, but Young et al. (31) used solar-simulating radiation (SSR), which is more relevant to natural solar exposure. Cartwright and Walter (36) published data that indicated that a commercially available oil-based 5-MOP sunscreen preparation enhanced phototumorigenesis in mouse skin. It has also been reported that a commercially available 5-MOP-containing sunscreen induced ODC activity in mouse skin exposed to UVB plus UVA (37). Theoretically, such data are more relevant to human use, but these experiments lacked vehicle (with and without sunscreen) controls, and it has been demonstrated that an oily vehicle enhances photocarcinogenesis (31,38). A more recent study, with the necessary controls, by Young et al. (39) has shown that low-concentration UVB sunscreens (Parsol MCX; 2-ethylyhexyl 4′-methoxycinnamate with Ultracyd; 1,7,7-tri-methol-3-benzylidene-bicyclo-[2.2.1]-2-heptanone) can have very marked inhibitory effects on 5-MOP-enhanced photocarcinogenesis. Animals were irradiated daily with SSR for about 45 weeks. Thereafter, half the animals were sacrificed and half were followed up without irradiation for 15 weeks. Comprehensive statistical analysis of the data showed that although 5-MOP was a photocarcinogen in the absence of UVB sunscreens, the addition of the sunscreens completely inhibited any enhancement by 5-MOP during the irradiation period of 45 weeks. Thus the groups treated with 5-MOP plus sunscreen were not significantly different from the groups treated with sunscreens alone. If the analyses were carried out for the total period (i.e., irradiation and follow-up), the majority of the statistical evaluations showed no difference between the 5-MOP plus sunscreen group and sunscreen only group, but a few analyses did indicate some evidence of increased risk in the 5-MOP groups. The main conclusions from these data were that the risks from the addition of 5-MOP to sunscreens were much lower than had been previously feared from earlier data.

The marked protective effect of Parsol MCX alone against 5-MOP phototumorigenesis was confirmed in another study (40) with Skh 1 hairless albino mice treated with preparations in a lotion base. However, in this study, 5-MOP plus sunscreen was significantly more phototumorigenic than sunscreen alone after 60 weeks of SSR exposure. These studies also compared Skh 1 mice with hairless pigmented mice of the same strain (Skh 2). Both types of mouse showed similar susceptibility to tumors, but it must be noted that epidermal melanogenesis did not occur in the dorsal skin of the Skh 2 mice; although this mouse shows a good pigmentation response in ear epidermis (41).

III. POSSIBLE MECHANISMS OF SUNSCREEN PROTECTION WITH 5-METHOXYPSORALEN

Psoralens are activated by UVA. The radiation source used in PUVA usually have an emission peak at about 365 nm, and the UVA sunscreens inhibit 5-MOP-induced erythema in the guinea pig (42). Thus, initially, it may seem surprising that UVB sunscreens should inhibit 5-MOP phototumorigenesis. It should not be forgotten that psoralen absorption spectra generally have maxima in the region of 300 nm. Most psoralen action spectrum studies in biological systems have been done with 8-MOP. These studies show that the waveband 320–340 nm is the most important for endpoints such as erythema (43), sunburn cell formation (44), inhibition of epidermal DNA synthesis (45), inhibition of mitogen-induced DNA synthesis in lymphocytes (46), and inhibition of growth in yeast (47). Young et al. have shown (48) that the 320 to 340-nm waveband is also the most significant in 8-MOP photocarcinogenesis. 5-MOP action spectra data are limited, but such that are available suggest that they are similar to 8-MOP (49,50). The only psoralen action spectrum for pigmentation appears to be with 4,5′-8-trimethylpsoralen in the guinea pig, in which a maximal plateau was seen at 330–350 nm (50).

The UVB sunscreens, such as Parsol MCX (2-ethylhexyl 4′-methoxycinnamate), show significant absorbance in the 320- to 335-nm region (39). Thus, it seems reasonable to suppose that the inhibition of photocarcinogenesis by Parsol MCX is due to optical filtering of relevant wavelengths of the action spectrum. However, Parsol MXC quenches the 5-MOP triplet state (51), which may be important in psoralen–DNA photobinding (52). Thus, protection may also be occurring at a more fundamental photophysical level.

IV. PHOTOCHEMOPROTECTION

Much of the data on the tanning efficacy of psoralens, especially 8-MOP, have been anecdotal, but efficacy has been confirmed (53), and PUVA is used as a means of treating vitiligo (54). One laboratory study indicated that 5-MOP-containing sunscreens did not enhance pigmentation (55), but it is possible that subjects were not observed for a long enough period. Field studies have provided evidence that 5-MOP-containing sunscreens do enhance tanning (56,57). Such preparations do enhance epidermal melanogenesis in the ear of the pigmented Skh 2 mouse (41), but not in dorsal skin (40).

Photochemoprotection has been demonstrated with 8-MOP in normal subjects using such endpoints as prevention of erythema (58,59), DNA damage, and sunburn cell formation (60). The underlying design of such experiments was to induce pigmentation with 8-MOP and UVR and then, at a later stage, challenge the tanned site with UVR alone and make comparisons with a non-8-MOP tanned site that had received a similar challenge. A PUVA tan is also used prophylactically in patients with sun-sensitive skins (3,61).

Until recently, evidence of photochemoprotection with 5-MOP has not been available. Sambuco et al. (62) showed that, under some conditions, a 5-MOP-containing sunscreen induced a tan that inhibited SSR-induced sunburn cell formation in the skin of the miniature pig. Sunburn cells are thought to be a consequence of keratinocyte DNA damage (63). Direct evidence of 5-MOP-induced protection against SSR-induced DNA damage in human keratinocytes in vivo has been shown (64,65). In these studies, previously untanned sites of the buttocks of skin type II subjects were tanned with sunscreens, with and without 5-MOP, and SSR alone. This was done by exposing the sites to 0.7 MED for 10 consecutive weekdays. One week after the final tanning treatment, the sites were challenged with 2 MED (determined on untanned skin) SSR.

Biopsies were taken and assessed for unscheduled DNA synthesis (UDS) as a marker of repair of SSR-induced DNA damage. The UDS comparisons of pretreated sites were made with a site that was challenged, but not previously tanned. The sites treated with sunscreen alone resulted in minimal tanning and afforded no protection from DNA damage. Protection from sites treated with sunscreen plus 5-MOP was 5-MOP dose-dependent and ranged from about 40 to 60%. Surprisingly, mean data (12 individuals; three experiments) from sites tanned with SSR alone did not show protection (65), although some protection was apparent in some individuals. The results from one experiment with skin type II volunteers are summarized in Table 1. Preliminary unpublished experiments by Potten and Young indicate a similar pattern of photoprotection in skin type I despite the poor tans observed in these subjects. A tan induced by SSR alone afforded no protection from sunburn cell damage in pig skin (62). In a separate human experiment, a 5-MOP-containing sunscreen tan afforded partial protection for up to 3 months. When such a tan (at about 3 months) was reexposed to SSR alone, protection was better than an SSR alone tan reexposed to SSR alone under the same conditions (65).

V. RISK-BENEFIT ANALYSIS

The data on 5-MOP photomutagenicity and photocarcinogenicity have prompted widespread concern and discussion about the use of 5-MOP in sunscreens (2,13,21,29,30,66–68). This concern is valid, given that 5-MOP and 8-MOP show similar photocarcinogenic potential in mice (31) and that there is good evidence that 8-MOP is photocarcinogenic in human skin (24).

Although the animal phototumorigenesis studies with 5-MOP and sunscreens (39,40) show differences in the degree of protection afforded by the sunscreen(s) from 5-MOP-enhanced phototumorigenesis, it is evident that UVB sunscreens are highly effective in this respect. Thus, the photocarcinogenic risk associated with the use of 5-MOP in sunscreens must be much lower than previously feared from photomutagenicity and photocarcinogenicity data without sunscreens. However, this risk has not been eliminated.

Also, the dorsal epidermis of mice in the 5-MOP photocarcinogenicity studies was incapable of tanning and was treated 5 days per week for about half the life span of a mouse, but human use of 5-MOP-containing sunscreens is likely to be intermittent. Despite the protective effects of the UVB filter and the likely differences in exposure patterns of laboratory mice and humans, it is apparent that the experimental carcinogenicity data indicate the possibility of enhanced skin cancer risk with 5-MOP. Thus, any advocation of the use of 5-MOP-containing sunscreens should be justified by benefits that outweigh long-term risks.

Table 1 Unscheduled DNA Synthesis (UDS), Tanning, and Skin Thickness in Four Skin Type II Volunteers

Treatment[a]/challenge[b]	UDS		Melanin histology[e]	Visual[f]	Skin thickening	
	Counts[c]	%[d]			Counts[f]	%[h]
None/none	0.6	0.08	1.0	0	14.7	95
None/SSR	7.6	100	1.7	0	15.5	100
SSR/SSR	6.4	84	5.0	3.5	27.5	177
S + SSR/SSR	7.4	97	2.0	1.0	18.0	116
S + 15 ppm 5-MOP + SSR/SSR	4.2	55	4.8	3.5	20.0	129
S + 30 ppm 5-MOP + SSR/SSR	2.7	36	5.5	4.5	25.1	162
S + 45 ppm 5-MOP + SSR/SSR	2.5	33	5.6	4.5	23.5	152

[a]10 consecutive weekdays; 0.7 MED SSR.
[b]2 MED SSR 1 week after last tanning treatment.
[c]Mean no gains/basal cell nucleus.
[d]As above, but expressed as percentage of no treatment/SSR challenge count.
[e]Graded scoring system 0–6.
[f]Graded scoring system 0–5.
[g]No cell layers in stratum corneum.
[h]As above, but expressed as percentage of no treatment/SSR challenge count.
Abbreviations: S, sunscreens; 2% Parsol MCX + 1% Eusolex 6300; SSR, solar-simulated radiation.
Source: Data modified from Ref. 64.

The human 5-MOP photochemoprotection data described may well tilt the risk/benefit analysis of 5-MOP sunscreen use in the direction of benefit. An important question here is whether the 5-MOP–DNA lesions, that may occur in obtaining protection, are of greater or lesser consequence (quantitatively and qualitatively) than solar UVR DNA lesions, such as thymine dimers, that may be inhibited by 5-MOP photochemoprotection. Such studies have not been done. Some investigations have been done with 8-MOP in human skin. There is indirect evidence that 8-MOP–DNA lesions may persist for up to 72 h (69), although the experimental design used did not permit the identification of the location of these lesions; namely, the epidermis or the dermis. More recently, monoclonal antibodies for 8-MOP–DNA lesions have shown the presence of 8-MOP nuclear lesions in keratinocytes in vitro and in vivo (70,71).

For the moment, the data suggest that there is a fine balance between inhibiting the photocarcinogenic potential of 5-MOP by sunscreens but, at the same time, maintaining its photochemoprotective potential. There would be no point in adding 5-MOP to a preparation if its protective efficacy was totally inhibited by the concentration of sunscreen present. It is unknown whether the sunscreen merely reduces the effective concentration of 5-MOP (e.g., is 25 ppm 5-MOP equivalent to 10 ppm in the presence of a sunscreen?) or whether the inhibitory effects are dependent on photobiological endpoint. Such would be expected if mechanisms, action spectra, threshold doses, time courses, interactions with other wavelengths, and the effects of exposure regimen vary with endpoint.

VI. CONCLUSIONS

The concern about the use of 5-MOP in sunscreens is justified. However, risk has been assessed on animal models incapable of showing potential benefit of pigmentation. It is clear that UVB sunscreens act as highly effective anticarcinogens in 5-MOP photocarcinogenicity. The results from the human photochemoprotection studies are encouraging and suggest that the judicious use of 5-MOP-containing sunscreens may have an important role in reducing solar radiation-induced DNA damage in skin. Such damage is widely believed to be the first step in skin cancer development. Given that the potential benefits of photochemoprotection are major, it is important that further research is carried out on the relation of risk to benefit of 5-MOP-containing sunscreens.

ACKNOWLEDGMENTS

The author's work on 5-MOP was funded by Laboratories Goupil (now Laboratoires Bergaderm).

REFERENCES

1. Roelandts, R., Vanhee, J., Bonamie, A., Kerkhofs, L., and Degreef, H. A survey of ultraviolet absorbers in commercially available sun products. *Int. J. Dermatol., 22*: 247–255 (1983).
2. Pathak, M. A. Sunscreens: Topical and systemic approaches for protection of human skin against harmful effects of solar radiation. *J. Am. Acad. Dermatol., 7*:285–312 (1982).
3. Hölzle, E., Plewig, G., Hofmann, C., and Roser-Maass, E. Polymorphous light eruption. *J. Am. Acad. Dermatol., 7*:111–125 (1982).
4. McFadden, N. UVA sensitivity and topical photoprotection in polymorphous light eruption. *Photodermatology, 1*:76–78 (1984).
5. Kligman, L. H., Akin, F. J., and Kligman, A. M. Sunscreens prevent ultraviolet photocarcinogenesis. *J. Am. Acad. Dermatol., 3*:30–35 (1980).
6. Kligman, L. H. and Kligman, A. M. The nature of photoageing: Its prevention and repair. *Photodermatology, 3*:215–227 (1986).
7. Plastow, S. R., Harrison, J. A., and Young, A. R. Early changes in dermal collagen of mice exposed to chronic UVB and the effects of a UVB sunscreen. *J. Invest. Dermatol., 91*:590–592 (1988).
8. Slaper, H. and Van der Leun, J. C. The use of sunscreens against long-term sun-damage, *Hazards of Light–Myths and Realities: Eye and Skin* (edited by J. Cronley-

Dillon, E. S. Rosen, and J. Marshall). Pergamon Press, New York, 1986, pp. 125–132.

9. MacKie, R. A., Elwood, J. M., and Hawk, J. L. M. Links between exposure to ultraviolet radiation and skin cancer. *J. R. Coll. Physicians*, *21*:91–96 (1987).
10. van Weelden, H. and van der Leun, J. C. UV-A induced tumors in pigmented and albino hairless-mice. *Human Exposure to Ultraviolet Radiation: Risks and Regulations* (edited by W. F. Passchier and B. F. M. Bosnjakovic). Elsevier Science Publishers, Amsterdam, 1987, pp. 45–52.
11. Ohme, C. Über die Zusamensetzung des Bergamottols. *Ann. Chem.*, *31*:316–321 (1839).
12. Cieri, U. R. Characterization of the steam nonvolatile residue of bergamot oil and some other essential oils. *J. Am. Off. Anal. Chem.*, *52*:719–728 (1969).
13. Forlot, P. Possible cancer hazard associated with 5-methoxypsoralen in suntan preparation. *Br. Med. J.*, *280*:648 (1980).
14. Freund, E. Über bischer noch nicht beschriebene kunstliche Hautverfarbungen. *Dermatol. Wochenschr.*, *63*:931–936 (1916).
15. Parrish, J. A. Fitzpatrick, T. B., Tanenbaum, L., and Pathak, M. A. Photochemotherapy of psoriasis with oral methoxsalen and longwave ultraviolet light. *N. Engl. J. Med.*, *291*:1207–1211 (1974).
16. Dall'Acqua, F., Terbojevich, M., Marciani, S., Vedaldi, D., and Recher, M. Investigations of the dark interaction between furocoumarins and DNA. *Chem. Biol. Interact.*, *21*:103–115 (1978).
17. Sa E Melo, T., Morliere, P., Santus, R., and Dubertret, L. Photoreactivity of 5-methoxypsoralen with calf thymus DNA upon excitation in the UVA. *Photobiochem. Photobiophys.*, *7*:121–131 (1984).
18. Dall'Acqua, F., Marciani, S., and Rodighiero, G. The action spectrum of xanthotoxin and bergapten for the photoreaction with native DNA. *Z. Naturforsch.*, *24b*: 667–671 (1969).
19. Dall'Acqua, F., Marciani, S., Veldaldi, D., and Rodighiero, G. Skin photosensitization and cross-linkings formation in native DNA by furocoumarins. *Z. Naturforsch.*, *29c*:635–636 (1974).
20. Averbeck, D. Photochemistry and photobiology of psoralens. *Proc. Jpn. Soc. Invest. Dermatol.*, *8*:52–73 (1984).
21. Ashwood-Smith, M. J., Poulton, G. A., Barker, M., and Mildenberger, M. 5-Methoxypsoralen, an ingredient in several suntan preparations, has lethal mutagenic and clastogenic properties. *Nature*, *285*:407–409 (1980).
22. Urbach, F. Modification of ultraviolet carcinogenesis by photoactive agents. *J. Invest. Dermatol.*, *32*:373–378 (1959).
23. Forbes, P. D., Davies, R. E., and Urbach, F. Phototoxicity and photocarcinogenesis: Comparative effects of anthracene and 8-methoxypsoralen in the skin of mice. *Food Cosmet. Toxicol.*, *14*:303–306 (1976).
24. Stern, R. S., Laird, N., Melski, J., Parrish, J. A., Fitzpatrick, T. B., and Bleich, H. L. Cutaneous squamous-cell carcinoma in patients treated with PUVA. *N. Engl. J. Med.*, *310*:1156–1161 (1984).
25. Henseler, T., Christophers, E., Hönigsmann, H., et al. Skin tumors in the European PUVA study. *J. Am. Acad. Dermatol.*, *16*:106–116 (1987).

26. Gibbs, N. K., Hönigsmann, H., and Young, A. R. PUVA treatment strategies and cancer risk. *Lancet*, *1*:150–151 (1986).
27. Hönigsmann, H., Jaschke, E., Gschnait, F., Brenner, W., Fritsch, P., and Wolff, K. 5-Methoxypsoralen (bergapten) in photochemotherapy of psoriasis. *Br. J. Dermatol.*, *101*:369–378 (1979).
28. Mezzadra, G., Guarneri, B., Grupper, C., and Forlot, P. Effects of chronic field exposure of humans to bergapten, *Psoralens in Cosmetics and Dermatology* (edited by J. Cahn, P. Forlot, C. Grupper, A. Meybeck, and F. Urbach). Pergamon Press, Paris, 1981, pp. 383–386.
29. 5-Methoxypsoralen. *IARC Monogr. 40*:327–347 (1986). [Evaluation of the carcinogenic risk of chemicals to humans].
30. Zajdela, F. and Bisagni, E. 5-Methoxypsoralen, the melanogenic additive in suntan preparations, is tumorigenic in mice exposed to 365 nm u.v. radiation. *Carcinogenesis*, *2*:121–127 (1981).
31. Young, A. R., Magnus, I. A., Davies, A. C., and Smith, N. P. A comparison of the phototumorigenic potential of 8-MOP and 5-MOP in hairless albino mice exposed to solar simulated radiation. *Br. J. Dermatol.*, *108*:507–518 (1983).
32. Abel, G. Chromosome damage induced in human lymphocytes by 5-methoxypsoralen and 8-methoxypsoralen plus UV-A. *Mutat. Res.*, *190*:63–68 (1987).
33. Connor, M. J. and Lowe, N. J. The induction of erythema, edema, and the polyamine synthesis enzymes ornithine decarboxylase and *S*-adenosyl-L-methionine decarboxylase in hairless mouse skin by psoralens and long-wave ultraviolet light. *Photochem. Photobiol.*, *39*:787–792 (1984).
34. O'Brien, T. G. The induction of ornithine decarboxylase as an early, possibly obligatory, event in mouse skin carcinogenesis. *Cancer Res.*, *38*:2644–2653 (1976).
35. Slaga, T. G. Mechanisms involved in two-stage carcinogenesis in mouse skin. *Mechanisms of Tumor Promotion*. Vol. 2: *Tumor Promotion and Skin Carcinogenesis* (edited by T. J. Slaga). CRC Press, Boca Raton, FL, 1984, pp. 1–16.
36. Cartwright, L. E. and Walter, J. F. Psoralen-containing sunscreen is tumorigenic in hairless mice. *J. Am. Acad. Dermatol.*, *8*:830–836 (1983).
37. Walter, J. F., Gange, R. W., and Mendelson, I. R. Psoralen-containing sunscreen induces phototoxicity and epidermal ornithine decarboxylase activity. *J. Am. Acad. Dermatol.*, *6*:1022–1027 (1982).
38. Gibbs, N. K., Young, A. R., and Magnus, I. A. Failure of UVR dose reciprocity for skin tumorigenesis in hairless mice treated with 8-methoxypsoralen. *Photochem. Photobiol.*, *42*:39–42 (1985).
39. Young, A. R., Gibbs, N. K., and Magnus, I. A. Modification of 5-methoxypsoralen phototumorigenesis by UVB sunscreens: A statistical and histologic study in the hairless albino mouse. *J. Invest. Dermatol.*, *89*:611–617 (1987).
40. Young, A. R. and Walker, S. L. Experimental photocarcinogenesis of psoralens. *Psoralens*: *Past, Present and Future of Photochemoprotection and Other Biological Activities* (edited by T. B. Fitzpatrick, P. Forlot, M. A. Pathak, and F. Urbach). John Libbey Eurotext, Montrouge, France, 1989, pp. 357–366.
41. Walker, S. L. and Young, A. R. A possible animal model for the role of induced pigmentation in photocarcinogenesis. *Br. J. Dermatol.*, *113*:787 (1985).

42. Saettone, M. F., Trambusti, M., and Giannaccini, B. Inhibition de l'action phototoxique du bergaptène par des filtre UV-A. *Int. J. Cosmet. Sci.*, *5*:201–213 (1983).
43. Cripps, D. J., Lowe, N. J., and Lerner, A. B. Action spectra of topical psoralens: A re-evaluation. *Br. J. Dermatol.*, *107*:77–82 (1982).
44. Young, A. R. and Magnus, I. A. An action spectrum for 8-MOP induced sunburn cells in mammalian epidermis. *Br. J. Dermatol.*, *104*:541–548 (1981).
45. Kaidbey, K. H. An action spectrum for 8-methoxypsoralen-sensitized inhibition of DNA synthesis in vivo. *J. Invest. Dermatol.*, *85*:98–101 (1985).
46. Gasparro, F. P., Berger, C. L., and Edelson, R. L. Effect of monochromatic UVA light and 8-methoxypsoralen on human lymphocyte to response to mitogen. *Photodermatology*, *1*:10–17 (1984).
47. Baydoun, S. A. and Young, A. R. An action spectrum for lethal photosensitization of *Candida albicans* by 8-MOP after low-dose broad-band UV-A irradiation; an action spectrum for 8-MOP 4′,5′-monoadducts? *Photochem. Photobiol.*, *46*:311–314 (1987).
48. Young, A. R., Walker, S. L., and Garmyn, M. G. A first approach to an action spectrum for 8-MOP phototumorigenesis in mice. *J. Invest. Dermatol.*, *90*:175–178 (1988).
49. Young, A. R. and Barth, J. Comparative studies on the photosensitizing potency of 5-methoxypsoralen and 8-methoxypsoralen as measured by cytolysis in *Paramecium caudantum* and *Tetrahymena pyriformis*, and growth inhibition and survival in *Candida albicans*. *Photochem. Photobiol.*, *35*:83–88 (1982).
50. Nakayama, Y., Morikawa, F., Fukuda, M., Hamano, M., Toda, K., and Pathak, M. A. Monochromatic radiation and its application—laboratory studies on the mechanism of erythema and pigmentation induced by psoralen. *Sunlight and Man* (edited by T. B. Fitzpatrick, M. A. Pathak, L. C. Harber, M. Seiji, and A. Kukita). University of Tokyo Press, Tokyo, 1974, pp. 591–611.
51. Morliére, P., Avice, O., Sa E Melo, T., Dubertret, L., Giraud, M., and Santus, R. A study of the photochemical properties of some cinnamate sunscreens by steady state and laser flash photolysis. *Photochem. Photobiol.*, *36*:395–399 (1982).
52. Bensasson, R. V., Land, E. J., and Salet, C. Triplet excited state of furocoumarins: Reaction with nucleic acid bases and amino acids. *Photochem. Photobiol.*, *27*:273–280 (1978).
53. Zaynoun, S., Konrad, K., Gschnait, F., and Wolff, K. The pigmentary response to photochemotherapy. *Acta Dermatvenereol.* *57*:431–440 (1977).
54. Kenny, J. A. Vitiligo treated by psoralens. *Arch. Dermatol.*, *103*:475–480 (1971).
55. Brenner, W. and Gschnait, F. The value of topical sunscreens containing psoralens. *Arch. Dermatol. Res.*, *267*:189–190 (1980).
56. Tronnier, H. and Agache, P. Field trial on suntan products containing bergamot oil in Tunisia. *Psoralens in Cosmetics and Dermatology* (edited by J. Cahn, P. Forlot, C. Grupper, A. Meybeck, and F. Urbach). Pergamon Press, Paris, 1981, pp. 411–417.
57. Sigafoes, R. Evaluation of the tanning potential of topical sunscreens containing psoralens. *Psoralens in Cosmetics and Dermatology* (edited by J. Cahn, P. Forlot, C. Grupper, A. Meybeck, and F. Urbach). Pergamon Press, Paris, 1981, 419–426.

58. Imbrie, J. D., Daniels, F., Bergeron, L., Hopkins, C. E., and Fitzpatrick, T. B. Increased erythema threshold six weeks after a single exposure to sunlight plus oral methoxsalen. *J. Invest. Dermatol.*, *32*:331–337 (1959).
59. Cripps, D. J. Natural and artificial photoprotection. *J. Invest. Dermatol.*, *76*:154–157 (1981).
60. Gschnait, F., Brenner, W., and Wolff, K. Photoprotective effect of a psoralen–UVA-induced tan. *Arch. Dermatol. Res.*, *263*:181–188 (1978).
61. Gschnait, F., Hönigsmann, H., Brenner, W., Fritsch, P., and Wolff, K. Induction of UV light tolerance by PUVA in patients with polymorphous light eruption. *Br. J. Dermatol.*, *99*:293–295 (1978).
62. Sambuco, C. P., Forbes, P. D., Davies, R. E., and Urbach, F. Protective value of skin tanning induced by ultraviolet radiation plus a sunscreen containing bergamot oil. *J. Soc. Cosmet. Chem.*, *38*:11–19 (1987).
63. Young, A. R. The sunburn cell. *Photodermatology*, *4*:127–134 (1987).
64. Young, A. R., Potten, C. S., Chadwick, C. A., Murphy, G. M., and Cohen, A. J. Inhibition of UV radiation-induced DNA damage by a 5-methoxypsoralen tan in human skin. *Pigment Cell Res.*, *1*:350–354 (1988).
65. Potten, C. S., Chadwick, C. A., Young, A. R., Murphy, G. M., and Cohen, A. J. A 5-methoxypsoralen-induced tan protects against DNA damage from a subsequent exposure to solar simulated radiation in human skin. *Psoralens*: *Past, Present and Future of Photochemoprotection and other Biological Activities* (edited by T. B. Fitzpatrick, P. Forlot, M. A. Pathak, and F. Urbach). John Libby Eurotext, Montrouge, France, 1989.
66. Ashwood-Smith, M. J. Possible cancer hazard associated with 5-methoxysoralen in suntan preparations. *Br. Med. J.*, *2*:1144 (1979).
67. Hook, I. Possible cancer hazard associated with 5-methoxypsoralen in suntan preparations. *Br. Med. J.*, *280*:1537–1538 (1980).
68. Natarajan, A. T., Verdegaal-Immerzeel, E. A. M., Ashwood-Smith, M. J., and Poulton, G. A. Chromosomal damage induced by furocoumarins and UVA in hamster and human cells including cells from patients with ataxia telangiectasia and xeroderma pigmentosum. *Mutat. Res.*, *84*:113–124 (1981).
69. Gange, R. W., Levins, P., Murray, J., Anderson, R., and Parrish, J. A. Prolonged skin photosensitization induced by methoxalen and subphototoxic UVA irradiation. *J. Invest. Dermatol.*, *82*:219–222 (1984).
70. Yang, X. Y., DeLeo, V., and Santella, R. M. Immunological detection and visualization of 8-methoxypsoralen–DNA photoadducts. *Cancer Res.*, *47*:2451–2455 (1987).
71. Yang, X. Y., Gasparro, F. P., DeLeo, V. A., and Santella, R. M. 8-Methoxypsoralen–DNA adducts in patients treated with 8-methoxypsoralen and ultraviolet A light. *J. Invest. Dermatol.*, *92*:59–63 (1989).

23

Tanning Accelerators

James M. Wilmott
Mary Kay Cosmetics, Inc., Dallas, Texas

Michele C. Duggan
Avon Products, Inc., Suffern, New York

Alexander P. Znaiden
Chesebrough-Ponds U.S.A., Trumbull, Connecticut

I. INTRODUCTION

What is a tan? How is it formed? What does it do? Why is it there? How is it controlled?

Is tanning simply an insignificant biological phenomenon of evolutionary specialization, a primordial leftover, or a yet to be understood fractal event? Has its total purpose been missed by the researcher's eye that often looks for singularity of function, specificity of action, and black and white boundaries for definition?

Is it a sunscreen? A regulator of vitamin D synthesis? A free radical trap? A heat absorption system? A hormone? A collection and disposal system for toxic, oxidative byproducts of cellular metabolism and photodamage? A defense mechanism to protect the cell's genetic materials from ultraviolet damage?

The complete picture of the mechanisms of tanning is still unknown, but it is slowly beginning to come into focus. Scientists have elucidated the pathways of synthesis, identified cause and effect relations, characterized the chemical composition, investigated the cells in which it is produced, and studied the consequences of unhealthy expression.

What is not well known is how to control the tanning process for aesthetic and therapeutic benefit. The following is presented to provide information on some of the agents that have demonstrated an ability to accelerate tanning and to offer suggestions that may lead to more effective materials in the future.

II. THE TANNING PROCESS

It has been suggested that the pigmentation response of skin may, in fact, be more than just the well-known "tanning" that occurs when one is exposed to ultraviolet (UV) light. Although most persons are aware of the relation between tanning and sunlight, not many appreciate the mechanism by which this occurs. To effectively enhance tanning, such knowledge is critical and will be briefly described.

Tanning may be simply defined as an enhancement in the production of pigment by the skin. This pigment is called melanin, and it can be either black (eumelanin) or red (phaeomelanin). The process by which melanin is formed is called melanogenesis. It takes place in cellular organelles known as melanosomes, which are produced by epidermal melanocytes. The melanosomes contain everything that is needed for melanin production. They are transported by melanocytic dendrites, to the epidermal keratinocytes. There are usually 36 keratinocytes associated with one melanocyte (1,2). This complex is known as the epidermal melanin unit (Fig. 1). Only when the melanin-containing melanosomes are in the keratinocytes is a tan visible.

The basic biochemical pathway of melanogenesis was elucidated by Mason and Raper, who later altered it to accommodate the contribution of all intermediates to the final product (Fig. 2). These investigators determined that the hydroxylation of the amino acid tyrosine to dehydroxyphenylalanine (dopa) and the oxidation of dopa to dopaquinone are under enzymatic control, whereas the remaining steps occurred spontaneously (3). However, subsequent investigations recognized that this scheme was too simplistic because it did not account for the different types of pigments that are observed in vivo. Figure 3 demonstrates the divergence of the pathway that is due to the combination of dopaquinone with either sulfur or oxygen to form the phaeomelanin or eumelanin monomer, respectively (2).

The sulfur-containing compounds of particular interest are the amino acid cysteine and the tripeptide glutathione (GSH). GSH or cysteine can be added to dopaquinone in various ways to produce four different forms of cysteinyldopa with 5-*S*-cysteinyldopa and 2-*S*-cysteinyldopa being prevalent.

The final step in this process is the polymerization of the eumelanin or phaeomelanin monomers within a peptide matrix to form the characteristic black and red pigments (4,5).

III. THE ENZYME

The generally recognized key steps in melanogenesis are the conversions of tyrosine to dopa and dopa to dopaquinone, which are both catalyzed by the enzyme tyrosinase. Tyrosinase is a copper-containing protein that is usually gly-

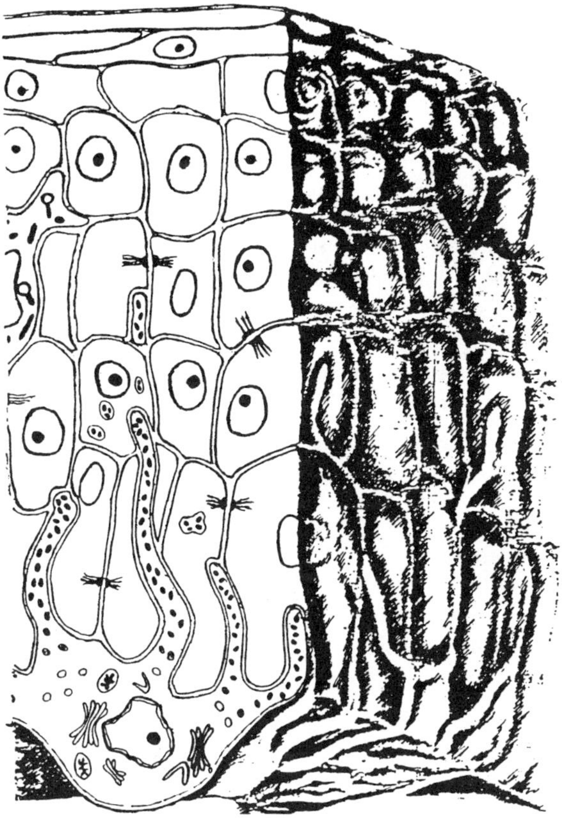

Figure 1 The epidermal melanin unit: The melanocyte (bottom) deposits pigment-filled melanosomes into neighboring keratinocytes.

cosylated with neuraminic acid (also known as sialic acid), galactose, and possibly mannose in the active state (1,2). The protein fraction is synthesized on ribosomes that are located on the rough endoplasmic reticulum. It then migrates to a site in the Golgi apparatus through the cytoplasm or the smooth endoplasmic reticulum. Once this site is cleaved from the Golgi apparatus, it is referred to as a premelanosome and exists for awhile in the cytoplasm of the melanocyte (6).

Tyrosinase, similar to many other enzymes, is synthesized in an inactive form, termed a proenzyme, which is a tetramer. McGuire et al. (2), in their work with frog epidermis, reported that activation occurs in cytoplasmic components before the formation of premelanosomes. Mishima et al. (7), in their research with B16

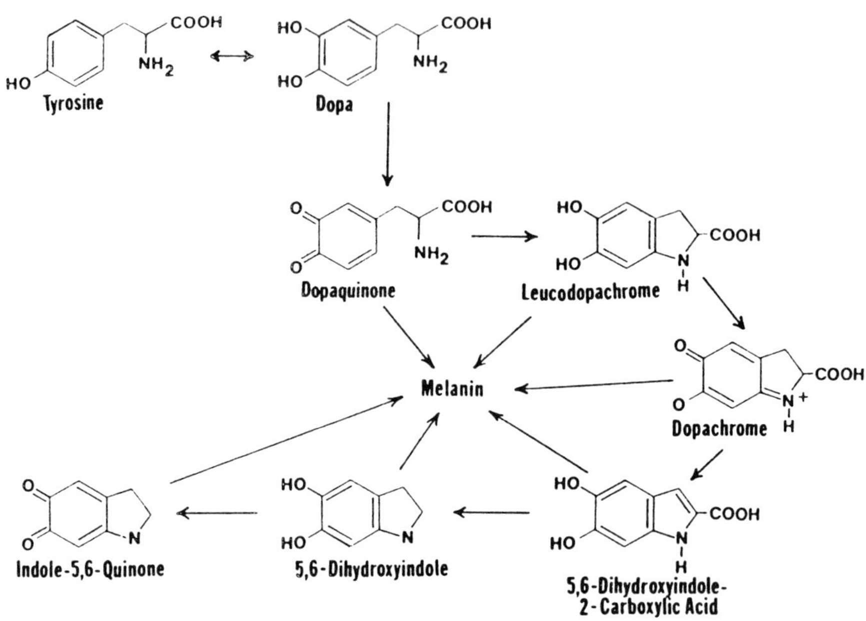

Figure 2 Mason–Raper pathway. (From Ref. 3.)

murine melanoma cells, confirmed that transport into the premelanosomes is necessary for melanogenic initiation (8).

In the early 1980s, Hearing et al. (8) discovered that tyrosinase exists in three slightly different forms (isoenzymes) that have been identified as T_1, T_3, and T_4. It has been speculated that a fourth isoenzyme, T_2, existed, but this proved to be an electrophoretic artifact of T_3. T_1 and T_3 are soluble and found in the cytoplasm. They have molecular weights of 66,000 and 56,700, respectively. Together, they are responsible for only 10% of all tyrosinase activity. Both have similar enzymatic properties and are believed to be associated with an inhibitory factor. This factor has been identified by Hearing et al. as 5,6-dihydroxyindole-blocking factor (IBF). As the name signifies, IBF inhibits the conversion of 5,6-dihydroxyindole to indole-5,6-quinone (2,8,9). T_4, found bound to the melanosomal membranes, has a molecular weight of 120,000 and is responsible for approximately 90% of tyrosinase activity. This high level of activity may be due to the presence of dopachrome conversion factor (DCF) and 5,6-dihydroxyindole conversion factor (ICF). The three isoenzymes also differ in carbohydrate content, T_1 and T_4 being glycosylated, whereas T_3 is nonglycosylated (8).

From their research with B16 murine melanoma cells and C57B1/6N murine hairbulbs, Hearing et al. (8) proposed the following pathway for tyrosinase ac-

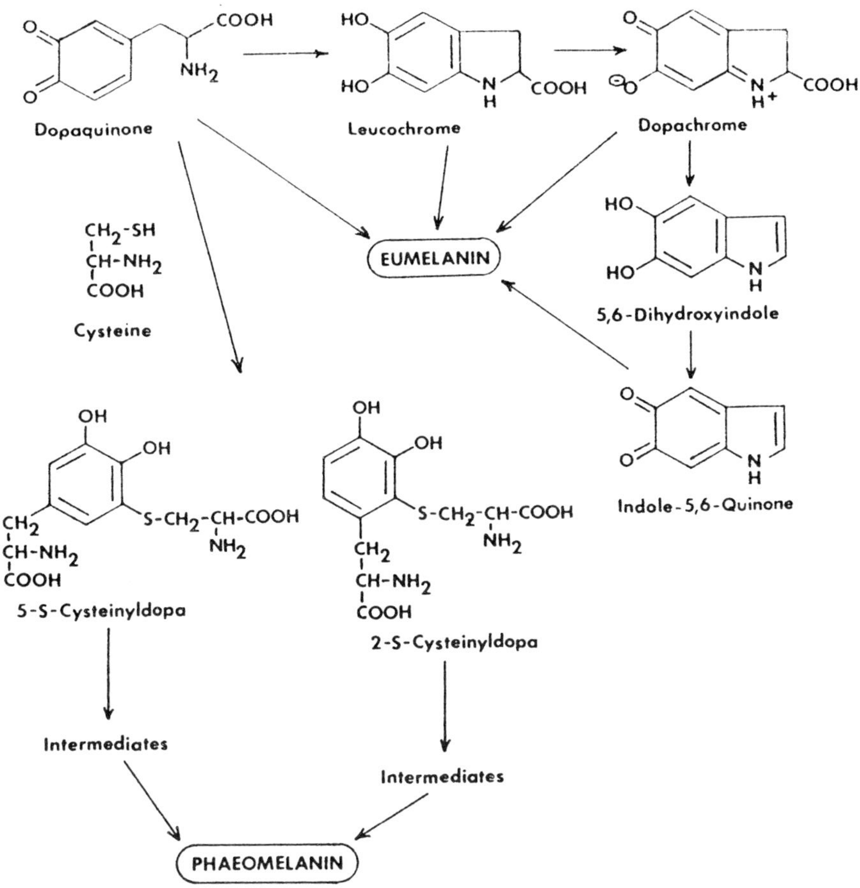

Figure 3 The biosynthetic pathway for eumelanin and phaeomelanin. (From Ref. 2.)

tivation. T_3 is newly synthesized by the ribosomes of the rough endoplasmic reticulum. It is then glycosylated, the sugars are added in the Golgi apparatus to form T_1. The T_1 isoenzyme is delivered to the premelanosome and is attached to the membrane to form a T_4 molecule. The authors supported this theory, in part, by incubating T_4 with the proteolytic enzyme trypsin. This cleaved the membrane fragments from T_4 and produced a molecule that was electrophoretically similar to T_1.

There was much debate over whether tyrosinase was one enzyme with two functions or two enzymes, tyrosinase and dopaoxidase, each with one function. However, in recent years, researchers have identified two binding sites on the tyrosinase enzyme, one for the oxidation of dopa and the other for tyrosine (2). In fact, it has been determined that the oxidation of dopa proceeds at a faster rate than the hydroxylation of tyrosine. Also, when dopa is added to a reaction mixture containing tryrosine and tyrosinase, the rate of hydroxylation is greatly increased, indicating that dopa functions as a cofactor for tyrosinase as well as a substrate (2). It is hypothesized that the increase in the reaction is due to a conformational change that enhances the efficiency of the enzyme.

Although it was believed that the reactions beyond dopaquinone were spontaneous, it has been determined that tyrosinase also catalyzes the conversion of additional intermediates farther along the melanogenic pathway (10). It has also been determined that these reactions are regulated by several biological factors. Pawelek's laboratory (11) discovered three of these regulatory factors in mouse melanoma cells. DCF increases the conversion of dopachrome to 5,6-dihydroxyindole 2-carboxylate, whereas ICF increases the conversion of 5,6-dihydroxyindole to the quinone and, subsequently, to melanin. IBF restricts melanogenesis at 5,6-dihydroxyindole and protects cells from the cytotoxic effects of melanin precursors (12). Experiments done by Hearing et al. (9) suggest that IBF is associated with tyrosinase isoenzymes T_1 and T_3, explaining their low level of activity. Both conversion factors, DCF and ICF, migrate with tyrosinase when chromatographed on both gel filatration and ion-exchange columns, suggesting that the conversion factors are complexed with the isoenzymes in the pigment cells (8). This same set of experiments also suggests that DCF is associated with T_4, the most active form of tyrosinase. Figure 4 represents a hypothesis developed by Wilmott and Duggan that describes the activation of the tyrosinase isoenzymes and their relation with the blocking and conversion factors.

From the foregoing, one can appreciate the intricacy involved in melanogenesis. Even though this complexity makes the understanding of tanning more difficult, it does provide the opportunity to better comprehend how one might influence melanogenic expression. This additional knowledge might provide the cosmetic or pharmaceutical formulator with the key to develop a product that produces a tan without the need for exposing the skin to high levels of damaging sunlight.

IV. SUPPORT SYSTEMS

As is true for all enzymatic reactions, melanogenesis does not exist alone, but rather, as a part of a complicated series of biochemical events. Any modification to these secondary systems can directly influence the rate and type of melanin produced. A chart identifying some of these secondary processes and their interaction with the primary melanogenic path is found in Figure 5.

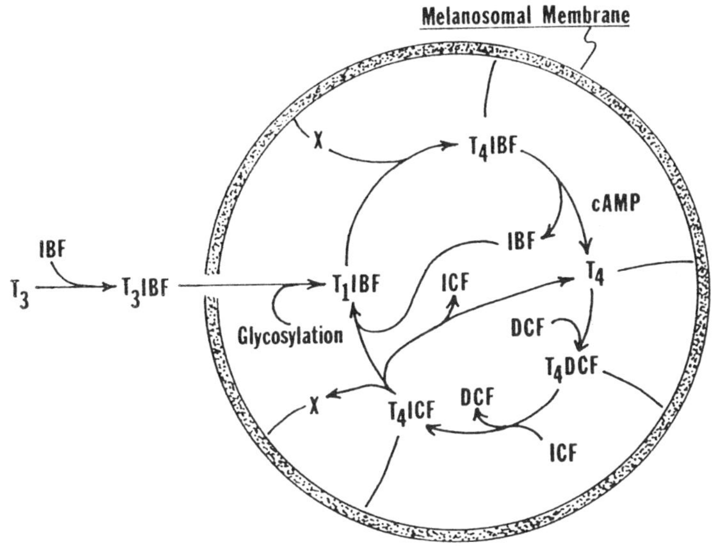

Figure 4 Proposed schematic for the activation of tyrosinase: X, membrane factor; IBF, 5,6-dihydroxyindole-blocking factor; DCF, dopachrome conversion factor; ICF, 5,6-dihydroxyindole conversion factor; T_1, T_3, T_4, tyrosinase isoenzymes.

A. Membrane Mediated

It is very apparent that the development of melanin is influenced by materials that bind to or complex with the membrane of the melanocyte. This "receptor" binding by an external factor produces an intracellular signal that ultimately stimulates melanogenesis.

Perhaps the most widely known materials purported to be involved with melanogenesis is melanocyte-stimulating hormone (α-MSH or α-melanotropin). It is produced by the anterior portion of the pituitary gland (9). Amino acid analysis has revealed the sequence of α-MSH (Fig. 6); two peptide fragments within the α-MSH have melanogenic activity (13).

In melanoma cells, α-MSH enhances tyrosinase activity by binding to a cell surface receptor on the melanocyte, which activates the enzyme adenylate cyclase. This catalyzes the conversion of adenosine triphosphate (ATP) to cyclic adenosine monophosphate (cAMP). Calcium is needed for signal transduction; that is, the movement of the signal from the cell surface to the adenylate cyclase (14). The cAMP is the true activator of tyrosinase. It is thought to bind to and remove an inhibitor of tyrosinase (15). This inhibitor, in fact, may be the IBF discussed by Pawelek et al., which is associated with the T_1 and T_3 isoenzymes.

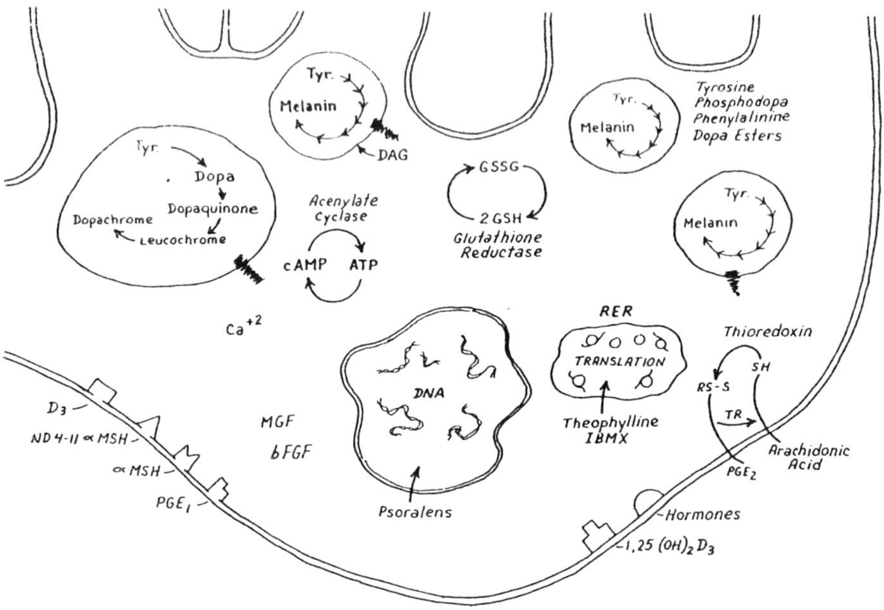

Figure 5 Support systems and agents that influence melanogenesis. DAG, diacylglycerol; GSH, glutathione; GSSG, oxidized glutathione; TR, thioredoxin reductase; RER, rough endoplasmic reticulum; IBMX, isobutylmethylxanthine; PGE_2, prostaglandin E_2; PGE_1, prostaglandin E_1; D_3, vitamin D_3; $1,25(OH)_2D_3$, 1α,25-dihydroxyvitamin D_3; αMSH, melanocyte-stimulating hormone; MGF, melanocyte growth factor; bFGF, basic fibroblast growth factor; ATP adenosine triphosphate; cAMP, cyclic adenosine monophosphate.

Acetyl-Ser-Tyr-Ser-Met

Glu-His-Phe-Arg-Trp

(Active Area I)

Gly-Lys-Pro-Val-NH_2

(Active Area II)

Figure 6 Amino acid sequence for MSH showing two active sites.

However, studies by Gilchrist (16) have demonstrated that diacylglycerol, and not AMP, may serve as the activating mechanism for nonmelanoma cells. Cyclic-AMP was also believed to be involved in the transfer of melanosomes into the keratinocyte as evidenced by the increased dendricity of melanocytes in the presence of MSH (2,17). Prostaglandin-E_2 (PGE_2) also increases melanocyte swelling and dendricity (18).

The MSH-induced activation may also occur at the genetic transcriptional or translational level through either cAMP or an alternative pathway. The presence of α-MSH increases the number of melanosomes. Furthermore, Hu et al. (19) have demonstrated an increase in tyrosinase synthesis in mouse melanoma cells that were treated with MSH.

In addition to the two active sites in the α-MSH molecule, researchers have identified α-MSH analogues that stimulate melanogenesis.The most potent analogue is Ac-[Nle 4, D-Phe 7] α-MSH 4-1 $1NH_2$ (ND4-11 α-MSH). Panasci et al. (20) demonstrated that this analogue was at least ten times more potent than α-MSH in their work with F1 variant B16 melanoma cells. They discovered that there are approximately 4500 receptors for this analogue per melanoma cell. The investigators theorize that once the analogue complexes with a receptor, the receptor-hormone complex is internalized into the cells, followed by intracellular degradation of the hormone (20). Hadley et al. (21) termed ND4-11 α-MSH "superpotent," having "ultraprolonged melanogenic effects" based on their work with Cloudman S91 cells. In this model, ND4-11 α-MSH increased tyrosinase activity 100-fold, with its melanogenic effects lasting up to 6 days after the analogue was removed from the cell culture medium (21).

Other materials also affect melanogenesis through a membrane receptor mechanism. Cholera toxin, PGE_1, and possibly cholecalciferol (vitamin D_3) are believed to stimulate melanogenesis by activating adenylate cyclase (1,9).

In contrast with vitamin D_3, the hormonal form of this vitamin (1α,25-dihydroxyvitamin D_3) increases tyrosinase activity, but not through a AMP intermediate (12). Finally, gonadal hormones, such as estrogen and progesterone, also enhance tyrosinase activity, but their intracellular mechanisms of action have not yet been elucidated (2).

Another more recently discovered membrane-bound reaction that influences pigmentation is the thioredoxin–thioredoxin reductase system. Thioredoxin is a flavoprotein present in the plasma membrane of both the melanocyte and keratinocyte. It is involved in the reduction of oxygen free radicals. It requires intracellular NADPH to accomplish reduction of superoxide anion radicals. Oxidized thioredoxin in the presence of tyrosinase, enhances the conversion of tyrosine to melanin. Reduced thioredoxin decreases the conversion of tyrosine to dopa even in the presence of tyrosinase (22). This finding is particularly significant since many of the mediators for reactions that cause enhanced pigmentation occur by a free radical mechanism. In fact, many of the processes involved in the inflammation processes generate free radicals of arachidonic acid (e.g., certain prostaglandins and leukotrienes). This process may represent the mechanism by which repeated irritation and sunlight achieve their melanogenic effect.

B. Intracellular Mechanisms

Within the cell, several peripheral events occur that influence the rate and manifestation of melanogenesis. Of particular importance is the role of DNA, which is at the core of all cellular activity because all molecules produced by cells are ultimately under DNA control. In melanogenesis, factors that direct the expression of DNA may enhance (or inhibit) either the availability of key melanogenic components or the proliferation of the melanocyte itself. In the former, either transcriptional or translational properties may be influenced. In the latter, the synthesis of DNA itself is enhanced, resulting in increased mitotic activity.

Materials that are theorized to promote tanning by their effect on DNA are psoralens. In 1974, Carter et al. (23) demonstrated in vitro stimulation of tyrosinase activity by trimethylpsoralen (TMP) and by 8-methoxypsoralen (8-MOP) in conjunction with UVA exposure. In 1983, Borkovic et al. (24), in their work with mouse melanoma cells, discovered that 8-MOP requires UVA exposure, whereas TMP functions in the absence of UV light. It has been purported that the different pigmentary effects of TMP and 8-MOP, along with different binding sites on the nuclear DNA, support the hypothesis of DNA level regulation (23).

Other materials, such as melanocyte growth factor and basic fibroblast growth factor, accelerate the growth cycle of the melanocyte in culture. This enhances the mitogenic activity of the cells (25).

Finally, many other intracellular biochemical reactions are required to support the melanogenic process. In fact, proper homeostasis is required for optimum cell function. Any modification of those systems (e.g., the selective stimulation or inhibition of a key enzyme, such as glutathione reductase) might have profound consequences on the development of melanin.

Evidence for the influential effect of the support systems in the development of pigmentation is found in the myriad chemicals and physiological disorders that result in hyperpigmentation (17,26,27). Most of these conditions act at a different site on or within the melanocyte (Table 1). This diversity of action has given rise to speculation that melanogenesis is much more than the development of color due to exposure of the skin to UV radiation. Rather, the conversion of tyrosine to melanin may be the mechanism for the ultimate removal of oxidative damage from the skin.

V. TANNING ACCELERATORS

Since the close of World War II, the possession of a tan has been associated with available leisure time and social success. The cosmetic value of pigmen-

Table 1 Factors Involved with Promoting Pigmentation

Source	Melanocytotic (increase in number of melanocytes)	Melanotic (increase in melanin)
Heritable or developmental	Lentigines	Café au lait macule
	Moynahan's syndrome	Neurofibromatosis
	Centrofacial neurodysraphic lentiginosis	Albright's syndrome
	Peutz-Jegher's syndrome	Silver-Russel syndrome
	PUVA	Westerhof syndrome
	Soto syndrome	Watson syndrome
		Bloom syndrome
		Gastrocutaneous syndrome
		Becker's melanosis
		Nevus spilus
		Ephelides (freckles)
		NAME or LAMB syndrome
		Ichthyosis nigricans
		Neurocutaneous melanosis
		Familial periorbital hyperpigmentation
		Familial progressive hyperpigmentation
		Dowling-Degos disease
		Dyskeratosis congenita
		Fanconi's syndrome
		Human chimera
		Acropigmentation of Dotti
		Reticulate acropigmentation of Kitamura
		Dermatopathia pigmentosa reticularis
		POEMS syndrome
		Carbon baby syndrome
Metabolic		Porphyria cutanea tarda
		Hemochromatosis
		Hepatolenticular degeneration
		Gaucher's disease
		Niemann-Pick disease
Endocrine		Melasma
		ACTH- and MSH-producing tumors
		Exogenous ACTH therapy
		Pregnancy
		Addison's disease
		Estrogen therapy

Table 1 Continued

Source	Melanocytotic (increase in number of melanocytes)	Melanotic (increase in melanin)
Chemical and drug		Arsenicals Busulfan Photochemical agents (psoralens, tar) Berlock dermatosis 5-Fluorouracil, systemic Cyclophosphamide Nitrogen mustard, topical Bleomycin
Physical	Lentigo, solar Ultraviolet radiation (tanning)	Ultraviolet radiation (suntanning) Thermal radiation Ionizing radiation: alpha, beta, gamma Trauma (e.g., chronic pruritus)
Inflammation and infection		Postinflammatory melanosis (exanthems, drug eruptions) Lichen planus Lupus erythematosus, discoid Lichen simplex chronicus Atopic dermatitis Psoriasis Tinea versicolor
Neoplastic		Melanoma Mastocytosis Acanthosis nigricans with adenocarcinoma and lymphoma
Miscellaneous	Lentigines, eruptive Lentigo, senile	Scleroderma, systemic Chronic hepatic insufficiency Whipple's syndrome Cronkhite-Canada syndrome

Source: Ref. 27.

tation has generated a proliferation of consumer products claiming to enhance or ''accelerate'' the tanning process. The early 1980s ushered in the era of tan accelerators that successfully created a niche in the sun products category. Today more than 15 product lines are marketed for the purpose of enhanced tanning. The rapid proliferation of tanning accelerators confirms the consumer desire for pro-pigmenting products. However, there is also a pharmacological need to develop these products to improve the outward appearance of pigmentary disorders, such as melasma, listed in Table 1 (27).

Pigmentation is a complex series of biochemical reactions that can be affected by a wide variety of materials and conditions. A broader understanding of the many mechanisms involved will help formulators to identify materials that can safely modify the tanning process.

A. Direct Manipulation of the Melanogenic Path

Most of the currently marketed tanning accelerators contain tyrosine, tyrosine derivatives, tyrosine–riboflavin (or ATP) complexes, amino acid blends, or some combination of these agents. Tyrosine is used to increase the substrate available for tyrosinase (28); however, this approach may have some limitations because

1. Tyrosine may not penetrate the skin.
2. Tyrosine has low water solubility; therefore, the amount in the skin may not be greatly increased.
3. The concentration of tyrosine in the skin may already be at the maximum level that the enzyme can handle.

Any of these factors could jeopardize the efficacy of the tanning accelerator (29.)

A glucose ester of tyrosine, glucose tyrosinate, has been developed to overcome the low water solubility of tyrosine, thereby permitting greater amounts of the amino acid to be present in the skin. Tyrosine was also complexed with riboflavin or ATP to accelerate its oxidation. Work by Okun et al. (30) also demonstrated that the enzyme peroxidase can rapidly hydroxylate tyrosine to dopa in melanoma cells. Finally, amino acid blends are added to provide precursors of tyrosine, in addition to increasing diffusability in the skin (31). Although a rationale for the use of these materials can be developed, there are questions concerning the magnitude of the effect that can be delivered.

Collaborative work done by Yale University and Plough (32) indicates that phosphodopa and its derivatives increase melanogenesis. This may be accomplished by increasing the amount of substrate–cofactor for the dopaoxidase function of tyrosinase. This approach is further supported by Yu and Van Scott (33), who patented the use of esters of dopa to enhance pigmentation in hair and skin.

B. Modification of the Melanogenic Support Systems

Release of Tyrosinase Inhibition

Hadley and associates' work (15) with vertebrate pigment cells indicates that tyrosinase inhibition can be released by increasing the level of cAMP, which is thought to bind to and remove an inhibitor of tyrosinase. This can be accomplished by the direct introduction of cAMP and derivatives of cAMP (e.g., dibutyryl cAMP), or by stimulating the activity of the enzyme adenylate cyclase. The latter can be achieved by the addition of α-MSH, cholera toxin, PGE_1, or vitamin D_3, all of which are believed to bind membrane-bound receptors (1,9). Of particular interest is the dramatic stimulatory effect that can be achieved through a series of MSH analogues developed by Hadley. The most potent analogue being the Ac-[Nle-4, D-Phe-7] α-MSH 4-11-NH_2 (ND 4-11 α-MSH).

An intriguing possibility for enhanced melanogenesis is the use of diacylglycerol (DAG). Of particular interest is 1,2-dicapryloylglycerol. It is speculated that the release of tyrosinase inhibitors in a normal nonmalignant melanocyte may occur through DAG and not through cAMP (16). An investigation of the promoting effect of various acyl groups appears warranted.

Regulation of DNA

By affecting the genetic expression of the melanocyte (DNA), one can alter melanogenesis. Psoralens are believed to function this way by binding to sites on the DNA molecule. Currently, the well-known psoralens (TMP and 8-MOP) are not available for use in cosmetic products owing to safety concerns. However, efficacious psoralen-like materials that possess lower toxicity may be possible. The resurgence in the investigation of natural products may help identify such materials.

Another way to increase melanogenesis through use of the genetic system is to increase the amount of tyrosinase present in the melanosome. Hu et al. (19) demonstrated that theophylline may directly increase the rate of tyrosinase synthesis. Because theophylline has demonstrated efficacy in this area, it may also be possible for other xanthine-based derivatives, such as caffeine, theobromine, and isobutyl methylxanthine (IBMX), to have a similar effect.

Finally, a logical way to enhance melanogenesis is to increase the number of melanocytes. Although work in this area is relatively new, experiments have successfully identified several key materials required for the proliferation of melanocytes in vitro. Cell culture studies conducted by Gilchrist et al. (25) have shown that materials, such as melanocyte growth factor, basic fibroblast growth factor, PGE_2, and other serum factors are needed for promoting cell growth. Other agents, such as choleratoxin, have also been employed to enhance the proliferation of melanocytes in culture.

Even if the efficacy of such materials appears inviting, the short-term use of them is remote. However, as research continues in this area, there is every hope that the judicious application of a carefully selected agent might enhance cell proliferation without the oncological consequences that might accompany the improper use of mitogens.

Use of Conversion Factors

An interesting possibility to control the development of pigment may be found by enhancing the apparent effect of the conversion factors DCF and ICF. It is possible that DCF and ICF function by removing IBF-related inhibition while structurally modifing the conformation of the enzyme to permit a more rapid conversion to 5,6-dihydroxyindole-2-carboxylic acid and indole-5,6-quinone, respectively. If this is true, then increasing the amount of the conversion factors would result in increased tyrosinase activity. Unfortunately, neither the conversion factors nor their mechanisms of action are well understood. However, greater knowledge in this area might provide a unique opportunity to enhance pigmentation.

Miscellaneous

Several membrane-mediated events produce a propigmentary effect without the involvement of cAMP. 1α,25-Dihydroxyvitamin D_3 and certain steroidal hormones (e.g., estrogen) have efficacy. These findings suggest that analogues of these agents might selectively promote tanning without the obvious toxicological consequences associated with the use of high levels of the parent compound.

Finally, the efforts of Woods' group (22), with their discovery of the role of the thioredoxin–thioredoxin reductase system in the release and inhibition of tyrosinase may offer unique opportunities to promote pigmentation. Materials that could selectively oxidize thioredoxin or inhibit the reductase enzyme would increase the level of available tyrosinase.

VI. CONCLUSION

Scientists appear to be on the verge of unraveling the total role of the melanocyte and the melanogenic process. Several have demonstrated that there is a close interaction between the melanocyte, the keratinocyte, and the Langerhans cell in the epidermis. Furthermore, communication from the dermal fibroblast as well as mediators circulating in the blood have shown the ubiquitous involvement of melanogenesis in maintaining the health of the body. These interactions confirm that "tanning" is a dynamic process, one that will require the investigator to have a broad perspective to moderate it safely.

VII. TANNING ACCELERATORS: AN UPDATE

Since our original document was published, an interesting phenomenon has occurred. There has been significant investigative activity in the field of melanogenesis that has led to the publication of myriad articles and patents. In fact, in the past 5 years there has been almost as much published on the topic as was found in the preceding 15 years. Ironically, there has been inversely proportional activity in the number of marketed products geared for pigmentation enhancement. A review of the current state of the market as well as of the possible causes for the situation will be discussed later. Initially, the latest findings concerning tanning enhancement will be addressed. Many are simply extensions of the work reviewed in the original document. However, several researchers have proposed some new approaches and developed some new hypothesis on the subject.

A. Direct Manipulation of the Melanogenic Path

Tur (34) has demonstrated by autoradiography that water-soluble α-tyrosine combined with vitamin B_2 and ATP as activators [Unipertan series] can penetrate down to the basal layer of the epidermis. Additionally, the application of these materials before UV exposure results in a tan that develops more rapidly and is more intense and long-lasting. Also, the addition of 21% collagen hydrolysate to a product containing 5% Unipertan enhanced tanning efficiency by as much as 40% versus a control (35). Engenito and Marchese (36) have patented the addition of a polyoxyethylene alkyl ether or polyoxyalkyl sorbitan fatty acid to a system containing tyrosine or a tyrosine derivative plus riboflavin. They claim that the presence of a surfactant increases the rate of penetration of active components into the skin, which ensures a deeper and more uniform tan. Dumas and Meybeck (37) claim that a composition containing L-tyrosine or sodium, calcium, methyl, ethyl, or stearyl tyrosinate which, when delivered to the skin by a liquid, lamellar phases, or liposomes, produces a significant increase in tanning following exposure to UV light.

Raab (38) conducted an independent evaluation of products containing what he called premelanins. These products mostly contained tyrosine or its derivatives, and all claimed a "speeding up" of the natural tanning process. He concluded that neither the topical nor oral administration of compounds from which melanogenesis is started will accelerate or deepen a tan.

An intriguing new approach to the direct manipulation of the melanogenic path is claimed by Chan and co-workers (39). Instead of stimulating the pathway by providing the initial precursor tyrosine, these inventors propose stimulation at the middle stage by the addition of an indole derivative plus at least one nitrobenzene derivative. Specifically, either 5.6-diacetoxy-*n*-methyl indole or 5.6-diacetoxyindole could be used in conjunction with *m*-nitroanisole.

B. Modification of the Melanogenic Support System

Release of Tyrosinase Inhibition

The use of these materials is an indirect approach to the simulation of melanogenesis. Instead of directly supplying the melanogenic path with higher levels of precursors, the approach manipulates the series of chemical reactions that support melanogenic activity. One approach that is being explored far more extensively than others is the use of diacylglycerols (DAGs), which are thought to bind tyrosinase inhibitors, thereby increasing enzymatic activity. Gilchrist and Gordon (40) have patented the use of diacylglycerol to induce melanin synthesis in the presence of or absence of UV light. Agin (41,42) has taken Gilchrist's work one step farther. She has determined that the addition of riboflavin or riboflavin phosphate to DAG or other protein kinase C activators such as triacylglycerols, lipopolysaccharides, unsaturated free fatty acids, short-chain free fatty acids, enzymes that hydrolyze glycophospholipids to DAGs, bryostatins, and glycerophospholipids, will promote a faster, darker, and safer tan when exposed to UV radiation.

An alternative method of releasing tyrosinase inhibition is accomplished by either an increase in cAMP or a stimulation of the enzyme adenylate cyclase. The focus of several researchers is arachidonic acid or its metabolites. Imokawa and Motegi (43) determined that PGE_2, leukotrienes (LT)C_4 and B_4 significantly stimulate melanogenesis in vitro, as indicated by enhanced tyrosinase activity. They also discovered that, of several known growth factors, only fibroblast growth factor stimulated melanogenesis. Reed and Warren (44) have patented the addition of up to 500 μg/ml of a tanning agent selected from PGE_1, PGE_2, and PGD_3. One or more of these agents are incorporated into a topical preparation that includes a diol or triol compound (e.g., 1, 2-propanediol or glycerol) and a polar lipid (e.g., oleyl alcohol or oleic acid) to enhance penetration through the stratum corneum. This combination is purported to enhance tanning, even in the absence of UV radiation.

An interesting French patent was issued to Cousse et al. (45). These inventors have identified that interleukin (IL)-1 and IL-6 contain a sequence of at least four amino acids derived from melanin-stimulating hormone. These materials, when chemically linked to a high molecular weight polysaccharide, stimulate melanin biosynthesis.

DNA Regulation

A large amount of investigative effort centers on the effect of certain materials to directly stimulate genetic activity within the melanocyte. By far, the preponderance of research centers on the effect of psoralens, which when used with selected UV radiation, induce tanning. Of particular interest are the 5-methoxy (5-MOP) and 8-methoxy (8-MOP) psoralens. Sambuco et al. (46) evaluated 8-MOP, along with other alleged tanning enhancers, on miniature pig skin UV irradiated daily. They determined that 8-MOP more clearly enhanced the UVR-induced tan and that this tan was effective in protecting the skin from further UV insult. This effect was confirmed by Imokawa and Motegi (10). Urbach and Sambuco (47) have reported that psoralens, in conjunction with a UVB sunscreen, enhance a tanning that was effective in protecting the skin from subsequent UV damage as assessed by sunburn cell formation.

Xanthine derivatives are also alleged to modulate the melanocyte's genetic composition by increasing the cell's translational capability to produce tyrosinase. Hubaud and Thirv (48) have patented the use of caffeine and theophylline with copper gluconate in a topical cosmetic composition. They claim that this composition, usually with the addition of an organic silicone, will accelerate the development of a suntan that protects the skin.

Analogously, Bonte and associates (49) have identified that a fairly complex composition featuring 0.01–2.0% of a xanthine base (preferably IBMX or theophilline) along with one or more extracts of *Coleus esquirolii*, *C. sculeltarioides*, or *C. xanthanthus* intensifies a tan and increases its natural defense against subsequent UV radiation. These inventors have taken a "cocktail" approach and have included the addition of tyrosine or its derivatives (i.e., a melanin precursor) as well as an activator selected from the B vitamins, quinine, rubefacients, keratin hydrolyzates, trace elements (e.g., Zn, Se, Cu), 5-α-reductase inhibitors (e.g., progesterone), cyproterine acetate, minoxodil, and azelaic acid.

New Approaches

Recent research has fostered some novel approaches in the field of melanogenesis. Donatien and co-workers (50) have demonstrated that bovine pituitary extract significantly enhanced melanocyte growth in vitro. Furthermore, the presence of keratinocyte-derived factors are essential for sustaining melanocyte dendricity and melanization. Schallreuter and Wood (51) have reported tanning acceleration from the application of a cream containing a sunscreen and what they describe as a pseudocatalase. Of particular note is the use of a 5–20 mmol copper I, iron II, or manganese II complex. Most preferably $MnCl_2$ with excess sodium bicarbonate is of interest.

Another unusual approach to achieving an enhanced tan is the use of melanin itself. Alene and Chedekel (52) have patented a composition combining the melanin with a color modification agent selected from oxidizing and reducing agents in an aqueous medium to initiate the color modification reaction then stopping the reaction once the desired color is achieved.

Some very interesting research has discovered that melanogenesis may be effected by some alternative routes. Rosen et al. (53) have reported that both UVA at 365 nm and UVB at 304 nm induce different pigmentation responses, but their effect on melanocyte number and function were indistinguishable. The UVA-induced tan was observed to develop more rapidly than that induced by UVB. This finding suggests that there are alternative biochemical pathways that stimulate melanogenesis and might be safely exploited for tanning enhancement.

Perhaps the most intriguing exploration of the tanning process is currently focused on the role of various specific enzymes in the melanogenic process.

Iizawa et al. (54) conducted studies that revealed that UVB radiation increased lipid peroxide levels and decreased or impaired the activity of reactive oxygen species (ROS) scavaging enzymes. After a 1–3 day delay, the activity of these enzymes (particularly superoxide dismutase and glutathione peroxidase) significantly increased, then slowly decelerated until reaching preirridation levels 3 weeks later. Either coincidentally or because of direct involvement, the melanogenic activity is enhanced from 1 to 2 weeks after the UV insult. These findings correlate well with some recent findings of Gilchrist et al. (55). They reported that the enzymatic response to the topical application of UV-damaged DNA fragments to guinea pig skin produced a tan which, after a 2-week induction period, lasted for 2 months. They reported that the DNA repair enzyme T4 endonuclease catalyzes the repair of pyrimidine dimers that are initiated by UV radiation. It is the presence of this DNA repair enzyme that speeds up melanogenesis. This hypothesis was confirmed in vitro when T4 endonuclease was delivered in carrier liposomes to enhance repair of UV-induced DNA damage in cultured human cells. This enhanced UV-induced melanogenesis, as measured by melanin content, increased tyrosinase activity and [^{14}C]dopa incorporation. The investigators concluded that individuals may differ in their ability to produce repair enzymes, which might underlie what is observed as different tanning abilities. Finally, they suggested that a safe, beneficial tan might be produced from the careful manipulation of the DNA repair enzyme mechanism (55).

The evolving work of Gilchrist and others clearly demonstrates the inter- and intradependence of the entire melanocyte cellular environment. Promising areas for future investigation include keratinocyte enzymes, the extracellular matrix, and the role of Merkel cells. The "primative" melanocyte can be very destructive (e.g., virulent melanomas) and is highly controlled by the entire cellular system.

C. Miscellaneous

They say everything old is new again, and this is true for the field of melanogenesis. Ironically, one of the earliest attempts to enhance tanning is once more becoming popular. Franco (26) has patented a composition comprising annatto and other naturally occurring oils (e.g., avocado oil, cod liver oil, and the like), which is purported to be a "highly effective cosmetic enhancement for suntanning." The mechanism for this approach is the change in the optical properties of the skin with the application of oil. This effectively allows a greater UV flux into the skin, with a concomitant increase in cellular damage and ultimately tanning.

D. The Market

Commercial tanning accelerators reached their height of popularity during the late 1980s. However, because of their claimed mechanism of action, these products often went beyond the clinical definition of a cosmetic and were under scrutiny from various regulatory agencies. In addition, several trade and popular journals questioned the effectiveness of such products.

Typically, a tanning enhancer will claim to either deepen a tan or accelerate its development. Unfortunately, the pigmentation process takes from 7–14 days to develop a perceptible "true" tan. Frequently, the customer is looking for immediate gratification and will often confuse immediate pigment darkening (IPD) with tanning. As a consequence, consumer expectations were not being met, and the products fell out of favor with the public. Tanning accelerators have since been replaced by a return to tanning oils and the more ubiquitous use of dihydroxyacetone sunless-tanning products.

Also, it is important to recognize that markets and marketing opportunities are very much dependent on "windows-of-opportunity," and as such tan accelerators are not the current hot interest of the cosmetic industry, as they were in the early and mid-1980s.

Today, we are seeking methods and ingredient solutions for ameliorating age spots and uneven hyperpigmentation resulting from accelerated, photodamaged, premature aging. Although this is the flip side of our earlier work, information derived from tan acceleration may be applied to broad skin lightening or discrete areas of depigmentation objectives, and visa versa. By better understanding how to shut down or retard the formulation of melanin, future researchers may identify elusive mechanisms to accelerate the tanning process—a goal that, for the most part, has yet to deliver in a truly meaningful consumer benefit.

A review of tanning accelerators currently being sold on the retail market follows:

Company	Product	Key ingredients
Clarins	Lait Solaire Bronzage Progressif	Tyrosine, phenylalanine
	Lait Solaire Bronzage Rapid	Tyrosine, phenylalanine

REFERENCES

1. Fitzpatric, T. B., Eisen, A. Z., Wolfe, K., Freedman, I. M., and Austen, K. F. *Dermatology in General Medicine*, 2nd ed. McGraw-Hill, New York, 1979.
2. McGuire, J., Newman, D., and Barisas, G. Tyrosinase–protyrosinase system in frog epidermis. *Yale J. Biol. Med.*, *46*:572–582 (1973).
3. Prota, G. Recent advances in the chemistry of melanogenesis in mammals. *J. Invest. Dermatol.*, *75*:122–127 (1980).

4. Prota, G. Some new aspects of eumelanin chemistry. *Advances in Pigment Cell Research* (edited by J. T. Bagnara). Alan R. Liss, New York, 1988, pp. 116–118.
5. Jimbrow, K., Yamana, K. Akutsu, Y., and Maeda, K. Nature and biosynthesis of structural matrix protein in melanosomes: Melanosomal structural protein as differentiation antigen for neoplastic melanocytes. *Advances in Pigment Cell Research* (edited by J. T. Bagnara). Alan R. Liss, New York, 1988, pp. 169–172.
6. Hearing, V. J. and Jimenez, M. Mammalian tyrosinase—the critical regulatory control point in melanocyte pigmentation. *Int. J. Biochem., 19*:1141–1147 (1987).
7. Mishima, Y. and Imokawa, G. Selective aberration and pigment loss in melanosomes of malignant melanoma cells in vitro by glycosylation inhibitors: Premelanosomes as glycoprotein. *J. Invest Dermatol., 81*:106–104 (1983).
8. Hearing, V. J., Korner, A. M., and Pawelek, J. M. New regulators of melanogenesis are associated with purified tyrosinase isoenzymes. *J. Invest. Dermatol. 79*:16–18 (1982).
9. Goldsmith, L. A., ed., *Biochemistry and Physiology of the Skin*. Oxford University Press, New York, 1983.
10. Korner, A. and Pawelek, J. Mammalian tyrosinase catalyses three reactions in the biosynthesis of melanin. *Science, 217*:1163–1165 (1982).
11. Pawelek, J., Korner, A., Bergstrom, A., and Bologna, J. New regulators of melanin biosynthesis and the autodestruction of melanoma cells. *Nature, 286*:617–619 (1980).
12. Hosoi, J., Abe, E., Suba, T., and Kuroki, T. Regulation of melanin synthesis of B-16 mouse melanoma cells by 1-alpha 25-dihydroxyvitamin D-3 and retinoic-acid. *Cancer Res., 45*:1474–1478 (1985).
13. Eberle, A. M. Studies on melanotropin (MSH) receptors of melanophores and melanoma cells. *Biochem. Soc. Trans., 9*:37–39 (1981).
14. Lucas, A. M., Thody, A. J., and Shuster, S. The role of calcium in MSH stimulated melanosome dispersion. *Peptides, 8*:955–960 (1987).
15. Hadley, M. E., Heward, C. B., Hruby, V. J., Sawyer, T. K., and Young, Y. C. S. Hormone receptors and vertebrate pigment cells. *Yale J. Biol. Med., 53*:385–448 (1980).
16. Gilchrist, B. Tufts University, Boston, MA, personal communication (1988).
17. Nordlund, J. J., Sober, A. J., and Hansen, T. W. Periodic synopsis on pigmentation. *J. Am. Acad. Dermatol., 12*:359–363 (1985).
18. Tomita, Y. Iwamoto, M., Masuda, T., and Tagami, H. Stimulatory effect of prostaglanelin E-2 on the configuration of normal human melanocytes in vitro. *J. Invest. Dermatol., 89*:299–301 (1987).
19. Hu, F., Mah, K., and Teramure, D. J. Electron microscopic and cytochemical observations of theophylline and melanocyte-stimulating hormone effects on melanoma cells in culture. *Cancer Res., 42*:2786–2791 (1982).
20. Panasci, L. C., McQuillan, A., and Kaufman, M. Biological activity, binding, and metabolic fate of Ac-[Nle4, D-Phe7] alpha-MSH4-11NH_2, with the Fl variant of B16 melanoma cells. *J. Cell. Physiol., 132*:97–103 (1987).
21. Hadley, M. E., Abdel Malek, A. Z., Marivan, M. M., Kreutzfeld, K. L., and Hruby, V. J. [Nle4, D-Phe7]-alpha-MSH: A superpotent melanotropin that irreversibly activates melanoma tyrosinase. *Endocr. Res., 11*:157–170 (1985).

22. Schalleuter, K. U., Pittelkow, M. R., and Wood, J. M. Free radical reduction by thioredoxin reductase at the surface of normal and vitiliginous human keratinocytes. *J. Invest. Dermatol.*, *87*:728–732 (1986).
23. Carter, D. M., Wolff, K., and Schnedl, W. 8-Methoxypsoralen and UVA promote sister chromatid exchanges. *J. Invest. Dermatol.*, *67*:548–551 (1976).
24. Borkovic, S. P., Alper, J. C., and McDonald, C. J. Stimulation of pigmentation in melanoma cells by trimethylpsoralen in the absence of UV irradiation. *Br. J. Dermatol.*, *108*:525–532 (1983).
25. Gilchrist, B. Tufts University, Boston, MA, personal communication (1987).
26. Ortonne, J. P. Chemical and drug induced hypermelanoses, *Advances in Pigment Cell Research* (edited by J.T. Bagnara). Alan R. Liss, New York, 1988, pp. 116–118.
27. Fitzpatrick, T. B., Eisen, A. Z., Wolfe, K., Freedberg, I. M., and Austen, K. F. *Dermatology in General Medicine*, 3rd ed. McGraw-Hill, New York, 1987, pp. 838–876.
28. Pawelek, J., Bolognia, J., McLane, J., Murray, M., Osber, M., and Slominski, A. A possible role for melanin precursors in regulating both pigementation and proliferation of melanocytes, *Advances in Pigment Cell Research* (edited by J. T. Bagnara). Alan R. Liss, New York, 1988, pp. 143–154.
29. Agin, P. P., Wilson, D. K., Shorter, G. G., and Sayre, R. M. Tyrosine does not enhance tanning. *Bioscience 33*:516–517 (1983).
30. Okun, M., Edelstein, L., Or, N., Hamada, G., Donnellan, B., and Burnett, J. Oxidation of tyrosine and dopa to melanin by mammalian peroxidase: The possible role of peroxidase in melanin and catecholamine synthesis in vivo. *Pigmentation: Its Genesis and Biological Control* (edited by V. Riley). Appleton-Century-Crofts, New York, 1972.
31. Pauly, M., European patent 0,010,483 (1980).
32. Plough, Inc., AU 8,424,243 (1984).
33. Yu, R. J. and Van Scott, E. J., U.S. patent 4,021,538 (1977).
34. Tur, W. Cosmetic active complexes for sunscreens causing accelerated, more intense, and more permanent skin pigmentation. *Seifen-Oele-Fette-Wachse*, *116*: 406–410.
35. Tur, W. Tanning accelerators: Increasing the effectiveness. *Cosmet. Toilet.*, *105*. (12):79–85 (1990).
36. Engenito, J. S. and Marchese, F. P. U.S. patent, 5,061,480 (1993).
37. Dumas, M. and Meybeck, A. U.S. patent 5,290,562 (1994).
38. Raab, W. The application of quick tanning substances. *Aerztl-Kosmetol.*, *18*:14–30 (1988).
39. Chan, A. C., Schultz, T. M., and Serban, G. U.S. patent 5,037,640 (1993).
40. Gilchrist, B. A. and Gordon, P. R. World patent, 9,404,122 (1994).
41. Agin, P. A. World patent, 9,107,168 (1993).
42. Agin, P. A. World patent, 9,107,167 (1993).
43. Imokawa, G. and Motegi, I. Skin organ culture model for examining epidermal melanization. *J. Invest. Dermatol.*, *100*:47–54 (1993).
44. Reed, J. C. and Warren, R. European patent 302,147 (1993).

45. Cousse, H., Dussourd, D. L., Martinez, J., and Pinel, A. French patent 2,691,465 (1994).
46. Sambuco, C. P., Cole, C. A., Forbes, P. D., and Urbach, F. Photobiologic effects of cosmetic tanning enhancers in miniature pigs. Presented at Symposium on Psoralens: Past, Present and Future, Paris, France, 1988.
47. Urbach, F. and Sambuco, C. P. Enhanced skin tanning by psoralens and a sunscreen–psoralen product in miniature pigs. Presented at Symposium on Psoralens: Past, Present and Future, Paris, France, 1988.
48. Hubaud, J. C. and Thirv, D. European patent 380,335 (1993).
49. Bonte, F., Dumas, M., and Meybeck, A. French patent 2,665,637 (1993).
50. Donatien, P., Surleve-Bazeille, J. E., Thody, A. J., and Taieb, A. Growth and differentiation of normal melanocytes in a TPA-free cholera toxin-free, low serum medium and influence of keratinocytes. *Arch. Dermatol. Res.*, *285*:385–392 (1993).
51. Schallreuter, K. U. and Wood, J. M. World patent 9,220,321 (1993).
52. Alene, A. B. and Chedekel, M. World patent 9,117,738 (1993).
53. Rosen, C. F., Seki, Y., Farinelli, W., Stern, R. S., Fitzpatrick, T. B., Pathak, M. A., and Gange, R. W. A comparison of the melanocytes response to narrow band UVA and UVB exposure in-vivo. *J. Invest. Dermatol.*, *88*:774–779 (1987).
54. Iizawa, O., Kato, T., Tagami, H., Akamatsu, H., and Niwa, Y. Long term follow-up study of changes in lipid peroxide levels and the activity of superoxide dismutase, catalase and glutalthine peroxidase in mouse skin after acute and chronic UV irradiation. *Arch. Dermatol. Res.*, *286*:47–52 (1994).
55. Gilchrist, B. A., Zhai, S., Eller, M. S., Yarosh, D. B., and Yaar, M. Treatment of human melanocytes and S91 melanoma cells with the DNA repair enzyme T4 endonuclease V enhances melanogenesis after ultraviolet irradiation. *J. Invest. Dermatol.*, *101*:666–672 (1993).
56. Franco, R. U.S. patent 5,279,817.

V

BIOLOGICAL EVALUATION OF SUNSCREEN PRODUCTS

24

Sun Protection Factor Determination in the United States

Molly E. Griffin and Teresa D. Bourget
Skin Research Foundation of California, Santa Monica, California

Nicholas J. Lowe
Skin Research Foundation of California, Santa Monica, and UCLA School of Medicine, Los Angeles, California

I. SUNSCREEN GUIDELINES IN THE UNITED STATES

A *Federal Register* published in 1978 (1) established guidelines for the formulation and evaluation of sunscreens marketed in the United States. These guidelines were reevaluated in 1988 and further revised in 1993 (2).

The 1978 publication suggested that sun protection factors (SPF) be categorized according to the level of protection afforded, ranging from minimal sun protection with an SPF of 2–4, up to an ultra-sun–protection maximum SPF of 15. These guidelines suggested that no product exceed a maximum SPF level of 15. As of the 1993 publication, the ultra-sun–protection maximum was changed to 30. The SPF numbers as high as 30 have been achieved mainly by increasing the concentration of sunscreening chemical to the maximum allowed under the monograph guidelines. Other products have improved their substantivity and SPF numbers by modification of the vehicle. Many products have claimed to have SPF numbers as high as 40 or 50.

II. SPF DETERMINATION IN THE UNITED STATES

The *Federal Register* gives the precise instructions on performance of SPF evaluation. The *sun protection* factor is defined as

$$\text{SPF} = \frac{\text{minimal erythema dose in sunscreen-protected skin}}{\text{minimal erythema dose in nonsunscreen-protected skin}}$$

Selection of subjects is of great importance for the determination of SPF. In general, skin types I to III are the ideal skin types to use for these investigations (Table 1). It is also suggested that test subjects should not have had sun exposure or exposure to tanning beds for at least 90 days before study enrollment.

Darker skin types require much longer irradiance times. In general, the longer the irradiance, the more likelihood of more variable minimal erythema dose (MED) determinations.

In my laboratory, for static SPF investigations, I prefer to use a lower back that previously has had no sun exposure (1,2). A minimal erythema dose (MED) is first determined. This is the minimum dose of ultraviolet (UV) radiation expressed as joules per square centimeter (J/cm^2) required to produce a minimal uniform skin erythema with clear margins at the borders of the exposure site. Light-proof adhesive metal foil is used to produce a template that contains 1-cm^2 areas. This is placed on the lower back's skin, and the different areas of skin are uncovered at precisely determined times. In this way an accurate MED is determined for each subject on the first day of the study.

The response to the ultraviolet radiation is determined usually 24 h following the irradiation. From previous experience, we know the approximate MED of each skin type with the solar simulator. The amounts of ultraviolet radiation are then delivered to approximate that anticipated MED time.

Table 1 Skin Types Determined by Historical Response to Sun[a]

Type	Definitions
I	Always burns easily; never tans (sensitive)
II	Always burns easily; tans minimally (sensitive)
III	Burns moderately; tans gradually (light brown)
IV	Burns minimally; always tans well (moderate brown)
V	Rarely burns; tans profusely (dark brown; insensitive)
VI	Never burns; deeply pigmented (insensitive)

[a]Based on 30 to 40 min of sun exposure in early summer without previous sun exposure that season.

The following day the sunscreens to be tested are applied in amounts of 2 mg/cm^2. These are measured carefully with a measuring pipette. The areas will have been previously marked using the prepared template. The measurement of 2 mg/cm^2 is usually used because most sunscreen products have a specific gravity of almost unity (1,2). It is most important that the sunscreen application be very uniform and the technician carefully trained in this technique. Large variations of SPF are obtained with irregularly applied amounts of sunscreen.

There have been recent studies to suggest that the average individual applies less than 2 mg/cm^2 of sunscreen in regular use (4,5; also see Chap. 30 for further detailed discussion of this subject). In the face of claims that the standard should, therefore, be changed to a density of less than 2 mg/cm^2, the 1993 *Federal Register* maintains that 2 mg/cm^2 is a reliable test density that has been in use for 10 years. A change in the standard for test density would make past scientific studies difficult to interpret because a comparison in application could not be made; therefore, this change would negate 10 years of data based on 2 mg/cm^2 density. In addition, consumers have adjusted their sunscreen use to today's SPF determinations, and any change in this system would lead to consumer confusion.

The sunscreens are allowed to dry on the skin for 15 min before radiation. Some sunscreens are highly protective as soon as they are applied. Further investigations are required to determine how rapidly and for how long sunscreens retain their SPF numbers.

After the sunscreens are applied, the volunteer is again irradiated at the sunscreen-protected sites with incremental amounts of irradiance based on a preliminary estimation of the SPF number for that sunscreen and a geometric formula recommended by the U.S. Food and Drug Administration (FDA). By this I mean we either obtain a spectrophotometric absorption estimate of the expected SPF, or determine the SPF with carefully measured incremental amounts of irradiance in a small number of volunteer patients. By this means of estimation, excessive ultraviolet burning of a volunteer's skin is avoided.

The volunteers are again examined 24 h after irradiation. The minimal erythema doses are then determined for both sunscreen-protected, as well as nonprotected skin, and the SPF is determined from the ratio of these results.

Table 2 summarizes the sunscreens allowed in the United States and their concentration range.

III. EUROPEAN TESTING

Differences existed between the U.S.- and European-specified amounts of application allowed. The U.S. monograph recommends 2 mg/cm^2. Australia, Japan, and Britain follow these FDA guidelines. Previously the German DIN recommended 1.5 mg/cm^2, with the belief that this is a much closer estimate of actual

Table 2 Sunscreens Allowed in United States Under FDA–OTC Panel, with Concentration Range

Chemicals	Approved %
UVA absorbers	
Oxybensone	2–6
Sulisobenzone	5–10
Dioxybenzone	3
Menthyl anthranilate	3.5–5
Avobenzone[a]	
UVB absorbers	
Aminobenzoic acid	5–15
Amyl dimethyl PABA	1–5
2-Ethoxyethyl *p*-methoxycinnamate	1–3
Diethanolamine *p*-methoxycinnamate	8–10
Digalloyl trioleate	2–5
Ethyl 4-bis(hydroxypropyl)aminobenzoate	1–5
2-Ethylhexyl-2-cyano-3,3-diphenylacrylate	7–10
Ethyl *p*-methoxycinnamate	2–7.5
Ethylhexyl salicylate	3–5
Glycerol aminobenzoate	2–3
Homomenthyl salicylate	4–25
Lawsone with dithydroxyacetone	0.25 with 3
Octyl dimethyl PABA	1.4–8
2-Phenylbenzimidazole-5-sulfonic acid	1–4
Triethanolamine salicylate	5–12
Physical	
Red petrolatum	30–100
Titanium dioxide	2–25

[a]Available in only UVA GUARD, as of July 1996.

sunscreen use in the general population. It is the belief of the U.S. FDA that the 2-mg/cm^2 guide is a more easily applied uniform density. In addition, any changes in this density would result in confusion, as previously discussed. It was also suggested that the differences in climate, population density, and geography allow the United States to include year-round sunbathing as a lifestyle. This would make the United States a more accurate model for sunbathing compared with Germany.

European guidelines for SPF testing were published recently under the COLIPA guidelines. The FDA and COLIPA guidelines are now very similar (see Chap. 26 for details on COLIPA testing guidelines).

Table 3 summarizes the sunscreen chemicals and their concentrations as used in Europe.

IV. TYPES OF ULTRAVIOLET APPARATUS AVAILABLE FOR SUNSCREEN TESTING

A variety of ultraviolet sources have been used worldwide for sunscreen evaluation. In the United States, the FDA monograph (1,2) recommends the use of appropriately filtered xenon arc solar simulators for the routine evaluation of sunscreens. However, this monograph does allow the use of alternative ultraviolet sources, providing correlation with a variety of sunscreens has been confirmed in comparison studies with an appropriately filtered xenon arc simulator. The FDA has set forth guidelines for appropriate solar simulator use. An appropriate solar simulator should provide a continuous emission spectrum from 290 to 400 nm; less than 1% of its total energy output should be contributed by nonsolar wavelengths shorter than 290 nm; and not more than 5% of its total energy output should be from wavelengths longer than 400 nm. An appropriate solar simulator should also have good beam uniformity and should have no time-related fluctuations in radiation emissions.

A. Xenon Arc Solar Simulators

The xenon arc sources are available with different lamp outputs. The xenon arc lamp power generally ranges between 150 and 5000 W. With the 5000-W sources, water cooling is usually required to reduce damage to reflecting mirrors within the lamp housing as well as to reduce the amount of heat energy delivered to the subject.

Sources with 1000–2500 W with appropriate filtration are available from Oriel and Kratos in the United States. Depending on the exact UV spectrum required, a variety of different UV filters can be employed in these machines. Figure 1 summarizes the spectral characteristics of these lamps.

Other less powerful units use 150-W xenon arc lamps. One example of this is the Berger Solar Simulator. The main advantage of these smaller units is their portability, but the main disadvantage is that only small areas of skin irradiance are possible; therefore, they are really suitable only for pilot or smaller sunscreen studies. In comparison with the smaller units, the 1000- to 2500-W xenon arc machines can irradiate an area with relatively uniform skin irradiance of approximately 10 × 15 cm. With xenon arc lamps, careful handling of the replacement lamps is most important, as these lamps can explode and are a potential source of injury. In addition, the housing of the lamps must be of sufficient strength to protect against such problems.

Table 3 Sunscreen Chemicals Used in Europe

European Economic Community chemical name	Maximum % concentration
4-Aminobenzoic acid	5
N,N,N,-Trimethyl-4-(2-oxoborn-3-ylidene)menthyl anilinium methyl sulphate	6
Homosalate	10
Oxybenzone	10
3-Imidazole-4-glycolic acid and ethyl ester	2
2-Phenylbenzimidazole-5-sulphonic acid and salts	8
Ethyl-4-bis(hydroxypropyl)aminobenzoate	5
Ethoxylated 4-aminobenzoic acid	10
Amyl 4-dimethylaminobenzoate	5
Glycerol 1-(4-aminobenzoate)	5
2-Ethylhexyl 4-dimethylaminobenzoate	8
2-Ethylhexyl salicylate	5
3,3,5-Trimethylcyclohexyl-2-acetaminido benzoate	2
Potassium cinnamate	2
4-Methoxycinnamic acid salts	8
Propyl 4-methoxycinnamate	3
Salicylic acid salts	2
Amyl 4-methoxycinnamate	10
2-Ethylhexyl 4-methoxycinnamate	10
Cinoxate	5
Digalloyl trioleate	4
Mexenone	4
Sulisobenzone	5
2-Ethylhexyl 2-(4-phenylbenzoyl)benzoate	10
5-Methyl-2-phenylbenzoxazone	4
Sodium 3,4-dimethoxyphenylglyoxylate	5
1,3-Bis(4-methoxyphenyl)propane-1,3-dione	6
5-(3,3-Dimethyl-2-norbonylidene)-3-penten-2-one	3
a′-(2-Oxoborn-3-ylidene)-*p*-xylene-2-sulphonic acid	6
a′-(2-Oxoborn-3-ylidene) toluene-4-sulphonic acid and salts	6
3-(4-Methylbenzylidene)bornan-2-one	6
3-Benzylidenebornana-2-one	6
a-Cyano-4-methoxycinnamic acid and it hexyl ester	5
1-*p*-Cumenyl-3-phenylpropane-1,3-dione	5
4-Isopropylbenzyl salicylate	4
Cyclohexyl 4-methoxycinnamate	1
1-(4-*tert*-Butylphenyl-3-(4-methoxyphenyl)propane-1,3-dione	5

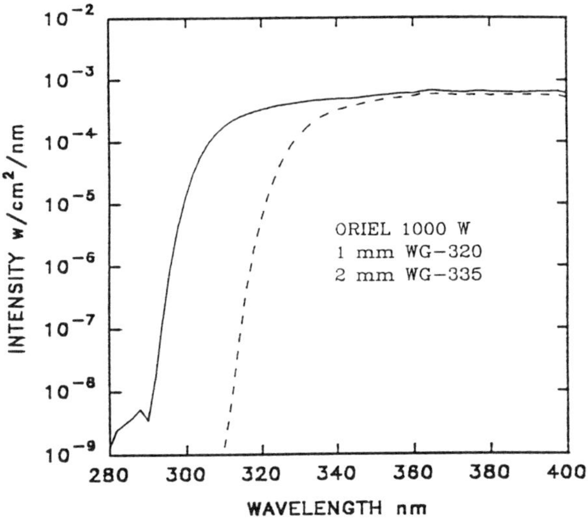

Figure 1 Xenon arc solar simulator.

B. High-Pressure Metal Halide Sources

Mercury vapor lamps have been used for some years as ultraviolet sources, but their solar simulation has been considered unsatisfactory because there are wide gaps between spectral lines in the ultraviolet range. The spectrum of the mercury vapor lamps can be improved using metal iodides to produce a more solar simulating spectrum (Figs. 2 and 3).

We have had the opportunity to study the use of a high-pressure metal halide lamp as an appropriate ultraviolet source for sunscreen evaluation (3). In a series of investigations, we compared the SPF numbers derived with a series of sunscreens comparing the filtered xenon arc lamp with the high-pressure metal halide simulator. Very similar SPF numbers were obtained with these two machines (Table 4). Therefore, it would seem feasible to use the high-pressure metal halide source as an alternative for sunscreen evaluation for products with SPFs less than 15. One of the major advantages is that this machine is very portable and has a higher energy output and larger skin irradiance area than does the 150-W xenon arc solar simulators.

C. Intermediate Mercury Vapor Sources

The West German DIN system formerly recommended the use of an Osram Vitalux Intermediate Pressure Mercury Vapor Source (6). This source was not solar simulating, as it has a mercury vapor line spectrum. In comparison studies

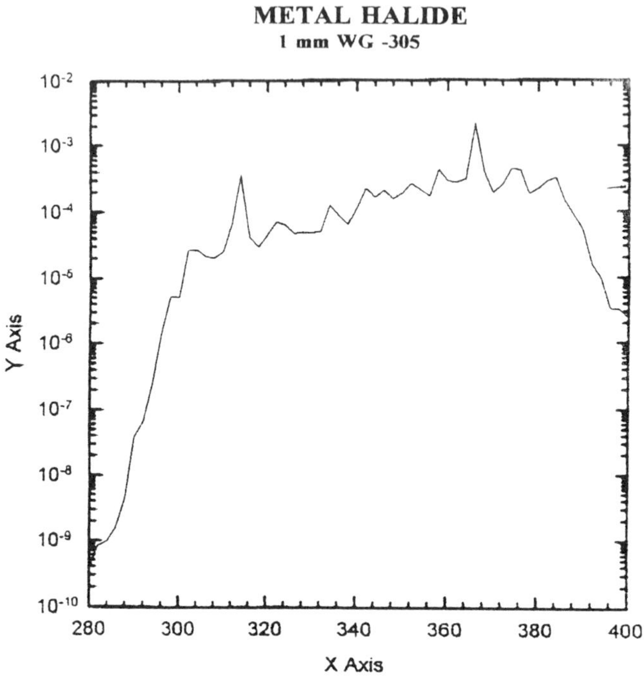

Figure 2 Metal halide simulator.

of predicted human SPF factors derived by using the in vitro mouse epidermal assay, Sayre and Agin showed that the Osram Vitalux Lamp gave very similar predicted protection factors with most sunscreens when compared with the xenon arc solar simulator (7). Again, these studies suggest that there are realistic alternatives to using filtered xenon arc solar simulators.

D. Other Sources Used for SPF Testing

Other ultraviolet sources previously used have included fluorescent sunlamp tubes. Although these have the advantage of ease of use and of a consistent ultraviolet spectrum, their major disadvantage is that they are not solar simulating (7), as they possess peak output at 313 nm, with a smaller peak at 365 nm. These sources also can produce erythema from nonsolar irradiation. Therefore, there is relatively little UVA in relation to UVB and, in addition, these units do contain contaminant UVC. Table 5 summarizes some differences between these UV sources.

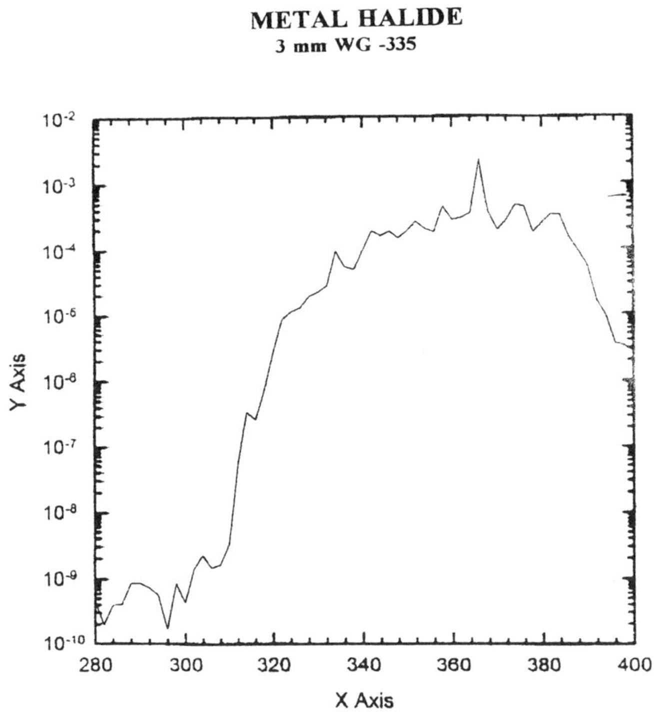

Figure 3 Metal halide simulator.

Table 4 Comparison of SPFs Using Two Different Solar Simulators Mean SPF ± SD

Sunscreen	Metal halide SPF	Xenon arc SPF
Homosalate	3.8 ± 0.5	4.3 ± 0.5
A	15.0 ± 2.1	16.2 ± 1.6
B	9.4 ± 2.0	9.6 ± 1.6
C	15.6 ± 2.5	16.5 ± 1.5
D	16.2 ± 1.6	16.8 ± 1.6

Table 5 Comparison Between Xenon Arc, Metal Halide Simulators, and Intermediate Pressure Mercury Vapor Sources

	Electrical power input	UV per W	Simulation of solar spectrum	Lamp life (h)	Ease of operation	Cost of solar simulator	Running costs	Portability
Xenon arc	150–5000	Low	Excellent	1500	Complex	Low High	Low High	150 W Yes 1000–5000 W No
Metal halide	400–1000	High	Adequate	2000	Easy	Low	Low	Yes
Intermediate pressure mercury vapor	300	Low	Poor	2000	Easy	Low	Low	Possible

E. Summary Comments on Ultraviolet Sources

Ideally a close simulation of the solar spectrum is desirable for human SPF testing. This is clearly most simulated using filtered xenon arc sources. However, as discussed in this chapter, other units, including a high-pressure metal halide simulator can be filtered to produce a relatively acceptable solar-simulating spectrum. Comparison of the different types of alternative ultraviolet apparatus with the filtered xenon arc solar simulators have been conducted and show similar SPF numbers with some sunscreens.

V. OUTDOOR TESTING OF SUNSCREENS

The importance of outdoor testing has been stressed by some authorities (8) because of the need to evaluate sunscreens in a setting more relevant to regular usage. In addition, the volunteers are moving and, therefore, frictional removal of sunscreen is much more likely than in a static laboratory solar simulator environment. There is the additional important factor that sunlight contains larger amounts of infrared radiation than does solar simulator-generated ultraviolet. In addition, there are other variables clearly relating to ambient humidity and the possibility of sweating by volunteers that hasten the removal of some sunscreens. All of these factors make it very difficult to standardize outdoor sunscreen testing.

A more practical solution may be to intentionally vary indoor laboratory factors, such as ambient humidity, the amount of infrared radiance in the solar simulator spectrum, and laboratory temperature. In this way, a more controlled indoor-reproducible evaluation of the sunscreens may be possible.

VI. WATER-RESISTANT AND VERY WATER-RESISTANT CLAIMS OF SUNSCREEN PROTECTIVENESS

The FDA monograph (1,2) outlines the procedures required to evaluate sunscreens in more rigorous conditions, under which the volunteers are exposed to different times of water exposure following sunscreen application. These claims are divided into two broad groups: water-resistant and very water-resistant (previously called waterproof; see also Chap. 7).

A water-resistance property may be claimed by evaluating the subject after two 20-min immersions with moderate activity in water. A very water-resistant claim requires four such immersions. Table 6 shows the types of protocols to be followed when evaluating either water-resistant or very water-resistant claims for sunscreens. In my laboratory, we routinely use a portable whirlpool jacuzzi bath, with steady water movement provided by a water agitator. The sunscreens

Table 6 Water-Resistance Claims Protocols

Action	Time (min)[a]
Water-resistant	
Apply sunscreen	0
In water	20
Out of water	40
In water	60
Out of water (air dry)	80
Irradiate	100
Very water-resistant	
Apply sunscreen	0
In water	20
Out of water	40
In water	60
Out of water	80
In water	100
Out of water	120
In water	140
Out of water (air dry)	160
Irradiate	180

[a]All erythema readings 24 h after irradiance.

were applied to lower backs immersed for the appropriate amounts of time in the whirlpool bath. Water jets were directed away from the sunscreen-treated skin, to avoid direct washoff. Water temperature, pH, and water movement can be easily standardized in this situation.

With this equipment it is clearly practical to alter the water temperature and the degree of water agitation, thereby simulating greater or lesser amounts of exercise. In addition, the degree of salinity can be altered to simulate either freshwater or seawater conditions. Few of these parameters have been studied as yet.

VII. RECENT TRENDS IN SUNSCREEN DEVELOPMENT

There are several recent important trends in sunscreen development. These include the increasing sun protection factor numbers, which have been mainly achieved using increasing amounts of sunscreen chemical. One concern has been a potential increased risk of irritancy and contact allergic sensitization reactions.

Some sunscreens are now claiming SPF numbers of 50, and it remains to be seen whether there will be an increase in topical toxicity from these sunscreens.

Further areas of development have included the development of micronized powders for use as physical reflecting sunscreens. This has great importance for patients with severely photosensitive skin diseases, but perhaps also as a means of reducing infrared damage. There have also been suggestions that infrared radiance may be important as an additional cause of particularly sun-induced skin aging (9–11). The current sunscreens absorb very poorly in infrared, but perhaps some of the new reflecting micronized powder sunscreens may afford some protection in this area of the solar spectrum.

The future potential risks of UVC irradiance if further ozone layer depletion occurs in the atmosphere, may lead to the requirement to shift the absorption spectrum of sunscreens down to UVC. This, fortunately, may be possible with some sunscreens, particularly with the *p*-aminobenzoic acid (PABA) esters. It has additionally been shown that modification of the vehicle, for example, the use of more polar solvents, may shift the absorption spectrum of several different sunscreens to the shorter wavelengths (12). Therefore, if there is more UVC irradiance reaching the earth's atmosphere in future years, it should be possible to modify the sunscreen to absorb at shorter ultraviolet wavelengths. Clearly, should such a change in the terrestrial solar spectrum occur, then modification of ultraviolet sources used for phototesting would be required.

In summary, significant advances in sunscreen technology and testing have been achieved in recent years. Future developments are required, in particular, more efficient and cosmetically acceptable reflecting sunscreens that may have an enhanced photoprotection against UVA and infrared radiation and new ways of evaluating these different wavelengths and protection.

REFERENCES

1. Sunscreen products for over-the-counter use. *Federal Register, 43*:28269 (1978).
2. Sunscreen drug products for over-the-counter human use; tentative final monograph; proposed rule. *Federal Register, 58*:28194–28302 (1993).
3. Lowe, N. J., Weingarten D., and Wortzman, M. Sunscreens and phototesting. *Clin. Dermatol., 6*:40–49 (1988).
4. Stenberg, C. and Larko, O. Sunscreen application and its importance for the sun protection factor. *Arch. Dermatol., 121*:1400–1402 (1985).
5. Sayre, R. M. Sunscreen applications: Flawed study? [letter]. *Arch. Dermatol., 122*: 745–746 (1986).
6. Deutsches Institut for Normung, Normenausschuss Litchttechnik, Berlin, W. Germany (1984).
7. Sayre, R. M. and Aging, P. P. Comparison of human sun protection factors to predict protection factors using different lamp spectra. *J. Soc. Cosmet. Chem., 35*:439–445 (1984).

8. Pathak, M. Sunscreens: Topical and systemic approaches for protection of human skin against harmful effects of solar radiation. *J. Am. Acad. Dermatol.*, *7*:285–312 (1982).
9. Kligman, L. H. Intensification of ultraviolet induced dermal damage by infra-red radiation. *Arch. Dermatol. Res.*, 227–229 (1982).
10. Freeman, R. G. and Knox, J. M. Influence of temperature on ultraviolet injury. *Arch. Dermatol.*, *89*:953–961 (1964).
11. Kligman, L. H. and Kligman, A. M. Reflections on heat. *Br. J. Dermatol.*, *110*: 376–377 (1984).
12. Agradis-Palyompis, L., Nash, R., and Shaath, N. The effect of solvents on the ultraviolet absorbance of sunscreens. *J. Soc. Cosmet. Chem.*, *38*:209–221 (1987).

25

European Guidelines (COLIPA) for Evaluation of Sun Protection Factors

Jack Ferguson
Oriflame International, Dublin, Ireland

I. INTRODUCTION

Sun products marketed in Europe to protect normal skin from the burning or aging effects of the sun are classed as cosmetics.

The European Union (EU, formerly EEC) has a cosmetic directive (1) that covers cosmetic product safety and gives a definitive list of ultraviolet (UV) filters and concentrations that may be included in a cosmetic product. The cosmetic directive does not require specific testing or proof of claims on the pack of any cosmetic product, including sun products, although certain national legislation such as the Trades Description Act in the United Kingdom could be applied if needed. However, the 6th Amendment to the directive (2), which will come into effect from January 1, 1997, requires an information file for each product and should include "proof of effect claimed for the cosmetic product, where justified by the nature of the effect or product." How to "prove the effect" and the effect on which products are not defined and are still open to interpretation, but it seems more than likely that proof of protection against sunburn will be required if a sun protection factor (SPF) is labeled on a product pack. Therefore, a standard SPF test for Europe would be desirable, if not necessary, to accommodate this change in legislation.

Germany is the only European Union country that has published an official SPF test, the DIN method (3). This method has been used by other countries, but has not been widely adopted, mainly because of the inadequacy of the recommended UV source (Osram Vitalux).

The European Cosmetics and Perfumery Trade Association (COLIPA) established a task force to investigate and recommend a method for determining sun protection factors for the purposes of European and international harmonization. The COLIPA Task Force "Sun Protection Measurement" first met in 1990, and its members included representatives from major European manufacturers of sun products and contract testing laboratories with practical knowledge of testing techniques. A critical appraisal was undertaken of existing SPF test methods to identify areas for improvement. A test protocol was established, and multicentered ring tests were carried out using four products representing a range of SPF values. The task force subsequently recommended a test protocol that has been officially adopted and published by COLIPA (4). The method has also been submitted to DIN for consideration, and it is hoped that they will adopt the method wholly or with slight modifications in the near future.

II. THE COLIPA SUN PROTECTIVE FACTOR TEST METHOD

The COLIPA SPF test method has all the main elements of SPF tests used throughout the world. It is a human volunteer test, performed in the laboratory, using a standardized solar simulator UV source. The quantity of product applied is 2 mg/cm^2 (see Table 4 for details of the test).

The COLIPA method is not just a regurgitation of the U.S. Food and Drug Administration (FDA) protocol or any other method. The task force attempted, and succeeded in many respects, to address the major perceived sources of experimental variation; consequently, an advanced approach to SPF testing has resulted. The main aspects of interest were the definition of the UV source, product weighing and application techniques, assessment of the minimum erythemal dose (MED), number of test volunteers, statistical criteria applied to test results, and the introduction of a high-SPF standard.

A. The Ultraviolet Source

Variation in the erythemal wavelengths of the ultraviolet source can cause large differences in the SPF result, particularly if a product contains predominantly UVB sunscreens. An overestimation of three or more SPF units may be observed if the UV source is too rich in the UVB waveband, compared with a standard sun at 35°N latitude. Alternatively, if the UV source is low in UVB wavebands, then the SPF would be underestimated.

Spectral output from a solar simulator cannot be defined simply by specifying the type of lamp (e.g., xenon arc) and the filter system (e.g., Schott WG 320 1 mm, and UG 11 2 mm). Variations in transmission properties of different glass filters of the same specification will mean that the output of such a defined system can vary significantly. Variation can also occur as a result of the age of the lamp. A lamp coming to the end of its life can have an output different from that when it was new.

The major cause of difference in UV content of solar simulator systems is the differences between filters from one batch to another, even from the same manufacturer (Fig. 1). It is important to define the quality and quantity of the UV output of the solar simulator, and this cannot reliably be done by simply defining the lamp and filters. Therefore, to reduce variability in the test, it is necessary to define UV output of the filtered lamp source and for it to be regularly monitored.

The variation in the measured SPF will be dependent on the relative intensity of each wavelength in the UV and also on the erythemal effectiveness of the different wavelengths. Any increase or decrease in the intensity of the highly erythematous wavelengths will have a greater effect on the SPF than similar changes in intensity of the wavelengths with a lower potential to induce erythema. Accordingly, the COLIPA method defines the solar simulator output by a cumulative erythemally effective spectral distribution (Table 1).

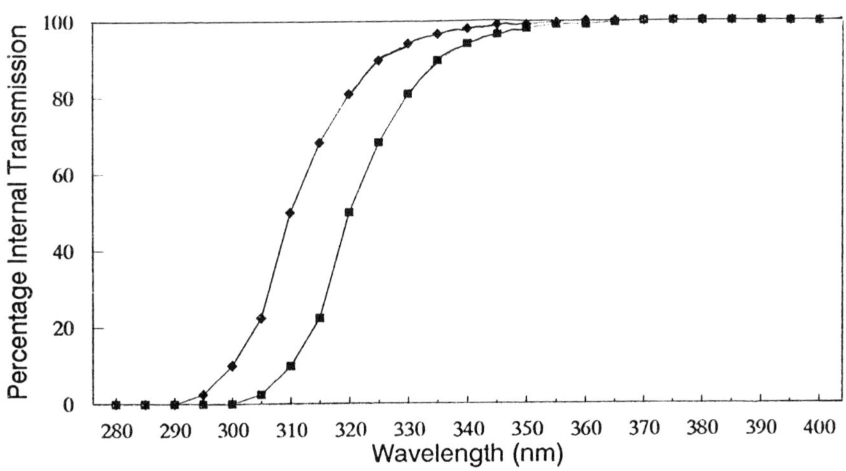

Figure 1 Transmission in characteristics of two Schott 2-mm WG320 glass filters supplied at different times. Although these products fulfill the manufacturer's quality control criteria, the UV transmission is sufficiently different to noticeably affect the SPF when used to test products. (Courtesy of Michael W. Brown, Boots Co. PLC, Nottingham, UK.)

Table 1 The Cumulative Erythemally Effective Spectral Distribution Used by COLIPA to Define a ''Standard Sun'' for SPF Testing of Sun Products

Wavelength	% Transmission 2-mm WG320	% Transmission minimum spec. 2-mm WG320
280	0.0	0.0
285	0.0	0.0
290	0.0	0.1
295	0.0	2.5
300	0.1	10.0
305	2.5	22.4
310	10.0	50.0
315	22.4	68.4
320	50.0	81.0
325	68.4	89.7
330	81.0	94.0
335	89.7	96.6
340	94.0	98.0
345	96.6	98.9
350	98.0	99.0
355	98.9	99.5
360	99.0	100.0
365	99.5	100.0
370	100.0	100.0
375	100.0	100.0
380	100.0	100.0
385	100.0	100.0
390	100.0	100.0
395	100.0	100.0
400	100.0	100.0

Source: Ref. 4.

B. Product Weighing and Application Techniques

It almost seems inappropriate to discuss the detail of such an elementary laboratory operation, but the fact is that the amount of product on the test site is fundamental to determining the SPF of the product (5). This is not just a consequence of the quantity that is applied, but also relates to the uniformity of the application of product over the test site.

Video recordings of the techniques used for weighing, applying, and spreading product on the test site by participating COLIPA laboratories indicated that, left to their own devices, technicians may apply product in a variety of different ways. This resulted in nonuniform distribution of product over the test site in several laboratories. To facilitate even distribution and accurate weighing of product, a small 1-mm syringe or micropipette is recommended and the amount of product dispensed using a "weight loss" technique and not by volume. With the aid of a syringe, product can be distributed on the volunteer's back by dotting small amounts over the entire test area before spreading with the finger (Fig. 2). A rubber finger-stall presaturated with product is recommended. Failure to presaturate can result in a loss of 5–10% of the weighed quantity.

C. Assessment of the Minimum Erythemal Dose

The subjectivity of MED selection together with the incremental nature of exposure of the skin to UV could be considered a significant source of error. The FDA, in a recent proposed modification of the tentative monograph, suggested a decrease in the 1.25 increment to 1.17 and that two observers should be used to determine the MED (6). We considered that a reduction in exposure increments to less than 1.25 would make it difficult to distinguish between sites and thus to select an MED level.

The 1.25 times incremental progression was used in a COLIPA study, and MEDs for protected and unprotected skin were assessed visually and by reflectance colorimetry (4,7) and the SPFs calculated. A threshold level of redness of $\Delta a^* = +2.5$ had previously been determined experimentally (4). Four laboratories participated in this study and tested four products with SPF values of 4, 12, 15, and 22. The correlation coefficient for visually and colorimetrically assessed SPF across all four products and all four centers was $R = 0.93$ ($n = 205$, $p = 0.0001$, SD = 4.0), with a slope of 0.994. The correlation coefficient for the mean SPF values determined visually and colorimetrically for each laboratory was $R = 0.999$ ($n = 1.6$, $p = 0.0001$, SD = 0.49), with a slope of 0.995. The very strong correlation between objective and subjective assessments of SPF suggests that MED does not represent a significant source of error in SPF determination.

One of the strengths of the SPF test is that each test subject is his own control. In this respect the MED determination can be strengthened by determining the MED of unprotected (u) skin and of protected (p) skin at the same time. This allows a better comparison and selection of the MED for treated skin. Clearly, an indication of the MED level of the subject has to be known in advance. The UV exposure of MEDu and MEDp sites should be done simultaneously. MEDu is determined 24 h later by selection in the usual way; MEDp is chosen by color matching to the chosen MEDu. Where a complete color-match is not present an

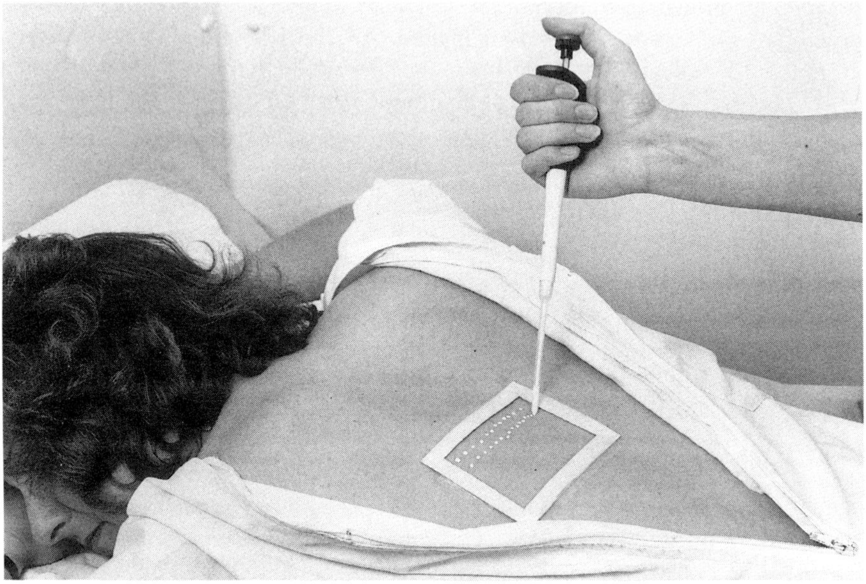

(a)

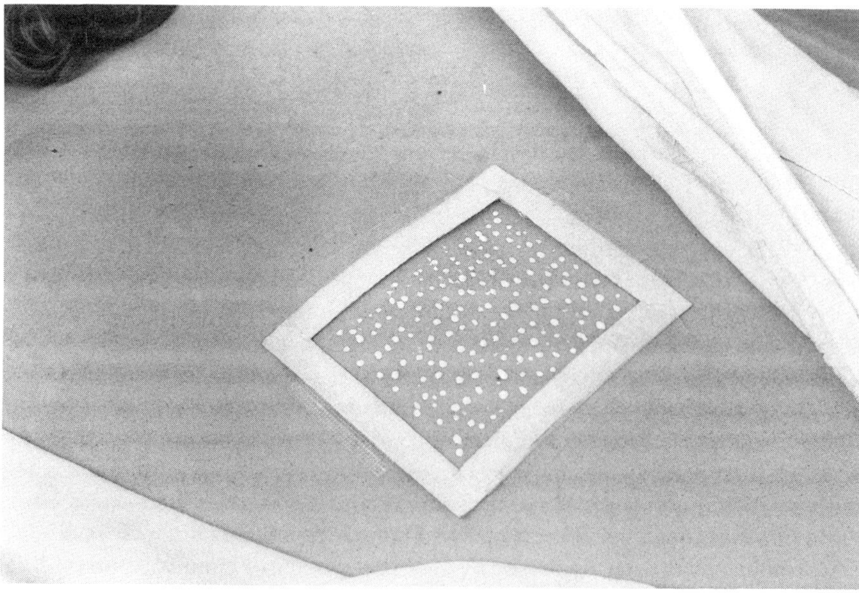

(b)

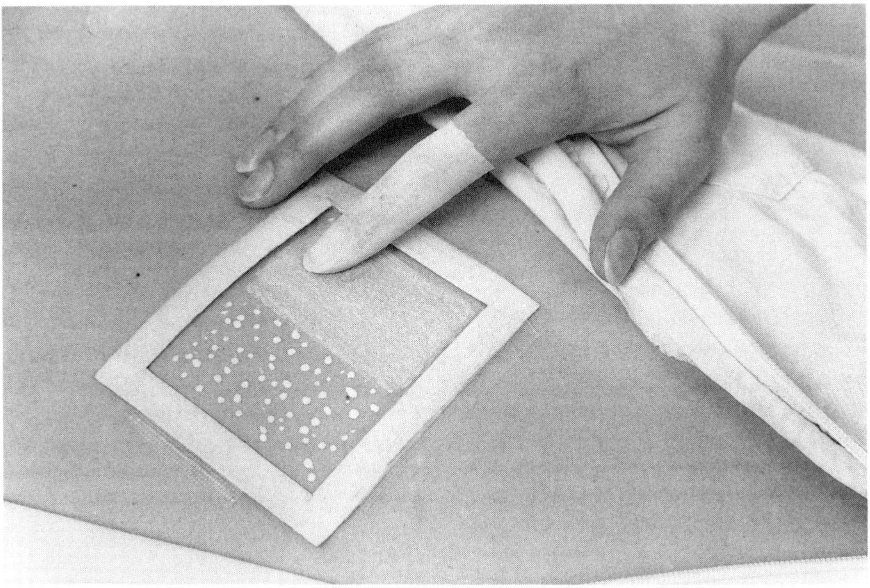

(c)

Figure 2 Product application techniques: (a) Application of product using micropipette to help distribute evenly, across test site. The quantity of product is determined using weight loss technique and not by volumes. (b) All 2 mg/cm^2 of product dotted over test site before spreading. (c) Product rubbed into test site using presaturated fingerstall. (Photographs courtesy of Mr. M.W. Brown and Mrs. M. Pykett, Boots Co., PLC, Nottingham, UK.)

interpolated MED energy value may be selected as, for example, the midpoint between two actual exposures.

D. Number of Test Subjects and Statistical Criteria of Tests

Previously published SPF standards have arbitrarily advocated 20 subjects per test. The Australian and Japanese standards (8,9) allow as few as 10 subjects, if statistical variation is of an acceptable level. This is recognition that it is not the absolute number of subjects in the test that is important, but the confidence in the experimental data and how predictive the SPF is of the true SPF.

This principle has been applied to the COLIPA test, and it is the level of statistical confidence in the data that determines actual panel size. The practical

Table 2 Formulas of Products Used in COLIPA SPF Interlaboratory Ring Tests

		% w/w
P1	Low-SPF standard (DIN K17N)	
	Propylene glycol stearate SE	1.0
	Mineral oil	5.0
	Stearic acid	1.5
	Octyl methoxycinnamate	2.7
	Cetearyl alcohol	0.4
	Propylparaben	0.1
	Methylparaben	0.1
	Triethanolomine	0.8
	Glycerine (85%)	4.0
	Carbomer	0.1
	Water	84.3
P2	High-SPF standard (CTFA/JCIA)	
	Lanolin	4.5
	Cocoa butter	2.0
	Glyceryl stearate	3.0
	Stearic acid	2.0
	Octyl dimethyl PABA	7.0
	Benzophenone-3	3.0
	Propylparaben	0.1
	Methylparaben	0.3
	Triethanolamine	1.0
	Sorbitol	5.0
	Water	71.6
P3	High-standard (COLIPA)	
	Cetearyl alcohol, PEG-40, castor oil, and sodium cetearyl sulfate	3.15
	Decyl oleate	15.0
	Octyl methoxydibenxoylmethane	0.5
	Butyl methoxydibenzoylmethane	0.1
	Propylparaben	0.1
	Methylparaben	0.3
	2-Phenyl-benzimidazole-8-sulfuric acid	2.78
	Carbomer	0.3
	Disodium EDTA	0.1
	Sodium hydroxide (45% solution)	1.2
	Water	73.57
P4	Commercial sunblock contained following sunscreens	
	4-*tert*-butyl-4′-methoxydibenzoylmethane	
	2-ethylhexyl-*p*-methoxycinnamate	
	Mexenone	
	Titanium dioxide	

application of this statistical approach requires that a minimum of ten volunteers be entered into the study initially. The mean SPF is calculated with the 95% confidence interval (CI) for the ten results. If the 95% CI falls within ±20% of the mean, then the test will be declared valid and complete. If the 95% CI does not meet the acceptance limits, then testing should be continued on further volunteers until the statistical criterion is met. If the variation of the test is so great that the statistical acceptance is not achieved at 20 or just over 20 subjects, then the test should be abandoned and repeated after evaluating experimental techniques.

This approach encourages good test procedures and practice, provides confidence in the test results, and can reduce unnecessary testing.

E. Interlaboratory Ring Tests and the Establishment of High-SPF Standard Products

A series of interlaboratory ring tests were performed. Seven European laboratories, all members of the task force, tested four products using the test protocol agreed by the task force.

The four test products were the P1, the DIN standard (SPF 4); P2, the suggested CTFA/JCIA high standard (SPF 15); P3, the COLIPA high standard (SPF 15); and P4, a commercially available sun block (SPF 20–25). All SPF values are approximate. P1, P2, and P3 were chosen as future test standards. P4 was added to extend the SPF range in the series. Formula details are shown in Table 2.

The ring tests were used to establish the practicability of the new test protocol, to make modifications to the test procedure where required, and to establish guideline values for the standard products. The final set of results are presented in Table 3.

Table 3 The Results of Four Products Tested in Seven Laboratories Using COLIPA SPF Test Method

		Mean SPF	Range	SD	Coefficient of variation	Range ± 1 SD
P1	DIN standard	4.2	3.9– 4.6	0.2	5.6	4.0– 4.4
P2	CTFA/JCIA high standard	12.7	11.1–14.1	1.2	9.6	11.5–13.9
P3	COLIPA high standard	15.5	14.2–18.2	1.5	9.4	14.0–17.0
P4	Commercial sunblock	22.4	11.3–33.9	5.8	26.0	16.6–28.2

Table 4 A Comparison of the Major Elements of the SPF Tests Published by COLIPA and FDA

Parameter	COLIPA test (1994)	FDA proposed test (1978)
UV source	Artificial lamp source. Output defined by distribution of erythemal effectiveness between 290 and 400 nm based on standard sun 40°N latitude.	Artificial lamp source. Continuous emission spectrum between 290–400 nm, similar to sunlight: sea level, 15° incidence UVC < 1% VIS > 400 nm < 5%
Test product standards	1. Low SPF standard (DIN K17N) SPF value approximately 4.0 at 2.0 mg/cm^2. 2. High SPF standards, tested at 2.0 mg/cm^2 a. COLIPA product approx. SPF 15 b. CTFA/JCIA product approx. SPF 13 The homosolate FDA standard may be added as an alternative to the DIN standard in the near future. 3. Neutral density filters may be used to check the system.	1. Low SPF standard Homosolate product approximate SPF 4
Product quantity and application	2 mg/cm^2, minimum area of 35 cm^2; prescribed method of weighing by loss and application technique for uniform coverage. Product drying time 15 min.	2 mg/cm^2, over an area of 50 cm^2. No prescribed method of weighing or application. Product drying time 15 min.

Test subjects	Minimum number of ten; actual number determined by statistical variation (no more than 20)	Minimum of 20
	Skin types I, II, and III according to Fitzpatrick, or ITA values greater than 28 as described by Chardon et al. (11)	Skin types I, II, and III
Test area	Region of the back between the scapula and the waist line	Same as COLIPA
	Irradiation sites minimum area of 0.4 cm^2; recommend 1 cm^2	Recommend 1 cm^2
MED determination	Constant flux variable time	Constant flux variable time
	Minimum of five sites	Five sites normally used
	MED read after 20 ± 4 h	MED read after 16–20 h
	1.25× increments between irradiation doses	1.25× incremental increase in radiation doses
	MED of protected skin (MEDu) determined simultaneously with unprotected skin (MEDp)	MEDu and MEDp may be determined sequentially
	Optional use of colorimetry to assess MED	
Calculation of SPF and determination of statistical acceptance	The calculated mean SPF is the arithmetic mean of the individual SPF values	Same as COLIPA
	95% confidence interval applied to the calculated mean ± 20%	Standard error of the mean ≤ 5 mean SPF

Source: Refs. 4 and 10.

The variability of the data between laboratories was acceptable and increased with increasing SPF value. The data in Table 3 can be used as guidance to the expected values of the standard products tested at 2 mg/cm^2. One standard product should be used in all tests, and the choice of standard for a given test remains with the testing laboratory. The high standard need not be used on every occasion when a high-SPF product is being tested, but test laboratories should test high standards on a regular basis.

III. COMPARISON OF THE COLIPA AND FDA SPF TESTS

The COLIPA SPF test is summarized in Table 4 and compared with the FDA test as described in the 1978 monograph (10). The main points of difference are described in the previous sections.

The question of whether or not the COLIPA and FDA methods are sufficiently similar to be interchangeable and acceptable to respective organizations and authorities has not yet been addressed.

The FDAs currently defined method would not meet COLIPA guidelines, in that the lamp output is insufficiently defined, advice on product application technique is inadequate, sequential assessment of MEDu and MEDp, different statistical criteria, and absence of a high-SPF standard.

The COLIPA method compares favorably with the FDA method relative to the lamp, product application rate (2 mg/cm^2), and rate incremental increases 1.25 $\times$), all areas for which the current German DIN method significantly varies from the FDA method. The points of differences are the statistical criteria, number of volunteers, and the absence of the FDA standard product. However, a further and recent international ring test has provided sufficient information for the use of the FDA homosolate standard in the COLIPA test (SPF 4.3) (unpublished data).

Worldwide improvements and greater harmonization in SPF testing, therefore, could be made if the FDA were to consider and recommend the improvements in SPF testing as described in the foregoing and in more detail in the COLIPA document (4).

The COLIPA SPF test is published (4) and has been recommended to, and adopted by, the European cosmetics industry as the preferred method of testing sun protection factors.

ACKNOWLEDGMENTS

I would like to acknowledge the contribution to the COLIPA method of my fellow members of the COLIPA Task Force 'SPM' without whom the method would not have been completed.

REFERENCES

1. Official Journal of European Communities No. L262/170, 27th September, 1976.
2. Official Journal of European Communities No. L151, 23rd June 1993.
3. Deutches Instit für Normung: Expermentelle dermatologische Bewertung des Erythemschutzes von externen Sonnenshutzmittlen für die menschliche Hauf. DIN 65701, 1984.
4. COLIPA Sun Protection Factor Test Method published by the European Cosmetic, Toiletry, and Perfumery Association (COLIPA), Brussels, Belgium, October 1994.
5. Brown, S. and Diffey, B. L. The effect of applied thickness on sunscreen protection: In vivo and in vitro studies. *Photochem. Photobiol.*, *44*:509–513 (1986).
6. Department of Health and Human Services FDA, USA. Sunscreen drug products for over the counter human use: Tentative final monograph: proposed rule: *Federal Register*, *58*(90):28194–28302 (1993).
7. Ferguson, J., et al. The COLIPA SPF test method. An advancement in SPF test methodologies. Paper presented at the IFSCC Congress, Venice, October 1994.
8. Standards Australia, Standards New Zealand. Sunscreen products evaluation and classification. AS/NZS, 2604 (1993).
9. Japanese Cosmetic Industry Association (JCIA). Standard SPF test method (1991).
10. Department of Health, Education and Welfare, FDA, USA. Sunscreen drug products for over-the-counter human use, proposed safety, effectiveness and labelling conditions. *Federal Register*, *43*(166):38206–38269 (1978).
11. Chardon, A., Cretois, I., and Hourseau, C. Comparative colorimetric follow-up on humans of the tannings induced by cumulative exposures to UVB, UVA and UVB + A radiations. 16th IFSCC Congress, New York, Preprint *1* 51–70 (1990).

26

Ultraviolet A Claims and Testing Procedures for OTC Sunscreens: A Summary and Review

Nicholas J. Lowe
Skin Research Foundation of California, Santa Monica, and
UCLA School of Medicine, Los Angeles, California

I. BACKGROUND

There is a well-documented need for effective ultraviolet (UV)B and UVA photoprotection by sunscreens for human skin (see Chap. 2). They reduce the number of precancerous sunlight-induced skin lesions (see Chap. 10), and this protection may be evidence for protection against predominantly UVB wavelengths.

Recent studies have shown that small suberythemogenic amounts of UVA radiation can cause significant photodamage to human skin. These studies revealed epidermal and dermal changes (1–3). Larger fluxes of UVA will lead to greater evidence of skin damage (4–7). Sunscreens, however, cannot be easily tested in these long-term studies or assays; therefore, short-term or "acute" assays remain important.

II. SPECIFIC METHODS

A. Sun Protection Factor Method

With a correctly filtered solar simulator or appropriate outdoor testing, the current sun protection factor (SPF) method accurately predicts the protectiveness of sunscreens against UV radiations mainly of between 290 and 340 nm, which are the most effective wavelengths for the induction of ultraviolet-induced cutaneous erythema (8). This is a well-proved effective method for predicting sunscreen effectiveness against these wavelengths in vivo on human skin and can be used to predict the static performance of a sunscreen. (Methods have also been developed that predict the protectiveness after water immersion.) This method should remain the established method for the evaluation of sunscreen protectiveness in the UVB and UVA-1 wavelengths. However, it does not predict the protectiveness against the longer UVA wavelengths which, in nondiseased skin, contribute little to UVA-induced erythema relative to the amount of expected long-wave UVA present in appropriately filtered solar simulators or natural sunlight (9).

B. Rationale for the Necessity to Evaluate Sunscreen Protectiveness Throughout the Ultraviolet A Spectrum

Skin photodamage occurs throughout the UVA region. It is possible that further research may provide data that show long-wave UVA to be particularly important in the pathogenesis of cutaneous aging and dyspigmentation. It has also been well established that some photosensitive diseases have their action spectra within the UVA spectrum and some in long-wave UVA region. Examples of this include many patients with polymorphic light eruption, chronic actinic dermatitis, actinic reticuloid, photosensitivity drug reactions, and topical photosensitivity reactions (10). This, therefore, establishes a rationale for the need of additional information about the sunscreen's performance throughout the UVA region. Suggestions for UVA phototherapy assays include use of an in vivo method that is (1) practical, (2) reproducible, and (3) has a defined biological endpoint.

III. IN VIVO HUMAN ULTRAVIOLET A ASSAYS

A. Immediate Pigment Darkening Assay

The major problem with the immediate pigment darkening (IPD) assay is that the response induced during skin exposure to UVA and assessed immediately after exposure is unstable (Table 1). The color change quickly fades during the first hour following irradiation (11). Because of this transient and changing effect, the final measurement of the biological endpoint is not stable. Consequently, the reciprocity law of the response can be questioned under certain experimental conditions. Another problem is that local oxygenation of skin (e.g., modified by the product layer or local pressure on the skin) may alter the intensity and appearance of the IPD reaction.

Visual endpoints are variable and difficult to determine, because the response observed immediately after exposure can also be distributed by thermal artifacts

Table 1 Immediate Pigment Darkening Reaction (IPD)

Advantages
Avoids photosensitizers
Medium irradiance times, average IPD between 5 and 15 J/cm^2
Problems
Transient and unstable effect
Imprecise visual endpoints

caused by the exposure and by the presence of colored protective products. Some subjects develop graying, whereas other subjects develop a brown discoloration. If there is any background pigmentation on the surrounding skin, this often makes visualization and grading difficult.

B. Persistent Pigment-Darkening (Equivalent to Delayed Pigment Darkening)

The persistent pigment color change following UVA radiation of humans' skin, classified as phototype II, III, or IV, observed at least 1 h after exposure (optimum 2 h), would appear to be a possible in vivo assay (12; Tables 2 and 3). The assay has the advantage of stability and persistence. It has high reproducibility, provided that an appropriate UV source is used with appropriate filters. The only concern for this assay is that the current UV sources may have UV energy outputs that are inadequate in the desired spectrum range to elicit the biological response endpoint within a practical timeframe (Figs. 1 and 2). Therefore, testing laboratories need to be advised that appropriately filtered high-energy sources are required (13).

Table 2 Levels of Irradiation Required to Product Minimal Pigment-Darkening Reactions in Skin Phototype 2 (n = 10)

Time after exposure	Xenon arc 345-nm filter UG11 (J/cm^2)	Halide 340-nm filter UG11 (J/cm^2)
0	5.3 ± 3.83	5.3 ± 3.2
2	27.7 ± 6.7	26.8 ± 5.0
24	28.2 ± 3.4	29.4 ± 1.7

For comparison: minimal phototoxic dose 0.48 ± 0.8 J/cm^2 (using 0.1% 8-MOP) 15 min before UVA using xenon arc 345-nm UG11 filter.

Table 3 Persistent Pigment-Darkening (PPD) at 2 or 24 h

Advantages
Avoids photosensitizers
Predicts response throughout UVA
Consistent response with established visual endpoint after 2 h
Problems
Prolonged irradiance times
Average DPD 20–25 J/cm^2

C. Ultraviolet A Erythema

The UVA erythema assay in nonsensitized skin is applicable mainly to skin phototype I. It requires the same UVA dose range as the persistent pigment-darkening (PPD) response and a delayed observation time, usually at 24 h (Table 4). The reaction is primarily induced by shorter UVA wavelengths and, by itself, fails to predict complete UVA spectrum effectiveness (12). In practice, when applied to skin with intermediate phototypes (II), both erythemal and pigmentation responses are present.

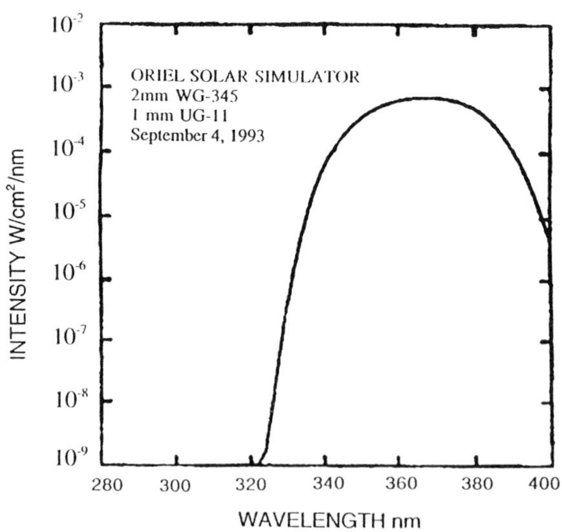

Figure 1 One "ideal" spectrum for UVA-induced delayed pigment, and erythema reactions: A xenon arc 1000-W solar simulator appropriately filtered. This source can also be used for phototoxicity reactions.

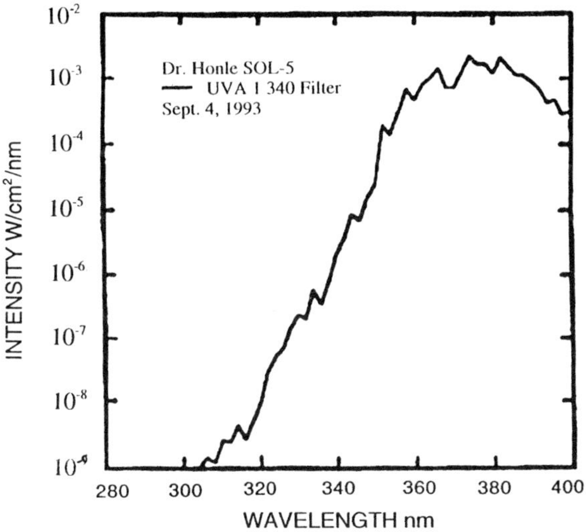

Figure 2 An alternative source for UVA-induced delayed erythema, and pigment reactions. A high-pressure metal halide/lamp appropriately filtered. This can also be used for evaluating phototoxicity reactions.

D. Phototoxic Protection Factor

There are practical advantages to the phototoxic protection factor (PPF) assay, with short irradiance times and clear erythemal visual gradings (10,14,15; Tables 5 and 6). Significant concern has been expressed that it is not relevant for nonphotosensitized human skin. It appears to overpredict the protectiveness of some sunscreens because of the relatively short irradiation times. Therefore, any effect of UVA on a photounstable product may be missed. There is persistent pigmentation, and there may be unwanted phototoxic erythema. It is not possible

Table 4 UVA Erythema Method

Advantages
Avoids the use of photosensitizers
Disadvantages
Requires skin types I or light skin type II
Requires prolonged irradiance (expected MED 25–30 J/cm^2)
Predicts mainly UVA up to 340 nm
Provides little information on long-wave UVA
Imprecise visual endpoints

Table 5 Comparison of Phototoxic Protection Factors (PPF) and Delayed Pigment-Darkening (DPD) Factors with Three Different Sunscreens

	Active agents	
UVA Guard	Octylmeth. cinn	7.5%
	Avobenzone	3%
	Oxybenzone	3%
Photoplex	Padimate O	7.0%
	Avobenzone	3%
"SPF 15" control standard	Octylmeth. cinn	7.5%
	Oxybenzone	3%

to use topical photosensitizers to predict water resistance; but this can be assessed using systemic psoralens. However, this assay does have relevance for some photosensitivity diseases and photosensitivity drug reactions.

IV. IN VITRO ULTRAVIOLET A METHODS

A simple solution absorption spectrum that reveals the UV absorption of the sunscreen would be adequate for gaining additional information to support the in vivo performance studies noted in Section III. Unfortunately, with pigment powders (e.g., titanium dioxide), effectiveness may not be revealed by this test. It is now felt that a method using a tape substrate, such as the Diffey method (Table 7), may be an option that has several advantages (9), as it addresses some of the influence of the vehicle. It is, however, not human skin.

Table 6 Phototoxic Protection Factor

Advantages
Short irradiance times (MPD 0.4–0.8 J/cm^2)
Clear erythema visual gradings
Greater range of PF (i.e., separates good from poor UVA filters)
Disadvantages
Use of photosensitizer
Occasional persistent pigmentation
Topical photosensitizer assay cannot be used for predicting water resistance
Systemic photosensitizer assay needed for predicting water resistance
Relevant for some photosensitivity diseases and photosensitivity drug reactions
Overpredicts some sunscreens (short irradiation times)

Table 7 In Vitro Methods (Diffey Method)

Advantages
Speed and low cost
Disadvantages
Poor correlation with some vehicles and actives
Does not predict the performance of a sunscreen on human skin
Will not reveal any problems with photoinstability

Table 8 Results in Different UVA Protection Factor Assays

Time after exposure (h)	Oxybenzone SPF 15 control oct. meth. cinn	Photoplex	UVA Guard
Pigment-darkening factors n = 12			
0	3.4 ± 1.1	5.1 ± 1.7	5.8 ± 2.1
2	2.9 ± 0.8	4.6 ± 1.6	4.9 ± 1.3
24	2.7 ± 0.8	4.3 ± 0.9	4.9 ± 0.8
Phototoxic protection factors n = 24			
48	3.03 ± 0.6	5.1 ± 1.0	9.08 ± 1.6

Table 9 Summary Suggestions for Predicting UV Protection Factors

For normal skin:
SPF assay for
UVB and UVA II
UVAPF or
Persistent pigment darkening or
An acceptable in vitro/in vivo method
For photosensitivity claims:
Phototoxic protection factors
Photosensitivity disease studies

Another method that needs further evaluation is the spectral reflectance method currently being developed by SPEX Company. Further information on this assay is required that compares it with in vivo method and with a variety of sunscreens.

V. SUMMARY SUGGESTIONS

1. The current human in vivo sun protection factor (SPF) should be retained to develop information on the protectiveness of a sunscreen in vivo primarily in the spectrum range between 290 and 340 nm.
2. An in vitro absorption assay could be used for preliminary screening of products.
3. Relative to the in vivo UVA methods that are proposed for assessing UVA efficacy of sunscreens in normal skin, the PPD response and the erythema response appear to be useful assays (Tables 8 and 9). One important difference between these two assays is selection of the appropriate skin types for the pigmentation and the erythemal responses. In addition, the assays have been demonstrated to give similar protection factors in testing a variety of products over a spectrum range of UVA protection.
4. In claims for photosensitivity protection, if the manufacturer of a sunscreen product wishes to make claims about its performance in photosensitivity diseases, an evaluation of protectiveness within that disease (i.e., polymorphic light eruption or photosensitivity) should be performed; the PPF method could then be used to establish those claims.
5. For UVA protection labeling, the level of UVA protectiveness needs to be expressed in an easily understandable form for the public. Preferably, it should avoid a numbering system such as the current SPF. The suggested labeling would contain the SPF up to a maximum of SPF 30. Its UVA protection would then be stated as being "low UVA protection," "moderate UVA protection," or "maximum UVA protection." This could be based on data determined by the persistent pigment-darkening–erythema reaction, or an acceptable in vitro assay.

REFERENCES

1. Lowe, N., Wieder, J., Myers, D., and Scott, I. Repetitive UVA induces major changes in previously non-sunexposed skin. *J. Invest. Dermatol.*, *102*:649 (1994).
2. Lafker, R., et al. UVA radiation effects on the skin. *J. Invest. Dermatol.*, *102*:647 (1994).
3. Staberg, B., Wulf, H. C., Poulsen, T., Klemp, P., and Brodhagen, H. The carcinogenic effect of sequential artificial sunlight + UVA irradiation in hairless mice. Consequences for solarium "therapy." *Arch. Dermatol.*, *119*:641–642 (1983).

4. Paul, B. S. and Parrish, J. A. The interaction of UVA and UVB in the production of threshold erythema. *J. Invest. Dermatol.*, *78*:371–374 (1982).
5. Gilchrest, B. A., Soter, N. A., Hawk, J. L. M., Barr, R. M., Black, A. K., Hensby, C. N., Mallet, A. I., Greaves, N. W., and Parrish, J. A. Histologic changes associated with ultraviolet A-induced erythema in normal human skin. *J. Am. Acad. Dermatol.*, *9*:213–219 (1983).
6. Gschnait, F., Schwarz, T., and Ladich, I. Treatment of polymorphous light eruption. *Arch. Dermatol. Res.*, *275*:379–382 (1983).
7. McFadden, N. UVA sensitivity and topical photoprotection in polymorphous light eruption. *Photodermatology*, *1*:76–78 (1984).
8. Sunscreen products for over-the-counter use. *Federal Register*, *43*:28206–28269 (1994).
9. Diffey, B. L. and Farr, P. M. An evaluation of sunscreens in patients with broad action-spectrum photosensitivity. *Br. J. Dermatol.*, *112*:83–86 (1985).
10. Lowe, N. J., Dromgoole, S. H., Sefton, J., Bourget, T., and Weingarten, D. Indoor and outdoor efficacy testing of a broad-spectrum sunscreen against ultraviolet. A radiation in psoralen-sensitized subjects. *J. Am. Acad. Dermatol.*, *17*:224–230 (1987).
11. Kaidbey, K. and Gange, R. W. Comparison of methods for assessing photoprotection against ultraviolet A in vivo. *J. Am. Acad. Dermatol.*, *16*:346–353 (1987).
12. Ruger, R., Holzle, E., Plewig, G., and Galosi, A. In-vivo UVA tests: Erythema, pigmentation, phototoxicity. *Skin Models* (Marks, R., Plewig, G., eds.), Springer-Verlag, Berlin, 1986, pp. 147–154.
13. Mutzhas, M. F., Hoizle, E., Hofmann, C., and Plewig, G. A new apparatus with high radiation energy between 320–460 nm: Physical description and dermatological applications. *J. Invest. Dermatol.*, *76*:42–47 (1981).
14. Lowe, N. J., Weingarten, D., and Wortzman, M. Sunscreens and phototesting. *Clin. Dermatol.*, *6*:40–49 (1988).
15. Gange, R. W., Soparkar, A., Matzinger, E., Dromgoole, S. H., Sefton, J., and DeGryse, R. Efficacy of a sunscreen containing butyl methoxydibenzoylmethane against ultraviolet A radiation in photosensitized subjects. *J. Am. Acad. Dermatol.*, *15*:494–499 (1986).

27

An Evaluation of Methods for Measuring Sunscreen Ultraviolet A Protection Factors

Joseph W. Stanfield, Susan H. Edmonds, and Patricia P. Agin
Schering-Plough Health Care Products, Memphis, Tennessee

I. INTRODUCTION

Increased awareness of the need for protection against the ultraviolet (UV)A portion of the solar spectrum (320–400 nm) has accelerated development of sunscreen products that offer improved UVA protection. The development of products that provide greater UVA protection and the labeling of such products require a reliable index of UVA protection. This is a formidable problem because sunlight is always a mixture of UVA and UVB, and the immediate and long-term effects of UVA are normally masked by the effects of energy in the more potent UVB wavelengths. Separation of UVA effects from those of full-spectrum UV radiation is difficult, and eliciting effects of UVA alone requires high doses of UV energy and relatively long exposure times.

The purpose of this chapter is to describe several methods for measuring sunscreen product UVA protection and to examine their validity and practicality, based on laboratory evaluations involving representative sunscreens with varying degrees of UVA protectiveness.

Sunburn, or erythema, is the most familiar effect of UV. It has a steep dose–response curve, which results in a well-defined endpoint, corresponding to a threshold dose, or minimal erythema dose (MED). The action spectrum for erythema, described by Parrish et al. (1), and more recently by McKinlay and Diffey (2) and Diffey (3) is similar to the action spectrum described by DeGrujil and Van der Leun (4) for photocarcinogenesis and represents a conservative estimate of prolonged connective tissue damage by UVA (5,6).

Protection against erythema from the full solar UV spectrum is measured and expressed as the sun protection factor (SPF). The SPF is the ratio of the sunscreen-protected MED to the unprotected MED expressed in energy units (e.g., J/cm^2) or seconds, assuming constant source irradiance. The mean SPF for a panel of human subjects determines the labeled SPF value for a sunscreen product (7). The SPF is an index of protection against the full solar UV spectrum, but UVA protection levels may vary widely for products having the same SPF. Therefore, a measure of sunscreen protection specific to the UVA portion of the solar spectrum (320–400 nm) and capable of differentiating among products with varying degrees of UVA protection is needed.

Several in vivo methods have been proposed for determining UVA protection. These include the immediate pigment-darkening (IPD) method, as described by Kaidbey and Barnes (8); the persistent pigment-darkening (PPD) method, as described by Chardon et al. (9); the protection factor in UVA (PFA) method, as described by Cole (10); and the phototoxic protection factor (PPF), as described by Lowe et al. (11).

The IPD method involves determining a UVA protection factor from the ratio of the sunscreen-protected minimal immediate pigment-darkening dose to the unprotected minimal immediate pigment-darkening dose, within 60 s after each UVA exposure. The IPD is a transient reaction occurring within preexisting melanin granules. Because it is an oxygen-dependent reaction, IPD is potentially a highly variable response (12,13). The action spectrum shows a broad peak UV effectiveness at 340–370 nm (14), which is substantially different from that for erythema (1).

The PPD method involves determining a UVA protection factor from the ratio of the sunscreen-protected minimal persistent pigment-darkening dose to the unprotected minimal persistent-pigment darkening dose, evaluated approximately 2 h after UVA exposure. The action spectrum for the PPD response has not yet been fully described.

The PFA method involves determining a UVA protection factor from the ratio of the sunscreen-protected minimal response dose (erythema or tanning) to the unprotected minimal response dose, approximately 24 h after UVA exposures. Action spectra for these responses are approximated by the erythemal effectiveness spectrum of McKinlay and Diffey (2).

The PPF method uses treatment with topical 8-methoxypsoralen (8-MOP) to increase subject sensitivity to UV by eliciting a phototoxic response, which is maximal in the UVA portion of the spectrum (15). Thus, the action spectrum differs substantially from that for erythema. The PPF is the ratio of the sunscreen-protected minimal phototoxic dose to the unprotected minimal phototoxic dose, measured 72 h after UVA exposures.

In addition to the foregoing in vivo methods, several in vitro methods for measuring sunscreen protection have been proposed (16). Most of these methods are similar, and only one will be addressed in this chapter: the UVA protection percentage (APP) method proposed by Sayre and Agin (17,18). The APP method uses an empirically determined UVA forward-scattering transmission spectrum of the sunscreen product determined on excised skin or a suitable artificial substrate, the McKinlay–Diffey erythemal effectiveness spectrum (2), and a standard solar UVA spectrum (19,20), to calculate the percentage of effective UVA energy blocked. APP is determined according to the following expression:

$$\text{APP} = \left[\frac{1 - \Sigma_\lambda S_\lambda E_\lambda T_\lambda}{\Sigma_\lambda S_\lambda E_\lambda}\right] 100\% \quad (1)$$

where

λ = each wavelength in the range from 320 to 400 nm
S_λ = solar spectral irradiance at wavelength λ
E_λ = erythemal effectiveness at wavelength λ
T_λ = product transmission at wavelength λ

The APP value may be converted to a UVA protection factor using the following formula:

$$\text{UVA protection factor} = \frac{1}{(1 - \text{APP}/100\%)} \quad (2)$$

For example, an APP of 80% could also be expressed as a UVA protection factor of 5.0. Thus, Eq. (1) becomes

$$\text{UVA protection factor} = \frac{\Sigma_\lambda S_\lambda E}{\Sigma_\lambda S_\lambda E_\lambda T_\lambda} \quad (3)$$

An estimate of the PFA of a sunscreen product may be calculated using Eq. (3) by replacing the solar spectral irradiance, S_λ, with spectral irradiance values for a solar simulator, which has been modified to deliver energy only in the UVA wavelengths, as demonstrated by Cole (10).

Diffuse reflectance spectroscopy, as described by Kollias et al. (21), can be used to determine the transmission spectrum of a sunscreen product directly from human skin. In this method, UV reflectance spectra are obtained rapidly and noninvasively from the skin, using a fiber optic bundle to transmit small amounts of UV energy to the skin and the reflected energy back to a photomultiplier detector, before and after sunscreen application. Because both incident and reflected radiation are attenuated by the layer of sunscreen on the skin, the square root of the ratio of the diffuse reflection spectrum from sunscreen-protected skin to that from unprotected skin at each wavelength yields the product transmission spectrum. This is shown schematically in Figure 1.

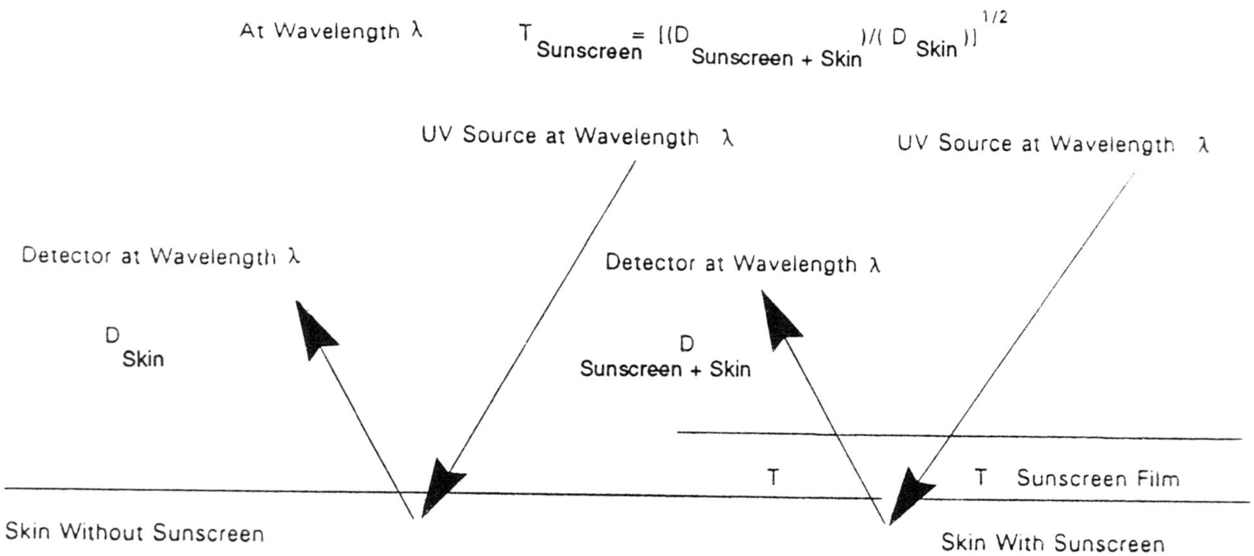

Figure 1 Sunscreen transmission obtained from diffuse reflectance measurements.

A disadvantage of the diffuse reflectance method is that sunscreen transmission is difficult to measure in the UVB range, owing to low UVB reflectance of skin and backscatter by the sunscreen product itself. However, the UVA portions of transmission spectra obtained by diffuse reflectance resemble UVA transmission spectra obtained using forward-scattering measurements, such as those presented by Sayre and Agin (17,18), and can be used to calculate an APP value directly from human skin.

Evidence that sunscreen UVA transmission spectra obtained by diffuse reflectance and forward-scattering methods can both be used to estimate UVA protection factors obtained by in vivo measurements with accuracy would support the APP concept as an appropriate index of sunscreen UVA protection. It would also demonstrate the potential value of diffuse reflectance as a rapid, noninvasive, in vivo method for determining the APP value as an index of product UVA protection, directly from human skin.

To evaluate in vivo UVA protection results for the IPD, PPD, and PFA methods, compared with APP values obtained by in vitro forward scattering and in vivo diffuse reflectance, a clinical study was conducted in 27 volunteer subjects treated with three sunscreen products with differing degrees of UVA protection.

Sunscreen transmission spectra obtained by in vitro forward scattering and in vivo diffuse reflectance were used to estimate product PFA values for comparison with those obtained by evaluating clinical endpoints.

II. METHODS AND MATERIALS

A. Subjects

Subjects included men and women with Fitzpatrick skin types I, II, III, and IV (22). All subjects provided written informed consent and agreed to avoid sunbathing and tanning bed use during the entire study.

B. Sunscreens

Sunscreen products included the Cosmetic, Toiletry, and Fragrance Association (CTFA) proposed SPF 15 sunscreen standard (octyl dimethyl PABA, 7.0% and oxybenzone, 3.0%); a commercial SPF 15 lotion, marketed by Schering-Plough [octyl methoxycinnamate, 7.5%; avobenzone (Parsol 1789, Givaudan), 3.0%; and oxybenzone, 3.0%] and a 5.0% oxybenzone lotion previously evaluated by Cole (10), which is an appropriate sunscreen standard for use in both in vivo and in vitro tests of the efficacy of sunscreen UVA protection.

C. Ultraviolet Sources

The UVA sources were 150-W xenon arc solar simulators, as described by Berger (23). Each solar simulator was equipped with a 1-mm WG320 UVC-blocking filter and a 3-mm WG335 UVB-blocking filter, in addition to a visible and infrared-blocking dichroic mirror, and a visible and infrared-blocking 1-mm UG-11 filter.

Spectral irradiance of the lamps was measured within 1 month of study initiation using a spectroradiometer (Optronic Model 742) calibrated with a UV source traceable to NIST standards.

From integrals of spectral irradiance curves and convolution with the McKinlay–Diffey erythemal effectiveness spectrum (2), the effective lamp power was approximately 0.8 mW/cm^2 when configured for full spectrum UV and 0.03 mW/cm^2 when the WG335 UVB-blocking filter was added.

The UVA spectrum of the lamp with the WG335 filter differs from that of the standard solar UVA spectrum in that the ratio of effective power in the UVA II region (320–340 nm) to that in the UVA I region (340–400 nm) is very low. The lamp UVA II/UVA I ratio is 0.2, whereas that for sunlight is approximately 1. This would tend to cause underestimation of the effects of UVA II, which is the most biologically important portion of the UVA spectrum.

D. Procedures

Test sites (50 cm^2) were selected on the back between the waistline and scapulae and lateral to the midline. Sites were observed with Woods lamp illumination to ensure uniform pigmentation and texture, and absence of warts, moles, nevi, scars, and blemishes. Excessive hair was shaved or clipped from the skin.

Subject skin type was estimated by sunburning and tanning history. To confirm subject skin type, the MED to full-spectrum UV (solar simulator with the WG335 filter removed) was determined by administering a series of UV doses to an unprotected site and grading the erythemal response 16–24 h later. Typical MED values for the lamps used in this study range from 6 to 8 s for untanned type I subjects to over 40 s for moderately tanned subjects with type IV skin.

In addition, colorimetry readings (Minolta Chromameter, Model CR300) were obtained from an adjacent untreated site. The value of total luminosity (L*) was used as an index of skin color (24). The L* scale extends from 0 (black) to 100 (white). Caucasian skin usually ranges in L* values from 75 or higher for very fair skin to 60 or less for moderately tanned skin. The Chromameter was calibrated daily using white and brown standards provided by Minolta.

Test products (100 ± 5 mg) were applied to designated 50-cm^2 areas of the back with the subject seated upright, and spread uniformly over the test site, using a finger cot. Evenness of application was verified by observation with a Woods lamp. The product was allowed to dry at least 15 min.

Each subject's protection factor for each type of evaluation was determined by exposing an untreated, unprotected skin site and an adjacent sunscreen-protected site to respective series of exposures from the solar simulator, with the WG335 filter in place.

The exposure doses for the unprotected sites were calculated using the geometric series expressed as follows:

$$\text{Unprotected dose} = \text{MRD}_E \times (1.25)^N, \tag{4}$$

where

MRD_E = expected minimal response dose and
N = −3, −2, −1, 0, 1, 2, and 3

Likewise, the exposure doses for the sunscreen-protected sites were calculated using the geometric series expressed as follows:

$$\text{Protected dose} = \text{PF}_E \times \text{MRD}_E \times (1.25)^N, \tag{5}$$

where

PF_E = expected protection factor
MRD_E = expected minimal response dose
N = −3, −2, −1, 0, 1, 2, and 3

If considered necessary, the series were extended to higher UV doses to ensure that threshold doses that produced defined endpoints were reached for PPD and PFA, as well as IPD.

Within 60 s after each exposure (IPD test), and again approximately 2 h after exposures (PPD test) and 16–24 h after exposures (PFA test), the irradiated sites were evaluated under bright, "warm white" illumination (approximately 1000 Lux at 6000 K) for pigmentation responses and erythema, using the following scales:

Pigmentation

0 = No response
0.5 = Equivocal response
1.0 = Unambiguous dark gray or brown pigmentation
1.5 = Well-defined dark gray or brown pigmentation with sharp borders
2.0 = Deep pigmentation

Erythema

0 = No response
0.5 = Equivocal response
1.0 = Unambiguous erythema
1.5 = Well-defined erythema with sharp borders
2.0 = Bright erythema

One test product was evaluated at each session (at a different skin site), and the procedure was performed for all three test products within 3 weeks of study enrollment.

E. Interpretation of Endpoints

The endpoint for the IPD test was pigmentation producing a grade ≥1 within 1 min after each UVA exposure, whereas the endpoint for the PPD test was pigmentation producing a grade ≥1 approximately 2 h following UVA exposures. The endpoint for the PFA test was erythema (erythema grade ≥1) or tanning (pigmentation grade ≥1), 16–24 h after UVA exposures.

Each protection factor was calculated as the ratio of the smallest dose that produced the endpoint for the sunscreen-protected site to that for the unprotected site. If, at a given evaluation time, an erythema endpoint occurred at a lower UVA dose than a pigmentation endpoint, any IPD or PPD response was not considered a valid endpoint. Likewise, an erythema response that occurred at a higher UVA dose than a pigmentation response was not considered a valid endpoint. Subjects for whom the appropriate endpoint could not be determined were considered unevaluable by that method.

Protection factors using the IPD, PPD, or PFA methods could not be determined if either the sunscreen-protected site or the unprotected site failed to yield a valid endpoint. Determination of a valid endpoint required at least one UVA exposure that produced a grade 0 or 0.5 response and progressively higher grades consistent with UVA dose.

III. RESULTS

A. Subjects

Subjects included 20 women and 7 men, ranging in age from 22 to 55 years (mean = 35.2, SD = 9.3). Subject demographic characteristics, skin type, L* and MED values are shown in Table 1.

B. Effect of Skin Type on Observed Endpoints

Mean values of UV doses that produced endpoints for unprotected sites, expressed in seconds, are shown in Figure 2. The full spectrum MED was progressively higher for skin types I through IV, and the coefficient of correlation (r^2) was 0.72 for MED versus L*. As expected, no IPD or PPD endpoint was observed for type I subjects, and the mean UV doses that produced the PPD endpoint were higher for types II and III than for type IV. In general, the endpoints for PPD, 24-h UVA erythema and 24-h UVA tanning occurred at UVA doses about twice those required to produce the IPD endpoint. Only one type

Table 1 Subject Demographic and Skin Type Data

Subject number		Age	Sex	MED	Type	L*
14		52	F	9	I	67.88
16		25	F	8	I	75.09
20		24	M	10	I	71.39
Type I	*N*	3		3		3.00
	Mean	34		9		71.45
	SD	16		1		3.61
	Max	52		10		75.09
	Min	24		8		67.88
5		47	F	16	II	67.76
13		30	F	12	II	72.08
15		23	M	9	II	70.14
17		36	F	13	II	70.72
18		26	F	9	II	73.50
21		55	F	12	II	70.78
22		32	F	10	II	71.57
23		28	F	10	II	76.67
24		30	M	7	II	72.90
25		30	F	10	II	73.52
26		27	F	10	II	70.71
27		44	F	10	II	72.39
Type II	*N*	12		12		12.00
	Mean	34		11		71.89
	SD	10		2		2.21
	Max	55		16		76.67
	Min	23		7		67.76
1		30	M	23	III	64.97
2		45	F	17	III	66.56
6		43	F	14	III	67.60
7		29	F	25	III	63.89
10		52	F	23	III	70.26
11		32	F	19	III	64.05
12		33	M	13	III	72.71
19		30	M	15	III	67.07
Type III	*N*	8		8		8.00
	Mean	37		19		67.14
	SD	9		5		3.07
	Max	52		25		72.71
	Min	29		13		63.89

Table 1 Continued

Subject number		Age	Sex	MED	Type	L*
3		37	F	20	IV	62.41
4		22	M	36	IV	54.99
8		36	F	20	IV	60.90
9		42	F	29	IV	64.39
Type IV	*N*	4		4		4.00
	Mean	34		26		60.67
	SD	9		8		4.05
	Max	42		36		64.39
	Min	22		20		54.99
All	*N*	27	M: 7	27		27.00
	Mean	35	F: 20	15		68.77
	SD	9		7		4.87
	Max	55		36		76.67
	Min	22		7		54.99

III subject and no type IV subject showed the 24-h UVA erythema endpoint. Conversely, no type I subject showed the 24-h UVA-tanning endpoint.

C. Protection Factors

With the method of Sayre and Agin (17,18), APP values of 63%, 87%, and 73% were determined for the CTFA standard SPF 15 sunscreen, commercial SPF 15 sunscreen, and the 5% oxybenzone lotion, respectively, using in vitro forward-scattering measurements. Corresponding protection factors based on APP are 2.7, 7.6, and 3.8, from Eq. (3). Figures 3 through 5 show mean in vivo UVA protection factors by skin type, for each of the methods evaluated, compared with protection factors based on APP determinations for the three sunscreen products. A complete listing of protection factor results is shown in Table 2.

In general, UVA protection factors determined by in vivo methods were higher than those predicted by APP. Thus APP appears to represent a conservative estimate of UVA protection. Protection factors based on IPD and PPD responses were not obtained from type I subjects because those subjects did not exhibit tanning responses. However, 24-h PFA values were obtained for all four skin types when either the erythema or tanning responses were used. For subjects with valid endpoints, the overall mean IPD protection factors were 3.3 for the CTFA SPF 15 standard, 3.8 for the 5% oxybenzone lotion, and 11.5 for the

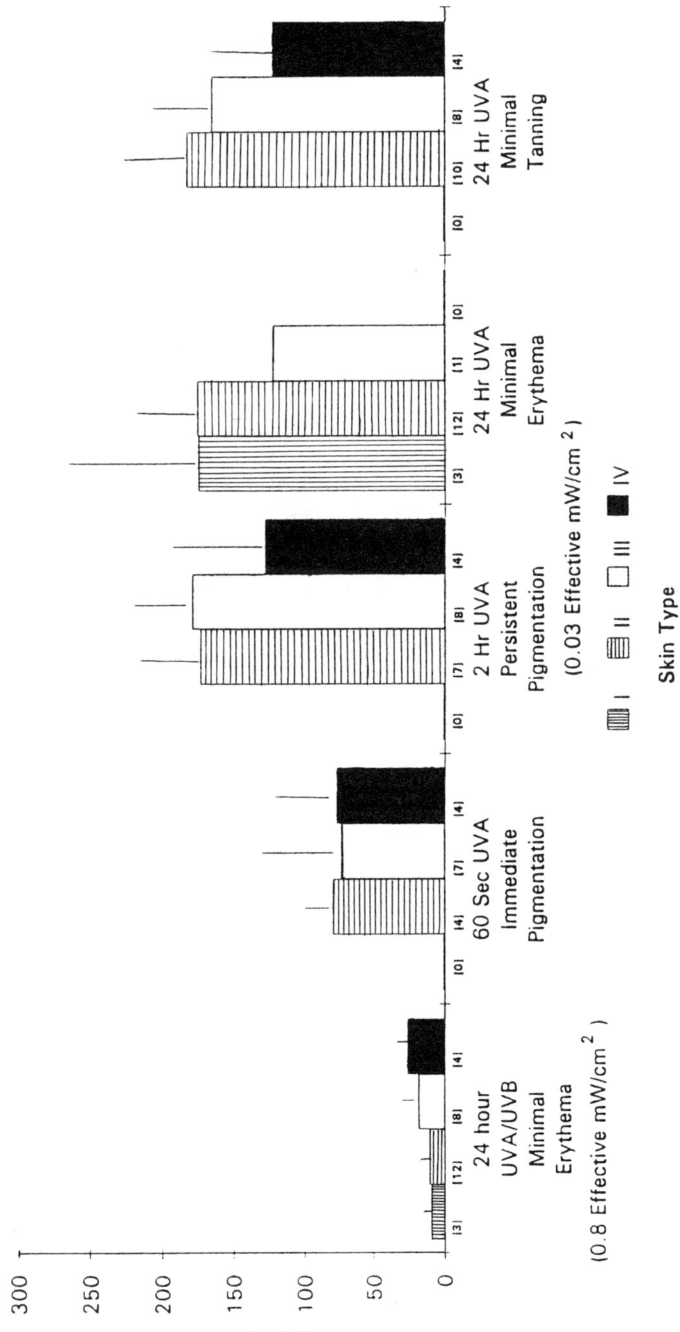

Figure 2 Mean UV endpoint doses (*N*).

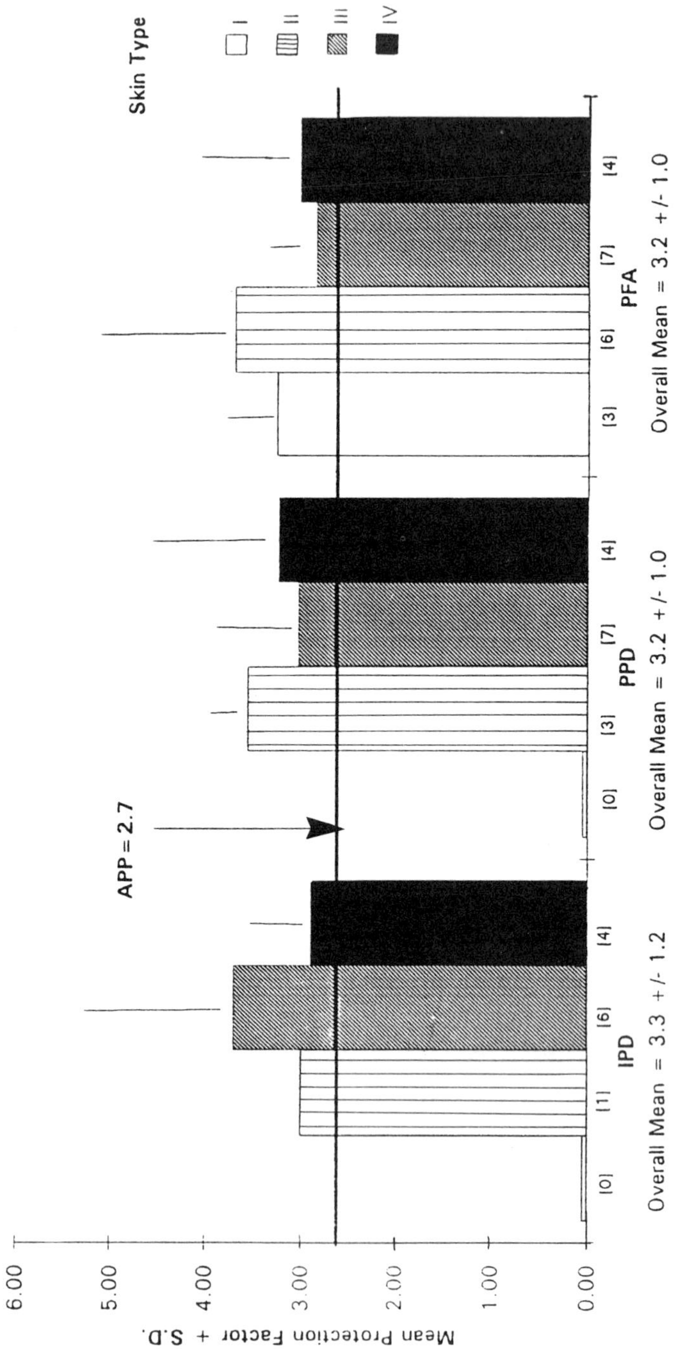

Figure 3 UVA protection factors for CTFA 15 standard (*N*).

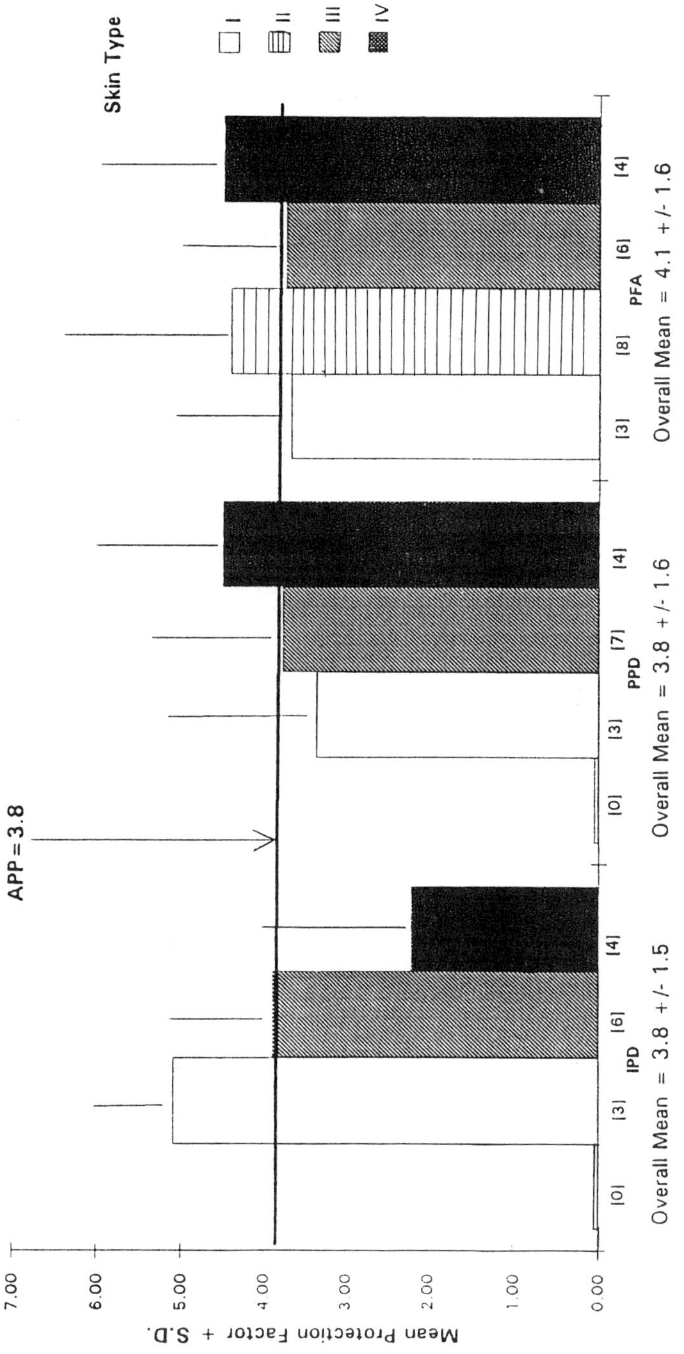

Figure 4 UVA protection factors for 5% oxybenzone lotion (*N*).

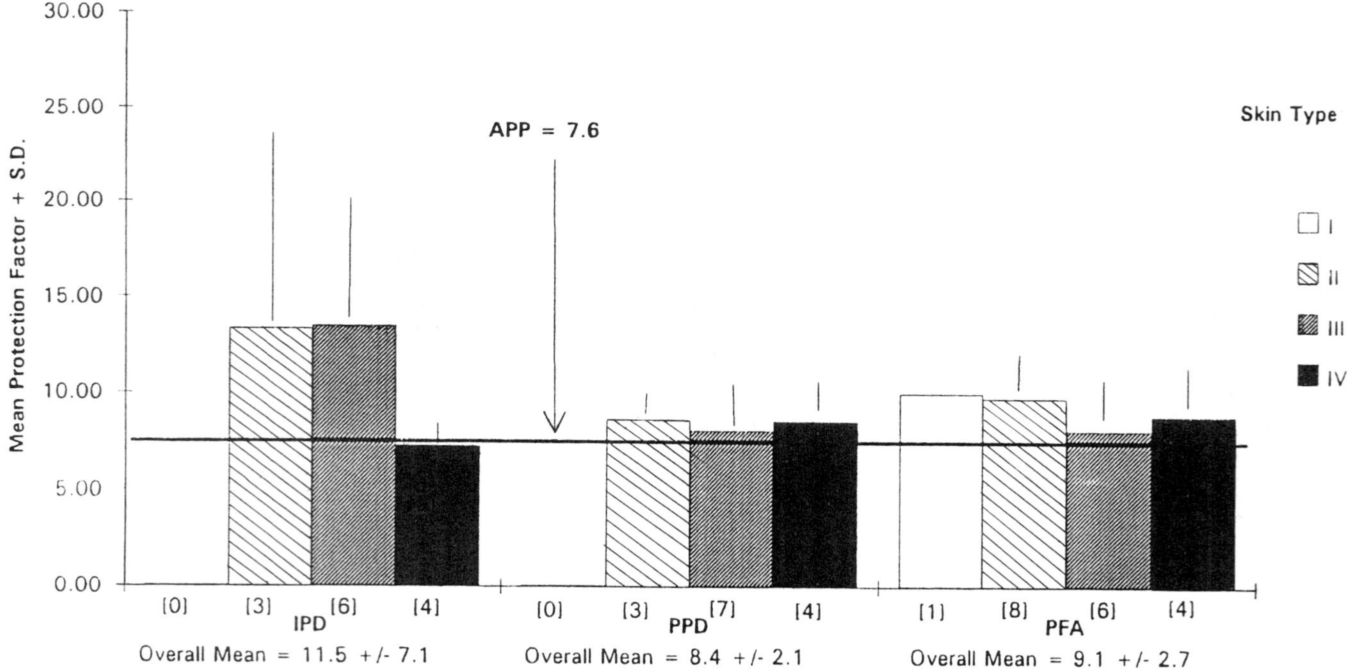

Figure 5 UVA protection factors for commercial SPF 15 sunscreen (*N*).

commercial SPF 15 sunscreen. Overall mean PPD protection factors were 3.2, 3.8, and 8.4, and overall mean PFA values were 3.2, 4.1, and 9.1, for the CTFA SPF 15 standard, the 5% oxybenzone lotion, and the commercial SPF 15 sunscreen, respectively.

An ANOVA was performed for the nine subjects who provided results for all three methods and all three products. For the IPD method, results for skin types III and IV were significantly different ($p = 0.004$). For all skin types, IPD results were significantly different from those obtained by the PPD method ($p = 0.009$) and the PFA method ($p = 0.006$). For PPD and PFA methods no significant differences in mean protection factors were found for method or skin type ($p > 0.05$).

For the 19 subjects with valid endpoints for the CTFA SPF 15 standard and the 5% oxybenzone lotion, product mean protection factors were significantly different for the PFA method ($p = 0.03$), but not for the IPD or PPD methods ($p > 0.05$).

Because of its substantially higher UVA protection, mean protection factors were significantly different for the commercial SPF 15 sunscreen and the other two products ($p < 0.05$) by all three methods).

For the 5% oxybenzone lotion, the effect of skin type was statistically significant ($p = 0.01$) for the IPD method, but not significant ($p > 0.05$) for the PPD and PFA methods. The mean PFA value of 4.1 was similar to the value of 4.0 reported by Cole (10).

Ultraviolet A transmission spectra were also obtained for each product by the in vivo diffuse reflectance method, and are shown, along with spectra obtained by in vitro forward scattering, in Figure 6. Although the general shapes of the transmission curves are similar, diffuse reflectance spectra yielded higher transmission values (lower apparent UVA protection) at the shorter wavelengths. For the 5% oxybenzone lotion and the commercial SPF 15 sunscreen, diffuse reflectance spectra showed lower transmission values (higher apparent UVA protection) at longer wavelengths.

Estimates of PFA calculated by Eq. (3) using transmission spectra obtained both by forward scattering and diffuse reflectance are shown in Figure 7. There was good agreement between the estimates, but both methods tended to underestimate in vivo PFA values determined using the endpoints of erythema or tanning.

The APP values calculated using diffuse reflectance measurements are compared with the original APP values obtained by forward scattering in Figure 8. There was excellent agreement among the estimates for all three products.

IV. DISCUSSION

The clinical study described in the foregoing has shown that simultaneous in vivo measurements of sunscreen UVA protection using the IPD, PPD, and PFA

Table 2 UVA Protection Factor Results

Subject no.	Skin type	CTFA 15 STD			5% Oxybenzone lotion			Commercial SPF 15 sunscreen		
		IPD	PPD	PFA	IPD	PPD	PFA	IPD	PPD	PFA
14	I			3.00			4.00			
16	I			3.76			2.04			
20	I			3.00			4.99			10.01
5	II		3.00		6.26	6.15	9.20	6.40	8.00	8.00
13	II		3.90	3.00	4.00	1.54	2.00	25.60	9.85	9.85
15	II			3.75		4.00	3.08			16.00
17	II				5.00	1.29	4.00	8.00	8.00	
18	II			1.92			4.00			10.40
21	II	3.00	3.75	4.69		3.20	4.00			
22	II									8.00
23	II									8.00
24	II			3.00			4.92			8.00
25	II			5.77						
26	II									10.00
27	II					4.00	4.00			
1	III	1.23	3.00	2.39	4.00	4.00	4.00	15.60	10.03	9.99
2	III		2.40	2.34	4.00	6.20	6.23		6.40	
6	III	3.75	3.00	3.00	6.24	4.00	4.00	10.01	10.01	6.39
7	III	3.00	4.72	3.00	4.00	5.35	4.00	10.00	6.43	8.00
10	III					4.00	3.19			
11	III	3.75	2.42	3.77	3.25	2.08	2.08	8.00	6.43	4.10
12	III	5.76	1.92	2.44	3.23	2.00	3.20	27.37	12.31	12.31
19	III	4.69	3.75	3.00	2.60	2.58	3.20	10.00	5.12	8.00

3	IV	2.40	3.77	3.77	3.20	6.29	6.29	8.00	8.00	10.03
4	IV	2.40	3.00	3.00	2.05	3.21	4.00	8.00	6.40	8.00
8	IV	3.77	4.68	3.76				6.46	9.96	12.49
9	IV	3.00	1.50	1.54	1.35	4.00	3.20	6.50	10.00	5.10
	Type I									
	Mean			3.25			3.68			10.01
	SD			0.44			1.50			
	N	0	0	3	0	0	3	0	0	1
	Type II									
	Mean	3.00	3.55	3.69	5.08	3.36	4.40	13.33	8.62	9.78
	SD		0.48	1.37	1.13	1.80	2.12	10.65	1.07	2.72
	N	1	3	6	3	6	8	3	3	8
	Type III									
	Mean	3.70	3.03	2.85	3.90	3.78	3.74	13.50	8.10	8.13
	SD	1.54	0.95	0.51	1.16	1.51	1.20	7.26	2.66	2.84
	N	6	7	7	7	8	8	6	7	6
	Type IV									
	Mean	2.89	3.24	3.02	2.20	4.50	4.50	7.24	8.59	8.91
	SD	0.65	1.35	1.05	0.94	1.60	1.60	0.88	1.73	3.13
	N	4	4	4	3	3	3	4	4	4
	All									
	Mean	3.34	3.20	3.20	3.78	3.76	4.07	11.53	8.35	9.09
	SD	1.22	0.96	0.95	1.45	1.58	1.60	7.07	2.05	2.71
	N	11	14	20	13	17	22	13	14	19

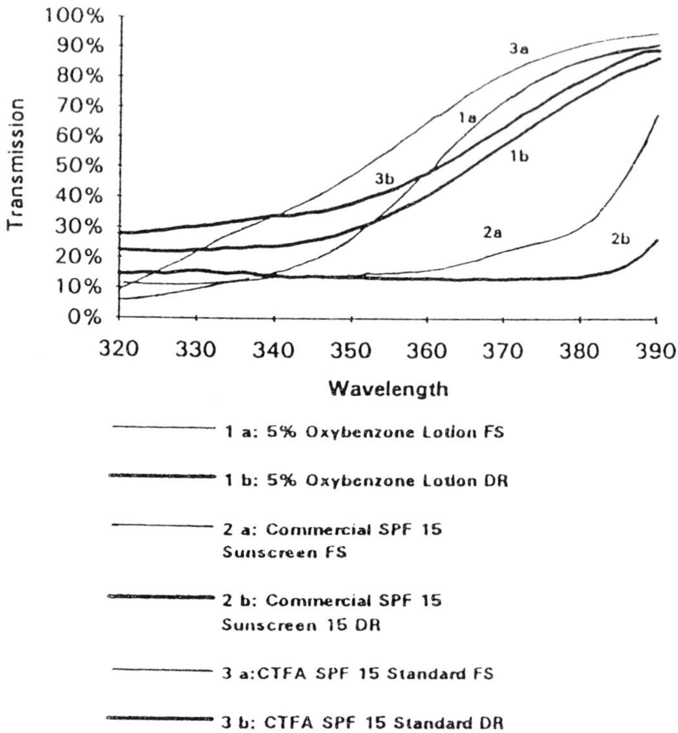

Figure 6 Comparison of UVA transmission spectra from forward scattering (FS) and diffuse reflectance (DR) measurements (mean of three spectra).

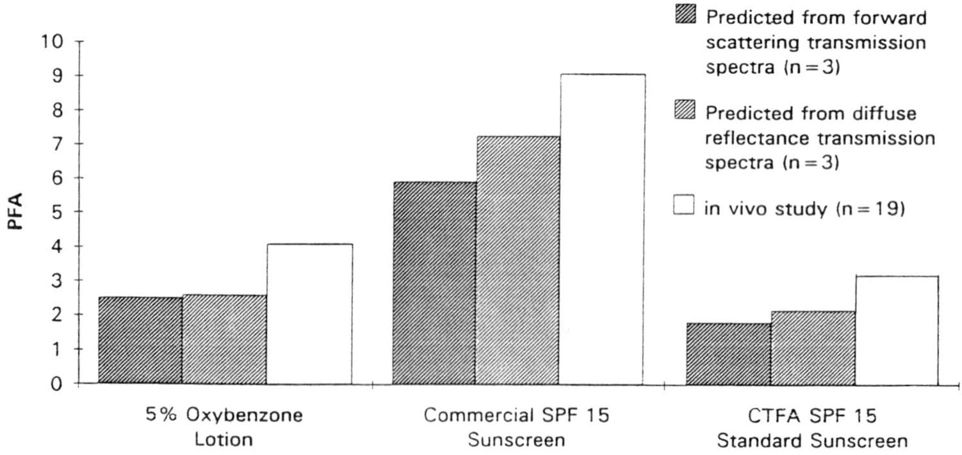

Figure 7 Predicted and measured UVA protection factor (PFA).

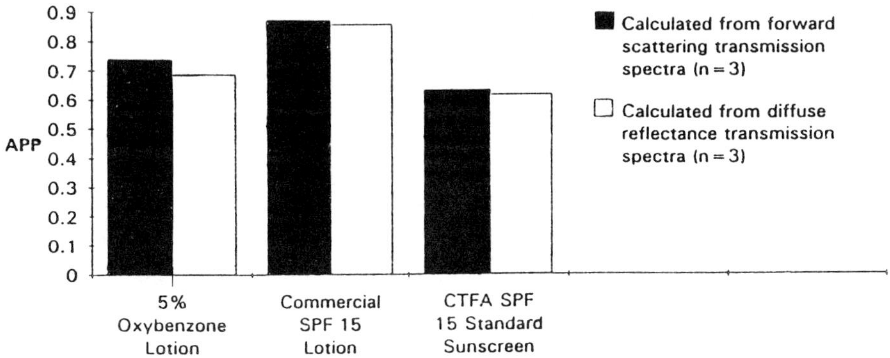

Figure 8 UVA protection percentage (APP) calculated from forward scattering and diffuse reflectance measurements.

methods are feasible, although a series of UVA exposures required up to 5 h to complete for a single product. This illustrates the difficulties associated with the isolation of UVA effects from those of full-spectrum UV.

The effects of comparable UVA doses on subjects of differing skin type were dramatically different: Type I subjects exhibited only erythema responses, type II exhibited predominately erythema responses, type III exhibited both erythema and tanning responses, and type IV exhibited tanning responses only. Thus, IPD and PPD tests are not practical for type I and II subjects, who are the individuals with the greatest need for UV protection. In addition, protection factors determined by the IPD method showed significant skin type effects. The only clinical method that yielded endpoints for all four skin types and discriminated among all three products with statistical significance was the PFA method.

Of the in vivo methods considered, only PFA uses responses that follow the erythemal effectiveness spectrum. Consequently, PFA is believed to represent protection against UV effects having effectiveness spectra similar to that of erythema, as discussed in the foregoing. The relevance of IPD to these effects is considered low, and the relevance of PPD to these effects is unknown. In addition, the PFA value of 4.1 for the proposed UVA standard agreed well with the value of 4.0 reported by Cole (10), which supports Cole's observation that PFA values are reproducible among different laboratories and suggests that the 5% oxybenzone lotion could be used routinely as a control in UVA studies, as the homosalate standard is used in SPF testing (7). However, the PFA test is subject to the limitation in UV source spectrum (i.e., deficient UVA II/UVA I ratio), as are the IPD and PPD tests.

The agreement in estimates of PFA using transmission spectra obtained by in vitro forward scattering and in vivo diffuse reflectance (see Fig. 7) demonstrated that transmission spectra from either measurement technique can be used, along with the UVA source spectrum, to produce UVA protection estimates which are conservative, and show reasonable agreement with an established, but more cumbersome, in vivo method. Although forward-scattering measurements require reduced sunscreen application density (0.5 mg/cm^2) (9), the application density used in diffuse reflectance measurements is the standard 2 mg/cm^2, which is the same as that used in sunscreen SPF testing (7).

The excellent agreement between APP values obtained by forward-scattering and diffuse reflectance measurements suggests that either method is appropriate for determining the APP of a sunscreen product.

Diffuse reflectance measurements required less than 1 min each to complete. Because the skin area measured is relatively small, a large number of sunscreen products could be evaluated per subject. In addition, the rapidity and ease of the method make it suitable for evaluating substantivity of UVA protection (e.g., resistance to water immersion, ruboff, and perspiration).

V. CONCLUSIONS

In a clinical study of 27 subjects treated with three sunscreen products with widely differing degrees of UVA protection, PFA measurements provided a more reliable and relevant UVA protection index than IPD or PPD. However, PFA suffers from the inherent spectral limitations of filtered xenon arc lamps and requires long exposure times and high UV doses.

The APP method had been shown earlier to provide an index of UVA protection that is independent of skin type and adaptable to any source spectrum (17,18). The results described in the foregoing suggest that APP is a conservative estimate of UVA protection, and that APP values may be determined directly from human skin using the noninvasive technique of diffuse reflectance spectroscopy, which can eliminate unnecessary, time-consuming exposure of subjects to UV radiation, while providing appropriate and reliable measurements of sunscreen product efficacy against UVA radiation.

The proposed 5% oxybenzone UVA standard appears to be appropriate for use as a control for in vivo and in vitro evaluations of sunscreen UVA protection.

ACKNOWLEDGMENTS

The authors wish to thank Mr. John Beard, Ms. Darlene Daniels, Ms. Phyllis Hinkle, Ms. Jo Ann Jackson, Dr. Robert J. Kaplan, and Ms. Lisa Sutton for their excellent technical and clinical assistance.

REFERENCES

1. Parrish, J., Jaenicke, K., and Anderson, R. Erythema and melanogenesis action spectra of normal human skin. *Photochem. Photobiol.*, *36*:187–191 (1982).
2. McKinlay, A. and Diffey, B. A reference spectrum for ultraviolet induced erythema in human skin. *CIEJ*, *6*:17–22 (1987).
3. Diffey, B. Observed and predicted minimal erythema doses: A comparative study. *Photochem. Photobiol.*, *60*:380–382 (1994).
4. deGruijl, F. and Van der Leun, J. Estimate of the wavelength dependency of ultraviolet carcinogenesis in humans and its relevance to the risk assessment of a stratospheric ozone depletion. *Health Phys.*, *67*:319–325 (1994).
5. Urbach, F. Ultraviolet A transmission by modern sunscreens: Is there a real risk? *Photodermatol. Photoimmunol. Photomed.*, *9*:237–241 (1993).
6. Kligman, L. H. UVA induced biochemical changes in hairless mouse skin collagen: A contrast to UVB effects. In: *Biological Responses to Ultraviolet A Radiation*. (Urbach, F., ed.), Valdenmar Publishing, Overland Park, KS, 1992, pp. 209–216.
7. U.S. Food and Drug Administration. Sunscreen drug products for over-the-counter human use; Tentative final monograph; proposed rule. *Federal Register*, *58*: 28194–28302 (1993).
8. Kaidbey, K. and Barnes, A. Determination of UVA protection factors by means of immediate pigment darkening in normal skin. *J. Am. Acad. Dermatol.*, *25*:262–266 (1991).
9. Chardon, A., Moyal, D., and Hourseau, C. Skin immediate pigment darkening applied to UVA protection assessment. Fourth Congress of the European Society for Photobiology, Amsterdam, Sept. 1–6, 1991.
10. Cole, C. Multicenter evaluation of sunscreen UVA protectiveness with the protection factor test method. *J. Am. Acad. Dermatol.*, *30*:729–736 (1994).
11. Lowe, N. J., Dromgoole, S. H., Sefton, J., et al. Indoor and outdoor efficacy testing of a broad spectrum sunscreen against ultraviolet A radiation in psoralen-sensitized subjects. *J. Am. Acad. Dermatol.*, *15*:224–230 (1987).
12. Tegner, E., Rorsman, H., and Rosengren, E. 5-5-Cysteinyldopa and pigment response to UVA light. *Acta Derm. Venerol.*, *63*:21–25 (1983).
13. Beitner, H. and Wennersten, G. A qualitative and quantitative transmission electron microscope study of the immediate pigment darkening reaction. *Photodermatology*, *2*:273–278 (1985).
14. Rosen, C., Jaques, S., Stuart, M., et al. Immediate pigment darkening: Visual and reflectance spectrophotometric analysis of action spectrum. *Photochem. Photobiol.*, *51*:583–588 (1990).
15. Cripps, D. J., Lowe, N. J., and Lerner, A. B. Action spectra of topical psoralens: A re-evaluation. *Br. J. Dermatol.*, *107*:77–82 (1982).
16. Sayre, R. Correlation of in vivo tests, in vitro SPF prediction, a survey of published studies. *Cosmet. Toilet.*, *108*:111–114 (1993).
17. Sayre, R. M. and Agin, P. P. A method for determination of UVA protection for normal skin. *J. Am. Acad. Dermatol.*, *23*:429–440 (1990).

18. Agin, P. P. UVA protection percent: A versatile method for determining UVA efficacy of sunscreens. *Biological Responses to Ultraviolet A Radiation* (F. Urbach, ed.). Valdenmar Publishing, Overland Park, KS, 1992, pp. 335–345.
19. Frederick, J. E. and Lubin, D. The budget of biologically at the ultraviolet radiation in the earth atmosphere system. *J. Geophys. Res.*, *93*:3825–3832 (1988).
20. Frederick, J. E. Solar ultraviolet irradiance at the earth's surface appropriate to the year 1992. Correspondence. June 1992.
21. Kollias, N., Gilles, R., and Anderson, R. R. The non-invasive determination of UVA sunscreen effectiveness in vivo. *Biological Responses to Ultraviolet A Radiation* (F. Urbach, ed.), Valdenmar Publishing, Overland Park, KS, 1992, pp. 335–345.
22. Fitzpatrick, T. The validity and practicality of sun-reactive skin types I through IV. *Arch. Dermatol.*, *124*:869–871 (1988).
23. Berger, D. S. Specification and design of solar ultraviolet solar simulators. *J. Invest. Dermatol.*, *53*:192–199 (1969).
24. Weatherall, I. L. and Coombs, B. D. Skin color measurement in terms of CIELAB color space values. *J. Invest. Dermatol.*, *99*:468–473 (1992).

28

Persistent Pigment-Darkening Response as a Method for Evaluation of Ultraviolet A Protection Assays

Alain Chardon, Dominique Moyal, and C. Hourseau
L'Oréal, Clichy, France

I. INTRODUCTION

Over the past few decades, the changing lifestyle and the development of leisure activities and holiday habits have led to an increase in daily exposure of the skin affecting an ever larger section of the population. At the same time, sun protective products have undergone improvements in their ability to protect skin against sunburn, caused mainly by ultraviolet B radiation (UVB) as reflected by continuous increase in sun protection factors (SPF).

Recently, the specific effects of the UVA part of sun radiation (UVA) have been more deeply investigated and shown to be biologically more active than previously thought, becoming a real concern to the scientific community and the health authorities, particularly for fair-skinned populations. As a result, protection from UVA has to be improved and the sunscreen industry has been challenged to comply with this requirement by developing and marketing new ingredients and appropriate formulations.

The UVA rays present a weak intrinsic erythemogenic activity, because the UVA minimal erythema dose (UVA-MED) varies with skin type from 20 to 80 J/cm^2, whereas UVB-MED ranges from 20 to 70 mJ/cm^2 (1,2). However, taking into account their relative high level in the sun flux at zenith angle, about 16 times that of UVB, their effective erythemal potential amounts to about one-fifth of that of UVB (erythemal effectiveness: UVB 84%, UVA 16%, UVB:UVA ratio 5.2:1), which is not negligible: an exposure time of 2 h in sun without protection can, in fact, represent more than one UVA-MED in fair skin (3). The SPF index remains the criterion of choice to substantiate sunscreen protection, because it already takes into account the whole UV erythemal effect and actually more than 80% of the potential harmful effects. But the participation of the UVA radiation in this erythemal index amounts to only about 10% of the total UV contribution, whereas it represents more than 90% of the sun UV radiant energy. Thus, the SPF does not convey a specific and accurate image of the product's effective UVA protection to the consumer. There appeared to be a need for a method specially dedicated to the assessment of sunscreen efficacy against UVA effects.

Persistent pigment darkening (PPD), or the stable portion of immediate pigment darkening (IPD), was proposed as a noninvasive, fast, readily induced and visible skin response offering a viable endpoint for such a method. To define accurate and reliable conditions of application, the characteristics of UVA-induced pigmentation were thoroughly studied in various skin types, using colorimetry as an objective tool to quantify and analyze the pigmented response. On the basis of this new appraisal, it was possible to establish a protocol, which was then calibrated and validated by testing physical standards and a variety of formulations.

II. ULTRAVIOLET A PROTECTION ASSESSMENT

A. Sun Radiation

In the spectrum of the sunlight that reaches our skin at ground level, the following wavelength bands can be distinguished: infrared (700–2000 nm); visible light (400–700 nm); ultraviolet A (320–400 nm), which is generally divided into UVA I, or long UVA (340–400 nm), and UVA II, or short UVA (320–340 nm); and the UVB wavelengths (290–320 nm) (4,5).

The maximal sun UVA irradiance at ground level, including direct flux from sun and atmosphere-scattered rays, is about 6 mW/cm^2, and maximal UVB sun irradiance is about 0.36–0.2 mW/cm^2 when sun is close to its zenith. Under these conditions, UVA and UVB together represent about 5% of the total radiant energy received (3). UVA radiation is less affected than UVB by such variables as sun elevation, altitude, or cloud cover (i.e., atmospheric filtration): its total flux is less attenuated in the morning and in the evening, when the sun is low over the horizon (4). Furthermore, the main portion of UVA radiation is not trapped by standard glass: automobile windows, verandas, conservatories, and windows in general fail to protect against UVA as they do for UVB, because the short wavelength cut-off of glass is about 320 nm.

B. Existing Methods

In contrast with usual UV protection assessment, for which erythema is the unique acute endpoint (i.e., determination of the SPF of sunscreens), UVA protection is not defined by a single specific endpoint. Development of in vivo evaluation methods for UVA is still in progress, and no one method has yet been standardized.

The proposed methods use guidelines similar to those for SPF assessment in human or animal models. They call on a variety of endpoints, such as erythema induced by moderate UVA doses and observed at 24 h (6–8); erythema induced by local skin photosensitization by psoralen or derivatives (9–12) or other sensitizers (13), using small UVA doses and observed at 48 h (maximal erythema is delayed when sensitization is involved); immediate pigmentation observed straight after exposure (11,14–16) or after a delay of 2 h, 24 h, or several days (17); sunburn-cell counts at 24 h in photosensitized mouse skin (17); or biochemical analysis (18,19).

An in vitro method has also been proposed using spectrophotometric measurement of UVA absorbance by the test product in solution (29), or by applying it on various substrates, such as excised epidermis (20,21), skin resin replica (22,23), or adhesive plastic tape (24). In this case, the spectral absorbance data are then transformed by calculation into various forms of protection factor, sometimes convoluted with a skin damage action spectrum (such as erythema).

Finally, physical measurements of the absorbance characteristics of a sunscreen film applied on live skin in the absence of UV exposure have recently been reported (25,26).

Some of the previous methods have been used comparatively to assess the efficacy of sunscreens (13,17).

C. The Need for a Specific Method

For enhanced efficacy, the absorption spectra of UVA filters used in sunscreen products should encompass the action spectrum of the damage from which one wishes skin protection. The action spectra of deep skin damage has not been totally established. To overcome this uncertainty, it would appear wise to take into account the widest UVA spectrum possible, particularly because long UVA radiation penetrates deeper into skin. Therefore, an in vivo method should be based on a measurable skin response, with an action spectrum extending throughout the whole UVA waveband.

D. Use of Persistent Pigment Darkening

From this point of view, and because of human ethical considerations, the in vivo method based on immediate pigment-darkening (IPD) skin response and, more specifically, on its lasting component, persistent pigment darkening (PPD) (27,28), has much to recommend it in the assessment of UVA protection, because it is noninvasive and relatively easy to perform in routine testing in humans.

Similar to the SPF determination using erythema as an endpoint (29–35), the UVA protection factor (UVA-PF) is obtained in this method by calculating the ratio of the UVA doses required to induce the same minimal pigmenting effects on human skin with and without protection by the test sample. Even though the IPD/PPD phenomenon has not been shown to constitute skin damage on its own, it constitutes a specific response to UVA and does effect a visible pigmentation, the intensity of which is related to the amount of UVA reaching the skin. It is thus a valid determinant in a method to evaluate a protection factor, provided certain other conditions are met in relation to the characteristics of the endpoint used.

The purpose of the studies exposed hereafter was to investigate these aspects of the IPD skin response: specific color components, setting-up and fading kinetics over 7 days, dose–response effects at various time lags, postexposure, time–flux reciprocity law, action–spectrum and efficacy spectrum, and skin typing. For each study, a short description of the protocol will be given with results and comments according to its consequence in establishing a UVA-PF method.

III. ULTRAVIOLET IMMEDIATE PIGMENTATION OF SKIN

A. Kinetics

The skins of Caucasian volunteers with very light, light, and intermediate color categories (36,37) were exposed to progressive doses (1–28 J/cm^2) of UVA (340–400 nm) supplied by a filtered metal halide source. The pigmentation induced was monitored visually and colorimetrically using the CIE (1987) L*a*b* color space (38). This system of quantitative assessment was selected because it is constructed to be in strict correlation with human vision and, over the years, has proved to be a particularly useful method of skin investigation (36,37,40–42).

During exposure, a bluish-gray skin coloration developed, reaching its maximum at the end of exposure (Fig. 1). Then a partial fading occurred rapidly within 1 h after the end of exposure, but for UVA doses higher than about 6–10 J/cm^2, a stable residual pigmentation remained, persisting for up to several days (43–46).

B. Color of Immediate Pigmentation

The pigmentation present at end of exposure was of an essentially bluish-gray hue. As it decreased, the pigmentation progressively lost its blue component so

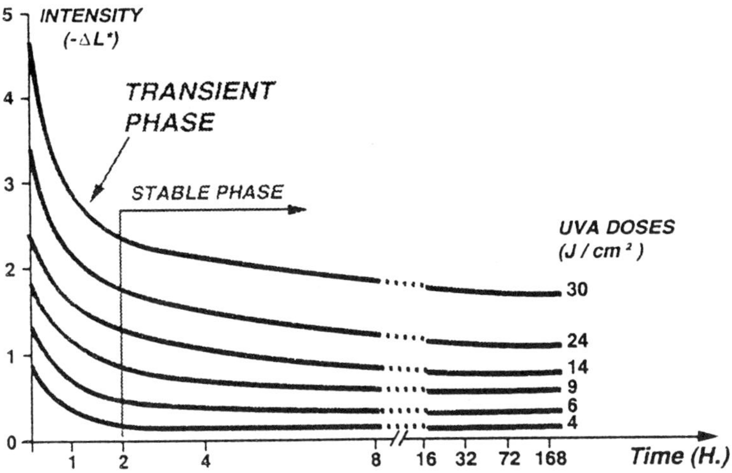

Figure 1 Kinetics of IPD skin response: The intensity of IPD skin response involves a transient phase during the first 2 h postexposure, then a more stable phase that can last several days.

that a gray residue was present about 2 h postexposure and beyond (Fig. 2). The color of this residual pigmentation could not be confused with that of a delayed pigmentation (neomelanization), because the latter occurs, with a typical brown (dark yellow) hue, only with high UVA doses or with cumulative sequential exposures, and with a delay of about 2–3 days (36,37). As a consequence of these colorimetric observations, for the purpose of developing a UVA-PF method, the intensity of immediate and persistent pigmentation induced by a single small UVA exposure was then colorimetrically quantified simply in terms of the difference in luminance ($-\Delta L^*$).

C. Dose–Response and Time Course

The dose–response curves (Fig. 3), recorded on the same volunteers immediately after exposure to progressive UVA doses, exhibited two distinct parts: with low UVA doses (up to about 4–6 J/cm^2) a rapid increase in pigmentation was observed and, with higher doses (6–30 J/cm^2 and beyond), a slow and practically linear rise took place. As time elapsed, the initial surge of the curve faded, whereby the higher part collapsed without significant change in slope so that, after about 2 h, the dose–response curves became practically *linear* and passed through the origin, *without significant threshold*, contrarily to that of the erythemal response. A small dose threshold appeared only afterward.

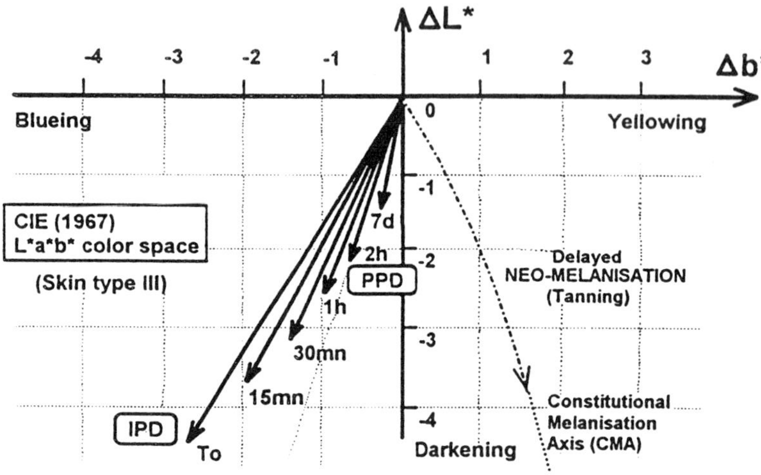

Figure 2 Color of IPD skin responses: Immediately postexposure the color of the UVA-induced pigmentation has a bluish-gray hue ($\Delta L^* < 0$, $\Delta b^* < 0$, $|\Delta L^*| > |\Delta b^*|$). The blue component fades away faster with time than the gray component so that, 2 h later and beyond, only the gray component remains, distinct from dark yellow hue of delayed neomelanization (tanning).

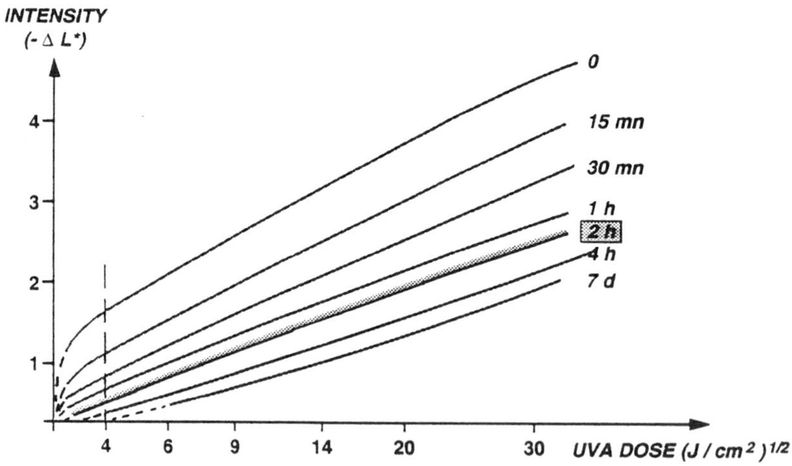

Figure 3 IPD: Dose–Response and Time Course: Immediately postexposure, the IPD dose–response curves are composed of two distinct sections corresponding to UVA doses below and above about 4 J/cm^2, respectively. About 2 h postexposure the residual response is quasilinear, without significant dose threshold. A small dose threshold appears afterward.

D. Composition of the IPD Response

Dose Effect

From these data, it may be deduced that the overall so-called IPD skin response likely includes two main subjacent, additive visible components (Fig. 4), as it can be inferred from the breakage and change in slope in the dose–response curves at near 5–6 J/cm^2: the first of these components, which shows an initial rapid rise, with no significant dose threshold, is unstable, and is saturated over about 6 J/cm^2; the second component appears with a small dose threshold, rises more progressively, and is stable. Although induced with different dose ranges, both phenomena result from an immediate effect of UVA exposure, for they can be observed immediately postexposure. In addition, the UVA dose (about 15 J/cm^2) required to induced a minimal persistent pigmentation (PPD) represents approximately the amount received in 1 h of sun exposure in low latitudes at noon.

As a whole, from these data and other studies (43–47), one could propose the hypothesis that three phases or phenomena may be discerned in ''melanic'' pigmentation, that are more or less intricate in acquired UVA-induced suntanning, but are distinct in their characteristics: that is transient IPD, persistent PPD, and neomelanization. However, the biological nature of these phenomena has

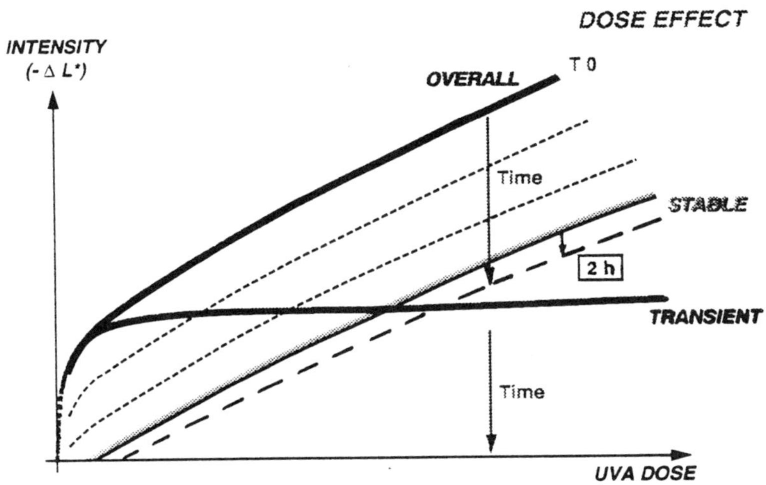

Figure 4 Composition of IPD response (dose effect): The skin response known as IPD appears to consist of two subjacent phenomena, the sum of which is observable immediately postexposure: a transient saturating response that fades rapidly (IPD proper) induced by small UVA doses, and a more stable pigmentation (PPD: persistent pigment darkening) induced at a more progressive rate by higher UVA doses.

not been fully explained. In rough terms, the first could result from an immediate and reversible photooxidative reaction; the second, from a partial polymerization of melanin precursors; whereas the third could result from UV-enhanced complete melanin synthesis.

Time Effect

When combining the pigmentation kinetics from both IPD subjacent phenomena, the resultant curves (Fig. 5) closely mimic the overall kinetic curves recorded, as described earlier (see Fig. 1). It was concluded from these findings that the selection of an appropriate time lag after exposure, once the transient component of IPD has faded, would leave the second, stable PPD phenomenon to serve as the endpoint in the UVA-PF method. This route has notable consequences on protection factor values, through the validity of the reciprocity law.

E. Reciprocity Law

The reciprocity law is the equivalence of exposure time and source flux on the intensity of the response. Its application to any skin response is intended to answer two questions relative to the method of UVA-PF determination. The first one is to determine how greatly the intensity of the response to a particular UVA dose can vary when this dose is applied under various conditions of source flux

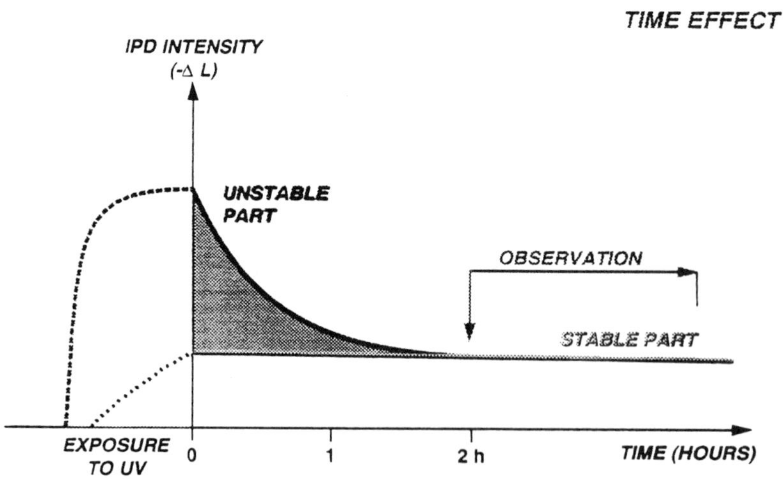

Figure 5 Composition of IPD response (time effect): Both components of IPD skin response are induced immediately during exposures, although with different UVA doses. The unstable portion fades away during the first 2 h postexposure, then a stable and weak portion remains that can last several days (PPD: persistent pigment darkening).

and exposure time. This question is related to the fact that different UVA sources, with a variety of UVA flux and exposure conditions, may be used in laboratories for UVA-PF determination (narrow- or wide-beam, low- or high-power, use of constant flux and variable time, or constant time and variable flux), and especially to the fact that protected skin is exposed with a flux, by definition, to UVA-PF times smaller than unprotected skin. The second question, arising if the reciprocity law does not appear valid is: What is the consequence to the accuracy of the UVA-PF results, or what kind of test conditions are needed to avoid or limit any influence on the results?

The reciprocity law of the IPD–PPD response was tested on eight volunteers (back) with skin types II and III, using doses of 4, 8, 16, and 32 J/cm^2 and flux values of 6, 12, 18, and 28 mW/cm^2. Change in flux was obtained by varying the distance from a UVA metal halide source to the skin. Observations and colorimetric measurements were performed immediately and then, 1 and 2 h after exposure (Fig. 6).

Probably because the UVA-induced IPD skin response is unstable during the first hour after exposure (transient pigmentation), the reciprocity law failed to appear valid during this time frame, as the higher the flux, the higher the response found, whatever the dose tested. Conversely, the intensities of response became very similar at 1 h postexposure and showed no significant difference (independently of source flux) for every dose tested at 2 h postexposure (i.e.,

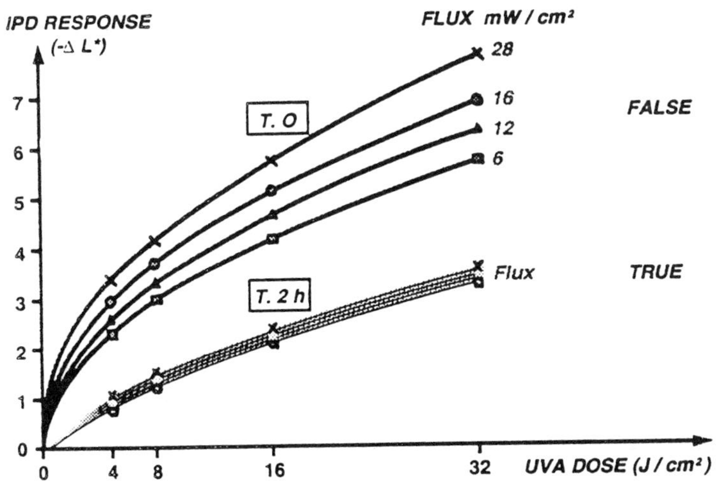

Figure 6 Reciprocity law for IPD skin response: Immediately postexposure, the pigmentation intensity is unstable and dependent on the UVA source flux. About 2 h postexposure, the intensity of the residual pigmentation is weaker, but stable and independent of source flux (reciprocity law is then complied).

when the pigmenting response had become stable: persistent pigmentation). From this finding it can be concluded that the UVA-PF method should rely on a stable response observed after a certain delay after exposure for the results to be independent of exposure conditions.

G. Linearity of Response

An added advantage in performing the observation 2 h postexposure is that the PPD dose–response curves became quasilinear (see Figs. 3 and 6), which elicits a clearer visual contrast between successive doses than with the response observed immediately. A consequence of the linearity on the UVA-PF method, when comparing of minimal-pigmenting doses on unprotected (MPDu) and protected (MPDp) skin, is that the ratio of MPDp to MPDu (UVA-PF) is practically independent of the pigmentation threshold used for their determination (the ratio of the respective colorimetric slopes could also be used, coupled with a simple dose transform).

The dose–responses curves of both IPD and PPD skin responses thus appear substantially different in shape from each other and from that of the erythemal response (35,39) used for SPF determination, and they should be treated accordingly: that is, because no actual dose threshold exists, a "minimal" pigmentation level will have to be clearly defined before beginning the determination of an action spectrum. Observations should be performed with some delay on the stable response, so that MPDu and MPDp can be determined *simultaneously*, by paired comparison of dose–response curves on protected and unprotected areas, which will allow the results to be accurately and reproducibly improved.

H. Action Spectra

Immediate Pigment Darkening

The back of 18 subjects with skin types II, III, and IV ("very light" to "tan" skin color categories; [36,37]), was subjected to continuous exposure to 320-, 330-, 350-, 375-, 400-, and 425-nm wavelengths, using a 1-kW xenon lamp provided with a single monochromator set at 20 nm (for 50% peak height) and a silica multifiber (ϕ 16 mm) light guide. With such a device, the UV flux recorded at skin level with a calibrated thermopile ranged from 4.3 mW/cm^2 at 320 nm to 6.8 mW/cm^2 at 425 nm (i.e., a maximal ratio of 1:1.6; wide differences in flux between the various wavelengths tested must be avoided because of direct influence of source flux on action spectrum owing to lack of validity of reciprocity law for unstable IPD). The colorimetric dose–response curves were recorded during exposure at 0.3, 0.5, 1, 2, 3, 4, 8 J/cm^2, and up to 16 and 32 J/cm^2 when necessary, according to wavelength, and were averaged over volunteers for each wavelength. At 320 nm, the minimal pigmentation dose could not be reached without inducing too much erythema. For various given

|ΔL*| values (1, 1.2, 1.5, and 2) used as the pigmentation threshold, the minimal pigmenting doses were then graphically determined, and the reciprocal MPDs were calculated allowing a set of action spectra to be plotted. By comparison with visual observations, the mean 1.2 value for |ΔL*| was retained as corresponding to a representative visual threshold for PPD and 1.5 for IPD.

These results confirmed previous studies (48–53) stating that IPD responses and action spectra, observed immediately after the end of exposure (within 1 min), develop throughout the UVA range with a continuous increase as the wavelengths shorten from 450 to 330 nm (Fig. 7), and they likely reach the UVB range. An increase in the preselected threshold level to a small extent reduced the threshold doses and lowered the action spectrum curve, but did not significantly change the relative effect of UVA wavelengths.

Persistent Pigment Darkening

Because delayed observation of the persistent pigmentation (PPD) was recommended as an endpoint, the action–spectrum of PPD observed at 2 h was also determined on the back of ten subjects with skin type III, using the same experimental equipment as used for IPD. A range of UVA doses, predetermined from the data of the IPD experiment described in the foregoing, were applied at 320-, 330-, 340-, 370-, and 400-nm wavelengths. The pigmentation induced was visually observed and colorimetrically monitored during fading from im-

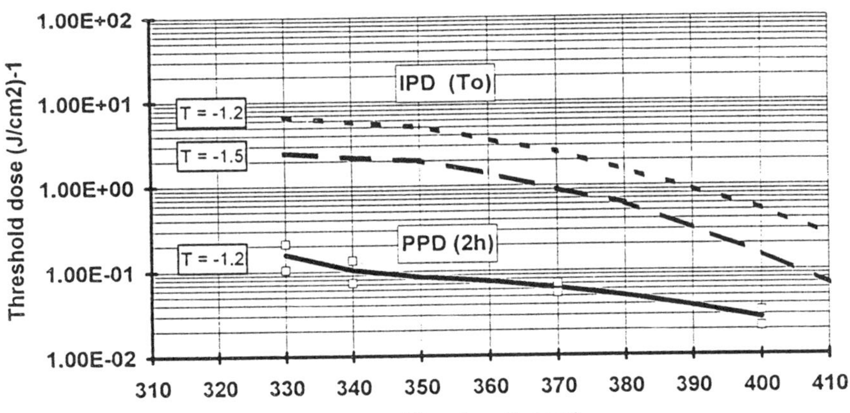

Figure 7 Action spectra of IPD and PPD responses: Action spectra of IPD at T_0 postexposure (given for two threshold values), and of PPD at 2 h are relatively flat and parallel and cover the whole UVA waveband, up to short visible. Doses for minimal IPD response (ΔL* = −1.5) are in practice five to ten times smaller than for PPD.

mediately until 2 h postexposure. The dose that induced a pigmentation threshold ($\Delta L^* = -1.2$) at 2 h was selected as the MPD. Reciprocal MPDs for each wavelength were calculated and averaged over volunteers, allowing the PPD action spectrum to be plotted (see Fig. 7).

At 320 nm, threshold doses could not be obtained because of erythemal response; hence, specific experimental conditions (narrower monochromator bandwidth) would be required at this wavelength. A delay in the observation time postexposure, in reducing the influence of small saturating effects, tends to linearize the action spectrum and to slightly increase the relative influence of long UVA wavelengths. Therefore, the PPD action spectrum appeared slightly flatter than the IPD action spectrum. Although awaiting new data, the action spectrum was supposedly flat between 330 and 320 nm. At similar visual threshold the doses required to induce minimal IPD responses were five to ten times lower than that for PPD.

I. Ultraviolet A Source Definition and Efficacy Spectrum

Thus, IPD and PPD responses cannot be considered as strictly specific of the UVA range of wavelengths and, consequently, the emission spectrum of the UVA sources should be restricted to the UVA range with suitable filters. In the present state of the art, a xenon arc lamp, already used for SPF determination, filtered with a specific Schott WG335 3-mm short cutoff filter and a UG11 1-mm to discard visible and IR rays, can be considered a valid source (54,55). According to our experience, the same filtration is also applicable to complex metal halide or combined sources with continuous emission spectrum. When the action spectrum of the PPD endpoint is multiplied by the emission spectrum of such a UVA source, the resulting efficacy spectrum (Fig. 8) appears quite suitable for the UVA protection assessment of sunscreens, as covering the whole UVA range. When using a xenon source in UVA-PF determination, UVB and short visible wavelengths contribute less than 1%, about 7% of the total PPD response is ascribed to UVAII and 92% to UVAI, whereas in SPF determination about 85, 10, and 5% of erythemal response are respectively due to UVB, UVAII, and UVAI. SPF and UVA-PF can thus be considered as quite complementary for the full evaluation of sunscreen effectiveness.

J. Skin Typing

The previous experiments on IPD–PPD responses were performed on very light, light, intermediate, and tan skin color categories. The choice of volunteers for the purpose concerned should depend only on a skin typology linked to the stable part of the IPD response. Apart from using a previous UVA irradiation test, it has not yet been possible to define more specific parameters that would

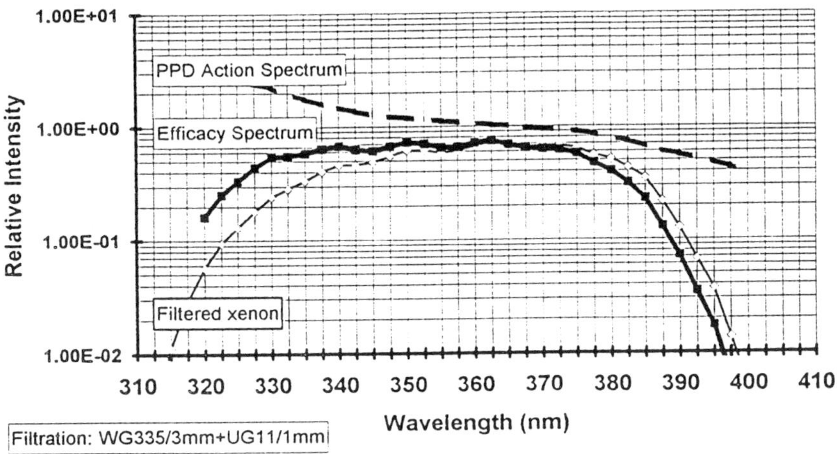

Figure 8 UVA source and efficacy spectrum: Coupling the emission spectrum of an adequately filtered xenon UVA source (320–400 nm) with PPD response results in a flat and broad efficacy spectrum, suitable for testing UVA-PF of sunscreen products.

allow a precise preselection of volunteers giving an expanded and linear response that is able to render the UVA-PF easier to determine. However, the general finding was that the panel of volunteers for UVA-PF assessment of sunscreens using PPD as endpoint could be drawn from roughly the same panel as that used for SPF determination, extended to phototype IV and separating some very fair skins (skin type I).

IV. DEFINITION OF A UVA-PF PROTOCOL

On the basis of this set of data about UVA-induced immediate skin pigmentation and of our experience in practicing the PPD method with formulations, the main parameters of a suitable protocol for the UVA protection assessment of sunscreen products could be defined as follows:

A. General Parameters

Conditions concerning the test area (back), size of test sites, product application (2 mg/cm^2 on at least 35 cm^2), selection of volunteers, inclusion and exclusion criteria, and such, follow the same current standards as those for the SPF determination (29–35) unless otherwise specified.

B. Endpoint

The PPD, or the lasting *stable* portion of the IPD skin response, is used as an endpoint.

C. Volunteers

At least 10 (maximum 20) subjects with skin types II, III, and IV are used.

D. Ultraviolet A Source

Xenon or metal halide sources, or a xenon–metal halide combination with continuous spectra, limited to the UVA spectrum (320–400 nm) by appropriate filters, such as Schott WG335 3 mm and UG11 1 mm for narrow-beam sources, or UG5 6 mm for wide-beam sources should be used.

E. Ultraviolet A Doses and Progression

Six UVA doses ranging from 4 to about 30 J/cm^2 are applied to unprotected skin of the volunteer's back in a 30–50% geometric progression (with potential interpolation, the dose progression is actually reduced to about 25%); a wide range of UVA doses allows the same dose range to be usable for any skin type. On protected skin the UVA doses are multiplied by the expected UVA-PF of the test product.

F. Time-Lag Postexposure

After exposure, the products, particularly those including pigments and dyes, are gently removed with a mild lotion; the volunteers are asked to rest for the time-lag period before observation and to avoid rubbing the test area. Observation of the skin response should be delayed for *at least* 1 h and typically 2 h after the end of the *last* exposure.

G. Determination of Minimal Pigmenting Doses

Visual observation of UVA-induced progressive pigmentation is performed under standard room and illumination conditions. The minimal pigmenting doses (MPD) are determined in a *simultaneous* observation of the protected (MPDp) and unprotected areas (MPDu) by paired matching of pigmentation thresholds, with possible free interpolation between successive doses. Colorimetry, using ΔL^* coordinate in CIE (1987) $L^*a^*b^*$ color space and plotting the dose-response curves (35,39) may also help determine MPD, defined with a difference in luminance of $\Delta L^* = -1.2$.

H. Ultraviolet A Protection Factor

The individual UVA protection factor (UVA-PFi) for a product is calculated as the ratio of MPDp to MPDu. The UVA protection factor of the product (UVA-PF) is the arithmetic mean of the individual UVA-PFi on the number n of volunteers selected. This number n may vary from 10 to 20, provided the standard error of the mean (SEM) is lower than, say, 10% of the mean.

V. EVALUATION AND RESULTS

A. Neutral-Density Physical Filters

The method was calibrated using neutral-density physical filters as standards with respective absorbances giving nominal UVA-PFs of about 4 and 8. Probably owing to invalidity of the reciprocity law under the given exposure conditions (wide-beam UVA source, 4–25-J/cm^2–dose range, with constant flux and variable exposure time) the results were significantly higher than expected and more widely dispersed at the short observation time (less than 1 min after the end of exposure of the highest dose). Then, when observed 1 and 2 h postexposure, the results tended toward the nominal values, with smaller interindividual variations (Fig. 9).

B. Cosmetic Formulations

Four UVA protective formulations (Table 1) were tested under various exposure conditions. A wide-beam UVA source was used to expose protected and unprotected areas in a single run, but the exposures of the various UVA doses were either started all at the same time and ended at intervals (mode 1, formulation F4), or started at intervals and all ended at the same time (mode 2, formulations F1–F3). The observations were performed after time lags of 0, 1/2, 1, and 2 h postexposure, beginning at the end of the longest exposure (in mode 1).

Depending on the exposure mode, initial UVA-PFs were too low or too high just after end of exposure (Fig. 10), because the intensity of the unstable response varied with time for short time lags, and minimal pigmenting dose varied

Table 1 SPF and (PPD) UVA-PF of Sunscreen Formulations

Formulation	SPF	UVA-PF (PPD)	SD	n
1	17	3.9	(0.9)	6
2	7	3.5	(1.4)	6
3	9	5.2	(0.7)	6
4	17	7.5	(1.4)	9

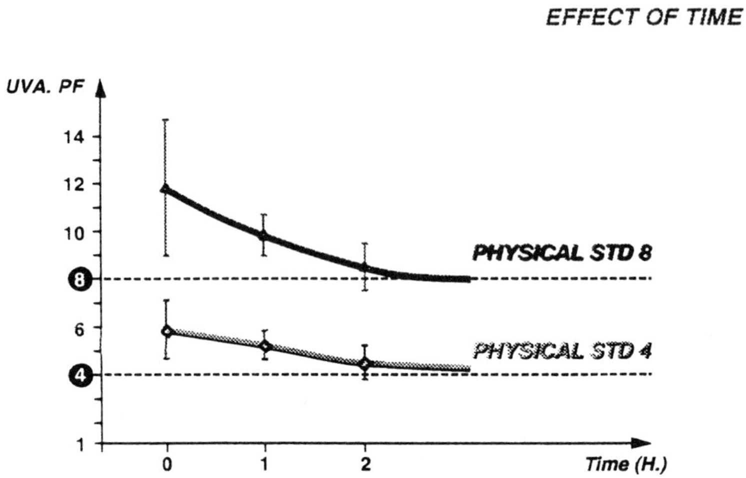

Figure 9 UVA-PF of neutral density physical filters: UVA-PF of neutral-density filters of known absorbances (PF 4 and 8) are higher than expected values and highly variable when determined on unstable response immediately postexposure. They decrease toward standard values with smaller variation at 2 h postexposure (wide-beam metal halide UVA source).

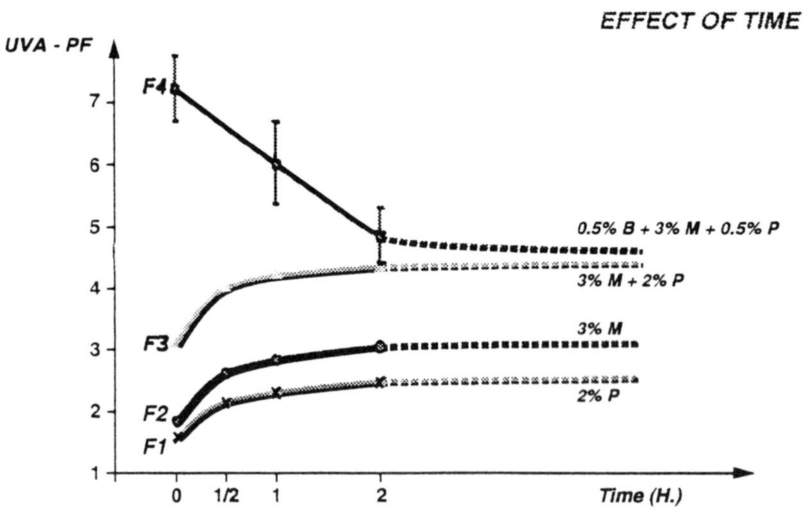

Figure 10 Effect of time lag postexposure with protective formulations: The UVA-PF of protective formulations varies with observation time lag postexposure and stabilizes at about 2 h postexposure. With too short time lags, variation of UVA-PF from expected values depends on exposure conditions; that is, all UVA doses starting (F4) or finishing at the same time (F1, F2, and F3).

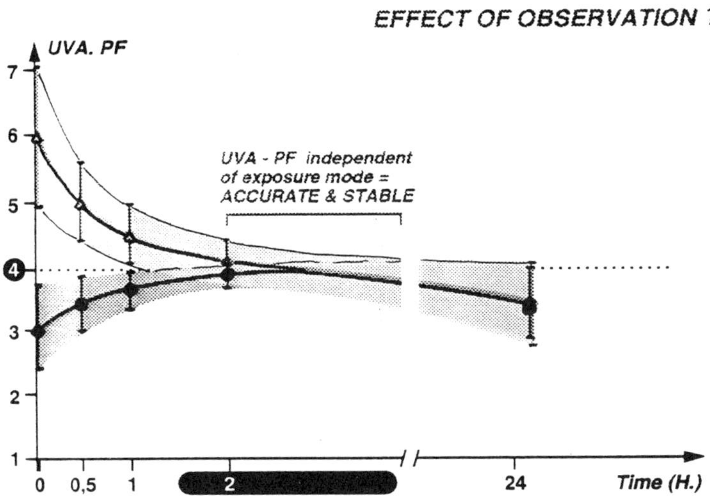

Figure 11 Effect of observation time lag postexposure: On the whole, UVA-PF results diverge at T_0 with a widening interindividual variability depending on exposure conditions, converge at 2 h and beyond with a reduced variability, then slightly decrease afterward at 24 h. Performing observations at about 2 h postexposure thus appears optimal for more accurate UVA-PF results.

accordingly. For longer time lags, beyond 1 h postexposure, the UVA-PFs tended to stabilize, which confirmed the trend found with the physical standards.

C. Effect of Observation Time Lag

The critical effect of observation time lag may be summarized by stating that results diverged at T_0, with widening interindividual variability (Fig. 11), then converge at 2 h and beyond with a reduced variability, and slightly decreased afterward at 24 h, with more variability: the mean variability among volunteers was about 18% at 2 h and 28% at 24 h. All these findings strongly support the recommendation of an observation time lag of about 2 h.

D. Ultraviolet A Filters

The dose effects of various UVA sunscreens in similar basic formulations were evaluated separately, according to the protocol defined in Section IV: 0, 2, and 3% butyl methoxydibenzoylmethane (Parsol® 1789); 2, 3, and 5% benzophenone-3; and 0, 2, 3, 5, and 10% terephthalylidene dibornanone sulfonic acid (Mexoryl® SX).

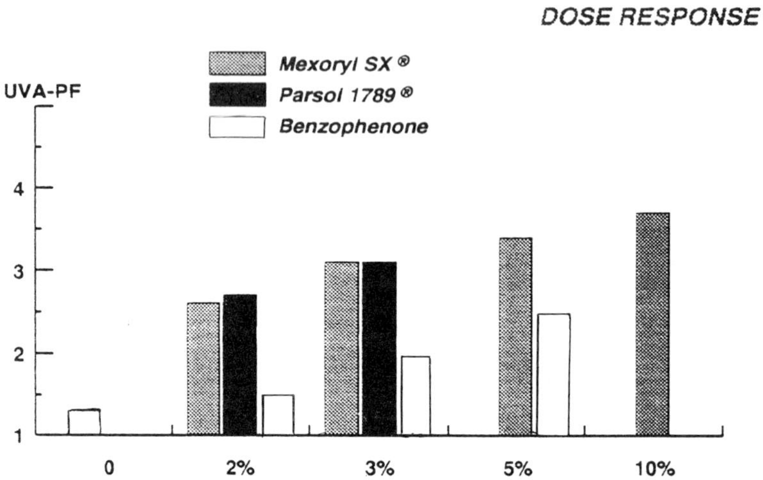

Figure 12 UVA-PF of UVA sunscreens: Increasing concentrations of various single UVA sunscreen formulations, tested with standard protocol based on PPD response, yield increasing UVA-PF values. Protective efficacy of vehicle is not significant. Mexoryl® SX and Parsol® 1789 appear significantly more efficient than benzophenone.

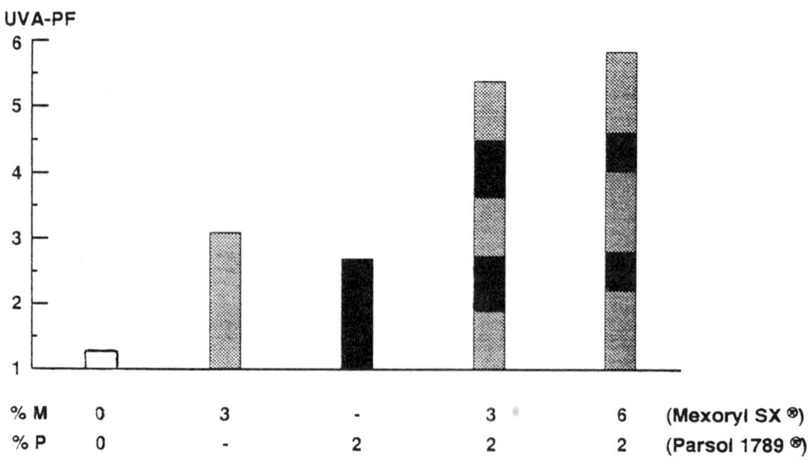

Figure 13 UVA-PF of combinations of UVA sunscreens: Combinations of UVA sunscreens in formulation led to UVA-PF values much higher than is possible with single, separate sunscreens, thereby optimizing the protective efficacy.

Although the vehicle (as well as a pure UVB filter) afforded a very small UVA-PF (about 1.3), benzophenone clearly yielded heightened concentration-dependent protection (Fig. 12), as did Parsol® 1789 and Mexoryl® SX, although in a significantly higher range of efficacy than for benzophenone.

E. Combination of Ultraviolet A Filters

With the same method, a strong increase in efficacy was found with various combinations of Mexoryl® SX and Parsol® 1789, which yielded a UVA-PF of 5.4 with only 5% active ingredients (Fig. 13).

Four realistic sunscreen formulations, with SPF values from 7 to 17 and highly effective UVA protection, based on various combinations of filters, were tested on volunteers with skin type III. They exhibited, as expected, UVA-PF values (see Fig. 1), correlated or not with their SPF values, showing that protection against UVA effects and against erythema may or may not be dependent according to the active ingredient content.

F. Reproducibility of Results

Some UVA protective formulations were evaluated several times in various tests over months. Although limited panels of volunteers were used, a good reproducibility of the results, even in a wide range of UVA-PF values, was a particular evident feature of this series of experiments (Fig. 14).

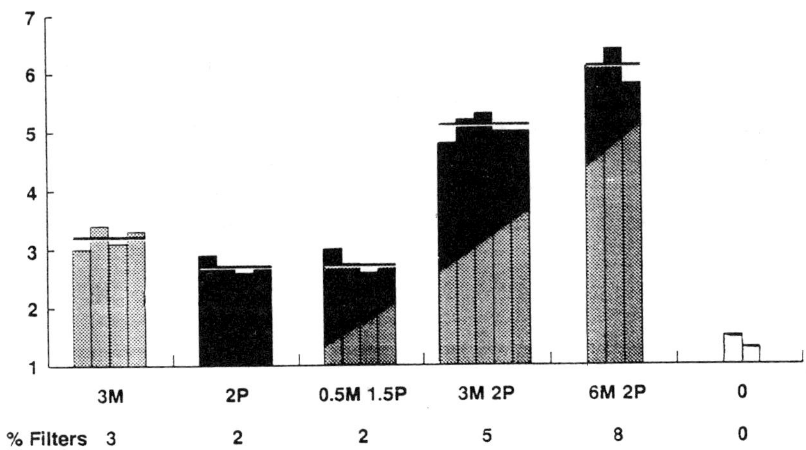

Figure 14 Reproducibility of the UVA-PF method over time: A variety of protective formulations and vehicles tested several times over a period of months yielded very close UVA-PF results, demonstrating the excellent reproducibility of the test method.

VI. VALIDATION

The PPD UVA-PF method has been proposed for validation to the Technical Committee TC6-24 ''UVA Sunscreen Testing'' of the Commission Internationale de l'Eclairage (CIE); to the European COLIPA Task Force ''Sun Protection Measurement,'' which is currently testing it in a multicentered test; and to the U.S. Food and Drug Administration (55–57).

The principle of the PPD method was selected and published in December 1995, by the Japan Cosmetic Industry Association (JCIA), to be officially applicable for the UVA protection efficacy of sunscreens in Japan from January 1, 1996 (58).

VII. CONCLUSION

The advantages of using the stable portion of the immediate skin pigmentation (persistent pigment darkening; PPD) as an endpoint for the assessment of UVA protection by sunscreens can be summarized as follows: It is a noninvasive, visible response, sensitive to the whole UVA waveband which, in combination with suitable sources, constitutes a suitable endpoint with adequate efficacy spectrum; the UVA protection factors can be determined after acceptable UVA exposures and a short time lag before observation, allowing repeated exposures on the same day; however, because lasting pigmentation gives a weak hue contrast on skin compared with erythema, specific conditions of observations by paired simultaneous determination of minimal pigmenting doses are recommended.

Delaying the observation for at least 1 h, typically 2 h, postexposure makes it possible to overcome the influence of various exposure conditions linked to UVA sources and to the default of the reciprocity law for the transient skin pigmentation. At the same time, this procedure also allows more comfortable and reliable conditions of observation, yielding more stable and accurate results.

The proposed protocol, elaborated through extensive colorimetric studies of the complex IPD phenomenon, calibrated with neutral density physical filters, and widely tested over recent years on a variety of formulations, appears to be a valuable method for UVA protection assessment of sunscreens.

REFERENCES

1. CIE. A reference action spectrum for ultraviolet induced erythema in human skin. CIE research note, 6 (1987).
2. Diffey, B., Farr, P., and Oakley, A. Quantitative studies on UVA induced erythema in human skin. *Br. J. Dermatol.*, *117*:57–66 (1987).
3. Diffey, B. and Larkö O. Clinical climatology. *Photodermatology*, *1*:30–37 (1984).

4. Robinson, N. *Solar Radiation*. Elsevier Publishing, New York, 1966.
5. Kromann, N., Wulf, H. C., Eriksen, P., and Brodthagen, H. Relative ultraviolet spectral intensity of direct solar radiation. Sky radiation and surface reflections. *Photodermatology*, *3*:73–82 (1986).
6. Stanfield, J. W., Feldt, P. A., Csortan, E. S., and Krochmal, L. Ultraviolet A sunscreen evaluations in normal subjects. *J. Am. Acad. Dermatol.*, *20*:744–748 (1989).
7. Mark, R., Gabriel, K. L., Affrime, A., Stoudemayer, T., Forbes, P. D., Davies, R. E., and Urbach, F. A method for evaluating the UVA protection of sunscreens in human volunteers. Poster. Dermal Clinical Evaluation Society, Elisabeth NJ, 1990.
8. Cole, C. and van Fossen, R. Measurement of sunscreen UVA protection: An unsensitized human model. *J. Am. Acad. Dermatol.*, *26*:178–184 (1992).
9. Diffey, B. and Farr, P. An evaluation of sunscreens in patients with broad action-spectrum photosensitivity. *Br. J. Dermatol.*, *112*:83–86 (1985).
10. Gange, R. W., Soparkar, A., Matzinger, E., Dromgoole, S., Sefton, J., and DeGryse, R. Efficacy of a sunscreen containing butylmethoxydibenzoylmethane against ultraviolet A radiation in photosensitized subjects. *J. Am. Acad. Dermatol.*, *15*:494–499 (1986).
11. Lowe, N. J., Dromgoole, S. H., Sefton, J., Bourget, T., and Weingarten, D. Indoor and outdoor efficacy testing of a broad-spectrum sunscreen against ultraviolet A radiation in psoralen-sensitized subjects. *J. Am. Acad. Dermatol.*, *17*:224–230 (1987).
12. Dalhen, R. F., Shapiro, S. I., Berry, C. Z., and Screiber, M. M. A method for evaluating sunscreen protection from longwave ultraviolet. *J. Invest. Dermatol.*, *55*: 164–168 (1970).
13. Kaidbey, K. and Gange, R. W. Comparison of methods for assessing photoprotection against UVA in vivo.
14. Gonzenbach H., Mascotti, R., Césarini, J-P. UVA sunscreen in vivo effectiveness measurement. *Cosmet. Toilet.*, *106*:79–84 (1991).
15. Poelman, M. C., Césarini, J. P., Ruse, F., and Binet, O. A non phototoxic method for testing in vivo the UVA photoprotection. Preprint of 14th IFSCC Congress, Barcelona, 1986.
16. Kaidbey, K. and Barnes, A. Determination of UVA protection factors by means of immediate pigment darkening in normal skin. *J. Am. Acad. Dermatol.*, *25*:262–266 (1991).
17. Roelandts, R., Sohrabvand, N., and Garmyn, M. Evaluating the UVA protection of sunscreens. *J. Am. Acad. Dermatol.*, *21*:56–62 (1989).
18. Garmyn, M. and Roelandts, R. Evaluating the UVA photoprotection of sunscreens with murine skin edema. *J. Invest. Dermatol.*, *8*:187–190 (1992).
19. Chew, S., DeLeo, V. A., and Harber, L. C. An animal model for evaluation of topical photoprotection against ultraviolet A (320–400-nm) radiation. *J. Invest. Dermatol.*, *89*:410–414 (1987).
20. Sayre, R. M. and Agin, P. A method for the determination of UVA protection for normal skin. *J. Am. Acad. Dermatol.*, *23*:429–440 (1990).
21. Poh Agin, P. UVA protection percent: A versatile method for determining UVA efficacy of sunscreens. *Biological Responses to Ultraviolet A Radiation* (F. Urbach, ed.), Valdenmar Publishing, Overland Park, KS, 1992, pp. 347–361.

22. Stockdale, M. A novel proposal for the assessment of sunscreen product efficacy against UVA. *Int. J. Cosmet. Sci.*, *9*:85–98 (1987).
23. Stockdale, M. UVA sunscreens. Methods for assessing their efficacy. *Cosmet. Toilet.*, *102*:111–115 (1987).
24. Diffey, B. L. and Robson, J. A new substrate to measure sunscreen protection factors throughout the ultraviolet spectrum. *J. Soc. Cosmet. Chem.*, *40*:127–133 (1989).
25. Sayre, R. In-vivo rapid non-invasive measurement of sunscreen UVA transmission. FDA workshop on UVA, Rockville MD, Proceedings 192–205, May 12, 1994.
26. Kollias, N. IPD reciprocity and optical methods for the evaluation of sunscreens. FDA workshop on UVA, Rockville MD, Proceedings 182–192, May 12, 1994.
27. Moyal, D., Chardon, A., and Hourseau, C. Colorimetric definition of the immediate pigment darkening: Its application for the determination of UVA sun-protection-factor. Poster B22. Proceedings of 16th IFSCC Congress, New York, 1990.
28. Chardon, A., Moyal, D., and Hourseau, C. Skin immediate pigment darkening applied to UVA protection assessment. Poster P12. Proceedings of ASP 19th Annual Meeting, San-Antonio, 1991; and Poster N3, 4th Congress of the ESP, Amsterdam, 1991.
29. Standard Association of Australia. Sunscreen products. Evaluation and classification. AS 2604, 1986.
30. Commission Internationale de l'Eclairage (CIE). Sunscreen testing (UVB)—technical report. Pub CIE 90. 1st ed., 1991.
31. Japan Cosmetic Industry Association (JCIA). Standard SPF test method. 1991.
32. Canada. Standardized proprietary medicine monograph, Sunburn protectants. June 25, 1991.
33. Southern African Bureau of Standards. Proposed specifications for sunscreen products. 541/50960 Draft, 1992.
34. U.S. Food and Drug Administration (FDA). Sunscreen drug products for over-the-counter human use; Tentative final monograph; proposed rule. *Federal Register*, *58*:28194–28302, 1993.
35. COLIPA. SPF test method. (A recommendation of the COLIPA Task Force ''Sun Protection Measurement''). COLIPA, Brussels, Belgium, 1994.
36. Chardon, A., Crétois, I., and Hourseau, C. Skin colour typology and suntanning pathways. *Int. J. Cosmet. Sci.*, *13*:191–208 (1991).
37. Chardon, A., Crétois, I., and Hourseau, C. Colour changes induced on various skin categories by repeated exposures to UV rays. *Biological Responses to Ultraviolet A Radiations* (F. Urbach ed.), Valdenmar Publishing, Overland Park, KS, 1992, pp. 159–175.
38. CIE. CIE 1976 uniform color spaces. Colorimetry. CIE Publication, 15.2:29–32 (1986).
39. Chardon, A., Dupont, G., Hourseau, C., and Grollier, G. F. Colorimetric determination of sun-protection-factor. Poster, Proceedings of 15th IFSCC Congress, London. Vol. A/A24, 1990, pp. 313–322.

40. Chardon, A., Moyal, D., Bories, M.-F., and Hourseau, C. Comparing suntans from actual sun using various SPF sunscreens. *Cosmet. Toilet.*, *108*:79–93 (1993).
41. Westerhof, W., van Hasselt, B. A., and Kammerijer, A. Quantification of UV-induced erythema with a portable computer controlled chromameter. *Photodermatology*, *3*:310–314 (1986).
42. Weatherall, J. and Coombs, B. Skin colour measurements in term of CIELab Color space values. *J. Invest. Dermatol.*, *99*:468–473 (1992).
43. Pathak, M. A., Riley, F. C., and Fitzpatrick, T. B. Melanogenesis in human skin following long wave ultraviolet and visible light. *J. Invest. Dermatol.*, *39*:435–443 (1962).
44. Kawada, A. UVB-induced erythema, delayed tanning, and UVA-induced immediate tanning in Japanese skin. *Photodermatology*, *3*:327–333 (1986).
45. Pathak, M. Immediate and delayed pigmentary and other cutaneous responses to solar UVA radiation (320–400 nm). *The Biological Effects of UVA Radiation* (F. Urbach and R. Gange, eds.), Praeger, New York, 1986, pp. 156–167.
46. Hönigsmann, H. Newer knowledge of immediate pigment darkening (IPD). *The Biological Effects of UVA Radiation* (F. Urbach and R. Gange, eds.), Praeger, New York, 1986, pp. 221–225.
47. Lavker, R. and Kaidbey, K. Redistribution of melanosomal complexes within keratinocytes following UVA irradiation: A possible mechanism for cutaneous darkening in man. *Arch. Dermatol. Res.*, *272*:215–228 (1982).
48. Rosen, C., Jacques, S., Stuart, M., and Gange, R. Immediate pigment darkening: visual and reflectance spectrophotometric analysis of action spectrum. *Photochem. Photobiol.*, *51*:583–588 (1990).
49. Parrish, J., Jaenicke, K., and Anderson, R. Erythema and melanogenesis action spectra of normal human skin. *Photochem. Photobiol.*, *36*:187–191 (1982).
50. Satoh, Y. and Kawada, A. Action spectrum for melanin pigmentation to ultraviolet light, and Japanese skin typing. *Brown Meloderma* (T. Fitzpatrick and M. Wick, eds.), University of Tokyo Press, Tokyo, 1986.
51. Gange, R., Park, Y.-K., Auletta, M., Kageetsu, N., Blackett, A., and Parrish, J. Action spectra for cutaneous responses to ultraviolet radiation. *The Biological Effects of UVA Radiations* (F. Urbach and R. Gange, eds.), Praeger, New York, 1986, pp. 57–66.
52. Ortel, B. and Gange, R. UV-A action spectra for erythema and pigmentation. *Biological Responses to Ultraviolet A Radiation* (F. Urbach, ed.), Valdenmar Publishing, Overland KS, 1992, pp. 79–82.
53. Irwin, C., Barnes, A., Veres, D., and Kaidbey, K. An ultraviolet radiation action spectrum for immediate pigment darkening. *Photochem. Photobiol.*, *57*:504–507 (1993).
54. Cole, C. Multicenter evaluation of sunscreens UVA protectiveness with the protection factor test method. *J. Am. Acad. Dermatol.*, *30*:729–736 (1994).
55. Chardon, A. Specifications for UVA sources for in-vivo UVA-PF determination. FDA workshop on UVA, Rockville MD. Proceedings, 90–92 (1994).

56. Chardon, A. Immediate pigment darkening persistent effect (PPD), reciprocity law, doses and observation time. FDA workshop on UVA, Rockville MD. Proceedings, 92–97 (1994).
57. Moyal, D. Persistent pigment darkening (PPD) as biological response for the UVA damaging effects. FDA workshop on UVA, Rockville MD. Proceedings, 97–105 (1994).
58. Japan Cosmetic Industry Association (JCIA). Measurement standards for UVA protection efficacy (1995).

29

Sunscreens: Effects of Amounts of Application of Sun Protection Factors

Annette Gottlieb
UCLA School of Medicine, Los Angeles, California

Teresa D. Bourget
Skin Research Foundation of California, Santa Monica, California

Nicholas J. Lowe
Skin Research Foundation of California, Santa Monica, and UCLA School of Medicine, Los Angeles, California

I. OVERVIEW

There has been controversy about the amount of sunscreen applied by individuals. At least one previous study on a group of volunteers suggested that approximately 1 mg/cm^2 was the average application by a participant. This contrasted with the amount (2 mg/cm^2) routinely advised for routine human evaluation of sun protection factors in the United States and presently in Europe. We have reevaluated different sunscreen formulations: a gel, an alcoholic lotion, a creamy lotion, and a cream used by a group of volunteers. The mean application amount was 1.3 mg/cm^2, with the gel applied in a slightly greater thickness overall, at 1.7 mg/cm^2, than the other vehicles. Sun protection factor (SPF) studies were then performed using a filtered xenon arc solar simulator and five different sunscreens.

Four of the sunscreens showed no significant reduction of SPF using sunscreen application of 1.3 mg/cm^2, compared with 2.0 mg/cm^2. However, one sunscreen with an SPF of 29 at 2.0 mg/cm^2 did show a significant decrease in protection at lower application amounts.

II. EVALUATING THE APPROPRIATE AMOUNT OF SUNSCREEN USAGE

Sunscreen use is now established as an important means of reducing the risks of photocarcinogenesis (1) and photoaging (2). The Food and Drug Administration (FDA) developed guidelines, originally in 1978 and then in 1994, for evaluating the protectiveness of a sunscreen (3). This monograph recommended that sunscreens be delivered at 2 mg or 2 $\mu L/cm^2$ amounts. Most of the claimed sun protection factor (SPF) numbers of sunscreens appropriately evaluated in the United States, therefore, have been determined using this quantity of sunscreen.

A previous study reported that sunscreens may, in fact, be delivered at much smaller amounts of approximately 1 mg/cm^2, resulting in lower SPF (4). This particular investigation was subsequently criticized (5) for some important reasons. Only a small sample of sunscreen was supplied (albeit per each area of the body), which may have artificially restricted the quantity of sunscreen applied. [A study by Lynfield and Schechter (6), found that individuals will use less than half the amount of topical medication supplied in a small tube compared with that in a large jar.] Another problem was that a nonsolar-simulating source was used, which may have resulted in relatively low SPF numbers. Also, the products used in the study are not available in the United States.

Therefore, we undertook a study to further investigate the average amount of sunscreens applied by individuals. In addition, we have determined the protection factors of several different sunscreens using the average amount applied by our volunteers. These SPF values have been compared with values for the same sunscreens at 2.0 and 1.0 mg/cm^2.

III. MATERIALS AND METHODS

For the application density studies, the following four sunscreen preparations were used:

Clear gel: padimate O (octyl dimethyl PABA), glyceryl PABA, with: alcohol (55%), water, PEG-6, carbomer-940, and hydroxyethylcellulose

Clear lotion: 5% aminobenzoic acid (PABA), 5% padimate O (octyl dimethyl PABA), 3% oxybenzone, 58% SD alcohol 40

Creamy lotion: octyl methoxycinnamate, oxybenzone, octyl salicylate

Cream: 7% ethylhexyl-*p*-methoxycinnamate, 4% oxybenzone, 2% titanium dioxide

A. Evaluation of Sunscreen Application Amounts

After informed consent was obtained, 20 healthy volunteers (17 women and 3 men) were randomly divided into four groups in which the individuals were asked to apply a particular kind of sunscreen as follows: (1) creamy lotion, (2) clear lotion, (3) clear gel, (4) cream. The active ingredients in the different sunscreens are those just described. The sunscreens were kept in their individual large-sized bottles as they would appear if bought in a store. The individuals were instructed to apply a particular sunscreen, using as much as they normally would if they were going to the beach, to six separate regions of the body: head and neck, chest, back, arms, thighs, and lower legs. Subjects wore sterile plastic gloves and applied the desired amount of sunscreen to one region at a time under the observation of one of the investigators. By weighing the sunscreen bottle and the gloves before and after each application and subtracting, the amount used per body region was determined, taking into account any residual sunscreen left on the hands. The height and weight of each volunteer were obtained and the body surface was calculated using a nomogram (7). The specific surface area for each body region was calculated using a modification of the "rule of 9s" for estimating burn injuries (8). In this way, we were able to determine the average layer thicknesses of four different sunscreens applied to various locations of the body.

B. Evaluation of Sun Protection Factors

For the studies of application densities, a filtered xenon arc solar simulator was used. Five sunscreens were used that had manufacturer-determined SPFs of 4, 8, 10, 15, and 29, respectively. (The sunscreen with the SPF of 4 was a standard formulation provided by the FDA, which was used as a control.) Controls were normal healthy volunteers who had skin types I or II, based on their history of skin responses to prior sun exposure (skin type I and II does not tan, skin type II sunburns and tans slightly). Lower back or buttock skin, which had not previously been sun-exposed was used for the studies. Informed consent was obtained after the procedure was fully explained.

On the first study day, the minimal erythema dose (MED) was determined for each volunteer by irradiating the unprotected skin with 25% increments of ultraviolet light at about the anticipated MED for that skin type. The MED was the shortest light exposure at which erythema was observed 24 h later and below which no visible change was noted. With a measuring micrometer syringe, sunscreens were carefully applied to achieve a uniform layer using a micropipette at 2.0, 1.3, and 1.0 $\mu m/cm^2$ over a precut template 1×6 cm with square holes 1 cm^2 placed on the skin. These sites were irradiated using a series of ultraviolet doses near the anticipated MED of the protected skin, as detailed by the recommended procedures in the FDA sunscreen monograph (3). A 2500-W filtered xenon arc solar simulator (Kratos) was used for irradiance. The skin was observed 24 h later and the actual SPF was determined by dividing the MED of the protected skin by that of the unprotected skin. In this way, the actual SPFs of five different sunscreens applied in three different layer thicknesses were determined.

IV. RESULTS

A. Amounts of Sunscreen Applied

There was variation in the amounts of sunscreen applied (Table 1) among the different body regions.

Overall, the average amount of sunscreen applied was 1.3 mg/cm^2. In general, the gel was applied at a greater density (average amount = 1.7 mg/cm^2) than the other vehicles in all areas of the body except the back.

B. Sun Protection Factors

There was no significant change in SPF for four of the tested sunscreens with applied amounts of 1.0 and 1.3 μm/cm^2 compared with 2.0 μm/cm^2. However, the fifth sunscreen (with a manufacturer-determined SPF of 29) did show a significant decrease in SPF both at 1.0 and 1.3 μm/cm^2. The results are summarized in Table 2.

V. DISCUSSION

The present study was undertaken to determine the average amount of sunscreen applied by most individuals, and to evaluate the subsequent effect on sun protection factors. We found the average sunscreen application amount to be 1.3 mg/cm^2, which was close to 1 mg/cm^2 found in a previous study by Stenberg and Larko (4). The gel was applied in a slightly greater amount when compared with that of the other vehicles (mean = 1.7 mg/cm^2) in all areas of the body

Table 1 Average Amount of Sunscreen Applied to Various Regions of the Body (mg/cm^2) (mean ± SD)

Sunscreen	Head and neck (n = 5)	Chest (n = 5)	Back (n = 5)	Arms (n = 5)	Thighs (n = 5)	Lower legs (n = 5)
Creamy lotion	1.58 ± 0.66	1.10 ± 0.27	0.95 ± 0.40	1.28 ± 0.45	1.16 ± 0.31	1.11 ± 0.41
Alcoholic lotion	1.54 ± 0.73	1.16 ± 0.78	1.45 ± 0.95	1.27 ± 0.56	1.07 ± 0.26	1.28 ± 0.36
Gel	2.23 ± 1.01	1.49 ± 0.50	1.26 ± 0.40	1.76 ± 0.64	1.97 ± 0.48	1.59 ± 0.85
Cream	1.03 ± 0.42	1.17 ± 0.58	0.93 ± 0.64	1.05 ± 0.53	1.38 ± 0.82	0.94 ± 0.77

Table 2 Sun Protection Factors for Five Sunscreens Applied in Three Different Amounts (mean ± SD)

Amount applied ($\mu L/cm^2$)	Sunscreens[a]				
	A	B	C	D	E
2.0	4.0 ± 0.37	8.7 ± 1.22	11.0 ± 1.74	16.0 ± 1.76	29.0 ± 1.73
1.3	4.2 ± 0.48	8.0 ± 1.03	11.1 ± 1.41	14.8 ± 1.80	25.0 ± 2.12[b]
1.0	4.3 ± 0.44	7.8 ± 0.44	10.0 ± 1.41	14.4 ± 1.34	17.4 ± 1.34[b]

[a]Manufacturer-determined SPFs are as follows: A = SPF 4; B = SPF 8; C = SPF 10; D = SPF 15; E = SPF 29.
[b]The mean SPF value is statistically different from the mean SPF for the same sunscreen applied at 2.0 $\mu L/cm^2$ ($p < 0.05$).

except the back. (Because the back is the most difficult area to reach, it is also probably the most unreliable area for application.) Otherwise, the varying consistencies of the different sunscreens did not appear to play a role in application density, which differs from the findings of Stenberg and Larko. It is likely that in routine use even lower amounts than those documented in this study may actually be delivered to the skin. This may occur when sunscreen is applied hurriedly or with application to both hairy skin and to areas of the body that are difficult to reach.

In the United States, to obtain SPF values, a sunscreen amount of 2 mg/cm^2 or 2 $\mu L/cm^2$ is recommended (3). Our results, in contrast with those of Stenberg and Larko (4), suggest that individuals may apply less than the 2 mg/cm^2 quantity without diminishing the SPF value, at least with the lower SPF sunscreens. An SPF 29 sunscreen retained an SPF of 25 with 1.3 $\mu L/cm^2$ and 17.4 with 1.0 $\mu L/cm^2$; this suggests that the use of high-SPF sunscreens (greater than 15) may still offer maximal protection, even if the sunscreen is applied in less than optimum quantities.

These results are not in agreement with those of Stenberg and Larko (4), who found that 1 mg/cm^2 gave only 50% of the protection achieved with the 2 mg/cm^2 application. This may be due to the different methods employed: They used a glass rod to apply a defined amount of sunscreen to a 10 × 30-cm area. We used a micropipette to apply microliter amounts to 1-cm^2 areas. Perhaps, by dealing with much smaller skin areas, we were able to achieve more uniform layers and thus retain more sunscreen protection.

We were careful to incorporate the suggestions of Sayre (5) including the use of a filtered xenon arc solar simulator (as stipulated in the current FDA monograph on evaluation of sunscreens), supplying large standardized purchased sunscreen containers to the volunteers, and using products that are available in the United States.

Despite these observations, the revised 1994 FDA sunscreen monograph and the 1995 COLIPA standards (see Chap. 25) recommended 2 mg/cm. It is obviously important to try to ensure that optimum quantities of sunscreen are applied by the public to protect their skin.

REFERENCES

1. Parrish, J. A., et al. UV-A. *Biological effects of Ultraviolet Radiation with Emphasis on Human Responses to Longwave Ultraviolet.* Plenum Press, New York, 1978.
2. Kligman, L. H. and Kligman, A. M. The nature of photoaging: Its prevention and repair. *Photodermatology*, *3*:215–227 (1986).
3. U.S. Department of Health, Education and Welfare. Sunscreen drug products for over-the-counter human use. *Federal Register*, *43*:38206–38269 (1978) and (1994).
4. Stenberg, C. and Larko, O. Sunscreen application and its importance for the sun protection factor. *Arch. Dermatol.*, *121*:1400–1402 (1985).
5. Sayre, R. M. Sunscreen application: Flawed study? [letter]. *Arch. Dermatol.*, *122*: 745–746 (1986).
6. Lynfield, Y. L. and Schechter, S. Choosing and using a vehicle. *J. Am. Acad. Dermatol.*, *10*:55–59 (1984).
7. Goraef, J. W. and Cone, T. E. Jr., eds. *Manual of Pediatric Therapeutics*, 3rd ed. Little, Brown & Co., Boston, 1985.
8. *Plastic and Reconstructive Surgery—Essentials for Students.* Plastic Surgery Educational Foundation, Chicago, 1986, p. 85.
9. Martini, M. C. Comparison des méthodes de détermination des SPF. *Int. J. Cosmet. Sci.*, *8*:215–224 (1986).

30

Indices of Protection from In Vitro Assay of Sunscreens

Brian L. Diffey
Dryburn Hospital, Durham, England

I. INTRODUCTION

The increasing concern about the deleterious effects in human skin of exposure to ultraviolet A (UVA) radiation (320–400 nm) has led to the development of topical sunscreens that attenuate much more effectively in this waveband than has previously been the case. As a consequence, there is considerable debate among manufacturers and regulatory agencies on appropriate ways of measuring and specifying UVA protection. Methods based on in vivo techniques (described elsewhere in this volume) have been proposed, although none has yet met with universal acceptance by manufacturers and governmental regulators, probably because all of the techniques have one or more limitations. Because of this, attention has been diverted to the development of reliable in vitro techniques that can provide a measure not only of UVA protection, but also, in many cases, an estimate of the sun protection factor (SPF), which is an indication of the product's ability to protect against sunburn induced by UVB or UVA radiation, or both. Yet, despite the development of in vitro methods, no index of UVA protection has been universally accepted. In this chapter I will describe means by which data from in vitro assays can be manipulated to yield indices of sun protection, UVA protection, and broad-spectrum protection.

The principle of in vitro assay is to measure the transmission of ultraviolet radiation (UVR) through a substrate, apply the sunscreen, and repeat the transmission measurement. The ratio of the transmission without and with the sunscreen applied gives a measure of photoprotection.

II. SUBSTRATES

A wide range of substrates have been used for in vitro assay. These include wool (1,2), pig skin (2), lyophilized pig epidermis (3), hairless mouse epidermis (4–7), human epidermis (8), human stratum corneum (9), synthetic skin casts (10,11), surgical tape (12–14), and a combination of a biomembrane barrier with a biomacromolecular matrix (15).

III. RADIOMETRIC SYSTEMS

Systems used for detection of the transmitted UVR employ either a broadband radiometer (3,6,7,10,11,13) or a scanning spectroradiometer (4,5,9,12,14).

IV. SUN PROTECTION FACTOR

A. Spectroradiometry

The spectral transmission of UVR through the substrate is measured on wavelength-by-wavelength basis both with and without the sunscreen applied. The source of UVR must have a continuous spectrum of radiation over the wavelength range of interest (normally 290–400 nm), but the shape of the spectrum is unimportant. That is, there is no need to simulate the solar spectrum. Indeed, this is undesirable, for the intensity of radiation in the UVB region (290–320 nm) will be so low that the signal/noise ratio can become compromised. Light sources of choice are either an unfiltered xenon arc lamp or a fluorescent sunlamp (e.g., Philips TL12 or Westinghouse FS sunlamp). Mercury arc or metal halide lamps are not suitable, as their emission is largely composed of discrete wavelengths.

The transmittance of the sunscreen at wavelength λ nm [$T(\lambda)$] is equal to the ratio of the photocurrent measured through the substrate with the sunscreen applied to that before the sunscreen is applied. Normally $T(\lambda)$ will be measured in 5-nm steps from 290 to 400 nm. The sun protection factor (SPF) is then calculated

$$\mathrm{SPF} = \frac{\sum_{290}^{400} E(\lambda)\varepsilon(\lambda)\Delta\lambda}{\sum_{290}^{400} E(\lambda)\varepsilon(\lambda)T(\lambda)\Delta\lambda} \tag{1}$$

where $E(\lambda)$ is the spectral irradiance of terrestrial sunlight under defined conditions, $\varepsilon(\lambda)$ is the erythema action spectrum (16), and $\Delta\lambda$ is the wavelength step (e.g., 5 nm). The exact numerical values of the SPF will depend on the choice of $E(\lambda)$. For example, choosing $E(\lambda)$ to represent midday winter sunlight at latitude 60°N will give a different SPF from that obtained when $E(\lambda)$ is selected to represent midday summer sunlight at latitude 20°N.

B. Broadband Radiometry

In this technique workers (6,7,10,13) have commonly used a xenon arc lamp optically filtered to resemble the UVB component of terrestrial sunlight, in combination with a radiometer, the spectral response of which is designed to match the action spectrum for erythema in human skin. They have then calculated the in vitro SPF as

$$\text{SPF} = \frac{\text{transmitted irradiance without sunscreen}}{\text{transmitted irradiance with sunscreen}} \tag{2}$$

However, this approach can lead to appreciable errors in the determination of the SPF, depending on the spectral transmittance of the substrate, as the following analysis shows.

The transmitted irradiance through the substrate alone (i.e., before the sunscreen is applied) is equal to

$$\int E(\lambda)S(\lambda)D(\lambda)d\lambda \tag{3}$$

where $E(\lambda)$ is the spectral irradiance from the solar simulator at wavelength λ nm; $S(\lambda)$ is the spectral transmittance of the substrate at wavelength λ nm; and $D(\lambda)$ is the spectral sensitivity of the detector at wavelength λ nm. The integration is over all wavelengths for which the detector has a nonzero spectral response.

With the sunscreen applied the transmitted irradiance will be:

$$\int E(\lambda)S(\lambda)D(\lambda)T(\lambda)d\lambda \tag{4}$$

where $T(\lambda)$ is the spectral transmittance of the sunscreen at wavelength λ nm. As a source of simulated sunlight is used, that is, $E(\lambda)$ is a good approximation of the solar spectral power distribution, it might be assumed that the SPF can be expressed by the ratio of the two irradiances:

$$\text{SPF} = \frac{\int E(\lambda)S(\lambda)D(\lambda)d\lambda}{\int E(\lambda)S(\lambda)D(\lambda)T(\lambda)d\lambda} \tag{5}$$

Equation (5) was evaluated numerically for four different substrates and four different ultraviolet radiation detectors, and for each was compared with the correct value of the SPF defined by Eq. (1). The substrates chosen were a neutral-density filter, human stratum corneum, human epidermis, and lyophilized porcine epidermis; the representative spectral transmittance values [$S(\lambda)$] of these substrates are shown in Figure 1. The four UVR detectors, for which spectral sensitivities are also shown in Figure 1, comprised a thermopile; two commercially available UVB radiometers, one with a peak spectral response at 310 nm and full width at half maximum sensitivity of 25 nm, and the other with peak spectral response at 297 nm and bandwidth of 7 nm; and a detector for which spectral sensitivity matched the erythema action spectrum (e.g., a Robertson–Berger sunburn meter).

The SPFs calculated from Eq. (5) for each substrate–detector combination are given in Table 1.

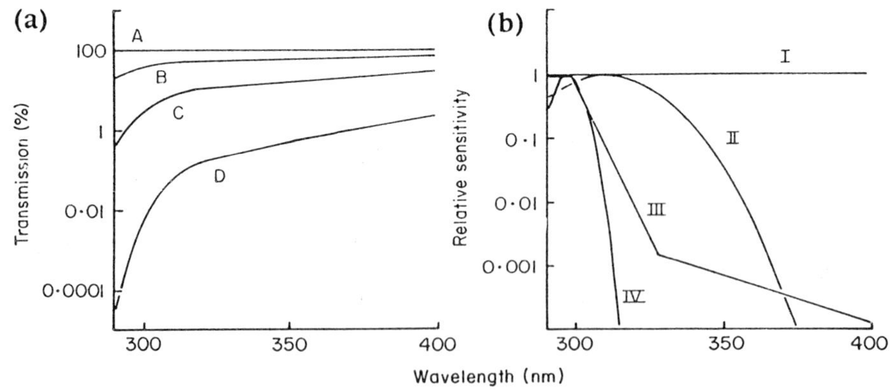

Figure 1 (a) Representative spectral transmittance of different substrates: A, neutral density filter; B, human stratum corneum; C, human full-thickness epidermis; D, lyophilized porcine epidermis. (b) Spectral sensitivity of different UVR detectors: I, thermopile; II, UVB radiometer (310 ± 25 nm); III, sunburn meter; IV, UVB radiometer (297 ± 7 nm).

Table 1 Sunscreen[a] SPFs Predicted from Different Substrate–Detector Combinations

	Substrate			
Detector	Neutral density filter	Human stratum corneum	Human epidermis	Porcine epidermis
Thermopile	1.1	1.1	1.0	1.0
UVB (310 ± 25 nm)	10.5	10.1	9.3	7.5
UVB (297 ± 7 nm)	30.9	30.9	30.8	30.7
Sunburn meter	15.0	13.0	9.0	3.9

[a]The sunscreen has a true SPF of 15.

In evaluating Eq. (5) for the thermopile (which has a wavelength-independent response) the spectral integration extended throughout the ultraviolet and visible regions into the near infrared, as these detectors exhibit a broad, flat spectral response. Clearly, virtually the whole signal produced by the detector will be a consequence of radiation that is not absorbed by the sunscreen. So applying the sunscreen to the substrate has a negligible effect on the detector signal, resulting in a SPF of approximately unity.

The UVB radiometer with the 25-nm bandwidth underestimates the correct SPF, whereas the UVB radiometer with the 7-nm bandwidth and peaking at 297 nm markedly overestimates the SPF. Even the "sunburn radiometer," the spectral sensitivity of which matches the erythema action spectrum, yields the correct SPF only when used in conjunction with a neutral-density filter. If a detector such as this is used with skin substrates, an SPF is obtained that is too low. The reason can be seen by examining Eqs. (1) and (5). For these equations to be identical, the product of the substrate transmission $S(\lambda)$ and the detector sensitivity $D(\lambda)$ should produce a function that has the same shape as the erythema action spectrum, $\varepsilon(\lambda)$. If this criterion is not met, then the determination of SPFs using simulated sources of sunlight becomes invalid.

V. ULTRAVIOLET A PROTECTION FACTORS

In this section different ways of manipulating in vitro spectroradiometric data are described.

A. Mean Ultraviolet Protection Factor

The simplest approach to defining a UVA protection factor (PF) from spectral transmittance data is to take the arithmetic average of the monochromatic protection factors, MPF(λ), from 320 to 400 nm. This is expressed as

$$\text{Mean UVA PF} = \frac{\sum_{320}^{400} \text{MPF}(\lambda)\Delta\lambda}{\sum_{320}^{400} \Delta\lambda} \tag{6}$$

where MPF(λ) is the reciprocal of the transmittance $T(\lambda)$ at wavelength λ nm. This index weights all UV wavelengths equally, and as the solar UVA spectrum increases with increasing wavelength, the mean UVA PF will be biased toward the protection provided by longer UVA wavelengths.

B. Erythemal Ultraviolet Protection Factor

Although UVA radiation constitutes about 95% of terrestrial UVR, it contributes only about 20% toward sunburn (17). Consequently, it is possible to define an erythemal UVA protection factor in the same way as a SPF, except that the lower limit of the summation begins at 320 nm, and not 290 nm; namely,

$$\text{Erythemal UVA PF} = \frac{\sum_{320}^{400} E(\lambda)\varepsilon(\lambda)\Delta\lambda}{\sum_{320}^{400} E(\lambda)\varepsilon(\lambda)T(\lambda)\Delta\lambda} \tag{7}$$

The justification for using this index might be that the erythema action spectrum is an appropriate surrogate for other photobiological effects, such as aging and skin cancer. However, there are no human data to confirm this.

C. Ultraviolet A Protection Percentage

The UVA protection percentage (APP) index, proposed by Sayre and Agin (18), is an extension of the erythemal UVA protection factor defined in Eq. (7). The UVA protection percentage is the fractional percentage of erythemally weighted solar UVA radiation absorbed by the sunscreen on the skin and is expressed as

$$\text{APP} = \left[1 - \frac{1}{\text{erythemal UVA PF}}\right] \times 100\% \tag{8}$$

D. Ultraviolet A/Ultraviolet B Absorbance Ratio

This index, defined as the ratio of the mean UVA absorbance to the mean UVB absorbance, was adopted by Boots the Chemist Limited in the United Kingdom (19). It is calculated as:

$$\frac{\int_{320}^{400} A(\lambda)d\lambda \Big/ \int_{320}^{400} d\lambda}{\int_{290}^{320} A(\lambda)d\lambda \Big/ \int_{290}^{320} d\lambda} \tag{9}$$

where $A(\lambda)$ is the effective absorbance and related to the sunscreen transmittance $T(\lambda)$ by $A(\lambda) = -\log[T(\lambda)]$. The absorbance ratio will range from zero for products exhibiting no protection against UVA radiation to unity for products exhibiting equal absorption at all wavelengths through the UV spectrum. This ratio then permits products to be classified into one of five categories:

Ratio mean UVA absorbance to mean UVB absorbance	UVA rating
<0.2	0
>0.2 and ≤0.4	1
>0.4 and ≤0.6	2
>0.6 and ≤0.8	3
>0.8	4

This method of defining UVA protection gives an indication of the physical properties of the sunscreen ingredients, and is independent of concentration and application thickness, unlike protection factors.

E. Broad-Spectrum Rating

It is pertinent to ask why there is so much concern about UVA protection per se? Figure 2 shows a typical spectrum of terrestrial UV radiation, together with the action spectrum for UV-induced erythema in human skin. A feature common to the two curves is that they span both the UVB and UVA regions. In other words, when we are outdoors we are not exposed to either UVB or UVA radiation, but to solar UVR, which always contains both. Also biological effects

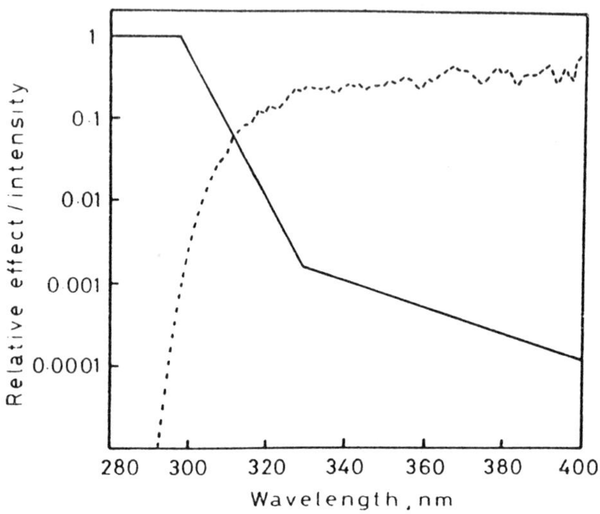

Figure 2 The action spectrum for ultraviolet-induced erythema in human skin (solid curve) and a typical spectrum of terrestrial solar ultraviolet radiation (broken curve).

that we may wish to protect against by using sunscreens are mediated by all wavelengths in the solar UV spectrum.

In terms of photoprotection any sunscreen has only one relevant attribute—its absorption spectrum. The two important features of an absorption spectrum are its amplitude and its width. The *amplitude* is a measure of how well the product absorbs radiation. In vivo determination of the SPF using a source of simulated sunlight is essentially a measure of the amplitude. The *width* of the absorption spectrum is a measure of how broadly across the UV spectrum the sunscreen is effective. A method proposed by Diffey (20) is to determine that wavelength λ_c for which the area under the absorption spectrum from 290 nm (the approximate lower wavelength limit of terrestrial sunlight) to λ_c is 90% of the integral of the absorption spectrum from 290 to 400 nm, that is

$$\int_{290}^{\lambda_c} A(\lambda)d\lambda = 0.9 \int_{290}^{400} A(\lambda)d\lambda \tag{10}$$

where $A(\lambda)$ is the absorbance of the product at wavelength λ nm. A five-point scale, analogous to the UVA rating, can be used to classify products as follows:

Table 2 A Comparison of UVA and Broad-Spectrum Indices Obtained in 35 Sunscreens by In Vitro Assay

Product	Quoted SPF	Mean UVA PF	Erythemal UVA PF	UVA protection percentage	UVA:UVB absorbance ratio	Broad-spectrum rating
1	4	2.2	2.7	62	0.25	2
2	4	1.8	2.1	53	0.17	1
3	4	1.9	2.4	58	0.30	3
4	4	1.9	2.4	59	0.37	3
5	4	2.2	2.7	63	0.34	2
6	4	1.6	2.0	49	0.13	1
7	4	1.7	2.1	52	0.26	2
8	4	2.4	3.0	67	0.36	3
9	4	2.3	2.8	65	0.33	2
10	4	1.9	2.4	59	0.34	3
11	4	1.9	2.4	58	0.26	2
12	4	2.4	2.9	66	0.56	4
13	4	3.9	4.6	78	0.72	4
14	4	2.6	3.1	68	0.34	3
15	8	3.4	3.6	72	0.30	2
16	8	3.2	3.7	73	0.34	3
17	8	3.0	2.9	65	0.23	2
18	8	2.8	3.4	71	0.38	3
19	8	4.0	4.6	78	0.49	3
20	8	4.1	4.4	77	0.44	3
21	8	2.3	2.9	66	0.26	2
22	8	6.1	7.0	86	0.71	4
23	8	5.3	6.2	84	0.70	4
24	8	4.4	5.4	81	0.61	4
25	8	3.9	4.6	78	0.39	3
26	15	6.2	5.6	82	0.36	3
27	15	5.8	6.9	86	0.53	4
28	15	6.5	7.2	86	0.48	3
29	15	6.1	6.4	85	0.43	3
30	15	6.0	5.5	82	0.42	3
31	15	4.6	4.6	78	0.29	3
32	15	6.9	5.0	80	0.32	2
33	15	6.9	6.9	86	0.49	3
34	15	4.7	5.7	83	0.49	4
35	15	6.5	6.9	85	0.48	3

Critical wavelength (nm)	Broad-spectrum rating
$\lambda_c < 325$	0
$325 \leq \lambda_c < 335$	1
$335 \leq \lambda_c < 350$	2
$350 \leq \lambda_c < 370$	3
$370 \leq \lambda_c$	4

The method makes no underlying assumptions about the form of action spectra for either immediate or long-term photobiological damage. However, acknowledgment is made of the fact that the efficiency of UVR to induce a given photobiological endpoint tends to decrease with increasing wavelength by adopting wavelength intervals for classifying the *broad-spectrum rating* which increase in an approximately logarithmic manner.

VI. COMPARISON OF PROTECTION FACTORS

To compare the classification of sunscreens by the methods outlined in section V.A to V.E, the absorption spectra of 35 different sunscreens were measured

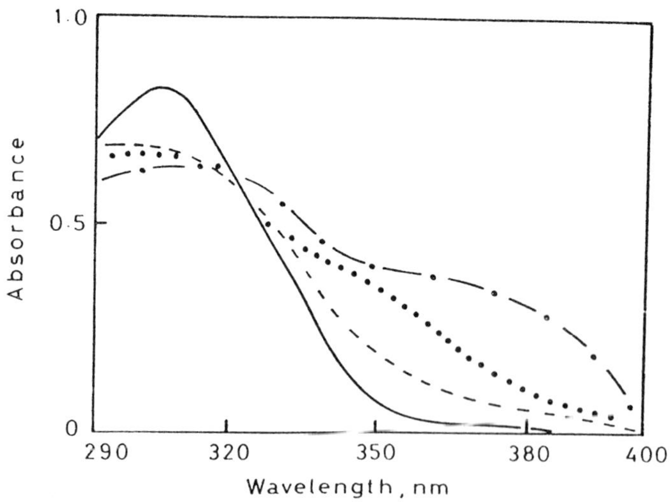

Figure 3 The absorbance of four sunscreens of SPF 4 measured using Transpore tape as the substrate: solid line, product 2; dashed line, product 7; dotted line, product 4; dash–dot line, product 12 (product numbers refer to those listed in Table 2).

using Transpore tape as a substrate (12). A comparison of the ratings obtained by each of the methods is given in Table 2. Examples of the spectral absorbance of four of these sunscreens are shown in Figure 3. Association analysis was applied to examine whether the SPF was independent of each of the UVA and broad-spectrum indices. Kendall's rank correlation test indicated strongly ($p < 0.0001$) that the null hypothesis of mutual independence between SPF and either the mean UVA protection factor, the erythemal UVA protection factor, or the UVA protection percentage (APP) should be rejected. The association between SPF and UVA/UVB absorbance ratio was equivocal ($p = 0.053$), whereas SPF was independent of the broad-spectrum rating ($p > 0.1$). Clearly, any rating additional to the SPF should be independent of the SPF, otherwise there is some redundancy in the additional index. On these grounds, a statement on any UVA protection factor is of limited value, because it is adding relatively little information about the product to what is already contained within the SPF. The broad-spectrum rating is preferred, as it provides additional, independent information to the SPF.

REFERENCES

1. Wurst, F., Prey, T., Washuttl, J., and Greiter, F. Studies on the adhesive power of several UV filter substances on sheep's wool before and after UV radiation. *Arz. Kosmetol.*, *8*:144–154 (1978).
2. Greiter, P., Bilek, P., Doskoczil, S., Washuttl, J., and Wurst, F. Methods for water resistance testing of sun protection products. *Int. J. Cosmet. Sci.*, *1*:147–157 (1979).
3. Stamper, B. Lyotropic liquid crystals as model vehicles for chemical sunscreens. PhD Thesis, Sunderland Polytechnic (1990).
4. Sayre, R. M., Pho Agin, P., LeVee, G., and Marlow, E. A comparison of in vivo and in vitro testing of sunscreening formulations. *Photochem. Photobiol.*, *29*:559–566 (1979).
5. Sayre, R. M., Pho Agin, P., Desrochers, D. L., and Marlowe, E. Sunscreen testing methods: In vitro predictions of effectiveness. *J. Soc. Cosmet. Chem.*, *31*:133–143 (1980).
6. Cole, C. and Van Fossen, R. Rapid in vitro evaluation of sunscreens: SPF and PFA. *Photochem. Photobiol.*, *47B*:73S (1988).
7. Cole, C. and Van Fossen, R. In vitro models for UVB and UVA photoprotection. *Sunscreens: Development, Evaluation, and Regulatory Aspects* (N. J. Lowe and N. A. Shaath, eds.). Marcel Dekker, New York, 1990, pp. 395–404.
8. Brown, S. and Diffey, B. L. The effect of applied thickness on sunscreen protection: In vivo and in vitro studies. *Photochem. Photobiol.*, *44*:509–513 (1986).
9. Pearse, A. D. and Edwards, C. Human stratum corneum as a substrate for in vitro sunscreen testing. *Int. J. Cosmet. Sci.*, *15*:234–244 (1993).
10. Stockdale, M. A novel assessment of sunscreen product efficacy against UVA. *Int. J. Cosmet. Sci.*, *9*:85–98 (1987).

11. Ferguson, J., Brown, M. W., Hubbard, A. W. and Shaw, M. I. Determination of sun protection factors. Correlation between in vivo human studies and an in vitro skin cast method. *Int. J. Cosmet. Sci.*, *10*:117–129 (1988).
12. Diffey, B. L. and Robson, J. A. A new substrate to measure sunscreen protection factors throughout the ultraviolet spectrum. *J. Soc. Cosmet. Chem.*, *40*:127–133 (1989).
13. Keeley, K. A., Laskar, P. A., Ewing, G. D., Dromgoole, S. H., Lichtin, J. L., and Sakr, A. A. In vitro sun protection factor evaluation of sunscreen products. *J. Soc. Cosmet. Chem.*, *44*:139–151 (1993).
14. Sellers, R. L. and Carpenter, F. G. An instrument for in-vitro determination of SPF. *Cosmet. Toilet.*, *107*:119–123 (1992).
15. Gordon, V. SPF evaluation via in vitro methodology. *Drug Cosmet. Ind.*, 30–75 (1992).
16. McKinlay, A. F. and Diffey, B. L. A reference action spectrum for ultraviolet induced erythema in human skin. *CIE J.*, *6*:17–22 (1987).
17. Diffey, B. L. The need for sunscreens with broad spectrum protection. *Biological Responses to Ultraviolet A Radiation* (F. Urbach, ed.). Valdenmar Publishing, Overland Park, KS, 1992, pp. 321–328.
18. Sayre, R. M. and Agin P. P. A method for the determination of UVA protection for normal skin., *J. Am. Acad. Dermatol.*, *23*:429–440 (1990).
19. Boots the Chemist Ltd. *The Guide to Practial Measurement of UVA/UVB Ratios.* The Boots CO PLC, Nottingham, England.
20. Diffey, B. L. A method for broad spectrum classification of sunscreens. *Int. J. Cosmet. Sci.*, *16*:47–52 (1994).

31

Noninvasive In Vivo Determination of Sunscreen Ultraviolet A Protection Factors Using Diffuse Reflectance Spectroscopy

Robert Gillies and Nikiforos Kollias
Wellman Laboratories of Photomedicine, Massachusetts General Hospital, Boston, Massachusetts

I. INTRODUCTION

In recent years, there has been an increased awareness of the potentially damaging effects of ultraviolet A (UVA; 320–400 nm) radiation to human skin. Long-term exposure to UVA has been associated with erythema (1,2), photoaging (3), damage to blood vessels (4), changes in the dermal connective tissue, and elastosis (5,6). The damage associated with UVA radiation occurs much deeper in the skin than the damage from UVB (280–320 nm) because of the deeper penetration of UVA. UVA exposure also causes phototoxic reactions with many commonly used drugs, such as the tetracycline antibiotics and 8-methoxypsoralen (8-MOP) used in PUVA treatment.

Each of these responses may be described by its UVA biological risk; that is, the ability of UVA to generate that biological response. The UVA biological risk may be estimated by convoluting the UVA source spectrum with the biological action spectrum of the response (i.e., the wavelength sensitivity for the particular response). The integrated value of the convoluted spectrum over the UVA wavelength range represents the UVA biological risk. For example, the biological risk for erythema from UVA in sunlight is calculated by the convolution of the CIE threshold erythema action spectrum (7) with the UVA portion of solar radiation. Although UVA is over 300 times less erythemogenic than UVB, the biological risk from UVA is only five times less at solar maximum than the biological risk from UVB. UVA is present in amounts 10–20 times greater in solar radiation than UVB, resulting in biological risk from UVA.

Topical sunscreens provide a major means of protecting the skin from biological risk by altering the optical properties of the skin. The efficacy of a sunscreen product (defined by its protection factor; PF) is assessed by measuring its ability to decrease biological response of the skin after exposure to radiation. Traditionally, the ability of a sunscreen to prevent a threshold erythema from solar radiation, defined as the solar protection factor (SPF) is measured using human phototesting. However, the protection measured in this manner is primarily protection from UVB and not UVA radiation. The generation of erythema, by using only UVA radiation that is produced by filtered xenon arc or high-pressure metal halide lamps, requires doses as high as 100–120 J/cm^2. In addition, the UVA irradiance of these lamps is 10–30 times higher than present in sunlight (30–100 mW/cm^2 for lamps, compared with about 3 mW/cm^2 for sunlight). Many UVA effects are also oxygen-dependent (3), and reciprocity may not hold. Thus, the extremely long exposure times and the question of reciprocity makes this method a time-consuming and an expensive process.

Attention has lately focused on developing laboratory methods for the routine assessment of the UVA protection factor (PFA). Several in vitro and in vivo methods using animals have been proposed, among which some employ photosensitizers such as topical 8-methoxypsoralen to reduce the exposure time required to elicit an erythemic response (8,9), changes in ear swelling in mouse models (10–12), or measurements of the transmission spectrum of products on excised mouse and human epidermis (13–15) and on tape products (16). The first two methods measure the sunscreen efficacy in preventing a biological response, whereas the latter techniques determine the sunscreen efficacy mathematically based on its transmission spectrum. Good correlation has been reported between these techniques and human phototesting results.

None of these methods measures the product's UVA efficacy on the biological system the sunscreen product was meant to protect; namely, in vivo human skin. The responses elicited on UVA exposure depend more on the source, the action spectrum, and the substrate of the test system. Each technique may determine a different efficacy for the same product. Moreover, these tests do not take into account in vivo effects, such as interaction of the sunscreen with human skin (e.g., chemical sunscreens such as iron oxides and dihydroxyacetone), skin topography, skin absorption, and sweating (17). Furthermore, the invasive nature of current in vivo techniques, either from previous exposures or the presence of photosensitizers, greatly alters the optical properties of the exposed site and renders it unfit for further testing.

It was felt that a noninvasive method that separates the optical properties of the skin from its biological responses, and was quick and inexpensive, was required for testing UVA efficacy. A more general approach is the noninvasive determination of the alterations of the UVA transmission properties of the skin following the application of a sunscreen product. Given a reliable and well-accepted action spectrum for the biological endpoints and the solar spectrum, the biological response may be predicted. The UVA efficacy of the sunscreen product may then be calculated once the transmission spectrum of the sunscreen on in vivo human skin has been measured.

Diffuse reflectance spectroscopy (DRS) is ideally suited for these measurements. DRS directly measures the optical properties of the skin decoupled from its biological responses. Application of a sunscreen product modifies the surface of the skin by providing an additional barrier, or filter, through which UVA radiation must penetrate before reaching the viable layers of the skin. The sunscreen product changes the surface of the skin and, thereby, changes the absorption properties of the skin. Measurement of the absorption properties of the skin before and after application of a sunscreen product permits calculation of the transmission spectrum of the product on in vivo human skin and, thus, the protection efficacy for the wavelength range of interest. This chapter presents measurements of the UVA protection factors of four Cosmetic, Toiletry, and Fragrance Association (CTFA) test sunscreens, compares these results with recently published human phototesting results from a multicenter study (15), and discusses the prospect of DRS for routine laboratory measurement of PFA values.

II. PFA PREDICTIONS USING DIFFUSE REFLECTANCE SPECTROSCOPY

The intensity of a biological response by an individual's skin depends on three major factors: (1) the action spectrum of the response; (2) the wavelength distribution and intensity of light incident on the surface of the skin, defined by a source spectrum; and (3) the transmission of the light into the skin, defined by a transmittance spectrum. The transmission spectrum of a sunscreen product on human skin is calculated as the ratio of the skin remittance spectrum with product (I_{with}) to the skin remittance spectrum without product (I_{without}). The collected backscattered light, of the same wavelength as the excitation light, is doubly attenuated as it passes through the product twice, as shown in Figure 1. The effective transmittance of the product at each wavelength is thus the square root of the ratio of the spectra:

$$T(\lambda) = \left[\frac{I_{\text{with}}(\lambda)}{I_{\text{without}}(\lambda)}\right]^{1/2} \tag{1}$$

where $I_{\text{with}}(\lambda)$ and $I_{\text{without}}(\lambda)$ are the remitted intensities at wavelength λ from the skin with and without application of the sunscreen product, respectively. The

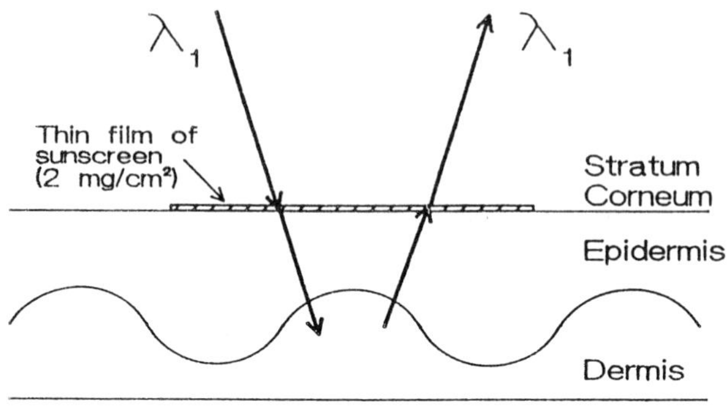

Figure 1 Schematic diagram of the light path through human skin during a DRS measurement.

square root arises from the double attenuation of the light as it passes through the sunscreen. The UVA protection factor of the product, defined by an PFA value, is

$$\text{PFA} = \frac{\sum_{320}^{400} S(\lambda)A(\lambda)}{\sum_{320}^{400} S(\lambda)A(\lambda)T(\lambda)} \tag{2}$$

where $S(\lambda)$ is the source spectrum, $A(\lambda)$ is the action spectrum of the biological response being investigated, and $T(\lambda)$ is the calculated transmission spectrum of the sunscreen product obtained on human skin in vivo.

III. MATERIALS AND METHODS

The DR spectra of the skin with and without product were measured with a Spex Fluorolog 2 fluorimeter (model 212, Spex Industries, Edison, NJ). The cuvette holder was replaced with a bifurcated fiber optic bundle, one arm of which was coupled to the excitation monochromator and the other arm coupled to the emission monochromator. The DR spectra were collected by scanning both monochromators synchronously at the same wavelength. The joined end of the fiber bundle was placed on the skin site to be measured, irradiated, and the collected backscattered excitation light passed to the emission monochromator. Both monochromators were set at 2-nm bandpass.

The location for measurement was first examined under Wood's light and a 4 × 4-cm area was selected that was relatively free of epidermal pigmented macules. The DRS spectra of the skin site without product were recorded. At least four different locations within the selected area were measured. A weighed amount (32 mg) of a sunscreen product was then applied to the selected area to produce a 2-mg/cm^2 coverage. This is the same coverage specified by the Food and Drug Administration (FDA) for human phototesting. The area was reexamined under Wood's light to check the homogeneity of the product application. Each product was allowed to air dry for 20 min, and DRS spectra of the skin with product were recorded. A UV-transmitting Teflon sheet (YSI No. 5775 oxygen membrane, YSI Inc., Yellow Springs, OH) was placed over the selected area just before each set of measurements. This sheet permitted placement of the fiber bundle tip in close contact with the sites being measured, but prevented product contamination of the fiber probe. The transmission spectrum of each sunscreen was calculated by Eq. (1). The PFA of each sunscreen for erythema was calculated by Eq. (2) using the CIE erythema action spectrum and the solar spectrum (18).

IV. RESULTS

Representative calculated transmission spectra of CTFA 060 (7% padimate O) and CTFA 057 (its vehicle), measured on a single volunteer, are shown in Figure 2. Each spectrum of each product represents the calculated transmission spectrum obtained from a single location within a 4 × 4-cm site. An average calculated transmission spectrum for each measured sunscreen product was obtained by averaging all the representative spectra. Figure 3 shows the average

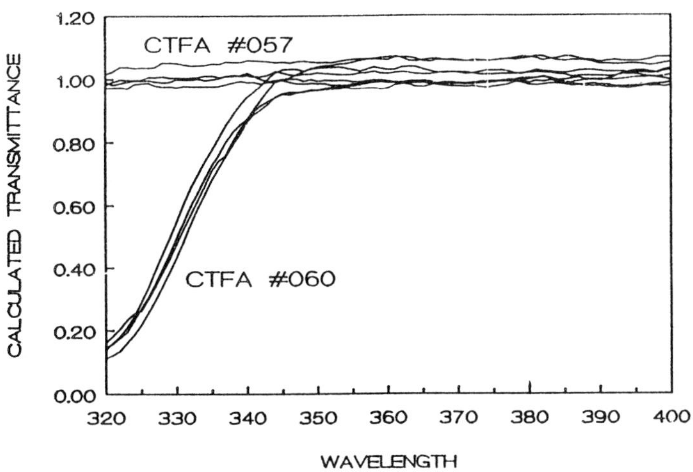

Figure 2 Representative calculated transmission spectra, obtained by diffuse reflectance spectroscopy, of CTFA 057 and CTFA 060 products. Each curve of the family of four curves represents the measurement of a single site within the 4 × 4-cm application area. Each formulation was applied at a 2-mg/cm^2 coverage.

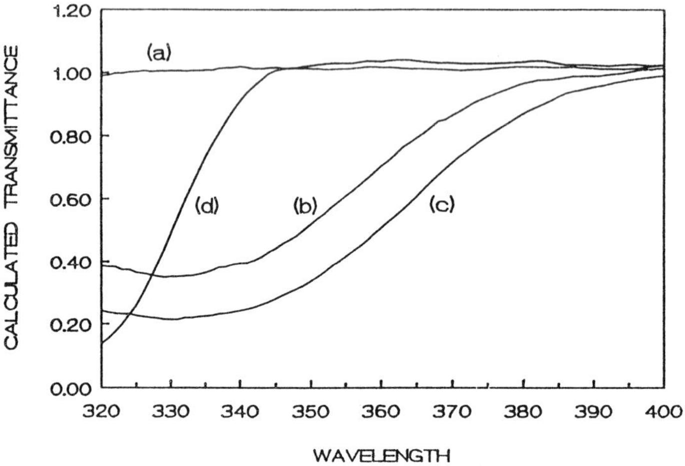

Figure 3 Average calculated transmission spectra, obtained by diffuse reflectance spectroscopy, of (a) CTFA 057, (b) CTFA 058, (c) CTFA 059 and (d) CTFA 060. Each curve is the average of the calculated transmission spectra of at least four sites with the 4 × 4-cm application area. Each product was applied at a 2-mg/cm^2 coverage.

calculated transmission spectra obtained for each product measured. All spectra resemble their absorption spectra obtained as thin films on a quartz flat.

An average PFA value for each product was calculated for each subject. The PFA values calculated for each subject were averaged (n = 10 for all products). These values and their standard deviations are listed in Table 1.

V. DISCUSSION

A routine laboratory method for the assessment of UVA photoprotection by sunscreen formulations must be quick, inexpensive, show good agreement with

Table 1 Calculated PFA Values of Sunscreen Formulations from Diffuse Reflectance Measurements

Active ingredient[a]	Calculated PFA value
CTFA 057	1.0 ± 0.1
CTFA 058	1.8 ± 0.2
CTFA 059	3.1 ± 0.5
CTFA 060	1.6 ± 0.1

[a]All sunscreen formulations were applied at 2-mg/cm^2 coverage.

phototesting results of human volunteers, and be reproducible. The time required for the PFA determination of four sunscreen formulations on a single volunteer, from start to finish, was approximately 1 h. The instrumentation consisted of a double-grating fluorimeter equipped with a bifurcated fiber bundle attachment. The high quality of the spectrometer was not required for these measurements, but was used because it was available in our laboratory. With faster-scanning speeds available with other spectrometers, the time required for the determination of PFA values may be reduced. A previous communication (19) that calculated the PFA values by diffuse reflectance of three commercial sunscreen products showed that the calculated values agreed closely with the PFA values stated by the manufacturers.

The results for the CTFA sunscreens shown in Table 1 demonstrate that the technique is reproducible from volunteer to volunteer. The standard deviation of the means for all measurements except 5% oxybenzone, expressed as a percentage, were within 10%. The large error for 5% oxybenzone arose from saturation of our particular spectrometer. However, because the PFA value is directly related to the film thickness (the thicker the film the higher the observed PFA value), thinner films of a product may be measured, and a PFA value for a standard 2-mg/cm^2 coverage calculated. The relation between the film thickness and the calculated PFA value lies in the calculation of the transmission spectrum of the product. In these measurements the sunscreen product acts essentially as a double-pass filter and, as such, may be treated as a filter. A simple film thickness correction to Eq. (1) may then be applied in the calculation of the transmission spectrum for standardization at a 2-mg/cm^2 coverage:

$$T(\lambda) = \left[\frac{I_{\text{with}}(\lambda)}{I_{\text{without}}(\lambda)}\right]^{(1/2)(2/c)} \tag{3}$$

where c is the sunscreen coverage in milligrams per square centimeter (mg/cm^2). Equation (3) reduces to Eq. (1) at a standard coverage of 2 mg/cm^2.

All the foregoing PFA calculations were performed using the CIE erythema action spectrum and a solar spectrum. These calculated values thus reflect the protection afforded by the sunscreen formulations to UVA-induced erythema from sunlight. This method, however, does not depend on any given biological endpoint nor source spectrum, and the same measurements may be used to predict a variety of biological endpoints. Identical calculations may be made for filtered solar simulators once the spectral irradiance of the source is measured. The same method may be used to calculate the protection of sunscreens to phototoxic responses provided that the action spectrum of the phototoxic response and the source spectrum are known. The only requirement for this technique is that the action spectrum of the specific endpoint and radiation source are known and the appropriate spectra are substituted into Eq. (2).

Application of a sunscreen product on skin alters its UVA transmission. The basic principle underlying the use of diffuse reflectance spectroscopy is that the optical properties of the sunscreen product are decoupled from the biological responses of the skin. With this technique, the optical properties of the sunscreen are measured directly, rather than any biological response of the skin. This avoids the problems generally associated with the use of exogenous photosensitizers, which measure the action spectrum of the photosensitizer, rather than the skin reaction of normal individuals.

The noninvasive nature of the technique does not alter the optical properties of either the skin site being measured or the optical properties of the sunscreen formulation being measured. Foremost, these measurements are performed on in vivo human skin. Interactions between the sunscreen formulation and human skin, such as sweating and absorption, are taken into account by the measurement. The noninvasive nature of the measurements is an important consideration for substantivity studies. This permits measurements of the same skin sites with time to determine the substantivity of the formulation (17) to human activities such as swimming or sweating. Short-term or long-term photodegradation of the sunscreen product may also be assessed with this technique. No restrictions on the type of sunscreen or wavelength range investigated are made. These measurements are equally as applicable to oil-based and fluorescent sunscreens as to water-based sunscreens. Replacing the solar spectrum in Eq. (2) with the appropriate irradiation spectrum permits other wavelength ranges to be investigated.

The PFA value of products other than sunscreens may also be investigated using this technique. Dihydroxyacetone (DHA) is used as an artificial tanning agent. It adheres to the stratum corneum and persists for approximately 1 week. One individual applied 2% DHA in eucerin on their skin over three successive days, resulting in a deep yellow coloration. Measurement of the transmission spectrum (Fig. 4) of DHA from this skin site shows that DHA has a sizable absorption in the UVA and may be a potential UVA sunscreen agent. The calculated PFA value for this application of DHA to human skin was 3.1.

The technique presented here seems ideal as a routine laboratory test for the efficacy of potential sunscreen products. It is quick, inexpensive, yields reproducible results, and the results are in good agreement with stated phototesting results. More importantly though is that these measurements are made in vivo on human skin, the material the products were designed to protect. PFA values for different biological endpoints may be calculated, provided a good action spectrum for the endpoint and the illumination spectrum is accurately known. The noninvasive nature of the technique permits multiple measurements of the same skin site without altering the optical properties of the skin, an important consideration for substantivity and sunscreen product photodegradation studies.

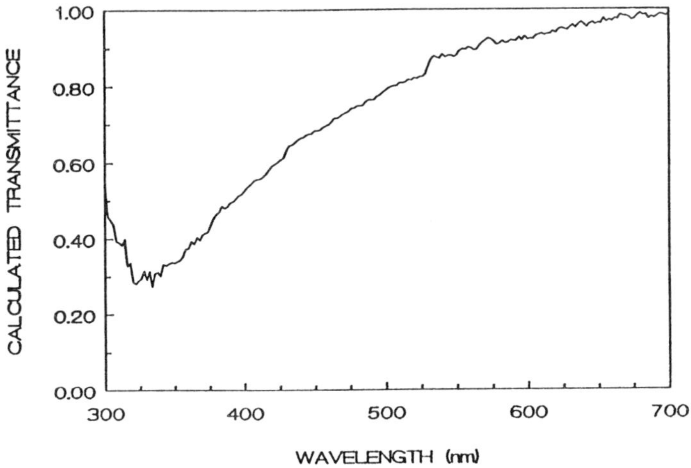

Figure 4 Calculated transmission spectrum obtained by diffuse reflectance spectroscopy of dihydroxyacetone (DHA) from in vivo human skin. A commercial product containing DHA was applied to human skin at a 2-mg/cm^2 coverage on three successive days. The measurement was performed on the fourth day after the skin site obtained a deep yellow coloration.

REFERENCES

1. Stanfield, J. W., Feldt, P. A., Csortan, E. S., and Krochmal, L. Ultraviolet A sunscreen evaluations in normal subjects. *J. Am. Acad. Dermatol.*, *20*:744 (1989).
2. Kaidbey, K. H. and Kligman, A. M. The acute effects of longwave ultraviolet radiation on human skin. *J. Invest. Dermatol.*, *72*:253 (1979).
3. Kaidbey, K. and Gange, R. W. Comparison of methods for assessing photoprotection against ultraviolet A in vivo. *J. Am. Acad. Dermatol.*, *16*:346 (1987).
4. Gilchrist, B. A., Soter, N. A., Hawk, J. L. M., Barr, R. M., Block, A. K., Hensby, C. N., Mallet, C. N., Greaves, A. I., and Parrish, J. A. Histologic changes associated with ultraviolet-A-induced erythema in normal human skin. *J. Am. Acad. Dermatol.*, *9*:213 (1983).
5. Chew, S., DeLeo, V. A., and Harber, L. C. An animal model for evaluation of topical photoprotection against ultraviolet A (320–380 nm) radiation. *J. Invest. Dermatol.*, *89*:410 (1987).
6. Kligman, L. H. and Kligman, A. M. Ultraviolet radiation-induced skin aging. *Sunscreens: Development, Evaluation and Regulatory Aspects* (N. J. Lowe and N. A. Shaath, eds.), Marcel Dekker, New York, 1990, p. 55.
7. McKinlay, A. F. and Diffey, B. L. A reference action spectrum for ultraviolet induced erythema in human skin. *CIE J.*, *6*:17 (1987).
8. Roelandts, R., Sohrabvand, N., and Garmyn, M. Evaluating the UVA protection of sunscreens. *J. Am. Acad. Dermatol.*, *21*:56 (1989).

9. Gange, R. W., Soparkar, A., Matzinger, E., Dromgoole, S. H., Sefton, J., and DeGryse, R. Efficacy of a sunscreen containing butyl methoxydibenzoylmethane against ultraviolet A radiation in photosensitized subjects. *J. Am. Acad. Dermatol.*, *15*:494 (1986).
10. Furukawa, R. D., Brown, W. R., Shivji, G. M., and Ramsay, C. A. Assessment by mouse model of the ultraviolet A protective effect of topical sunscreens. *J. Am. Acad. Dermatol.*, *20*:1031 (1989).
11. Cole, C. A., Forbes, P. D., and Ludwigsen, K. Sunscreen testing using the mouse ear model. *Photodermatology*, *6*:131 (1989).
12. Bissett, D. L., Hillebrand, G. G., and Hannon, D. P. The hairless mouse as a model of skin photoaging: Its use to evaluate photoprotective materials. *Photodermatology*, *6*:228 (1986).
13. Sayre, R. M., Agin, P. P., LeVee, G. J., and Marlowe, E. A comparison of in vivo and in vitro testing of sunscreen formulas. *Photochem. Photobiol.*, *29*:559 (1979).
14. Cole, C. A. and VanFossen R. Rapid in vitro evaluation of sunscreens: SPF and PFA. *Photochem. Photobiol.*, *47*:73S (1988).
15. Cole, C. A. and VanFossen, R. L. In vitro models for UVB and UVA photoprotection. *Sunscreens: Development, Evaluation and Regulatory Aspects* (N. J. Lowe and N. A. Shaath, ed.), Marcel Dekker, New York, 1990, p. 395.
16. Diffey, B. L. and Robson, J. A new substrate to measure sunscreen protection factors throughout the ultraviolet spectrum. *J. Soc. Cosmet. Chem.*, *40*:127 (1989).
17. Sayre, R. M., Marlowe, E., Agin, P. P., LeVee, G. J., and Rosenberg, E. W. Performance of six sunscreen formulations on human skin. *Arch Dermatol.*, *15*:46 (1979).
18. Sayre, R. M. and Agin, P. P. Comparison of human sun protection factors to predicted protection factors using different lamp spectra. *J. Soc. Cosmet. Chem.*, *35*: 439 (1984).
19. Kollias, N., Gillies, R., and Anderson, R. R. The non-invasive determination of UVA sunscreen effectiveness in vivo. *Biological Responses to UVA Radiation* (F. Urback, ed.), Valdenmar Publishing, Overland Park, KS, 1992, p. 371.

32

Polymorphous Light Eruption: Its Reproduction and Prevention by Sunscreens

Dominique Moyal
L'Oréal, Clichy, France

O. Binet
Fondation A. de Rothschild, Paris, France

I. INTRODUCTION

Polymorphous light eruption (PLE) is an idiopathic photodermatosis frequently found in humans originating from northern parts of Europe (1–3). The condition of PLE can cause great discomfort; therefore, it is desirable to gain greater knowledge of this photodermatosis and to be able to reproduce it under controlled conditions in real and simulated sunshine, to evaluate the efficacy of potential forms of preventive treatment. The percentage of individuals at potential risk is approximately 10–20%, of which the majority are females.

Onset of the condition occurs within the first three decades of life, and its incidence tends to diminish at about the age of 50. A family history has been reported (2). Affected patients are drawn from all Caucasian phototypes (types I–V). The eruption is triggered by initial exposure to sufficiently strong sunlight in spring or summer, generally after two sessions.

The cutaneous lesions take various forms. Classically, they consist of papules, reticulated erythema, and vesicles, and may be accompanied by pruritus. They appear only on the areas exposed to sunlight. They are principally restricted to the chest, arms, abdomen, and legs, but in the most severe cases, can extend further, affecting particularly the face.

The eruptions recur year after year. Severity varies from one individual to another and lesions may show a tendency to subside after a few days, provided that the subject avoids exposure to the sun.

The solar wavelengths incriminated in the triggering of PLE apparently span the entire ultraviolet (UV) spectrum. Some studies incriminate principally UVA (320–400 nm) (4–7), others also demonstrate sensitivity to UVB (8). According to Farr and Diffey (9), it is possible that the use of certain sunscreens that mainly protect the skin against UVB cause an increase in the incidence of PLE.

Various modes of treatment have been proposed to prevent the onset of PLE: Psoralen-UVA (PUVA) therapy and UVB phototherapy (5,10–13) afford protection in 65–90% of cases. However, these types of treatment, in particular PUVA therapy, are tedious. Other forms of treatment have been studied, such as oral β-carotene (14,15) or hydroxychloroquine (16) and have had limited success. Protection by the topical route involving uniquely UVB filters or the use of too low a level of protection against UVA, has been ineffective (6).

The aim of this chapter is to present the experimental work carried out to reproduce PLE under simulated and real sunlight and to evaluate the efficacy of sunscreen products with a high protective capacity, particularly in the overall UVA range, to prevent the eruption.

II. REPRODUCTION OF POLYMORPHOUS LIGHT ERUPTION UNDER AN ULTRAVIOLET A SIMULATOR

The studies were carried out in women of Caucasian origin, aged from 18 to 50 years, with phototypes II to IV. All subjects professed to have suffered from PLE and were selected after consideration of their medical history; antinuclear factors were negative. None of these women had experienced PLE on their face. They were included in the studies following clinical examination after having signed an informed consent form. They had not been exposed to UV for at least 3 months.

The first study consisted of verifying the role of UVA in the onset of PLE. The subjects were exposed solely on the chest, the predominant site of the eruption. A total of 52 subjects were included and exposed without photoprotection under a UVASUN 5000 metal halide lamp equipped with a Schott WG 335 filter, 3 mm in thickness, and a black filter to eliminate wavelengths longer than 440 nm and shorter than 320 nm. The UVA flux was measured using a Centra Osram UV meter equipped with a UVA-sensitive cell. The subjects were exposed daily for a maximum of 5 days, at a dose of 25 J/cm^2 on days 1 and 2, corresponding to 2 h of Mediterranean sunlight in June; 30 J/cm^2 on the next 2 days; and 35 J/cm^2 on the fifth day. Exposure was stopped if any evidence of an eruption was detected. A clinical examination was performed by one of us (OB) approximately 5 h after each exposure session, as the eruption was not immediately visible on completion of exposure. The objective signs of erythema, papules, vesicles, and edema were accorded a score, as were subjective signs of pruritus and burning experienced during and after exposure. Of the 52 women 49 experienced induction of PLE lesions under UVA alone (Fig. 1): 13 after two exposures (50 J/cm^2) (25% of subjects); 19 after three exposures (80 J/cm^2), giving a cumulative 61% of subjects; 7 after four exposures (110 J/cm^2), giving a cumulative 75%, and 10 after five exposures (145 J/cm^2), amounting overall to 94% of subjects. It was thus demonstrated that relatively low UVA doses, corresponding to a few hours in the sun, are able to induce PLE lesions in women prone to suffer from the disease.

III. PREVENTION OF PLE BY TOPICAL SUNSCREEN PRODUCTS

A. Products Studied

All products contained photostable sunscreen combinations. The various classes of products tested are displayed in Table 1. These classes were defined in terms of the sun protection factor (SPF) and the UVA protection factor (UVA PF).

The SPF was determined in accordance with the methodology recommended by the U.S. Food and Drug Administration (FDA) (17) and the UVA protection factor by the method based on persistent pigment darkening (PPD; see Chap. 29) (18).

B. Indoor Studies

A UVA simulator was used under a condition sufficient to trigger the onset of PLE (total UVA dose of 145 J/cm^2 over a maximum period of 5 days). Forty-eight women who were prone to PLE were enrolled in the study. One-half of their chest received no application and the other half was pretreated with a

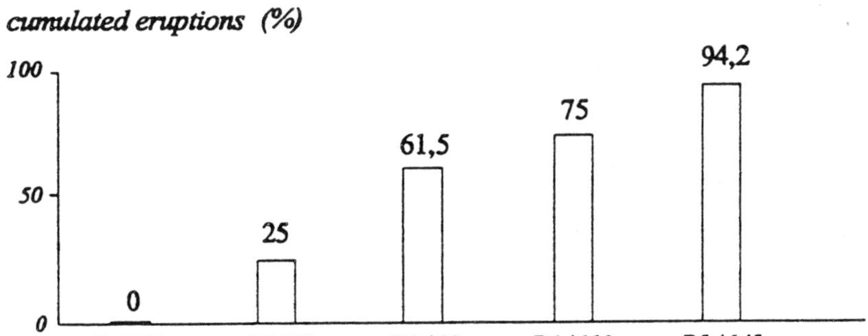

Figure 1 Percentages of PLE eruption under UVA simulator on 52 female patients.

Table 1 Description of Product Classes Tested, Defined in Terms of SPF and UVA PF: Sunscreen Combinations of the Products[a]

Product class	P1	P2	P3	P4
SPF	15	20	40	> 40
UVA PF (PPD)	1	4	6–9	> 10

[a]P1: UVB filters (19.5%): 4-methylbenzylidene camphor; octyl methoxycinnamate; homosalate. P2: UVB filter (5%): 4-methylbenzylidene camphor; UVA filters (3%): MEXORYL® SX (hydrosoluble, λ max = 345 nm); butyl methoxydibenzoylmethane (liposoluble, λ max = 358 nm); titanium dioxide (5%). P3: UVB filters: 4-methylbenzylidene camphor (5%); octocrylene (10%); phenylbenzimidazole sulfonic acid (3%); UVA filters (4%–6%): MEXORYL® SX; butyl methoxydibensoylmethane; titanium dioxide (5%). P4: UVB filter (10%): octocrylene; UVA filters (7%): MEXORYL® SX; butyl methoxydibensoylmethane; titanium dioxide (5%). (Note: MEXORYL® SX = Terephtalylidene dibornanone sulfonic acid).

protective product. Seventeen subjects applied the P2-type products and 31 subjects applied those of P3 type. PLE was elicited on the unprotected half-chest of the 48 women.

With P2, only two cases of PLE were observed (i.e., 12% of subjects), and in these two cases, the onset of PLE was delayed and less intense by comparison with the unprotected half-chest. PLE was totally prevented with P3.

In conclusion, products containing adequate UVA sunscreen absorbers can prevent the triggering of PLE by UVA rays. These studies were complemented under natural sunlight using realistic conditions of exposure.

C. Outdoor Studies

The first study involved 20 PLE-prone women under the sunlight of Guadelupe (latitude 16°15N, longitude 15°35W). Comparison was made per half-body of a protectant between a P1 type that filters only in the UVB spectrum and, hence, protects from actinic erythema and a protectant of P2 type that possesses good UVB-filtering capacity and a UVA protection factor of 4.

The results (Fig. 2) show that, when a sunscreen product that protects only in the UVB range was applied, the onset of PLE was triggered very rapidly, whereas, with a product of P2 type, no incidence of PLE occurred on 30% of subjects (6/20), and in the others, the onset was delayed and the eruption less severe.

Further studies were carried out under natural sunlight in Djerba, Tunisia (latitude 33°50N, longitude 10°48E), where the efficacy of products with higher UVA protection factor was compared. Thirty-five women were protected on half-body by a product containing solely UVB filters (P1) and on the other half by a product of class P3 (UVA PF between 6 and 9). On eight other women, whose

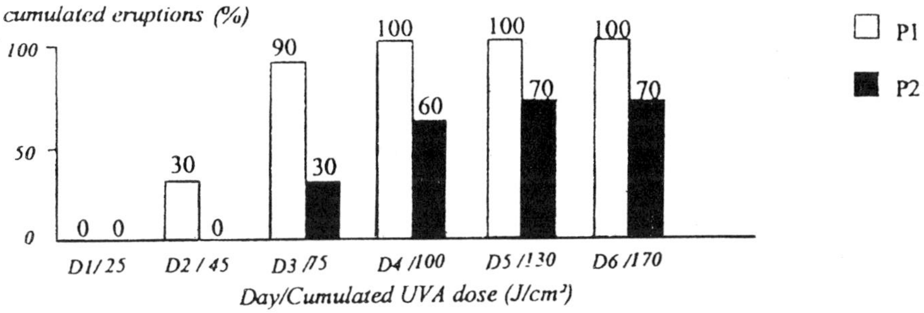

Figure 2 Percentages of PLE eruption in the sunlight of Guadelupe on 20 PLE-prone women: Comparison of two types of sunscreen products: P1 filtering solely in the UVB spectrum and P2 possessing a UVA PF of 4.

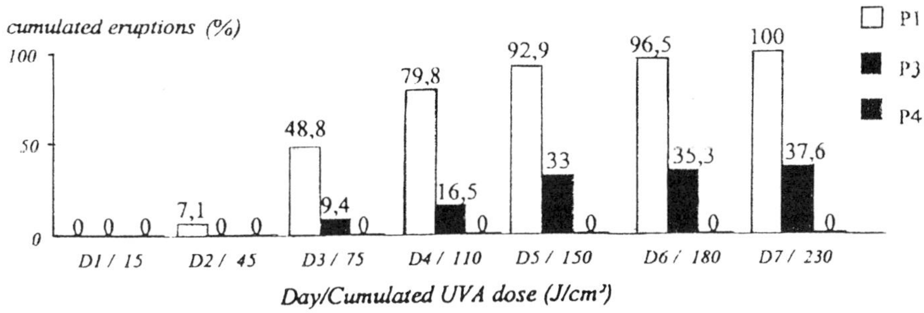

Figure 3 Percentages of PLE eruption in the sunlight of Tunisia on PLE-prone women: Comparison of three types of sunscreen products. P1 and P3 possess a UVA PF between 6 and 9, and P4 has a UVA PF higher than 10.

susceptibility to PLE in the sun when protected only by a P1 product was well known, a comparison was made per half-body of a product of class P3 PF between 6 and 9 and another of class P4 the UVA protection factor of which was higher than 10. The results obtained are shown in Figure 3.

Among the 43 subjects who received a P3-type product on one half-body, 27 cases of complete prevention were recorded and the other 16 had the onset of their PLE delayed or the severity who reduced relative to the condition observed using a sunscreen product filtering UVB only (type P1). In the 8 subjects who received a product of P4 type, no incidence of PLE was observed.

IV. CONCLUSION

The studies carried out gave evidence for the important role of UVA in the induction of PLE. The findings demonstrate that applying adequate sunscreen products may significantly prevent the onset of this photodermatosis. Maximal prevention was achieved by using sunscreen products providing a high protection throughout the UV spectrum and, in particular, the UVA spectrum. The best results were obtained by using a combination of two complementary UVA filters Parsol 1789 and Mexoryl SX; which allow a high level of UVA absorbance and fully prevents PLE.

REFERENCES

1. Morison, W. L. and Stern, R. S. Polymorphous light eruption: A common reaction uncommonly recognised. *Acta Derm. Venereol. (Stockh.)*, *62*:237–240 (1982).
2. Ros, A. and Wennersten, G. Current aspects of polymorphous light eruptions in Sweden. *Photodermatology*, *3*:298–302 (1986).
3. Pao, C., Norris, P. G., Corbett, M., and Hawk, J. L. M. Polymorphic light eruption: Prevalence in Australia and England, *Br. J. Dermatol.*, *130*:62–64 (1994).
4. Hölze, E., Plewig, G., Hofmann, C., and Røser-Maass E. Polymorphous light eruption: experimental reproduction of lesions. *J. Am. Acad. Dermatol.*, *7*:111–125 (1982).
5. Ortel, B., Tanew, A., Wolff, K., and Hönisgman, H. Polymorphous light eruption: Action spectrum and photoprotection. *J. Am. Acad. Dermatol.*, *14*:748–753 (1986).
6. McFadden, N. UVA sensitivity and topical photoprotection in polymorphous light eruption. *Photodermatology*, *1*:76–78 (1984).
7. Neumann, R. A., Pohl-Markl, H., and Knobler, R. M. Polymorphous light eruption: Experimental reproduction of skin lesions by whole body UVA irradiation. *Photodermatology*, *4*:252–256 (1987).
8. Miyamoto, C. Polymorphous light eruption: Successful reproduction of skin lesions, including papulovesicular light eruption, with ultraviolet B. *Photodermatology*, *6*: 69–79 (1989).
9. Farr, P. M. and Diffey, B. L. Adverse effects of sunscreens in photosensitive patients. *Lancet* *1*:429–431 (1989).
10. Gschnait, F., Hönigsmann, H., Brenner, W., Fritsch, P., and Wolff, K. Induction of UV light tolerance by PUVA in patients with polymorphous light eruption. *Br. J. Dermatol.*, *99*:293–295 (1978).
11. Murphy, G. M., Logan, R. A., Lovell, C. R., Morris, R. W., Hawk, J. L. M., and Magnus, I. A. Prophylactic PUVA and UVB therapy in polymorphic light eruption: A controlled trial. *Br. J. Dermatol.*, *116*:531–538 (1987).
12. Addo, H. A. and Snarma, S. C. UVB phototherapy and photochemotherapy (PUVA) in the treatment of polymorphic light eruption and solar urticaria. *Br. J. Dermatol.*, *116*:539–547 (1987).
13. Bisland, D., George, S. A., Gibbs, N. K., Aitchison, T., Johnson, B. E., and Ferguson, J. A comparison of narrow band phototherapy (TL-01) and photochemotherapy (PUVA) in the management of polymorphic light eruption. *Br. J. Dermatol.*, *129*:708–712 (1983).

14. Corbett, M. F., Hawk, J. L. M., Herxheimer, A., and Magnus, I. A. Controlled therapeutic trials in polymorphic light eruption. *Br. J. Dermatol.*, *107*:571–581 (1982).
15. Parrish, J. A., Levine, M. J., Morison, W. L., Gonzalez, E., and Fitzpatrick, T. B. Comparison of PUVA and beta-carotene in the treatment of polymorphous light eruption. *Br. J. Dermatol.*, *100*:187–191 (1979).
16. Murphy, G. M. Hydroxychloroquine in polymorphic light eruption: A controlled trial with drug and visual sensitivity monitoring. *Br. J. Dermatol.*, *116*:379–386 (1987).
17. U.S. Food and Drug Administration (FDA). Sunscreen drug products for over-the-counter human use; proposed safety, effective and labeling conditions. *Federal Register*, *43/166*:38206–38209 (1978).
18. Chardon, A., Moyal, D., and Hourseau, C. Persistent pigment darkening (PPD) skin response as a method for the UVA protection assessment of sunscreens. *Med. Staff*, *27*:2–5 (1995).

33

Ultraviolet Protection Offered by Clothing: An In Vitro and In Vivo Assessment of Clothing Fabrics

Nicholas J. Lowe
Skin Research Foundation of California, Santa Monica, and UCLA School of Medicine, Los Angeles, California

Teresa D. Bourget
Skin Research Foundation of California, Santa Monica, California

Shaun Hughes
Solumbra, Inc., Seattle, Washington

Robert M. Sayre
University of Tennessee Center for Health Sciences, Memphis, Tennessee

I. INTRODUCTION

Clothing has long been considered a tool in the arsenal of sun protective agents for the dermatological patient, but little guidance has been provided either for the practicing physician in his or her recommendations, or for the patient suffering sun sensitivity illnesses, about what constitutes proper protective clothing. In general, the medical advice has been that long-sleeved shirts, long pants, and wide-brimmed hats, all made of tightly woven fabrics offer sufficient sunlight protection (1,2). Furthermore, medical literature has often considered topical sunscreens to be superior to clothing for photoprotection.

Today, medical professionals and patients alike face a choice in determining proper clothing as a first-line sun protective tool. Historically, tightly woven natural-fiber summer clothing has been popular. Since the 1920s, the trend in clothing has been to lighter-weight, looser-woven fabrics. Summer clothing designs have emphasized light, airy appearances, with the tendency toward more exposure of the skin. As a result, individuals with normal responses to sunlight can experience phototrauma and photodamaging injuries even through clothing

(3). Special clothing is now available that claims to offer UV protection to photosensitive patients. This clothing, both by design and fabric construction, is intended to serve as a mechanical UV block, shielding the skin.

Little medical research has focused on sun protection offered by fabrics, clothing styles, or benefits of proper clothing for the patient with sun-sensitive conditions. The research that has been performed has often been purely in vitro estimates of a fabrics capabilities (4–6). Sayre and Hughes (7) used in vitro methods to analyze UV protection offered by normal summer-clothing fabrics and those that claim to offer sun protection. Other recent studies have appeared that stress clothing design and examine both in vitro test procedures and test fabrics on normal individuals (8–10). Strangely enough, although many papers urge the use of clothing, there has been only one published trial reporting benefits to one patient with xeroderma pigmentosum whose condition improved by wearing sun-protective garments (11).

Prolonged exposure to sunlight is a causal factor in the development of human skin cancer, in particular basal and squamous cell carcinoma. It is also a putative cause for the rapidly increasing risk of malignant melanoma. The rate of skin cancer in equatorial regions is seven to eight times that in the Arctic region, which correlates well with a greater ultraviolet (UV) irradiance in the equitorial regions. It is also affected by recreational and outdoor activities, daily sunlight exposure, and dressing habits. In tropical regions, clothing designs and fabrics are less likely to offer effective sun protection. This chapter reports research that shows the highly variable sun protectiveness of different articles of clothing intended for outdoor activities.

In addition to the skin cancer risk, there are individuals who have significant risk of a variety of photodermatoses. These can range from the relatively common problem of polymorphic light eruption, affecting approximately 11% of the population, to the more severe and persistent problem of chronic actinic dermatitis, which is less frequent, but can be severe and disabling when it occurs. Many of the current sunscreens are relatively inefficient at protecting individuals against these forms of photosensitivity.

II. METHODS AND MATERIAL

A. Transmittance Measurements

In vitro transmittance measurements were performed on selected fabrics and garments (7). The measurements were made using an Optronic model 752 spectroradiometer in transmittance mode (Fig. 1). Forward-scattered UV radiation (280–400 nm) was captured from radiation passing through the fabric and spectrally analyzed for both dry and dampened fabrics. The transmittance of three adjacent areas of the fabric was measured both when dry and after moistened with tap water.

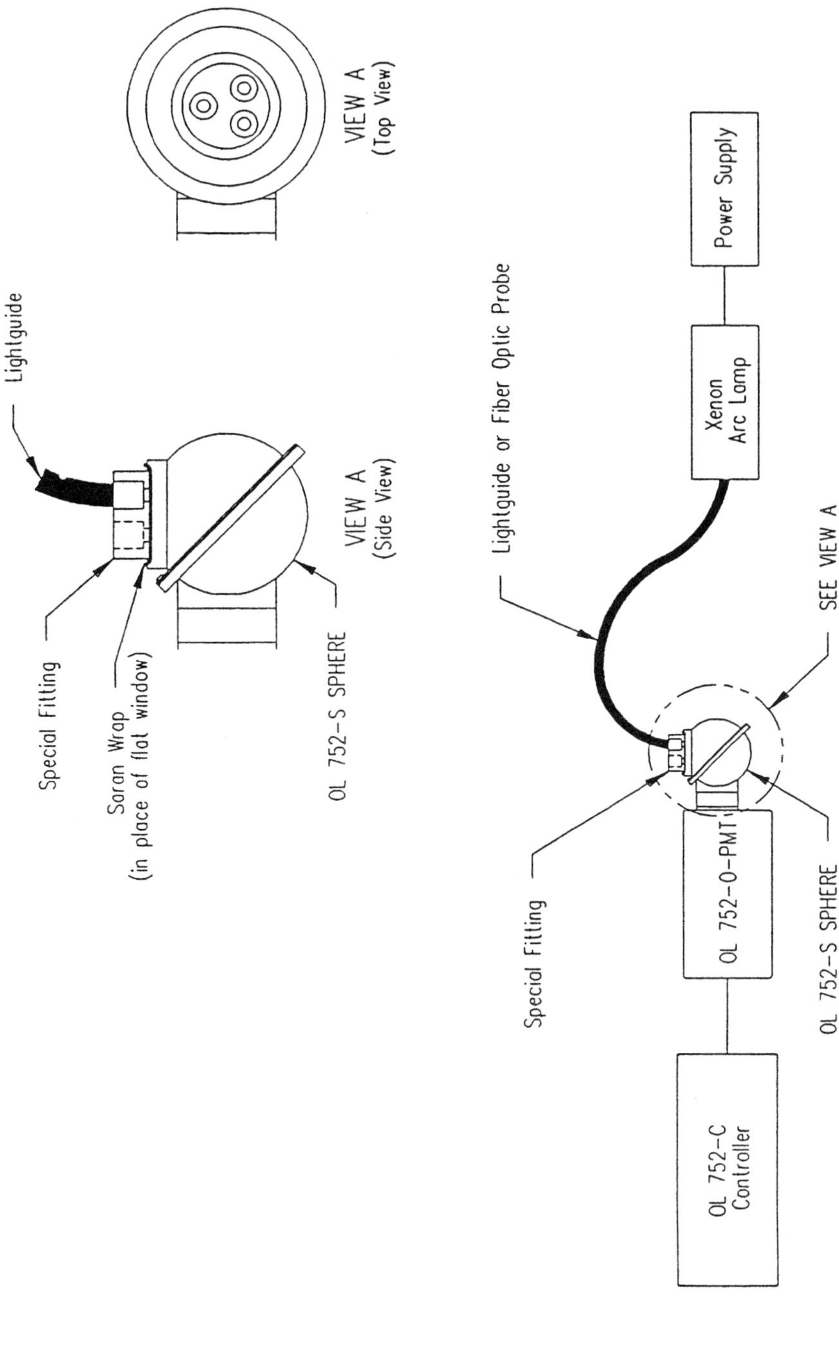

Figure 1 Instrumental setup to measure UV transmittance of fabrics.

B. Analysis of Transmittance Measurement

The transmittance measurements were analyzed to predict the sun protective factors (SPF) of each fabric wet and dry by mathematically integrating the transmittance, erythemal response spectrum, and a solar spectrum (Fig. 2). The *risk of injury* to a specific source spectrum for unprotected skin was simply the risk spectrum multiplied by the source spectrum:

Risk of erythema to unprotected skin:

$$\text{Unprotected risk} = \int_{\lambda\,=\,290\text{ nm}}^{400\text{ nm}} E_\lambda * S_\lambda \, d\lambda \tag{1}$$

where

E = CIE erythemal action spectrum
S = noonday summer sunlight spectrum
λ = wavelength of light in nanometers (nm)

Risk of erythema to fabric-protected skin:

$$\text{Fabric-protected risk} = \int_{\lambda\,=\,290\text{ nm}}^{400\text{ nm}} E_\lambda * S_\lambda * T\lambda \, d\lambda \tag{2}$$

where

E = CIE erythemal action spectrum
S = noonday summer sunlight spectrum
T = transmittance of fabric
λ = wavelength of light in nanometers (nm)

Estimation of the sun protection factor (SPF) fabric-protected skin

$$\text{Fabric SPF} = \frac{\displaystyle\int_{\lambda\,=\,290\text{ nm}}^{400\text{ nm}} E_\lambda * S_\lambda \, d\lambda}{\displaystyle\int_{\lambda\,=\,290\text{ nm}}^{400\text{ nm}} E_\lambda * S_\lambda * T_\lambda \, d\lambda} \tag{3}$$

where

E = CIE erythemal action spectrum
S = noonday summer sunlight spectrum

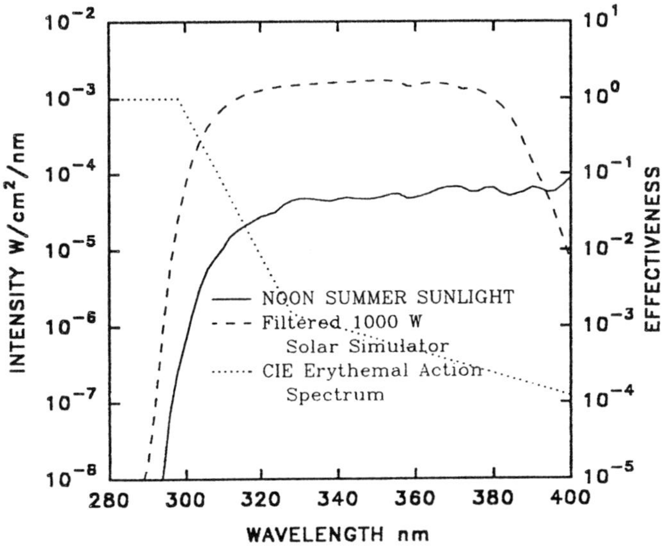

Figure 2 UV noonday summer sunlight spectrum from Albuquerque NM, July 3, 1986 (solid line). 1000-W xenon arc solar simulator filtered with a 1-mm Schott WG-320 filter (dashed line). CIE erythema risk spectrum (dotted line).

T = transmittance of fabric
λ = wavelength of light in nanometers (nm)

C. In Vivo Efficacy Validation

In vivo UV-testing methods were similar to those used in sunscreen SPF testing except a simple pass–fail method was adopted (14). Unlike topical sunscreens for which a given product can provide a range of protection depending on formula application and environmental conditions, patients who wear clothing should expect maximum protection all the time. A 1000 W xenon arc solar simulator was filtered to meet proposed spectral standards (see Fig. 2). The UV output was measured with an international light radiometer fitted with a sensor that absorbed at a peak of 311 nm. The unprotected control minimal erythema dose (MED) was determined for each volunteer by irradiating 25% incremental UV exposure close to the anticipated MED for the individual. The MED was determined 24 h after UV irradiation. A single exposure of 10, 15, or 31 MEDs was administered through the test clothing materials, based on their estimated in vitro protection. This avoided putting the volunteers at any risk of severe UV burns.

III. RESULTS

Any erythema of the skin protected by a fabric and exposed to UV radiation was considered an unacceptable injury and caused that fabric to fail the validation test. Such changes in skin color may be due to erythema, immediate and delayed pigmentation, or even vascular injury. For the sun-sensitive patient and persons at high risk for skin carcinogenesis, all injuries should be considered unacceptable.

A. Materials Tested

Table 1 shows the six fabrics and garments tested in the study. Clothing fabrics were selected based on three criteria: normal summer-clothing fabrics found in the wardrobe of a malignant melanoma patient or in widespread use in North America; fabrics claiming to offer sun protection without any numerical-labeling claims; and fabrics used in North America for clothing using numerical sun protection claims. Only clothing that has been reviewed under medical device regulations is permitted to make numerical fabric sun protection factor claims. Most normal summer-clothing fabrics are fabricated from cotton fibers. These

Table 1 Testing Results of UV Protection Afforded by Different Garments: In Vitro and In Vivo Fabric Test Results

Fabric	Garment	In vitro predicted SPF		In vivo human exposure tests	
		Dry	Wet	MED[a] exposures	Pass–fail[b]
1	White t-shirt	7.4	5.5	15	Fail
2	White polo shirt	8.7	6.8	15	Fail
3	California life-guard blue t-shirt	7.0	5.2	15	Fail
4	White gardening shirt	5.1	2.6	10	Fail
				15	Fail
5	Solumbra sun-protective shirt				
	White	58.6	55.6	31	Pass
	Blue	92.1	89.7	31	Pass
	Black	250.4	317.7	31	Pass
6	Solumbra II sun-protective shirt	204.8	242.8	31	Pass

[a]MED, minimal erythema dose.

[b]Pass, no visible erythema; Fail, erythema.

fibers are knitted or woven into a lightweight fabric. Shirts from leading manufacturers of sportswear or active wear were used, and are widely available in the market place.

One shirt was tested in different areas to test whether fabric UV protection was uniform over the entire garment. Most of these garments appeared to be tightly woven or knit. One long-sleeved T-shirt was allegedly chosen by a dermatologist to protect lifeguards from sun exposure while on the job. All garments had been laundered and worn. Knit fabrics have apertures that increase in size with tension; these fabrics were not stretched when tested.

The proprietary Solumbra SPF fabrics are constructed of man-made fibers and were woven. No coatings, which could affect long-term UV performance of the fabrics, have been applied to these fabrics.

Results from both in vitro and in vivo tests are found in Table 1. Calculated in vitro SPF predictions and pass–fail in vivo results are listed. Figures 3 and 4 show some of the in vitro transmittance measurements obtained. Specifically note in Figure 3 that both garments transmit more UV radiation when dampened than when dry.

Figure 4 compares the transmittance of a white Solumbra fabric both wet and dry with the transmittance of a proposed high-SPF 15 sunscreen standard. In this figure note the high UVA transmittance of the topical sunscreen. At wavelengths longer than 340 nm, this sunscreen formula provides almost negligible

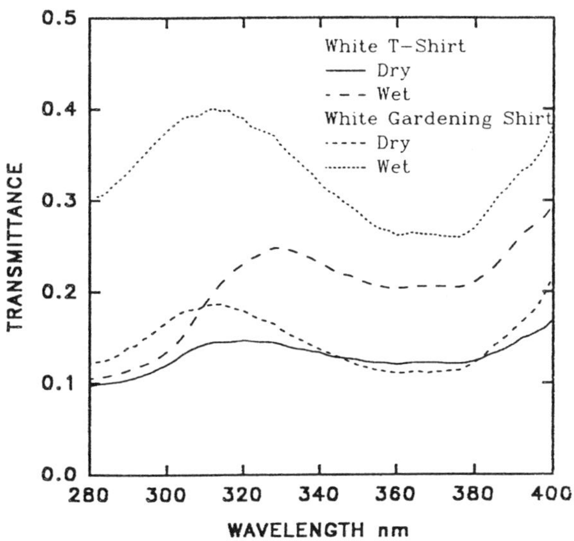

Figure 3 Transmittance spectrum of cotton T-shirts, wet and dry.

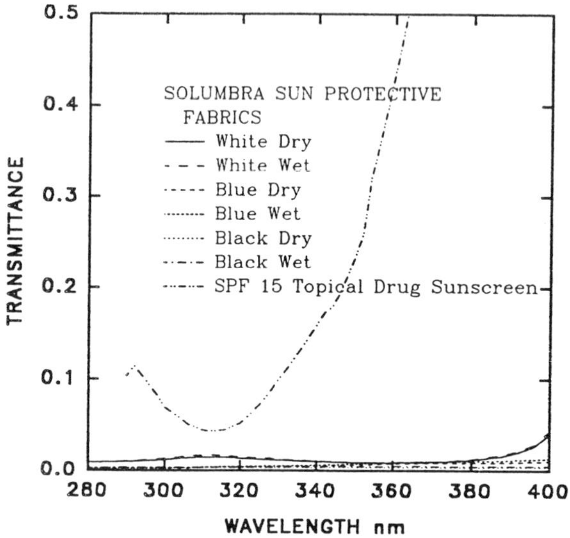

Figure 4 Transmittance spectrum of Solumbra white fabric. Also shows transmittance spectrum of SPF 15 standard sunscreen formula.

protection. On the other hand, the white Solumbra fabric exhibits almost complete protection throughout the UVA spectrum. Compare this level of UVA protection with the cotton shirt and T-shirts shown in Figure 3. Neither shirt in Figure 4 exhibits adequate UVA protection for dermatological patients.

Figure 5 shows the results of testing 24 h after exposure. In this figure, the results of control 1-, 2-, and 3-MED exposures are shown. The lower left sample receiving 15 MEDs through a garment, shows that the fabric failed to protect at this level. In high-intensity sunlight this volunteer would have been injured by even a lower exposure. The lower right sample (Solumbra I) shows a fabric that is effective after a 31-MED exposure. Notice the absence of indication that exposure has occurred.

IV. DISCUSSION

Our research has focused on UV protection offered both by normal summer clothing fabrics and new fabrics specifically designed to offer UV protection. Both in vivo and in vitro testing methods were used to measure UV protection. Results from these two methods can be compared. A comparison of in vitro SPF predications with in vivo validation of performance indicates consistent responses. Fabrics that were shown through in vitro testing to be ineffective in providing SPF 15 protection, failed to protect against erythema in the in vivo validation tests. Fabrics that were shown through in vitro testing to be effective in providing SPF 31+ protection protected against erythema in the in vivo validation tests.

Only Menzies and associates' recent report (10) compares in vitro SPF predictions for fabrics with in vivo clinical testing. In that study they report poor agreement between in vitro protection predictions and clinical tests: The in vitro predictions consistently overestimated the protection provided. In our study, the in vitro predictions were consistent with the results of the in vivo clinical tests reported. We used different equipment in making in vitro measurements than the equipment reportedly used by Menzies. The results show that our equipment detected erythemic radiation passing through the fabrics, and they correlate the in vitro with the in vivo human skin responses to UV radiation.

We have found that a summer clothing fabric's opaqueness to visible light may have little relation to its ability to block UV radiation, although a loose weave may allow more visible and UV light through. A high level of UV radiation may still be transmitted through opaque cotton fabrics. An assortment of these fabrics appears to offer inadequate (less than 15 SPF) sun protection for UV-sensitive patients. Physicians should be aware that these shirts tested were selected for use in high UV exposure activities or were marketed under UV protective claims.

The fabrics specifically designed for sun protective clothing offered high UV protection. Neither dampness nor color seemed to degrade UV protection. In fact, the darker-colored fabrics provide a greater protection. But the levels of UV protection provided by the white Solumbra fabrics is so high that most dermatological patients would be completely protected during normal activities.

It must be remembered from the transmittance of the SPF 15 sunscreen standard that even poorly protective clothing provides broader and better UVA protection. Patients need to be warned that, although clothing can help protect their bodies, non–clothing-protected body areas, even when protected by the "best" topical sunscreens, cannot tolerate extended exposures to sunlight. Furthermore, the broad protective capabilities of fabrics exceed the UVA protective capabilities of all topical sunscreen products. A highly UV-sensitive dermatological patient is provided high, broad-spectrum UV protection by sun-protective apparel, but still should consider limiting exposure of nonclothed skin because topical products are unable to provide equivalent protection.

V. CONCLUSIONS

New fabrics provide sun-sensitive patients with high-SPF clothing which also provides broad-spectrum UVA and UVB protection. Summer-weight cotton clothing sometimes provides minimal protection and cannot be recommended for light-sensitive patients or for high-exposure activities. Our research indicates that many summer-weight fabrics when wet provide less protection. Fabric protection, as well as clothing design, is important in providing UV protection to patients. Broad-spectrum high UV-protective fabrics may be useful for patients who have experienced both acute and chronic photodamage. The superior UV protection may be especially important to individuals with lupus, porphyrias, vitiligo, albinism, sun allergies, and most importantly, a history of skin cancers. Individuals taking medications that make them photosensitive could also benefit from UV-protective apparel.

ACKNOWLEDGMENT OF CONFLICT OF INTEREST

Nicholas J. Lowe, M.D. and Teresa Bourget, B.S., received a research grant from Solumbra, Inc. to the Skin Research Foundation of California. Robert Sayre, Ph.D. received a research grant from Solumbra, Inc. Shaun Hughes is the founder and President of Solumbra, Inc., Seattle, Washington.

REFERENCES

1. Berne, B. and Fisher, T. Protective effects of various types of clothing against UV radiation. *Arch. Dermatol.*, *60*:459–460 (1980).
2. Keeling, J. H., Draus, E. W., Pathak, M., et al. Design and protection from ultraviolet radiation. *Mil. Med.*, *154*:250–255 (1989).
3. Ream, G. P. and Devillez, R. L. An unusual case of suntanning. *Arch. Dermatol.*, *114*:287 (1978).
4. Welsh, C. and Diffey, B. The protection against solar actinic radiation afforded by common clothing fabrics. *Clin. Exp. Dermatol.*, *6*:577–582 (1981).
5. Robson, J. and Diffey, B. Textiles and sun protection. *Photodermatol. Photoimmunol. Photomed.*, *7*:32–34 (1990).
6. Sliney, D. H., Benton, R. E., Cole, H. M., et al. Transmission of potentially hazardous actinic ultraviolet radiation through fabrics. *Appl. Ind. Hyg.*, *2*:36–44 (1987).
7. Sayre, R. M. and Hughes, S. Sun protective apparel: Advancements in sun protection. *Skin Cancer*, *8*:41–47 (1993).
8. Diffey, L. and Cheesman, J. Sun protection with hats. *Br. J. Dermatol.*, *127*:10–12 (1992).
9. Jevtic, A. P. The sun protective effect of clothing, including beachwear. *Aust. J. Dermatol.*, *31*:5–7 (1990).
10. Menzies, S. W., Lukins, P. B., Greenoak, G. E., et al. A comparative study of fabric protection against ultraviolet-induced erythema determined by spectrophotometric and human skin measurements. *Photodermatol. Photoimmunol. Photomed.*, *8*:157–163 (1992).

11. Bech-Thomsen, N., Wulf, H. C., and Ullman, S. Xeroderma pigmentosum lesions related to ultraviolet transmittance by clothes. *J. Am. Acad. Dermatol.*, *24*:365–368 (1991).
12. McKinlay, A. F. and Diffey, B. L. A reference action spectrum for ultraviolet induced erythema in human skin. *Human Exposure to Ultraviolet Radiation: Risks and Regulations* (W. F. Passchier and B. F. M. Bosnjakovic, eds.), Ecerpta Medica, Amsterdam, 1987, pp. 83–87.
13. Sayre, R. M., Cole, C., Billhimer, W., et al. Spectral comparison between solar simulators and sunlight. *Photodermatol. Photoimmunol. Photomed.*, 7:153–165 (1990).
14. Food and Drug Administration. Sunscreen drug products for over-the-counter human use: tentative final monographs; proposed rule. *Federal Register*, *58*(90): 28193–28301 (1993). 21 CFR 352 et al.

34

Contact Sensitization and Photocontact Sensitization of Sunscreening Agents

Jens O. Funk
Fred Hutchinson Cancer Research Center, Seattle, Washington

Sydney H. Dromgoole
Oclassen Pharmaceuticals, Inc., San Rafael, California

Howard I. Maibach
University of California Medical Center, San Francisco, California

I. INTRODUCTION

Sunscreens can be classified into two major types: chemical and physical sunscreens (1). Physical sunscreens, such as titanium dioxide and zinc oxide, reduce the amount of light penetrating the skin by creating a physical barrier that reflects, scatters, or physically blocks the ultraviolet light reaching the skin surface. Chemical sunscreens reduce the amount of light reaching the stratum corneum by absorbing the radiation. Examples of chemical sunscreens include *p*-aminobenzoic acid (PABA) and PABA derivatives such as padimate O, cinnamates, benzophenones, salicylate derivatives, and dibenzoylmethane derivatives.

Because chemical sunscreens are applied topically to the skin in relatively high concentrations (up to 26%), contact sensitization can occur. Similarly, because these chemicals absorb radiation, they have the potential to cause photosensitization. Both types of sensitization can occur not only with the various sunscreening agents, but also with excipients, such as emulsifiers, antioxidants, and preservatives, that are included in the various hydroalcoholic lotions, ointments, and oil-in-water emulsions.

Modified with permission from the *Journal of the American Academy of Dermatology*, Mosby Year Book, 1990.

Despite extensive sunscreen use, infrequent reports of sunscreen-induced adverse effects have been published, including allergic or photoallergic reactions. Because of the lack of a well-developed adverse reaction reporting system, we have inadequate data to accurately predict the degree of hypersensitivity to sunscreening agents.

This chapter reviews the reports of contact sensitization and photocontact sensitization induced by various sunscreening agents currently in use in the United States. Strategies for patch and photopatch testing, as well as immediate-type testing, are discussed.

II. LITERATURE REPORTS

Published reports of contact and photocontact sensitization and contact urticaria induced by sunscreening agents are listed in Table 1. Representatives of all major

Table 1 Contact and Photocontact Sensitization to Sunscreening Agents

Chemical	Number of subjects with contact dermatitis	Number of subjects with photocontact dermatitis	Total
p-Amino benzoic acid (PABA)	60	63	123
Isopropyl-methoxydibenzoylmethane (Eusolex 8020)[a]	34	18	52
Oxybenzone (Eusolex 4360)	12	30	42
Glyceryl aminobenzoate (glyceryl PABA)	30	7	37
Amyl-*p*-dimethylaminobenzoate (padimate A)	12	16	28
Octyl-dimethyl PABA (padimate O)	19	3	22
3-(4-Methyl-benzylidene)camphor (Eusolex 6300)[a]	17	5	22
Ethoxyethylmethoxycinnamate (Cinoxate)	6	12	18
Butylmethoxydibenzoylmethane (avobenzone)[b]	7	11	18
Hydroxymethoxy methylbenzophenone (mexenone)[a]	9	1	10
2-Phenylbenzimidazole-5-sulfonic acid	5	3	8
Sulisobenzone	3	2	5
Homosalate	2	2	4
Dioxybenzone	2	0	2
Bornelone	2	0	2
4-Isopropyldibenzoylmethane	2	0	2
Digalloyl trioleate	1	0	1

[a]Not an FDA-approved sunscreening agent.
[b]Avobenzone is the U.S. Adopted Name (USAN) of this chemical. It is also known as Parsol 1789, the Cosmetics, Toiletry and Fragrances Association (CTFA) name.
Source: Refs. 2–62.

sunscreen categories, including PABA derivatives, anthranilates, salicylates, cinnamates, and benzophenones, have caused allergic reactions (2–62).

A. *p*-Aminobenzoic Acid

In 1976, Willis (63) suggested that the sensitization potential of PABA was minimal. Wennersten et al. (7) reported that a total of 78 of 1883 (3.9%) subjects tested with 5% PABA in alcohol in the Scandinavian Standard Photopatch Tray had either allergic or photoallergic responses to PABA. The use of PABA as a sunscreening agent in Europe and the United States has decreased significantly in recent years. PABA has been replaced by ester derivatives, such as padimate O that, unlike PABA, are not water-soluble and tend to remain on the surface layer, with less than 10% penetrating the corneum even after 24 h (47). These PABA esters appear to be less sensitizing than PABA; however, this impression remains to be proved.

B. *p*-Aminobenzoic Acid Derivatives

Sensitization to glyceryl PABA was first reported by Sulzberger et al. (64), and during the next 30 years additional reports appeared (3,15–19,21,50,60). Many of the cases of glyceryl PABA sensitization showed uniformly strong reactions to benzocaine, suggesting that the sensitization may be due to the presence of impurities in the glyceryl PABA. This suggestion was first made by Fisher (19) and has since been confirmed (20,22,42,65,66). Benzocaine impurities (1–18%) occurred in many commercial sources of glyceryl PABA. Thus, many of the early reports of contact allergy to glyceryl PABA may have falsely implicated glyceryl PABA as the sensitizer. Thune (8) reported two cases of allergic or photoallergic reactions to glyceryl PABA in which there was no reaction to benzocaine, suggesting true allergy to the PABA derivative. However, no allergic responses were observed when these subjects were patched with glyceryl PABA that had been purified by high-pressure liquid chromatography (HPLC) (66). This suggests the presence of an, as yet, unknown impurity (impurities) other than benzocaine as the sensitization source. The glyceryl PABA scenario underscores the importance of using purified raw materials in the manufacture of consumer products, such as sunscreening agents, and the need for careful interpretation of patch test results.

Other PABA derivatives that have caused sensitization or photocontact sensitization include octyl dimethyl PABA (padimate O), amyl dimethyl PABA (padimate A), and ethyl dihydroxy PABA. The number of case reports of sensitization or photocontact sensitization with padimate O is fewer than that reported with PABA and glyceryl PABA, suggesting a lower sensitization potential with this derivative. This may be because padimate O is not a true PABA ester, because it does not contain the NH_2 grouping present in glyceryl PABA, PABA, and benzocaine (20).

Padimate A also lacks the NH_2 grouping and, although it was included in the proposed sunscreen monograph (1) as a safe and effective sunscreening agent, this derivative can cause phototoxicity (45,67). It may have accounted for the erythemal response observed by Katz 30 min after sun exposure (46). This compound is no longer used in sunscreens in the United States. In addition to finding benzocaine impurities in the glyceryl PABA raw materials, Bruze et al. (65) found that some PABA esters contain 0.2–4.5% PABA. Therefore, it is possible that these PABA impurities may account for some of the reports of sensitization to the PABA derivatives.

C. Salicylates

There are only two cases of contact allergy and two reports of photocontact allergy to homomenthyl salicylate in the literature (10,25) and no reports of sensitization to octyl salicylate, the major salicylate derivative incorporated in many sunscreens.

D. Cinnamates

Cinnamates are chemically related to or are found in balsam of Peru, balsam of Tolu, coca leaves, cinnamic acid, cinnamic aldehyde, and cinnamon oil, ingredients used in perfumes, topical medications, cosmetics, and flavoring. Thune reported eight cases of sensitivity to cinnamates, two cases of photoallergy to 2-ethoxyethyl-*p*-cinnamate, and six subjects with contact allergy to other cinnamates, such as amyl cinnamaldehyde, amyl cinnamic acid, and cinnamon oil (8). The final report of the Scandinavian multicenter photopatch study 1980–1985 contains one case of contact allergy and two cases of photocontact allergy to cinnoxate (60). Cross-sensitization among cinnamon derivatives has been reported (68).

E. Benzophenones

There have been 30 reported cases of photocontact allergy (8,10,13,38,43,44, 58,60,62) and 12 cases of contact allergy (9,25,58,60,62) to oxybenzone, and three reports of contact allergy (5,29) and two reports of photocontact allergy to sulisobenzone (10). Benzophenone-10 (mexenone), a benzophenone derivative not used in sunscreens in the United States, can also cause contact and photocontact dermatitis (8,30,36,49).

F. Dibenzoylmethanes

Dibenzoylmethane derivatives, such as isopropyldibenzoylmethane (Eusolex 8020) and butyl dibenzoylmethane (avobenzone) have been incorporated in European sunscreens as UVA absorbers since 1980 (67). Instances of contact allergy or photoallergy to the dibenzoylmethanes, or sunscreens and lipsticks containing these derivatives, have been reported, although most of these reports have been associated with the isopropyl derivative (11,27,28,31–33,51,52). As a consequence of these adverse reactions to Eusolex 8020, several manufacturers of products that contained this ingredient have stopped incorporating this derivative into their products (51,52). The manufacturers of Eusolex 8020 have voluntarily withdrawn this UV filter from the market (69,70).

There have been fewer reports of contact allergy or photoallergy to the butyl dibenzoylmethane derivative, avobenzone (11,27,28,30,31). It is possible that some of these reactions to avobenzone may have been cross-reactions resulting from prior exposure to the isopropyl derivative (27,28,30). Greater use of these compounds with appropriate testing should help clarify their relative sensitization potential.

G. Camphor Derivatives

3-(4-Methylbenzylidene) camphor (Eusolex 6300) is a sunscreening agent used extensively in Europe, often in combination with Eusolex 8020, but it is not approved for use in the United States. There have been several reports of allergic and photoallergic reactions to sunscreens containing this agent (27,30,32, 35,37,52,60).

H. Miscellaneous

Other chemical sunscreens that have caused allergic reactions include diagalloyl trioleate (12), the glycerol ester of *o*-amino-*m*-(2,3 dihydroxyproxy)benzoic acid (71), a dioxane derivative (6), and 2-phenyl-5-methylbenzoazol (witisol) (72). None of these ingredients are approved for use in sunscreens in the United States.

I. Titanium Dioxide

Physical blockers such as titanium dioxide and zinc oxide have the advantage of not being sensitizers, but such agents may be so occlusive that they can cause miliaria (73). Kaminester reported that the inclusion of titanium dioxide in a PABA sunscreen blocked the appearance of photoallergy (74). It is possible that the reflection and scattering of light by titanium dioxide reduced the amount of UV light that penetrated the skin and elicited the photoallergic response.

J. Excipients

Note that contact allergy can also be caused by excipients included in the formulations. These chemicals include mineral oil, petrolatum, isopropyl esters, lanolin derivatives, aliphatic alcohols, triglycerides, fatty acids, waxes, propylene glycol, emulsifiers, thickeners, preservatives, and fragrances. Table 2 lists some of the excipients used in sunscreen formulations that have caused allergic reactions (75–87). A more extensive list of vehicle constituents in cosmetics that can cause allergic responses has been published (75). De Groot et al. (85) indicated that preservatives, fragrances, and emulsifiers are the main classes of ingredients responsible for cosmetic allergy, with Kathon CG (a mixture of methylisothiazolinone–methylchloroisothiazolinone) producing contact allergic reactions in 27.7% of subjects tested. In view of the potential to cause sensitization, the Cosmetic Ingredient Review (CIR) Expert Panel in the United States has recently requested additional patch testing of Kathon CG (83).

Many sunscreens available in the United States provide a complete list of ingredients including the excipients. The listing of all ingredients in sunscreens should be encouraged so that consumers, especially those with known sensitivities to chemicals, or patients suffering from light sensitivity disorders (88,89)

Table 2 Excipients That Can Cause Contact or Photocontact Dermatitis

Avocado oil
t-Butyl alcohol
Methylparaben
Phenyl dimethicone
Solvent red 1
Solvent red 3
Triethanolamine stearate
Benzyl alcohol
Cetylstearyl alcohol
Sorbitan sesquiolate
Imidazolidinyl urea (Germall 115)
Methylisothiazolinone–methylchloroisothiazolinone (Kathon CG)
Glyceryl monostearate
6-Acetoxy-2,4-dimethyl-*m*-dioxane
Carbowaxes
Ethyl alcohol
Glycerol
Isopropyl alcohol
Isopropyl myristate
Petrolatum
Stearyl alcohol

Source: Refs. 73–87.

are fully informed about the composition of the formulation before the purchase and application of the product to the skin.

III. CROSS-SENSITIZATION REACTIONS

Cross-sensitization reactions are defined as additional allergic responses that occur in one or more compounds (90,91). The allergen causing the primary reaction is the primary sensitizer, whereas secondary allergens are those compounds that can cause cross-sensitization. In general, these compounds have similar chemical structures or are metabolized to chemical structures similar to the primary allergen. However, in many instances, it is difficult to determine which allergen is the primary allergen.

Cross-sensitization reactions with other sunscreens as well as with chemicals with similar chemical structures have been reported, but are uncommon. Table 3 summarizes the published reports of cross-sensitization among the sunscreening agents. Many sunscreens containing PABA or PABA derivatives in the United States include a statement in the labeling that the use of the product is contraindicated in subjects who have known sensitivities to *p*-phenyldiamine, aniline, procaine, benzocaine, sulfonamides, and several azo dyes. These contraindications were based on the early studies by Meltzer and Baer (16), and Curtis and Crawford (17) with glyceryl PABA. It is established by Fisher (19,20,35) and by Bruze et al. (65,66) that benzocaine, and possibly other impurities, were responsible for some of the reports of cross-sensitization associated with glyceryl PABA. Therefore, it is possible that some of the other reported cross-sensitizations to glyceryl PABA may also have been due to cross-reactions to benzocaine and not to glyceryl PABA. This possibility underscores the need to ensure that the reported sensitization of various chemicals is due to the chemical and not to impurities. This is important for two reasons (66): (1) the subjects may be advised to avoid the wrong substances; (2) effective and safe compounds may be prematurely withdrawn from the market place.

Note that the number of cross-sensitizations to octyl dimethyl PABA (padimate O), the most common and widely used UVB absorber in sunscreens, is small. Camarasa and Serra-Baldrich (9) reported one subject sensitive to PABA who also reacted to padimate O and oxybenzone, whereas one subject sensitive to padimate O also reacted to PABA. Rietschel and Lewis (25) reported that two subjects who had contact dermatitis to homosalate were not sensitive to PABA, menthyl anthranilate, cinnamates, or PABA esters. Fisher noted that four patients with sensitivity to benzocaine and glyceryl PABA did not show cross-reactions to padimate O or padimate A (20), whereas Thune (8) reported that cross-sensitization did not occur between padimate O and padimate A, but all subjects sensitive to the PABA derivatives were also sensitive to PABA.

Table 3 Cross-Sensitization to Other Chemicals and Sunscreens

Sunscreen	Cross-sensitizer(s)	Refs.
p-Aminobenzoic acid (PABA)	Padimate O, oxybenzone	4, 9
	Glyceryl PABA	3, 23
	Sulfonamides	64
	p-Phenylenediamine (PPD)	8, 89
	Benzocaine	3, 33
	Padimate A	3, 8
	Padimate O	3, 8, 9
	Glyceryl PABA	3
Cinoxate	Cinnamaldehyde	6
	p-Methoxyisoamylcinnamate	6
	Balsam of Peru	4
	Benzylcinnamate	4
	Methylcinnamate	4
	Cinnamyl alcohol	4
Oxybenzone	Dioxybenzone	13, 14
	p-Phenylenediamine	44
Padimate O	*p*-Aminobenzoic acid	3, 8, 9
	Cinoxate	33
	Isobutyl PABA	47
Padimate A	*p*-Aminobenzoic acid	8
Isopropyl-methoxydibenzoylmethane (Eusolex 8020)	*t*-Butyl-methoxydibenzoylmethane (avobenzone)	27, 28, 30, 31
Glyceryl PABA	Sulfonamides	15, 16, 50
	Benzocaine	16, 17, 19–21, 42
	Procaine	15, 19, 20
	Aniline	15, 17, 19
	p-Diphenyldiamine	15, 17, 19, 20
	PABA	8, 15, 17, 19, 20
	Picric acid	16
	Azodye A	16
	Butyl aminobenzoate (butesin)	16, 17
	Saccharin	16
Sulisobenzone	Oxybenzone	29

It is of paramount clinical relevance to know if PABA and its esters cross-react in either allergic or photoallergic dermatitis. Adequate quantitative data are needed; it appears that true cross-reactions (not caused by contaminants) do not routinely occur. Patch and photopatch testing with each of the sunscreens could be used to define this problem efficiently. Even if such testing is negative, we suggest cautious application to an exposed noncosmetic area for 5 days before widespread application.

Cross-sensitization to the cinnamate derivative Cinoxate includes cinnamic aldehyde, *p*-methoxyisoamylcinnamate, balsam of Peru, benzylcinnamate, cinnamyl alcohol, and methylcinnamate (4,6).

De Groot et al. (30,31) reported on the cross-sensitization among the dibenzoylmethane derivatives used in European sunscreens. Several subjects sensitive to Eusolex 8020 were also sensitive to avobenzone. Schauder and Ippen (28) have reported that, in Europe, (photo)allergic reactions to Eusolex 8020 are more frequent than to the *t*-butyl derivative, avobenzone. The authors also noted that there were no subjects who reacted to avobenzone but not to Eusolex 8020, whereas the reverse situation is not infrequently observed.

Contact dermatitis and photoallergic contact dermatitis to different sunscreens can occur simultaneously (28). This phenomenon can be explained by a combination of cross-reaction and coupling allergy.

IV. SUNSCREEN SENSITIZATION IN PHOTODERMATOSES

MacLeod and Frain-Bell have emphasized the importance of using a sunscreen that is appropriate for the action spectrum of the patients' photodermatoses (92). However, there is increasing evidence that some subjects with photodermatoses, such as chronic actinic dermatitis, polymorphous light eruption, or persistent light reactions, not only have increased sensitivity to light, but also increased sensitivity to chemicals (83–85,88,89). Thune and Eeg-Larsen (93) found that of 18 males with persistent light reactivity, 17 showed allergic reactions. Of these, 12 were positive photopatch test reactions and 11 were contact allergic reactions. Four of these subjects had positive patch tests to sunscreens: 2 to PABA, 1 to oxybenzone, and 1 to padimate A. In addition, 10 patients with polymorphous light eruption exhibited positive photopatch or contact allergic reactions. Two subjects exhibited photosensitivity to PABA and 1 subject was sensitive to two cinnamate derivatives.

Many of the reports included in Table 1 involved subjects with various photosensitivities. These photosensitivities are listed in Table 4. Photosensitive persons who require protection from the sun need to be patch and photopatch tested to the various sunscreening agents to determine the specific sensitizers or photosensitizers. Once the offending agents have been identified, the person can

Table 4 Photodermatoses That May Be Aggravated by Sunscreening Agents

Atopic dermatitis
Lichen rubeo planus
Polymorphous light eruption (PMLE)
Chronic actinic dermatitis (chronic photosensitive eczema, actinic reticuloid)
Xeroderma pigmentosa (XP)
Persistent light reaction (PLR)
Herpes labialis
Musk ambrette photosensitivity
Idiopathic photosensitivities
Lupus erythematosus (LE)
Phytophotodermatitis
Seborrheic dermatitis

Source: Refs. 2, 4, 5, 7, 8, 14, 16, 17, 27, 28, 30, 31, 33, 36, 41, 48, 52, 93.

eliminate the sunscreens that contain these chemicals, and their presumably endogenous photosensitivity may be easier to manage (96).

Some persons may not be sensitive to the active sunscreening agents, but to one or more of the many excipients included in sunscreen formulations. In these cases, the selection of a suitable sunscreen is more difficult, and customized sunscreens eliminating the sensitizing compounds may be necessary.

V. SUNSCREENS IN THE UNITED STATES

Numerous sunscreening preparations are currently sold in the United States. Only 14 of the 21 sunscreen agents included in the proposed sunscreen monograph are commonly used in sunscreens in the United States (Table 5). One additional sunscreening agent used in sunscreens that is not included in the proposed FDA sunscreen monograph is *t*-butyl dimethoxydibenzoylmethane (avobenzone). This chemical has been used extensively in sunscreens in Europe for several years. Zinc oxide, included as an inactive ingredient in the sunscreen monograph, is commonly used as a sunscreening agent.

Table 6 lists examples of the sunscreen formulations sold in the United States together with the active ingredients and their sun protection factors (SPF), a measure of sunscreen protection against sunburn. Most of the formulations that are combination sunscreens contain one or more UVB absorbers and an UVA absorber to provide protection against the damaging effects of UVA. Several sunscreens also include physical blockers, such as titanium dioxide.

Table 5 Approved Sunscreening Agents Used in United States

Aminobenzoic acide (PABA)
2-Ethoxyethyl-*p*-methoxycinnamate (Cinoxate)
Dioxybenzone
Ethyl 4-[bis(hydroxypropyl)]aminobenzoate
2-Ethylhexyl 2-cyano-3,3-diphenylacrylate (octocrylene)
Ethylhexyl *p*-methoxycinnamate (Parsol MCX)
Octyl salicylate
Glyceryl PABA
Homomenthyl salicylate (homosalate)
Menthyl anthranilate
Oxybenzone
Octyl dimethyl PABA (padimate O)
2-Phenylbenzimidazole-5-sulfonic acid
Titanium dioxide
t-Butyl dimethoxydibenzoylmethane (avobenzone)[a]

[a]Not included in the proposed sunscreen monograph (1).
Source: Ref. 1.

In general, as the SPF of the sunscreen formulation increases, the number of active ingredients increases to three or four, and occasionally the total amount of active sunscreens increases up to 26%. Not all formulations list the specific concentrations of the active ingredients, and it is possible that some may contain higher concentrations. As with many chemicals, increasing the concentrations of the active ingredients may increase the likelihood of sensitization (14). The majority of sunscreens contain octyl dimethyl PABA (padimate O) as the main UVB absorber. The cinnamate derivative, octyl methoxycinnamate (Parsol MCX), is also commonly used as a UVB absorber.

Most sunscreens with more than one ingredient contain oxybenzone as the additional ingredient. The absorption peak of this compound lies in the UVB region and extends partially into the UVA. Other agents that absorb in the UVA region include sulisobenzone, dioxybenzone, menthyl anthranilate, and avobenzone. The latter chemical, which has an absorption maximum in the middle of the UVA region (358 nm), has been approved in the United States in one sunscreen formulation (Photoplex, Herbert Laboratories, Santa Ana CA) (see Table 6).

A meticulous and thoughtful review of sunscreen composition of virtually all available formulations was published by Schauder (97). Although in German, the tables are self-explanatory.

Table 6 Selected Sunscreen Formulations Available in the United States

Trade name	SPF	Active ingredients
Four sunscreening ingredients		
Sundown-30 (Johnson & Johnson)	30	Parsol MCX, octyl salicylate, oxybenzone, titanium dioxide
Cancer Garde-33 (Eclipse Labs)	33	Padimate O, Parsol MCX, oxybenzone, titanium dioxide
Three sunscreening ingredients		
Solar Plus 15 (Person & Covey)	15	Padimate O, dioxybenzone, oxybenzone
Bain de Soleil (Bain de Soleil)	15	Octylmethyl PABA, oxybenzone, dioxybenzone
Sundown-15 (Johnson & Johnson)	15	Padimate O, Parsol MCX, oxybenzone
Total Eclipse-15 (Eclipse Labs)	15	Padimate O, octyl salicylate, oxybenzone
Pre Sun 15/29 (Westwood-Squibb)	15/29	Parsol MCX, octyl salicylate, oxybenzone
Solbar-50 (Person & Covey)	50	Parsol MCX, octocrylene, oxybenzone
Two sunscreening ingredients		
Nivea Visage (Beiersdorf)	5	Octyl methoxycinnamate, oxybenzone
Sundown Extra/Maximal/Ultra Protection	6/8/15	Padimate O, oxybenzone
Maxafil (Genderm)	6–8	Cinoxate, menthyl anthranilate
A-Fil (Genderm)	8–15	Menthyl anthranilate, titanium dioxide
Solbar PF 15 (Person & Covey)	15	Octyl methoxycinnamate, oxybenzone
Solar SPF 15 (Doak)	15	Aminobenzoic acid, titanium dioxide
Ti-Screen (Pharmaceuticals)	15	Octyl methoxycinnamate, oxybenzone
PreSun 15 Facial Sunscreen (Westwood-Squibb)	15	Padimate O, oxybenzone
Photoplex (Herbert Labs)	15	Padimate O, avobenzone
One sunscreening ingredient		
Eclipse Gel (Eclipse Labs)	11	Octyldimethyl PABA
Pabagel, Pabanol	12/13	PABA
PreSun Gel, PreSun Lotion (Westwood-Squibb)	16/19	PABA
Supershade by Coppertone (Plough)	20	PABA

VI. PATCH AND PHOTOPATCH TESTING: CONTACT VERSUS PHOTOCONTACT DERMATITIS

The criteria for separating allergic contact and allergic photocontact dermatitis using patch-testing techniques are imprecise. General criteria and their interpretation are listed in Table 7. Often, the results are not all or none, as implied in the table. Frequently, there is a difference in the response intensity, with either the contact or photocontact response being greater. Infrequently, serial dilutions are performed with either the putative antigen or the amount of ultraviolet light employed. Until a significant number of patients are studied in this way, it will be unclear how many of them represent contact versus photocontact sensitization.

A. Light Source

Wennersten et al. (38) recommend that patients with suspected photocontact allergy be phototested before implementation of patch testing. The aim of this preliminary light testing is to detect any abnormal sensitivity to UVA and UVB wavebands. It is generally agreed that UVA sources are adequate and sufficient to elicit responses, an important convenience as UVA does not produce erythema in normal fair-skinned persons until a dose of 20–30 J/cm^2 is delivered. Most photoallergies will be defined with a far smaller dose. Although some data on the dose of light required to elicit a response exist, this remains incomplete and must be studied in context with the dose of antigen and the vehicle. Until the light and antigenic intensities are more completely defined, most physicians use a PUVA unit, a bank of UVA bulbs in a diagnostic unit, or a hot quartz (Kromayer) unit, with an appropriate filter to remove light with wavelengths below 320 nm (UVB).

The effect of UV irradiation on photopatch test substances in vitro has been reported by Bruze et al. (98). It appears that 5 J of UVA is almost always adequate. All 13 photoactive compounds formed photoproducts after UVA irradiation; 8 substances were decomposed by both UVA and UVB radiation; 5

Table 7 Patch and Photopatch Testing[a]

Contact test site response	Photocontact test site response	Interpretation
Positive	Positive	Allergic contact dermatitis
Negative	Positive	Photoallergic contact dermatitis
Negative	Negative	Not sensitized

[a]See text for details.

by UVA alone. It is also possible that some patients may require UVB to elicit photoallergic dermatitis. However, since UVB testing is not done routinely, it may be some time before this problem is clarified.

B. Masked Patch Test

Epstein (99) observed that many patients are so sensitive to light that the dose delivered under an ordinary patch will elicit reactions. He provided details of testing the nonexposed site, using a large light-impermeable black patch applied in a dimly lit room. This precaution remains essential for those decisions of separating the two mechanisms.

C. Antigen Source

Commercial sources of appropriately diluted sunscreen antigens are not presently available in the United States. On request, many manufacturers provide patch-test kits of individual ingredients for their products. A ''standard'' series of sunscreen antigens has been proposed by the International Contact Dermatitis Research Group (ICDRG). In Europe, these test kits are commercially available (Fig. 1). These sunscreen antigens are available in 2% concentrations, although the maximum nonirritating doses of putative antigens in a given vehicle have not been defined. Test concentrations and vehicles for the dermatological testing of many other cosmetic ingredients in sunscreen formulations have been published elsewhere (100). We currently lack adequate virgin controls for the high concentrations used in contemporary formulations. When high concentrations are required to elicit allergic contact dermatitis, an impurity or a photoproduct may be the actual allergen.

D. Vehicle

The specific vehicle in which the allergens are dissolved or suspended is important (101). The ICDRG list (see Fig. 1) employs petrolatum as diluent. This vehicle appears to be adequate to elicit reactions in many patients. However, the bioavailability of the antigen may be too limited in some cases. Thus, Mathias et al. (23) required ethanol to demonstrate PABA sensitivity; Schauder and Ippen (28) noted more pronounced test reactions to avobenzone in isopropylmyristate compared with petrolatum. The optimal vehicle and concentration for each ingredient remains an area for study.

E. Enigmas

Some patients develop sunscreen-induced dermatitis reactions that appear allergic or photoallergic in a morphological and historic sense, yet fail to demonstrate a positive patch or photopatch test despite appropriate testing. Such false-neg-

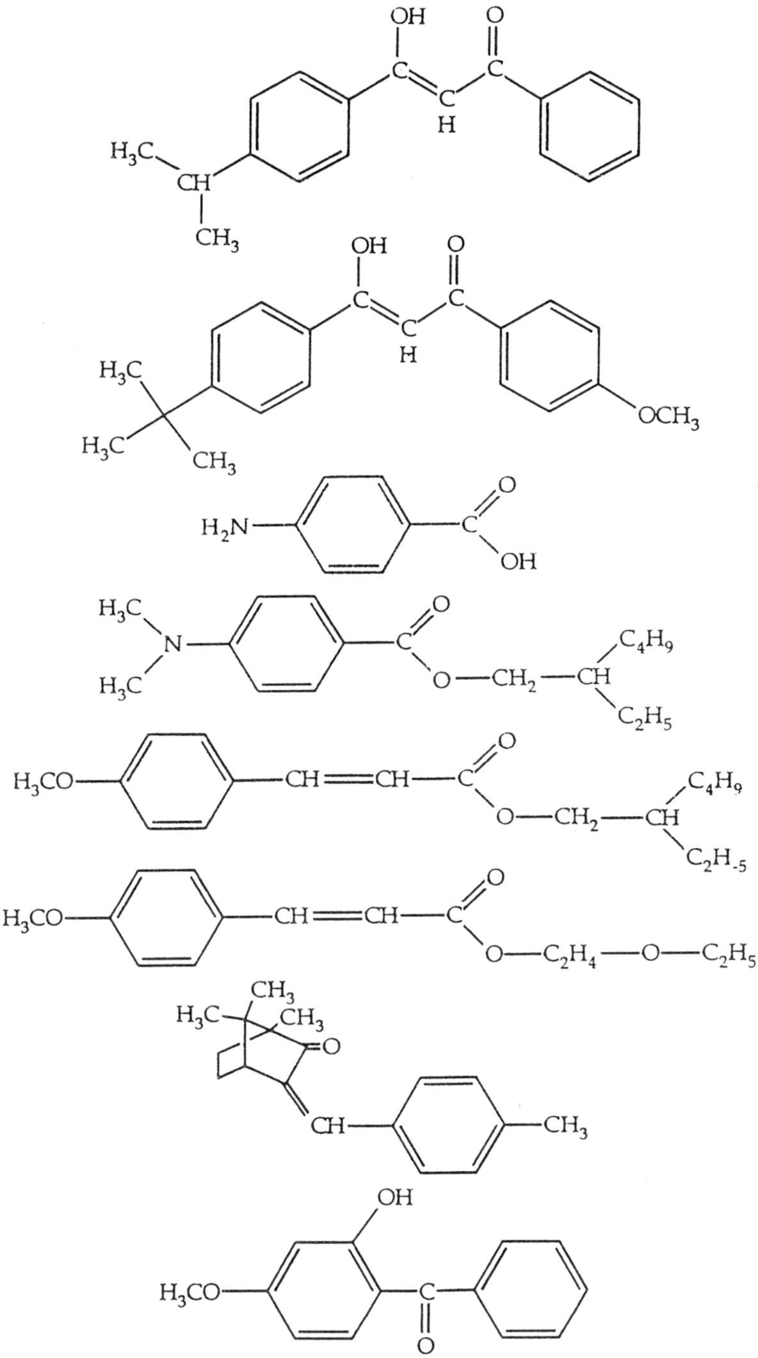

Figure 1 Sunscreen antigens available in Europe, 2% concentration in petrolatum. Names of compounds (from top to bottom): 1-(4-isopropylphenyl)-3-phenyl-1,3-propanediol; 4-t-butyl-4′-methoxy-dibenzoylmethane; p-aminobenzoic acid; 2-ethylhexyl-p-dimethylaminobenzoate; 2-ethylhexyl-p-methoxycinnamate; 2-ethoxyethyl-p-methoxycinnamate; 3-(4-methylbenzylidene)-camphor; and oxybenzone. (Note: Sunscreen antigens available from Trolab Company, Hermal Chemie, Hamburg, Germany.)

ative reactions are more difficult to identify than false-positive reactions (102). Table 8 provides the basic strategy employed in attempting to elucidate the nature of the reaction. Unfortunately, in some patients, even these extensive maneuvers fail to elicit the culprit.

F. Excited Skin Syndrome

Many of the reported positives, and especially the cross-reaction studies, may represent false-positives owing to the excited skin syndrome. This state of skin hyperirritability, often induced by a concomitant dermatitis, is responsible for many nonreproducible patch tests. Bruynzeel and Maibach (103) detail strategies for minimizing false-positive reactions.

G. Burning and Stinging: Contact Urticaria and Subjective (Sensory) Irritation

Many products, when applied to the face, burn, itch, or sting (29,104–107). The mechanism may be either subjective (sensory) irritation or contact urticaria. The later describes a wheal and flare response that occurs within 30–60 min after cutaneous exposure. This response may be allergic or nonallergic and has been reported with benzophenone (29). It is most likely that the diagnosis is generally missed, as only rarely is immediate type (wheal and flare) testing performed. Von Krogh and Maibach (108) review diagnostic criteria and testing methods. Subjective irritation remains a major factor in decreased use of sunscreens (i.e., patients discontinue use because of discomfort). Much work remains to be done to resolve this complication (109).

Table 8 Strategy for Identifying the Cause When Routine Patch Testing Is Negative

Intervention	Comment
Increase UVA dose	Avoid UVA erythema; utilize UVA control
Increase concentration of sunscreen	Upper limit of nonirritating dose not completely defined
Alter vehicle	Ethanol has been found to be effective
Test other components	Sunscreen manufacturers often helpful in providing test kits
Add suberythemogenic doses of UVB	
Perform provocative use test on final formulation	
Consider ‘‘compound’’ allergy	

VII. CONCLUSIONS

The potential of sunscreening agents to cause sensitization if theoretically high because of the high concentrations of these active ingredients in the formulations. However, despite the widespread use of sunscreens, the small number of published reports of contact or photocontact sensitization to these agents suggests that such sensitization is a minimal problem in dermatology. Thune et al. (60) stated that photocontact dermatitis was confirmed in 10.9% and contact dermatitis in 9.5% of all persons tested within the Scandinavian multicenter photopatch study. Wennersten et al. (38) reported that the frequency of hypersensitivity to sunscreens in their patient population is about 0.1%, while Nater and De Groot (75) report that the incidence of adverse effects caused by sunscreens is about 1–2%, with a low- to medium-risk index. The latter figure was based on surveys of cosmetics including the FDA pilot study (110) and the U.S. manufacturers' file (111). A valid criticism, as pointed out by Davies et al. (40), is that these reports, mainly single case reports, do not accurately reflect the true sensitization rates of these agents. Many mild reactions that occur in individuals may not be reported or may be interpreted as failures of sunscreen protection. The establishment of a registry for adverse reporting of sunscreening agents, similar to that developed for the nonsteroidal anti-inflammatory agents by Stern and Bigby (112), would help characterize the sensitization rates of these agents. In addition, much work needs to be done both in epidemiology and in developing appropriate testing techniques.

REFERENCES

1. Food and Drug Administration. Sunscreen drug products for over-the-counter human drugs: proposed safety, effective and labeling conditions. *Federal Register*, *43*:38206 (1978).
2. Horio, T. and Higuchi, T. Photocontact dermatitis from *p*-aminobenzoic acid. *Dermatologica*, *156*:124 (1978).
3. Marmelzat, J. and Rapaport, M. J. Photodermatitis with PABA. *Contact Dermatitis*, *6*:230 (1980).
4. Cronin, F. *Contact Dermatitis*. Churchill Livingstone, London, 1980, pp. 450–454.
5. Adams, R. M., Maibach, H. I., Clendenning, W. F., et al. A five-year study of cosmetic reactions. *J. Am. Acad. Dermatol.*, *13*:1062 (1985).
6. Fagerlund, V.-L., Kalimo, K., and Jansen, C. Valonsuoja-aineet fotokontaktiallergian aiheuttajina, *Duodecim*, *99*:146 (1983).
7. Wennersten, G., Thune, P., Brodthagen, H., et al. The Scandinavian multicenter photopatch study: Preliminary results. *Contact Dermatitis*, *10*:305 (1984).
8. Thune, P. Contact and photocontact allergy to sunscreens. *Photodermatology*, *1*:5 (1984).

9. Camarasa, J. G. and Serra-Baldrich, E. Allergic contact dermatitis to sunscreens. *Contact Dermatitis*, *15*:253 (1986).
10. Menz, J., Muller, S. A., and Connoly, S. M. Photopatch testing: A six-year experience. *J. Am. Acad. Dermatol.*, *18*:1044 (1988).
11. English, J. S. C. and White, I. R. Allergic contact dermatitis from isopropyl dibenzoylmethane. *Contact Dermatitis*, *15*:94 (1986).
12. Sams, W. M. Contact photodermatitis. *Arch. Dermatol.*, *73*:142 (1956).
13. Pariser, R. J. Contact dermatitis to dioxybenzone. *Contact Dermatitis*, *3*:172 (1977).
14. Thompson, G., Maibach, H., and Epstein, J. Allergic contact dermatitis from sunscreen preparations complicating photodermatitis. *Arch. Dermatol.*, *113*:1252 (1977).
15. Baer, R. L. and Meltzer, L. Sensitization to monoglyceryl *para*-aminobenzoate. *J. Invest. Dermatol.*, *11:5* (1948).
16. Meltzer, L. and Baer, R. L. Sensitization to monoglycerol *para*-aminobenzoate: A case report. *J. Invest. Dermatol.*, *12*:31 (1949).
17. Curtis, G. H. and Crawford, P. F. Cutaneous sensitivity to monoglycerol *para*-aminobenzoate. *Cleve. Clin. Q.*, *18*:35 (1951).
18. Goldman, G. C. and Epstein, E. Contact photosensitivity dermatitis from sun-protective agent. *Arch. Dermatol.*, *100*:447 (1969).
19. Fisher, A. A. Sunscreen dermatitis due to glyceryl PABA: Significance of cross-reactions to this PABA ester. *Curr. Contact News*, *18*:495 (1976).
20. Fisher, A. A. Dermatitis due to benzocaine present in sunscreens containing glyceryl PABA (Escalol 106). *Contact Dermatitis*, *3*:170 (1977).
21. Caro, I. Contact allergy/photoallergy to glyceryl PABA and benzocaine. *Contact Dermatitis*, *4*:381 (1978).
22. Hjorth, N., Wilkinson, D., Magnusson, B., Bandmann, H.-J., and Maibach, H. Glyceryl *p*-aminobenzoate patch testing in benzocaine-sensitive subjects. *Contact Dermatitis*, *4*:46 (1978).
23. Mathias, C. G. T., Maibach, H. I., and Epstein, J. Allergic contact photodermatitis to *para*-aminobenzoic acid. *Arch. Dermatol.*, *114*:1665 (1978).
24. Kroon, S. Standard photopatch testing with Waxtar, *para*-aminobenzoic acid, potassium chromate and balsam of Peru. *Contact Dermatitis*, *18*:35 (1983).
25. Rietschel, R. L. and Lewis, C. W. Contact dermatitis to homomenthyl salicylate. *Arch. Dermatol.*, *114*:442 (1978).
26. Fowler, J. F. Allergic cheilitis due to benzophenone-3. Presented to the Patch Test Clinic Symposium at the American Academy of Dermatology Annual Meeting, San Antonio, 1987.
27. Schauder, S. and Ippen, H. Photoallergic and allergic contact dermatitis from dibenzoylmethanes. *Photodermatology*, *3*:140 (1986).
28. Schauder, S. and Ippen, H. Photoallergic and allergic contact dermatitis from dibenzoylmethane compounds and other sunscreens. *Hautarzt*, *39*:435 (1988).
29. Ramsay, D. L., Cohen, H. J., and Baer, R. L. Allergic reaction to benzophenone. *Arch. Dermatol.*, *105*:906 (1972).
30. De Groot, A. C. and Weyland, J. W. Contact allergy to butyl methoxydibenzoylmethane. *Contact Dermatitis*, *16*:278 (1987).

31. De Groot, A. C., van der Walle, H. B., Jagtman, B. A., and Weyland, J. W. Contact allergy to 4-isopropyl-dibenzoylmethane and 3-(4-methylbenzylidene) camphor in sunscreen Eusolex 8021. *Contact Dermatitis, 16*:249 (1987).
32. Woods, B. Dermatitis from Eusolex 8021 sunscreen agent in a cosmetic. *Contact Dermatitis, 7*:168 (1981).
33. English, J. S. C., White, I. R., and Cronin, E. Sensitivity to sunscreens. *Contact Dermatitis, 17*:159 (1987).
34. Haussman, A. and Kleinhaus, D. Allergisches Kontaktekzem durch UV-Strahlenfilter in Sonnenschutzcremes—zwei Fallbeobachtungen. *Z. Hautkr., 61*:1654 (1986).
35. Fisher, A. A. The presence of benzocaine in sunscreens containing glyceryl PABA (Escalol 106). *Arch. Dermatol., 113*:1299 (1977).
36. Millard, L. G. and Barrett, P. L. Contact allergy from mexenone masquerading as an exacerbation of light sensitivity. *Contact Dermatitis, 3*:222 (1977).
37. Hunloh, W. and Goerz, G. Contact dermatitis from Eusolex 6300. *Contact Dermatitis, 9*:333 (1983).
38. Wennersten, G., Thune, P., Jansen, C. T., and Brodhagen, H. Photocontact dermatitis: current status with emphasis on allergic contact photosensitivity (CPS) occurrence, allergens, and practical phototesting. *Semin. Dermatol., 5*:273 (1986).
39. Goodman, T. Photodermatitis from a sunscreening agent. *Arch. Dermatol., 102*: 563 (1970).
40. Davies, M. G., Hawk, J. L. M., and Rycroft, R. J. G. Acute photosensitivity from the sunscreen 2-ethoxyethyl-*p*-methoxycinnamate. *Contact Dermatitis, 8*:190 (1982).
41. Murphy, G. M. and White, I. R. Photoallergic contact dermatitis to ethoxyethyl-*p*-methoxycinnamate. *Contact Dermatitis, 16*:296 (1987).
42. Kaidbey, K. H. and Allen, H. Photocontact dermatitis to benzocaine. *Arch. Dermatol., 117*:77 (1981).
43. Holzle, E. and Plewig, G. Photoallergishe Kontaktdermatitis durch benzophenonhaltige Sonnenschutzpräparate. *Hautarzt, 33*:391 (1982).
44. Knobler, E., Almeida, L., Ruzkowski, A. M., Held, J., Harber, L., and DeLeo, V. Photoallergy to benzophenone. *Arch. Dermatol., 125*:801 (1989).
45. Emmett, E. A., Taphorn, B. R., and Kominsky, J. R. Phototoxicity occurring during the manufacture of ultraviolet-cured ink. *Arch. Dermatol., 113*:770 (1977).
46. Katz, S. I. Relative effectiveness of selected sunscreens. *Arch. Dermatol., 101*:466 (1970).
47. Weller, P. and Ereeman, S. Photocontact allergy to octyldimethyl PABA. *Aust. J. Dermatol., 25*:73 (1984).
48. Kalimo, K., Eagerlund, V.-L., and Jansen, C. Concomitant photocontact allergy to a benzophenone derivative and a sunscreen preservative, 6-acetoxy-2,4-dimethyl-*m*-dioxane. *Photodermatology, 1*:315 (1984).
49. Bury, J. N. Photoallergies from benzophenones and beta carotene in sunscreens. *Contact Dermatitis, 6*:211 (1980).
50. Satulsky, E. M. Photosensitization induced by monoglycerol paraaminobenzoate: A case report. *Arch. Dermatol., 62*:711 (1950).

51. Roberts, D. L. Contact allergy to Eusolex 8021. *Contact Dermatitis*, *8*:302 (1988).
52. Alomar, A. and Cerda, M. T. Contact allergy to Eusolex 8021. *Contact Dermatitis*, *20*:74 (1989).
53. De Groot, A. C. and Weyland, J. W. Cosmetic allergy to the UV-absorber bornelone. *Dermatosen*, *37*:13 (1989).
54. Foussereau, J., Cavelier, C., and Protois, J. C. Contact dermatitis from Eusolex 8020 elucidated by chemical analysis. *Contact Dermatitis*, *20*:311 (1989).
55. Garioch, J. J. and Forsyth, A. Allergic contact dermatitis from 4-isopropyldibenzoylmethane in a light moisturising cream. *Contact Dermatitis*, *20*:312 (1989).
56. Marguery, M. C. and Bazex, J. Photocontact dermatitis due to 2-hydroxy, 4-methoxybenzophenone (oxybenzone): Report of four cases. *Br. J. Dermatol. Suppl. 34*, *121*:59 (1989).
57. Motley, R. J. and Reynolds, A. J. Photocontact dermatitis due to isopropyl and butyl methoxy dibenzoylmethanes (Eusolex 8020 and Parsol 1789). *Contact Dermatitis*, *21*:109 (1989).
58. Lenique, P., Machet, L., Vaillant, L., Bensaid, P., Muller, C., Khallouf, R., and Lorette, G. Contact and photocontact allergy to oxybenzone. *Contact Dermatitis*, *26*:177 (1992).
59. Crowe, M. J., Banks, S. L., Guin, J. D. Photoallergic and allergic contact dermatitis to butyl-methoxydibenzoylmethane. *Am. J. Contact Dermatitis*, *3*:33 (1992).
60. Thune, P., Jansen, C., Wennersten, G., Rystedt, I., Brodthagen H., and McFadden, N. The Scandinavian multicenter photopatch study 1980–1985: Final report. *Photodermatology*, *5*:261 (1988).
61. DeLeo, V. A., Suarez, S. M., and Maso, M. J. Photoallergic contact dermatitis. Results of photopatch testing in New York, 1985 to 1990. *Arch. Dermatol.*, *128*: 1513 (1992).
62. Fischer, T. and Bergstrom, K. Evaluation of customers' complaints about sunscreen cosmetics sold by the Swedish pharmaceutical company. *Contact Dermatitis*, *25*:319 (1991).
63. Willis, I. Sensitization potential of *para*-aminobenzoic acid. *Cosmet. Toilet.*, *91*: 63 (1976).
64. Sulzberger, M. B., Kanof, A., Baer, R. L., and Lowenberg, C. Sensitization by topical application of sulfonamides. *J. Allergy*, *18*:92 (1947).
65. Bruze, M., Fregert, S., and Gruvberger, B. Occurrence of *para*-aminobenzoic acid and benzocaine as contaminants in sunscreen agents of *para*-aminobenzoic acid type. *Photodermatology*, *1*:277 (1984).
66. Bruze, M., Gruvberger, B., and Thune, P. Contact and photocontact allergy to glyceryl *para*-aminobenzoate. *Photodermatology*, *5*:162 (1988).
67. Kaidbey, K. H. and Kligman, A. M. Phototoxicity to a sunscreen ingredient: Padimate A. *Arch. Dermatol.*, *114*:547 (1978).
68. Calnan, C. D. Cinnamon dermatitis from an ointment. *Contact Dermatitis*, *2*:167 (1976).
69. White, I. R. Risk of contact dermatitis from UV-A sunscreens. Reply. *Contact Dermatitis*, *29*:221 (1993).

70. Weyland, J. W., Rooselaar, J., Assink-Generinick, J. M. B., and Hartog, B. J. *Sun Cosmetics* 1984. Market Survey, Cosmetics Report No. 36. Food Inspection Service, Enschede, The Netherlands, 1985.
71. Van Ketel, W. G. Allergic contact dermatitis from an aminobenzoic acid compound used in sunscreens. *Contact Dermatitis, 3*:283 (1977).
72. Mork, N.-J. and Austad, J. Contact dermatitis from witisol, a sunscreen agent. *Contact Dermatitis, 10*:122 (1984).
73. Fisher, A. A. *Contact Dermatitis*, 2nd ed. Lea & Febiger, Philadelphia, 1973, p. 209.
74. Kaminester, L. H. Allergic reaction to sunscreen products. *Arch. Dermatol., 117*: 66 (1981).
75. Nater, J. P. and De Groot, A. C. *Unwanted Effects of Cosmetics and Drugs Used in Dermatology*, 2nd ed. Elsevier, Amsterdam, 1985, p. 360.
76. De Groot, A. C., Van der Meeren, H. L. M., and Weyland, J. W. Contact allergy to avocado oil in a sunscreen. *Contact Dermatitis, 16*:108 (1987).
77. Edwards, E. K. and Edwards, E. K. Allergic reaction to *t*-butyl alcohol in a sunscreen. *Cutis, 29*:476 (1982).
78. Edwards, E. K. and Edwards, E. K. Allergic reaction of triethanolamine stearate. *Cutis, 31*:195 (1983).
79. Edwards, E. K. Allergic reactions to benzyl alcohol in a sunscreen. *Cutis, 28*:332 (1981).
80. Hannuksela, M. Skin contact allergy to emulsifiers. *Int. J. Cosmet. Sci., 10*:9 (1988).
81. De Groot, A. C. and Weyland, J. W. Hidden contact allergy to formaldehyde in imidazolidinyl urea. *Contact Dermatitis, 17*:124 (1987).
82. Dooms-Goossens, A., De Boulle, K., Doom, M., and De Groot, H. Imidazolidinyl urea dermatitis. *Contact Dermatitis, 14*:322 (1986).
83. Methylisothiazolinone: Additional skin retention data. *Rose Sheet*, April 25:7 (1988).
84. Ford, G. P. and Beck, M. H. Reactions to quarternium 15, bronopol and Germall 115 in a standard series. *Contact Dermatitis, 14*:271 (1986).
85. De Groot, A. C., Bruynzeel, D. P., Bos, J. D., et al. The allergens in cosmetics. *Arch. Dermatol., 124*:1525 (1988).
86. De Groot, A. C., Weyland, J. W., Bos, J. D., and Jagtman, B. A. Contact allergy to preservatives (I). *Contact Dermatitis, 14*:120 (1986).
87. Hannuksela, M., Kousa, M., and Pirila, V. Allergy ingredients of vehicles. *Contact Dermatitis, 2*:105 (1976).
88. Green, C., Catterall, M., and Hawk, J. L. M. Chronis actinic dermatitis and sunscreen allergy. *Clin. Exp. Dermatol., 16*:70 (1991).
89. Eransway, A. F. Sensitivity to Kathon CG: Findings in 365 consecutive patients. *Contact Dermatitis, 19*:342 (1988).
90. Baer, R. L. Cross-sensitization phenomenon. *Modern Trends in Dermatology* (R. M. B. McKenna, ed.), Butterworths, London, 1954, p. 232.
91. Dupuis, G. and Benezra, C. Cross-sensitization. *Allergic Contact Dermatitis to Simple Chemicals. A Molecular Approach*, Marcel Dekker, New York, 1982, pp. 87–126.

92. MacLeod, T. M. and Frain-Bell, W. The study of the efficacy of some agents used for the protection of the skin from exposure to light. *Br. J. Dermatol.*, *84*:266 (1971).
93. Thune, P. and Eeg-Larsen, T. Contact and photocontact allergy in persistent light reactivity. *Contact Dermatitis*, *11*:98 (1984).
94. Addo, H. A. and Frain-Bell, W. Persistence of allergic contact sensitivity in subjects with photosensitivity dermatitis and actinic reticuloid syndrome. *Br. J. Dermatol.*, *117*:555 (1987).
95. Addo, H. H. A., Sharma, S. C., Eierguson J., et al. A study of composite plant extract reactions in photosensitivity dermatitis. *Photodermatology*, *2*:68 (1985).
96. Fisher, A. A. The role of nonsensitizing alternatives in the management of allergic contact dermatitis. *Semin. Dermatol.*, *5*:263 (1986).
97. Schauder, S. *Göttinger Liste.* Lichtfilterhaltige Hautpflegepräparate in der Bundesrepublik Deutschland. Grosse, Berlin, 1990.
98. Bruze, M., Fregert, S., and Luggren, B. Effect of ultraviolet irradiation of photopatch test substances in vitro. *Photodermatology*, *2*:32 (1985).
99. Epstein, S. Masked photopatch tests. *Contact Dermatitis*, *41*:369 (1963).
100. Maibach, H. I., Akerson, J. M., Marzulli, F. N., et al. Test concentrations and vehicles for dermatological testing of cosmetic ingredients. *Contact Dermatitis*, *6*: 369 (1980).
101. Fischer, T. and Maibach, H. I. Patch testing in allergic contact dermatitis: An update. *Semin. Dermatol.*, *5*:214 (1980).
102. Rycroft, R. J. G. False reactions to non-standard patch tests. *Semin. Dermatol.*, *5*: 225 (1986).
103. Bruynzeel, D. P. and Maibach, H. I. Excited skin syndrome (angry back). *Arch. Dermatol.*, *122*:323 (1986).
104. Parrish, J. A., Pathak, M. A., and Fitzpatrick, T. B. Facial irritation due to sunscreen products. *Arch. Dermatol.*, *111*:525 (1975).
105. Calnan, C. D. Stinging sensation from ethoxyethyl-methoxycinnamate. *Contact Dermatitis*, *4*:294 (1978).
106. Cullen, S. I. Safety and efficacy of two sunscreens in tretinoin-treated acne vulgaris. *Curr. Ther. Res.*, *26*:625 (1979).
107. Forbes, M. A., Brannen, M., and King, W. C. Benzophenone as a sunscreen. *South. Med. J.*, *59*:321 (1966).
108. Von Krogh, G. and Maibach, H. I. The contact urticaria syndrome, *Dermatotoxicology*, 3rd ed. (F. N. Marzulli and H. I. Maibach, eds.). Hemisphere Publishing Company, Washington DC, 1987, Chap. 15.
109. Lammintausta, K., Maibach, H. I., and Wilson, D. Mechanisms of subjective (sensory) irritation. Propensity to non-immunologic contact urticaria and objective irritation in stingers. *Derm. Beruf Umwelt.*, *36*:45 (1988).
110. *(PB-242 480) An investigation of Consumers' perception of adverse reactions to cosmetic products.* (1975). Westat Inc., prepared for the Food and Drug Administration, National Technical Information Service U.S. Department of Commerce, Springfield, Virginia.

111. *Tabulation of Cosmetic Product Experience Report, Submitted to the Food and Drug Administration under Voluntary Cosmetic Regulatory Program (Jan 1974–June 1975)*. Food and Drug Administration, Division of Cosmetic Technology, Washington, DC.
112. Stern, R. S. and Bigby, M. An expanded profile of cutaneous reactions to nonsteroidal anti-inflammatory drugs. *JAMA*, *252*:1433 (1984).

VI

CHEMICAL EVALUATION OF SUNSCREEN PRODUCTS

35

Quality Control of Sunscreens

Nadim A. Shaath
KATO Worldwide, Ltd./Nickstadt-Moeller, Mount Vernon, New York

I. INTRODUCTION

The cosmetic industry is incorporating sunscreen agents into a variety of cosmetic and toiletry products. This has complicated the task of the quality control chemist, who is responsible for the analysis of the sunscreen raw materials, as well as products containing mixtures of sunscreen chemicals.

Owing to the diversity and number of the analytical procedures required for the successful analysis of a sunscreen chemical's raw materials, as contrasted with those procedures designed to separate and identify sunscreen chemicals in cosmetic formulations, it was decided to address each problem separately. This chapter deals with quality control procedures required for the analysis of sunscreen's raw materials as provided by suppliers (1). Chapter 36 deals with the separation and identification of sunscreen chemicals and mixtures in formulations of cosmetic and toiletry products.

Of the 19 chemical sunscreen active ingredients allowed by the Food and Drug Administration (FDA) for use in the United States (2), all are aromatic acids or their esters or salts with the exceptions of the benzophenones (Table 1). In fact, the esters are generally either *ortho-* or *para-*disubstituted aromatic compounds as shown in Table 1.

The same is true for the 37 chemicals on the European Economic Community (EEC) and COLIPA list for ultraviolet (UV) filters used in Europe. The excep-

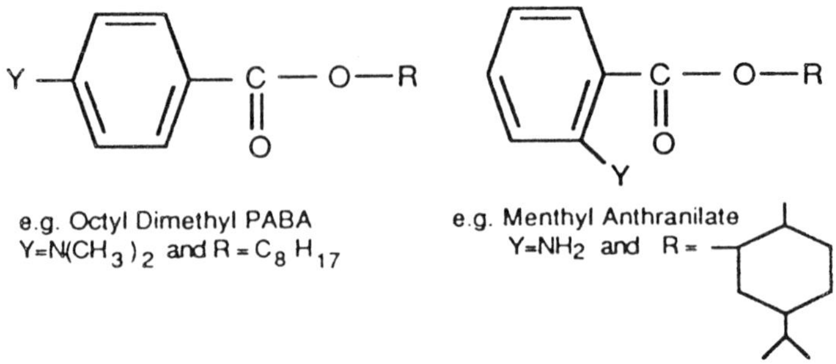

Table 1 FDA–OTC Panel Category 1[a] Sunscreens and Corresponding EEC Designations

Sunscreen	Approved (%)	λ_{max} (ethanol)	Extinction coefficient (ethanol)	EEC no.[b]	EEC (% conc.)
A. Chemical					
UVA Absorbers					
Oxybenzone	2–6	325	9400	1.4	10
Sulisobenzone	5–10	324	8400	2.17	5
Dioxybenzone	3	327	10440	NA	NA
Menthyl anthranilate	3.5–5	336	5600	NA	NA
UVB Absorbers					
Aminobenzoic acid	5–15	283	15300	1.1	5
Amyl dimethyl PABA	1–5	NA	NA	2.3	5
2-Ethoxyethyl-*p*-methoxycinnamate	1–3	NA	NA	NA	NA
DEA-*p*-methoxycinnamate	8–10	290	24930	2.9	8
Digalloyl trioleate	2–5	NA	NA	2.15	4
Ethyl 4-bis(hydroxypropyl) aminobenzoate	1–5	312	27000	2.1	5
2-Ethyl-hexyl-2-cyano-3,3-diphenylacrylate	7–10	303	12600	NA	NA
Ethylhexyl-*p*-methoxycinnamate	2–7.5	311	23300	2.13	10
2-Ethylhexyl salicylate	3–5	307	4900	2.6	5
Glyceryl aminobenzoate	2–3	297	18700	2.4	5
Homomenthyl salicylate	4–15	306	4300	1.3	10
Lawsone with dihydroxyacetone	0.25 with 33	NA	NA	NA	NA
Octyl dimethyl PABA	1.4–8	311	27300	2.5	8
2-Phenylbenzimidazole-5-sulfonic acid	1.4	310	28250	1.6	8
Triethanolamine salicylate	5–12	298	3000	2.11	2
B. Physical					
Red petrolatum	30–100	NA	NA	NA	NA
Titanium dioxide	2–25	NA	NA	NA	NA

[a]From Ref. 2.

[b]For conditions and complete EEC (European Economic Committee) sunscreen chemical list see encyclopedia and UV absorbers for sunscreen products in *Cosmetics & Toiletries*, March 1987.

tions are the benzophenones (5.38, 5.39, 5.40), the diketones (5.52, 5.64, 5.66), and the ketones (5.56, 5.60, 5.61). The compound 5-(3,3-dimethyl-2-norbornylidene)-3-penten-2-one (5.56) is the only nonaromatic sunscreen chemical approved for use globally.

With few exceptions, these compounds can be easily analyzed using standard chromatographic and spectroscopic techniques. Other standard physical and chemical tests should be undertaken to ensure satisfactory control over the purity and consistency of the sunscreen chemical being investigated.

II. QUALITY CONTROL PROCEDURES

Quality control procedures can be subdivided into four types of analyses:

A. Physical analyses
B. Chemical analyses
C. Chromatographic analyses
D. Spectroscopic analyses

For a thorough analysis of sunscreen actives, several of the procedures listed in Table 2 may be necessary (3). The analyst should, however, consult the sunscreen supplier for the selection of the tests that are appropriate.

A. Physical Analyses

Odor

Sunscreen chemicals generally exhibit little or no odor. They should be compared with a standard that is kept in a cool dry place, in an amber bottle void of headspace, and away from heat and light. Request a new standard from the sunscreen supplier every 6–9 months. Odors should be smelled on blotters and not directly from the containers. The sunscreen chemical should not be stored next to highly odoriferous material because cross-contamination is possible for sunscreens that are otherwise odorless, especially if they are viscous liquids or solids.

Color

The color of the sunscreen should be noted on the specification sheet. Caution is to be exercised when comparing the color of a new batch with a standard stored over a time. The color of the standard may have changed on standing for more than 6 months. The chemistry of the sunscreen may be indicative of the potential color problem. For example, cinnamates and *p*-aminobutyric acid (PABA) derivates may darken on standing because air oxidation is possible. Phenolic derivatives, such as salicylates, can interact with iron contaminants and

Table 2 Quality Control Procedures

A. Physical analyses	C. Chromatographic analyses
Odor description	TLC pattern
Color	GLC pattern
Physical appearance	a. Column
Melting point (°C)	b. Temperature conditions
Refractive index at 20°C	c. Detector
Specific gravity at 25°C	HPLC pattern
Optical rotation	a. Solvent
Solubility	b. Column
Moisture determination	c. Detector
Viscosity	
pH determination	D. Spectroscopic analyses
Flash point (°F)	Ultraviolet spectrum
	UV pattern
B. Chemical analyses	λ_{max}
Saponification value	Molar absorptivity (ϵ)
Acid value	K-value
Functional group analysis	Percentage purity
Metal content	NMR spectrum
	IR spectrum
	MS spectrum

form pinkish colorations. Generally, unlined metal drums should be avoided. Restoration of original colors may be possible in some sunscreen liquids through simple chemical treatments, distillations, or otherwise. Sunscreen suppliers should be aware of such procedures and will recommend the most appropriate remedy.

Physical Appearance

Without resorting to any instrumentation, recording the physical appearance alone may be a critical factor in acceptance or rejection of the sunscreen chemical. Observe the nature of the product as to whether it is a solid or a liquid, clear or cloudy, viscous or free-flowing. Check for particulates, sediments, or foreign substances in the product.

Melting Point

This property is applicable to solid sunscreens such as benzophenones, DEA cinnamate, or PABA. Record the temperature and the melting range in degrees Celsius (C°). Observe the behavior during the melting process. Report any swelling, sweating, charring, or other characteristics. This may reveal information on the purity of the sunscreen raw material.

Refractive Index

The refractive index $[n]_{20}^{D}$ is the change in direction of a light ray passing from one medium to another of different density. The index of refraction of a substance may also be expressed as the ratio of the velocity of light in a vacuum to its velocity in the substance. It varies with the wavelength of the incident light, temperature, and pressure. The usual light source is the D line of sodium and the standard temperature is 20°C. This method applies to liquid sunscreens that are not too viscous to analyze.

Specific Gravity

The specific gravity is the ratio of the density of a substance to the density of a reference substance (water). The temperature should be specified at all times. In the absence of a gas chromatograph, this property is helpful in determining the purity of the sunscreen chemical.

Optical Rotation

The optical rotation $[n]_{D}^{25}$ is the change in direction of the plane of polarized light either to the right or to the left as it passes through a molecule containing one or more asymmetric (chiral) carbon atoms. The optical rotation is measured by the use of a polarimeter. This method applies only to sunscreens that are optically active.

Solubility

The supplier should recommend a procedure for monitoring the solubility behavior of the sunscreen in various solvents. Depending on the application, the sunscreen may be tested at several concentrations in different solvents of varying polarities. Usually a 5% solution of the sunscreen is tested in water, ethanol, hexane, mineral oil, isopropyl myristate, corn oil, acetone, isopropyl palmitate, or glycerine.

Moisture Determination

The presence of moisture in the sunscreen product may be detrimental to certain applications. A Karl Fisher titration (manual or automatic) for the determination of moisture content should be performed if it is suspected that the product contains over 0.5% water. The Karl Fisher reagent is a mixture of iodine, sulfur dioxide, and pyridine in a solvent system containing methanol and chloroform. The moisture in the sample is titrated with the reagent until the potentiometric endpoint is reached. The percentage moisture content is then measured.

Viscosity

This is a measure of the resistance of the fluid sunscreen to flow, expressed in dyne second per square centimeter or poises. Centipoises, which are 0.01 poises, are also used in the industry.

pH Determination

The pH is a value that represents the acidity or alkalinity of an aqueous solution. It is defined as the logarithm of the reciprocal of the hydrogen-ion concentration of a solution pH = ln $1/H^+$. The pH of a number of sunscreen acids and their salts should be routinely determined in the quality control laboratory.

Flash Point

A closed cup determination of the flash point in degrees Fahrenheit of the sunscreen is required by the Department of Transportation (DOT) whenever the chemical is to be transported outside of the laboratory facilities (4). A flash point below 140°F is considered to be flammable and possibly combustible. If the sunscreen has a low flash point, care must be exercised in handling. Store in a manner appropriate to the potential hazard it poses.

B. Chemical Analyses

Saponification Value

This procedure is applicable to the analysis of sunscreen chemicals that are esters. Esters can be converted to carboxylic acids and alcohols by alkaline hydrolysis. The acids can be detected and quantitatively determined by neutralization. The saponification value is defined as the number of milligrams of KOH required for the saponification of the ester and neutralization of free acids in 1 g of sample (5). Typical values range from 160 to 240.

$$\underset{\text{ester}}{\mathrm{RCOOR'}} + \underset{\text{water}}{\mathrm{H_2O}} \xrightarrow{\mathrm{OH^-}} \underset{\text{alcohol}}{\mathrm{R'OH}} + \underset{\text{acid}}{\mathrm{RCOOH}}$$

$$\underset{\text{acid}}{\mathrm{RCOOH}} + \underset{\text{base}}{\mathrm{KOH}} \longrightarrow \underset{\text{salt}}{\mathrm{RCOOK}} + \underset{\text{water}}{\mathrm{H_2O}} \tag{2}$$

Acid Value

Some sunscreen products contain carboxylic acids that may be present as unreacted products from the synthesis of the sunscreen (e.g., esterification), as degradation products (e.g., aldehydes), or as part of the compound (e.g., PABA). The amount of acid present is determined by titrating with a standardized base (6). If the acid value is performed to determine impurity, values lower than 5 should be observed. Some products require a percentage acid as opposed to an acid value.

$$\underset{\text{acid}}{R{-}\overset{\overset{\displaystyle O}{\|}}{C}{-}OH} + \underset{\text{base}}{NaOH} \longrightarrow \underset{\text{acid salt}}{R{-}\overset{\overset{\displaystyle O}{\|}}{C}{-}ONa} + \underset{\text{water}}{H_2O} \tag{3}$$

Functional Group Analysis

In the absence of chromatographic and spectroscopic techniques, many sunscreens containing functional groups may be analyzed with conventional chemical methods. Acetals, acids, aldehydes, esters, ketones, or phenols are analyzed both qualitatively and quantitatively through functional group analyses. For example, hydroxylamine hydrochloride will react with the carbonyl group in ketones, forming oximes and liberating HCl. This reaction occurs at alcoholic reflux temperatures at a pH of 6. The amount of ketone can be determined from the amount of HCl liberated. For more details consult the many books and monographs written on the subject (7).

Metal Content

The presence of metal contaminants, such as iron or chromium, can affect the color, odor, and occasionally the performance of the sunscreen chemical. The analysis of these metals by either atomic absorption spectroscopy (8) or ion chromatographic procedures (9) is recommended.

C. Chromatographic Analyses

Chromatography may be defined as the science of separation techniques involving a mobile phase (the solvent in thin-layer chromatography [TLC] and high-performance liquid chromatograph [HPLC], or the helium/nitrogen gas in gas liquid chromatography [GLC]) passing through a stationary phase (alumina or silica in TLC, C-18 packing in HPLC, silicones or carbowax in GLC) (10). The ability to separate various components in a mixture of sunscreen chemicals (owing to the presence of impurities or isomers) is based on selective and preferential partitioning of these components between the mobile phase and the stationary phase.

Thin-Layer Chromatography

In the absence of a gas chromatograph, TLC can be used effectively in assaying sunscreens to develop a solvent system suited for the specific application. Plates, reagents, and sprays are now commercially available, making TLC determinations fast, reliable, and inexpensive (11).

Gas–Liquid Chromatography

The last 20 years have witnessed an explosion in the use of gas–liquid chromatography (also known as gas chromatography [GC] or vapor phase chromatography [VPC]) in the analysis of sufficiently volatile and thermally stable organic chemicals. The supplier should recommend suitable conditions for the analysis by GLC. The equipment available, however, may be the determining factor. The following parameters are crucial in any GLC sunscreen analysis (12).

1. *Column*: The type of column (packed or capillary), the packing material (polar or nonpolar), and the length and diameter of the columns must be specified.
2. *Temperature conditions*: The injector, detector, and oven temperatures should be noted. Any variations could lead to erroneous results. If temperature programming is available, record the initial and final temperature, as well as the temperature ramp.
3. *Detector*: From the many detectors commercially available, thermoconductivity (TC) detectors and flame ionization detectors (FID) are adequate for most analyses.

Other factors that may be important are the carrier gas, flow rate, split ratio, and undercoating of columns. However, as long as the standard and the new batch are analyzed back to back, any condition mentioned in the foregoing may be adequate for the analysis. Caution must be exercised in comparing data run on two different machines or on two different days without ensuring that the conditions are identical. The use of internal or external GLC standards may be helpful in those circumstances.

A typical set of GLC conditions for the analysis of sunscreen chemicals follows:

Sample size: 0.1 μl
Oven temp: 65°–250°C
Program rate: 4°C/min
Injection port temp: 210°C
Detector: FID
Carrier flow: 1 ml/min
Split ratio: 200:1
Attenuation: 0
Threshold: 0
Detector temp: 250°C
Carrier gas: Helium

High-Performance Liquid Chromatography

Liquid chromatography is quite similar to GLC except for the mobile phase, which is a liquid instead of a gas (13). Many laboratories are currently equipped with HPLC machines and are routinely analyzing their products with this procedure. The following conditions have to be specified:

1. *Solvent*: State whether the analysis requires gradient elution or isocratic conditions. If gradient elution is used, the mixing ratio of the binary or ternary solvents must be specified. The use of ultrapure-grade solvents is highly recommended.
2. *Column*: Other than the length, width, and such, the most important factor is the type of packing in the column. Specify whether it is a reverse-phase column or a normal-phase column.
3. *Detector*: The use of either an ultraviolet (UV) or refractive index (RI) detector is recommended. If a UV detector is used, the exact wavelength should be specified. Other detectors, such as conductivity, fluorescence, or electrochemical, may be suitable for a particular application (14).

Unlike GLC, HPLC may require extensive preworkup, the use of pure reagents, and generally longer analysis time. However, it is the method of choice if the sunscreen is either thermally unstable or is insufficiently volatile (e.g., PABA or sunscreen salts).

D. Spectroscopic Analyses

Spectroscopic methods of analysis are excellent methods for the identification of sunscreen chemicals and for the elucidation of their molecular structure (15). This is accomplished by recording the energy absorbed or emitted by the sunscreen chemical in any of the wavelengths of the electromagnetic spectrum in response to excitation by an external energy source.

Ultraviolet Spectrum

Ultraviolet spectroscopy is used in the characterization of sunscreen chemicals which, by definition, are UV radiation absorbers. UV spectroscopy involves the absorption behavior of the chemical and can be either qualitative or quantitative. The qualitative application of absorption spectroscopy depends on the fact that a given molecular species absorbs light in specific regions of the spectrum. Such a display is called an absorption spectrum and serves as a fingerprint for identification purposes. The quantitative application is for determining an unknown concentration of a given species.

Several useful data may be extracted from the UV spectrum to aid in the identification and characterization of sunscreen active chemicals, namely the UV pattern, the λ_{max}, the molar absorptivity (ϵ), the K-value, and percentage purity.

1. *UV pattern*: Pattern recognition of the UV spectrum may provide information for the identification of sunscreen chemicals. Care must be exercised in selecting the appropriate solvent for the analysis at an extremely high purity (ultrapure or spectroscopy-grade solvents). The concentration and dilution tech-

niques have to be carried out carefully because the UV pattern is directly affected by the concentration, as shown in the Beer-Lambert law, which governs the UV absorption (17). Use of second and higher derivatives on instruments so equipped may enhance the apparent difference between similar spectra.

The basic law governing the absorption of radiant energy is expressed by the Beer-Lambert relation:

$$\epsilon = \frac{AM}{bc}$$

where ϵ is the molar absorptivity in moles per centimeter (mol/cm) gram and is a constant, characteristic of the solute. The molar absorptivity is directly proportional to the chemical's ability to absorb UV radiation. Therefore, the greater the absorptivity, the more UV radiation the chemical absorbs; A is the absorbance at the wavelength maximum; M is the molecular weight of the solute in moles per cubic centimeter (mol/cm^3); b is the cell width in centimeters; c is the concentration of the solute in grams per cubic centimeter (g/cm^3).

2. λ_{Max}: The wavelength of maximum absorption, a λ_{max}, is characteristic of the sunscreen chemical under investigation, provided that the solvent is clearly defined. The λ_{max} for several sunscreen chemicals used in the United States is listed in Table 1.

3. *Molar absorptivity* (ϵ): The molar absorptivity (ϵ), or molar extinction coefficient, is a measure of the radiation's attenuation in an absorbant medium. It is characteristic of the sunscreen chemical and dependent on the wavelength of light, nature of the solvent, concentration and the temperature. Typical molar absorptivity values are reported in Table 3.

The following procedure is used to determine molar absorptivity (ϵ):

a. Weigh accurately on an analytical balance 50 mg of the sunscreen chemical to be analyzed into a 100-ml volumetric flask.
b. Dilute with isopropanol or 95% ethanol and mix.

Table 3 Molar Absorptivity Values for Several Sunscreens in Ethanol

Sunscreen chemical	Molar absorptivity $\times 10^3$
p-Aminobenzoic acid (PABA)	13
Octyl *p*-methoxycinnamate	23
Octyl dimethyl PABA	28
Ethyl dihydroxypropyl PABA	26
2-Ethylhexyl-2-cyano-3,3-diphenyl acrylate	13
Homomenthyl salicylate	5

c. Transfer quantitatively 1 ml of the solution prepared to a 100-ml volumetric flask, dilute with ethanol, and mix.
d. Using a 1-cm–path length UV cuvette, record the differential absorption spectrum of the final dilution against an optical reference of 95% ethanol or isopropanol. Scan and record from 400 to 200 nm.
e. Measure the maximum absorbance (A) for the UV absorber at its characteristic wavelength maximum (λ_{max}).
f. Knowing the molecular weight of the absorber and the cell path (b) as 1 cm, calculate the molar extinction coefficient using the formula given on foregoing paragraph 1.

For example: Sunarome WMO (octyl salicylate) had an absorbance value of 0.104, a molecular weight of 250.33, a concentration of 0.00502 g/L, and a cell path length of 1 cm. Thus,

$$\epsilon = \frac{0.104 \times 250.33}{0.00502 \times 1} = 5186$$

4. *K-value*: The K-value is used in the sunscreen industry to rate the effectiveness of a sunscreen chemical. It is the ratio of its maximum absorbance to its concentration in grams per liter measured in a cell of 1-cm–path length.

$$\text{K-value} = \frac{\text{UV absorbance}}{\text{concentration of sunscreen (g/L)}}$$

The K-values for a 1-g/L solution of several sunscreens are listed in Table 4.

The experimental method involves accurate weighing of the sunscreen chemical under investigation, dissolving it in a known volume of ethanol or isopropanol and, after a series of dilutions, recording its absorbance in UV spectrophotometer at its λ_{max}. The spectrophotometer parameters are:

Ordinate (min/max): 0.1
Wavelength (min/max): 200.0–400.0
Scan speed: 120

Table 4 K-Values for 1 g/L Solution of Several Sunscreens

Sunscreen chemical	K-value
Octyl dimethyl PABA	106
Triethanolamine salicylate	12
Menthyl anthranilate	28
Octyl methoxycinnamate	83

Slit: 1
Response: 3

For example: Sunarome OMC (octyl methoxycinnamate) had an absorbance value at 309.9 nm of 0.4382 at a concentration of 0.00526 g/L.

$$\text{K-value} = \frac{0.4382}{0.00526} = 83.47$$

Note that the accuracy of this procedure is ± 2 K-values.

5. *Percentage Purity*: Although GLC is the best method to determine the purity of a sunscreen chemical, UV spectroscopy could also be used for that purpose, provided accurate weighings are performed on an analytical balance and a reference material of known purity is available to calculate molar absorptivity.

The following procedure may be used to calculate the percentage purity of a sunscreen (e.g., octyl salicylate):

a. Weigh by difference 0.1000–0.1300 g of sample into a 200-ml volumetric flask. Dilute to mark with chloroform or methylene chloride and shake.
b. Transfer 10 ml to a 100-ml volumetric flask. Dilute with chloroform or methylene chloride and shake.
c. Using 1-cm–path length UV cells, record the differential absorption spectrum of the prepared test solution versus an optical reference of chloroform. The instrument is referenced at 350 nm. Scan and record, beginning at 400–250 nm, measuring absorbance values.
d. Draw a line on the recorder chart between the points on the absorption curve at 332.5 nm and 282.5 nm, and determine the intercept of a line drawn perpendicular to the absorbance maximum at 307 nm. Subtract the absorbance at the intercept from the absorbance value of 307 nm to obtain the absorbance change (ΔA).

Using the following simplified formula:

$$\%\ (\text{w/w})\ \text{of sunscreen} = \frac{(\Delta A) \times (13.806)}{\text{weight in grams}}$$

$$\Delta A \text{ at } 307\ \text{nm} = 1.150 - 0.225 = 0.925$$

$$\text{weight of octyl salicylate} = 0.128\ \text{g}$$

therefore,

$$\%\ (\text{w/w})\ \text{of octyl salicylate} = \frac{0.925 \times 13.806}{0.128} = 99.7\% \pm 2\%$$

The reasons for the rather large error (±2%) in calculations involving UV spectrophotometry exist even though these experiments may be done with the maximum accuracy possible. The sources of error other than deviations from Beers-Lambert law include pH, temperature, stray light, instrumental error, and human error (18). Experimental conditions may have a dramatic effect on the UV spectrum, rendering some of the data unusable or dubious. These include inexperience in operating the spectrophotometer, fingerprints on the cuvettes, dirty glassware, air bubbles in the UV cutette, errors in the precision of weighing and pipetting, the use of a different grade of a particular reagent or solvent, and not calibrating the analytical balance or the spectrophotometer.

Solvent effects may have a dramatic effect on the UV spectrum (19). Figure 2 illustrates the effect of solvents (polar vs. nonpolar) on the ultraviolet spectrum of dioxybenzone.

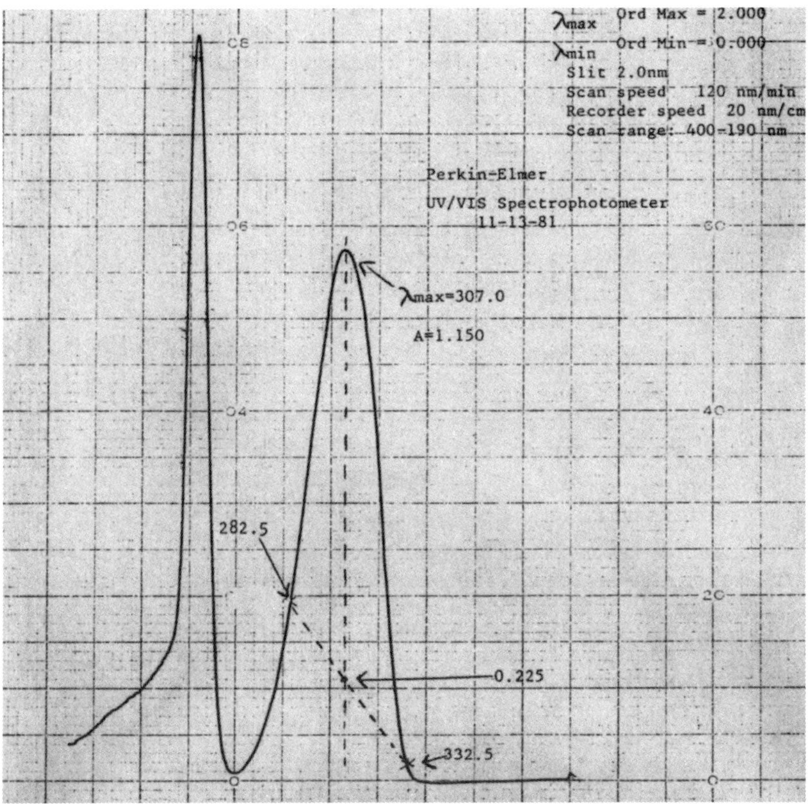

Figure 1 The UV spectrum of octyl salicylate in chloroform.

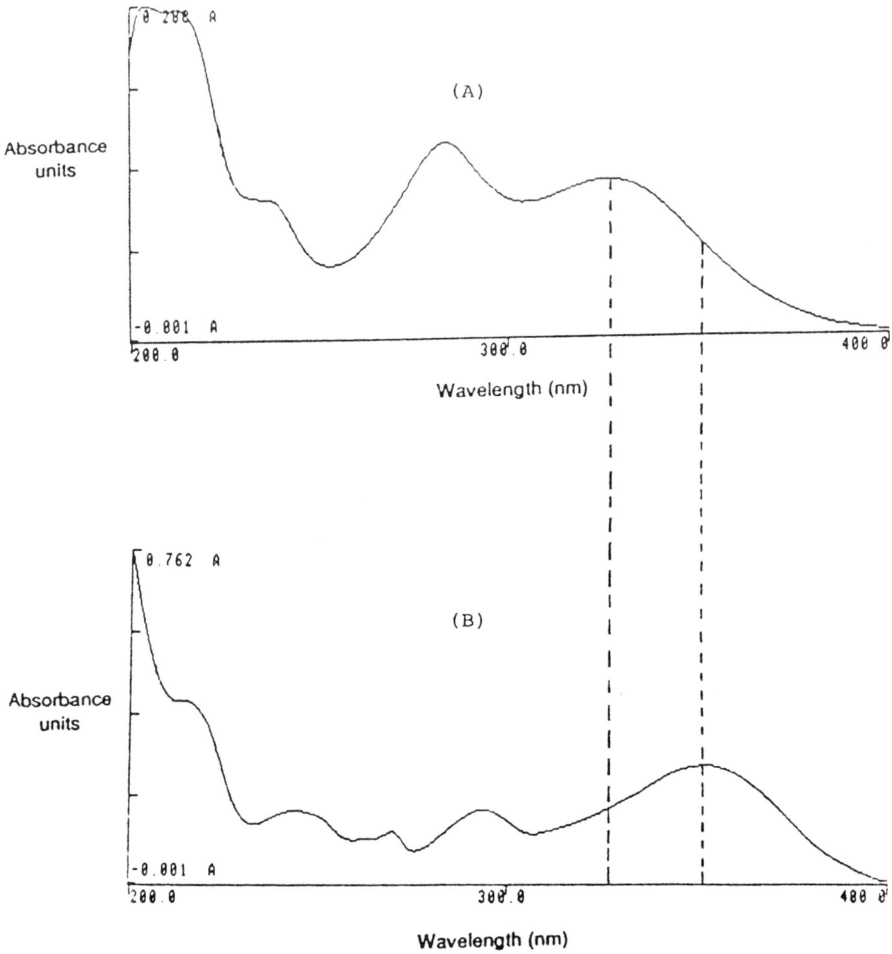

Figure 2 Ultraviolet absorption spectra of dioxybenzone in (A) 100% ethyl alcohol (0.00349 g/L) and (B) hexane (0.00494 g/L).

Nuclear Magnetic Resonance (NMR) Spectroscopy

Nuclear magnetic resonance is a powerful technique used to examine the environment of the nuclei of the atoms which make up the sunscreen chemical (20). The nuclear environment depends strongly on the nature of the chemical bonds that hold the atoms together in the molecule and is also dependent on the number of types of other atomic nuclei in the immediate vicinity. Thus, in the

NMR experiment the hydrogens (protons) are "seen" and their environment characterized. Figure 3 illustrates the NMR spectrum of octyl dimethyl PABA.

Infrared Spectrum

When the sunscreen chemical interacts with infrared (IR) radiation, portions of the incident radiation are absorbed at particular wavelengths. This absorbed energy causes the atoms in the chemical to undergo a series of twisting, bending, rotational, and vibrational motions. These motions, occurring simultaneously, produce a highly complex absorption spectrum that is uniquely characteristic of the functional groups (e.g., carbonyl group, phenolic group) that the sunscreen molecule comprises and of the overall configuration of the atoms as well. The use of the IR spectrum as a "fingerprint" for the sunscreen chemical provides the analyst with an extremely powerful technique for identification and characterization (21). The IR spectrum of menthyl anthranilate is shown in Figure 4.

Mass Spectrometry

The main advantage of mass spectrometry (MS) is its increased sensitivity over other analytical techniques and its specificity in identifying unknowns and for

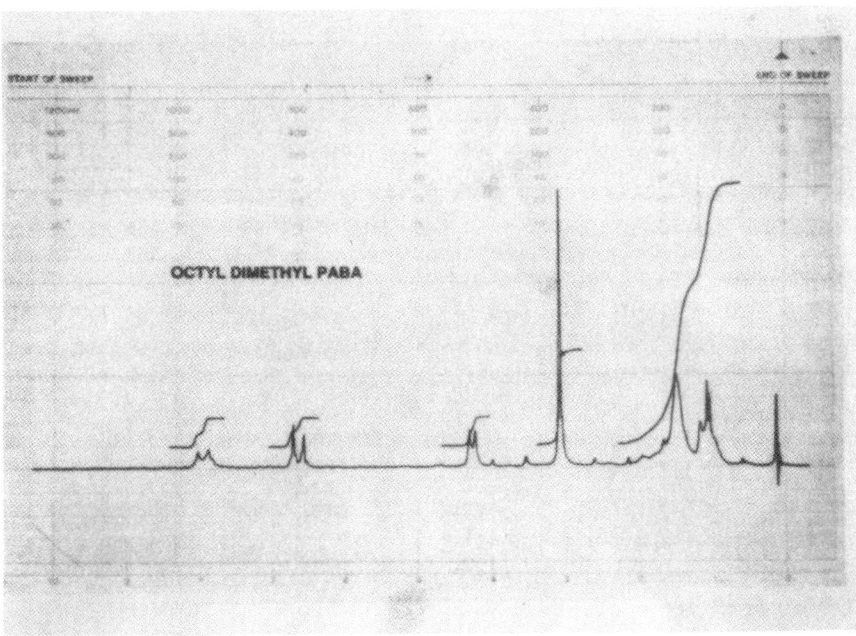

Figure 3 NMR spectrum of octyl dimethyl PABA.

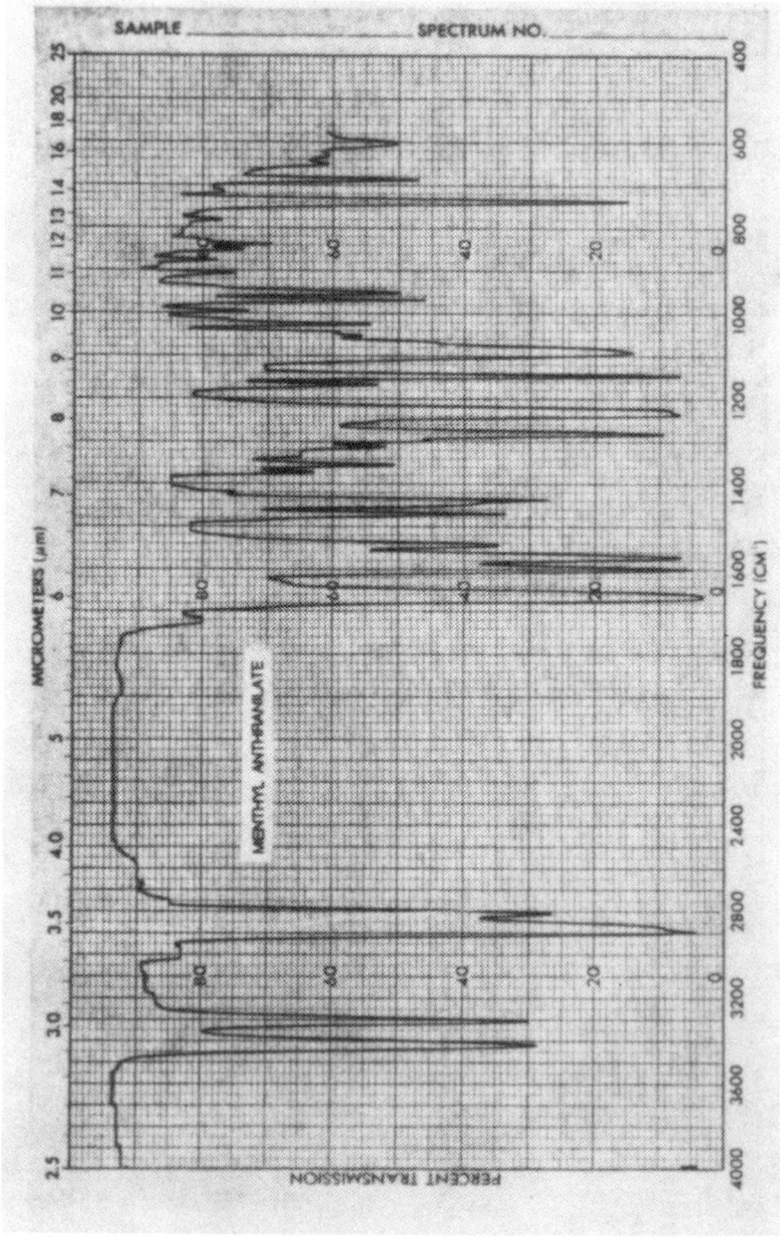

Figure 4 The IR spectrum of menthyl anthranilate.

confirming the presence or absence of suspected compounds (22). The excellent specificity results from characteristic fragmentation patterns which can give information about molecular weight and molecular structure. The mass spectrum of octyl *p*-methoxycinnamate is shown in Figure 5.

The interfacing of two powerful analytical techniques, namely the separation technique of capillary GLC and the identification technique of MS, has produced the single most powerful analytical tool to date: the GC–MS. the automation and the computerization currently available in the state-of-the-art instrumentation of GC–MS has had a major effect on the flavor and fragrance industry; the food industry; the petroleum industry; and the cosmetic, toiletries, and household products industry (23). With the use of a selective ion-monitoring (SIM) technique in GC–MS, for example, subparts per million of an impurity can be detected in sunscreen chemicals (24). However, the importance as to the absence or presence of those low-level impurities in the cosmetic and toiletry product is currently considered insignificant and is not measured. Nevertheless, the use of this advanced technique, if available, facilitates analysis tremendously and decreases the time needed to complete even the most complicated tasks. Table 5 lists typical GC–MS conditions utilizing both a quadrupole instrument and an ion-trap mass spectrometer.

III. CONCLUSION

With the rapid growth of the sunscreen industry in both sales and the number of new product introductions, UV filters are incorporated into a multitude of products used in our daily lives. It has thus become necessary for the quality control chemist to know much more about analytical procedures, the chemical structure of sunscreens and their potential interaction with solvents, impurities, the environment (heat, light, oxygen, or other), and other UV filters. The pro-

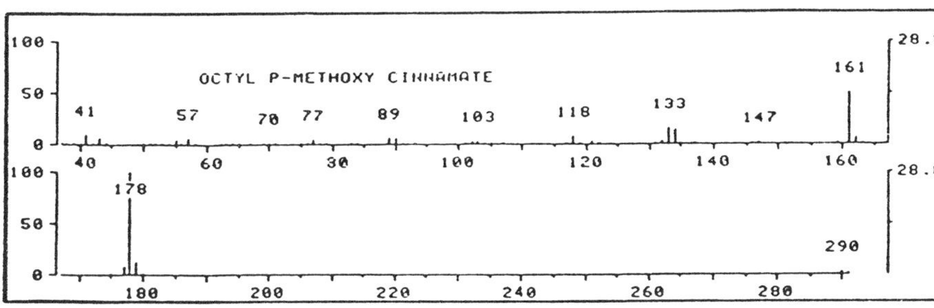

Figure 5 The mass spectrum of octyl *p*-methoxy cinnamate.

Table 5 GC–MS Conditions

GC–MS conditions	Quadrupole	Ion trap
Sample size	0.5 μl	0.05 μl
Oven temperature	65–210°C	65–210°C
Program rate	2°C	2°C
Injection port temperature	210°C	210°C
Detector	Total ion chromatogram	Total ion chromatogram
Carrier flow	1 ml/min	1 ml/min
Split ratio	200:1	300:1
Peak threshold	30	2
MS ionization voltage	70 eV	70 eV
Scan time	108.3 Amu/s	1 scan/s
Scan start delay	0 s	180 s
Start, stop masses	35–400	35–400
Electron multiplier voltage	2400 eV	1700 eV

The GC–MS column used is a 50-m Ultra (#1) Hewlett Packard Capillary column cross-linked methyl silicone, with a film thickness of 0.33 μm and an internal diameter of 0.20 mm.

cedures outlined in this chapter are meant to provide the analyst with the general methods for the analysis of sunscreen chemicals. Sunscreen suppliers should be consulted for the selection of tests appropriate to the sunscreen chemicals they offer. The analytical chemist should also research and develop new techniques that can identify impurities, isomers, and other ingredients more accurately. Thus, this increased activity in the sunscreen industry will only increase the burden on the analytical, research, and quality control chemists to meet the new challenges of purity, consistency, accuracy, and, more importantly, claim substantiation.

REFERENCES

1. Shaath, N. A. The analysis of sunscreen chemicals, Part 1. Quality control procedures for sunscreen chemicals. *Cosmet. Toilet.*, *3*:69–81 (1987).
2. *Federal Register*, *266*:38206–38269 (1978).
3. (a) Kateman, G. and Pijpers, F. W. *Quality Control in Analytical Chemistry*. John Wiley & Sons, New York, 1981; (b) Kuwana, T. *Physical Methods in Modern Chemical Analysis*. Academic Press, New York, 1978.
4. Shugar, G., Shugar, R., Brauman, L., and Bauman, R. *Chemical Technicians Ready Reference Book*, McGraw Hill, New York, 1981.

5. (a) Association of Official Analytical Chemists. *Official Methods of Analysis*. Procedure 28.028, 1984; (b) *U.S. Pharmacopeia National Formulary*. 20th Revision, 1980, p. 920; (c) *Handbook of Chemistry and Physics*. CRC Press, New York, 1985–6, p. 218.
6. (a) Association of Official Analytical Chemists. *Official Methods of Analysis*. Procedure 16.237, 1984; (b) *Handbook of Chemistry and Physics*, CRC Press, New York; 1985–6, p. C710.
7. (a) Shriner, R., Fuson, R., Curtin, D., and Morrill, T. *The Systematic Identification of Organic Compounds*. John Wiley & Sons, New York, 1980; (b) Cheronis, N. and Entrikin, J. *Semimicro Qualitative Organic Analysis*. Interscience Publishers, New York, 1957; (c) Fieser, L. and Williamson, K. *Organic Experiments*, D.C. Heath & Co., Toronto, 1983; (d) Siggia, S. *Instrumental Methods of Organic Functional Group Analysis*. John Wiley & Sons, New York, 1972.
8. Willard, H., Merrit, L., Dean, J., and Settle, F. *Instrumental Methods of Analysis*. Wadsworth Publishing, Belmont, CA, 1981.
9. Freiser, H. *Ion Selective Electrodes in Analytical Chemistry*. Plenum Press, New York, 1980.
10. Poole, C. and Schuette, S. *Contemporary Practice of Chromatography*. Elsevier, New York, 1984.
11. Stahl, E. *Thin Layer Chromatography*. Springer Verlag, New York, 1969.
12. (a) Jennings, W. *Gas Chromatography with Glass Capillary Columns*. Academic Press, New York, 1980; (b) Freeman, R. *High Resolution Gas Chromatography*. Hewlett Packard, New York, 1981; (c) Horvath, C. *High Performance Liquid Chromatography, Advances and Perspectives*. Academic Press, New York, 1980.
13. Snyder, L. and Kirkland, J. *Introduction to Modern Liquid Chromatography*. John Wiley & Sons, New York, 1979.
14. Vickrey, T. *Liquid Chromatography Detectors*. Marcel Dekker, New York, 1983.
15. Silverstein, R., Bassler, and Morill, T. *Spectrometric Identification to Organic Compounds*. John Wiley & Sons, New York, 1963.
16. (a) Scott, A. *Interpretation of Ultraviolet Spectra of Natural Products*. Pergamon Press, New York, 1964; (b) Rao, C. *Ultraviolet and Visible Spectroscopy*. Butterworths, New York, 1975.
17. Jaffe, H. and Orchin, M. *Theory and Application of Ultraviolet Spectroscopy*. John Wiley & Sons, New York, 1962.
18. Willard, H. H., Merritt, L. L., Dean, J. A., and Settle, F. A. *Instrumental Methods of Analysis*. Wadsworth Publishing, Belmont CA, 1981, pp. 85–94.
19. (a) Shaath, N. A. On the theory of ultraviolet absorption by sunscreen chemicals. *J. Soc. Cosmet. Chem.*, *82*:193 (1987); (b) Agrapidis-Paloympis, L. E., Nash, R. A., and Shaath, N. A. The effects of solvents on the ultraviolet absorbance of sunscreens. *Cosmet. Toilet.*, *38*:209–221 (1987); (c) Agrapidis-Paloympis, L. E. *The Influence of Solvent on the UV Absorbance of Sunscreens*. M.S. thesis, St John's University, New York, 1986.
20. Casy, A. F. *PMR Spectroscopy in Medicinal and Biological Chemistry*. Academic Press, New York, 1971.
21. Colthup, N. B., Daly, L. H., and Wiberley, S. E. *Introduction to Infrared and Raman Spectroscopy*. Academic Press, New York, 1974.

22. (a) Budzikiewicz, H., Djerassi, C., and Williams, D. *Mass Spectrometry of Organic Compounds*. Holden Day, San Francisco, 1967; (b) McLafferty, M. *Interpretation of Mass Spectra*. Benjamin, New York, 1967; (c) Jennings, W. and Shibamoto, T. *Qualitative Analysis of Flavor and Fragrance Volatiles by Glass Capillary Chromatography*. Academic Press, New York, 1980.
23. Chapman, J. *Computers in Mass Spectrometry*. Academic Press, New York, 1978.
24. Hewlett Packard 5993 GC/MS System Users Manual, Volume 1, 4–23, Palo Alto.

36

Modern Analytical Techniques in the Sunscreen Industry

Nadim A. Shaath and Gideon I. Andemicael
KATO Worldwide, Ltd./Nickstadt-Moeller, Mount Vernon, New York

Peter M. Griffin
Ungerer and Company, Lincoln Park, New Jersey

I. INTRODUCTION

The recent growth of the sunscreen market with the introduction of sunscreens into a variety of cosmetic and toiletry products increases the burden on both the quality control chemist and the research chemist to analyze raw materials and products containing mixtures of sunscreen chemicals.

With the present advanced technology of analytical techniques and concomitant increased knowledge of the relation between the physicochemical properties and the structure of organic molecules, chemists are employing an ever-widening variety of spectroscopic techniques to aid in the identification, characterization, and analysis of the purity of sunscreen ingredients. Comparison of the infrared, ultraviolet, nuclear magnetic resonance, and mass spectra as well as the gas and liquid chromatographic patterns of the targeted sunscreen ingredients with those of known reference compounds is currently the best criteria for purity assessment. Thus, the use of advanced instrumentation and computerization in laboratories undertaking research or analysis of sunscreen products has become a basic routine.

The principal areas in which sunscreen chemicals and sunscreen cosmetic formulations are analyzed follow:

A. Quality Control and Assurance

Sunscreen chemicals and formulations have to be routinely scrutinized for purity and for those spectroscopic characteristics that impart to each their unique properties as sunscreening agents. For a review of the procedures in the quality control laboratory, the reader should refer to Chapter 35.

B. Research Purposes

Advanced analytical techniques are routine in the characterization of new and novel sunscreening agents. The synthesis of any new sunscreen has to be fully characterized and assessed by spectroscopic and chromatographic techniques for purity and quality.

C. Technical Purposes

The stability of sunscreens as well as the functional properties of their ingredients has to be determined both spectroscopically and chromatographically. The analysis of competitive products for matching or duplication work is successfully accomplished by these procedures. The solution of the many problems are encountered in sunscreen formulations, such as base changes because of color or odor, as well as phase and efficacy changes are routinely accomplished by these advanced techniques.

D. Regulatory Compliance and Enforcement

The sunscreen industry is extremely well regulated both through self-regulation and through monitoring by the Food and Drug Administration (FDA) in the United States (1), COLIPA in the European Economic Community (EEC) countries (2), and governmental agencies in Japan, Asia, Australia, Canada, the non-EEC countries, and South American countries. The presence or absence of ingredients as well as the quantity present of a sunscreen in a cosmetic formulation in the United States, for example, is determined by the OTC–FDA monograph and thus has to be closely monitored and enforced.

II. PROBLEMS IN THE ANALYSIS OF SUNSCREENS

The analysis of sunscreening agents (see Table 1 for U.S.-approved chemicals and Table 2 for chemicals approved in Europe, in Chap. 15) present in complex and diverse cosmetic formulations poses unique and challenging problems to the analyst. These problems are twofold, the first deals with the cosmetic bases and the ingredients present therein, and the second deals with the chromatographic problems inherent in the analysis of such complicated formulations containing sensitive ingredients.

A. Analysis of Cosmetic Formulations

Cosmetic formulations are extremely complex mixtures of ingredients and the sunscreen active chemical. The sunscreen is generally present at levels varying from 2 to 20% depending on the desired sun protection factor (SPF), and is incorporated into a variety of everyday products, from moisturizers, creams, lotions, gels, to lipsticks, makeup foundations, shampoos, mousses, and other hair and skin preparations (3). Thus diversity poses potential problems to the chemist analyzing these products including:

1. The ingredients are found in many different physical forms and difficult to extract matrices. Thus, the method of choice for the separation has to be considered carefully. Separation techniques are either chemical, physical, or chromatographic. These will be briefly discussed at a later stage.
2. The ingredients may be difficult to analyze owing to nonvolatile components that do not lend themselves to routine analytical procedures. For example, sunscreen salts as 4-diethanolamine *p*-methoxycinnamate, or *N,N,N*-trimethyl-4-(2-oxoborn-3-ylidene methyl) anilinium methyl sulfate, tend to decompose if analyzed by gas chromatography. They are best analyzed by high-performance liquid chromatography (HPLC) or other specific techniques outlined later in this chapter. Other salts such as potassium cinnamate, 4-methoxycinnamic acid salts, salicylic acid salts, and sodium-3,4-dimethoxyphenyl glyoxylate are also nonelutable in the GC and, therefore, are analyzed by HPLC or are chemically converted to a GC-elutable derivative.
3. Because of the complexity of cosmetic formulations containing sunscreens, artifacts arising from the interaction of the chemicals with one another, or from environmental factors, tend to complicate the analysis. Artifacts that may be formed in a cosmetic formulation are due to one or more of the following potential interactions.
   a. The reaction of the sunscreen ingredient with other cosmetic or fragrance ingredients, for example, *p*-aminobenzoic acid (PABA) may undergo a number of potential reactions giving rise to additional artifact chemicals that appear on the chromatogram and complicate the analysis. Such reactions are shown as follows:

(1) The reaction of PABA with carboxylic acids yielding amides of PABA:

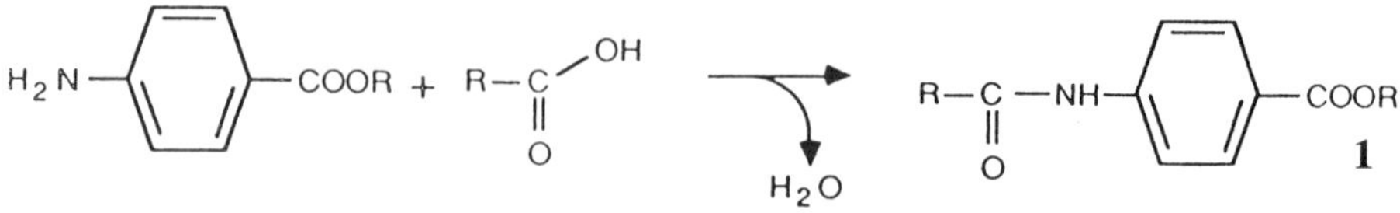

1

where R=H, glyceryl, ethyl

(2) The reaction of aldehydes with the amine portion of PABA yielding Schiff bases:

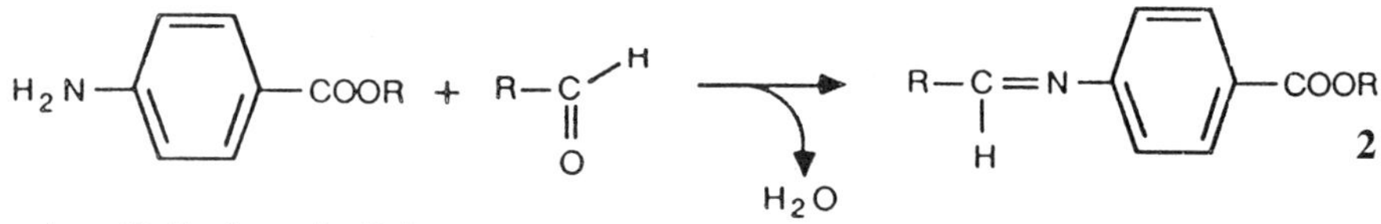

2

where R=H, glyceryl, ethyl

b. The reaction of the sunscreen ingredient with solvents such as glycerine, ethyl, or isopropyl alcohol:

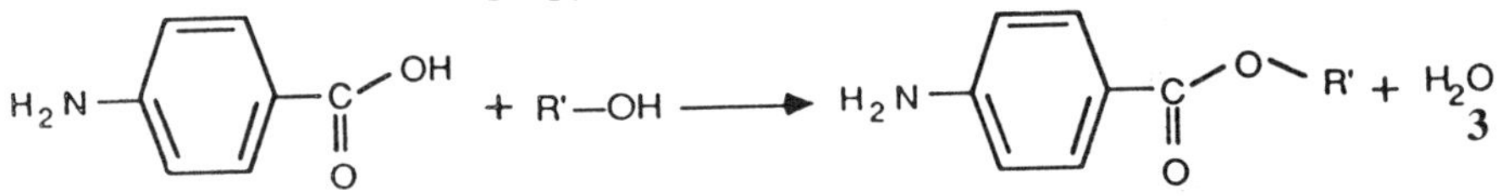

3

c. Reactions caused by environmental factors, which include heat, moisture, light, metals, and air oxidation. For example, the salts will tend to hydrolyze, forming the corresponding carboxylic acid. Amines will undergo air oxidation, yielding colored components. Phenols, such as oxybenzone, may form salts with metals (M^+) as shown below.

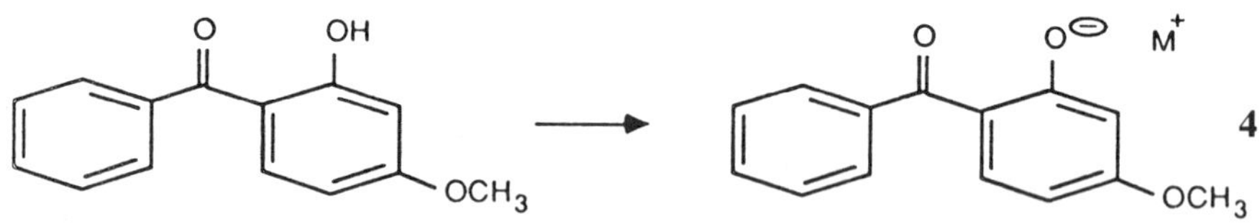

4

Esters of PABA may undergo hydrolysis at pH levels below 3, yielding the free acid and free alcohol.

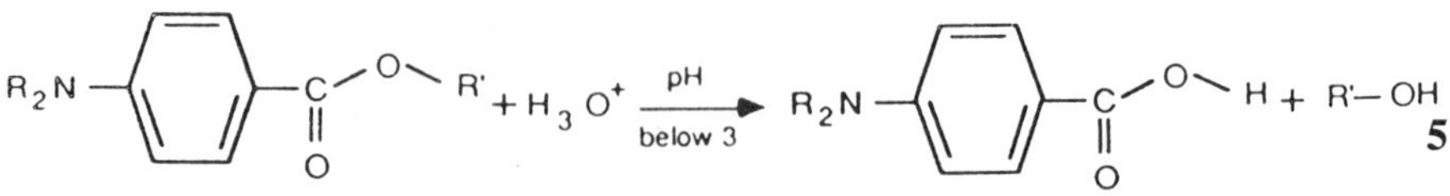

5

Other cosmetic ingredients may air-oxidize, yielding more reactive byproducts. For example, benzyl alcohol may yield benzaldehyde or benzoic acid that tends to react further with sunscreen or cosmetic ingredients.

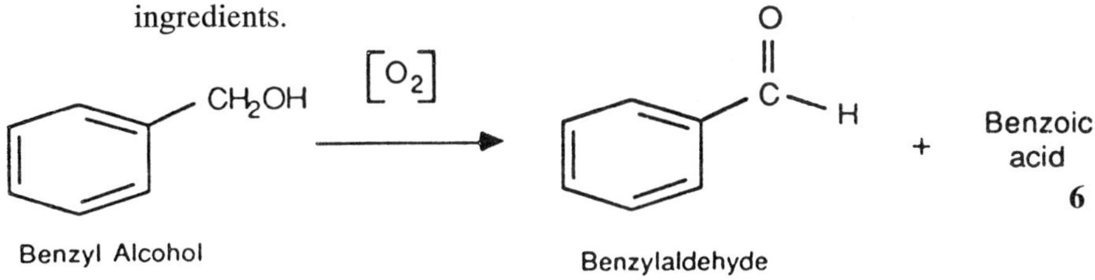

6

d. Reaction of the sunscreen ingredient with the packaging has to be considered. Acidic ingredients may react with the metal-lined containers, alcohol-containing ingredients may react with the metal-lined containers, alcohol-containing ingredients may react with aluminum packaging, and esters tend to react with plastic containers. For example, polystyrene tends to react with esters, aldehydes, ketones, and hydrocarbons; polyvinyl chloride reacts with esters, aldehydes, and ketones; whereas the polyolefins tend to be stable with most ingredients except high terpene content oils.

B. Problems in Chromatographic Analysis

Several problems in GC analyses cause peak distortion and irregularities in the chromatograms that often lead to erroneous identification and complicate the analysis.

Overloading

Polar molecules such as the free fatty acids, or the polyhydroxylated compounds such as monopropylene glycol, tend to show distorted peaks typically reminiscent of overloading the columns. This leads to errors both in quantitation and qualitative identification of the unknowns.

Chemisorption

A number of polar molecules, especially the sulfonic acid derivatives, such as sulisobenzone, will exhibit peaks in the GC indicative of chemisorption. This problem affects both the threshold (the minimum perceptible quantity capable of being analyzed) or distortions in the peak shapes.

Reactions in the Gas Chromatograph

The presence of air or oxygen as well as high temperatures coupled with excessive contact with metals leads to a number of reactions in the interfaces, including:

1. Heated injection ports or MS interfaces tend to degrade sensitive or unstable chemicals such as Schiff bases.
2. The presence of metals in the liner or the interface causes several acidic molecules, such as phenols or acids, to potentially interact.
3. Amines chemically degrade in the GC to yield either oxidized products, deaminated products, or *N*-demethylated byproducts.
4. There are a number of chemicals that are considered ''problem'' chemicals which include, H_2O, H_2S, H_3PO_4, CS_2, and HCOOH.
5. The presence of oxygen in the carrier gas or system causes some alcohols to oxidize as shown in **6**.

III. SOLUTIONS IN SUNSCREEN ANALYSIS

Combining modern analytical techniques with standard chemical and physical laboratory practices enables the analytical chemist to tackle most of the problems encountered in the characterization, structural elucidation, and identification of sunscreen active ingredients in many cosmetic formulations.

Table 1 outlines the two general approaches that must be undertaken in tandem for the characterization and analysis of any sunscreen ingredient in isolation or in the presence of other ingredients in a cosmetic formulation.

The following is a description of the separation and identification techniques used in an analytical laboratory to analyze sunscreen products.

A. Separation Techniques

The identification of the components of a mixture involves first a separation into its individual compounds followed by their identification. It is rarely pos-

Table 1 Outline of Separation and Identification Techniques

Separation	Identification
Chemical/physical	Chemical
Chromatographic	Chromatographic
CC and TLC	GC–RT
GC	GC–MS–DS
HPLC	HPLC–RT
Head space (HS; purge and trap)	HS–GC–RT
Ion chromatography	Specialized techniques
	Spectroscopic
	MS
	IR
	UV
	NMR

sible to identify the constituents of a complicated mixture without a separation. The separation of the components in a mixture should be as nearly quantitative as possible to assure the validity of the data.

The method of separation chosen should be such that the compounds are obtained as they exist in the original mixture without any change or deterioration. There are two types of separation techniques: chemical–physical methods and chromatographic methods.

Chemical–Physical Methods

1. *Extraction*: The goal of an extraction procedure is to obtain the original sunscreen from within whatever matrix it is incorporated (4) (see Appendix A). This matrix may be hydroalcoholic, oil-based, emulsion, mousse, gel, or shampoo base. The proper selection of solvents is crucial in ensuring the proper and complete extraction desired (Table 2).
2. *Distillation*: This is a separation process in which a liquid is converted to vapor, depending on the boiling point of each component (5). The vapor then condenses and each component in the mixture is individually collected. The purpose of distillation is purification, concentration, or separation of the components of a mixture.
3. *Sublimation*: This is the direct passage of a substance from its solid state to its vapor state without passing through the intermediate (liquid) state (6). This method is used in the isolation and purification of select sunscreens in certain emulsions.

Chromatographic Methods

Chromatography may be defined as the science of separation techniques involving a mobile phase (e.g., the solvent in column chromatography) passing

Table 2 Advantages and Disadvantages in Selecting Solvents for Extractions

Solvent	Advantage	Disadvantage
Methanol	Removes sunscreen without extracting lipids or carbohydrates	Will take out water and has a high potential for artifact formation
Methylene chloride	Best general purpose solvent, nonreactive, easily evaporated, does not extract water	Has tendency to form emulsion
Chloroform	Similar to methylene chloride, but more easily separated owing to high density	Has higher boiling point which may interfere with GC
Toluene	Best solvent for high molecular weight, does not absorb water	Has high boiling point, may interfere with GC analysis

through a stationary phase (e.g., the alumina in column chromatography). The ability to separate various components of a mixture of organic compounds is based on selective partitioning of these components between the mobile phase and the stationary phase (7).

The sunscreen industry uses at least five major classes of chromatographic procedures.

1. *Column chromatography and thin-layer chromatography* (*CC*). This method is directly applicable to preparative-scale separation and purifications because one can simply choose the size of the column and its contents to fit the dimensions of the sample to be fractionated (7,8).
2. *Gas chromatography* (*GC*). GC is useful for relatively volatile and thermally stable organic compounds; this method involves a gaseous mobile phase (helium or nitrogen) and a liquid stationary phase (7,9). The sample to be analyzed is injected at a temperature sufficient to vaporize it. The gaseous sample is then partitioned between the stationary liquid phase and the mobile phase, causing the separation of the components in the mixture. The separation is a function of both the polarity and volatility of the components of the sample.
3. *High-performance liquid chromatography* (*HPLC*). Liquid chromatography is quite similar to GC, except the mobile phase is a liquid instead of a gas (10). Pumps delivering thousands of pounds per square inch push the mobile phase through columns of very fine particles. HPLC offers advantages of high speed, reusable columns, automatic and continuous solvent addition, reproducible programmed gradients of solvents, and automatic and continuous monitoring of the samples eluted. As the chemicals separate, they are detected by both the refractive index detector and UV absorption detector. HPLC is primarily used when samples contain components not volatile enough to be detected by gas chromatography.
4. *Head space* (*purge and trap*). This method is used when the volatile components to be analyzed are incorporated into a matrix of nonvolatile components (11). The volatile components are separated from the nonvolatile matrix by purging with helium, then collecting this ''head space'' fraction of gas on Tenax traps followed by eluting the traps onto the GC for analysis. This method or other suitable methods can be employed to separate volatile materials from both liquid and solid matrices, such as purging sunscreens from a plastic matrix.
5. *Ion chromatography*. With the advent of the modern ion chromatographs (e.g., Dionex 4000i), separations of highly ionic compounds, such as salts, anions, cations, sugars, and phenols, has become routine in the instrumentation laboratories (12). The ion chromatographs use several types of highly sophisticated detectors, such as the suppressed ion conductivity and

the pulsed amperometric detectors, which are especially suited for the detection of ionic polar compounds.

B. Identification Procedures

After a satisfactory separation is achieved, the following identification methods are employed. Most sunscreen chemicals are readily identified by the chemical and spectroscopic techniques used in research laboratories.

Chemical

Various chemical tests can be extremely helpful in fully characterizing all the components being investigated. Details of many of the procedures can be found in Chapter 35 or by consulting the many monographs on the subject (13). The procedures include: alcohol analysis, ester content, aldehyde or ketone content, acid value, saponification number, analysis of phenols, derivative formation, and identification.

Chromatographic

1. *GC–RT*: Gas chromatography is used as an identification technique. A wealth of information has been accumulated that allows the identification of chemicals by using a relative retention time library (14). The information is best when used on the state-of-the-art gas chromatographic capillary columns. The procedure for determining the relative retention indices is found in Appendix B.
2. *GC–MS–DS*: The gas chromatograph interfaced with the mass spectrometer data system is the primary instrument involved with sunscreen identification (15). A capillary gas chromatography column is employed that separates the sample. As each chemical is separated, it is detected and analyzed by the mass spectrometer. All GC–MS data are automatically stored in the data system computer. After the run is completed, the computer is programmed to search for identifications in the mass spectrometry library. Libraries can be formed that contain all available information on sunscreen active chemicals.
3. *HPLC–RT*: High-performance liquid chromatography is generally a separation technique, but as gas chromatography can be an effective identification technique. Initially, a standard is run and the retention time (RT) is measured. The unknown compound can then be identified by comparing its chromatogram with that of the standard (10a).
4. *HS–GC–RT*: The coupling of two separation techniques, namely, head space (HS) and gas chromatography (GC), permits the analysis of complex matrices from which the volatiles are difficult to extract for direct gas chromatographic analysis (11b). With the use of a retention index (RT)

library of chemicals, identification of sunscreens in normally difficult to analyze products is readily accomplished.

5. *Specialized techniques*: There are several specialized techniques that offer the analyst additional insights into any analytical problem on hand. It is beyond the scope of this chapter to describe the details of the procedures. The reader, however, should consult the many excellent monographs on the topic of interest (15–17).

   The specialized chromatographic techniques can be subdivided into three categories:

   (1) *Use of Special Detectors in GC* (15):

   Thermal conductivity (TC)
   Flame ionization detectors (FID)
   Nitrogen–phosphorus detector (NPD)
   Mass-selective detectors (MSD)
   Electron capture detectors (ECD)
   Infrared detectors (IR)
   Flame photometric detectors (FPD)
   Halls detectors

   (2) *Specialized Detectors in HPLC–Ion Chromatography* (16):

   Refractive index detectors
   Ultraviolet detectors
   Infrared detectors
   Mass spectrometer detectors
   Fluorescence detectors
   Coulometric detectors
   Suppressed conductivity detectors
   Pulsed amperometric detectors

   (3) *Supplementary Chromatographic Techniques* (7):
   Heart-cutting
   Trapping
   Head space–purge and trap
   Split–splitless techniques
   On-column injection
   Dual capillary channels
   New column technology
   Precolumn derivatization
   Postcolumn derivatization

Spectroscopy

Spectroscopy is a branch of analytical chemistry devoted to the identification of elements and elucidation of atomic and molecular structure. This is accomplished by measurement of the radiant energy absorbed or emitted by a substance in any of the wavelengths of the electromagnetic spectrum in response to excitation by an external energy source (17). Chapter 37 deals exclusively with the spectroscopic data and their interpretation of the most common sunscreen chemicals.

1. *Mass spectrometry* (*MS*): The mass spectrometer produces charged particles consisting of the parent ion and ionic fragments of the original molecule and sorts these ions according to their mass/charge ratio (18). The mass spectrum is a record of the numbers of different kinds of ions; the relative numbers of each are characteristic for every compound, including isomers. The main advantages of mass spectrometry as an analytical technique are its increased sensitivity over other analytical procedures and its specificity in identifying unknowns or for confirming the presence of suspected compounds. The excellent specificity results from characteristic fragmentation patterns, which can give information about molecular weight and molecular structure. The tandem technique of MS–MS offers further refinements and improvements in sensitivity and specificity over conventional GC–MS (18e).
2. *Infrared spectroscopy* (*IR*): Infrared spectroscopy involves the twisting, bending, rotational and vibrational motions of atoms in a molecular (19). On interaction with infrared radiation, portions of the incident radiation are absorbed at particular wavelengths. The multiplicity of vibrations occurring simultaneously produces a highly complex absorption spectrum that is uniquely characteristic of the functional groups that compose the molecule and the overall configuration of the atoms as well. Routinely, chemicals can be separated and collected from cosmetic bases and an IR is performed to assist in the identification of the unknown. With the advent of the more modern GC–FTIR (Fourier transform IR) instruments, separation of a complex mixture followed by rapid (Fourier transform) infrared analysis can be accomplished in minutes with extremely small (ng or mg) samples (19c).
3. *Ultraviolet spectroscopy* (*UV*): Ultraviolet spectroscopy involves the absorption behavior of substances and can be either qualitative or quantitative (20). The qualitative application of absorption spectrometry depends on the fact that a given molecular species absorbs light in specific regions of the spectrum, and in varying degrees characteristic of that particular sample. Such a display is called an absorption spectrum of that molecular species, and serves as a fingerprint for identification purposes. The quantitative application is for determining an unknown concentration of a given species. The data derived from a UV spectrum include the λ_{max} (maximum wavelength of absorption), the extinction coefficient (ϵ), and the K-value (see Chap. 36 for details).

4. *Nuclear magnetic resonance spectroscopy* (*NMR*): Nuclear magnetic resonance can be used to examine the environment of the nuclei of the atoms that make up a molecule (21). A given nuclear environment depends strongly on the nature of the chemical bonds that hold the atom to the molecule and is also dependent on the number and types of other atomic nuclei in the immediate vicinity. Proton or hydrogen magnetic resonance (PMR), is the most common NMR technique. In PMR experiments, the hydrogens in the molecule are "seen" and their environment characterized. A powerful diagnosis of molecular structure can, therefore, be obtained from an NMR experiment. The coupling of this technique with powerful computer manipulation (Fourier transform) permits the analysis of extremely small quantities of chemicals.

IV. SPECIFIC SOLUTIONS IN SUNSCREEN ANALYSIS

A. TLC Techniques

Thin layer chromatography (TLC) has been widely used in the analysis of sunscreens (22). After separation of the individual components by the migration of a suitable solvent over the plate, quantitation is achieved by scraping the spot from the plate and measuring its UV absorption. TLC is also a useful method to quickly screen solvent mixtures for normal-phase HPLC when using silica gel columns. This is particularly important because silica columns are extremely slow in reaching equilibrium, especially when the solvent's polarity is decreasing. Finally, TLC equipment is simple and inexpensive and, therefore, available to any laboratory interested in sunscreen analysis.

Several factors, however, have contributed to the decline of the use of TLC in sunscreen analysis relative to the newer forms of chromatography. TLC is a very labor-intensive technique compared with HPLC or GC. This is often a decisive disadvantage in the setting of an industrial or regulatory laboratory.

Because the range of polarities encountered in sunscreens is wide, no single solvent system can given satisfactory separation for the entire range. Therefore, each sample must be run by three different solvent systems. Furthermore, although several systems of standardizing R_f values have been proposed (23), the running of a known reference standard is still the most reliable method of identification. Both of these problems have been overcome in HPLC and GC. The techniques of solvent gradient and temperature programming, respectively, overcome the general elution problem of performance over a wide range of retention times.

The ability to accurately measure retention volumes and capacity factors in HPLC and retention indices in GC eliminates the necessity of running authentic markers to establish identity. Finally, the greater separation capability of HPLC, and especially GC, reduces the problem of components that coelute.

Liem and Hilderink (2a) analyzed all the sunscreens on the European market by thin-layer chromatography. This TLC data in addition to GC and HPLC data obtained from our laboratory are presented in Table 3. Clearly, many of the components are not separated by the TLC plate. As with the HPLC on silica columns, the various esters of salicylic acid are not separated from each other. This is a result of intramolecular hydrogen bonding in the *ortho*-substituted esters. There is, therefore little interaction with the polar surface of the silica and all salicylates travel rapidly through the chromatographic system near the solvent front. All the other sunscreens, however, have R_f values reflecting the polarity of the molecule.

Table 3 GC, TLC and HPLC Data of Sunscreen Chemicals

No.	Chemical name	TLC[a]	GC[b]	HPLC[c]
1	Benzophenone 3	74	16.09	17.70
2	Benzophenone 4	0	DNE[d]	8.99
3	Benzophenone 8	51	17.67	15.60
4	DEA methoxycinnamate	NA[e]	DNE	12.60
5	Ethyl dihydroxypropyl PABA	6	MP[f]	14.65
6	Glyceryl PABA	29	MP[g]	MP[g]
7	Homosalate	90	MP[h]	MP[h]
8	Menthyl anthranilate	NA	17.11	22.90
9	Octocrylene	NA	[i]	22.90
10	Octyl dimethyl PABA	71	18.49	24.54
11	Octyl methoxycinnamate	76	18.94	24.20
12	Octyl salicylate	91	14.11	25.60
13	PABA	NA	11.25	3.43
14	2-Phenylbenzimidazole-5-sulfonic acid	0	DNE	7.83
15	TEA salicylate	NA	DNE	10.28
16	Butylmethoxydibenzoyl methane	NA	[j]	26.70

[a]TLC data are $R_f \times 100$; refer to Ref. 2a.
[b]Ethyl ester retention index (I_E); refer to Appendix B.
[c]HPLC data in minutes; refer to Appendix F.
[d]Does not elute under present conditions.
[e]Data not available.
[f]Multiple peaks owing to impurities at 16.11, 19.19, and 19.24.
[g]Multiple peaks owing to many impurities.
[h]Multiple peaks owing to the two isomers of homosalate. In GC Ultra no. 1, their retention index is 14.68 and 14.95; in CBXs at 18.04 and 18.35 and in HPLC at 24.90 and 25.80.
[i]Elutes beyond our current ethyl ester marker.
[j]Under present conditions, butylmethoxydibenzoyl methane does not elute. A GC analysis may be obtained using the manufacturer's suggested conditions (see Givaudan Corporation's Technical Bulletin for Parsol 1789).

B. Gas Chromatographic Techniques

General Techniques

The actual method used to quantitate sunscreen levels by GC is the internal standard method. The external standard methodology which is commonly used in HPLC requires a sample volume that is large enough to measure accurately. On both methyl silicone and carbowax columns we have found diethylphthalate to be a suitable internal standard. The equation by which sunscreen concentration may be calculated is:

$$\% = \frac{\text{wt sample}}{\text{wt injected}} \times \text{rel resp. factor} \times \frac{\text{area sample}}{\text{area int std}} \times \text{wt int std} \times 100$$

The setting of integration parameters is critical. The greatest source of error is from delayed start of peak integration because of front tailing. Back tailing from interactions between the analyte and the support (the column wall) is far less common in capillary chromatography than it is when using packed columns owning to the higher coating efficiency. The column-operating parameters will also have a profound effect on peak shapes, with slow program rates and large injection size contributing to front tailing and small sample size contributing to back tailing.

The removal of the sunscreen from the cosmetic base may be accomplished in several ways. The simplest is by the use of disposable columns with a bonded reversed phase (see Appendix C). The only precaution to be observed is to be certain that the sample capacity of the column is not exceeded. Sunscreens may also be extracted from a cosmetic base with organic solvents. Some solvents that are useful for this purpose are listed in Table 2.

As a specific example of the quantitative analysis of sunscreens in a commercial product, we will discuss a lotion analyzed by capillary GC. Approximately 0.1 g of diethylphthalate (DEP) is weighed accurately and added to 8 g of sample. They are then added to 25 g of water in a separatory funnel. This is then extracted three times with 25-ml portions of methylene chloride, and the bottom layer is collected. After most of the methylene chloride has been evaporated, a sufficient amount of acetone is added back to the sample to allow injection with an automatic sampler. Acetone will assure that all components extracted will remain in the solution and is sufficiently high boiling that it will not evaporate in a vial with a septum cap. Because the needle residence time and injection size have a great influence on splitter performance, quantitative precision can be achieved only with automatic injection. The sample is then injected three times and the results are averaged.

In a typical sunscreen lotion, a great number of peaks will be observed by capillary GC in addition to the sunscreen agents. They result from the other volatile components of the sunscreen base, as well as volatile impurities in the high molecular weight constituents of the base. Typical examples of the former are perfume components, cyclic siloxanes, and mineral oils. Examples of the latter are straight chain alcohols from alkyl sulfates, free fatty acids from vegetable oils and low oligomers from ethylene oxide adducts. Before the actual analysis, a test run should be made using the same column and chromatographic conditions to assure that there is no interference at the retention time of the sunscreen or the internal standard (see Appendix D for a procedure using an internal standard method of analysis).

Packed Column Techniques

Most sunscreen chemicals have sufficient volatility to be analyzed within the traditional limits of gas chromatography (24). However, their boiling points and polarities are high enough that care is required to obtain accurate quantitative results. Because they are often accompanied in the functional product by other GC elutables, such as fragrance chemicals, solvents, and emollients, the high-resolving power of capillary columns is desirable. The solubility and volatility characteristics of these chemicals make it impractical to separate them by extraction or distillation before chromatographic analysis. Finally, the boiling range of sunscreen chemicals and other elutables likely to accompany them require the use of temperature programming.

Although they are not as efficient as modern capillary columns in terms of separation efficiency, conventional, packed columns offer the analyst several distinct advantages. First, they are rugged, inexpensive, and compatible with almost any GC instrument that may be available. Second, they are more suitable for quantitative analysis as a result of their high sample capacity. These properties make them well suited to GC applications and routine monitoring.

The high sample capacity and high carrier flow rate eliminates the need for a splitter in the injection system. Thus, the problem of splitter discrimination among analytes of different boiling points is eliminated. Because the range of retention indices on a nonpolar column varies from 12.40 to 34.10, peak width can be a serious source of error in sunscreen analysis (Table 3).

The splitter also makes the effect of variable injection times or injector temperature much more significant. Because a packed column has only one path to atmosphere, pressure surges and concentration gradients from the injection process are less prone to distort the peak areas in the chromatogram.

The greater sample capacity of packed columns also overcomes the problems of incorrect integration resulting from abnormal peak shapes. When polar compounds are analyzed on nonpolar (methyl silicone) columns at slow program rates, asymmetric peaks result. The peaks exhibit front tailing and result in peaks that resemble right triangles. Some modern integrators use the second derivative of the front slope of the peak (the curvature) to activate area integration. This prevents area integration from being initiated by upward baseline drift. However, with the use of very thin film thickness in capillary columns (0.2–0.3 μm) and the necessity to use nonpolar stationary phases for the elution of high-boiling sunscreens, this "overloaded" peak shape occurs at concentration ranges required to yield a suitable response with the flame ionization detector. As a result, peak integration is delayed until much of the area at the beginning of the peak has been lost.

In terms of difficulties encountered with the integration of tailing peaks, thin-film capillary columns give superior performance to packed columns. In this case, tailing results from interaction between the analyte and the support, rather than with the stationary phase. Therefore, the greater film thickness of the packed column does not improve the peak shape, whereas the greater inertness of fused silica and the higher coating efficiency of the capillary column does.

The greater sample capacity of packed columns also allows the use of thermal conductivity detectors (TCD). The TCD has several advantages over the flame ionization detector. The range of response factors on the TCD is much smaller than encountered with other detectors. It is also a universal detector responding to noncombustible compounds, such as water, which might otherwise require a separate analysis. In addition, the TCD is nondestructive. When sample identification requires trapping of a GC peak to generate IR, UV, or NMR data, the high sample capacity of a packed column, in combination with a TCD, which does not require effluent splitting, yields an ideal system for preparative GC. If the area percentage of the unknown peak is greater than 5%, enough material may be collected in one or two runs to yield UV or IR data. If the concentration is low or if NMR data are required, repetitive GC runs may be necessary.

Many methods for trapping GC peaks for spectroscopic analysis have been described in the literature (7,9). We have found a very simple technique, which uses standard 18-gauge needles with Luer fittings, to be the most effective. A piece of glass tubing, approximately 1/4″ longer than the needle, is cut from soft glass stock and sealed at one end with a propane burner. The male Luer fitting from a glass syringe may be removed and attached to a swagelock fitting with silver solder. The swagelock fitting may then be connected to the outlet from the TC detector. As the appropriate peak elutes from the GC, a needle with its glass "vial" may be connected by the Luer fitting to the GC. After the peak has been collected, the needle and vial are transferred to a clinical centrifuge where the liquid is pulled from the walls of the needle and vial to a single droplet at the bottom. A glass capillary pipette is then used to transfer the liquid to the spectroscopic cell.

The trap, of course, must be maintained below the boiling point of the analyte. This may be done by dropping a volatile liquid, such as acetone or methylene chloride, on the glass trapping vial. However, the mistake most commonly made in attempting to trap a GC peak is to cool the trap too much. This will result in the formation of fine aerosol droplets that will be blown through the trap by the carrier gas and yield low recoveries of the analyte. For fairly high-boiling material, such as the sunscreens, it is necessary to heat the trap to between 40° and 70°C to prevent aerosol formation. At these temperatures the traps will still be sufficiently cold to condense the sunscreens.

The extinction coefficients of sunscreens are sufficiently high to record the UV spectrum in a conventional 1-cm or 1-mm–path length quartz cuvette. The IR spectra are considerably weaker and should be run in a microcell specifically designed for this purpose. If a commercial microcell is unavailable, a satisfactory one can be fabricated from a plain NaCl plate by applying nail polish to a small (1 × 4-mm) area in the center of the plate. The unprotected area is then etched away by placing the plate face down in water for about 1 min. After any adhering water has been removed by blotting away with a paper towel, the protective nail polish is removed with methylene chloride. When the sample is applied to the raised portion of the cell and a plain salt plate is placed on top, the sample is held in the small raised area in the center of the cell by capillary action. Currently, the use of the coupling technique of GC–FTIR has eliminated the need of separately trapping on a GC followed by analyzing by infrared spectrophotometers. GC–FTIR instruments, however, are still quite expensive and are not readily available in most laboratories dealing with the analysis of sunscreens.

The major limiting factor in the use of packed columns is the relatively small number of theoretical plates available. When dealing with the pure sunscreens themselves, this is not a problem, and packed column GC is the best way to obtain rapid and quantitative accurate results. When dealing with sunscreens as part of a complex matrix, such as are available in commercial preparations, difficulty may be encountered in separating the peaks of interest from other volatile components.

Capillary Column Techniques

Capillary columns have achieved their current popularity for two fundamental reasons. First, they afford up to 5000 theoretical plates per meter in 50-m column lengths. Second, the cross-linked capillary stationary phase yields excellent, reproducible retention times, stable baselines, and long working lives. In addition, the gas velocities used with capillary columns are ideal for interfacing with the mass spectrometer, as well as selective detectors such as the IR, FPD, TID, and ECD. The one negative feature of capillary columns is that the low sample capacity of thin film stationary phases require complicated sample introduction techniques, which make accurate quantitation more difficult.

The separating power of capillary columns is sufficiently high that it is not necessary to have columns with a wide array of stationary phases available. Generally, a cross-linked methyl silicone phase and a bonded polyethylene glycol phase are all that is required. If the same percentage is obtained on these two phases, it is highly unlikely that another component of the mixture is coeluting with the peak of interest. Additionally, because the two different columns can be connected to the same injection port by a two-hole ferule, the analysis on two dissimilar phases can be carried out simultaneously.

Before the advent of cross-linked stationary phases, the temperature limitations of capillary columns would have made the analysis of sunscreens difficult. Extremely long retention times and broad peaks would be the best that could be expected with nonpolar phases. Cross-linked methyl silicone capillary columns can, by contrast, be programmed up to 300°C with a flat base line at zero attenuation. This allows all the nonionic sunscreens to be eluted as sharp peaks. Bonded polyethylene glycol can be programmed up to 250°C with moderate bleed, but the late peaks do show significant broadening.

The cross-linked stationary phases also require far less stringent control in sample preparation than conventional columns. Should the column eventually become contaminated from the build-up of nonvolatile impurities, it may be regenerated by simply cutting several feet from the head end of the column. If this is insufficient, it may be rinsed with a suitable solvent, such as methylene chloride, without in any way degrading the stationary phase. Likewise, extraction solvents need not be evaporated because the cross-linked phases can withstand the splitless injection of several microliters of solvent.

The 1-ml/min flow of helium that a capillary column produces is within the capacity of the MS pumping system without the need to enrich the sample concentration by the use of a selective interface. This eliminates problems associated with sample adsorption on the interface. The sample comes in contact with only the inert fused silica column since once it leaves the injection port.

Cross-linked methyl silicone capillary columns are most commonly used for GC–MS. On conventional quadrupole and magnetic sector instruments, the amount of material required to yield satisfactory mass spectra is often great enough to result in peak overloading with the relatively polar sunscreens. The Perkin Elmer/Finnigan Ion Trap Detector is sufficiently sensitive that spectra can be readily generated on sunscreen mixtures without peak shape distortion. Typical injection size on the ITD would be 0.5 μl of a 1% solution of sunscreen, using a split ratio of 200:1 (see conditions in Appendix E).

The detection limits obtained by the FID can also be lowered for some sunscreens by use of the thermal ionic detector (TID). The detector will respond to nitrogen-containing sunscreens, such as derivatives of anthranilic acid and *p*-aminobenzoic acid. The TID also confirms the identity of a sunscreen when it is combined with a retention index.

Qualitative analysis can also be accomplished with FTIR. Sensitivity is not quite as good as the preceding methods, and the use of wide-bore capillary columns is advisable. A further problem with the FTIR method is that the vapor-phase spectra recorded at elevated temperatures cannot be directly compared with literature spectra run at room temperature in the liquid phase. Consequently, it is necessary to generate a library of spectra on the GC–FTR instrument.

To confirm the identity of a GC peak, it is necessary to obtain the retention index on another column of dissimilar polarity. The most widely used pair of stationary phases is polyethylene glycol and methyl silicone. With capillary columns, two are easily attached to the same injection port, allowing the polar and nonpolar retention indices to be determined simultaneously. These two column types are particularly useful because their retention behavior toward various classes of chemicals is quite different.

Although the original method developed by Kovats (25) was based on the use of straight-chain hydrocarbons during isothermal runs, we have adopted the method of Van den Dool and Krantz (14), which uses ethyl esters and temperature programming. The wide range of boiling points and polarities encountered is sunscreen analysis make the Van den Dool method more practical. In addition, the elimination of the logarithmic relation simplifies calculations. In the Van den Dool system the retention index of a substance is simply proportional to its difference from the retention time of the two adjacent markers (see Appendix B).

$$I_{\mathrm{E}} = I_{\text{low marker}} + \frac{(t \text{ sample} - t \text{ low marker})}{(t \text{ high marker} - t \text{ low marker})}$$

where I_{E} = retention index calibrated with ethyl esters

Qualitative information about an unknown can also be obtained from capillary GC by the use of retention indices. In fact, the difference in retention indices for the two phases is characteristic of the functionality of the analyte. The behavior of methyl silicone is simpler and has often been described as following boiling point (bp) order. A more accurate description would be that retention time on methyl silicone is a function of molecular weight. Taking as an example two volatiles the boiling points of which are known with great accuracy, methanol (MW 32 bp 65°C) elutes well before acetone (MW 58 bp 56°C).

The retention behavior of polyethylene glycol is much more complex, with molecular weight and polarity both playing a part, but with hydrogen bonding being the decisive factor. So strong is this effect that one cannot even predict that a member of a homologous series will elute in the order of carbon chain length. Thus, the first three carboxylic acids elute in the following order: acetic, formic, and propionic. The effect of intramolecular hydrogen bonding is particularly evident in *ortho*-substituted aromatics that contain adjacent donor and acceptor groups. For sunscreens such as anthranilates, salicylates, and *ortho*-hydroxybenzophenones, retention indices will be much lower than expected (see Table 3). For a specific procedure refer to Appendix E.

C. Liquid Chromatographic Techniques

Liquid-phase chromatography has the great advantage of being applicable to the separation of all sunscreens, even those that are salts. In addition, the injection process is simple and free from the discrimination problems encountered in GC. On the negative side, there are far fewer theoretical plates with which to resolve the sunscreens from each other and from the other constituents of the cosmetic base. Furthermore, the linear range of the most important detectors (UV) is limited and difficult to reproduce.

Normal-phase chromatography on silica gel columns produces a good separation of the sunscreen chemical from one another. However, problems with the retention time reproducibility, particularly when cosmetic base ingredients are also present, have limited silica's use in routine work. Because separation on silica is based on adsorption, retention behavior is profoundly affected by the samples that have been previously run. Obviously, all samples run on silica must be free of water because it permanently changes the surface composition of silica.

Silica separates compounds in the order of their ability to bond to the polar surface of the stationary phase. Each chemical has to be displaced on the active surface by a solvent molecule with a higher affinity for it. Because the salicylates have two functional groups, ester and phenol, they would be expected to be strongly retained by the silica column. In fact, they are eluted very quickly with only the hydrocarbons being more easily displaced. This is due to the *ortho*-substitution pattern of the hydroxyl and carbonyl groups, resulting in intramolecular hydrogen bonding. As a result, these groups are unavailable for bonding with the surface of the silica and elute very rapidly. A further result of this lack of affinity of salicylates for silica is that isomers of homomenthyl salicylate and octyl salicylate elute together. The parabens (*p*-hydroxybenzoates) on the other hand, interact rather strongly with silica and are retained longer on the column, allowing complete separation from their closely related isomers, the salicylates.

By far, the most widely used mode of liquid chromatograph for the separation of sunscreens is reversed-phase (26). With stationary phases of the "C-18" type, the problems of column memory effects from both the analyte and the mobile phase, and the resulting retention time instability are completely overcome (10). Furthermore, solvent gradients to cover the wide range of polarities encountered in sunscreens may be readily used.

Mobile phases used in the HPLC separation of sunscreens are generally the conventional methanol or acetonitrile mixtures with water. Both organic solvents have excellent transmittance in the 270- to 350-nm region where sunscreens have their maximum absorbance. After they have been properly degased, both mix smoothly with water if a gradient is desired.

As many of the sunscreens have either acidic or basic properties, it is necessary to control the pH of the mobile phase at a level that locks the analyte into one of its alternate forms to obtain sharp chromatographic peaks. Because most reversed-phased columns have a silica backbone that is unstable above pH 7, acids or acid buffers are normally used to suppress ionization equilibria. The acid or buffer used will also have an effect on the solute retention time, presumably owing to ion pair formation.

The one major limitation of HPLC as compared with GC is the limited number of theoretical plates available. This results in difficulties in separating the entire range of sunscreens that are commercially available in a single run. An example of this is given by Gagliardi et al. (26a), who describe the analysis of sunscreens available on the European market. The 15 sunscreen chemicals studied were divided into two classes: group A, with 6 chemicals that were soluble in acidified water, and group B, with 9 chemicals that were not soluble. All the chemicals in group A were either aromatic amines that were converted to their ammonium salts, or sulfonic acid derivatives, the water solubility of which was sufficiently great that the 10^{-3}-M sulfuric acid used would not suppress it. All the chemicals in group B were either neutral or weak acids and, therefore, were easily extractable into chloroform. Using an RP-18 column under ion-pairing conditions with group A and partition conditions with group B, a complete separation was achieved for all 15 chemicals.

In our laboratory, we have encountered a similar problem when attempting to separate the ten sunscreens that are of commercial importance in the U.S. market. By using a methanol/water gradient and phosphoric acid to suppress the acid base equilibria of the analytes, we were able to separate all components on a C-18 reversed-phase column except the first isomer of homomenthyl salicylate and octyl *p*-methoxycinnamate (see Appendix F). Because the ratio of these two isomers is fairly constant in commercial samples and the second peak is resolved (but not to the baseline), an approximate analysis of all ten sunscreens could be obtained in one run (see Table 3). If the salicylates are first separated out as a group from the other sunscreens, both fractions can then be completely separated with baseline resolution. Salicylates can be extracted as their phenolate ions using aqueous KOH. Salicylates are unusually stable toward saponification and there will be no loss of ester if the prescribed concentrations and extraction times are observed.

V. CONCLUSION

A survey of the published literature revealed the lack of advanced analytical procedures for the assaying of sunscreen ingredients present in a complex cosmetic formulation. We have presented several procedures that use both capillary gas chromatography and high-performance liquid chromatography for the qualitative and quantitative determinations of sunscreen active ingredients in cosmetic formulations. In HPLC we have shown two procedures, one using normal-phase silica gel separations and the other reverse-phase bonded C-18 column separations and identifications. A table of retention indices for capillary GC in two phases, one polar and the other nonpolar, along with the retention indices on thin-layer chromatography and high-performance liquid chromatography was presented.

With a group of substances as diverse as sunscreens that include both ionic and nonionic compounds, a wide range of volatility and molecular weights, and a variety of isomers, we will probably not attain the ideal single method that can determine any or all of them in one matrix. The occurrence, however, of more than four sunscreens in any cosmetic formulation is at present highly unlikely. But with the tools that are available, the analyst can identify and quantitate any combination of sunscreens he or she may encounter, although the task might be laborious and time-consuming. We hope this review will be helpful to those engaged in sunscreen analysis and that it will point out to researchers those areas where opportunities exist for faster, more sensitive, and reliable methods.

APPENDIX A: SHAMPOO AND CONDITIONER EXTRACTION PROCEDURE*

A new method was developed for the extraction of sunscreen components from shampoo and conditioners. The procedure is as follows:

1. Take 50 g (approx.) of shampoo and mix with enough anhydrous sodium sulfate to absorb all the water. Let this dry into a solid block.
2. When dry, take the solid sodium sulfate/shampoo block and grind in a mortar and pestle. Extract with methylene chloride and filter.
3. Evaporate and inject into capillary GC. When a good curve is achieved, take the extract and again dissolve in methylene chloride and filter.
4. In comparing the two extracts, some differences will be noticed. In the first extract, some shampoo base components will still be present in the extract, as well as any straight-chain alcohols and ester. By adding urea to the extract, the shampoo base components will be complexed and removed, as well as any straight-chain alcohols or esters in the fragrance. This is why two extracts are necessary for comparing the chromatographs.
5. Fragrance components will also be extracted by this procedure.

---

*Felton Worldwide, Inc., Research and Development Laboratories Technical Bulletin 199.

APPENDIX B: DETERMINATION OF RELATIVE RETENTION INDICES BY ETHYL ESTERS*

The use of relative retention indices is a valuable tool in aiding in the identification of individual components of a complex mixture by GC. For a given column, the indices are constant and reproducible within experimental error.

When running samples, the instrument should be properly warmed up, checking to make sure that the column is void of contaminants. The ethyl ester mixture is injected first and the sample is run immediately after. (For multiple sampling, the ethyl esters should be reinjected every fifth run—1, 6, 11, ...)

The ethyl esters consist of the C_2, C_3, C_4, C_5, C_6, C_7, C_8, C_9, C_{10}, C_{11}, C_{12}, C_{14}, C_{16}, C_{18}, and C_{20} of the homologous series where R represents the remaining carbon chain: RCOO CH_2CH_3.

The mixture is prepared by adding approximately equal amounts of each pure ethyl ester. Fifty milliliters of this mixture should last a very long time.

Before injecting the ethyl ester mixture, a sample of isobutane is run by placing a syringe into a cigarette lighter and withdrawing some vapor. This serves as a starting point for and can be stopped as soon as the peak comes out. For noncapillary column runs, air can be used instead of isobutane by pulling back on the syringe containing the ethyl ester sample to let some air in and then injecting.

Injection sizes should be about the same for all samples, and the run times should all be equal (about 60 min). The pattern obtained (but not necessarily the retention times) should be similar to the accompanying example.

To calculate the relative retention index of any given peak, the following procedure is used:

1. For each sample peak to be analyzed, sandwich is actual retention time between the nearest two ethyl ester peaks (next highest and nearest lower marker).
2. Take the actual difference between the two ethyl ester peaks needed.
3. Take the difference between the sample peak and the lower ethyl ester.
4. Divide step 3 by step 2. Note that when using ethyl ester markers where the lower value ≥ 12, divide the difference by 2 as there is a two-carbon difference between the markers.
5. Add step 4 to the lower ethyl ester marker value.

This is the relative retention index of each of these sample peaks. For a given column, it should be repetitious, regardless of the machine and conditions.

---

*Felton Worldwide, Inc., Research and Development Laboratories Technical Bulletin 130.

APPENDIX C: PREP SEP EXTRACTION PROCEDURE*

A method was developed to extract organics from small sample volumes. This involves using C-18 Prep Sep extraction columns from Fisher Scientific. The procedure is as follows:

1. First condition the column, starting with methanol, then with water, buffer, or solvent. In most samples, water will be used. To do this, pipette a small amount of solvent into the top of the extraction column, place a column in a vial, and centrifuge until the solvent passes through the column into the vial.
2. If the solution is highly concentrated or contains a large amount of organic material (e.g., ethanol), dilute sample first with several times its own volume of water. Next load the sample onto the column and the centrifuge into a vial. Organoleptically significant material will remain on the column.
3. Now elute the organics with a nonpolar solvent, such as methylene chloride, using centrifugation as before. This volume is kept relatively small to expedite recovery by evaporation.
4. The organics are now ready to be analyzed by GC.
5. Discard the Sep-pack.

APPENDIX D: A METHOD FOR INTERNAL STANDARD ANALYSIS*

A method of internal standard analysis was developed for use with capillary CG. It enables the user to determine quantitative values for various peaks. The procedure is as follows:

1. Determine which peaks you want to quantitate and separate them as best you can.
2. Choose an internal standard that will not interfere with any unknown peaks, and will also show a linear response to the GC.
3. Make solutions of exact amounts of the unknowns plus exact amounts of the internal standard. Everything must be soluble, volatile, and not so concentrated that the GC will be overloaded.
4. Make the known solutions at various concentrations and inject each one into the GC. These are your standards.

---

*Felton Worldwide, Inc., Research and Development Laboratories Technical Bulletin 201.
*Felton Worldwide, Inc., Research and Development Laboratories Technical Bulletin 197.

5. Make a solution of exact weights of unknown plus internal standard and inject into GC.
6. Record exact area counts of unknown peaks, internal standard peaks, and corresponding counts from the peaks in the unknown samples.
7. The calculations are as follows:

$$\text{Response factor} = \frac{\text{weight of sample}}{\text{weight of int std}} \times \frac{\text{area count of int std}}{\text{area count of sample}} \tag{1}$$

Obtain response factors for each individual sample on as many columns as needed. Average the separate concentrations to obtain one average response factor per known constituent per column.

Weight of sample peak in unknown =

$$\text{Wt of int std in unknown} \times \text{AV resp fact} \times \frac{\text{area count of sample}}{\text{area count of int std}} \tag{2}$$

An example of this calculation using menthanol follows:

	Menthol (g)	Area on OV1	Area on CBX	Int Std (g)	Area on OV1	Area on CBX
Standard 1	0.595	1110.2	2243.1	0.0994	1521.0	2412.9
Standard 2	0.0629	1121.8	2500.1	0.998	1363.9	2451.3
Standard 3	0.0612	1066.3	2403.8	0.1013	1324.8	2471.0
Standard 4	0.0534	926.7	2103.5	0.1074	1377.9	2577.8
Unknown	—	92.7	206.0	0.1069	1231.7	2430.0

Response factor for menthol in standard 1 on OV1 [Eq. (2)]

$$\frac{0.0595}{0.0994} \times \frac{1521.0}{1110.2} = 0.820$$

Determine the remaining response factors and get their average:

Average response factor for menthol on OV1 = 0.7691

Average response factor for menthol on CBX = 0.6231

Determine the weight of menthol in the unknown sample [Eq. (2)]:

Weight of menthol in unknown in OV1 =

$$0.1069 \times 0.769 \times \frac{92.74}{1231.7} = 0.00619 \text{ g}$$

Weight of menthol in unknown on CBX =

$$0.1069 \times 0.6231 \times \frac{206.0}{2430.0} = 0.00565 \text{ g}$$

APPENDIX E: GC ANALYSIS OF SUNSCREENS IN COSMETIC PRODUCTS*

A. Preparation of calibration sample for GC analysis
   1. Weigh 1 g of each of the following:
      a. Homomenthyl salicylate
      b. Homomenthyl salicylate
      c. Oxybenzone
      d. Menthyl anthranilate
      e. Octyl dimethyl PABA
      f. Octyl *p*-methoxycinnamate
      g. Diethylphthalate (DEP) (internal standard—''ISTD'')
   2. Add 21 g of acetone and mix well.

B. Preparation of sunscreen lotion for GC analysis
   1. Weigh 7–8 g of sunscreen lotion and 0.09–0.13 g of DEP in a beaker. Add about 20–30 g of water and mix well until it is homogeneous.
   2. Pour into a separatory funnel and add approximately 20–30 g of methylene chloride. Mix it gently for about 30 s, releasing the pressure two or three times in between.
   3. Let the mixture settle down for a few minutes and collect the methylene chloride (bottom layer) in a beaker. Evaporate methylene chloride completely.
   4. Add 65% of acetone into the extracted sample.
   5. Pour the calibration sample into three different GC vials. Run the extracted sample and the calibration samples back to back on the GC.

C. Calculation
   1. Response factor calculation for the calibration sample:

$$\frac{\text{weight of octyl sal}}{\text{weight of ISTD (DEP)}} \times \frac{\text{area count of ISTD}}{\text{area count of octyl sal}} = \text{RF}$$

   Follow this calculation for each sunscreen in the calibration sample. Do the same for all three calibration samples and obtain the average response factor for each sunscreen.

   2. Calculation of the extracted sample:

$$\begin{aligned}&\text{Weight of sunscreen in cosmetic product}\\&\quad = \text{average response factor}\\&\quad \times \text{weight of ISTD added into sunscreen lotion}\\&\quad \times \frac{\text{area of extracted sunscreen sample}}{\text{area of ISTD}}\end{aligned}$$

*Felton Worldwide, Inc., Research and Development Laboratories Technical Bulletin 244.

Following is a typical set of GLC conditions for the analysis of sunscreen chemicals:

Sample size:	0.1 μl	Split ratio:	200:1
Oven temp:	100–250°C	Attenuation:	0
Program rate:	6°C/min	Threshold:	0
Injection port temp:	210°C	Detector temp:	250°C
Detector:	FID	Carrier gas:	Helium
Carrier flow:	1 ml/min		

GLC Columns: 50-m Ultra No. 1 Hewlett-Packard capillary cross-linked methyl silicone; film thickness: 0.33 μm, internal diameter: 0.20 mm and 60-m cross-linked carbowax Supelco Wax Film thickness: 0.25 μm; internal diameter: 0.25 mm.

APPENDIX F: ANALYSIS OF SUNSCREENS IN COSMETIC PRODUCTS BY HPLC*

A. Reverse-Phase Procedure

1. PRINCIPLE: Reverse-phase HPLC is used with external standard analysis to determine the percentage of sunscreen(s) in finished products.
2. CONDITIONS:
   a. General conditions: See Table 1
   b. Column: AS1 25 cm × 3.9 mm C-18 10-μm irregularly packed column Waters 37- to 50-μm C-18 (corasil packed) guard column
   c. Solvent: 87% acetonitrile, 10% water

   Solvents should be HPLC grade and must be filtered and degased by vacuum filtration before use.
3. PROCEDURE:
   Standard: Weigh 120–200 mg of sunscreen(s) into a 100-ml volumetric flask. Dilute to volume with methanol and mix thoroughly. Dilute this solution by pipetting a 1- to 10-ml aliquot into a 100-ml volumetric flask. Fill to volume with methanol and mix thoroughly.

   Sample: Weigh accurately 300 mg of sample into a 100-ml volumetric flask. Fill to mark with methanol and mix thoroughly. Centrifuge at 2500 rpm for 5–10 min. Filter supernatant to remove insoluble particles.

---

*Felton Worldwide, Inc., Research and Development Laboratories Technical Bulletin 211 A and B.

Equilibrate HPLC for 15 min. Inject standard. Calibrate the integrator for external standard analysis and inject the sample.

4. CALCULATIONS: External standard analysis calculations can be performed with or without the data module integrator.

   For a standard containing sunscreen A the formula is:

$$\frac{\text{mg}}{\text{ml}} \text{ of A} = \frac{\text{weight of A}}{\text{volume of solvent}} \times \text{dilution factor}$$

   For a sample containing sunscreen A

$$\% \text{ A} = \frac{\text{mg}}{\text{ml}} \text{ of A} \times \frac{\text{area of A in sample}}{\text{area of A in standard}} \times \frac{100 \text{ ml}}{\text{weight of sample}} \times 100$$

Example: Brand (X) with an SPF 8 contained both octyldimethyl PABA and oxybenzone. The quantity for oxybenzone is calculated as

$$\frac{\text{mg}}{\text{ml}} \text{ oxybenzone} = \frac{120 \text{ mg}}{100 \text{ ml}} \times \frac{10 \text{ ml}}{100 \text{ ml}} = 0.12$$

$$\text{Therefore: } \% \text{ oxybenzone} = 0.120 \times \frac{7819000}{15046000} \times \frac{100 \text{ ml}}{0.315 \text{ mg}} \times 100 = 1.98\%$$

B. Normal-Phase Procedure

1. PRINCIPLE: Normal-phase HPLC is used with external standard analysis to determine the percentage of sunscreen(s) in finished products.
2. CONDITIONS:
   a. General conditions: See Table 1
   b. Column: Waters resolve column with 5-μm spherical silica packing. Waters LC-precolumn filter with Perkin-Elmer 37- to 44-μm silica packing as guard column
   c. Solvent: 90% hexane, 10% methylene chloride, 0.25% (20 drops) glacial acetic acid
3. PROCEDURE:

   Standard: Weigh 100–500 mg of standard sunscreen(s) into a 100-ml volumetric flask. Fill to the mark with hexane and mix thoroughly.

   Water-based sample: Weigh 0.5 g of sample into a 100-ml beaker. Dissolve with 100 ml of water. Add 5.0 g of sodium chloride to break the emulsion. Drop in a magnetic stirrer and mix for 10 min. Transfer the

solution to a 250-ml separatory funnel. Pipette in 50 ml of methylene chlo-
ride and extract. Collect bottom layer and add anhydrous sodium sulfate to dry.

Oil-based sample: Dissolve 0.5 g of sample in 150 ml IPA. Add 0.5 g sodium chloride. Extract in a 1000-ml separatory funnel with 600–700 ml water and 50 ml methylene chloride. Collect the bottom layer and add anhydrous sodium sulfate to dry.

Allow HPLC to equilibrate for 45 min. Inject standard. Calibrate the integrator for external standard analysis and inject the sample.

Note: Because column packing is silica, all samples must be water-free.

4. CALCULATIONS: External standard quantitation can be done with or without data module integrator. The percentage of a sunscreen chemical A in a finished product is given by the formula:

$$\% = \frac{\text{injection vol. A}}{\text{injection vol. sample}} \times \frac{\text{peak area sample}}{\text{peak area A}} \times \text{conc. A} \times \frac{50}{\text{wt. of sample}}$$

Sample calculation: Lotion brand Y (SPF 6) containing octyl dimethyl PABA

$$\%\ \text{ODPABA} = \frac{3}{3} \times \frac{22437200}{12095900} \times 0.0315 \times \frac{50}{0.6076} = 4.81\%$$

Parameter	Normal-phase	Reverse-phase
Wavelength	*See Table 2	*See Table 2
Flow	2.0 ml/min	0.8 ml/min
Chart speed	2.0 cm/min	1.0 cm/min
Injection volume	10.0 μl	10 μl
Run time	15 min	15 min

*Because of the difference in the λmax and molar absorptivity between sunscreens the wavelength at which the samples were run was the λmax of the individual sunscreens.

HPLC system: Waters Associates
720 System controller
730 Data module integrator
WISP 710B auto sampler
450 Variable wavelength detector
6000A Solvent delivery pump

λmax and Molar Absorptivity of Some Commonly Used Sunscreens

Sunscreen	λ_{max}	Molar absorptivity
Octyl salicylate	310	4,100
Menthyl anthranilate	334	5,400
Oxybenzone	328	8,800
Octyl methoxycinnamate	289	22,900
Octyl dimethyl PABA	300	25,000

REFERENCES

1. *Federal Register*, *43*(266):38206-38269 (1978).
2. (a) Liem, P. H. and Hilderink, T. H. U.V. Absorbers in sun cosmetics 1978. *Int. J. Cosmet. Sci.*, 341–361 (1979); (b) Janousek, A. Current status of UV filters in the EEC. *Cosmet. & Toilet.*, *7*:55 (1987).
3. Butts, S. T. Use of sunscreen agents no longer confined to sun care products. *Soap Cosmet. Chem. Special.*, *10*:32 (1984).
4. Kok, M. F., Yong, F. M., and Lim, G. Rapid extraction method for reproducible analysis of aroma volatiles, *J. Agric. Food Chem.*, *37*:779–781 (1987).
5. (a) Maarse, H. and Belz, R. *Isolation, Separation and Identification of Volatile Compounds in Aroma Research*. Akademie-Verlag, Berlin, 1981; (b) Gabriel, D. M. *J. Soc. Cosmet. Chem.*, *25*:33–48 (1974).
6. Pavia, D. L., Lampman, G. M., and Kriz, G. S., Jr. *Introduction to Organic Laboratory Techniques*. W. B. Saunders, Philadelphia, 1976.
7. (a) Poole, C. and Schuette, S. *Contemporary Practice of Chromatography*. Elsevier, New York, 1984; (b) Heftmann, E. *Chromatography*. Van Nostrand Reinhold, New York, 1975.
8. Landgrebe, J. A. *Theory and Practice in the Organic Laboratory*. D. C. Heath & Co., 1977.
9. (a) Jennings, W. and Shibamoto, T. *Qualitative Analysis of Flavors & Fragrance Volatiles by Capillary Gas Chromatography*. Academic Press, New York, 1980; (b) Lipsky, S. R. *LC-GC*, *4*:432–441 (1986); (c) Lipsky, S. R. *LC-GC*, *4*:898–906 (1986); (d) Freeman, R. *High Resolution Gas Chromatography*, 2nd ed. Hewlett Packard, California (1981).
10. (a) Snyder, L. and Kirkland, J. *Introduction to Modern Liquid Chromatography*, 2nd ed. Wiley, New York, 1979; (b) Ross, M. *J. Chromatogr.*, *160*:199 (1978); (c) Latz, H. and Ernes, D. *J. Chromatogr.*, *166*:189 (1978); (d) Horvath, C. *High Performance Liquid Chromatography, Advances and Perspectives*. Academic Press, New York, 1980; (e) Runner, P. J. *Maintaining and Troubleshooting HPLC Systems*. Wiley, New York, 1981.
11. (a) Charalambous, G. *Analysis of Food and Beverages: Headspace Techniques*. Academic Press, New York, 1978; (b) Hachenberg, H. and Schmidt, A. P. *Gas Chromatographic Headspace Analysis*. Heyden, London, 1977; (c) Kolb, B. *Applied Headspace Gas Chromatography*. Heyden, London, 1980.

12. Weiss, J. *Handbook of Ion Chromatography*. Dionex Corporation, Sunnyvale CA, 1986.
13. (a) Association of Official Analytical Chemists. *Official Methods of Analysis*, 14th ed., 1984; (b) *Handbook of Chemistry and Physics*, 66th ed. CRC Press, Boca Raton FL, 1985–6; (c) Shriner, R., Fuson, R., Curtin, D., and Morrill, T. *The Systematic Identification of Organic Compounds*, 6th ed. John Wiley & Sons, New York, 1980.
14. Van Den Dool, H. and Krantz, R. *J. Chromatog.*, *11*:463 (1963).
15. David, D. J. *Gas Chromatographic Detectors*. John Wiley & Sons, New York, 1974.
16. Vickrey, M. *Liquid Chromatography Detectors*. Marcel Dekker, New York, 1983.
17. (a) Dixon R. N. *Spectroscopy & Structure*. John Wiley & Sons, New York, 1965; (b) Silverstein, R., Bassler, G., and Morill, T. *Spectrometric Identification of Organic Compounds*. 3rd ed. John Wiley & Sons, New York, 1963.
18. (a) McLafferty, F. *Interpretation of Mass Spectra*. Benjamin, New York, 1967; (b) Frigerio, A. *Essential Aspects of Mass Spectrometry*. Spectrum Publications, Flushing, NY, 1974; (c) Thomas, A. F. and Willhalm, B. *Organic Mass Spectrometry*. *11*:831–834 (1976); (d) Budzekiewicz, H., Djeraassi, C., and Williams, D. *Mass Spectrometry of Organic Compounds*. Holden Day, San Francisco, 1967; (e) Smith, S. L. *Capillary Gas Chromatography in Essential Oil Analysis* (P. Sandra and C. Bicchi, eds.). Springer-Verlag, Heidelberg, 1987, pp. 367–384.
19. (a) Colthup, N. B., Daly, L. H., and Wiberley, S. E. *Introduction to Infrared and Raman Spectroscopy*. Academic Press, New York, 1974; (b) Welti, D. *Laboratory Methods in Infrared Spectroscopy*. Heyden, London, 1972; (c) Startin, J. R. *Applications of Mass Spectrometry in Food Science* (J. Gilbert, ed.). Elsevier Applied Science, New York, 1987, pp. 289–335.
20. (a) Scott, A. *Interpretation of the Ultraviolet Spectra of Natural Products*, Vol. 7. Pergamon Press, New York, 1964; (b) Rao, C. *Ultraviolet and Visible Spectroscopy*, 3rd ed. Butterworths, London, 1975; (c) Jaffe, H. and Orchin, M. *Theory and Application of Ultraviolet Spectroscopy*. John Wiley & Sons, New York, 1962.
21. Casy, A. F. *NMR Spectroscopy in Medicinal & Biological Chemistry*. Academic Press, New York, 1971.
22. Stahl, E. *Thin Layer Chromatography*. Springer Verlag, New York, 1969.
23. Masse, M. O., Herpol, Borremans, M., Grimee, R., and Gleviczky, S. *Int. J. Cosmet. Sci.*, *4*:235–246 (1982).
24. Cumpelik, B. M. *Cosmet. Toilet.*, *5*:67–75 (1982).
25. Kovats, E. *Adv. Chromatogr.*, *1*:229 (1965).
26. (a) Gagliardi, L., Amato, H., Basili, A., Carazzutti, G., and Tonelli, D. *J. Chromatogr.*, *408*:409–415 (1987); (b) Gagliardi, L., Amato, A., Basili, A., Carazzutti, G., Gattavecchia, F., and Tonelli, D. *J. Chromatogr.*, *362*:450–454 (1986); (c) Tan, H. S. I., Sih, R., Moseley, S. E., and Lichtin, J. L. *J. Chromatogr.*, *291*:275–282 (1984); (d) Konig, H. and Ryschka, R. *Fresenius Anal. Chem.*, *315*:434–437 (1983).

37

Interpretation and Evaluation: Spectroscopic Data from Sunscreens

Nadim A. Shaath, Gideon I. Andemicael, and Lisa E. Paloympis
KATO Worldwide, Ltd./Nickstadt-Moeller, Mt. Vernon, New York

Peter M. Griffin
Ungerer and Company, Lincoln Park, New Jersey

I. INTRODUCTION

A substantial amount of spectroscopic, chemical, and physical data on sunscreen chemicals (1) is available in the literature, but it is incomplete, extremely fragmented, and often presented in incompatible formats. Because spectroscopic data is essential for the analysis and identification of sunscreen chemicals, we have prepared this chapter to serve as a single reference source for sunscreen analysts. For the analytical methods used to generate the following data, see Chapter 36.

The 17 chemicals analyzed here are among the most widely used worldwide (2). If the sunscreen under investigation is an isomeric mixture, the major isomer is described. For example, commercially available glyceryl PABA contains several impurities, including benzocaine and PABA itself (3). Another example, ethyl dihydroxypropyl PABA, commercially sold under the trade name Amerscreen P, contains four identifiable isomers (4).

The spectroscopic data compiled are the sunscreen's ultraviolet (UV), mass (MS), infrared (IR), and nuclear magnetic resonance (NMR) spectra. Table 3, Chapter 36, contains chromatographic data, including gas chromatography (GC) on both polar and nonpolar stationary phases, high-performance liquid chromatography (HPLC), and thin-layer chromatography (TLC).

The instruments used to obtain the spectra are: Perkin Elmer Lambda 4B UV Spectrophotometer, Hewlett Packard 5995 Gas Chromatograph/Mass Spectrom-

BENZOPHENONE-3

Molecular Formula: $C_{14}H_{12}O_3$
Molecular Weight: 228

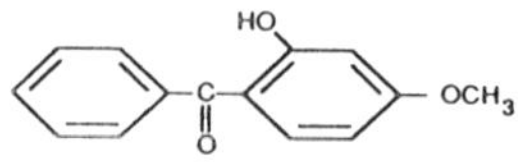

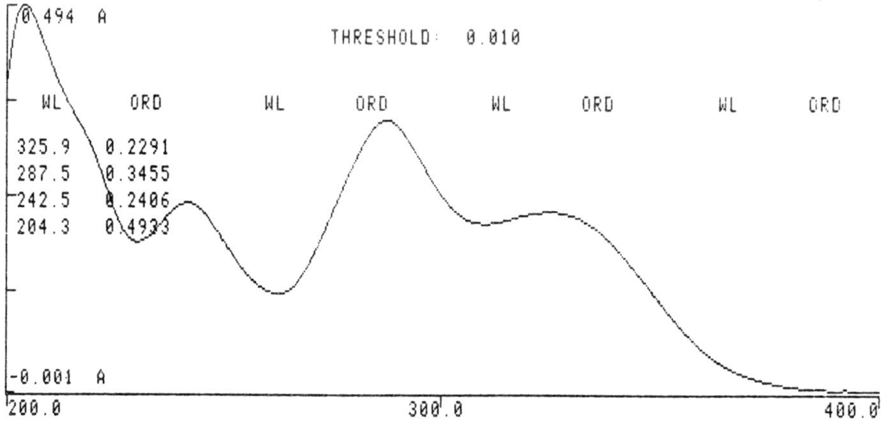

UV Spectrum. Concentration: 4.67 mg/L in ethyl alcohol; λ max: 288/325 nm; ϵ: 14,000/9400. Two absorption maxima appear for this compound, one at 288 nm and the other at 325 nm; the first with an extinction coefficient of 14,000 and 9400 for the second.

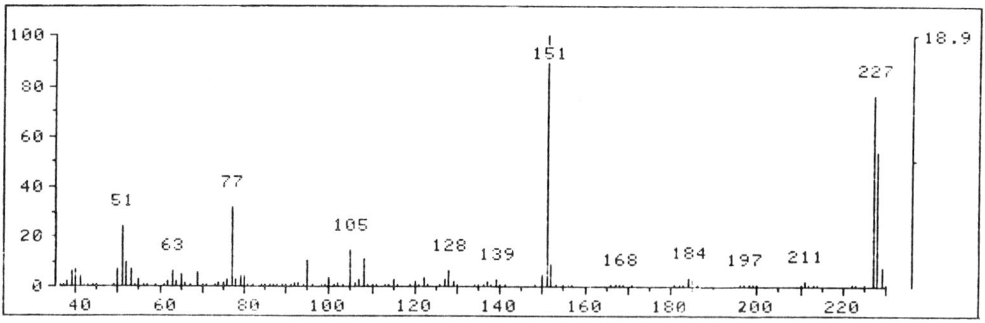

Mass Spectrum. The base peak at m/e 157 is generated by cleavage of the bond between the carbonyl group and the unsubstituted benzene ring. Cleavage of this bond with retention of the charge by the benzene ring yields the typical aromatic fragment at m/e 77. The substituted benzoyl cation is stabilized by the electron-withdrawing effect of the hydroxyl and methoxyl groups, whereas the unsubstituted benzoyl fragment at m/e 105 which results from the alternative cleavage adjacent to the carbonyl group is much weaker (18%). As expected for this type of aromatic structure, the molecular ion at m/e 228 results in an intense peak while the peak at m/e 227 is even more intense because of the aromatic ether.

BENZOPHENONE-3 (Cont.)

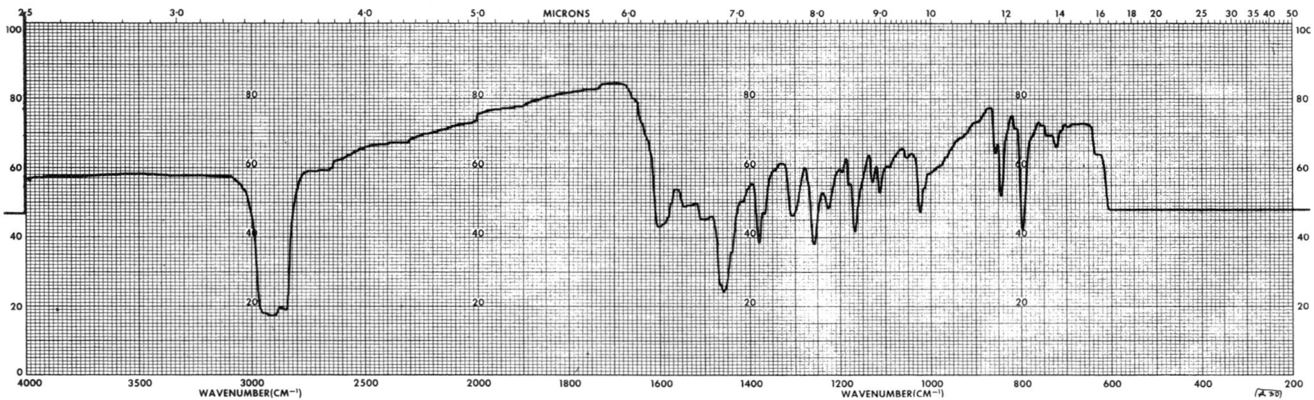

IR Spectrum. Solvent: mineral oil; NaCl plates. The carbonyl frequency is at 1590 cm^{-1} as a result of conjugation with the two benzene rings and intramolecular hydrogen bonding with the ortho hydroxyl group. Bands associated with the monosubstituted benzene ring are observed at 695 cm^{-1}. The aromatic carbon-oxygen bond is at 1230 cm^{-1}, while the aliphatic C—O is at 1085 cm^{-1}.

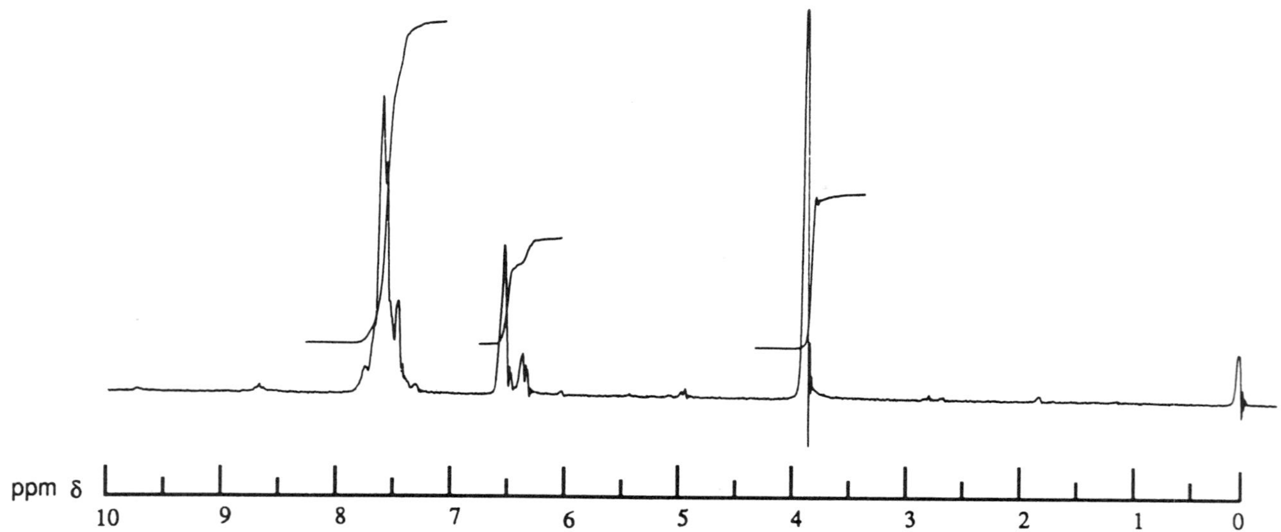

NMR Spectrum. Solvent: $CDCl_3$; 60 MHZ. The O—CH_3 resonance appears as a singlet at δ of 3.8 ppm. The balance of the aromatic signals, along with the phenolic proton, appear in the area of 6.5 to 7.6 ppm and integrate for eight protons. The carbonyl group and the hydroxyl proton are characterized more conclusively via the IR and MS spectra.

BENZOPHENONE-4

Molecular Formula: $C_{14}H_{12}O_6S$
Molecular Weight: 308

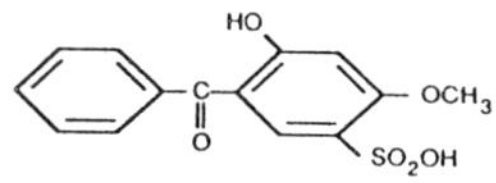

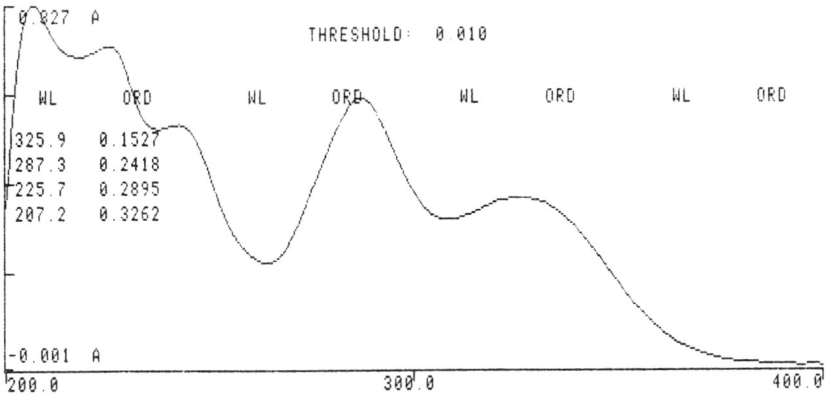

UV Spectrum. Concentration: 6.10 mg/L in ethyl alcohol; λ max: 286/325 nm; ϵ: 13,400/8,400. Here, again, as in all benzophenone derivatives, two absorption maxima are observed; at 286 nm with an ϵ of 13,400 and 325 nm with an ϵ of 8400.

(Data unavailable)

Mass Spectrum. Because the compound is a sulfonic acid, it is too polar to analyze by capillary GLC. We have produced all MS spectra using GC–MS, rather than a direct inlet to assure sample purity. Many mass spectra that have entered the literature have been contaminated with starting materials or by-products. Additionally, spectra generated under GC–MS conditions will more closely resemble those encountered by the analyst when working with real samples.

BENZOPHENONE-4 (Cont.)

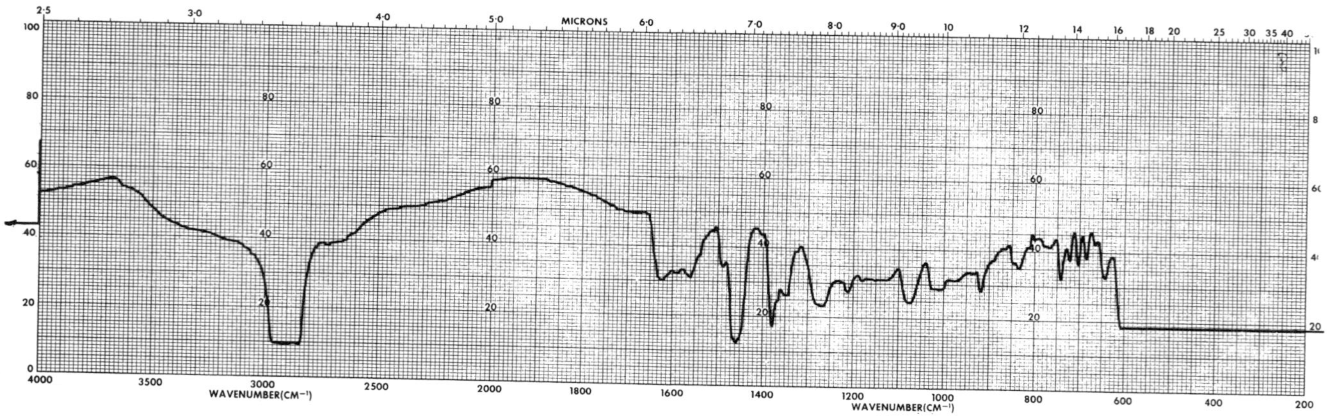

IR Spectrum. Solvent: mineral oil; NaCl plates. The conjugated, internally hydrogen-bonded ketone absorbs at 1625 cm^{-1}. The ring frequencies are at 1490, 1560, and 1590 cm^{-1}. There is a band at 1270 cm^{-1} from the aromatic ether and at 1075 cm^{-1} from the methoxyl group. The hydrated sulfonic acid absorbs at 1210 and 1015 cm^{-1}. The aromatic hydrogen frequencies absorb at 685, 722, and 742 cm^{-1}.

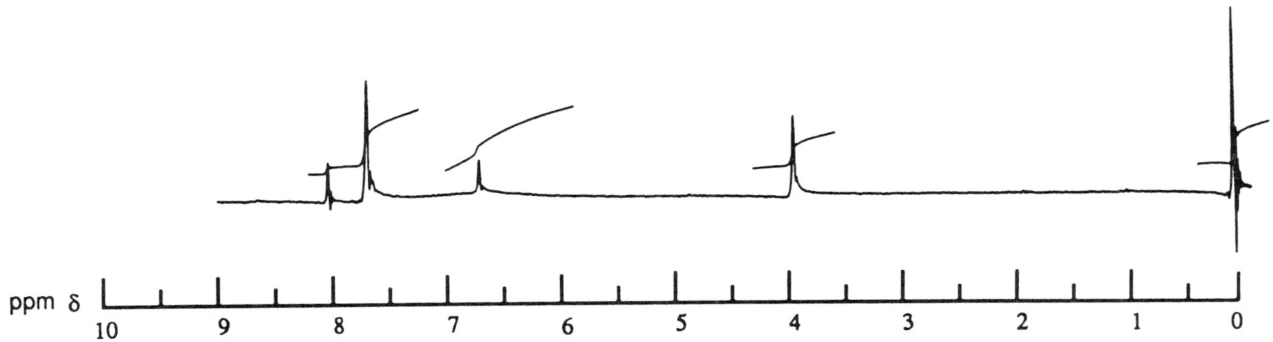

NMR Spectrum Solvent: DMSO-d_6; 60 MHz. Due to the insolubility of sulisobenzone in the usual NMR solvent ($CDCl_3$), the spectrum was run in DMSO-d_6. The seven aromatic protons, as well as the hydroxyl and sulfonic acid proton, appear in the area of 6.6 to 8.0 ppm. The OCH_3 resonates as a singlet at δ of 3.9 ppm.

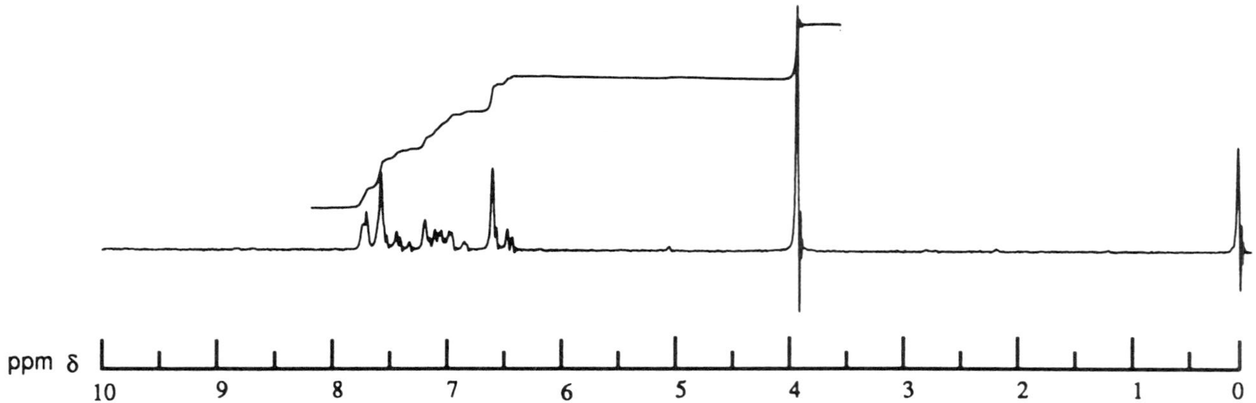

NMR Spectrum. Solvent: $CDCl_3$; 60 MHz. The two phenolic protons and the seven aromatic protons resonate at 6.4 to 7.8 ppm. The OCH_3 appears as a sharp singlet at 3.9 ppm.

DEA METHOXY CINNAMATE

Molecular Formula: $C_{14}H_{21}NO_5$
Molecular Weight: 283

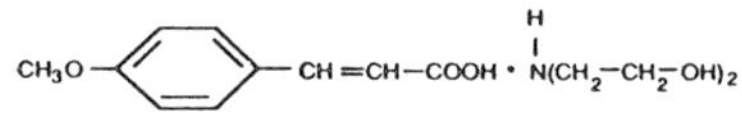

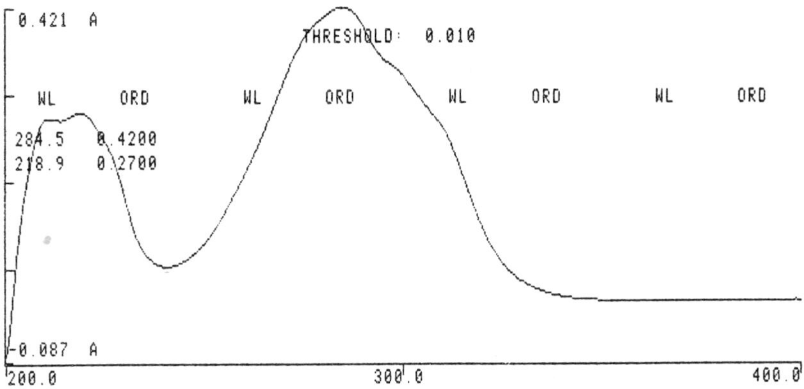

UV Spectrum. Concentration: 5.34 mg/L in ethyl alcohol; λ max: 290 nm; ε: 24,930. The λ max for this compound appears below that which would normally be expected for cinnamates since this compound is a salt and is subject to solvent effects in ethanol. Thus the λ max is shifted to 290 nm. It has, however, an excellent extinction coefficient of 24,930.

(Data unavailable)

Mass Spectrum. This compound is a salt and therefore unsuitable for GC analysis. A mass spectrum was not produced for the same reasons as benzophenone-4.

DEA METHOXY CINNAMATE (Cont.)

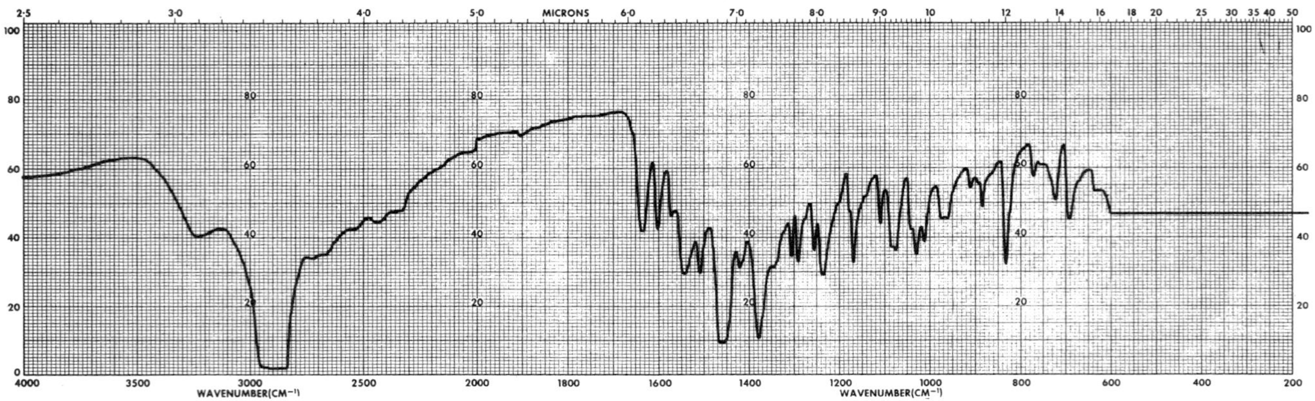

IR Spectrum. Solvent: mineral oil; NaCl plates. The carbonyl frequency of the conjugated ester salt is at 1635 cm^{-1}. The conjugated olefinic bond absorbs at 1602 cm^{-1} while the aromatic double bonds absorb at 1672, 1645, and 1608 cm^{-1}. The aromatic and aliphatic C—O bonds absorb at 1235 cm^{-1} and 1080 cm^{-1}, respectively. The hydrogen deformation bands for the para disubstituted benzene ring is at 832 cm^{-1}. The hydroxyl groups of the diethanol ammonium cation give rise to a broad bond at 3240 cm^{-1}.

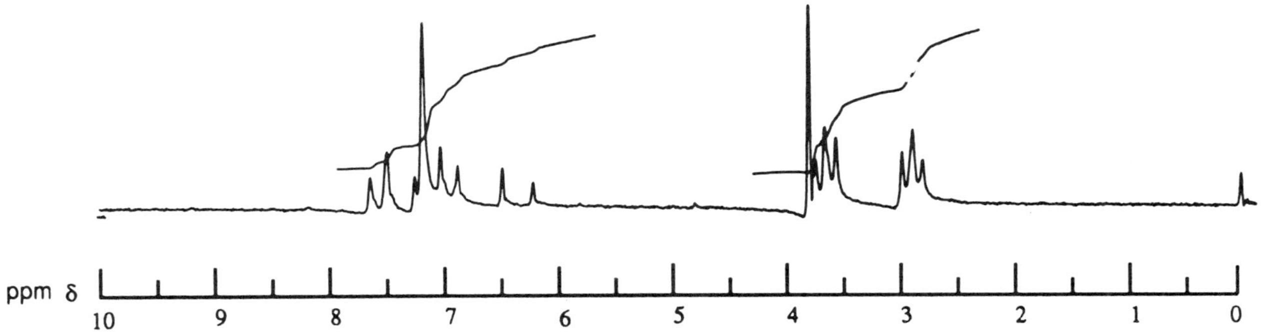

NMR Spectrum. Solvent: DMSO-d_6; 60 MHz. Being a salt, the NMR spectrum was run in DMSO-d_6 with the four aromatic protons, the two olefinic protons, the N—H and the two O—H protons appearing as a multiplet from 6.2 to 7.6 ppm. The $(CH_2—OH)_2$ appear as a triplet integrating for four protons due to their proximity to two CH_2 protons and they resonate at 3.6 ppm. The (HN—CH_2—) also appear as a triplet integrating for 4 protons due to their proximity to two —CH_2 protons and they resonate at 2.80 ppm. Finally, the OCH_3 protons integrate for three protons and they resonate at 3.70 ppm.

ETHYL DIHYDROXYPROPYL PABA

Molecular Formula: Mixture
Molecular Weight: 273 (average)

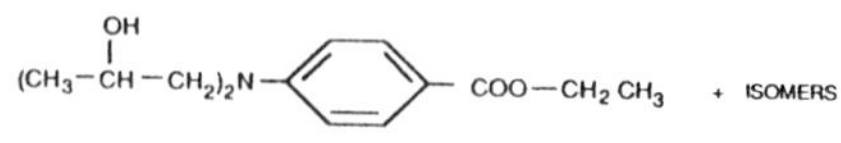

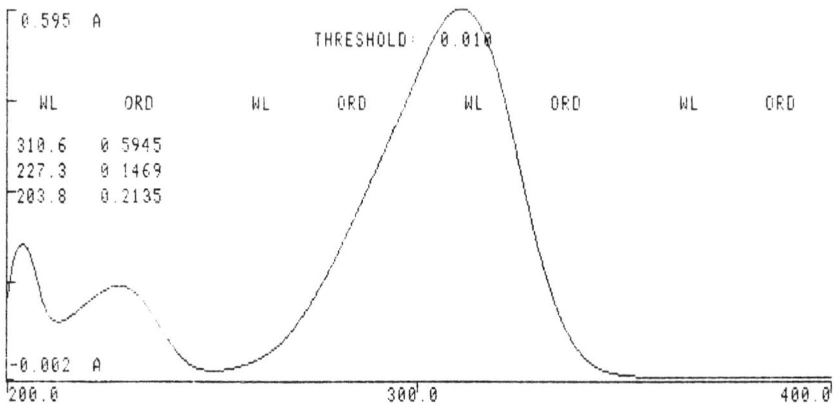

UV Spectrum. Concentration: 4.49 mg/L in ethyl alcohol; λ max: 312 nm; ϵ: 27,000. This compound exhibits a λ max at 312 nm where most PABA derivatives absorb. The extinction coefficient (27,000) is almost comparable to octyl dimethyl PABA itself.

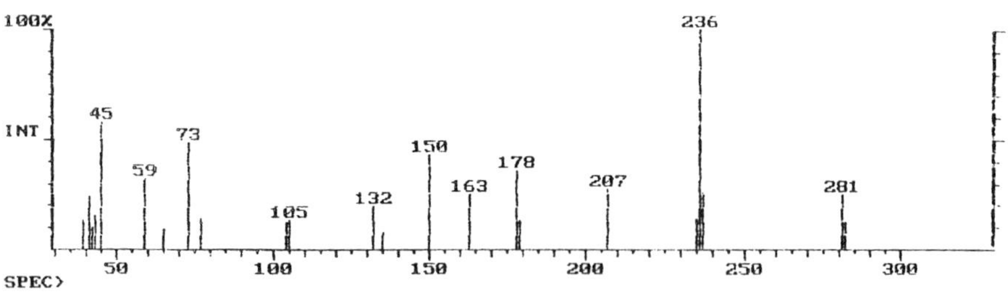

Mass Spectrum. The molecular ion at m/e 281 is clearly visible. The base peak at 236 m/e results from the loss of 45 amu (OCH_2CH_3) to yield the substituted benzoyl ion. The next major fragment at m/e 178 represents the loss of one of the two equivalent β hydroxypropyl substituents with the retention of a proton by the amine. Finally, this fragment loses 28 amu (CO) from the carbonyl group to yield m/e 150. Fragmentation of the secondary alcohols produces a strong ion at m/e 45 ($H-O^+=CH-CH_3$).

ETHYL DIHYDROXYPROPYL PABA (Cont.)

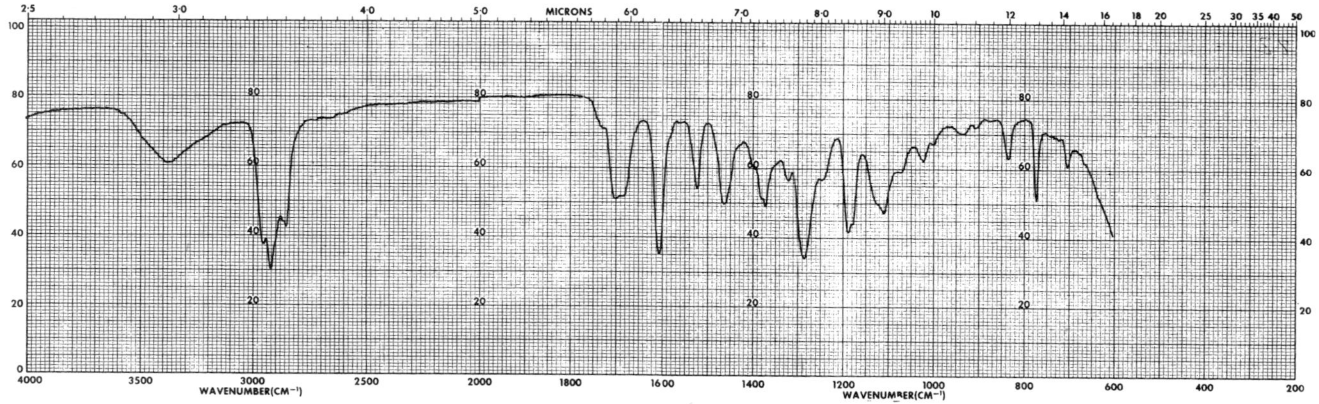

IR Spectrum. Solvent: mineral oil; NaCl plates. The two secondary hydroxyl groups yield a broad band centered at 3350 cm^{-1}. The carbonyl band is also broadened and centered around 1690 cm^{-1}. The intense aromatic bands are at 1605 cm^{-1} and 1522 cm^{-1}. The aromatic ester produces an intense band at 1283 cm^{-1} and the aromatic hydrogen deformation band occurs at 835 cm^{-1} and 772 cm^{-1}.

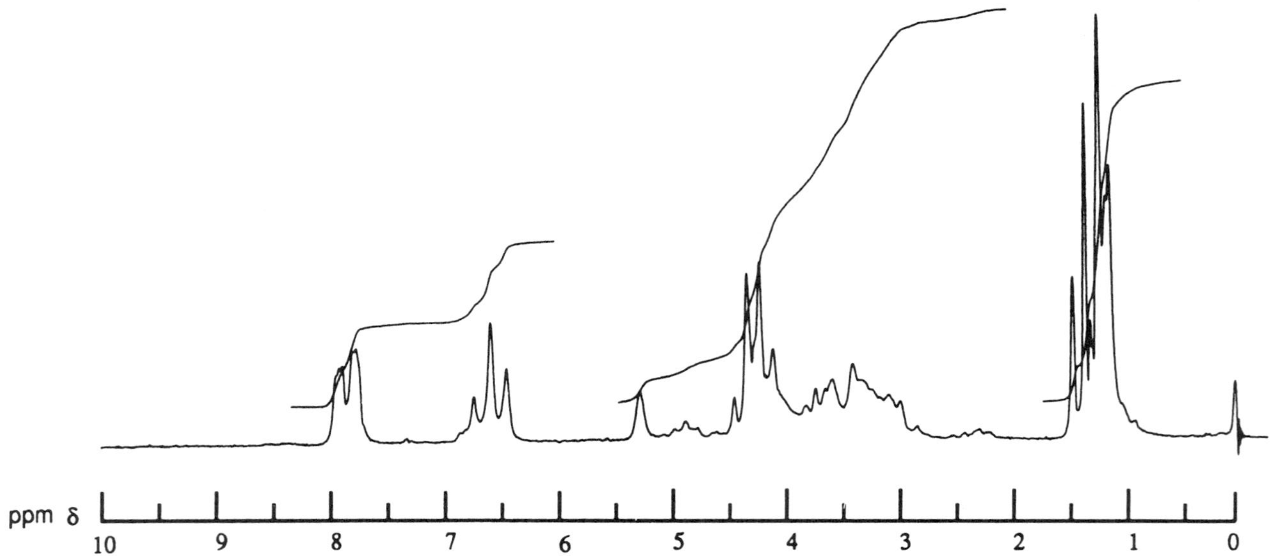

NMR Spectrum. Solvent: $CDCl_3$; 60 MHz. Since this compound is commercially unavailable in pure form, but rather is available as a mixture (see GC data), it is quite difficult to assess the character of this product by NMR spectroscopy. Nevertheless, evidence to its structure can be ascertained through a few features in the NMR spectrum. The aromatic protons appear as an AB quartet with the two protons next to the carbonyl resonance at 6.6 ppm and the two next to the nitrogen atom at 7.8 ppm. The total aliphatic proton signal is present as compared to the aromatic proton signals. The mass spectrum provides conclusive evidence for its structure and the gas chromatographic pattern reveals the number of isomers and impurities present.

GLYCERYL PABA

Molecular Formula: $C_{10}H_{13}O_4$
Molecular Weight: 211

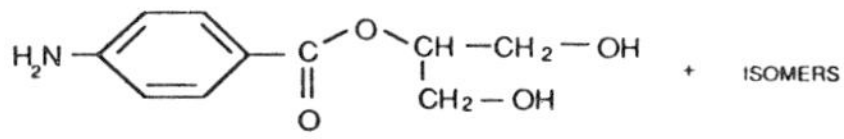

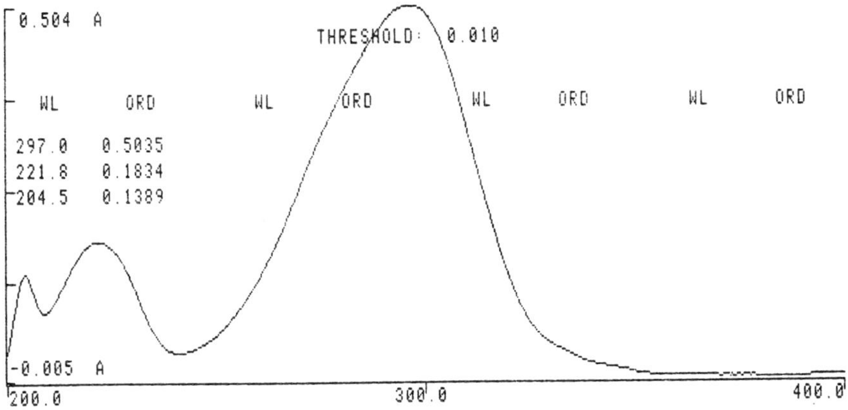

UV Spectrum. Concentration: 5.67 mg/L in ethyl alcohol; λ max: 297 nm; ϵ: 18,700. Since this compound is unusually polar in character, it will experience a shift in its spectrum due to solvent effects. Thus the λ max is 297 nm with a lower extinction coefficient of 18,700, as compared to octyl dimethyl PABA. Whenever excessive solvent interactions occur in the ground state (e.g., PABA has an ϵ of 15,300 as compared to octyl dimethyl PABA's 27,300), a lowering of the absorption is observed since symmetry is somewhat perturbed and resonance delocalizations diminish.

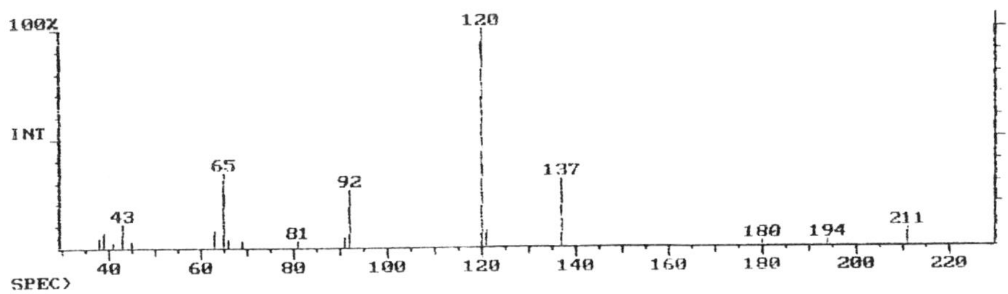

Mass Spectrum. This glycerol ester yields a very weak molecular ion at m/e 211. The ester cleaves with the transfer of one proton from the alcohol to produce the free acid at m/e 137. The *p*-aminobenzoyl fragment at m/e 120 is the base peak. Elimination of CO yields the fragment at m/e 92.

GLYCERYL PABA (Cont.)

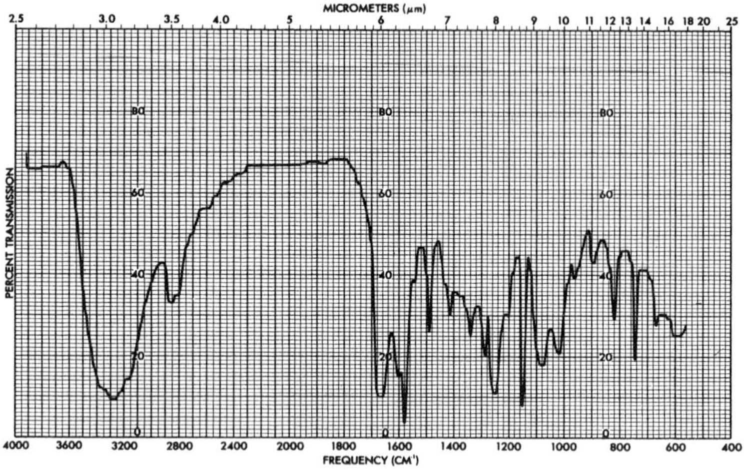

IR Spectrum. Solvent: neat; NaCl plates. The strongest band in its IR spectrum comes from the two unesterified hydroxyl groups in the glycerol portion of the molecule. It is a broad band centered at 3350 cm^{-1} and is superimposed on the weaker symmetrical and unsymmetrical N—H stretching frequencies which are only visible as shoulders at 3460 cm^{-1} and 3215 cm^{-1}. The double bonds of the benzene ring absorb at 1605 cm^{-1} and 1515 cm^{-1}. The carbonyl of the ester absorbs at 1690 cm^{-1} while the corresponding C—O single band is at 1170 cm^{-1}. The C—O band of the primary alcohol occurs at 1040 cm^{-1} and that of the secondary alcohol is at 1115 cm^{-1}. The hydrogen deformation band corresponding to para disubstitution occurs at 840 cm^{-1}. An indication of the overall aromatic character of this molecule is the weakness of the aliphatic C—H stretching at 2940 cm^{-1}.

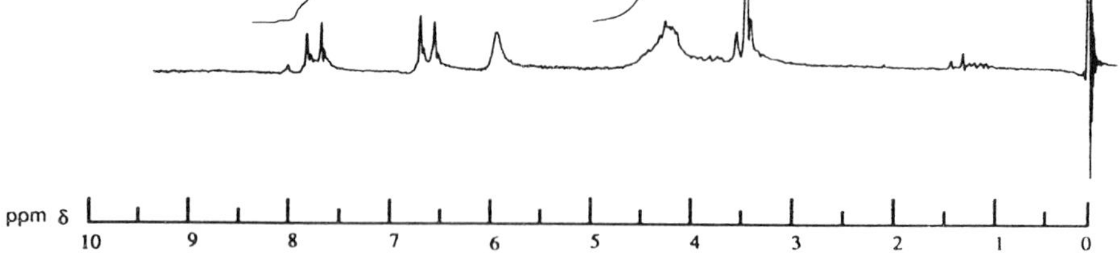

NMR Spectrum. Solvent: DMSO-d_6; 60 MHz. This product exists in several isomeric forms and hence was not pure enough to yield a satisfactory analysis of the NMR spectrum.

HOMOSALATE

Molecular Formula: $C_{16}H_{22}O_3$
Molecular Weight: 262

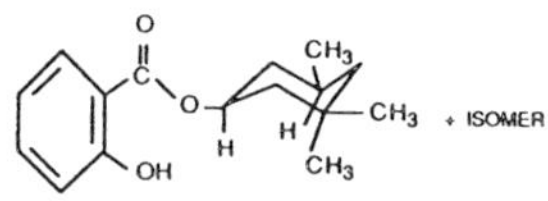

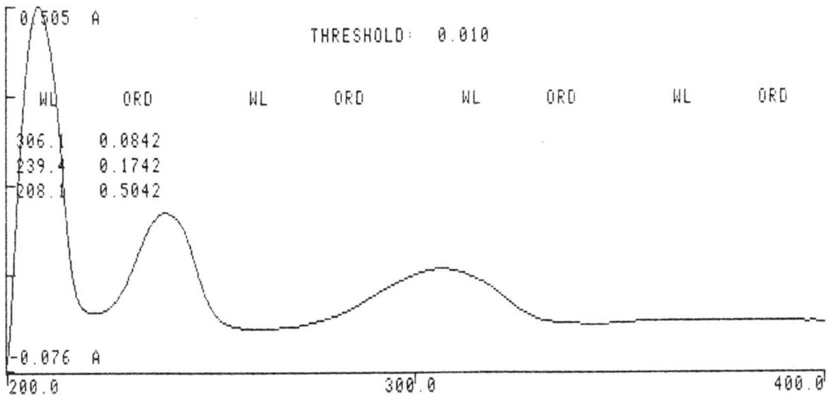

UV Spectrum. Concentration: 7.59 mg/L in ethyl alcohol; λ max: 306 nm; ϵ: 4,500. Salicylates have the ideal λ max which corresponds to the maximal sun flux of 308 nm. Homosalate absorbs at 306 nm. However, the ortho relationship in the salicylate moiety causes a deviation of the planarity of the groups substituted on the benzene ring, which in turn diminishes the flow of electrons during resonance, resulting in lower extinction coefficients (a hypochromic shift). The ϵ is 4,500.

Mass Spectrum. Since the fragmentation is directed almost exclusively by the salicylate moiety, all the major fragments (m/e 120, 138, 121) are the same as octyl salicylate. Only the weak molecular ion, m/e 264, is different.

HOMOSALATE (Cont.)

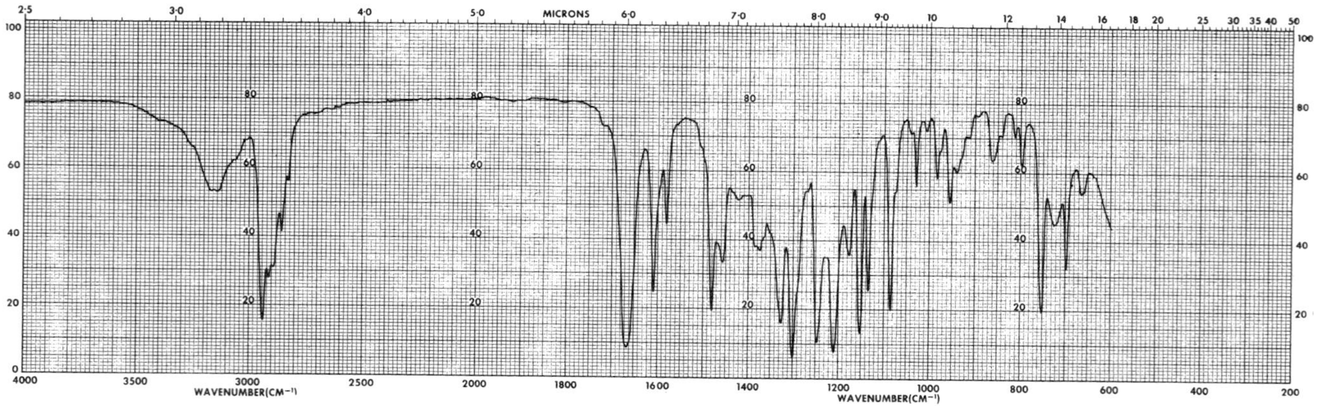

IR Spectrum. Solvent: neat; NaCl plates. The infrared spectrum of homomenthyl salicylate has all the typical salicylate bands discussed under octyl salicylate. The only differences in the IR spectra are in the C—H stretching region between 3000 and 2800 cm^{-1}, and in the fingerprint region between 1000 and 800 cm^{-1}.

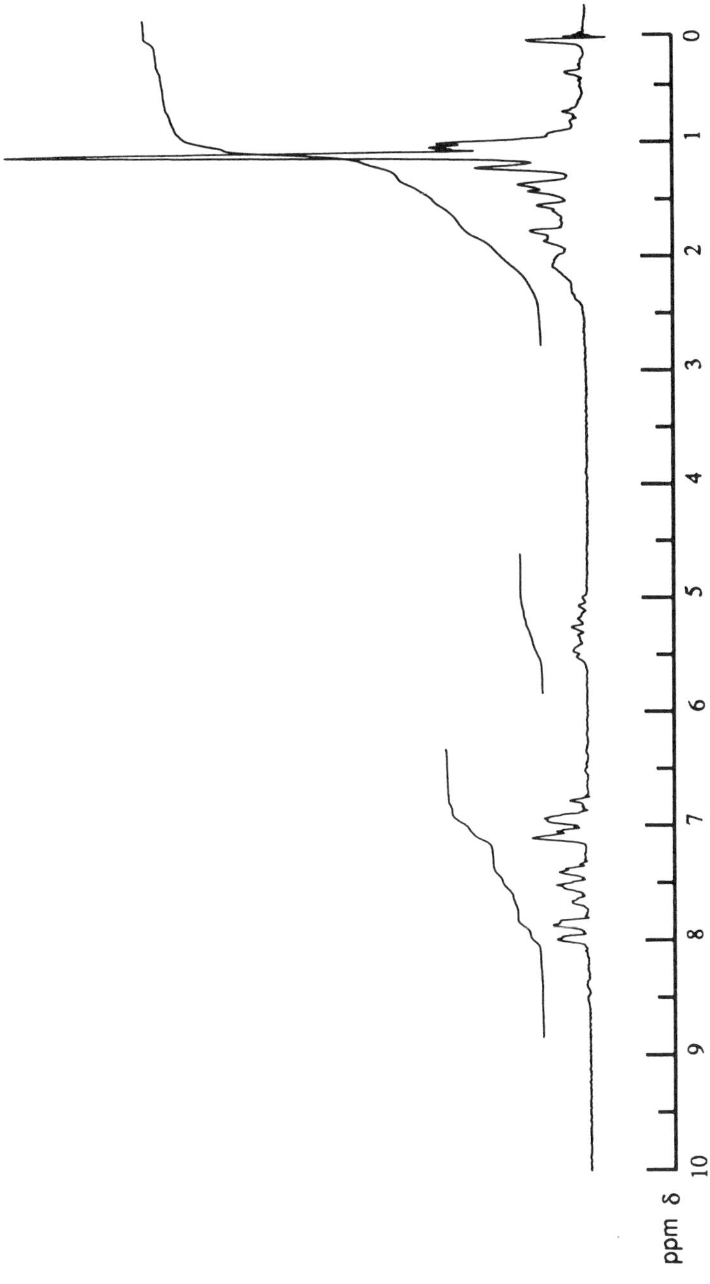

NMR Spectrum. Solvent: $CDCl_3$; 60 MHz. This product is commercially available as a cis/trans mixture. Hence the NMR spectrum is a composite of the two possible stereoisomers. The total proton count reveals a ratio of five protons resonating between 6.7 and 7.9 ppm for the four aromatic and the phenolic proton. A sharp singlet appears at 1.0 ppm integrating for six protons for the gem-dimethyl protons present on the cyclohexyl ring. Ten additional aliphatic protons, part of the cyclohexyl ring, give rise to a resonance at 0.85–2.20 ppm. The most characteristic and diagnostic resonance belongs to the *HC*—O resonance which is shifted downfield to a multiplet centered around 5.25 ppm and integrates for one proton.

MENTHYL ANTHRANILATE

Molecular Formula: $C_{17}H_{25}NO_2$
Molecular Weight: 275

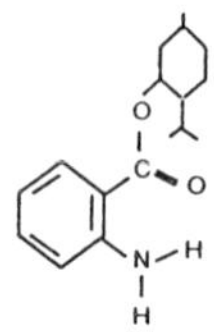

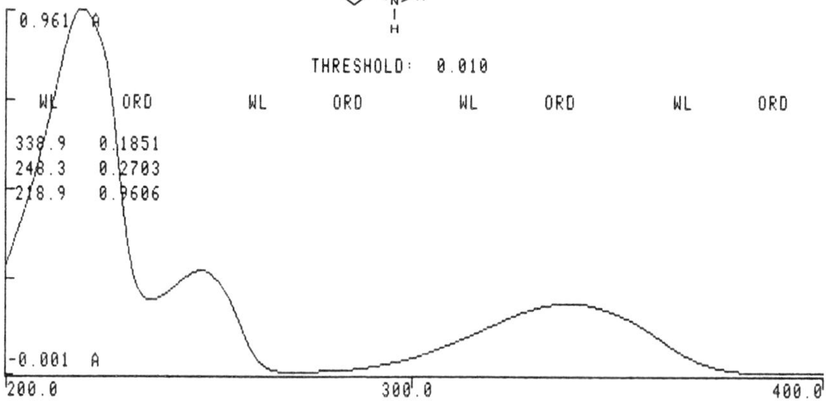

UV Spectrum. Concentration: 4.75 mg/L in ethyl alcohol; λ max: 338 nm; ϵ: 5,600. The ortho relationship of the amine group to the carboxylic ester grouping on the ring has two major effects on the UV absorbance. On the positive side, hydrogen bonding occurs through space, links the flow of electrons not only internally through resonance, but externally through this hydrogen-bonding interaction, resulting in the extension of conjugation. This increased conjugation lowers the energy requirements for the electronic transition between the ground state and the excited state which will be expressed in a higher λ max of 338 nm. But on the negative side, interaction of the two reasonably bulky groups outside the ring causes the planarity to deviate, reducing the electron flow during resonance to give a hypochromic shift.

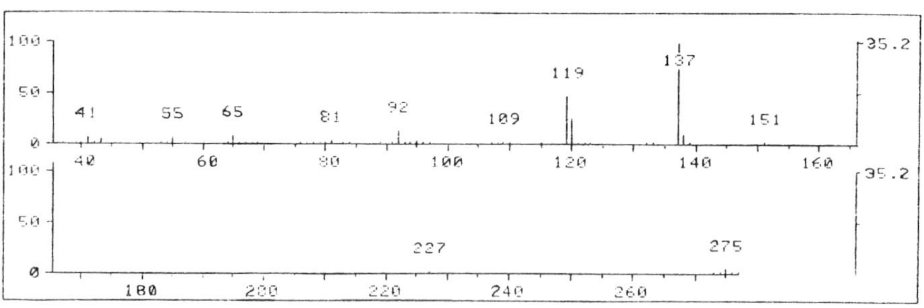

Mass Spectrum. The mass spectrum is also greatly affected by the ortho substitution pattern. The expected amino phenyl-substituted acylium ion at 120 m/e is only one-half as intense as the ion at m/e 119 resulting from the transfer of one amino hydrogen to the lost neutral alkyl fragment. The base peak at m/e 137 is also a rearrangement to yield an amino benzoic acid. The M^+ ion at 275 m/e is weak but observable.

MENTHYL ANTHRANILATE (Cont.)

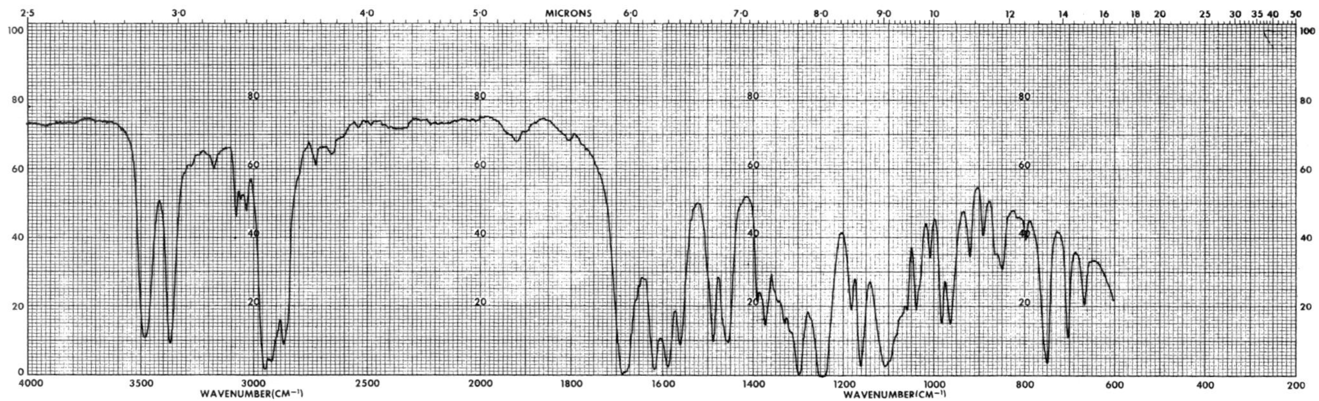

IR Spectrum. Solvent: neat; NaCl plates. The symmetric and asymmetric stretching modes of the primary aromatic amine at 3482 cm^{-1} and 3370 cm^{-1} are very conspicuous. The aliphatic C—H stretching bands between 3000 cm^{-1} and 2850 cm^{-1} are strong. The position of the C═O bands at 1685 cm^{-1} (compared to 1702 cm^{-1} for octyl dimethyl PABA) shows the effect of strong intramolecular hydrogen bonding with the ortho substituted amino group. The C—O bands for the ester are complex series with maxima at 1295 cm^{-1}, 1246 cm^{-1}, 1161 cm^{-1}, and 1108 cm^{-1}. The ortho hydrogen deformation frequency is at 750 cm^{-1}.

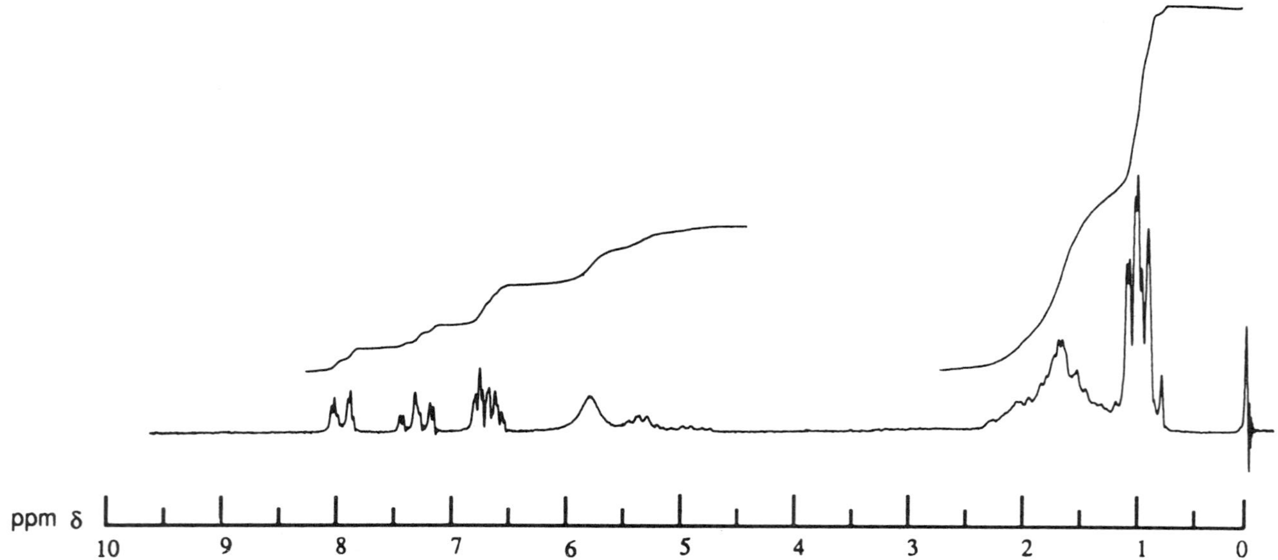

NMR Spectrum. Solvent: $CDCl_3$; 60 MHz. A number of menthol isomers may be linked to this anthranilic acid moiety giving rise to a complicated spectrum in the NMR. Nevertheless, the four aromatic protons appear as a multiplet in the region of 6.6 to 7.9 ppm. The NH_2 signal is a broad singlet at 5.8 ppm. The 18 aliphatic protons resonate as a multiplet from 0.7 to 2.2 ppm. Here, again, the 19th aliphatic proton, the (—C*H*—O) appears shifted downfield as a multiplet centered at 5.0 ppm.

OCTOCRYLENE

Molecular Formula: $C_{24}H_{27}NO_2$
Molecular Weight: 361

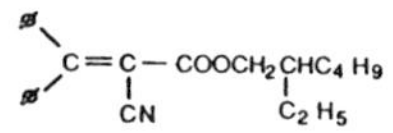

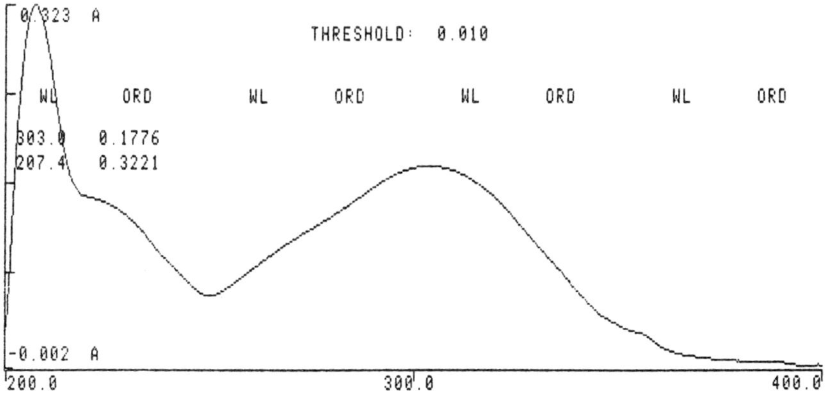

UV Spectrum. Concentration: 5.07 mg/L in ethyl alcohol; λ max: 303 nm; ϵ: 12,600. The λ max for this compound appears at 303 nm, but, in reality, a number of absorptions occur in the vicinity, giving rise to a considerably broadened spectrum in the UV. The ϵ = 12,600 which is considered above average.

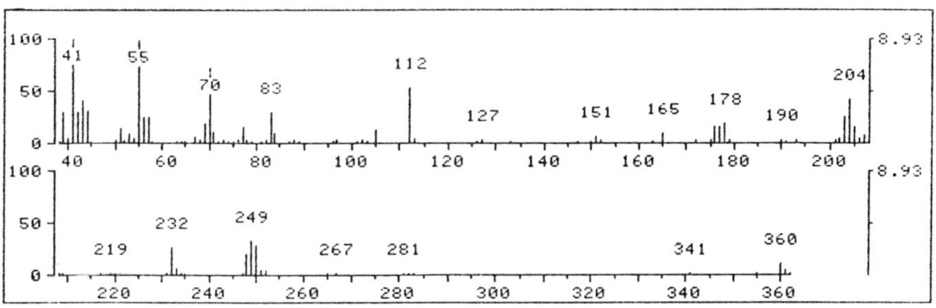

Mass Spectrum. Octocrylene has a weak molecular ion at m/e 361 and a somewhat stronger M^+-1 ion at 360. The major fragments are those expected of an ester. Fragments corresponding to the free acid and the protonated acid occur at 249 and 250, respectively. Additionally, an unusual fragment, one amu lower than the free acid, also occurs. This may correspond to a cyclic structure with a band between the carboxylate and nitrile groups. The next fragment is an acylium ion at m/e 232 which then eliminates CO to yield 204. All the other major fragments originate in the alcohol portion of the ester. Thus m/e 112 in the alkene portion results from the elimination of the acid moiety from the alcohol. The three low mass fragments m/e 70 (C_5H_{10}), 55 (C_4H_7), and 41 (C_3H_5) represent the typical hydrocarbon breakdown pattern for the 2-ethyl hexanol portion of the structure.

OCTOCRYLENE (Cont.)

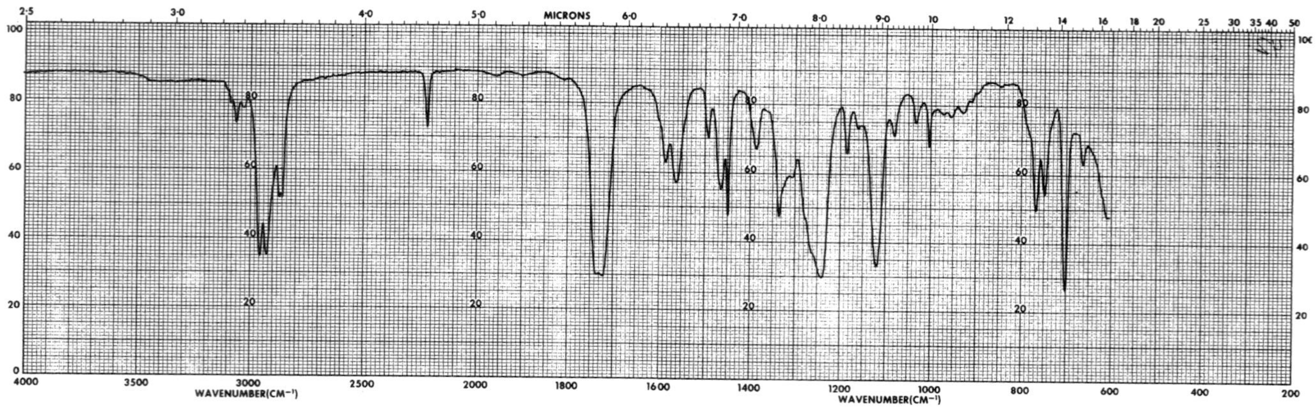

IR Spectrum. Solvent: neat; NaCl plates. The most characteristic band in this spectra is the moderate absorption at 2220 cm^{-1} from the C≡N stretching mode. The C=O band of the ester is 1730 cm^{-1}, while the C—O is at 1238 cm^{-1}. The alkene C=C is at 1588 cm^{-1}. The aromatic C=C bands are at 1563 cm^{-1} and 1490, 1465, and 1450 cm^{-1}. The aromatic hydrogen deformation frequencies for the two monosubstituted rings are at 702, 750, and 769 cm^{-1}.

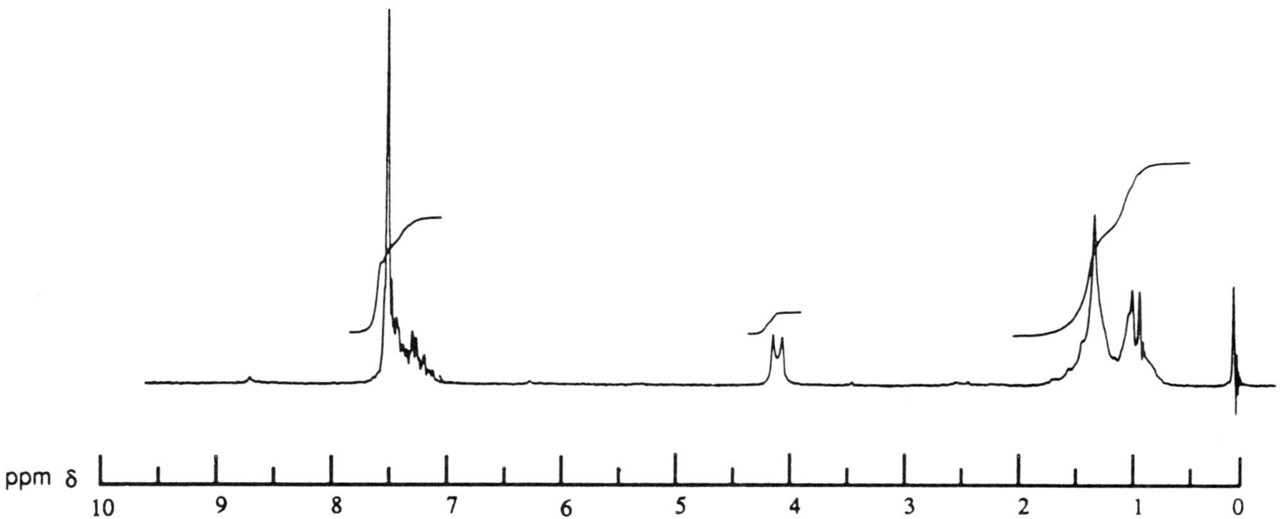

NMR Spectrum. Solvent: $CDCl_3$; 60 MHz. With 27 protons in this compound the NMR spectrum as expected is rather complicated. However, the 10 aromatic protons appear as a multiplet at 7.5 ppm. The (O—CH_2) protons are shifted, as expected, downfield to 4.1 ppm and since they are coupled to one proton, they are split into a doublet. The balance of the aliphatic protons (15 H) appear as a multiplet from 0.8 to 1.6 ppm.

OCTYL DIMETHYL PABA

Molecular Formula: $C_{17}H_{27}NO_2$
Molecular Weight: 277

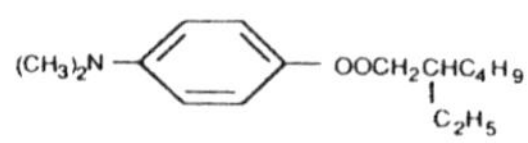

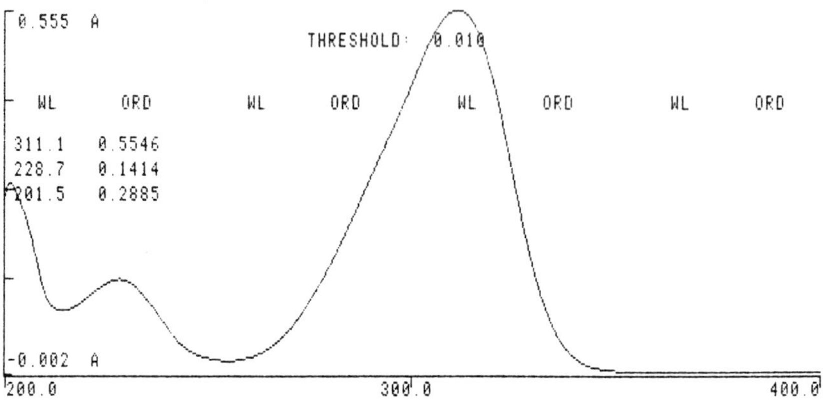

UV Spectrum. Concentration: 6.11 mg/L in ethyl alcohol; λ max: 311 nm; ϵ: 27,300. This compound has perhaps the most ideal λ max of 311 nm and a strong absorption of ϵ = 27,300, making it one of the best sunscreens available, yielding protection in the UVB region.

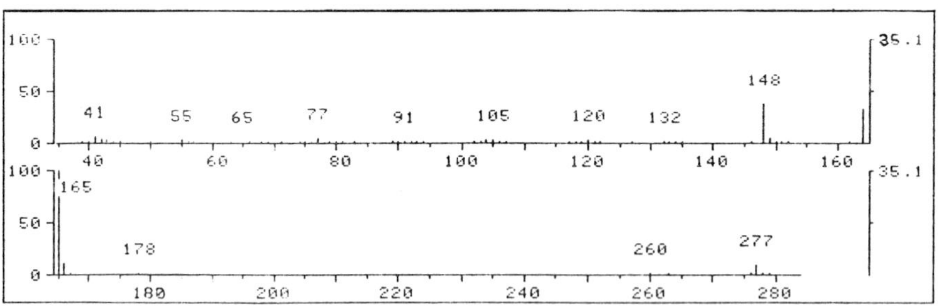

Mass Spectrum. The base peak of the spectrum is a rearrangement product at m/e 165 and corresponds to *p*-dimethyl amino benzoic acid. The next most intense peak is at m/e 140 and corresponds to cleavage of the C—O bond adjacent to the aromatic carbonyl groups. Its high intensity is typical of carbonyl groups conjugated to benzene rings. The molecular ion at m/e 277 is fairly intense as would be expected from an aromatic molecule. Most other fragments are of minor intensity and can be attributed to common fragmentation processes.

OCTYL DIMETHYL PABA (Cont.)

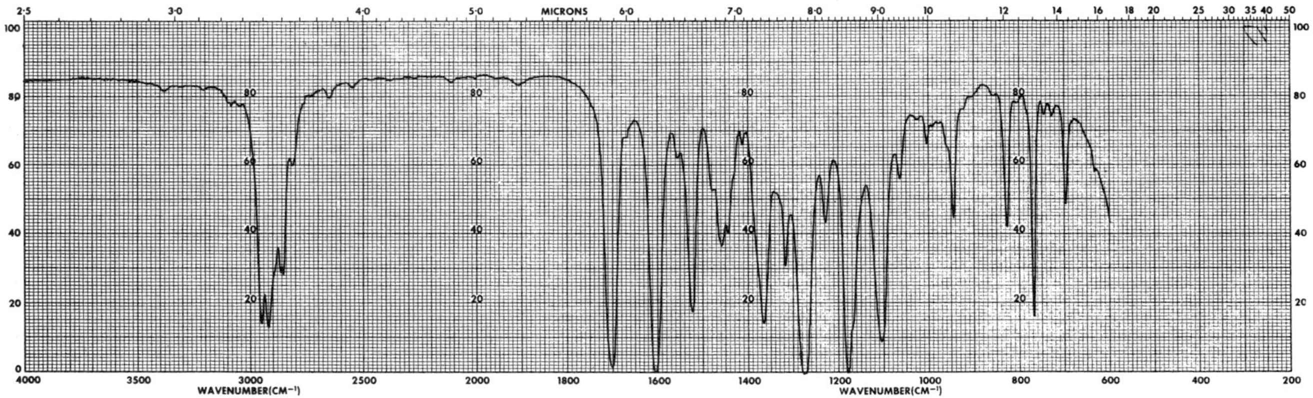

IR Spectrum. Solvent: neat; NaCl plates. The most prominent bands in the spectrum are from the ester group (C═O stretching 1702, C—O 1275, 1178, and 1106 cm^{-1}) and the benzene ring (C═C stretching 1605 and 1525 cm^{-1}, C—H deformation 769 cm^{-1}).

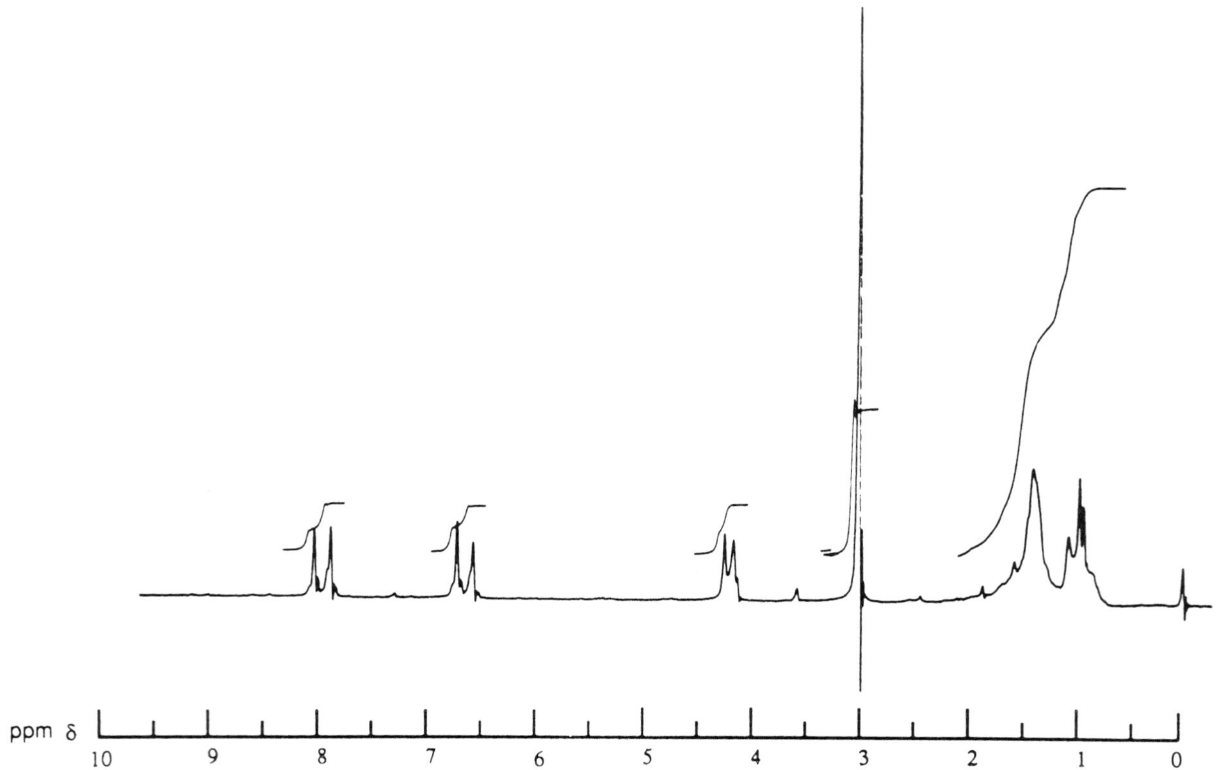

NMR Spectrum. Solvent: $CDCl_3$; 60 MHz. This compound lends itself extremely well for characterization and identification through NMR spectroscopy. The four aromatic protons appear as an AB quartet with the two next to the carbonyl group resonating around 6.6 ppm, and the two next to the nitrogen element at 7.9 ppm. The 6 gem-dimethyl protons on the nitrogen appear as a sharp singlet at 3.0 ppm. The (O—CH_2) protons are again shifted downfield to 9.2 ppm and appear as a doublet since they are coupled only to one adjacent proton. Finally, the remaining 15 protons appear as a multiplet from 0.8 to 1.7 ppm.

OCTYL METHOXY CINNAMATE

Molecular Formula: $C_{18}H_{26}O_3$
Molecular Weight: 290

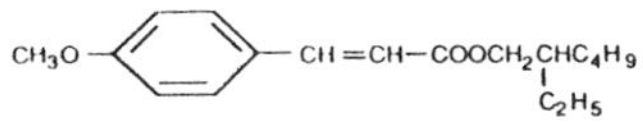

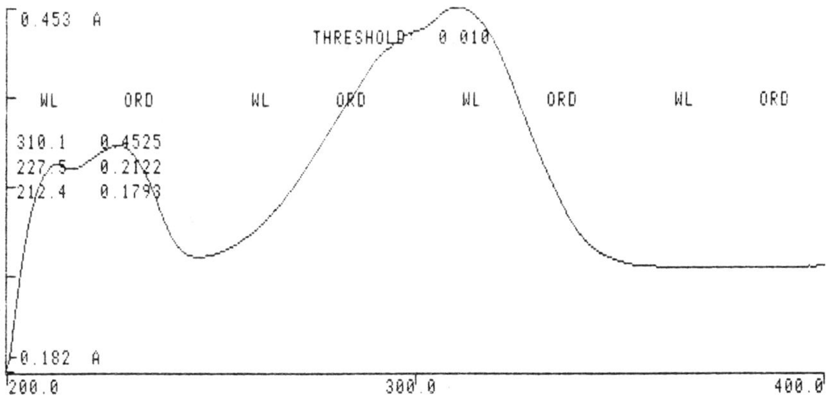

UV Spectrum. Concentration: 8.27 mg/L in ethyl alcohol; λ max: 311 nm; ϵ: 23,300. The additional conjugation present in this compound due to the double bond between the aromatic ring and the carbonyl group gives rise to a λ max at 311 nm with a shoulder absorbance at 290 nm. The extinction coefficient is considered high at ϵ = 23,300. Since the methoxyl group is not as efficient in delocalizing electrons as the amine group in the PABA derivatives, two effects result. The λ max would be expected to be lower but it is counterbalanced by the previously mentioned additional conjugation due to the presence of the double bond. However, this very same double bond which gives rise to cis-trans isomerism will cause a slight disruption of the resonance delocalization thereby resulting in a lower extinction coefficient than in PABA derivatives (23,300 vs. 27,300).

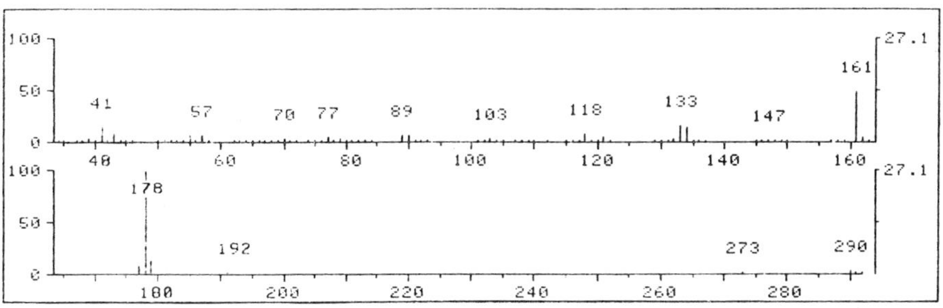

Mass Spectrum. As with most of the higher esters of aromatic acids, the base peak is a rearrangement to the free acid itself, m/e 178. The next most intense peak is at m/e 160 and represents the acylium ion derived from the parent ester by cleavage at the C—O bond adjacent to the carbonyl. A weak M^+ ion at 290 is observable.

OCTYL METHOXY CINNAMATE (Cont.)

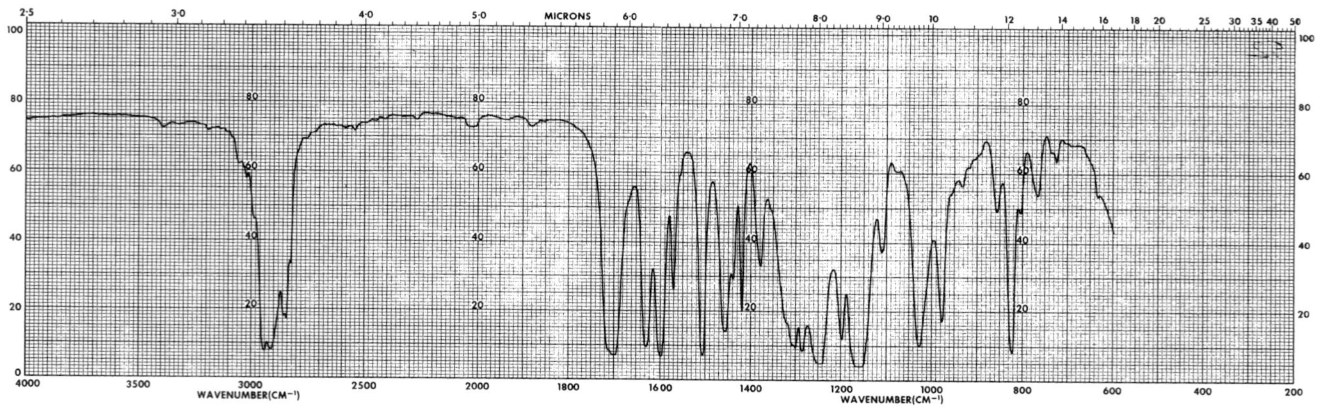

IR Spectrum. Solvent: neat; NaCl plates. The IR of OMC has an unusually large number of strongly absorbing functional groups. The conjugate ester C=O stretching band is at 1700 cm^{-1} while the conjugated C=C stretching band is at 1628 cm^{-1}. The aromatic ring frequencies appear at 1508, 1569, and 1598 cm^{-1}. The C—O bond from the para methoxyl group on the ring is at 1245 cm^{-1} while the ester C—O frequency is at 1160 cm^{-1}. The characteristic hydrogen deformation bands are at 978 cm^{-1} for the trans alkene and 822 cm^{-1} for the ring hydrogens with para substituents.

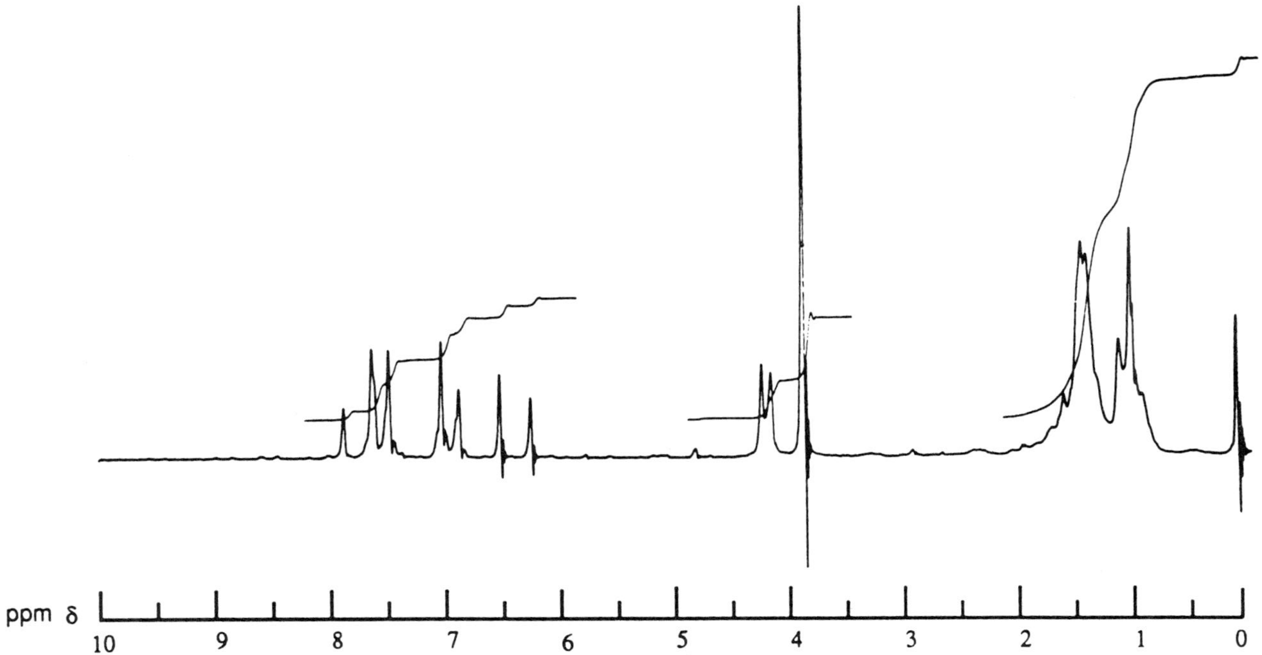

NMR Spectrum. Solvent: $CDCl_3$; 60 MHz. The four aromatic protons, along with the two alkene protons, appear as a multiplet from 6.2 ppm to 7.8 ppm. The three methoxyl protons resonate as a sharp singlet at 3.8 ppm. The balance of the octyl group gives rise to the characteristic resonance signals with the O—CH_2 appearing as a doublet at 4.3 ppm and the remaining 15 protons resonating as a multiplet from 0.8 to 1.8 ppm.

OCTYL SALICYLATE

Molecular Formula: $C_{15}H_{22}O_3$
Molecular Weight: 250

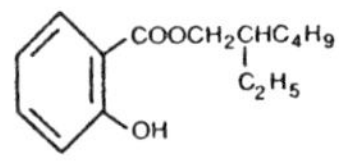

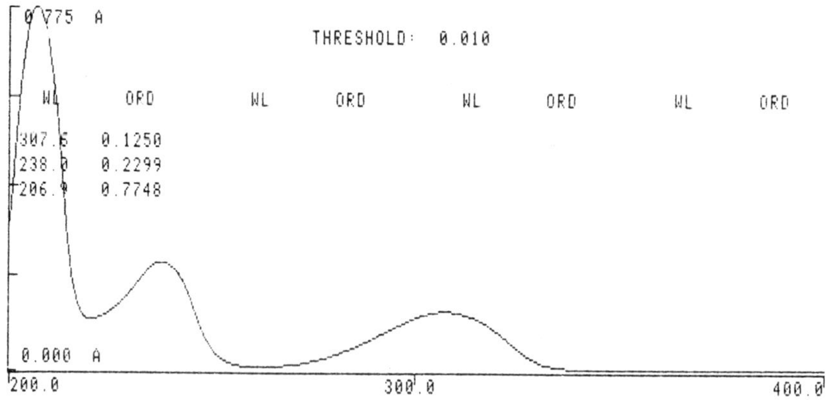

UV Spectrum. Concentration: 6.16 mg/L in ethyl alcohol; λ max: 307 nm; ϵ: 5,000. The UV spectrum is similar to homosalate with a λ max of 307 nm. However, due to the presence of a single isomer, its extinction coefficient is ϵ = 4900, slightly higher than in homosalate (ϵ = 4500).

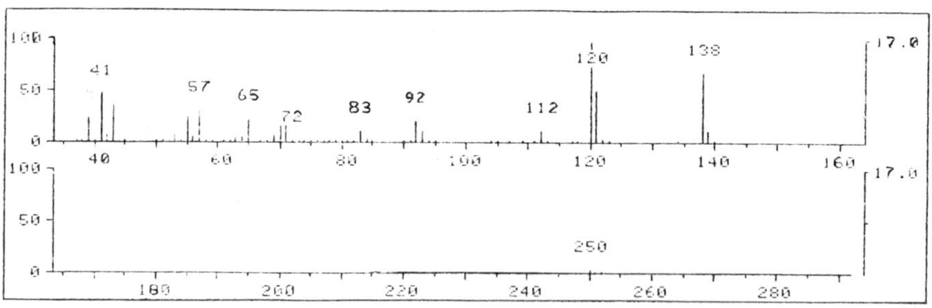

Mass Spectrum. As a result of the ortho hydroxyl group, transferring a hydrogen to the eliminated alkyl group, the base peak for octyl salicylate is at 120 m/e. The expected ortho hydroxyl acylium ion at m/e 121 is present at half the intensity. The second most intense peak is at m/e 138 and represents a rearrangement to salicylic acid. A weak molecular ion is present at m/e 250. There is a fragment-by-fragment analogy between the MS of salicylates and anthranilates, two ortho-disubstituted aromatic compounds.

OCTYL SALICYLATE (Cont.)

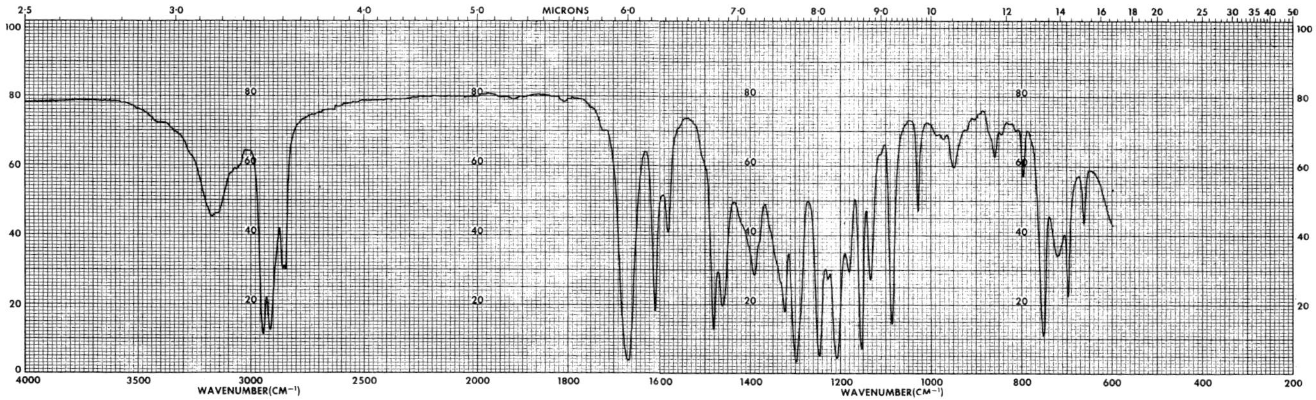

IR Spectrum. Solvent: neat; NaCl plates. The most unusual band in the spectrum of octyl salicylate is the internally bonded O—H at 3170 cm^{-1}. This band is at a lower frequency, broader, and independent of solvent concentration due to the intramolecular hydrogen bonding between the ortho substituted hydroxyl group and the carbonyl of the ester group. The conjugated carbonyl group absorbs at 1668 cm^{-1} which is also abnormally low due to this strong internal hydrogen bond. The aromatic ring has bands at 1608, 1581, and 1480 cm^{-1}. All salicylate esters have a characteristic complex pattern of bands at 1323, 1297, 1245, 1209, 1153, 1132, and 1087 cm^{-1}, resulting from the C—O absorption of the ester and the phenolic hydroxyl. Also characteristic of all salicylate esters is a pattern of bands at 751, 720, and 698 cm^{-1}.

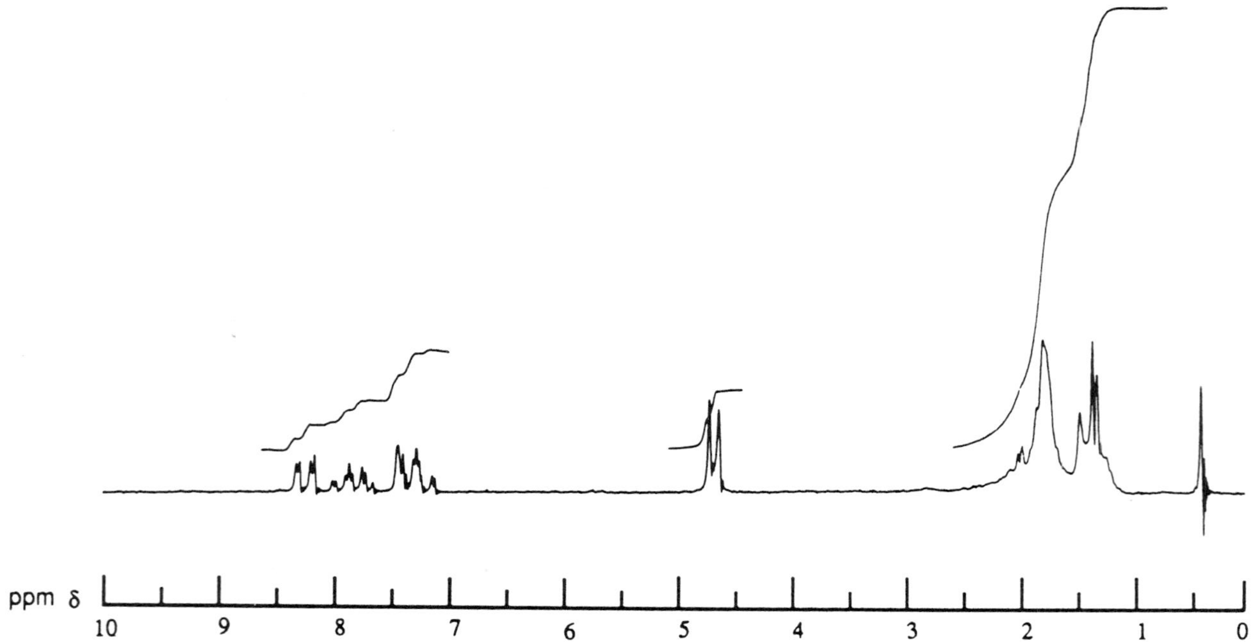

NMR Spectrum. Solvent: $CDCl_3$; 60 MHz. The four aromatic and the phenolic protons give rise to a characteristic multiplet from 6.70 ppm to 7.9 ppm integrating for five protons. The rest of the signals in the spectrum are the usual resonances of the octyl group, with the two O—CH_2 protons appearing as a doublet at 4.3 ppm and the remaining 15 protons as a characteristic multiplet from 0.8 to 1.8 ppm.

PABA

Molecular Formula: $C_7H_7NO_2$
Molecular Weight: 137

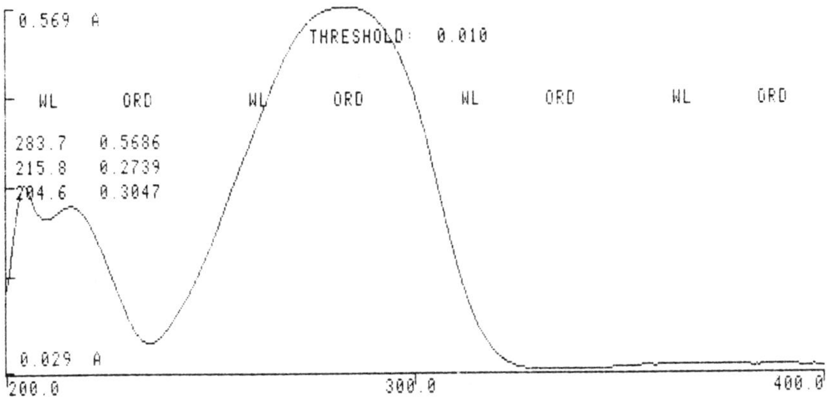

UV Spectrum. Concentration: 5.09 mg/L in ethyl alcohol; λ max: 283 nm; ϵ: 15,300. As discussed earlier, in the case of the glyceryl PABA the hydrogen bonding in this compound affects the overall symmetry which causes a lowering of the extinction coefficient to ϵ = 15,300. The polarity of this molecule will also cause a considerable hypochromic solvent shift to 283 nm.

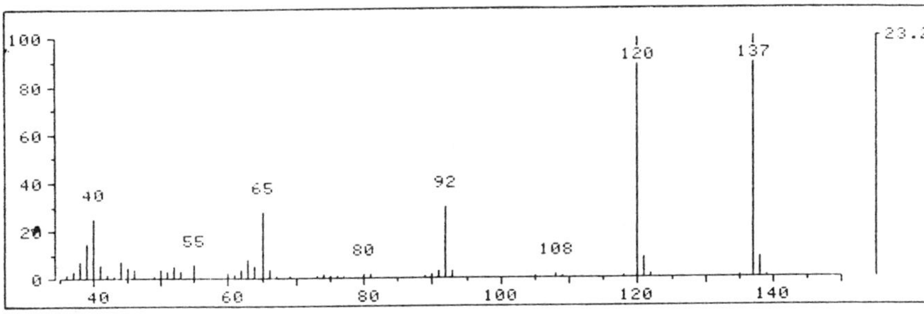

Mass Spectrum. The base peak of PABA is the molecular ion at m/e 137 due to the stability of aromatic molecules with small substituents. The fragment at 120 amu is of almost equal intensity and results from the loss of a hydroxyl group to yield a substituted benzoyl group. This in turn eliminates CO to form m/e 92.

PABA (Cont.)

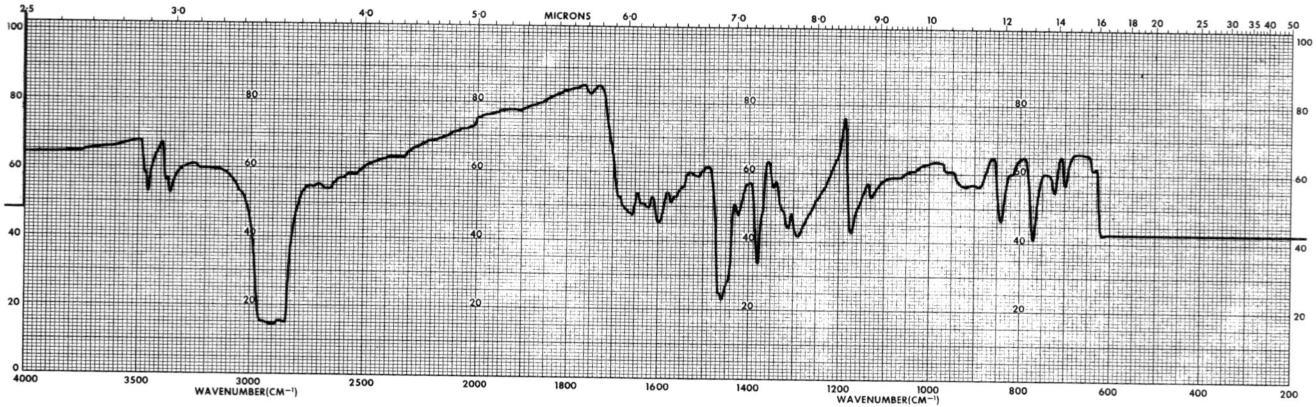

IR Spectrum. Solvent: mineral oil; NaCl plates. The symmetric and unsymmetric stretching frequencies of the primary amine are present at 3455 and 3360 cm^{-1}. The conjugated acid carbonyl band occurs at 1660 cm^{-1}. A complex series of bands between 1650 and 1500 cm^{-1} represent the aromatic C=C bands and the band from nitrogen to the ring. The para substitution band is at 840 cm^{-1}.

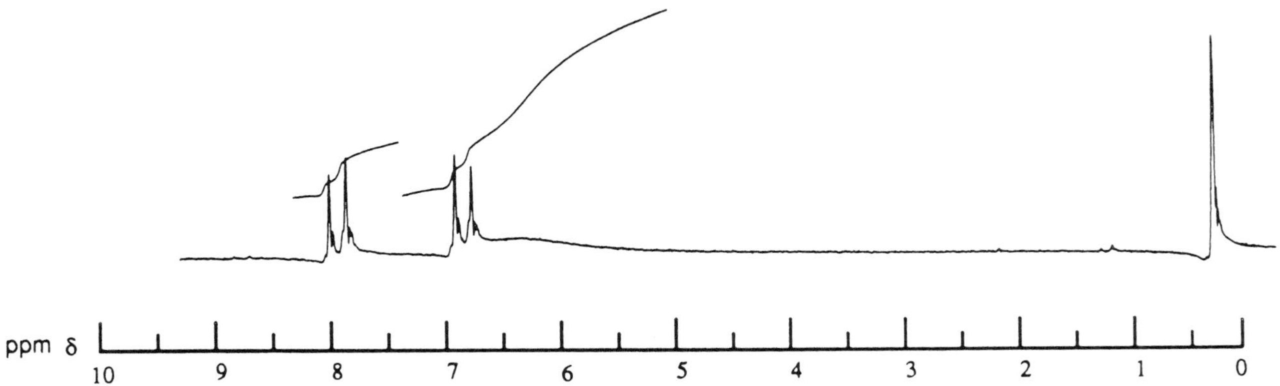

NMR Spectrum. Solvent: DMSO-d_6; 60 MHz. Due to the poor solubility characteristics of this compound and complicated further by the extreme intermolecular hydrogen bonding between the carboxyl proton on one PABA and the amine protons of another PABA, the NMR spectrum of this compound reveals an extremely broad signal for the two amine and the carboxylic protons. Under certain concentrations these signals may completely disappear. On the other hand, the characteristic AB quartet for all PABA derivatives appears reasonably well, with the two aromatic protons in the vicinity of the carboxylic acid resonating as a doublet centered at 6.5 ppm, and the remaining two protons in the vicinity of the amine moiety resonating as the other half of the quartet at 7.7 ppm. This compound can be better characterized through its IR and MS spectrum.

TEA SALICYLATE

Molecular Formula: $C_{13}H_{21}NO_6$
Molecular Weight: 287

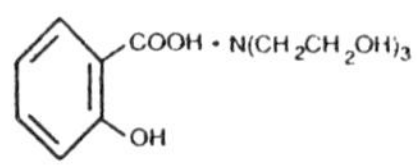

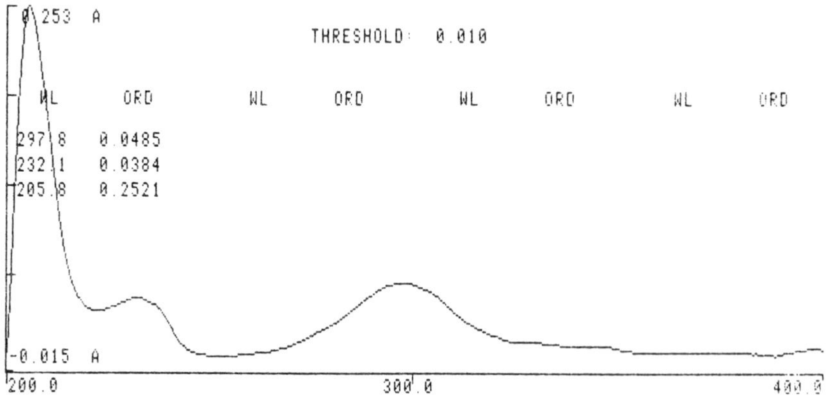

UV Spectrum. Concentration: 9.90 mg/L in ethyl alcohol; λ max: 298 nm; ϵ: 3,200. As a salicylic acid salt it, too, will experience excessive solvent interactions, lowering the λ max to 298 nm and its extinction coefficient to 3200.

(Data unavailable)

Mass Spectrum. This compound is a salt and as such is not suitable for analysis by GLC. Therefore, we have not produced mass spectral data for the reasons listed in entry 2 (Benzophenone-4).

TEA SALICYLATE (Cont.)

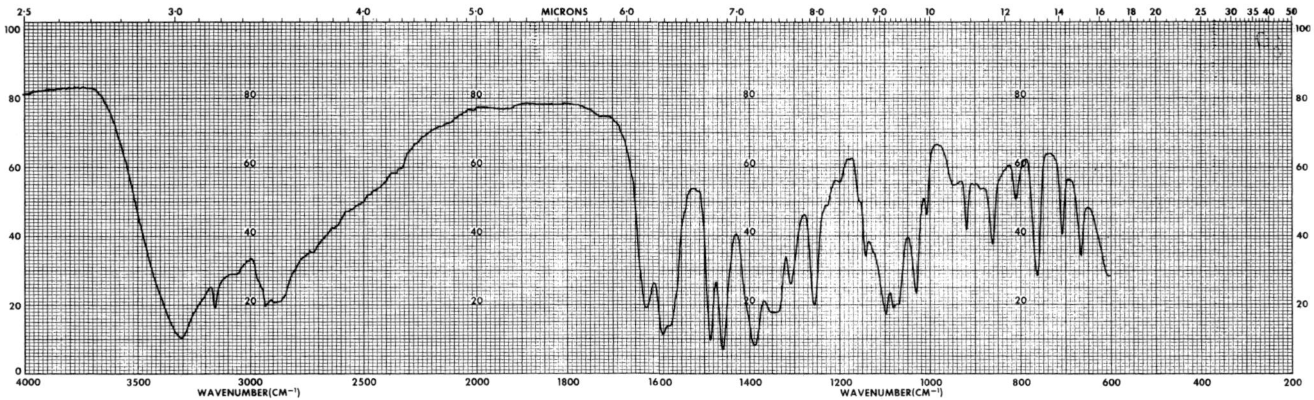

IR Spectrum. Solvent: neat; NaCl plates. A very intense and broad band centered at 3350 cm^{-1} represents the three hydroxyl groups present in the cation. The carbonyl group of the acid salt absorbs at 1687 cm^{-1}. The aromatic ring absorbs at 1605 and 1518 cm^{-1}. The C—O bond of the carboxylate absorbs at 1280 cm^{-1} while the three C—O bonds in the cation absorb at 1115 and 1045 cm^{-1}. The aromatic hydrogens absorb at 772 and 835 cm^{-1}.

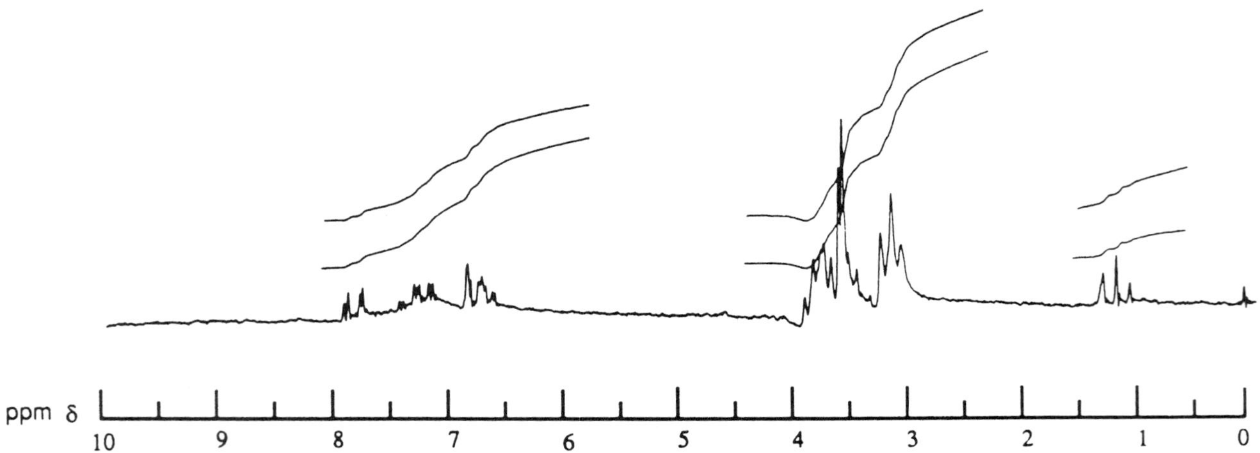

NMR Spectrum. Solvent: DMSO-d_6; 60 MHz. Since this compound, as a few other sunscreen chemicals (e.g., DEA cinnamate) are salts, they are best characterized in D_2O or DMSD-d_6. Here, again, the 4 aromatic and phenolic protons appear as a characteristic multiplet centered around 7.2 ppm with the balance of the aliphatic protons resonating as a multiplet around 3 ppm.

BUTYL METHOXY DIBENZOYL METHANE

Molecular Formula: $C_{20}H_{22}O_3$
Molecular Weight: 310

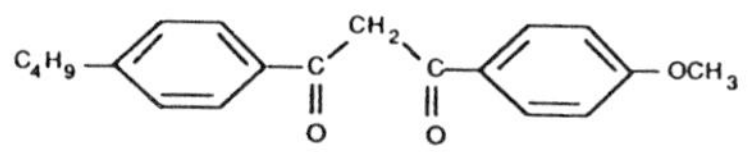

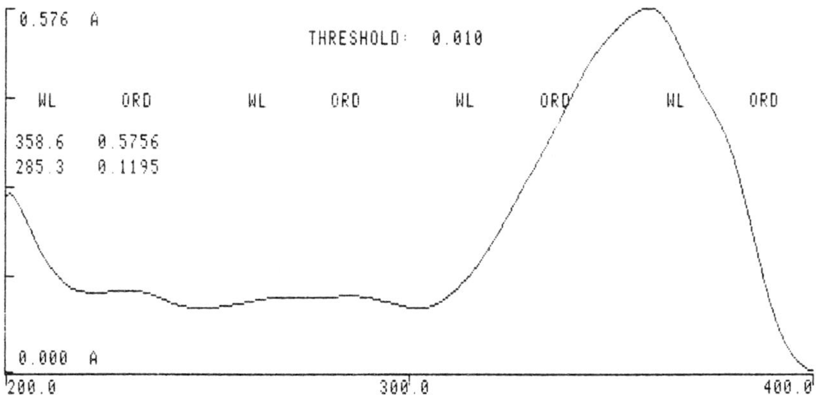

UV Spectrum. Concentration: 5.14 mg/L in ethyl alcohol; λ max: 358 nm; ϵ: 34,720. This molecule undergoes a keto-enol tautomerism which allows for the extended conjugation throughout the whole molecule. The net result is an extreme lowering of the energy requirements for the electronic transition with a λ max of 358 nm. Due to symmetry considerations, however, this molecule possesses an unusually high extinction coefficient of ϵ = 34,720.

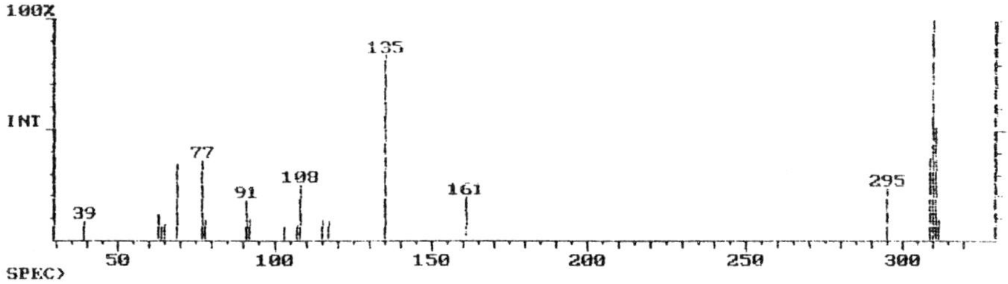

Mass Spectrum. The molecular ion m/e 310 is the base peak of the mass spectrum. There are two different para-substituted benzoyl fragments that may be formed from this structure. The one with the electron-withdrawing methoxyl group is by far the stronger, yielding a fragment at m/e 135 (80%). The benzoyl fragment with an electron-withdrawing tert butyl group yields a much smaller fragment at m/e 161.

BUTYL METHOXY DIBENZOYL METHANE (Cont.)

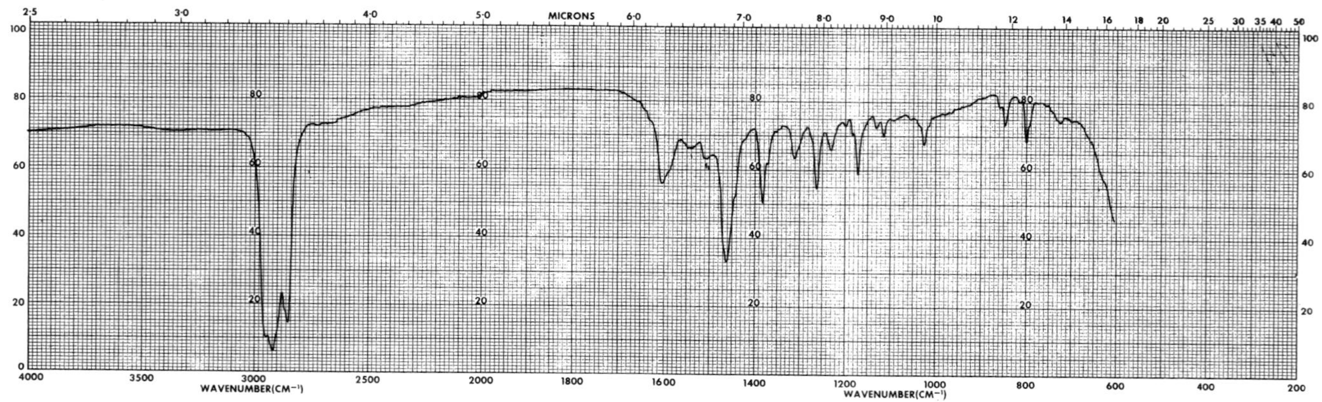

IR Spectrum. Solvent: mineral oil; NaCl plates. The carbonyl group of this diketone absorbs at the unusually low frequency of 1605 cm^{-1}. Although conjugation to the benzene ring contributes to the lowering of this frequency, the hydrogen bonding of the enol form of the β diketone is the primary cause. There are bands from the benzene rings at 1505 and 1545 cm^{-1}. The aromatic C—O of the ether absorbs at 1263 cm^{-1} while the aliphatic C—O bond absorbs at 1172 cm^{-1}. Aromatic hydrogens in the two para-substituted benzene rings absorb at 800 and 847 cm^{-1}.

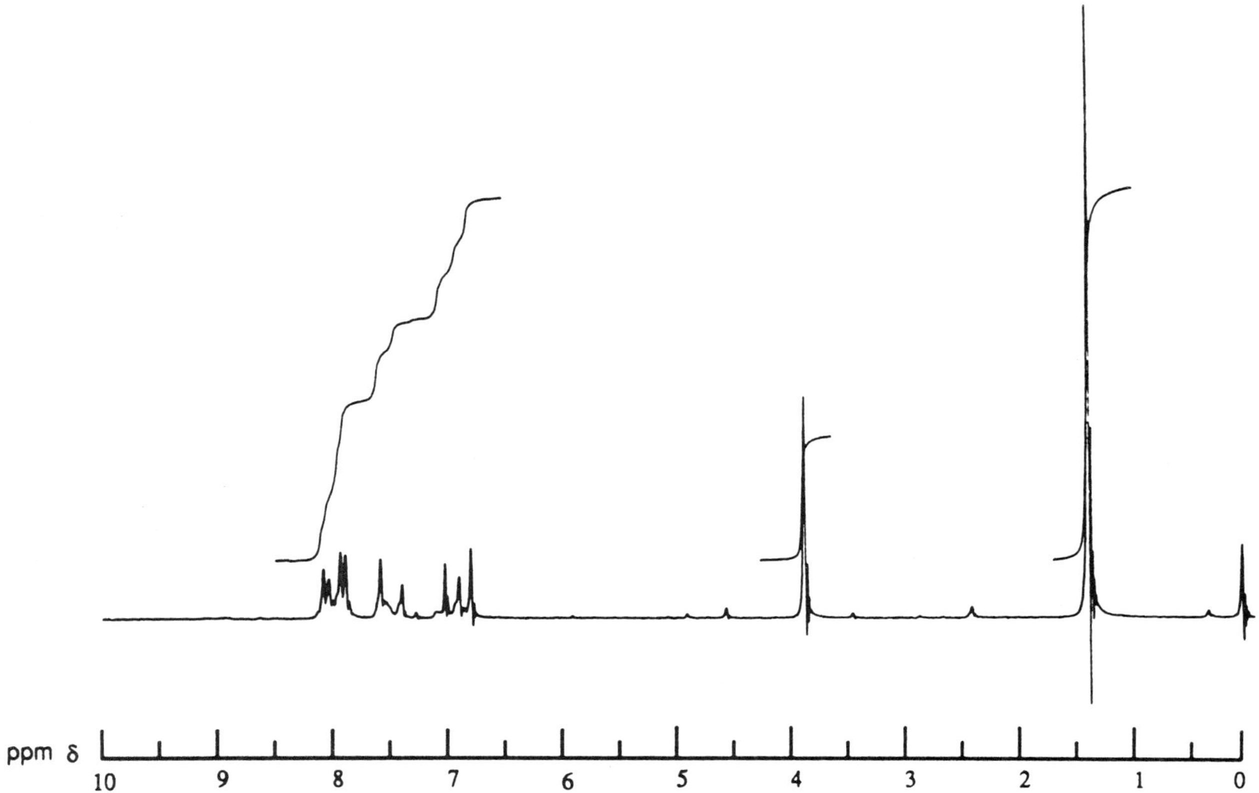

NMR Spectrum. Solvent: $CDCl_3$; 60 MHz. Despite the large number of protons (22) in this compound, the NMR spectrum is definitive in its characterization. The tertiary butyl group (9H) appears as a sharp singlet at 1.4 ppm. The methoxyl protons also appear as a singlet at 3.8 ppm as expected. The OC—CH_2—CO is shifted downfield in the 7-8 ppm region along with the eight aromatic protons giving rise to a huge but characteristic multiplet from 6.8 to 8.1 ppm.

ETOCRYLENE

Molecular Formula: $C_{18}H_{15}NO_2$
Molecular Weight: 277

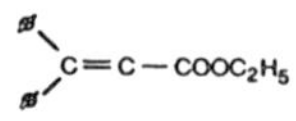

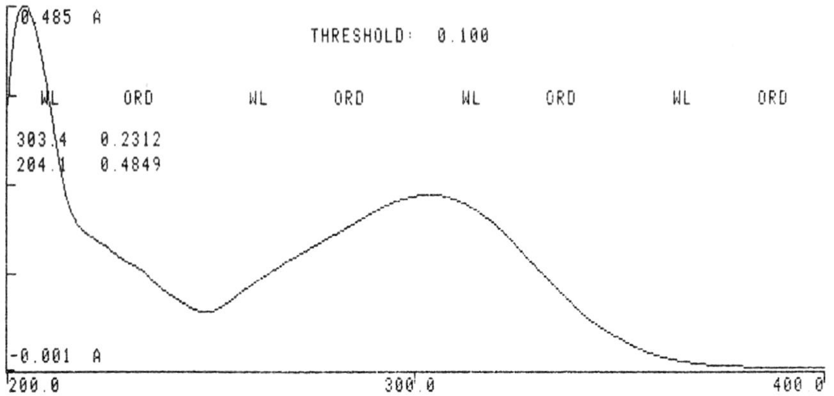

UV Spectrum. Concentration: 5.30 mg/L in ethyl alcohol; λ max: 303 nm; ϵ: 12,600. Here, as in octocrylene, the absorption is rather broad with a λ max of 303 nm. The extinction coefficient is ϵ = 12,600.

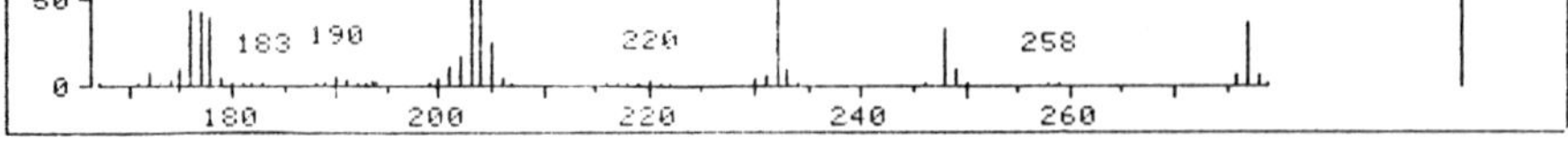

Mass Spectrum. Due to the short aklyl chain, the molecular ion at m/e 277 is much more intense (45%) than the corresponding ion in its larger homologue, octocrylene. The unusual fragment at 1 amu less than the molecular weight of the acid (m/e 248 is present). The typical ester acylium ion at m/e 232 and the product of CO elimination at m/e 204 are very prominent. Cleavage of either of the two bonds between the phenyl group and the alkene yields a strong fragment at m/e 77.

ETOCRYLENE (Cont.)

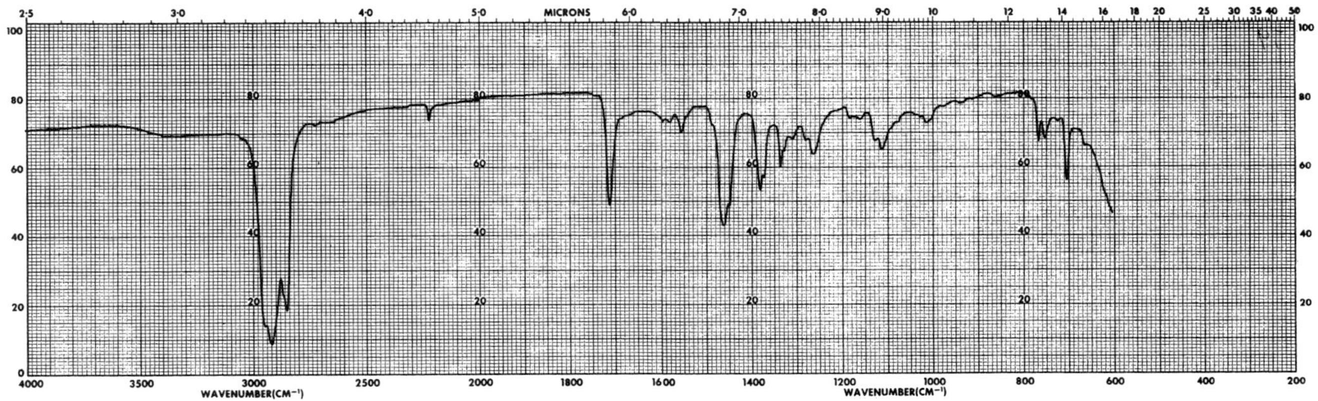

IR Spectrum. Solvent: mineral oil; NaCl plates. Since they differ only by the nature of the aklyl group substituted on the ester, etocrylene has the same IR frequencies as octocrylene. Only the relative intensities of absorptions, such as C—H stretch between 3000 and 2880 cm^{-1}, show significant differences.

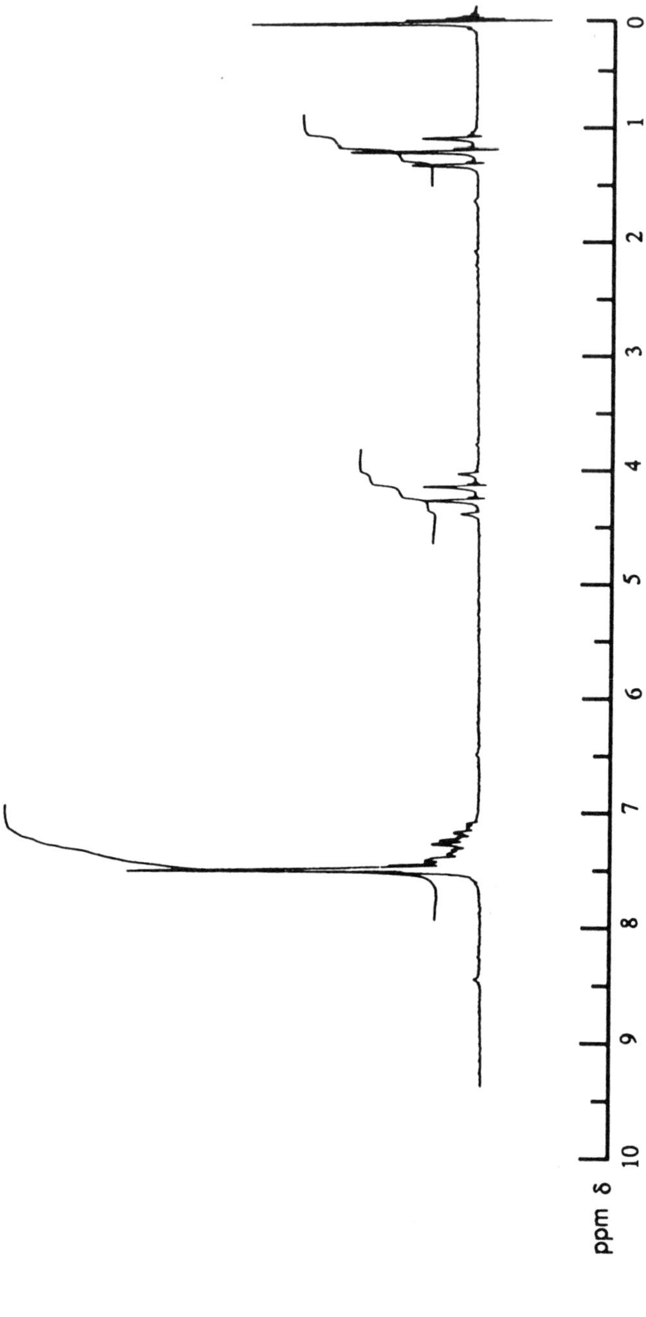

NMR Spectrum. Solvent: $CDCl_3$; 60 MHz. As in octocrylene, the 10 aromatic protons appear as a multiplet at 7.5 ppm. The ethyl group here is extremely simple to characterize with the usual quartet appearing at 4.2 ppm integrating for two protons, and the triplet contered at 1.2 ppm integrating for three protons.

eter, Perkin Elmer 1320 Infrared Spectrophotometer, and Varian EM-360L 60 MHz NMR. (For operating conditions see the Appendix at the end of the chapter.)

The molecular structure, molecular formula, molecular weight, and operating conditions used to obtain the spectra are listed with the spectroscopic data, followed by our interpretation of the spectra. Although fingerprinting is an excellent tool of identification, analyzing the spectral characteristics is essential for enabling the chemist to identify both the sunscreen and any impurities present.

This chapter has summarized the interpretation of the spectroscopic properties of sunscreens that are commercially available in the United States as of 1989. We await the work of our colleagues in other areas, such as the EEC, Japan, and Australia, which will describe the properties of sunscreen materials that were not available to us.

We hope this chapter will assist chemists working in both government and industry to devise more accurate and reliable analytical techniques. Because the analysis of sunscreens directly affects the health and safety of the public, considerable attention must be devoted by the analytical chemist to ensure the efficacy and purity of the sunscreen formulations.

APPENDIX: ANALYTICAL INSTRUMENT CONDITIONS

UV Conditions

Instrument	Perkin Elmer Lambda 48 UV/VIS Spectrophotometer
Cuvette	Quartz, 1-cm path length
Ordinate min/max	0–1
Wavelength min/max	200.0–400.0
Scan speed	120
Slit	1
Response	3

MS Conditions

Sample size	1 μl
Oven temperature	100°–210°C
Program rate	6°C
Injection port temperature	210°C
Detector	Total ion chromatogram
Carrier flow	1 ml/min

Split ratio	200:1
Threshold	30
MS ionization	70 eV
Scan time	108.3 amu/s
Scan start delay	0 min
Start, stop masses	35–400
Electron multiplier voltage	2600 eV
GC/MS column	50-m Ultra (No. 1) Hewlett Packard capillary column cross-linked methyl silicone
Film thickness	0.33 μm
Internal diameter	0.20 mm

IR Conditions

Instrument	Perkin Elmer 1320 Infrared Spectrophotometer
Chart expansion	1
Scan time/split program	12 min
Range	4000–600 cm^{-1}

NMR Conditions

Instrument	Varian EM-360L 60 MHZ NMR
Filter	0.05 s
RF power	0.05 mg
Sweep time	5 min
Sweep width	10 ppm
End of sweep	0 ppm
Nucleus	proton
Sample temperature	25°C

REFERENCES

1. Shaath, N. A. Encylopedia of UV absorbers for sunscreen products. *Cosmet Toilet., 102(2)*:21 (1987).
2. (a) Stenhagen, E., Abrahamson, S., and McLafferty, F. W. *Registration of Mass Spectral Data*, Vols. 1–4. John Wiley & Sons, New York, 1974; (b) Pouchert, C. J. *The Aldrich Library of NMR Spectra*, 2nd ed. Vols 1 and 2. Aldrich Chemical Company, Milwaukee, 1983; (c) Pouchert, C. J. *The Aldrich Library of Infrared Spectra*, 3rd ed. Aldrich Chemicals Comapny, Milwaukee, 1981; (d) Grasselli, J. G. and Ritchey, W. M. *Atlas of Spectral Data and Physical Constants for Organic Compounds*, Vols. 1–4. CRC Press, Cleveland, 1975; (e) *Ultraviolet Spectral Data.*

Chemical and Petroleum Research Laboratory, Carnegie Institute of Technology, Pittsburgh, 1959.

3. Fisher, A. A. Dermatitis due to benzocaine present in sunscreens containing glyceryl PABA (Escalol 106). *Contact Dermatitis*, *3*:170–171 (1977).
4. Masse, M. O., Herpol-Borremans, M., Grimee, R., and Glevuyky, S. *Int. J. Cosmet. Chem.*, *4*:235–236 (1982).

Index